MICROPROCESSOR SYSTEMS DESIGN

PWS-KENT RELATED TITLES OF INTEREST

SECOND EDITION

MICROPROCESSOR SYSTEMS DESIGN

68000 HARDWARE, SOFTWARE, AND INTERFACING

ALAN CLEMENTS

TEESSIDE POLYTECHNIC

PWS-KENT PUBLISHING COMPANY

BOSTON

PWS–KENT
Publishing Company

20 Park Plaza
Boston, Massachusetts 02116

PWS-KENT Publishing Company is a division of Wadsworth, Inc.

Library of Congress Cataloging-in-Publication Data

Clements, Alan
    Microprocessor systems design: 68000 hardware, software, and
interfacing / Alan Clements.—2nd ed.
        p.   cm.
    Includes bibliographical references and index.
    ISBN 0-534-92568-5
        1. Motorola 68000 (Microprocessor) 2. System design.   I. Title.
QA76.8.M6895C54   1992   91-39492
004.165—dc20   CIP

International Student Edition ISBN 0-534-98356-1

 This book is printed on recycled, acid-free paper.

Sponsoring Editor: Jonathan Plant
Assistant Editor: Mary Thomas
Production Coordinator: Robine Andrau
Interior and Cover Designer: Robine Andrau
Manufacturing Coordinator: Marcia Locke
Interior Illustrator/Typesetter: Santype International Limited
Cover Photo: Dominique Sarraute/The Image Bank
Cover Printer: New England Book Components, Inc.
Printer/Binder: R.R. Donnelley & Sons Company

Printed in the United States of America

92 93 94 95 96 — 10 9 8 7 6 5 4 3 2 1

For my colleagues:
Brien Dummigan, Steve Dunne,
Simon Lynch, Sharon Morgan,
Alex Siday, Colin Wheeler

CONTENTS

PREFACE

The microprocessor revolution, which placed a central processing unit (CPU) on a single chip, has turned the computer into a handful of chips and has handed these chips over to the electronic engineer—that is, an electronic engineer can now take a microprocessor and design it into a system whose complexity may vary from the trivial to the sophisticated. *Microprocessor Systems Design: 68000 Hardware, Software, and Interfacing*, Second Edition shows how a microprocessor is transformed into a system capable of performing its intended task. It concentrates on the interface between the microprocessor and the other components of a microprocessor system.

Microprocessor Systems Design is written for two audiences. The first is engineering and computer science students who have to study the design of microprocessor systems in order to get the qualification that frees them to design real systems in industry. For them, therefore, this book is written to support a typical course in microprocessor systems design. The second audience is practicing engineers who do not have exams to pass but who must produce an actual system to their employer's specification.

To address both audiences, I have included more practical information than is normally found in textbooks on microprocessor systems design. In particular, I have placed great emphasis on the timing diagram and on the analysis of microprocessor read/write cycles. An understanding of the timing requirements of a microprocessor is vital to the engineer.

Although I am writing about the design of microprocessor systems, I have included a section on programming a microprocessor in assembly language. This inclusion has been made because many of the small educational microprocessor systems can be programmed only in assembly language and because the peripherals appearing in this book are described at the physical level and may be programmed in assembly language.

Anyone writing a book about the design of microprocessor systems must choose an actual microprocessor to illustrate the techniques involved. The 68000 family, which is one of several high-performance microprocessor families currently available, has been chosen because of its powerful but relatively simple instruction set, its sophisticated interfacing capabilities, and its ability to support multitasking. Moreover, the 68000 family is probably the most commonly used vehicle to teach microprocessor architecture and systems design in both the United States and Europe. We speak of the 68000 family rather than the 68000 because the 68000 is just one of a family of closely related microprocessors. If you understand the 68000, you will have no difficulty in understanding other members of its family.

The 68000 was introduced to the world as a 16-bit machine because it has a 16-bit data bus (and at that time it was competing with 8-bit processors). In spite of its external 16-bit bus, it has 32-bit internal registers and executes operations on 32-bit values, so it is not unreasonable to refer to it as a 32-bit machine. The later

68020 and 68030 processors, which evolved from the 68000, have a full 32-bit data bus and can accurately be called 32-bit machines. Since it would be awkward to keep changing terminology, the term 32-bit microprocessors is used throughout this book and includes the 68000 in this category.

Do not think that 8-bit microprocessors are now superfluous. Because they are cheaper than 32-bit microprocessors, 8-bit microprocessors will be around for many years and can often be used in simpler systems. However, 32-bit microprocessors, such as the 68000, are more relevant to teaching because they have all the features of 8-bit devices plus many new features not found on the older chips.

The following paragraphs briefly describe the contents of *Microprocessor Systems Design.* Chapter 1 introduces the basic building blocks of a microprocessor system and provides a framework within which a real system can be discussed. Chapter 2 provides a general introduction to the architecture and programming of the 68000 microprocessor, an introduction to some of the 68020's new instructions, and an overview of the 68020's very powerful addressing modes. Sufficient detail of the 68000's assembly language is given to enable the reader to follow the fragments of assembly language appearing elsewhere in the text.

Chapter 3 is devoted to program design techniques, including an overview of some of the topics closely related to the design of software for embedded microprocessor systems. Chapter 4 examines the basic hardware characteristics of the 68000, including the concepts of the timing diagram and the relationship between the 68000 and random access memory.

Chapter 5 looks at the memory subsystem and is divided into two distinct sections: address decoding and the design of static memory arrays. Chapter 6, which covers exception handling, is especially valuable, given the 68000's wealth of exception-handling facilities.

In Chapter 7, we are concerned with the design of more sophisticated microprocessor systems. Multimicroprocessor systems, dynamic memories, memories capable of detecting and automatically correcting errors, and memory management systems are all discussed. Chapter 7 provides an opportunity to discuss some of the 68020's special features, such as its coprocessor interface and its on-chip instruction cache (both the 68030 and 68040 caches are also described). This chapter ends with an introduction to the revolutionary 68040.

The operation of the parallel interface and a description of the 68230 PI/T are outlined in Chapter 8, together with a discussion of the direct memory access (DMA) controller.

Chapter 9 deals with the serial interface between a microcomputer and a peripheral. This chapter also covers the international standards used to define the electrical characteristics of a serial data-link. Chapter 10 examines the buses linking together the functional modules of a microprocessor system. The characteristics of buses are dealt with under various headings, including timing and protocols, electrical properties, and bus control in microprocessor systems. The popular VMEbus is covered. The final chapter, chapter 11, presents a practical discussion of some issues involved in microprocessor systems design. Beginning with adequate specifications and moving through design, construction, testability, and maintainability, this chapter concludes with a worked example of a 68000-based microcomputer.

A complete *Instructor's Manual* to accompany the book is now available from PWS-KENT Publishing Company. This supplement includes complete solutions to text problems, an annotated version of the 68000 instruction set, and documentation for the bound-in MONITOR/EMULATOR diskette. The diskette and documentation allow readers to use more extensively the TS2MON monitor program, which appears at the end of chapter 11. The EMULATOR software includes both a 68000 emulator and a cross-assembler, which can run and test 68000 assembly language programs on non-68000-based hardware, such as IBM PCs and compatibles. The *Instructor's Manual* with diskette is available free to adopters of *Microprocessor Systems Design*, Second Edition.

Acknowledgments

I would like to thank all those who helped me with the production of this book. Indeed, if it were not for Karl Amatneek, Nick Panos, and Lance Leventhal in San Diego, this book would not have been written. I had originally intended to write a book on the 6809 microprocessor. These three assaulted me verbally until I agreed to write about the 68000. In particular, I would like to thank Karl for his help in organizing my visits to the United States, Nick for the time he spent with me discussing the 68000, and Lance for his constructive criticism of my manuscript.

During my visits to the United States, I met many of the reviewers of my manuscript as I traveled from Boston to San Diego. I would like to thank the following reviewers:

A.B. Bonds
Vanderbilt University

George Brown
Rochester Institute of Technology

Ray Doskocil
Motorola, Phoenix, Arizona

Frederick Edwards
University of Massachusetts—Amherst

Anura P. Jayasumana
Colorado State University

Joseph E. Lang
University of Dayton

Lance Leventhal
Emulative Systems Co., Inc.

Mark Manwaring
Washington State University

Zane C. Motteler
California Polytechnic State University

Elizabeth O'Neil
University of Massachusetts—Boston

David Poplawski
Michigan Technological University

Clara Serrano
Motorola, Austin, Texas

Peter G. von Glahn
Villanova University

Marvin Woodfill
Arizona State University

I must single out two of the reviewers for special praise. Clara Serrano at Motorola in Austin and Ray Doskocil at Motorola in Phoenix are both very busy engineers who spent considerable time working on my manuscript and who gave up even more time to chat with me about my book.

I would also like to thank those at PWS-KENT Publishing Company who helped and guided me through the many stages in the production of this book—specifically Jonathan Plant, Mary Thomas, and Robine Andrau. And, finally, a special thanks to my wife, Sue, who read the draft manuscript and provided much useful feedback.

Technical data relating to the MC68000 and other associated products described in this publication is reproduced by kind permission of Motorola Semiconductors. Motorola reserves the right to make any changes to the product herein to improve reliability, function, or design. Motorola does not assume any liability arising out of the application or use of any product or circuit described herein; neither does it convey any license under its patent rights nor the rights of others.

I encourage readers to send me suggestions and comments regarding this book through PWS-KENT Publishing Company in Boston or to contact me directly at Teesside Polytechnic (Middlesbrough, England) by means of E-mail. My address is ACT012@UK.TP.PA which is on the UK JANET network.

THE MICROCOMPUTER

In this chapter we introduce the microcomputer and identify the characteristics of its major component parts. Once we have introduced these parts and described the functions they perform in a microcomputer, we are in a better position to deal with them in detail in later chapters. However, before we can continue, we need to determine exactly what we mean by the terms *microcomputer* and *microprocessor*.

The microprocessor is a central processing unit (CPU) on a single chip and is entirely useless on its own. To create a viable computer requires memory components, interface components, timing and control circuits, a power supply, and a cabinet or other enclosure. This book shows how these other components can be connected to a microprocessor to produce a microcomputer.

A microcomputer is defined as a stand-alone system based on a microprocessor. A stand-alone system is one that is able to operate without additional equipment. It should be distinguished from the individual functional parts of a microcomputer (the memory, the CPU, the interfaces, and the power supply). Therefore, an understanding of how a microprocessor is interfaced to other components is necessary to the engineer who wishes to design a microcomputer. The term *microprocessor system* should be regarded as meaning the same as *microcomputer* throughout this book.

The applications of a microcomputer are legion and hardly need elaborating on today. Broadly speaking, the microcomputer falls into one of two categories: the general-purpose digital computer and the embedded computer. The general-purpose digital computer is what most people understand by the word *computer*. It has all the necessary memory and peripherals required by a user to execute a wide range of applications programs.

An embedded computer is one dedicated to a specific application and is normally transparent (or "invisible") to the user of the system in which it is located. A typical embedded microcomputer lies at the heart of an automatic bank teller. A customer inserts a credit card in a slot and the microcomputer reads the relevant details from its magnetic strip. The customer keys in an identity code; the microcomputer validates the code and then invites the customer to perform a transaction. Once the transaction has been completed, the microcomputer updates the data in the magnetic strip and operates the mechanical subsystem that returns the card to its owner. The user of the automatic teller is entirely unaware of the embedded computer and of its function.

The fundamental difference between an embedded computer and a general-purpose computer is the optimization of the former to suit a single function. Typically, an embedded computer contains only the components strictly necessary for it to execute its allocated task. In contrast, a general-purpose computer is highly likely to have sufficient memory and peripherals for it to be able to handle a broad range of tasks. In particular, the general-purpose computer almost always has facilities for expansion, thereby enabling the user to add more memory or peripherals at a later date.

1.1   MICROPROCESSOR SYSTEMS

In this book, we examine the design of two types of microcomputer: the single-board computer (SBC) and the modular computer, which is composed of a number of separate units linked by a bus. The single-board computer is usually associated with the embedded computer, where it is designed to execute a single, fixed task.

The modular computer based on a bus is frequently a general-purpose digital computer and has a degree of sophistication not normally found on an SBC. To be fair, the introduction of high-density memories, high-performance peripherals, and programmable logic elements has now made it possible to produce very powerful SBCs. When we come to design modular microcomputers, we will discover that their design must be flexible so that they can be adapted to a wide number of different applications.

The block diagram of a possible modular, general-purpose, digital computer based on a 16/32-bit microcomputer is given in figure 1.1. This system consists of a number of modules linked together by a bus. Figure 1.2 provides a photograph of a typical rack-mounted modular system. Although the words *card* and *module* are frequently interchangeable, a card may not necessarily constitute a module in the same sense as in the concept of "modular design and modularity."

Information is moved between the cards by means of a bus, labeled *system bus* in figure 1.1, to distinguish it from buses within the individual cards. A bus existing only within a card is termed a *local bus*. A system bus allows cards to be added to or removed from a microprocessor system. A microcomputer can, therefore, be built out of standard building blocks: the cards. The advantage of a modular approach is that a manufacturer can produce a sufficiently large range of cards to permit any user to construct a microcomputer to his or her own specifications. Furthermore, once this trend begins, a number of independent manufacturers soon start to sell special-purpose cards. This process can most readily be seen in the case of some popular personal computers.

Of course, manufacturers begin to make their own cards for sale only when the market is sufficiently large. The size of the market depends on the widespread acceptance of a particular system bus and of standard size cards that can be plugged into it. Standards for buses were once generated in an ad hoc fashion by some manufacturer or entrepreneur and later adopted by an international committee. In the case of

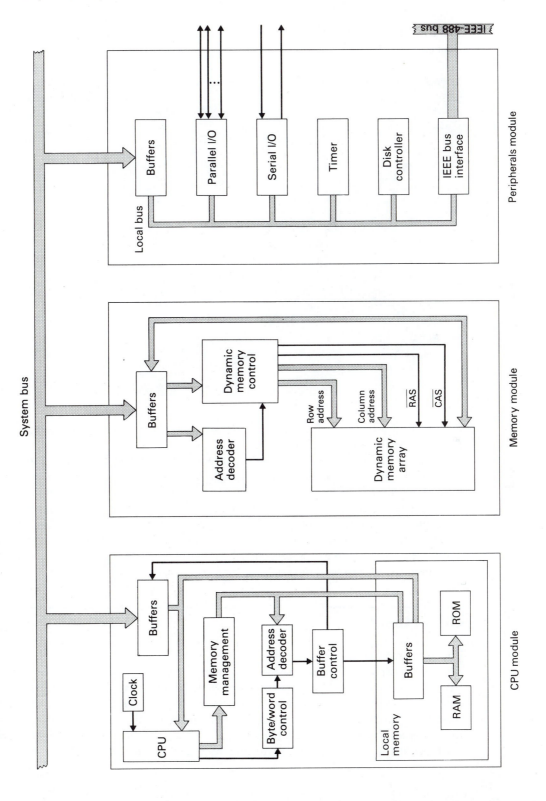

FIGURE 1.1 Block diagram of a possible modular microcomputer

FIGURE 1.2   A modular microprocessor system. Photograph supplied courtesy of BICC-Vero.

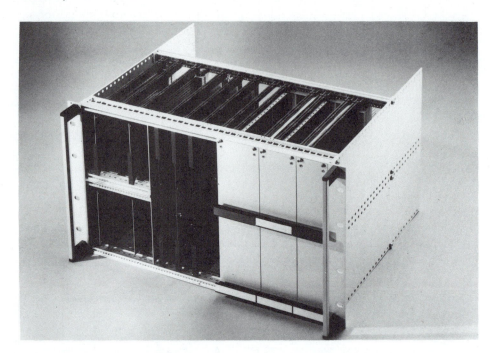

the 68000, Motorola first created the Versabus, which was later modified by committees to become the VMEbus.

In the block diagram of figure 1.1, three modules are connected to the system bus: a CPU module, a memory module, and a peripherals module. In practice, there is no reason why these functions cannot be distributed among the various modules of a system. For example, it is quite normal to locate a block of memory on a card containing one or more peripherals. However, the functions have been divided between the modules here in order to provide a reasonable sequence for the teaching of microprocessor systems design. The three modules comprising this system are the CPU module, the memory module, and the peripherals module.

CPU Module

The CPU module contains the microprocessor and its associated control circuitry. The control circuitry includes all the components (excluding memory and peripherals) that must be connected to a microprocessor to enable it to function. Most microprocessors do not contain sufficient circuitry on-chip to allow them to be connected directly to memory components. Generally speaking, the more sophisticated the microprocessor, the greater the demand for additional control circuitry. The following functions are some of those performed by the CPU module.

Clock and CPU Control Circuits

The clock provides the sequencing information needed by the microprocessor to control its internal operations. Modern 32-bit microprocessors require a very simple clock circuit generating a continuous square wave with a frequency of between 4 and 40 MHz. Some very high performance systems have clock frequencies of 50 MHz and even beyond. This clock is a master clock and is normally distributed to all modules in the system. Other control circuitry includes the power-on-reset circuit, which forces the microprocessor to execute a start-up routine shortly after the system has been powered up.

Address Decoder

The primary function of the address decoder is to divide the memory space of the CPU into smaller units and to allocate these to individual memory components. The CPU module is usually provided with some local memory that is not accessed from the system bus. This feature enables the CPU module to be tested independently of the rest of the microprocessor system. The address decoder also distinguishes between a memory access to local memory and one to memory situated on another module.

Address and Data Bus Buffers

The microprocessor or any other device distributing information throughout the system cannot normally provide sufficient electrical drive to cater for the loading on the bus from other elements. Special integrated circuits, called *buffers* or *bus drivers*, act as an interface between the microprocessor and the bus. Additional circuitry is needed to control the bus drivers, as only one device may drive the bus at any instant.

Buffer Control

The buffer control logic determines the mode of operation of the bus drivers and receivers on the CPU module. For example, during an access to local memory, the buffers driving the system bus must be disabled.

Although 16-bit microprocessors are designed to operate on 16-bit words, there are occasions when the microprocessor wishes to operate on an 8-bit entity. If memory were organized as 16-bit words, the CPU could not deal with transfer of a single 8-bit word between itself and memory. Most memories in 16-bit micro-computers are byte organized, so that a 16-bit word is stored as two 8-bit words, each independently accessible. Byte/word control logic is required to enable the microprocessor to access either a single byte or both bytes of a 16-bit word. Today's 32-bit microprocessors are extensions of, or enhancements to, earlier 16-bit micro-processors and can, generally, use their 32-bit data buses to access a byte, word, or longword (i.e., 32 bits) in a single bus cycle.

Bus Arbitration Control

In some microcomputers, the microprocessor has sole control of the system bus and determines the nature of each and every bus access, that is, a read from memory or a write to memory. In other microcomputers, more than one device may access the bus. It may be a DMA (direct memory access) controller, which is used to transfer data directly between a peripheral and memory without the active intervention of the CPU, or it may be another microprocessor. The latter case arises in a multiprocessor environment in which two or more microprocessors can operate independently on separate data. Clearly, if more than one device is able to control the system bus, a set of rules (i.e., a protocol) is needed to determine which can access the bus at any instant and to ensure that every bus controller gets fair access to the bus. The action of determining which device gets access to the bus is called arbitration and is normally carried out by hardware, as software arbitration would be unacceptably slow. The term *master* is generally used to describe the device that is currently controlling the bus and that may initiate read and write cycles to other devices. Similarly, the term *slave* is employed to refer to devices that are accessed by a master.

Memory Management

If there is one thing that separates the world of the 16/32-bit microprocessor from that of the 8-bit microprocessor, it is memory management. Memory management is a generic term and is applied to all those techniques that translate the address of information generated by the computer (i.e., an address that may fall anywhere within its address space) into the address of that information within the available system memory. Most 8-bit and many 16/32-bit microcomputers do not use memory management techniques. The address generated by the CPU corresponds exactly to the actual location of information in the microprocessor's random access memory. One form of memory management is associated with the term *virtual memory* and can be employed to make a small random access memory appear to the CPU as if it were much larger than it really is. This form of memory management is performed by holding part of the program or data in high-speed random access memory and the rest of it on disk. Whenever the CPU cannot find the data it requires in the high-speed memory, it moves more data from the disk to the high-speed memory.

Memory management is a rather complex topic and is best introduced by an analogy. Many newspapers have a page containing classified advertisements. The reader replies to an advertisement by writing to the address provided in the paper. This is the physical address of the advertiser. Sometimes the advertisements have box numbers instead of actual (physical) addresses. The reader writes to the newspaper quoting the box number, and the paper forwards the letter to the appropriate advertiser. The paper is performing the action of memory management. It is translating a box number, which we can call a logical address, into the physical address of an advertiser.

The advantage of the preceding arrangement is that the reader does not care where the advertiser actually lives. The advertiser may live in the street next to the newspaper offices or in another country. The advertiser may move from one office to another. All this is irrelevant to the reader, who simply writes to the box number, leaving it up to the newspaper to perform the address translation.

In computer terms, the address generated when a program is executed on a microcomputer is called a *logical address* and corresponds to the box number in the preceding analogy. The address of information in memory is called its physical address and corresponds to the address of the advertiser. One of the purposes of memory management is to free programmers from all worries about where their programs and associated data are to lie in memory. As mentioned above, virtual memory techniques allow data to be moved between fast RAM and slower disk-based memory with the memory management unit keeping track of the data automatically. Another function is to monitor all memory accesses and to provide a user program with protection from modification by another program.

Although all first-generation microprocessors and many of the early 16/32-bit microprocessors employed external memory management units (MMUs), there has been a trend to integrate MMUs into the CPU chip itself. For example, both the 68030 and the 68040 include sophisticated MMUs on-chip.

Memory Module

The memory module of a microcomputer contains the bulk of the random access memory accessible from the system buses. The term *bulk* is used because there may well be small quantities of memory on modules other than the memory module. The memory on the memory module may be read/write random access memory or read-only memory. Read/write memory can be read from or written to and is frequently called simply RAM in computer literature. Read-only memory (ROM) can be read from but not written to under normal conditions. Consequently, ROM is used to hold programs and data that change infrequently, if at all. Typically, ROMs hold interpreters for languages, such as BASIC or Forth, operating systems, and bootstrap programs. A bootstrap program is the first program run when a computer is switched on and is used to start up a system. It is a minimal program designed to read the operating system from mass storage, transfer it to the system's random access memory, and then initiate the execution of the operating system.

When we come to chapter 5, dealing with memories, we will find that a range of memory components is available to the designer. The actual memory component chosen for any given application represents a trade-off between the desirable characteristics of the component and its cost.

The component selected for the memory module in figure 1.1 is called dynamic memory. Dynamic memory is the most cost effective form of read/write random access memory available and is frequently chosen as the means of implementing large memories. Unfortunately, dynamic memory is somewhat more complex to use than other forms of read/write memory, and a dynamic memory module must be carefully designed. Chapter 7 deals with the design of memory systems using dynamic memory components.

Peripherals Module

The peripherals module (see figure 1.1) contains the circuits that form an interface between the microcomputer and the rest of the world. For example, a serial input/output interface enables the microcomputer to communicate with any of the CRT terminals currently available. The serial port moves data from point to point one bit at a time. A parallel port simultaneously moves several bits of data between the microcomputer and an external device, such as a printer. Many parallel interfaces transfer 8 bits at a time, but some can be programmed to move fewer than 8 bits or as many as 16 bits. Note that a parallel port may perform sundry other tasks, such as checking that the printer has accepted the data transmitted to it.

A timer is the general name for an integrated circuit that performs a variety of functions associated with the measurement of time and the generation of pulses. The actual facilities provided by individual timer chips vary from manufacturer to manufacturer. In general, the timer is able to generate single or repetitive sequences of pulses. It can measure the period or frequency of incoming pulses and is able to interrupt the microprocessor at fixed intervals. The latter operation permits the implementation of multitasking systems in which the computer switches from one job (task or program) to another each time it is interrupted.

The disk controller forms an interface between the microcomputer and a mass storage device, which may be a floppy disk drive or a hard disk drive. Most disk controllers are exceedingly sophisticated devices and often rival microprocessors themselves in complexity. The principal function of the disk controller to translate the data from the microcomputer into a format suitable for storing on the disk, and vice versa.

The IEEE-bus controller forms an interface between the microcomputer and the popular IEEE-488 bus. Conceptually, the IEEE-488 bus behaves in a way very similar to the system bus. Hewlett-Packard originally devised it and intended it to link together programmable instruments in a laboratory or industrial environment. By controlling test equipment and measuring devices from the IEEE-488 bus, implementing an automatic testing station is possible. A system under test is connected to the test equipment and measuring devices. The computer configures (i.e., sets up) all the equipment via the IEEE-488 bus and then reads the test results from the same bus. Today, the IEEE-488 bus is also used to link peripherals, such as printers or disk drives, with microprocessor systems. The advantage of this bus is that it is now an international standard and is (theoretically) device and manufacturer independent.

The peripherals just described represent some of the most popular functions obtainable in the form of a single chip. In a real system, the various peripheral interfaces are likely to be distributed between the modules of the microcomputer.

1.2   EXAMPLES OF MICROPROCESSOR SYSTEMS

Before we begin to examine the design of actual microprocessor systems, looking at two generic applications is instructive and will give us an idea of some of the factors involved in microprocessor systems design.

Example 1. Morse Code Transmitter

All radio amateurs once transmitted Morse code by tapping out the dots and dashes on a Morse key (i.e., a simple on/off switch). It is now relatively easy to design a circuit with a conventional keyboard that generates a single Morse character each time a key is depressed.

Figure 1.3 gives the block diagram of a possible Morse code generator. This is a truly basic computer and has the absolute minimum number of components needed to execute its *single* function. The CPU is connected to three major components: a parallel I/O port that detects a keystroke and generates the Morse code output, a read/write memory that holds temporary variables, and a read-only memory that contains the program to generate the Morse code. In the vast majority of systems, a few gates and other components are needed to perform certain system functions and to "glue" the CPU to its memory and peripherals.

Before leaving this example, we must make an observation. The majority of manufacturers would probably not employ a general-purpose microprocessor like the 68000 to build a simple system such as a text to Morse code translator. Semiconductor manufacturers produce a large range of low-cost microcontroller chips for such applications. These chips are devices that are usually based on a standard CPU and often include on-chip ROM, scratch pad read/write memory, and several peripherals. Microcontrollers make it possible to construct true one-chip computers for

FIGURE 1.3 Microprocessor-controlled Morse code generator

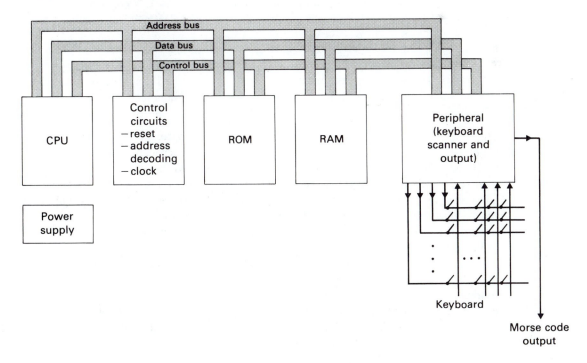

embedded control systems. Until recently, microcontrollers were based on 8-bit architectures because the CPU plus the memory and peripherals took up quite a large chunk of silicon real estate. However, the end of the 1980s has witnessed the introduction of 16/32-bit microcontrollers.

Figure 1.3 represents the simplest form of microcomputer. It is simple because it has a low component count and does not employ many of the powerful features found on some microprocessors. It is, of course, a single-board computer and may be embedded within a radio transmitter. There are relatively few major design decisions in the production of the type of system of figure 1.3, as there are so few components to deal with. In general, the design decisions are often largely economic and depend of the scale of production of the system.

Example 2. Personal Computer

Figure 1.4 provides the block diagram of a possible general-purpose personal computer. In this type of computer there are more functional blocks than in the basic computer presented in figure 1.3 because the general-purpose personal microcomputer is fairly complex and has many different functions to perform. It must be able to input data from several sources (e.g., a keyboard or an external data-link), be able to store and retrieve data from some mass storage device, and be capable of outputting data to a CRT terminal or to its own TV display.

Most of the functional units in figure 1.4 are as complex as the whole system of figure 1.3. The designer has to juggle with the conflicting requirements of each module and has to produce a compromise. Much of this book is concerned with the type of computer of figure 1.4 and with the decisions the designer must make.

SUMMARY

This chapter has set the stage for our course in microprocessor systems design. We have introduced the microprocessor around which the microcomputer is to be built. Later chapters show how the microprocessor is connected (i.e., interfaced) to external memory and to input/output devices. Although a microcomputer can be built from a handful of components on a single board, it can also be a very complex arrangement of modules that communicate with each other by means of a bus. When the basic principles of microcomputer design have been presented, we will return to the design of buses for microcomputers.

One of the most important concepts introduced in chapter 1 is that of *modularity*. Modern systems are invariably designed as a number of largely independent modules that work together to achieve the desired effect. Only by decomposing the design of a complex microcomputer into the design of a collection of less complex subsystems can we create highly reliable and sophisticated products. The electrical

FIGURE 1.4 Block diagram of a personal computer

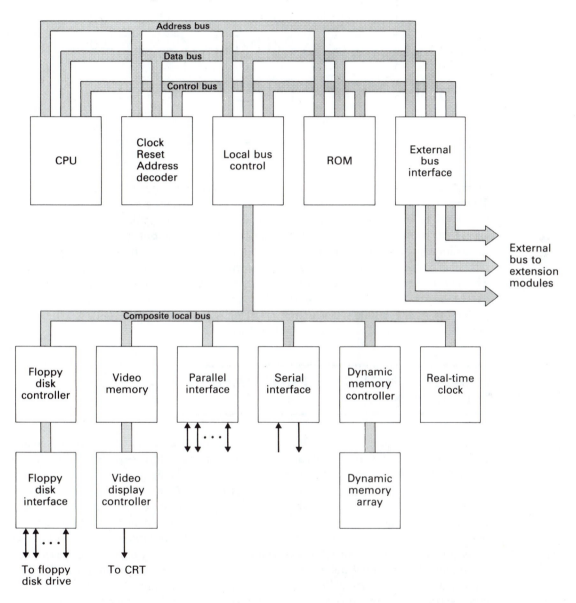

highway, or bus, which is introduced in this chapter, is one of the main tools of the engineer producing modular subsystems. The bus provides the means of linking modules together. Modularity is not limited to the realm of hardware. When we come to the design of software in chapter 3, we will see that this statement is as true for the design of software as it is for the design of hardware.

PROGRAMMING THE 68000 FAMILY

In this chapter we describe the programming model of the 68000 microprocessor. The instruction set and the addressing modes of the 68000 are treated at an elementary level to enable the student to understand the assembly language programs in later chapters. A prerequisite is a basic knowledge of computer architecture so that, for example, the concepts of a program counter, a Boolean operation, a conditional test, and a stack are familiar. We begin with a look at typical assembler directives. This is followed by an introduction to one of the 68000's most powerful features— its wealth of addressing modes. We then provide an introduction to some of the highlights of the 68000's instruction set. Finally, we look at the 68020, which is a powerful extension of the 68000's architecture.

2.1 ASSEMBLY LANGUAGE PROGRAMMING AND THE 68000

Assembly language is a form of the native language of a computer in which machine code instructions are represented by mnemonics and addresses and constants are written in symbolic form (just as they are in high-level languages).

Before we can discuss the instruction set of the 68000, we must define some of the conventions adopted in writing assembly language programs. An assembly language is made up of two types of statement: the executable instruction and the assembler directive. An executable instruction is one of the processor's valid instructions and is translated into the appropriate machine code form by the assembler. Assembler directives tell the assembler something about the program. Basically, they link symbolic names to values, allocate storage, set up predefined constants, and control the assembly process. The assembler directives to be dealt with here are: TTL, EQU, DC, DS, ORG, SECTION, and END. Real assemblers have very many more directives, but most of these often control the format of the output and are not relevant here.

Since the principal purpose of this chapter is to introduce the 68000's instruction set, we will not concentrate on assemblers. It is worth mentioning, however, that a powerful processor like the 68000 usually has a large range of sophisticated development tools. For example, many assemblers are employed in conjunction with linkers and libraries. The linker permits a program to be developed as a series of separate (i.e., independent) modules that can then be combined, or "linked." A module can be written and the linker told about all the variables and names that are declared outside the module. During the assembly phase, real (i.e., actual) addresses of operands are replaced by dummy addresses. When the modules are linked, the linker is able to put together all the various modules, and any dummy addresses are replaced by the actual addresses.

Another tool often available to the assembly language programmer is the *macro-assembler*, which permits the programmer to define macros. A macro is a unit of in-line code that is given a name by the programmer and that can be called rather like a subroutine. Unlike a subroutine, the full code of the macro is embedded in the program each time it is used—that is, the macro is a form of shorthand used by the programmer to avoid writing certain sequences of instructions over and over again. You can also employ the macro to *rename* one (or more) instructions. For example, if you do not like using MOVE.W D0,—(A7) to push data on the stack, you can define the macro PUSH D0 to replace the less clear MOVE.W D0,—(A7). Here's a simple example of a macro. Suppose you want to employ the three instructions

```
PEA        (A0)
MOVE.W     D3, — (A7)
BSR        Sub1
```

repeatedly in your program. If your assembler supports macros, you can give the instructions the name PushAndCall and then write PushAndCall in your program whenever you wish to execute them. The assembler replaces the PushAndCall mnemonic by the three instructions when the assembly is carried out.

Because assembly language is concerned with the "native processor" itself, it is, therefore, strongly dependent on the architecture of the processor. This situation contrasts strongly with high level languages that are, theoretically, independent of the system on which they run.

First of all, it is necessary to define the size (in bits) of the units of data manipulated by the 68000. The 68000 permits 8-, 16-, and 32-bit operations. A 32-bit entity is called a longword, a 16-bit entity a word, and an 8-bit entity a byte. To avoid confusion, longword, word, and byte will always refer to 32-, 16-, and 8-bit values, respectively, throughout this book.

The sizes of operands are indicated by .B, .W, and .L extensions to instructions, which indicate byte, word, and longword values, respectively. By the way, some microprocessors support even longer operands. The next step up after the longword is the *quadword*, which is 64 bits, or 8 bytes, long. Although the 68000 does not support quadword operations, the 68020 implements multiplication and division instructions that take quadword operands.

To provide the reader with an idea of what one looks like, an example of an assembly language program fragment follows. Note that the symbol $ indicates that the number following is a hexadecimal value. For example, $004A = 74_{10}$. The symbol # indicates that the operand following it is a literal or immediate value and not an address.

```
                    TTL             Program to input text into a buffer
BACK_SPACE          EQU             $08                    ASCII code for back space
DELETE              EQU             $7F                    ASCII code for delete
CARRIAGE_RET        EQU             $0D                    ASCII code for carriage return
*
                    ORG             $004000                Data origin
LINE_BUFFER         DS.B            64                     Reserve 64 bytes for line buffer
*
*   This procedure inputs a character and stores it in a buffer
*
                    ORG             $001000                Program origin
                    LEA. L          LINE_BUFFER,A2         A2 points to line buffer
NEXT                BSR             GET_DATA               Call subroutine to get input
                    CMP.B           #BACK_SPACE,D0         Test for back space
                    BEQ             MOVE_LEFT              If back space then deal with it
                    CMP.B           #DELETE,D0             Test for delete
                    BEQ             CANCEL                 If delete then deal with it
                    CMP.B           #CARRIAGE_RET,D0       Test for carriage return
                    BEQ             EXIT                   If carriage return then exit
                    MOVE.B          D0,(A2) +              Else store input in memory
                    BRA             NEXT                   Repeat
                                                           Remainder of program
                     :
                    END
```

Don't worry about understanding the preceding fragment of a 68000 assembly language program. Its only purpose is to show what an assembly program looks like. A line is composed of three parts: a label, an instruction or assembler directive, and a comment. The first and third components are optional. The assembler regards a word beginning in the leftmost column (e.g., 'DELETE' or 'NEXT') as a label that can be used to refer to the line it labels. Labels are chosen by the programmer and may consist of a string of up to 8 alphanumeric characters, as long as the first character is a letter.

Any text not starting in the first column is regarded as either a 68000 instruction (e.g., BSR GET_DATA) or as an assembler directive (e.g., ORG $001000). Instructions take the form of a mnemonic and an optional parameter (or parameters separated by commas). Further text following an instruction or an assembler directive is regarded as a comment and is ignored. A few assemblers require the programmer to precede a comment by a delimiter (e.g., a colon or a semicolon).

If the first character on any line is an asterisk, the entire line is regarded as a comment and is ignored by the assembler.

Assembler Directives

TTL TTL means "title" and gives the program a user-defined name that it puts at the top of each page of the listing when the program is printed.

EQU The equate directive simply links a name to a value, making programs much easier to read. For example, it is better to equate the name CARRIAGE_RETURN to $0D and use this name in a program than to write $0D and leave it to the reader to figure out that $0D is the ASCII value for carriage return. Consider the following example of its use:

STACK_FRAME	EQU 128	Define a stack frame of 128 bytes
	⋮	
	LINK A1,# −STACK_FRAME	Reserve 128 bytes for local storage

Not only is STACK_FRAME more meaningful to the programmer than its numerical value of 128, but it is easy to modify the numeric value by changing the EQU 128 that appears at the head of the program.

DC The define a constant assembler directive permits the programmer to specify a constant that will be loaded into memory before the program is executed. This assembler directive is qualified by .B, .W, or .L to specify constants of 8 bits, 16 bits, or 32 bits respectively. For example, the assembler directive DC.B 3 loads the byte $03 into memory, whereas DC.L 3 loads the longword $00000003 into memory. The constant specified by the directive is loaded in memory at the "current" location and the location counter is incremented by the size of the constant (by 1, 2, or 4 for .B, .W, or .L constants). Note that it is usual to precede the DC directive by a label to enable the programmer to refer to the constant. A number (i.e., constant) without a prefix is treated as a decimal value. Prefixing it with a $ indicates a hexadecimal value; prefixing it with a % indicates a binary value. Enclosing a text string in single quotes indicates a sequence of ASCII/ISO characters. The following example demonstrates the syntax of the DC directive.

	ORG $001000	Start of data region
First	DC.B 10,66	The values 10 and 66 are stored in consecutive bytes
	DC.L $0A1234	The value $000A1234 is stored as a longword
Date	DC.B 'April 8 1985'	The ASCII characters as stored as a sequence of 12 bytes
	DC.L 1,2	Two longwords are set up with the values 1 and 2

The effect of these directives can be illustrated by a memory map, as shown in figure 2.1. Note in figure 2.1 that successive *words* have even addresses that are called "word boundaries."

Constants are automatically aligned on a word boundary if they are either a word or a longword (.W or .L). Note that it is possible to employ a *symbolic value* or even an *expression* in a DC assembler directive. Consider the following example.

BASE	DC.W	57	Store the word $0039 in memory
OFFSET	DC.W	3	Store the word $0003 in memory
	DC.L	3*BASE + OFFSET	Store the longword 3*39 + 3 in memory

DS The define storage directive, DS, reserves storage locations. Its effect is similar to that of DC, but no information is stored in memory—that is, DC is used to set up constants that are to be loaded into memory before the program is executed, whereas DS reserves memory space for variables that will be generated by the program at run-time. DS is also qualified by .B, .W, or .L and its operand defines the size of the storage area. Consider the following examples:

	DS.B	4	Reserve 4 bytes of memory
	DS.B	$80	Reserve 128 bytes of memory
	DS.L	16	Reserve 16 longwords (64 bytes)
VOLTS	DS.W	1	Reserve 1 word (2 bytes)
TABLE	DS.W	256	Reserve 256 words

FIGURE 2.1 Use of DC directive

Address	Memory contents		
001000	0A	42	DC. B 10, 66
001002	00	0A	DC. L $0A1234
001004	12	34	
001006	41	70	
001008	72	69	
00100A	6C	20	DC. B 'April 8 1985'
00100C	38	20	
00100E	31	39	
001010	38	35	
001012	00	00	
001014	00	01	DC. L 1, 2
001016	00	00	
001018	00	02	

A label in the left-hand column equates the label with the first address of the defined storage. The label references the lowest address of the defined storage value. In the above example, TABLE refers to the location of the first of the 256 words reserved (or allocated) by the DS directive.

   ORG   The origin assembler directive defines the value of the location counter that keeps track of where the next item is to be located in the target processor's memory. The operand following ORG is the absolute value of the origin. An ORG directive can be located at any point in a program, as the following example illustrates:

```
              ORG       $001000         Origin for data
  TABLE       DS.W      256             Save 256 words for "TABLE"
  POINTER_1   DS.L      1               Save one longword for "POINTER_1"
  POINTER_2   DS.L      1               Save one longword for "POINTER_2"
  VECTOR_1    DS.L      1               Save one longword for "VECTOR_1"
  INIT        DC.W      0, $FFFF        Store two constants ($0000, $FFFF)
  SETUP1      EQU       $03             Equate "SETUP1" to the value 3
  SETUP2      EQU       $55             Equate "SETUP2" to the value $55
  ACIAC       EQU       $008000         Equate "ACIAC" to the value $8000
  RDRF        EQU       0               RDRF = Receiver Data Register Full = 0
  PIA         EQU       ACIAC + 4       Equate "PIA" to the value $8004
  *
              ORG       $018000         Origin for program
  ENTRY       LEA       ACIAC,A0        A0 points to the ACIA
              MOVE.B    #SETUP,(A0)     Write initialization constant into ACIA
  *
  GET_DATA    BTST.B    #RDRF,(A0)      Any data received?
              BNE       GET_DATA        Repeat until data ready
              MOVE.B    2(A0),D0        Read data from ACIA
  *
```

The preceding fragment of code has the memory map shown in figure 2.2. The first occurrence of ORG (i.e., ORG $001000) defines the point at which the following instructions and directives are to be loaded. The four lines after ORG define four named storage allocations of 262 words in all. Following these, two words, $0000 and $FFFF, are loaded into memory. The address of the next free location is $001210. The three EQUs define constants for use in the rest of the program. Thus, whenever the name SETUP1 is used, the assembler replaces it by its defined value, 3. Note that it is possible to write an expression at any point in the assembler where a numeric value must be provided. For example, PIA EQU ACIA + 4 causes the word PIA to be equated to the value ACIA + 4 (= $008000 + 4 = $008004).

The second ORG (i.e., ORG $018000) defines the origin from which the following instructions are loaded. It is not necessary to allocate separate regions of

FIGURE 2.2 Memory map demonstrating application of DS, DC, and ORG directives

Address	Value	Label
001000		TABLE
⋮		
0011FF		
001200		POINTER_1
001201		
001202		
001203		
001204		POINTER_2
001205		
001206		
001207		
001208		VECTOR_1
001209		
00120A		
00120B		
00120C	00	INIT
00120D	00	
00120E	FF	
00120F	FF	
001210		
⋮		
018000	41	LEA ACIAC, A0
018001	F9	

memory for data and instructions in this way: the first instruction would have been located at $001210 had the second ORG not been used.

SECTION   The section directive sets the program counter to zero as if it were equal to ORG $000000. The purpose of SECTION is to force the assembler to generate program counter relative addressing modes in order to make the resulting code position independent. The meaning of these terms is made clear later in the text. Note that this definition of SECTION is a simplified version of that found in 68000 assemblers.

END   The end directive simply tells the assembler that the end of a program has been reached and that there are no further instructions or directives to be assembled.

2.2 PROGRAMMER'S MODEL OF THE 68000

The first step in examining a microprocessor is to look at its on-chip (i.e., internal) storage. Programmer-accessible registers fall into three groups (data, address, and special-purpose registers) and are dealt with in turn. Figure 2.3 shows the arrangement of the 68000's internal storage. This diagram is slightly simplified as there are really two A7 registers (we will discuss this concept more fully later).

Because the 68000 has address and data registers and address and data pins, there is some danger of confusing registers and pins when writing about them. To avoid this confusion, data and address registers are always denoted by a letter and a single digit (i.e., D0 to D7 and A0 to A7) and data and address pins are always denoted by a letter and two digits (i.e., D_{00} to D_{31} and A_{00} to A_{31}).

In this section we concentrate only on the 68000; we will introduce the 68020 later. However, we should make it clear that the architecture of the 68000 forms a subset of the 68020's architecture, and 68000 code will run on a 68020 with no modification (apart from one tiny and insignificant exception). From the point of view of the writer or user of *applications* programs, there is no difference between the register sets of the 68000 and the 68020. As we shall discover later, the differences between the 68000 and the 68020 (or 68030) register sets are relevant only to the systems programmer who is interested in operating system facilities.

Data Registers

From an information storage point of view, the 68000 is internally organized as a 32-bit machine; its registers are, therefore, 32 bits wide. The 68000 has eight general-purpose data registers, D0 to D7, and most operations involving the manipulation of data act on the contents of these registers.

The eight data registers are entirely general in the sense that any operation permitted on Di is also permitted on Dj. Some microprocessors restrict certain oper-

FIGURE 2.3   Programming model of the 68000

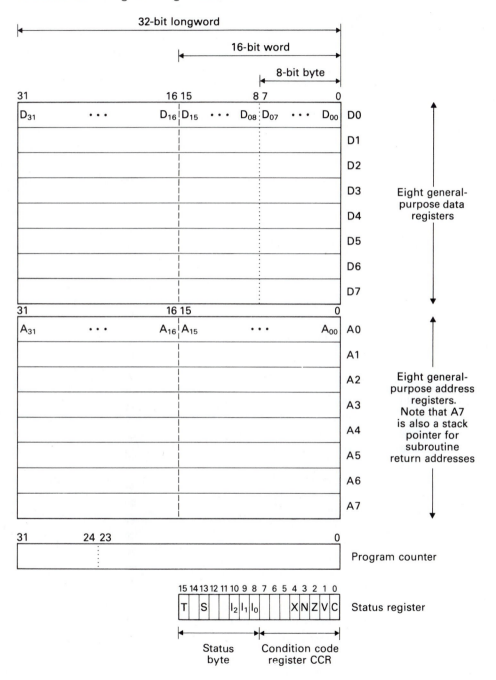

ations to specific registers. This restriction is undesirable because it forces the programmer to remember "what can be done to what."

As computers with 16- or 32-bit data wordlengths are less than ideal for the manipulation of text with its byte-oriented characters, microprocessor manufacturers have attempted to improve the efficiency of a 32-bit machine by allowing 8-bit operations to take place on part of the contents of 16- or 32-bit registers. In figure 2.1, the data registers are split into two words by a line of long broken dots and the lower-order words are in turn divided into two bytes by a line of short dots.

Operations on longwords, words, and bytes are denoted by the addition of .L, .W, and .B, respectively, to mnemonics. For example, the operation ADD.L D0,D1 adds the 32-bit contents of data register D0 to the 32-bit contents of D1 and puts the 32-bit result in D1. The operation ADD.B D0,D1 adds the least significant 8 bits of D0 to the corresponding 8 bits of D1 and puts the result in D1. Note that when a subsection of a data register is operated on, the remainder of the register is unaffected. For example, ADD.B D0,D1 does not affect bits 8 through 31 of register D0 or D1. It should be appreciated that, when a subsection of a data register takes part in an operation, the subsection is always the lowest-order unit of the register—that is, bits 0 through 7 or bits 0 through 15. Note that, throughout this text, the notation [Di] is interpreted as meaning "the contents of register Di."

Address Registers

The 68000 has eight address registers, designated A0 to A7 in figure 2.3. These registers should be regarded as pointer registers, because they hold addresses specifying the location of data in memory. All address registers are 32 bits wide, and operations performed on their contents affect the whole longword. Byte operations on bits 0 through 7 of an address register are not permitted.

The contents of an address register represent a single entity and the idea of separate address fields has no meaning. Therefore, an operation on the low-order word of an address register always affects the entire contents of that register—that is, if the low-order word of an address register is loaded with a 16-bit operand, the sign bit of the operand is extended into bits 16 through 31 of the high-order word. This happens because the contents of address registers behave like signed two's complement values. For example, if the low-order word of an address register is loaded with %1010 0000 0000 0111 ($A007), the address is sign extended to give the longword %1111 1111 1111 1111 1010 0000 0000 0111 ($FFFFA007). Programmers must be aware of this fact.

Do not be alarmed by the idea of negative (i.e., two's complement) addresses! Suppose A1 contains the value $FFFFFFFA (representing −6) and A2 contains the value $00001000. If we add A1 to A2 we get $FFFFFFFA + $00001000 = $00000FFA (in 32-bit arithmetic), which is six locations *back* from the address pointed at by A2. In other words, a negative address means "backward from the current location."

As in the case of the data registers, an operation on Ai can also be applied to Aj. However, A7 is a special-purpose address register and has an additional role to those of A0 to A6. It acts as the stack pointer used by subroutines to store return addresses in memory. Figure 2.3 is a simplified diagram.

One of the simplifications we introduced in figure 2.3 is the omission of an address register. The 68000 has, in fact, *two* A7 registers. We do not dwell on this point here, as it is of no immediate importance. We can say that the 68000 operates in one of two modes: *user* and *supervisor*. The operating system runs in the supervisor mode and programs controlled by the operating system run in the user mode. Each of these modes has its own A7. Consequently, if a user program corrupts its stack pointer, the entire system does not crash because the operating system's own stack pointer remains unaffected. After a hardware reset (e.g., on initial power-up), the 68000 runs in the supervisor mode and the supervisor stack pointer is automatically selected. The assembly programmer may write either A7 or SP (i.e., stack pointer). Moreover, the programmer may write USP (user stack pointer) or SSP (supervisor stack pointer) to label explicitly the A7 in use.

The designers of the 68000 family have, in effect, said, "Let there be address registers and data registers." Such a view has committed them to a philosophy that treats addresses and data separately. Some microprocessors have registers whose contents may contain data or the address of data. These microprocessors allow all arithmetic and logical operations to take place on addresses and data values alike.

However, addresses and data are used in entirely different ways and, therefore, should not be treated in the same way. Consequently, designers of the 68000 family have created two sets of registers, each with its own rules. One rule is that a word or byte operation on a data register does not in any way affect the bits of the register not taking part in the operation. The same rule does not, of course, apply to the contents of an address register.

When the 68000 was first introduced, there was some debate about whether it was a 16-bit or a 32-bit computer. As stated earlier, the 68000 has 32-bit internal registers and can carry out 32-bit operations on data or addresses. It is, however, interfaced to external systems by a 16-bit data bus, thereby forcing all 32-bit accesses to be implemented as two consecutive 16-bit accesses. Moreover, the external address bus of the 68000 is only 24 bits wide and address bits A_{24} through A_{31} have no effect on the address leaving the chip. Consequently, addresses (and the contents of address registers) are frequently written as six hexadecimal characters (i.e., 24 bits), as the 8 most significant bits of an address have no meaning as far as the system connected to a 68000 is concerned.

To be precise, the 68000 has a 23-bit address bus, A_{01} to A_{23}, that is employed to specify one of 2^{23} possible words. Two signals from the 68000, UDS* and LDS*, are used to specify whether the processor is accessing a word or an upper byte or a lower byte of a word. For this reason, we can think of UDS* and LDS* as acting as a pseudo A_{00} and talk about a 24-bit address bus.

The 68020 has full 32-bit address and data buses (incidentally, the 68020 has a true A_{00} pin and does not employ UDS* and LDS* pins to distinguish between the upper and lower bytes of a data word).

It should be appreciated that, although the 68000 is a word-oriented (16-bit) device with certain 32-bit facilities, its memory is byte addressable—that is, it can address both 16-bit and 8-bit quantities with equal ease. A byte address may be odd

FIGURE 2.4 Manner in which the 68000 stores bytes, words, and longwords in its memory space

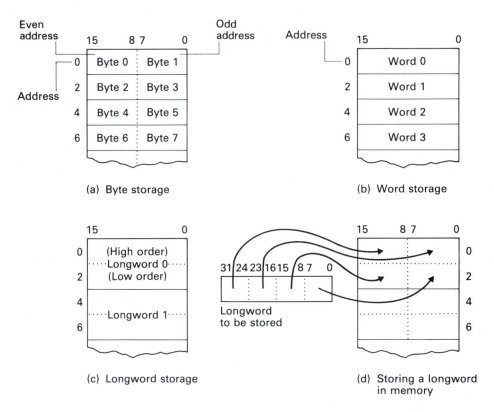

(a) Byte storage

(b) Word storage

(c) Longword storage

(d) Storing a longword in memory

or even. In byte addressing, the byte represented by bits D_{00} to D_{07} is the odd byte at the odd address, and the byte represented by bits D_{08} to D_{15} is the even byte at the even address. When the 68000 programmer refers to a longword, its address is defined as the address of the high-order 16 bits of the longword. The next even address holds the low-order 16 bits of the longword. Figure 2.4 illustrates the way in which the 68000's memory is arranged.

Special-Purpose Registers

The 68000 has two special-purpose registers: the program counter (PC) and the status register (SR). The program counter is 32 bits wide and contains the address of the next instruction to be executed. (This description is a simplification of the true state of affairs. The 68000 is a processor that is able to "look ahead" and to fetch instructions before they are needed. Therefore, the 68000 program counter does not always point to the *next* instruction to be executed.) It is a conventional PC with only one quirk. In order to fit the 68000 into a 64-pin package, the external address bus is restricted to 24 bits, which gives an addressing range of 2^{24} bytes (16M bytes or 8M words). Bits 24 through 31 of the program counter are not directly accessible in the 68000.

The 16-bit status register (SR) is divided into two logical fields. The 8 most significant bits are called the *system byte* and control the operating mode of the 68000. The importance of the system byte is dealt with in later chapters. All that need be done here is to introduce its 5 bits: T (the trace mode bit), S (the user/ supervisor mode bit), and I_0, I_1, and I_2 (the interrupt mask bits). The system byte cannot be modified by the programmer when the 68000 is running in the user mode.

The least significant 8 bits of the SR constitute the condition code register (CCR) and indicate certain things about the nature of each arithmetic and logical operation executed by the 68000. The position of the bits of the CCR is defined in figure 2.5.

The carry bit is conventional, as are the V, Z, and N bits, and represents the carry-out of the most significant bit of an operand during an arithmetic or logical operation. The word *operand* rather than *word* is used because the 68000 allows longword, word, and byte operations; therefore, the carry bit represents the carry-out from bits 31, 15, or 7, respectively. The following example should clarify the picture.

An add operation such as ADD.B D0,D1 affects only the 8 low-order bits of D1. The carry resulting from the addition is copied into the carry flag, and bits 8 through 31 of D1 remain unchanged. If [D0] = $12345678 and [D1] = $13579BDF, the action of ADD.B D0,D1 results in [D1] = $13579B57 and the carry flag is set to 1. Had we performed a word operation with ADD.W D0,D1, the carry bit would have been set to the carry-out resulting from bit 15.

Under certain circumstances the X bit, or extend bit, is identical to the carry bit. During an addition, subtraction, negation, or shifting operation, the X bit is made to reflect the state of the carry bit, C. The X bit is included in the condition code byte because it is a "pure extension bit" and is used only when a byte, word, or longword is extended beyond 8, 16, or 32 bits, respectively.

The X bit has been provided because the carry bit is often employed by programmers as a multipurpose test flag. For example, the carry bit is occasionally used to transfer information between subroutines. If C is set following a return from a subroutine, it may indicate that an error occurred in the subroutine. Sometimes the conflict between the use of the C bit as a carry bit acting as the extension bit of the operand during an arithmetic operation and its use as a general-purpose semaphore is inconvenient. For this reason, the X bit is provided exclusively for use in arithmetic operations that generate a true carry-out. Instructions such as CMP, MOVE, AND, OR, CLR, TST, MULU, and DIVU affect the state of the carry bit but have no effect on the state of the X bit.

The overflow flag, V, is set if the result of an arithmetic operation (longword, word, or byte), when interpreted as a two's complement value, yields an incorrect sign bit. The zero flag, Z, is set if a result (longword, word, or byte) is zero. The negative flag is a copy of the most-significant bit of a result (longword, word, or byte).

FIGURE 2.5 68000 status word

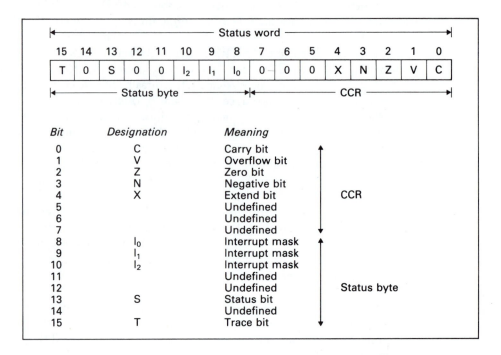

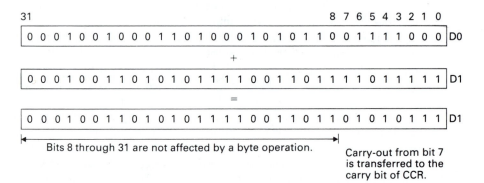

Bits 8 through 31 are not affected by a byte operation.

Carry-out from bit 7 is transferred to the carry bit of CCR.

2.3 ADDRESSING MODES OF THE 68000

In this section, we look at the ways in which the 68000 specifies the address of operands. Most microprocessor instructions have two or three fields. These include the nature of the instruction, the address of the source operand, and the address of the destination operand. The address of an operand can be specified in a number of different ways, collectively called *addressing modes*. The type of sophisticated addressing modes associated with the 68000 provide programmers with a tool for accessing data structures such as tables, arrays, and vectors.

Notation to Define Addressing Modes

Before continuing, we need to develop an unambiguous notation to help us describe information manipulation within the 68000. Such a notation is called *register transfer language* (RTL). Registers are denoted by their names. Square brackets mean "the contents of," so that [D0] means the contents of data register D0. We can therefore write [D4] = 50 to mean that the contents of register D4 is equal to 50.

The base of a number is denoted by a prefix. No prefix indicates decimal, a percent sign (%), binary, and a dollar sign ($), hexadecimal. A 32-bit number is represented by eight hexadecimal characters and a data register holds unsigned integer values in the range $0000 0000 to $FFFF FFFF. For purposes of style, I have left a space between the four least-significant hexadecimal digits of a number and the remaining most-significant digits—except in assembly language programs where numbers must not contain embedded spaces. As stated earlier, addresses are frequently represented by six hexadecimal characters, because A_{24} to A_{31} have no meaning in the system connected to the 68000.

The backwards arrow (←) indicates a transfer of information. Let us consider some examples.

[D4] ← 50	Put 50 into register D4
[D4] ← $1234	Put $1234 into register D4
[D3] ← $FE 1234	Put $FE 1234 into register D3

The following symbols are used in the definition of addressing modes.

SYMBOL	MEANING
M	Location (i.e., address) M in the main store
Ai	Address register (i = 0 to 7)
Di	Data register (i = 0 to 7)
Xi	General register i (Xi may be an address or a data register)
[M]	The contents of memory location M (M is used only to represent a general memory location)
[X]	The contents of register X (X may be an address or a data register)
$[D_i(0:7)]$	Bits zero to seven of register D_i
⟨ ⟩	Enclose a parameter required by an expression
ea	The effective address of an operand
[M(ea)]	The contents of a memory location whose effective address is ea
d8	An 8-bit signed offset (a constant in the range −128 to 127)
d16	A 16-bit signed offset (a constant in the range −32K to 32K)
d32	A 32-bit signed offset (a constant in the range −2G to 2G)

In order to illustrate the action of addressing modes, we introduce two fundamental instructions: ADD and MOVE. The assembly language form of the ADD instruction is ADD ⟨source⟩,⟨destination⟩, and it has the effect

[destination] ← [destination] + [source]

In RTL terms, the assembly language instruction ADD ⟨s⟩,⟨d⟩ is defined as [M(d)] ← [M(s)] + [M(d)]. That is, the contents of "source" are added to the contents of "destination," and the result of their addition becomes the new contents of "destination." Destination and source are both effective addresses. Addressing modes are concerned with the way in which the source and destination addresses of the operands are determined, as we shall soon see.

The second instruction we introduce is MOVE, which has the assembly language form MOVE ⟨source⟩,⟨destination⟩, and has the effect

[destination] ← [source]

The MOVE instruction copies the contents of the effective address specified by "source" into the location specified by "destination." Note that the 68000's MOVE instruction is its most general instruction (in terms of the range of addressing modes it supports) and replaces a whole host of instructions. For example, a typical 8-bit microprocessor has the instructions LDA (load accumulator), STA (store accumulator), and PHA (push accumulator), all of which are replaced by the 68000's MOVE.

Immediate Addressing

In the immediate (or literal) addressing mode, the actual operand follows the instruction and allows constants to be set up at the time the program is written. For example, the instruction ADD.L #9,D0 adds the number 9 to the contents of data register D0. The "9" is an immediate operand because it forms part of the instruction and the CPU does not have to carry out a further memory access to obtain it. Once a constant is defined in the literal mode, it cannot be changed when the program is running. (The constant can be changed if the technique of using self-modifying code is adopted, but wise programmers should never resort to this practice.)

The hash symbol (#) precedes the operand and indicates to the assembler that the following value is to be used with the immediate addressing mode. The 68000 permits byte, word, and longword immediate operands. The following examples illustrate addressing modes in terms of their assembly language form and define them in RTL.

ASSEMBLY LANGUAGE FORM	RTL FORM	ACTION
MOVE.W #$8123,D3	[D3(0:15)] ← $8123	The hexadecimal value $8123 is transferred into the lower-order word of register D3.
MOVE.L #$8123,D3	[D3(0:31)] ← $8123	The hexadecimal value $0000 8123 is transferred into D3.

A typical application of immediate addressing is in setting up control loops. The following example uses immediate addressing several times to preset all the elements of an array of 128 bytes to $FF.

```
        MOVEA.L   #$001000,A0    Load A0 with the address of the array
        MOVE.B    #128,D0        D0 is the element counter (preset to 128)
LOOP    MOVE.B    #$FF,(A0)+     Store $FF in this element and increment pointer
        SUBQ.B    #1,D0          Decrement element counter
        BNE       LOOP           Repeat until all the elements are set to $FF
```

Absolute Addressing

Absolute addressing means that the instruction contains the operand's address and is sometimes called *direct addressing*. Absolute addressing is so called because the actual (i.e., absolute) address of an operand is specified at the time the program is written; this address is constant and is not modified in any way by the processor.

The 68000 provides two variants of absolute addressing: absolute short addressing and absolute long addressing. In absolute short addressing, the address of the operand is a 16-bit word following the instruction. This word is sign extended to 32 bits before it is used to access the operand. Consequently, absolute short addresses in the range $0000 to $7FFF are sign extended to $0000 0000 through $0000 7FFF, while absolute short addresses in the range $8000 through $FFFF are sign extended through $FFFF 8000 to $FFFF FFFF; that is, the programmer can use absolute short addressing to access only the top and the bottom 32K bytes of memory space.

Absolute long addressing requires two 16-bit words following an instruction to generate a 32-bit absolute address. This procedure allows the whole of memory to be accessed. The programmer does not have to worry about long and short forms of addressing modes, as the assembler automatically selects the appropriate version. Some examples of absolute addressing are now provided.

ASSEMBLY LANGUAGE FORM	RTL FORM	ACTION
MOVE.L D3,$1234	$[M(\$1234)] \leftarrow [D3(16{:}31)]$ $[M(\$1236)] \leftarrow [D3(0{:}15)]$	The contents of register D3 are copied into memory location $1234. Because the 68000's memory is byte organized, locations $1234 to $1237 are required.
MOVE.W $1234,D3	$[D3(0{:}15)] \leftarrow [M(\$1234)]$	The contents of memory location $1234 are moved into the lower-order word of D3.

```
PTM  EQU    $FFFFC120
     MOVE.L PTM,D2            [D2] ← [M($FFFFC120)]
```

The contents of memory location $FFFF C120 are moved into register D2. Note that the address $FFFF C120 is stored as $C120 and is automatically sign extended to 32 bits ($FFFF C120).

Absolute addressing is employed when the address of an operand is known at the time of writing the program. This happens under two circumstances. The first corresponds to memory-mapped input/output, where a given memory address is used as an input or output port address. The second is in programs that will never be relocated. That is, when the program and its associated data always occupy the same addresses in memory, irrespective of the machine on which the program is run (or of the operating system or whether the system uses memory management). Whenever possible, 68000 programmers should avoid absolute addressing in order to produce position-independent code (PIC). Position-independent code avoids the use of absolute addressing and can be placed anywhere in memory *without* recomputing the address of operands.

Register Direct Addressing

The register direct addressing mode specifies operands (source or destination) that are the contents of the 68000's internal registers. Register direct addressing does not involve a memory access. The effective address of an operand is given by the name of the register (address or data) specified in the instruction. For example, the instruction MOVE.L D0,D3 copies the entire contents of data register D0 into register D3.

Consider the following examples of register direct addressing.

```
MOVE.L   D0,D3   = [D3] ← [D0]
MOVE.W   D0,D3   = [D3(0:15)] ← [D0(0:15)]
MOVE.B   D0,D3   = [D3(0:7)] ← [D0(0:7)]
MOVE.L   A1,D0   = [D0] ← [A1]
ADD.L    D1,D2   = [D2] ← [D2] + [D1]
ADD.L    #12,D2  = [D2] ← [D2] + 12
```

Note that the general MOVE instruction does not allow the transfer of the contents of a data register into an address register. Thus, although MOVE.L A1,D0 is legal, the inverse operation MOVE.L D1,A0 is not. This philosophy of segregating addresses and data makes it difficult for the programmer to carelessly corrupt an address. We shall soon see that a special instruction, MOVEA, is available for the transfer of information to address registers.

Information within the computer is stored in either memory locations or in registers inside the CPU. Theoretically, it does not matter where data is held, as long

as it is manipulated according to the appropriate algorithm. In practice, the storage of data in on-chip registers is preferred because registers can be accessed faster than memory locations.

Address Register Indirect Addressing

Register indirect addressing means that the address of an operand is in a register. This register is called a "pointer register" and is one of the 68000's eight address registers. In RTL terms, the effective address of an operand is specified by: ea = [Ai]. The following examples illustrate the effect of address register indirect addressing. Address register indirect addressing is specified in assembly language form by enclosing the address register in parentheses. The expression [M([Ai])] is read as "the contents of the memory location whose address is in address register Ai."

ASSEMBLY LANGUAGE FORM	RTL DEFINITION	ACTION
MOVE.L (A0),D3	$[D3] \leftarrow [M([A0])]$	The contents of the memory location whose address is in A0 are copied into register D3.
MOVE.W D4,(A6)	$[M([A6])] \leftarrow [D4(0:15)]$	The lower-order word of D4 is copied into the memory location whose address is in A6.

For example, in figure 2.6 MOVE.W (A0),D0 moves the word $17AB_{16}$ from memory location 1234 into data register D0. As we have already noted, the contents of the address registers cannot themselves be modified by a MOVE operation, because MOVE is explicitly forbidden from acting on the contents of an address register. The

FIGURE 2.6 Address register indirect addressing

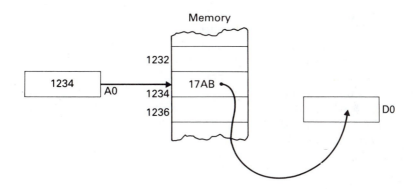

instruction MOVEA is provided to move information into an address register. MOVEA ⟨ea⟩,Ai is defined as

MOVEA: [Ai] ← [ea]

For example, MOVEA.L D4,A3 copies the 32-bit contents of D4 into address register A3. Note that MOVEA.W D4,A3 copies the low-order word of D4 into the low-order word of A3 and then copies the sign bit (i.e., bit 15) into bits 16 through 31 of A3. This instruction is almost the same as MOVE, except that the destination address of the operand must be an address register. There are two other differences between MOVE and MOVEA. The more general operation, MOVE, updates the contents of the condition code register. As MOVEA is used only to generate an address, the contents of the CCR are not affected by MOVEA. Another difference is that MOVE permits a byte operation, while MOVEA is defined only for longwords and words (i.e., MOVEA.L, MOVEA.W).

As stated earlier, the two instructions, MOVE and MOVEA, are provided to force the programmer to appreciate the distinction between addresses and data. However, some assembler writers have circumvented this by allowing the programmer to write MOVE for both MOVE and MOVEA. The assembler itself automatically chooses the appropriate operation code. The philosophy of segregating addresses and data arouses the same passions in some people as seat-belt legislation did in others. Some programmers have told me that they resent being forced to write programs in a "certain" way.

Register indirect addressing provides an efficient method of accessing data because the address of the operand does not have to be read from memory by the instruction that accesses the operand. The address is, of course, already in the CPU in an address register. Consider the following example.

```
              MOVEA.L    #ACIA,A0     Load A0 with the address of
*                                     the ACIA (ACIA is an I/O device)
READ_STATUS   MOVE.B     (A0),D0      Place the contents of the
*                                     location pointed at by A0 in D0
              BTST.B     #0,D0        Test bit 0 of D0 for data ready
              BEQ        READ_STATUS  Repeat until ACIA ready
              MOVE.B     2(A0),D0     Read the input
```

Note that this addressing mode is efficient only if instruction MOVE.B (A0),D0 is executed several times. MOVE.B ACIA,D0 could have been written, but such an instruction would require a memory access to read the address of ACIA every time the instruction is executed. Furthermore, this addressing mode provides a method of generating addresses dynamically during the execution of a program. For example, if MOVEA.L #ACIA,A0 loads the address register A0 with the address of ACIA, then ADDA.L #16,A0 sets up an address in A0 sixteen byte locations onward.

A good example of address register indirect addressing is provided by the *linked list*. The simplest type of singly linked list is composed of a chain of units, each

FIGURE 2.7 A linked list

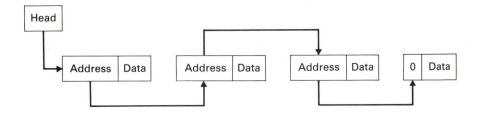

linked to its successor by an address (figure 2.7). The last element in the list has a null (i.e., zero) address.

Suppose we wish to add a new element to the list and insert it at the end. All we have to do is read the address field of the first element to locate the next element. Then we read the address field of this element in order to move to the next element. The list can be traversed in this way until its end is reached. We know that the end of the list has been located when the address of the next element is zero. The following fragment of code is designed to insert a new item into the list. Initially, the longword variable HEAD points to the first item in the list, and the longword variable NEW contains the address of the new item to be inserted.

```
          MOVEA.L    #HEAD,A0    A0 initially points to the start of the list
LOOP      TST.L      (A0)        IF the address field = 0
          BEQ.S      EXIT            THEN exit
          MOVEA.L    (A0),A0         ELSE read the address of the next element
          BRA        LOOP        Continue
EXIT      MOVEA.L    #NEW,A1     Pick up address of new element
          MOVE.L     A1,(A0)     Add new entry to end
          CLR.L      (A1)        Add new terminator
```

In the preceding example, the work is done by the instruction MOVEA.L (A0),A0. Initially address register A0 is pointing to the next element in the list and the source operand effective address (A0) means read the longword pointed at by A0. This longword is, of course, the next address in the chain.

Address Register Indirect with Postincrement Addressing

Register indirect with postincrement addressing is a variation of address register indirect addressing and is also called address register indirect with autoincrementing addressing. The basic operation is the same as address register indirect, except that the contents of the address register from which the operand address is derived are incremented by 1, 2, or 4 after the instruction has been executed. A byte operand causes an increment by 1, a word operand by 2, and a longword by 4. An exception to this rule occurs when the stack pointer, A7, is used with byte addressing. The contents of A7 are then automatically incremented by 2 rather than by 1. In this way, the stack pointer always points to an address on a word boundary. Some examples should make this clear.

ASSEMBLY LANGUAGE FORM	RTL DEFINITION	ACTION
MOVE.L (A0)+,D3	[D3] ← [M([A0])] [A0] ← [A0]+4	The contents of the memory location whose address is in A0 are copied into register D3. The contents of A0 are then increased by 4.
MOVE.W (A7)+,D4	[D4(0: 15)] ← [M([A7])] [A7] ← [A7]+2	The 16-bit contents of the memory location whose address is in A7 are copied into the lower-order 16 bits of D4. The contents of A7 are then increased by 2.
MOVE.B (A7)+,D4	[D4(0:7)] ← [M([A7])] [A7] ← [A7]+2	The 8-bit contents of the memory location whose address is in A7 are copied into the lower-order 8 bits of D4. The contents of A7 are then increased by 2, rather than by 1, because A7 is the stack pointer.

One application of this addressing mode is accessing a data structure where the individual elements are stored consecutively. For example, consider the following fragment of a program designed to fill a 16-element array of longwords (called BUFFER) with zeros.

```
        MOVE.B    #16,D0      Set up a counter for sixteen elements
        MOVEA.L   #BUFFER,A0  A0 points to the first element of the array
LOOP    CLR.L     (A0)+       Clear element and move pointer to next element
        SUBQ.B    #1,D0       Decrement the element counter
        BNE       LOOP        Repeat until the count is zero
```

Address Register Indirect with Predecrement Addressing

This variant of address register indirect addressing is similar to the preceding one except that the specified address register is decremented *before* the instruction is carried out. The decrement is also by 4, 2, or 1, depending on whether the operand is a longword, a word, or a byte, respectively. The two following examples demonstrate how this address mode differs from address register indirect addressing with postincrement.

ASSEMBLY LANGUAGE FORM	RTL DEFINITION	ACTION
MOVE.L — (A0),D3	$[A0] \leftarrow [A0] - 4$ $[D3] \leftarrow [M([A0])]$	The contents of address register A0 are first decremented by 4. The contents of the memory location pointed at by A0 are then moved into register D3.
MOVE.W — (A7),D4	$[A7] \leftarrow [A7] - 2$ $[D4(0:15)] \leftarrow [M([A7])]$	The contents of address register A7 are first decremented by 2. The contents of the memory location pointed at by A7 are moved into register D4.

By means of its autoincrementing and autodecrementing modes, the 68000 is endowed with eight stack pointers—A0 through A7. If the stack is considered to grow toward lower addresses, the PUSH operation is implemented by predecrementing and storing and the PULL operation by reading and postincrementing.

The operations necessary to push the entire contents of D0 and the lower-order word of D1 onto the stack pointed at by A4 are

```
MOVE.L      D0, — (A4)
MOVE.W      D1, — (A4)
```

If the contents of the 16-bit word on the top of this stack are to be pulled and stored in D6, the following operation may be used:

```
MOVE.W      (A4) + ,D6.
```

Autoincrementing and autodecrementing addressing modes are widely used to deal with data in tabular form (lists or arrays). For example, if sixteen words of data are stored consecutively in memory with their starting address in address register A0, they can be added together by executing the instruction ADD.W (A0)+,D0 sixteen times.

Let us suppose that there are two tables, each N bytes long. The following program compares the contents of the tables, element by element, to determine whether they are identical. The "work" is done by a CMPM (A0)+,(A1)+ instruction that compares the contents of the location pointed at by A0 with the contents of the location pointed at by A1 and then increments both pointers. CMPM stands for "compare memory with memory."

```
TABLE_1      EQU      $002000      Location of Table 1
TABLE_2      EQU      $003000      Location of Table 2
N            EQU      $30          Forty-eight elements in each table
               .
               .
               .
             MOVEA.L  #TABLE_1,A0  A0 points to the top of Table 1
             MOVEA.L  #TABLE_2,A1  A1 points to the top of Table 2
             MOVE.B   #N,D0        D0 is the element counter
```

NEXT_ELEMENT	CMPM.B	(A0)+,(A1)+	Compare a pair of elements
	BNE	FAIL	If not the same then exit to FAIL
	SUBQ.B	#1,D0	Else decrement element counter
	BNE	NEXT_ELEMENT	Repeat until all done
SUCCESS	.		Deal with success (all matched)
FAIL	.		Deal with fail (not all matched)

Register Indirect with Displacement Addressing

In the register indirect addressing with displacement mode, the effective address of an operand is calculated by adding the contents of an address register to the sign-extended 16-bit displacement word forming part of the instruction. Remember that "sign-extended" means that the 16-bit two's complement displacement is internally transformed into a 32-bit two's complement number, so that it can be added to the 32-bit contents of an address register. The assembly language form of this addressing mode is d16(An). In RTL form, the effective address of an operand is given by ea = d16 + [Ai]. Two examples should make this clear.

ASSEMBLY LANGUAGE FORM	RTL DEFINITION	ACTION
MOVE.L 12(A4),D3	$[D3] \leftarrow [M(12 + [A4])]$	The contents of the memory location whose address is given by the contents of register A4 plus 12 are moved into register D3.
MOVE.W − $04(A1),D0	$[D0] \leftarrow [M([-4 + [A1])]$	The contents of the memory location whose address is given by the contents of register A1 minus 4 are moved to register D0. The offset, −4, is stored as the two's complement value $FFFC.

For example, in figure 2.8 the instruction MOVE.W 8(A0),D0 loads data register D0 with the contents of the memory location whose address is 8 bytes higher than the value of A0. This addressing mode is roughly equivalent to indexed addressing in 8-bit microprocessors. However, its range is limited because the displacement is only 16 bits, rather than the 32 bits required to provide a comprehensive indexed addressing mode; that is, the offset can specify a location +32,767 bytes ahead or −32,768 bytes back from the contents of an address register.

Register indirect addressing with displacement is a useful tool for writing position-independent code. An address register is loaded with the starting position of the data in memory. All data accesses are then made with the effective operand

FIGURE 2.8   Register indirect with displacement addressing

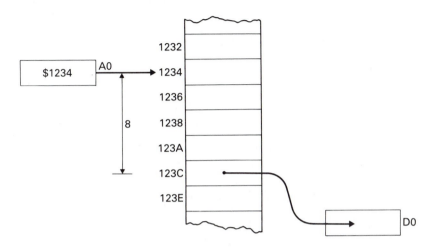

address "offset,(Ai)," where Ai points to the data area and "offset" indicates the location of the operand with respect to the start of the table. The monitor in chapter 11 makes extensive use of register indirect addressing to achieve position independence.

Consider the following use of register indirect addressing to access a look-up table. We can use two techniques to convert a 4-bit hexadecimal digit into its ASCII character form. One is to take an algorithmic approach:

```
HEX:=HEX+$30
IF HEX>$39 THEN HEX:=HEX+7
```

The other conversion technique is to resort to a simple look-up table. In the following example, the longword contents of data register D2 are to be printed as a string of eight ASCII-encoded characters using a look-up table to perform the translation. The subroutine PRINT_CHAR prints the contents of D0.B as an ASCII character.

```
*   D2.L contains the longword to be printed as 8 hex characters
*   D1.W is used as a counter to count the 8 characters
*   D0.B is used to send a character to the print subroutine
*   D3.L is used as a temporary register
*   TRANS is the address of the translation table
*
            MOVE.B    #8,D1            Eight hexadecimal characters to print
LOOP        ROL.L     #4,D2            Move next nibble to least significant position
            MOVE.B    D2,D3            Copy least significant byte to D3
            ANDI.L    #$0000000F,D3    Clear all D3 apart from least significant nibble
            MOVEA.L   D3,A0            A0 now contains least significant nibble
            MOVE.B    TRANS(A0),D0     Read the ASCII character from the table
            BSR       PRINT_CHAR       Display it
            SUB.B     #1,D1            Subtract 1 from loop counter
            BNE       LOOP             Repeat until all eight characters printed
            .
            .
TRANS       DC.B      '0123456789ABCDEF'   Translation table as ASCII string
```

FIGURE 2.9 Using register indirect addressing to access a table

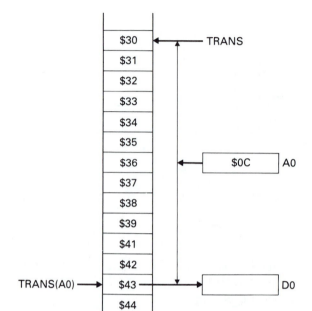

In this example, the ROL.L #4,D2 instruction nondestructively shifts the next nibble into the least significant position. This is copied to D3, masked with $0000 000F to strip it to the 4 least significant bits, and then copied to address register A0. Then it can be used as an offset into the character translation table starting at location TRANS. The appropriate ASCII character is read from the table and sent to the printer. In figure 2.9 A0 contains the value $0C, which is used to index into the table to access location TRANS+$0C. This location contains the hexadecimal value $43, which is the ASCII representation of the character C. Since the 68000's address register indirect addressing mode permits only a 16-bit offset, it follows that the table TRANS must lie in the region of memory from $0000 to $7FFF (or from $FF 8000 to $FF FFFF).

Register Indirect with Index Addressing

Register indirect with index addressing takes the register indirect addressing mode one step further. To form the effective address of an operand the contents of the specified address register are added to the contents of a general register, together with an 8-bit signed displacement. The general register may be an address register or a data register and is termed an index register. The assembly language form of this addressing mode is d8(An,Xn.W) or d8(An,Xn.L). In RTL form, the effective address of an operand is given by ea = d8 + [Ai] + [Xj]. The following example shows how an indexed address is computed.

ASSEMBLY LANGUAGE FORM	RTL DEFINITION	ACTION
MOVE.L 9(A1,D0.W),D3	[D3] ← [M(9+[A1]+[D0(0:15)])]	The contents of the memory location whose effective address is given by the contents of A1 plus the sign-extended contents of the low word of D0 plus a constant, 9, are moved into register D3.

For example, in figure 2.10, the instruction MOVE $18(A0,D0),D2 loads the contents of data register D2 with the contents of the memory location whose effective address is "$12(A0,D0)". This address is given by the contents of A0 plus the contents of D0.W (sign extended to 32 bits) plus the constant 12_{16}.

This is the most general and most complex form of addressing so far encountered. Some notes are needed to bring out its special features.

1. The 8-bit displacement is a signed, two's complement value, offering a range of -128 to $+127$. An 8-bit displacement is permitted simply because

FIGURE 2.10 Indexed addressing

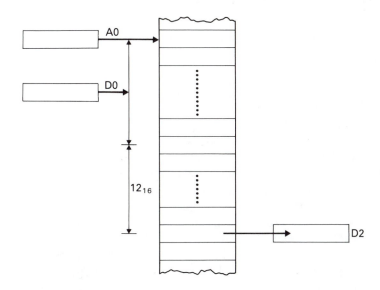

there are only 8 bits in the instruction code left for this purpose after the operation has been specified by the other bits.

2.   The contents of the 32-bit index register may be treated as a 32-bit long-word or a 16-bit word. For example, MOVE.L $12(A1,D0.L),D3 forms the effective address of the source operand by adding the entire contents of D0 to A1 plus $12. The instruction MOVE.L $12(A1,D0.W),D3 forms the effective address of the operand by adding the lower-order contents of D0 (i.e., bits 0 through 15, sign extended to 32 bits) to the contents of A1 plus $12.

This addressing mode is used to handle two-dimensional tables. Let us suppose that we need to access the seventh item in a table of records. If the head of the table is pointed at by A0 and the location of the record (from the start of the table) is in D6, the operation MOVE.L 6(A0,D6.L),D0 will access the required item.

Program Counter Relative Addressing

Program counter relative addressing is very similar to register indirect addressing, except that the address of an operand is specified with respect to the contents of the program counter (PC) rather than with respect to the contents of an address register. Two forms of program counter relative addressing are implemented on the 68000: program counter with displacement and program counter with index. The effective addresses generated by these modes are as follows:

Program counter with displacement: ea = [PC] + d16
Program counter with index: ea = [PC] + [Xn] + d8

The assembly language form of these instructions is LABEL(PC) and LABEL(PC,Xi), respectively. Consider the following application of this addressing mode (see figure 2.11):

```
            MOVE.B     TABLE(PC),D2
                .
                .
                .
   TABLE    DC.B       Value1
            DC.B       Value2
```

The assembler uses the offset TABLE in the instruction MOVE.B TABLE(PC),D2 to calculate the difference between the contents of the program counter and the address of the memory location TABLE. The result gives the 16-bit signed offset, d16, required by the instruction MOVE.B TABLE(PC),D2. When the instruction is executed, the offset is added to the contents of the PC to give the address of TABLE and Value1 is loaded into the lower-order byte of D2.

The power of this addressing mode lies in the fact that it allows the programmer to specify the address of an operand with respect to the program counter; that is, if the program is moved (i.e., relocated) in memory, the address of the operand does not have to be recalculated. Therefore, program counter relative

FIGURE 2.11 Program counter relative addressing

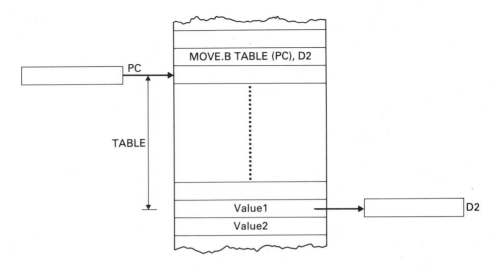

addressing enables the programmer to write position-independent code because the operand, Value1, is always d16 locations onward from the instruction that accesses it. An advantage of this addressing mode is that the resulting position-independent code can be placed in read-only memory and located anywhere in a processor's address space.

However, the 68000 permits only source operands to be specified by program counter relative addressing! Consequently, MOVE LIST(PC),D2 is a legal operation, whereas MOVE D2,LIST(PC) is illegal. It has been argued that program counter relative addressing should not be allowed to *modify* a source operand because this would make self-modifying code easy to write. Therefore, program counter relative addressing can only be used to read constants. We will soon discover how the LEA instruction can be used to generate destination operands with program counter relative addresses.

Permitted Addressing Modes

The 68000 has a very regular architecture in the sense that it has no special-purpose data or address registers that can be used only in conjunction with certain instructions (apart from the status register), although it does not have regular addressing modes. Some instructions can be used with almost all the possible addressing modes, while other instructions are limited to one addressing mode. This situation arises as a result of the limited number of op-code/addressing mode combinations possible with a 16-bit instruction. The chip's designers have attempted to provide the most frequently used instructions with the greatest number of addressing modes. There is no simple way to learn which instruction can be used with what addressing modes. The appendix provides a list of legal addressing modes for each instruction.

The Stack

Because of the 68000's autoincrementing and autodecrementing addressing modes, this CPU has, effectively, eight stack pointers—A0 through A7. Therefore, up to eight stacks can be active at any time. However, when a jump to a subroutine is executed, the return address is saved on the stack pointed at by A7.

The 68000 stack is arranged so that the stack pointer contains the address of the element at the top of the stack. Some processors point to the next free element above the stack. The 68000 stack grows from high to low memory when data is pushed onto it—that is, the stack pointer is decremented before each push. Similarly, the stack pointer is incremented after data has been pulled off the stack. A7 is automatically adjusted by 2 or 4 for word or longword operations, respectively (see figure 2.12).

A word is pushed on the stack by, for example, MOVE.W Dn, −(A7) and pulled off the stack by MOVE.W (A7)+,Dn. The 68000 also permits a word to be pulled from one stack and pushed on to another in one operation by MOVE.W (A3)+, −(A4). In chapter 3 we show how the stack is used in implementing modules and in passing data between modules.

Earlier we explained that the 68000 has *two* A7 registers: the supervisor stack pointer (SSP) and the user stack pointer (USP). The actual stack pointer (i.e., A7) active at any given instant is determined by the operating mode (supervisor or user) of the 68000. When the 68000 is executing a user task, it runs in the user mode and employs the user stack pointer. Any exception, trap, or interrupt (i.e., a call to the operating system) forces the 68000 to switch from user to supervisor mode and to employ the supervisor stack pointer. Consequently, the return from exception

FIGURE 2.12   The 68000's stack

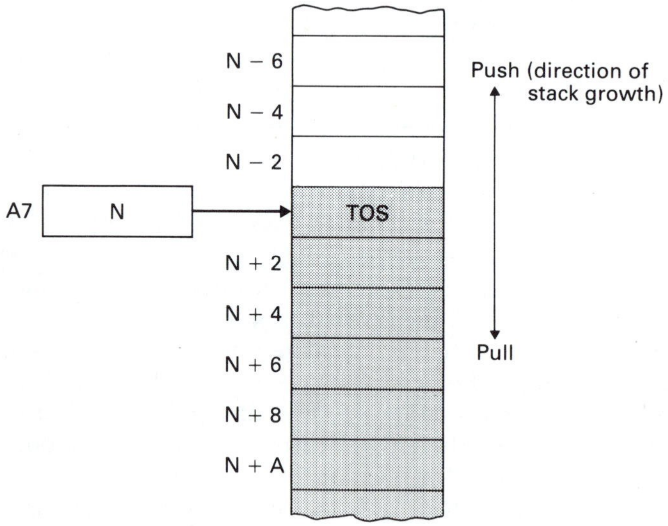

address is stored on the supervisor stack rather than the user stack. This aspect of the 68000 is explained fully in chapter 6.

Using Address Register Indirect Addressing to Access Array Elements

We are now going to look at an application of address register indirect addressing. Some readers may wish to omit this section until after they have read section 2.4. Address register indirect addressing provides a particularly useful tool for the accessing of elements in multidimensional arrays or matrices. The m-row by n-column matrix A can be written in the form

$$
\begin{matrix}
a_{1,1} & a_{1,2} & a_{1,3} & a_{1,4} & \cdots & a_{1,n} \\
a_{2,1} & a_{2,2} & a_{2,3} & a_{2,4} & \cdots & a_{2,n} \\
. & & & & \cdots & . \\
. & & & & \cdots & . \\
. & & & & \cdots & . \\
a_{m,1} & a_{m,2} & a_{m,3} & a_{m,4} & \cdots & a_{m,n}
\end{matrix}
$$

Since memory is essentially a one-dimensional array, the two-dimensional matrix must be mapped onto the memory array. We do this by storing a matrix as a series of rows (or columns), one after the other. If the matrix is stored as sequential rows, we speak of *row order*. If the matrix is stored in row order, its memory map might look like figure 2.13.

Consider the location of element $a_{i,j}$ in matrix A. Assume that the first element (i.e., $a_{1,1}$) is located in memory at address A. Row i starts at location $A + (i - 1)n$. Element $a_{i,j}$ has the address $A + (i - 1)n + j - 1$. The subscripts i and j appear as $(i - 1)$ and $(j - 1)$, respectively, because the array starts at element $a_{1,1}$ rather than $a_{0,0}$.

Suppose we wish to calculate the matrix sum $C = A + B$, where A, B, and C are m-row by n-column matrices and each element is a byte value. The element $c_{i,j}$ is defined as

$$c_{i,j} = a_{i,j} + b_{i,j}$$

We have taken the row and column numbers from 1 to m and 1 to n, respectively, for the sake of simplicity, although it is probably better to use 0 to $m - 1$ and 0 to $n - 1$.

If we use an address register to point to the start of an array, the element offset can be loaded into a data register and indexed addressing can be employed to access the desired element. Consider the following fragment of code designed to calculate the sum $C = A + B$, where A, B, and C are $m \times n$–byte matrices.

```
m    EQU    ⟨m⟩           Number of rows (1 to m)
n    EQU    ⟨n⟩           Number of columns (1 to n)
A    EQU    ⟨address⟩     Start address of array A
B    EQU    ⟨address⟩     Start address of array B
```

```
C         EQU        <address>        Start address of array C
          MOVEA.L    #A,A0            A0 points to base of matrix A
          MOVEA.L    #B,A1            A1 points to base of matrix B
          MOVEA.L    #C,A2            A2 points to base of matrix C
          CLR.W      D2               Clear the element offset
          MOVE.W     #m,D0            D0 is the row counter
L2        MOVE.W     #n,D1            D1 is the column counter
L1        MOVE.B     (A0,D2.W),D6     Get element from A
          ADD.B      (A1,D2.W),D6     Add element from B
          MOVE.B     D6,(A2,D2.W)     Store sum in C
          ADDQ.W     #1,D2            Increment element pointer
          SUB.W      #1,D1            Repeat until n columns added
          BNE        L1
          SUB.W      #1,D0            Repeat until m rows added
          BNE        L2
```

FIGURE 2.13 Storing a matrix in memory

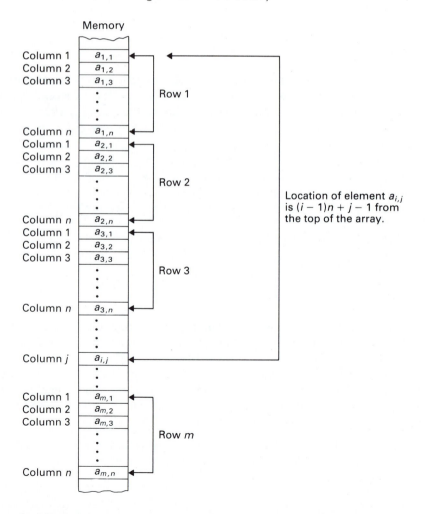

Location of element $a_{i,j}$ is $(i-1)n + j - 1$ from the top of the array.

In the preceding example, it is easy to step through the three arrays element by element simply by incrementing an offset to a pointer (i.e., D2.W). Consider a more general example. Suppose we have to calculate the effective address of element $A_{i,j}$ of m-row by n-column matrix A, where A0 points to the first element of the array, D0 contains i, and D1 contains j. We can use the following code:

```
MOVEA.L    #A,A0      A0 points to base of array A
SUBQ.L     #1,D0      Rows are numbered from 1
MULU       #n,D0      D0 now contains the row offset
SUBQ.L     #1,D1      Columns are numbered from 1
ADD.L      D1,D0      D0 now contains column + row offset
ADDA.L     D0,A0      A0 now points to the i,jth element
```

Note that we have to subtract 1 from the row and column pointers, since the array subscripts are numbered from 1 rather than from 0. If the elements were larger than byte values (e.g., word, longword, or quadword), it would be necessary to scale the values in D0 and D1 accordingly (by 2, 4, or 16, respectively). We have chosen .L operations on data registers D0 and D1 since we later add D0.L to A0.

2.4 INTRODUCTION TO THE 68000 FAMILY INSTRUCTION SET

Any one writing about the 68000 family is faced with a dilemma. Should the writer integrate the 68000 and later members of the family or should he or she cover the 68000 first and then provide details of the 68020 etc. at the end of the chapter? In general, we have chosen the latter approach to avoid burdening the student who, initially, needs to know only about the 68000.

It would be impossible to do justice to the power of the 68000 family without turning this book into an assembly language manual. Therefore, we have attempted to give an overview of the 68000's instruction set but have omitted much of the fine detail. In particular, greater emphasis is placed on the more interesting or unusual aspects of the 68000, as we assume that most readers are already familiar with basic microprocessor architectures. Definitions of the 68000 family instructions are given in the Appendix. As stated earlier, most instructions operate on byte, word, or longword operands and the operand size is specified by a .B, .W, or .L, respectively, after the mnemonic. However, if no size is specified, a .W (word) is taken as the default value.

Instructions can be divided into a relatively small number of groups. One possible grouping is

Data movement
Arithmetic operations
Logical operations
Shift operations
Bit manipulation
Program control

Other groupings are equally acceptable: Some combine logical and shift operations into a single logical group, others split logical operations into a Boolean logical group and a separate group including all shift operations (as we have done), and others include bit manipulation operations within the logical group.

Instructions and the Condition Code Register

After the execution of an instruction, the contents of the condition code byte of the status register are updated. Table 2.1 shows how the condition code is affected by the 68000's instructions.

Data Movement

There are 13 data movement operations in the 68000's instruction set. A data movement instruction simply copies information from one place to another. Such an instruction is hardly exciting, but it has been reported that 70 percent of the average program consists of data movement operations. The 68000 implements the following data movement instructions: MOVE, MOVEA, MOVE to CCR, MOVE to SR, MOVE from SR, MOVE USP, MOVEM, MOVEQ, MOVEP, LEA, PEA, EXG, SWAP. Some of these instructions have already been encountered in the section on addressing modes, but are included here for completeness. Note that certain instructions that affect the status byte of the SR may not be executed when the 68000 is operating in the user mode.

MOVE, MOVEA   The MOVE instruction copies an 8-, 16-, or 32-bit value from one memory location or register to another memory location or register. All the addressing modes discussed so far can be used to specify the source of the data or its destination, with three exceptions: immediate addressing, address register direct addressing, and program counter relative addressing cannot be used to specify a destination. The V and C bits of the condition code register (CCR) are cleared by a MOVE; the N and Z bits are updated according to the value of the destination operand; the X bit is unaffected. The MOVEA instruction permits all source addressing modes but only the address register direct destination addressing mode. The MOVEA instruction, like all other instructions operating on the contents of an address register, does not affect the CCR.

MOVE to CCR   The MOVE to CCR instruction moves data into the condition code register and is a *word operation*. A MOVE ⟨ea⟩,CCR copies the lower-order byte of the operand at the specified effective address into the CCR. The high-order byte of the operand is ignored. This instruction allows the programmer to preset the CCR.

MOVE to SR, MOVE from SR   The first of these MOVE instructions copies a word to the status register. It is a privileged instruction and can be executed only when the 68000 is operating in its supervisor mode. This topic is covered in

TABLE 2.1 Relationship between instructions and the CCR. (Reprinted by permission of Motorola Limited)

MNEMONIC	DESCRIPTION	OPERATION	X	N	Z	V	C
				CONDITION CODES			
ABCD	Add decimal with extend	$(Destination)_{10} + (source)_{10} + x \rightarrow destination$	•	U	•	U	•
ADD	Add binary	$(Destination) + (source) \rightarrow destination$	•	•	•	•	•
ADDA	Add address	$(Destination) + (source) \rightarrow destination$	—	—	—	—	—
ADDI	Add immediate	$(Destination) + immediate\ data \rightarrow destination$	•	•	•	•	•
ADDQ	Add quick	$(Destination) + immediate\ data \rightarrow destination$	•	•	•	•	•
ADDX	Add extended	$(Destination) + (source) + x \rightarrow destination$	•	•	•	•	•
AND	AND logical	$(Destination) \wedge (source) \rightarrow destination$	—	•	•	0	0
ANDI	AND immediate	$(Destination) \wedge immediate\ data \rightarrow destination$	—	•	•	0	0
ASL, ASR	Arithmetic shift	$(Destination)\ shifted\ by\ \langle count \rangle \rightarrow destination$	•	•	•	•	•
B$_{cc}$	Branch conditionally	If $_{cc}$ then PC + d → PC	—	—	—	—	—
BCHG	Test a bit and change	$\sim(\langle bit\ number \rangle)$ OF destination → Z; $\sim(\langle bit\ number \rangle)$ OF destination → $\langle bit\ number \rangle$ OF destination	—	—	•	—	—
BCLR	Test a bit and clear	$\sim(\langle bit\ number \rangle)$ OF destination → Z; 0 → $\langle bit\ number \rangle$ OF destination	—	—	•	—	—
BRA	Branch always	PC + displacement → PC	—	—	—	—	—
BSET	Test a bit and set	$\sim(\langle bit\ number \rangle)$ OF destination → Z; 1 → $\langle bit\ number \rangle$ OF destination	—	—	•	—	—
BSR	Branch to subroutine	PC → − (SP), PC + d → PC	—	—	—	—	—
BTST	Test a bit	$\sim(\langle bit\ number \rangle)$ OF destination → Z	—	—	•	—	—
CHK	Check register against bounds	If Dn $\langle 0$ or Dn$\rangle\langle ea \rangle$ then TRAP	—	•	U	U	U
CLR	Clear an operand	0 → Destination	—	0	1	0	0
CMP	Compare	(Destination) − (source)	—	•	•	•	•
CMPA	Compare address	(Destination) − (source)	—	•	•	•	•
CMPI	Compare immediate	(Destination) − immediate data	—	•	•	•	•
CMPM	Compare memory	(Destination) − (source)	—	•	•	•	•
DB$_{cc}$	Test condition, decrement, and branch	If $\sim$cc then Dn − 1 → Dn; if Dn ≠ −1 then PC + d → PC	—	—	—	—	—

Continued

45

TABLE 2.1 (*Continued*)

MNEMONIC	DESCRIPTION	OPERATION	CONDITION CODES				
			X	N	Z	V	C
DIVS	Signed divide	(Destination)/(source) → destination	—	●	●	●	0
DIVU	Unsigned divide	(Destination)/(source) → destination	—	●	●	●	0
EOR	Exclusive OR logical	(Destination) ⊕ (source) → destination	—	●	●	0	0
EORI	Exclusive OR immediate	(Destination) ⊕ immediate data → destination	—	●	●	0	0
EXG	Exchange register	Rx ⟷ Ry	—	—	—	—	—
EXT	Sign extend	(Destination) sign-extended → destination	—	●	●	0	0
JMP	Jump	Destination → PC	—	—	—	—	—
JSR	Jump to subroutine	PC → − (SP); destination → PC	—	—	—	—	—
LEA	Load effective address	Destination → An	—	—	—	—	—
LINK	Link and allocate	An → − (SP); SP → An; SP + displacement → SP	—	—	—	—	—
LSL, LSR	Logical shift	(Destination) shifted by ⟨count⟩ → destination	●	●	●	0	●
MOVE	Move data from source to destination	(Source) → destination	—	●	●	0	0
MOVE to CCR	Move to condition code	(Source) → CCR	—	●	●	●	●
MOVE to SR	Move to the status register	(Source) → SR	—	●	●	●	●
MOVE from SR	Move from the status register	SR → destination	—	—	—	—	—
MOVE USP	Move user stack pointer	USP → An or An → USP	—	—	—	—	—
MOVEA	Move address	(Source) → destination	—	—	—	—	—
MOVEM	Move multiple registers	Registers → destination; (Source) → registers	—	—	—	—	—
MOVEP	Move peripheral data	(Source) → destination	—	—	—	—	—
MOVEQ	Move quick	Immediate data → destination	—	●	●	0	0
MULS	Signed multiply	(Destination) × (source) → destination	—	●	●	0	0
MULU	Unsigned multiply	(Destination) × (source) → destination	—	●	●	0	0
NBCD	Negate decimal with extend	$0 - (Destination)_{10} - x \rightarrow$ destination	●	U	●	U	●
NEG	Negate	0 − (Destination) → destination	●	●	●	●	●
NEGX	Negate with extend	0 − (Destination) − x → destination	●	●	●	●	●
NOP	No operation		—	—	—	—	—
NOT	Logical complement	~(Destination) → destination	—	●	●	0	0
OR	Inclusive OR logical	(Destination) V (source) → destination	—	●	●	0	0
ORI	Inclusive OR immediate	(Destination) V immediate data → destination	—	●	●	0	0

Mnemonic	Description	Operation	X	N	Z	V	C
PEA	Push effective address	Destination → − (SP)	—	—	—	—	—
RESET	Reset external devices		—	—	—	—	—
ROL, ROR	Rotate (without extend)	(Destination) rotated by ⟨count⟩ → destination	—	●	●	0	●
ROXL, ROXR	Rotate with extend	(Destination) rotated by ⟨count⟩ → destination	●	●	●	0	●
RTE	Return from exception	(SP) + → SR; (SP) + → PC	●	●	●	●	●
RTR	Return and restore condition codes	(SP) + → CC; (SP) + → PC	●	●	●	●	●
RTS	Return from subroutine	(SP) + → PC	—	—	—	—	—
SBCD	Subtract decimal with extend	$(Destination_{10})$ − $(source)_{10}$ − x → destination	●	U	●	U	●
Scc	Set according to condition	If cc then 1's → destination else 0's → destination	—	—	—	—	—
STOP	Load status register and stop	Immediate data → SR; STOP	●	●	●	●	●
SUB	Subtract binary	(Destination) − (source) → destination	●	●	●	●	●
SUBA	Subtract address	(Destination) − (source) → destination	—	—	—	—	—
SUBI	Subtract immediate	(Destination) − immediate data → destination	●	●	●	●	●
SUBQ	Subtract quick	(Destination) − immediate data → destination	●	●	●	●	●
SUBX	Subtract with extend	(Destination) − (source) − x → destination	●	●	●	●	●
SWAP	Swap register halves	Register (31:16) ←→ register (15:0)	—	●	●	0	0
TAS	Test and set an operand	(Destination) tested → CC; 1 → [7] OF destination	—	●	●	0	0
TRAP	Trap	PC → − (SSP); SR → − (SSP); (vector) → PC	—	—	—	—	—
TRAPV	Trap on overflow	If V then TRAP	—	—	—	—	—
TST	Test an operand	(Destination) tested → CC	—	●	●	0	0
UNLK	Unlink	An → SP; (SP) + → An	—	—	—	—	—

⊕ Logical exclusive OR   ● Affected
∧ Logical AND   — Unaffected
∨ Logical OR   0 Cleared
~ Logical complement   1 Set
              U Undefined

NOTE: The terminology in this table is essentially that of Motorola and sometimes differs from the conventions adopted elsewhere in the text.

chapter 6 and is entirely irrelevant to the basic operation of the 68000. The MOVE from SR instruction allows the contents of the processor status word to be examined. Note that this is a privileged instruction in the 68010/20/30's instruction set but not in the 68000's. These two instructions are executed only by operating system software and are not required by user (i.e., applications) programmers. Their assembly language forms are MOVE ⟨ea⟩,SR and MOVE SR,⟨ea⟩.

MOVE USP   As already stated, the 68000 has two A7 registers: one associated with the user mode and one with the supervisor mode. These A7s are called USP (user stack pointer) and SSP (supervisor stack pointer), respectively, whenever it is necessary to distinguish between them. When the 68000 is operating in the supervisor mode, the instructions MOVE.L USP,An and MOVE.L An,USP transfer the USP to address register An, and vice versa, thereby allowing the operating system to manipulate the user stack. When the 68000 is in the user mode, the SSP is entirely hidden from the user and cannot be accessed.

MOVEM   The move multiple registers instruction offers the programmer a very simple way of transferring a group of the 68000's registers to or from memory with a single instruction. Its assembly language form is MOVEM ⟨register list⟩,⟨ea⟩ or MOVEM ⟨ea⟩,⟨register list⟩. MOVEM operates only on words or longwords. The effect of MOVEM is to transfer the contents of the group of registers specified by the "register list" (described later) to consecutive memory locations or to restore them from consecutive memory locations. Programmers use this instruction to save working registers on entering a subroutine and to retrieve them at the end of the subroutine. The contents of the CCR are not affected by a MOVEM.

The register list is defined as Ai–Aj/Dp–Dq. For example, A0–A4/D3–D7 specifies address registers A0 through A4 and data registers D3 through D7, inclusive.

The instruction MOVEM.L D0–D7/A0–A6, –(SP) pushes all the data registers and address registers A0 through A6 onto the stack pointed at by A7. The instruction MOVEM.L (SP)+,D0–D7/A0–A6 has the reverse effect and pulls the registers off the stack. The autodecrementing addressing mode is used to specify the destinations of the register and the autoincrementing addressing mode is used to specify the source of the registers.

MOVEQ   The MOVEQ (move quick) instruction is intended to move a 32-bit literal value in the range −128 to +127 to one of the eight data registers. The data moved is a byte which is sign extended to 32 bits. Therefore, although this instruction moves an 8-bit value, it yields a 32-bit result. For example, the operation MOVEQ # −3,D2 has the effect of loading the value $FFFF FFFD into data register D2.

MOVEP   The move peripheral instruction copies words or longwords to an 8-bit peripheral. Byte-oriented peripherals are connected to the 68000's data bus in such a way that *consecutive* bytes in the peripheral are mapped onto successive odd (or even) addresses in the 68000's memory space. For example, a peripheral with four internal registers may have the memory map in figure 2.14.

FIGURE 2.14 Mapping an 8-bit peripheral onto 16-bit memory

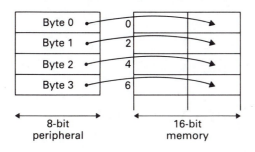

It is impossible to move more than a byte at a time to an 8-bit peripheral by means of a conventional MOVE.W or a MOVE.L because these move a word or a longword to consecutive bytes in memory. The MOVEP instruction is designed to move a word or a longword between a data register in the 68000 and a byte-wide, memory-mapped peripheral. The contents of the chosen register are moved to consecutive *even* (or odd) byte addresses. For example, MOVEP.L D2,0(A0) copies the four bytes in D2 to the addresses given by [A0 + 0], [A0 + 2], [A0 + 4], [A0 + 6]. Only register indirect with displacement addressing is permitted with this instruction. MOVEP does not affect the CCR. The assembler form of MOVEP is

        MOVEP   Dx,d16(Ay)

or

        MOVEP   d16(Ay),Dx

Data bytes are transferred between a data register and alternate bytes of memory, starting at the location specified and incrementing by two. The high-order byte of the data register is transferred first and the low-order byte is transferred last. If the address is even, all transfers are made on the high-order half of the data bus; if the address is odd, all the transfers are made on the low-order half of the data bus. Figure 2.15 shows an example of the instruction MOVEP.L D0,(A0).

LEA The load effective address (LEA) instruction calculates an effective address and loads it into an address register. This instruction can only be used with

FIGURE 2.15 Action of the MOVEP instruction

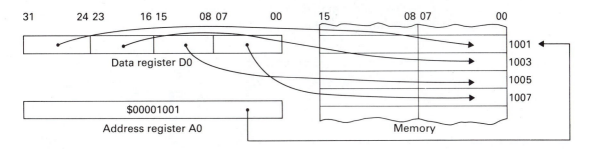

32-bit operands. The LEA instruction is one of the most powerful instructions provided by the 68000. Two examples of its use are now given.

ASSEMBLY LANGUAGE FORM	RTL FORM	ACTION
LEA $0010FFFF,A5	[A5] ← $0010FFFF	Load the address $0010FFFF into register A5.
LEA $12(A0,D4.L),A5	[A5] ← $12 + [A0] + [D4]	The contents of A0 plus the contents of D4 plus $12 are loaded into A5.

In the second example, LEA $12(A0,D4.L),A5, the address evaluated from the expression $12 + [A0] + [D4] is deposited in A5. If the instruction MOVEA.L $12(A0,D4),A5 had been used, the *contents* of that address would have been deposited in A5. The load effective address has been provided to avoid the repeated and time-consuming calculation of effective addresses by the CPU. It is clearly more efficient to put the effective address into an address register by means of a LEA ⟨ea⟩,An instruction than to recalculate the address every time it is used. For example, if the operation ADD.W $1C(A3,D2),D0 is to be repeated many times, it is better to execute a LEA $1C(A3,D2),A5 once and then to repeat ADD.W (A5),D0.

A realistic example of the application of the LEA instruction is given in chapter 3, where we consider the design of a command line interpreter. In that example, a table in memory is pointed to by an address in address register A2. The length of each entry in the table is 6 bytes plus the contents of the first element in the entry. We wish to calculate the address of the next entry. By executing MOVE.B (A2),D1, we can store the length of the current entry (less 6) in D1. Executing a LEA 6(A2,D1.W),A3 then adds the contents of A2 to the contents of the lower-order word of D1 plus 6 and deposits the result in A3—that is, [A3] ← [A2] + [D1(0:15)] + 6. The contents of A3 becomes, of course, the address of the next entry in the table. Note that D1(8:15) must be cleared by CLR.W D1 before carrying out the LEA.

The real importance of the LEA instruction lies in the support it offers for position-independent programming. Remember that program counter relative addressing permits us to use source operands but not destination operands. If we use LEA with program counter relative addressing to generate the address of a distinction operand, this *relative* address is loaded into an address register and can be used with register indirect addressing to achieve position-independent code. The following example shows how this is done.

CASE I		CASE 2	
MOVE.B	TABLE(PC),D0	LEA	TABLE(PC),A0
		MOVE.B	D0,(A0)
.		.	
.		.	
.		.	
TABLE	...	TABLE	...

These examples both generate position-independent code. In case 1, program counter relative addressing is permitted because it is not used to modify a destination operand. In case 2, program counter relative addressing is achieved by loading the relative address into A0.

The LEA ⟨ea⟩,An instruction calculates the effective address ⟨ea⟩ and deposits it in address register An without affecting the contents of the condition code register. Human ingenuity being what it is, some programmers employ the LEA instruction to perform certain arithmetic operations. Consider the effective address d8(Ai,Xj.L) and the instruction LEA d8(Ai,Xj.L),An. This instruction carries out the operation

$$[An] \leftarrow [Ai] + [Xj] + d8$$

In other words, the load effective address instruction permits the addition of the contents of two registers plus an 8-bit signed constant. Moreover, the result is put in a different register (without overwriting the contents of either source register). As you would expect, the contents of the CCR are not modified by an LEA operation.

The use of such techniques to squeeze performance out of a machine is somewhat dubious. On the one hand, the operation does a lot of computation in a single instruction. On the other hand, someone encountering an LEA instruction in a program would expect it to be generating an address (for use as a pointer) and not simply to be used as a vehicle for calculation. Unless it is a life-or-death matter, readability of code always takes precedence over speed.

PEA The push effective address (PEA) instruction calculates an effective address and pushes it onto the stack pointed at by address register A7 (i.e., the stack pointer). The only difference between PEA and LEA is that LEA deposits an effective address in an address register, whereas PEA pushes it onto the stack. Thus, PEA ⟨ea⟩ is equivalent to LEA ⟨ea⟩,Ai followed by MOVE.L Ai, −(SP).

PEA is used to push the address of an operand onto the stack prior to calling a subroutine. The subroutine can read this address from the stack, load it into an address register, and access the actual data by means of address register indirect addressing. Consider the following example. (We will look at subroutines in more detail in Chapter 3.)

```
P1      DS.W       1                Parameter P1 (one word)
        .
        PEA        P1               Push the address of P1 on the stack
        BSR        ABC              Call subroutine 'ABC'
        LEA        4(SP),SP         Increment stack pointer to remove address of P1
        :
ABC     LEA        4(SP),A0         Load the address of P1 into A0
        MOVE.W     (A0),D0          Now read the value of P1
        :
```

The offset 4 in the instruction LEA 4(SP),A0 is necessary because the address of P1 is "buried" on the stack under the return address.

EXG The EXG instruction exchanges the entire 32-bit contents of two registers. In RTL terms, [Xi] ← [Xj]; [Xj] ← [Xi], where Xi and Xj represent any data or address registers. Although EXG allows the contents of two data registers to be exchanged, its main application is in transferring a value calculated in a data register to an address register. The CCR is not affected by an EXG.

SWAP   The SWAP instruction has the assembler form SWAP Dn and exchanges the upper- and lower-order words of a data register. In RTL terms SWAP is expressed as $[Di(16:31)] \leftarrow [Di(0:15)]; [Di(0:15)] \leftarrow [Di(16:31)]$. This instruction has been provided because all operations on 16-bit words act only on bits 0 to 15 of a register. By using a SWAP, the high-order word Di(16:31) can be moved to the low-order word and then operated on by a word-mode instruction. The SWAP instruction affects the CCR in the same way as the MOVE instruction—V and C are cleared, N and Z are updated, and X is unaffected. Note that a byte swap is executed by means of a rotate instruction (defined later). ROL.W #8,Dn exchanges the upper and lower bytes of the lower-order word of D0.

Integer Arithmetic Operations

The 68000 family has a conventional set of arithmetic operations, all of which are integer operations. Except for division, multiplication, and operations whose destination is an address register, all arithmetic operations act on 8-, 16-, or 32-bit entities. The arithmetic group of instructions includes: ADD, ADDA, ADDQ, ADDI, ADDX, CLR, DIVS, DIVU, MULS, MULU, SUB, SUBA, SUBQ, SUBI, SUBX, NEG, NEGX, EXT.

ADD   The ADD instruction adds the contents of a source location to the contents of a destination location and deposits the result in the destination location. Either the source or the destination must be a data register. Memory-to-memory additions are not permitted with this instruction. The ADD instruction cannot be used to modify the contents of an address register.

ADDA   The ADDA instruction is almost identical to the ADD instruction and is necessary whenever the destination of the result is an address register. ADDA must, of course, be used only with word and longword operands and has no effect on the contents of the condition code register. For example, ADDA.L D3,A4 adds the entire contents of D3 to A4 and deposits the results in A4.

ADDQ   The ADDQ (add quick) instruction is designed to add a literal (i.e., constant) value, in the range 1 to 8, to the contents of a memory location or a register. ADDQ may be used with byte, word, and longword operands. Note that ADDQ can also be applied to the contents of an address register. Some readers might find this a little inconsistent. The term "quick" is employed because the instruction format of ADDQ includes the 3-bit constant to be added to the destination operand. Therefore, an ADDQ #4,D1 is executed faster than the corresponding ADD #4,D1.

ADDI   The add immediate (ADDI) instruction adds a literal value of a byte, word, or longword to the contents of a destination operand and then stores the result in the destination. The destination may be a memory location or a data register. Although ADD #$1234,D4 and ADDI #$1234,D4 are almost equivalent, the ADDI and ADD # instructions are coded differently because ADDI permits a literal to be added to the contents of a memory location. For example, ADDI.W #$1234,(A0) adds the constant $1234 to the memory location pointed at by A0.

Some assemblers permit only the mnemonic ADD and then automatically chose the appropriate op-code for ADD, ADDI, ADDA, or ADDQ. For example, if the literal is in the range 1 to 8, the code for ADDQ is selected; otherwise the code for ADD is selected.

ADDX The add extended instruction adds the contents of a source location to the contents of a destination location plus the contents of the X bit of the condition code register and deposits the result in the destination location. Only two addressing modes are permitted by ADDX. Both source and destination must be data registers or they must be memory locations accessed by the address register indirect addressing mode with predecrement. The following three examples should make the operation of ADDX easier to understand.

ASSEMBLY LANGUAGE FORM		RTL FORM
ADDX.L	D3,D4	$[D4] \leftarrow [D3] + [D4] + [X]$
ADDX.B	D3,D4	$[D4(0:7)] \leftarrow [D3(0:7)] + [D4(0:7)] + [X]$
ADDX.L	$-(A3),-(A4)$	$[A3] \leftarrow [A3] - 4; [A4] \leftarrow [A4] - 4$ $[M([A4])] \leftarrow [M([A3])] + [M([A4])] + [X]$

The ADDX instruction is used to perform multiprecision addition. For example, if a 64-bit integer is stored in D0 and D1 (with D1 holding the most significant 32 bits) and another 64-bit integer is held in D2 and D3 (with D3 holding the most-significant 32 bits), the following operations perform the 64-bit addition.

ADD.L	D0,D2	$[D2] \leftarrow [D2] + [D0]$	Add low-order longwords
ADDX.L	D1,D3	$[D3] \leftarrow [D3] + [D1] + [X]$	Add high-order longwords together with a carry-in

The first operation, ADD.L D0,D2, adds together the two lower-order 32-bit longwords. Any carry-out from the most-significant bit position (i.e., into bit 32) is stored in the X bit. When the two higher-order 32-bit longwords are added together by ADDX.L D1,D3, the carry-out recorded by the X bit is added to their sum.

CLR The clear instruction loads the contents of the specified data register or memory location with zero. As no explicit instruction clears the contents of an address register, SUBA.L An,An will clear the contents of An.

DIVS, DIVU The two operations DIVS and DIVU carry out integer division. The assembly language form is DIVU ⟨ea⟩,Dn (or DIVS ⟨ea⟩,Dn). The 32-bit longword in data register Dn is divided by the 16-bit word at the effective address given in the instruction. The quotient is a 16-bit value and is deposited in the lower-order word of Dn. The remainder is stored in the upper-order word of Dn. DIVU performs unsigned division and DIVS operates on two's complement numbers.

For example, the operation DIVU #$24,D4 (assuming that [D4] = $0002881E) would yield the result [D4] = $001E1200. The 32-bit contents of D4 are divided by the 16-bit literal $0024 to give $001E1200 in D4, which is interpreted as a 16-bit quotient $1200 in D4(0:15) and a 16-bit remainder $001E in D4(16:31). Since the programmer frequently does not require the remainder, the programmer must remember to clear bits 16 to 31 of the destination register after a division (if, of course, the destination register is later going to take part in 32-bit arithmetic).

Consider the following example of the DIVU instruction in the conversion of a binary value in the range 0 to 255 into three BCD characters. For example, the binary value 11001100_2 would be converted to 0000 0010 0000 0100 (i.e., 204_{10}). The source is D0.B and the result is in D0.W.

```
CLR.L     D1            Clear D1 as DIVU requires a 32-bit dividend.
MOVE.B    D0,D1         Copy the source to D1.
DIVU      #100,D1       Divide D1 by 100 to get 100s digit in D1(0:15).
MOVE.W    D1,D0         Save the 100s digit in D0 (in least significant position).
SWAP      D1            Move remainder in D1(15:31) to low-order word.
AND.L     #$FFFF,D1     Clear most significant word of D1 before division.
DIVU      #10,D1        Divide remainder by 10 to get 10s digit in D1(0:15).
LSL.W     #4,D0         Move 100s digit in result one place left.
OR.W      D1,D0         Insert 10s digit into result.
LSL.W     #4,D0         Move both 100s and 10s digits one place left.
SWAP      D1            Move remainder in D1(15:31) to low-order word.
OR.W      D1,D0         Insert 1s digit in least significant place.
```

If we use the value of 1100110 for D0, we can trace the execution of the preceding code. Note that the values of D0 and D1 in the comment field are the values after the execution of the operation.

```
CLR.L     D1            D0 = 000000CC    D1 = 00000000
MOVE.B    D0,D1         D0 = 000000CC    D1 = 000000CC
DIVU      #100,D1       D0 = 000000CC    D1 = 00040002
MOVE.W    D1,D0         D0 = 00000002    D1 = 00040002
SWAP      D1            D0 = 00000002    D1 = 00020004
AND.L     #$FFFF,D1     D0 = 00000002    D1 = 00000004
DIVU      #10,D1        D0 = 00000002    D1 = 00040000
LSL.W     #4,D0         D0 = 00000020    D1 = 00040000
OR.W      D1,D0         D0 = 00000020    D1 = 00040000
LSL.W     #4,D0         D0 = 00000200    D1 = 00040000
SWAP      D1            D0 = 00000200    D1 = 00000004
OR.W      D1,D0         D0 = 00000204    D1 = 00000004
```

MULS, MULU   As with division operations, two multiplication instructions are available. MULS forms the product of two signed (two's complement integers) and MULU forms the product of two unsigned integers. The assembly language forms of these instructions are MULS ⟨ea⟩,Dn and MULU ⟨ea⟩,Dn. Multiplication is a 16-bit operation that multiplies the low-order 16-bit word in Dn by the 16-bit word at the effective address in the operand. The 32-bit longword product is deposited in Dn.

For example, the operation MULU #$1234,D4 (assuming that [D4] = $12345678) would yield the result [D4] = $06260060, because the 68000 multiplies the literal $1234 by the lower-order word of D4 (i.e., $5678) to give the 32-bit result $06260060. Note that the simple operation MULU.W Dn,Dn will square the low order 16-bit contents of Dn to give a 32-bit result in Dn.

SUB, SUBA, SUBQ, SUBI, SUBX   The operations SUB, SUBA, SUBQ, SUBI, and SUBX are the subtraction equivalents of ADD, ADDA, ADDQ, ADDI, and ADDX, respectively. Each instruction subtracts the source operand from the destination operand and places the result in the destination operand. For example, SUBI.B #$30,D0 is interpreted as $[D0(0:7)] \leftarrow [D0(0:7)] - \$30$.

NEG   The negate instruction subtracts the destination operand from zero and deposits the result at the destination address. NEG is a monadic operation and has the assembly language form NEG ⟨ea⟩. The operand address may be a memory location or a data register but not an address register. This instruction simply forms the two's complement of an operand.

NEGX   The negate with extend instruction forms the two's complement of an operand minus the X bit.

EXT   The sign extend instruction has the assembly language form EXT.W Dn or EXT.L Dn. The former instruction sign extends the low-order byte in Dn to 16 bits by copying Dn(7) to bits Dn(8:15). Similarly, the latter instruction sign extends the low-order word in Dn to 32 bits by copying Dn(15) to bits Dn(16:31).

BCD Arithmetic

Like all conventional computers, the 68000 family uses binary arithmetic and represents signed integers in two's complement form. Since pure "8421-weighted" binary arithmetic is not always convenient in a decimal world, many processors support a limited form of BCD arithmetic. BCD arithmetic avoids the need to convert from decimal form to binary form before carrying out binary arithmetic and then the need to convert from binary to decimal form after the arithmetic.

The 68000's instruction set includes three instructions that support BCD arithmetic: ABCD, SBCD, and NBCD. These three instructions are add, subtract, and negate, respectively, a packed BCD byte (i.e., two BCD digits at a time). As the 68000's BCD instructions are fairly specialized, they do not support a wide range of addressing modes, and ABCD and SBCD can be used only with the data register direct and address register indirect with predecrementing modes—that is, the only two valid addressing modes are

ABCD Di,Dj   and   ABCD   −(Ai),−(Aj)

The ABCD (add BCD with extend) instruction adds two BCD digits packed into a byte together with the X-bit from the CCR. Similarly, the SBCD performs the subtraction of the source operand together with the X-bit from the destination operand (i.e., [destination] ← [destination] − [source] − [X]). The NBCD ⟨ea⟩ instruction subtracts the specified operand from zero together with the X-bit and forms the ten's complement of the operand if X = 0 or the nine's complement if X = 1. Figure 2.16 illustrates the action of these three BCD instructions. In each case the effect of the corresponding instruction is illustrated with numeric values.

Each of these BCD instructions employs the X-bit of the CCR, because they are intended to be used in chained calculations (i.e., operations on a string of BCD

FIGURE 2.16 The effect of the 68000's BCD instructions

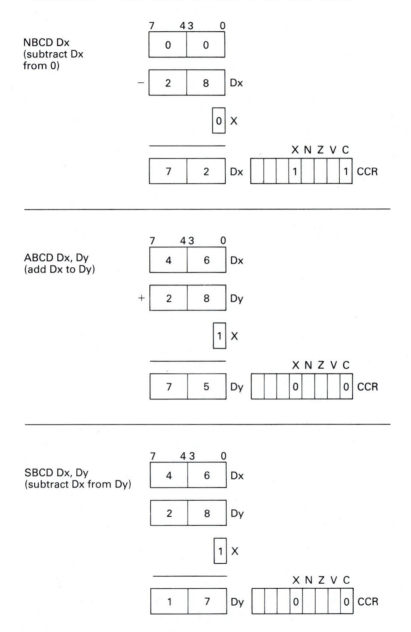

digits). As each pair of digits is added (subtracted), the X-bit records the carry (borrow) and can be used in an operation on the next pair of digits. The BCD instructions operate on the Z-bit (zero bit) of the CCR in a special way. The result of a BCD operation clears the Z-bit if the result is nonzero. If, however, the result is zero, the Z-bit is unaffected. The reason for this behavior is easy to understand if you consider the addition of, say, three pairs of BCD digits:

122056
248023
——————
370079

As you can see, the addition of the digits in the hundreds and thousands columns (i.e., 20 + 80) yields a zero result in these columns, even though the whole result (370079) is not zero.

A simple example will suffice to demonstrate the application of BCD instructions. Assume that two strings of BCD digits are stored in memory. One string starts at location String1 and the other at location String2. When the strings of BCD digits are added together, the destination location is to be String2 (the original String2 is overwritten). Both strings are made up of 12 BCD digits, which requires 6 bytes (since the BCD digits are packed two to a byte). Note also that the strings must be stored so that their least significant byte is at the highest address, because the auto-decrementing mode starts at a high address and moves towards lower addresses. For example, if String1 is 123456123456 and String2 is 001122334455, the result will be String2 = 124578457911.

	MOVE.W	#5,D0	Six bytes in the string to be added
	MOVE.W	#4,CCR	Clear X-bit of CCR and set Z-bit
	LEA	String1+1,A0	A0 points at source string
	LEA	String2+1,A1	A1 points at destination string
LOOP	ABCD	−(A0),−(A1)	Add pair of digits (with any carry-in)
	DBRA	D0,LOOP	Repeat until 6 bytes (12 digits) added

Each time the loop is executed, a pair of digits is added, and any carry is recorded by the X-bit. The X-bit is added to the running total on the next pass around the loop. Note that when the addresses of the two strings are loaded into address registers, the values loaded are the addresses of the strings plus 1 (e.g., LEA String1+1,A0), since the autodecrementing addressing will subtract 1 before it is first used.

Logical Operations

The 68000 implements four Boolean operations: AND, OR, EOR, and NOT. All logical operations can be applied to longword, word, and byte operands. Additionally, logical operations can, with immediate addressing, be applied to the contents of the status register or the condition code register. Operations on the SR are carried out to alter the mode of operation of the 68000 and are privileged. In general, logical operations are used to modify one or more fields of an operand. A logical AND masks out bits (i.e., clears them), an OR sets bits, and EOR toggles them (i.e., causes them to change state). The following instructions illustrate the effect of these logical operations. In each case an immediate operand is used. The low-order byte of D0 before each operation is %11110000. Note that the immediate operands of the above instructions use the mnemonics ANDI, ORI, and EORI. The immediate forms of these instructions are able to specify a data register as an operand or a memory location. Logical operations affect the CCR in exactly the same way as MOVE instructions.

ANDI.B	#%10100110,D0	[D0] ← 10100110 . 11110000
		[D0] ← 10100000
ORI.B	#%10100110,D0	[D0] ← 10100110 + 11110000
		[D0] ← 11110110
EORI.B	#%10100110,D0	[D0] ← 10100110 ⊕ 11110000
		[D0] ← 01010110

The actual logical operations supported by the 68000 in terms of their assembly language forms are as follows.

AND	⟨ea⟩,Dn	
AND	Dn,⟨ea⟩	
ANDI	#⟨data⟩,⟨ea⟩	
ANDI	#⟨data⟩,CCR	
ANDI	#⟨data⟩,SR	(privileged)
EOR	Dn,⟨ea⟩	
EORI	#⟨data⟩,⟨ea⟩	
EORI	#⟨data⟩,CCR	
EORI	#⟨data⟩,SR	(privileged)
NOT	⟨ea⟩	
OR	⟨ea⟩,Dn	
OR	Dn,⟨ea⟩	
ORI	#⟨data⟩,⟨ea⟩	
ORI	#⟨data⟩,CCR	
ORI	#⟨data⟩,SR	(privileged)

An operation labeled "privileged" can be executed only when the 68000 is operating in its supervisor mode (see chapter 6). Note that logical instructions are not entirely symmetric. For example, the operation EOR ⟨ea⟩,Dn is not permitted!

Shift Operations

In a shift operation, all bits of the operand are moved one or more places to the left or right, subject to the variations described below. The 68000 is particularly well endowed with shift operations.

All shifts can be categorized as logical, arithmetic, or circular. In a logical shift, a 0 enters at the input of the shifter and the bit shifted out is clocked into the carry flip-flop. An arithmetic shift left is identical to a logical shift left, but an arithmetic shift right causes the most significant bit, the sign bit, to be propagated right. This action preserves the correct sign of a two's complement value. For example, if the bytes 00101010 and 10101010 are shifted one place to the right (arithmetically), the results are 00010101 and 11010101, respectively. In a circular shift, the bit shifted out is moved to the position of the bit shifted in. No bit is lost during a circular shift.

In figure 2.17 note that an arithmetic shift left and a logical shift left operation are virtually identical. In each case, all the bits are shifted one place left. The bit

FIGURE 2.17 68000's shift and rotate instructions

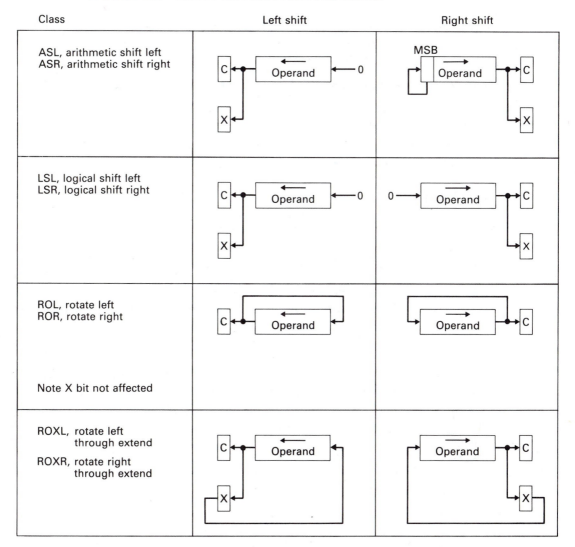

shifted out enters the carry bit and extend bit of the CCR and a 0 enters the vacated position (the least significant bit). There is, however, one small difference between an ASL and an LSL. Since an arithmetic left shift multiplies a number by 2, it is possible for the most significant bit of the value being shifted to change sign and therefore generate an arithmetic overflow. The V bit of the CCR is set if this event occurs during an ASL. Since logical operations are applied to strings of bits, an LSL instruction clears the V bit (since arithmetic overflow is meaningless).

The 68000 has eight shift operations in its instruction set, illustrated in figure 2.17. The symbol C denotes the carry bit of the condition code register; X means the extend bit of the CCR.

Arithmetic shifts update *all* bits of the CCR. The N and Z bits are set or cleared as we would expect. The V bit is set if the most significant bit of the operand is changed at any time during the shift operation. The C and X bits are set according to the last bit shifted out of the operand. However, if the shift count is zero, C is cleared and X is unaffected. Logical shifts and rotates clear the V bit.

Assembly Language Form of Shift Operations

All eight shift instructions are expressed in one of three ways. These are illustrated by the ASL (arithmetic shift left) instruction.

Mode 1.	ASL Dx,Dy	Shift Dy by Dx bits
Mode 2.	ASL #⟨data⟩,Dy	Shift Dy by #data bits
Mode 3.	ASL ⟨ea⟩	Shift the contents of ea by one place

A shift instruction can be applied to a byte, word, or longword operand, with the exception of mode 3 shifts, which act only on words.

In mode 1, the "source" operand, Dx, specifies how many places the destination operand, Dy, needs to be shifted. Dy may be shifted by 1 to 32 bits. In mode 2, the literal, #⟨data⟩, specifies how many places Dy needs to be shifted; this must be in the range 1 to 8. In mode 3, the memory location specified by the effective address, ⟨ea⟩, is shifted one place. Many microprocessors permit only the *static* shifts of modes 2 and 3. The 68000 permits *dynamic* shifts (i.e., mode 1) because the number of bits to be shifted is computed at run-time.

Bit Manipulation

The 68000 provides four instructions that act on a single bit of an operand rather than on the entire operand. In each of the bit manipulation instructions, the complement of the selected bit is moved to the Z bit of the CCR and then the bit is either unchanged, set, cleared, or toggled. The N, V, C, and X bits of the CCR are not affected by bit operations. Bit manipulations may be applied to a byte or to a longword. Even though bit manipulation instructions act on just one bit, they act on either one bit of a specified byte in memory or on one bit of a longword in a data register. The four instructions in this group are BTST, BSET, BCLR, and BCHG.

BTST   The BTST instruction tests a specified bit of an operand. If the bit is zero, set the Z bit of the condition code register. A bit test does not affect the value of the operand under test in any way.

BSET   The bit test and set instruction causes the Z bit of the CCR to be set if the specified bit is zero and then forces the specified bit of the operand to be set to 1.

BCLR   The bit test and clear instruction works exactly like BSET, except that the specified bit is cleared after it has been tested.

BCHG   The bit test and change instruction causes the value of the specified bit to be reflected in the Z bit of the CCR and then toggles the state of the specified bit.

In order to apply these four instructions, two items of information are needed by the assembler: the effective address of the operand and the position of the bit to be tested. All addressing modes are available to these four instructions for the destination operand except immediate addressing, address register direct, and program counter relative addressing. The location of the bit to be tested is specified in one of two ways: it may be provided in an absolute (i.e., constant or static) form in the instruction or given as the contents of a data register. In the following example, BTST is used as an illustration, but any of the members of this group could have been chosen.

The assembly language forms of these instructions are

```
BTST      Dn,<ea>           e.g., BTST D3,$1234
BTST      #<data>,<ea>      e.g., BTST #4,D6
```

If the destination address is a memory location, the source operand is treated as a modulo-8 value. If the destination address is a data register, the source operand is treated as a modulo-32 value. As an example of bit manipulation, consider a subroutine to count the number of ones in a byte. On entry to the subroutine, D0(0:7) contains the byte to be tested and on exit it contains the number of ones in the byte. No other registers must be modified by the subroutine.

```
*   D0 = input/output register (only D0.B modified)
*   D1 = one's counter (not modified by subroutine)
*   D2 = pointer to bit of D0 to be tested (not modified by subroutine)
*
ONES_COUNT   MOVEM.L   D1-D2, -(A7)    Save working registers
             CLR.B     D1              Clear one's counter
             MOVEQ     #7,D2           D2 initially points at msb of D0.B
NEXT_BIT     BTST      D2,D0           Test the D2. th bit of D0
             BEQ.S     LOOP_TEST       If zero then nothing more to do
             ADDQ.B    #1,D1           Else increment one's count
LOOP_TEST    SUBQ.B    #1,D2           Decrement bit pointer
             BGE       NEXT_BIT        Repeat until count negative
             MOVE.B    D1,D0           Transfer one's count to D0
             MOVEM.L   (A7)+,D1-D2     Restore working registers
             RTS                       Return
*
```

2.5 PROGRAM CONTROL AND THE 68000 FAMILY

The computational power of all computers lies in their ability to choose between two or more courses of action on the basis of the available information. Without such powers of decision, the computer would be almost worthless.

Compare Instructions

A computer chooses between two courses of action by examining the state of one or more bits in its CCR and associating one action with one outcome of the test and another action with the other outcome. Although the CCR bits are updated after certain instructions have been executed, two instructions can be used to explicitly update the CCR. These are the bit test (defined previously) and the compare.

CMP, CMPA, CMPI, CMPM   The compare group of instructions subtract the contents of one register (or memory location) from another register (or memory location) and update the contents of the condition code register accordingly. The N, Z, V, and C bits of the CCR are all updated and the X bit remains unaffected. The difference between the subtract and compare operation is that a subtraction evaluates $P = R - Q$ and replaces R with P, while the compare operation merely evaluates $R - Q$ and does not modify R. All instructions in this group, except CMPA, take byte, word, or longword operands.

The basic compare instruction, CMP, compares the source operand with the destination operand. The destination operand must be a data register and the source operand may be specified by any of the 68000's addressing modes. For example:

CMP.L	D0,D1	Evaluates [D1] — [D0]
CMP.B	TEMP1,D3	Evaluates [D3(0:7)] — [TEMP1]
CMP.L	TEMP1,D3	Evaluates [D3(0:31)] — [TEMP1]
CMP.W	(A3),D2	Evaluates [D2(0:15)] — [M([A3])]

Note that CMP ⟨ea1⟩,⟨ea2⟩ evaluates [ea2] — [ea1] so that the *first* operand is subtracted from the *second*. Some other microprocessors perform this operation in the reverse order. CMP has three variations on its basic form: CMPI, CMPA, and CMPM. CMPI is used to execute a comparison with a literal and is written CMPI #⟨data⟩,⟨ea⟩. Like the corresponding ADDI, the CMPI instruction allows a literal operation to be performed with the contents of a memory location. Consider two examples of CMPI operations.

CMPI.B	#$07,D3	Evaluates [D3(0:7)] — 7
CMPI.W	#$07,TEMP	Evaluates [TEMP] — 7

CMPA means "compare address" and is necessary whenever an address register is compared with the contents of an effective address. Its assembly language form is CMPA ⟨ea⟩,An and it operates only on word and longword values.

CMPM means "compare memory with memory" and is one of the few instructions that permits a memory-to-memory operation. Only one addressing mode is allowed with CMPM: register indirect with autoincrementing. The assembler form of this instruction is, therefore, CMPM (Ai)+,(Aj)+. This instruction is used to compare the contents of two tables, element by element.

A typical application of the CMPM instruction is in comparing two blocks of memory for equality, as the following example demonstrates.

BLOCK1	EQU	⟨address 1⟩	
BLOCK2	EQU	⟨address 2⟩	
SIZE	EQU	⟨number of words in block⟩	
*			
	LEA	BLOCK1,A0	A0 points to first block
	LEA	BLOCK2,A1	A1 points to second block
	MOVE.W	#SIZE-1,D0	D0 is the word counter
LOOP	CMPM.W	(A0)+,(A1)+	Compare a pair of words
	BNE	NOT_SAME	IF not same THEN exit
	DBRA	D0,LOOP	ELSE repeat until all done
ALL_SAME		exit here if they are the same	
	⋮		
NOT_SAME		exit here if they are different	

Branch Instructions

The 68000 provides the programmer with a toolkit containing three instructions for the implementation of all conditional control structures. These instructions are

Bcc ⟨label⟩	Branch to label on condition cc true
BRA ⟨label⟩	Branch to label unconditionally
DBcc Dn,⟨label⟩	Test condition cc, decrement, and branch

Of these three instructions, the first two are entirely conventional and are found on all microprocessors. The last instruction is more unusual and is not provided by most microprocessors.

Branch Conditionally Fourteen versions of the Bcc d8 or Bcc d16 instruction exist, where cc stands for one of fourteen logical conditions. If the specified condition cc is true, a branch is made to the instruction whose address is d8 or d16 locations onward from the start of the next instruction. The number of locations branched is specified by either an 8-bit value, d8, or a 16-bit value, d16. The displacement, d8 or d16, forms part of the instruction and is an 8- or 16-bit signed two's complement value, permitting a branch forward or backward from the location following the instruction Bcc d8 or Bcc d16. An 8-bit signed offset allows a branch of $+127$ bytes forward or -128 bytes backward and a 16-bit offset provides a range of $+32,767$ bytes forward and $-32,768$ bytes backward.

The assembler sometimes automatically selects the short 8-bit displacement or the long 16-bit displacement according to the distance to be branched. Otherwise the programmer must write Bcc.S d8 to force a short 8-bit branch, as the extension .S selects the 8-bit displacement. Note that the value of d8 or d16 is automatically calculated by the assembler, as the displacement is invariably in the form of a label rather than an address.

Table 2.2 defines the fourteen possible values of cc. After an arithmetic or logical operation is carried out (together with certain other operations), the values of the Z, N, C, and C flags in the condition code register are updated accordingly. These flag bits are then used to determine whether the appropriate logical condition is true or false. For example, BCS LABEL causes the state of the carry bit to be tested. If the bit is set (i.e., 1), a branch to LABEL (i.e., the point in the program labeled LABEL) is made. Otherwise, the instruction immediately following the BCS LABEL is executed.

You should take care with the conditions PL (plus) and MI (minus) that test the sign of the N-bit. If a two's complement operation is performed, the sign bit will change if arithmetic overflow occurs. The PL and MI conditions are best reserved for testing the most significant bit of an operand when it is used as a logical flag.

Information stored and manipulated by the computer is often in an unsigned integer form or in two's complement form. Consequently, some conditional tests are intended to be applied after operations on two's complement values, while others are applied after operations on integer (or any other non–two's complement) values. To illustrate this point, consider the following two examples.

TABLE 2.2 Conditional tests and the 68000

MNEMONIC (cc)	CONDITION	FLAGS TESTED	BRANCH TAKEN IF:
CC	Carry clear	C	$C = 0$
CS	Carry set	C	$C = 1$
NE	Not equal	Z	$Z = 0$
EQ	Equal	Z	$Z = 1$
PL	Plus	N	$N = 0$
MI	Minus	N	$N = 1$
HI	Higher than	C, Z	$\bar{C}.\bar{Z} = 1$
LS	Lower than or same as	C, Z	$C + Z = 1$
GT	Greater than	Z, N, V	$N.V\bar{Z} + \bar{N}\bar{V}\bar{Z} = 1$
LT	Less than	N, V	$N.\bar{V} + \bar{N}.V = 1$
GE	Greater than or equal to	N, V	$N.\bar{V} + \bar{N}.V = 0$
LE	Less than or equal to	Z, N, V	$Z + (\bar{N}.V + N.\bar{V}) = 1$
VC	Overflow clear	V	$V = 0$
VS	Overflow set	V	$V = 1$
T	Always true	None	Always
F	Always false	None	Never

NOTE: Some of these tests are designed to operate on unsigned values (HI, LS) and some on signed, two's complement values (PL, MI, GT, LT, GE, LE).

CASE 1		CASE 2	
ADD.L	D0,D1	ADD.L	D0,D1
BCS	ERROR	BVS	ERROR
ERROR...		ERROR...	

Both of these cases add the contents of D0 to D1 and deposit the result in D1. However, in case 1 the numbers are interpreted as being in unsigned integer form. If, when adding two 32-bit integers, a carry is generated out of the most-significant bit position, the carry flag is set. The instruction BCS ERROR causes a branch to ERROR to be made if a carry-out occurred. The part of the program labeled ERROR can deal with (i.e., recover from) the out-of-range condition.

Case 2 considers both numbers to be in two's complement form. After the addition has been completed, the state of the overflow flag is tested and a branch to ERROR is made if overflow occurred during the addition.

Example of Use of Conditional Instructions

As an example of the application of conditional branch instructions, consider the conversion of hexadecimal values to their ASCII character equivalents. Table 9.1 in chapter 9 gives the relationship between ASCII-encoded characters and their binary or hexadecimal equivalents. An excerpt from this table is provided for reference:

ASCII CHARACTER	HEXADECIMAL CODE
0	30
1	31
2	32
3	33
4	34
5	35
6	36
7	37
8	38
9	39
A	41
B	42
C	43
D	44
E	45
F	46

The algorithm for the conversion of a hexadecimal value into its ASCII-encoded equivalent can readily be derived from this table. In the following, HEX represents a single hexadecimal number and CHAR the ASCII-encoded character equivalent. Thus, if HEX = $0A, the corresponding value of CHAR is $41. By inspecting the preceding table, we can derive a relationship between CHAR and HEX:

```
CHAR: = HEX + $30
IF HEX > $39 THEN CHAR: = CHAR + $7.
```

In terms of 68000 assembly language, this algorithm can now be written as follows.

```
          MOVE.B    HEX,D0      Get HEX value to be converted into D0
          ADDI.B    #$30,D0     Add $30 to it
          CMPI.B    #$39,D0     Test for hexadecimal values in the range $0A to $0F
          BLS.S     EXIT        If not in range $0A to $0F then exit
          ADDQ.B    #$07,D0     Else add 7
EXIT      MOVE.B    D0,CHAR     Save result in CHAR
```

Branch Unconditionally   The branch unconditional instruction, BRA, causes a branch to the instruction whose address is marked by the label following the BRA mnemonic. An 8-bit or 16-bit signed offset follows the op-code for BRA, providing a branching range of up to 32K bytes. The unconditional branch is equivalent to the GOTO instruction in high-level languages. The 68000 also has a jump instruction, JMP, which is functionally equivalent to the branch instruction BRA. The only difference between the two instructions is that BRA uses relative addressing while JMP uses the following addressing modes:

```
JMP     (An)
JMP     d16(An)
JMP     d8(An, Xi)
```

```
JMP       Absolute_address
JMP       d16(PC)
JMP       d8(PC, Xi)
```

As an example of the application of an unconditional branch, consider the implementation of the CASE statement that is found in many high-level languages. In the program below, the variable TEST contains the integer used to determine which of three courses of action (labeled ACT1, ACT2, ACT3) is to be carried out. If TEST contains a value greater than 2, an exception is raised.

```
CASE         MOVE.B    TEST,D0    Put the value of TEST in D0
             BEQ.S     ACT1       If zero then carry out ACT1
             SUBQ.B    #1,D0      Decrement TEST
             BEQ.S     ACT2       If zero then carry out ACT2
             SUBQ.B    #1,D0      Decrement TEST
             BEQ.S     ACT3       If zero then carry out ACT3
EXCEPTION    . . .                Else deal with the exception
             BRA.S     EXIT       Leave CASE

ACT1         . . .                Execute action 1
             BRA.S     EXIT       Leave CASE

ACT2         . . .                Execute action 2
             BRA.S     EXIT       Leave CASE

ACT3         . . .                Execute action 3
             . . .
             . . .
                                  (Fall through to exit)
EXIT         . . .                Single exit point for CASE
```

This method of implementing a CASE statement is not unique and would not be used if there were many more possible values of TEST. A better method is to use a JMP with a computed address such as JMP d8(A0, D3), where D3 contains a value that is a function of TEST.

Test Condition, Decrement, and Branch   The DBcc instruction is not found in 8-bit microprocessors and provides a powerful way of implementing loop mechanisms. As in the case of the Bcc instruction, there are fourteen possible computed values of cc plus the two static (i.e., constant) values, cc = T and cc = F. When cc = T (i.e., DBT), the tested condition is always true, and when cc = F (i.e., DBF), the tested condition is always false.

The DBcc instruction has the assembly language form DBcc Dn,⟨label⟩, where Dn is one of the eight data registers and ⟨label⟩ is a label used to specify a branch address. When the 68000 encounters a DBcc instruction, it first carries out the test defined by the cc field. If the result of the test is true, the branch is not taken and the next instruction in sequence is executed. Note that this has the opposite effect to a Bcc instruction. The branch is limited to a 16-bit displacement.

If the specified condition, cc, is not true, the low-order 16 bits of Dn are decremented by 1. If the resulting contents of Dn are equal to -1, the next instruction in sequence is executed. Otherwise a branch to ⟨label⟩ is made. The DBcc instruction can be defined as follows.

```
DBcc Dn,⟨label⟩:        If cc TRUE      THEN EXIT
                                        ELSE
                                        BEGIN
                                        [Dn]: = [Dn] − 1
                                        IF [Dn] = −1 THEN EXIT
                                                        ELSE [PC] ← label

                                        END_IF
                                        END

                        END_IF
                        EXIT
```

Unlike the Bcc instruction, DBcc allows the condition F (i.e., false) to be specified by cc. For example, DBF Dn,⟨label⟩ always cause Dn to be decremented and a branch made to ⟨label⟩ until the contents of Dn are −1. Some assemblers permit the use of the mnemonic DBRA instead of DBF. The simplest application of DBcc is the mechanization of a loop. Let us suppose that a loop must be executed N times. The following program achieves this.

```
        MOVE.W    #200,D0    Load D0 with 200
NEXT    ...                  Start of body of loop
                             Body of loop
        DBF       D0,NEXT    Decrement D0 and branch if not −1
```

Register D0 is preloaded with 200 and the D0 loop is entered. The DBF D0,NEXT instruction causes D0 to be decremented by 1 to yield 199. A branch is then made to NEXT, the start of the body of the loop. When D0 contains 0, the next execution of DBF D0,NEXT yields −1 and the loop is terminated. Note that the loop is repeated $[Dn] + 1$ times (i.e., 201). Interestingly, DBcc Dn,⟨label⟩ works only with 16-bit values in Dn; that is, loops greater than 65,536 cannot be achieved directly by this instruction. This procedure speeds up the operation of the DBcc, as a 32-bit decrement and test would take longer than a 16-bit decrement and test because (for some operations) the 68000 is internally organized as a 16-bit machine and two operations have to be carried out to implement a 32-bit operation.

The DBcc instruction is designed for applications in which one of two conditions may terminate a loop. One condition is the loop count in the specified data register, and the other is the condition specified by the test. A typical computer application of DBcc concerns the input of a block of data.

Data is received by an application program and processed as a block of 256 words. If the word $FFFF occurs in the input stream, the processing is terminated. The data elements to be input are stored in a memory location INPUT by some external device. Another memory location, READY, has its least significant bit cleared to zero if there is no data in INPUT waiting to be read. If the least significant bit of READY is true, data can be read from INPUT. The act of reading from INPUT automatically clears the least significant bit of READY. This behavior corresponds closely to real input mechanisms.

```
        MOVE.W    #255,D1      Set up D1 as a counter with maximum block size 256
WAIT    BTST.B    #0,READY     Test bit 0 of READY
        BEQ.S     WAIT         Repeat test until not zero
```

```
MOVE.W      INPUT,D0          Get input and move it to D0
   ⋮                          }Process input

CMPI.W      #$FFFF,D0         Test input for terminator
DBEQ        D1,WAIT           Continue for 256 cycles or until true
```

Note that a mistake is very easy to make with the DBcc instruction! The lower-order word of the data register specified by the DBcc is decremented, as explained earlier. Therefore, this register must be set up by a *word* operation. Sometimes, it is easy to think that a .B operation is sufficient if the loop count is less than 255!

2.6   MISCELLANEOUS INSTRUCTIONS

We now briefly introduce some of the 68000's instructions that do not fall neatly into any of the groups described earlier.

Set Byte Conditionally   Set byte conditionally is a somewhat unusual instruction not found in most other microprocessors. The assembly language form is Scc ⟨ea⟩, where cc is one of the fourteen logical tests in table 2.2 and ⟨ea⟩ is an effective address. When Scc is encountered by the 68000, it evaluates the condition specified by cc and, if true, sets all the bits of the byte specified by ⟨ea⟩. If the condition is false, all the bits specified by ⟨ea⟩ are cleared. After Scc ⟨ea⟩ has been executed, the contents of ⟨ea⟩ are, therefore, either $00 or $FF.

The best way of looking at Scc is to regard it as doing the groundwork for a deferred test. Let us suppose that an operation is carried out and we need to note, say, whether the result was positive, for later processing. One way of doing this is as follows.

```
        BSR         GET_DATA      Get input to be tested in D0
        CLR.B       FLAG          Clear FLAG
        TST.L       D0            Test result
        BMI.S       NEXT          If negative then exit with FLAG = 0
        MOVE.B      #$FF,FLAG     Else set all the bits of FLAG
NEXT ...                          Continue
```

In the preceding example, the longword in D0 returned by GET_DATA is to be tested. If the longword is negative, zero is stored in FLAG; otherwise $FF is stored. This operation requires four instructions. By using Scc we can simplify it to the following:

```
        BSR         GET_DATA      Get input to be tested in D0
        TST.L       D0            Test result
        SPL.B       FLAG          If negative then FLAG = 0
                                  else FLAG = $FF
```

The use of SPL saves two instructions and also requires less time to run than the version of the program using a BMI instruction.

CHK   The CHK instruction has the assembly language form CHK ⟨ea⟩,Dn and checks the lower-order word of data register Dn against two bounds. If

$Dn(0:15) < 0$ or if $Dn(0:15) > $ [ea] then a call to the operating system is made; otherwise the next instruction in sequence is executed. Operating system calls are covered in chapter 6.

NOP The "no operation" instruction has no effect on the CPU other than to advance the program counter to the next instruction. A NOP wastes time and memory space—time because it must be read from memory, interpreted, and executed, and space because it takes up a word of memory space. Some programmers use a NOP to generate a defined time delay. Others use it to patch a program; they insert several NOPs when writing a program, and, if the program has bugs, they replace the NOPs by a jump to the code that fixes the bug. Of course, we know of such programmers—but this is something we would never do ourselves, would we?

RESET RESET is a privileged instruction that, when executed, forces the 68000's RESET* output to active-low for 124 clock periods. The instruction resets any device connected to the RESET* pin but has no effect on the 68000 itself.

RTE The return from exception instruction is privileged and is used to terminate an exception handling routine in the same way that an RTS terminates a subroutine call.

STOP The assembly language form of the STOP instruction is STOP #n, where n is a 16-bit word. When a STOP is encountered, the value of n is loaded into the status register and the processor ceases to execute further instructions. Normal processing continues only when a trace, interrupt, or reset exception occurs. STOP is, of course, a privileged instruction.

TAS The test and set instruction has the assembly language form TAS ⟨ea⟩ and tests the byte specified by the effective address. If the byte is zero or negative, the N and Z flags of the CCR are set accordingly. The V and C flags are cleared and the X flag is unaffected. Bit 7 (the MSB) of the operand is set to one, that is, [ea(7)] ← 1. This instruction requires a read-modify-write cycle because the operand must first be read to carry out the test and then written to in order to set bit 7. The TAS instruction is indivisible because the read-modify-write cycle is always executed to completion and cannot be interrupted by another processor requesting the bus. The purpose of this instruction is to facilitate the synchronization of processors in a multiprocessor system.

TRAPV If the overflow bit, V, of the CCR is set, executing a TRAPV instruction causes the TRAPV exception to be raised and a call to the operating system to be made. If $V = 0$, a TRAPV instruction has no effect other than to advance the PC to the start of the next instruction.

2.7 SUBROUTINES AND THE 68000

Like almost all microprocessors, the 68000 has an on-chip hardware facility to implement subroutine call and return mechanisms. Such a facility is, of course, based on the CPU's supervisor/user stack (A7). The 68000 has a conventional jump to subroutine (JSR) instruction. Executing JSR ⟨ea⟩ causes the address of the next instruction (i.e., the return address) to be pushed on to the stack pointed at by A7 and a jump to be made to the effective address. The addressing modes supported by JSR are register indirect, register indirect with displacement, indexed, absolute, and program counter relative. Note that register indirect with predecrementing or postincrementing is not permitted! (I leave it to you to work out why the previous sentence ends with an exclamation mark.) Thus, JSR LABEL means jump to the subroutine whose absolute address is given by LABEL and JSR (A3) means jump to the address found in address register A3.

In addition to JSR, the 68000 also offers a BSR (branch to subroutine) instruction. The effects of JSR and BSR are the same; the only difference lies in their addressing modes. BSR uses an 8-bit or a 16-bit displacement following the op-code, which is added to the contents of the program counter to create a relative address. In general, programmers choose BSR rather than JSR because its addressing mode is always program counter relative: the displacement following a BSR does not depend on the location of the program in memory. Consequently, the relative branch provided by BSR makes the design of relocatable and reentrant programs very easy. A reentrant program is one that may be interrupted and used by the interrupting routine without corrupting any of the data required by the first user.

Subroutines are normally terminated by RTS (return from subroutine), which loads the return address into the program counter from the top of the stack. Occasionally, the programmer may wish to restore the contents of the condition code register to its presubroutine value after returning from the subroutine. Saving the condition code register on the stack after the subroutine call can be achieved by MOVE CCR, −(A7). At the end of the subroutine, the instruction RTR (return and restore condition codes) is executed to pull (pop) the contents of the condition code register and then the program counter off the stack.

The following example illustrates the use of RTR.

```
              .
              BSR      GET_DATA
              :
GET_DATA      MOVE.W   CCR,−(A7)            Save CCR on stack
              MOVEM.L  D1–D7/A0–A6,−(A7)    Save working registers on stack
              :
              MOVEM.L  (A7)+,D1–D7/A0–A6    Restore working registers
              RTR                           Restore CCR and return
```

We return to the topic of subroutines in chapter 3 when we look at how parameters are passed to and from them.

2.8   INTRODUCTION TO THE 68020'S ARCHITECTURE

The differences between the 68000 and the 68020 (or the 68030) are, at the same time, both subtle and radical. This seemingly contradictory statement can be better understood by saying that we can regard the 68020 as just a faster version of the 68000 or as a super 68000 with some very powerful additions. For the purpose of this introduction, the 68020 and 68030 are identical, since the 68030 is essentially a 68020 with the addition of a sophisticated memory-management unit. This section assumes a greater knowledge of assembly language programming than the previous sections in this chapter, because the reader who is interested in the more powerful features of the 68020 will already be familiar with the 68000. Other readers may omit this material until after they have read chapters 3 and 6.

The 68020 is compatible with 68000 object code and executes all 68000 instructions (except MOVE SR,⟨ea⟩, which is privileged on the 68020 but not on the 68000). It is not necessary to employ the 68020's special instructions and new addressing modes to achieve a significant increase in its performance with respect to the 68000, since the 68020 represents an entirely new and more efficient implementation of the 68000 *core machine*. However, an even greater improvement is possible if all the 68020's facilities are implemented.

An immediately obvious difference between the 68020 and the 68000 is the 68020's address and data bus. The 68020's address bus is 32 bits wide, which extends the logical address space to 4G bytes, and the data bus is 32 bits wide, which makes it possible to access a longword in a single bus cycle. Unlike the 68000, the 68020 has internal 32-bit buses and a 32-bit execution unit plus an instruction cache. These enhancements applied to a 68020 with a 16-MHz clock provide a four- to sixfold increase in speed with respect to a 68000 clocked at 8 MHz.

One of the 68000's attributes that make it so much more sophisticated than its predecessors is its ability to support more operating system functions. Before the advent of the 68000 family, microprocessors were designed to execute a particular instruction set with little regard to needs of operating systems. To be fair, this approach to microprocessor design was not unreasonable, since before the mid-1980s most microprocessors were employed in systems with no operating system or in systems with minimal operating systems.

Why is an operating system different from an applications program (i.e., a user program running under an operating system)? A simple answer to this question is that the operating system must control the computer and protect it from certain types of damage that a user program may cause. Suppose a program is being tested and that the programmer intends to clear a block of 1024 bytes of memory. The programmer makes a mistake in the loop terminator (perhaps by writing BEQ instead of BNE) and the program does not terminate. Instead, it proceeds to clear all memory. A good operating system will prevent this from happening by making certain that user programs cannot access memory not allocated to them. That is, a user program is prevented from interfering with either the operating system or with other user programs. Chapters 6 and 7 have more to say about these aspects of a processor.

Some microprocessors, such as the 68000, provide only limited help to an operating system, whereas others, such as the 68020 and 68030, provide a much greater degree of help. In this introduction we are interested only in the user mode of the 68000 and the 68020. Operating systems run in the supervisor mode, which is described later when we cover exceptions. Here we are going to describe the differences between the 68000 and the 68020 from the point of view of the user programmer.

The 68020 and the 68030 have more registers than the 68000. However, all these new registers (including the cache control register) are dedicated to operating system functions. As far as we are concerned, there are still eight data and eight address registers. Even the condition code register of the 68020 is the same as the 68000's.

The 68020 is housed in a larger package than the 68000 (figure 2.18) and has a full 32-bit data bus and address bus. The 32-bit data bus enables the 68020 to run faster because it can access a longword in a single machine cycle. The 32-bit address bus means that each of the 32 bits of an address register or the program counter are connected to pins, and therefore the 68020 can address 2^{32} bytes (i.e., 4G bytes). You might think that these modifications have no effect on someone who wishes to run an existing 68000 program on a 68020 microprocessor (apart from causing the program to run faster). Some 68000 programmers, however, used the following argument.

If the 68000 employs only address bits A_{00} to A_{23} to specify a byte location (since the 68000 lacks address pins A_{24} to A_{31}, although it does have a 32-bit PC), it does not matter what we do with address bits A_{24} to A_{31}. For example, the addresses $0012\ 3456 and $4012\ 3456 access the same physical location, since the only difference between them is that state of A_{30} that is not connected to an address pin. Consequently, we can employ address bits A_{24} to A_{31} as *tag* bits that define the type of address—that is, we can "label" an address as pointing to a byte or a word, a vector or a matrix, or to any other type of object. Such a facility is useful in a language such as LISP. These tag bits may be used by software to check the type of the address but have no effect on the actual address leaving the 68000.

Of course, programs written for the 68000 relying on this convention will not run on a 68020 system that makes use of address pins A_{24} to A_{31}. This point is made here because it demonstrates how even apparently harmless differences between two processors can cause problems.

In this book, we are going to discuss the differences between the 68000 and the 68020 from three aspects: the user programmer, the hardware designer (chapter 4), and the systems or operating systems programmer (chapter 6). We begin with the differences at the user programmer level. These differences fall into two groups: new and enhanced instructions and new and enhanced addressing modes. In this context, *enhanced* implies a modest improvement to an existing 68000 facility.

Enhanced 68000 Instructions

Designers of the 68020 have not "gone to town" with new instructions but rather have gently extended the 68000's instruction set. Sometimes they have improved existing instructions by making them more flexible and sometimes they have added entirely new classes of instructions. We will look at some of the enhanced instructions before tackling the new instructions.

Multiplication and Division

The 68000's multiplication and division instructions (MULU, MULS, DIVU, and DIVS) have been extended to handle both 16-bit and 32-bit operations. We will

FIGURE 2.18 The 68020 and the 68000

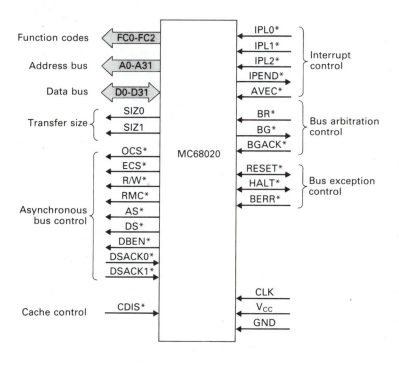

consider only the 68020's unsigned multiply and division here, as the corresponding signed instructions perform corresponding operations on signed two's complement numbers.

The 68000 has just a single multiply instruction, MULU ⟨ea⟩,Dn, which multiplies the 16-bit word at the specified effective address with the low-order word of Dn and deposits the 32-bit result in Dn. The assembly language format of the 68020's multiplication instructions is as follows:

ASSEMBLER SYNTAX	MULTIPLIER	MULTIPLICAND	RESULT
MULU.W ⟨ea⟩,Dn	16	16	32
MULU.L ⟨ea⟩,Dn	32	32	32
MULU.L ⟨ea⟩,Dh:Di	32	32	64

These operations can be expressed in RTL form by

MULU.W ⟨ea⟩,Dn	$[Dn(0:31)] \leftarrow [Dn(0:15)] \times [ea(0:15)]$
MULU.L ⟨ea⟩,Dn	$[Dn(0:31)] \leftarrow [Dn(0:31)] \times [ea(0:31)]$
MULU.L ⟨ea⟩,Dh:Dl	$[Dh(0:31),Dl(0:31)] \leftarrow [Dl(0:31)] \times [ea(0:31)]$

The 68020's word multiplication instruction, MULU.W ⟨ea⟩,Dn, is identical to the corresponding 68000 multiplication. The 68020 extends the 68000 multiplication instruction by the inclusion of the two longword multiplications. Both longword multiplications perform a 32-bit by 32-bit multiplication. One yields a 32-bit result (i.e., the most significant 32 bits of the 64-bit product), and the other yields a true 64-bit (i.e., quadword) product. In order to do this, two registers must be specified by the instruction (i.e., Dh:Dl), where Dh is the most significant 32 bits of the product and Dl is the least significant. For example, the operation MULU.L #\$12345678,D0:D1 multiplies the 32-bit contents of D1 by the literal \$1234 5678 and deposits the 64-bit result in the register pair D0,D1.

The 68020 has four unsigned division operations that carry out the operation dividend/divisor = quotient + remainder. As in the case of the 68000's MULU.W instruction, the 68000's DIVU.W instruction, which yields a 16-bit quotient and a 16-bit remainder, is extended to cater for 32-bit quotients and remainders.

ASSEMBLER SYNTAX	DIVIDEND	DIVISOR	QUOTIENT	REMAINDER
DIVU.W ⟨ea⟩,Dn	32	16	16	16
DIVU.L ⟨ea⟩,Dq	32	32	32	None stored
DIVU.L ⟨ea⟩,Dr:Dq	64	32	32	32
DIVUL.L ⟨ea⟩,Dr:Dq	32	32	32	32

These operations can be expressed in RTL form by

DIVU.W ⟨ea⟩,Dn	$[Dn(0:15)] \leftarrow [Dn(0:31)]/[ea(0:15)]$ $[Dn(16:31)] \leftarrow$ remainder.
DIVU.L ⟨ea⟩,Dq	$[Dq(0:31)] \leftarrow [Dq(0:31)]/[ea(0:31)]$
DIVU.L ⟨ea⟩,Dr:Dq	$[Dq(0:31)] \leftarrow [Dr(0:31),Dq(0:31)]/[ea(0:31)]$ $[Dr(0:31)] \leftarrow$ remainder.
DIVUL.L ⟨ea⟩,Dr:Dq	$[Dq(0:31)] \leftarrow [Dr(0:31)]/[ea(0:31)]$ $[Dr(0:31)] \leftarrow$ remainder.

In this case there are three .L variations in addition to the 68000's .W division. Note that in contrast to the multiplication instructions, which permit maximum source operands of 32 bits, the operation DIVU.L ⟨ea⟩,Dr:Dq specifies a 64-bit quadword dividend. The most significant 32 bits are in Dr and the least significant bits are in Dq.

Branch Instructions

The 68000's Bcc (branch on condition cc = true) instruction is extended by the 68020 to cater for 32-bit displacements. Instead of being able to jump a maximum of plus or minus 32K bytes from the current instruction, the 68020 programmer can now execute a relative jump of up to ±2G bytes on either side of the current instruction. The same extension is applied to unconditional branch, BRA, and branch to subroutine, BSR, instructions. The LINK instruction (to be described in the next chapter) also caters for 32-bit displacements.

Check and Compare Instructions

The 68000 has a special check instruction, (CHK ⟨ea⟩,Dn), that compares the value in data register Dn to zero and to an upper bound specified by ⟨ea⟩. The upper bound is a two's complement integer, X, which means that the check can be carried out either on a negative number in the range −X to 0 or on a positive number in the range 0 to +X. If the value in register Dn is less than zero or greater than the upper bound, a software interrupt (i.e., exception or trap) occurs. The term *software interrupt* describes a call to some facility provided by the operating system. Chapter 6 deals with exceptions in more detail.

The 68020 extends the 68000's CHK instruction by including two variations: CHK2 and CMP2 (i.e., check 2 and compare 2). We will describe only the CHK2 instruction here, since the CMP2 instruction differs in one detail only. The CHK2 ⟨ea⟩,Rn instruction compares the value in address/data register Rn with upper and lower bounds, just like the CHK instruction. The principal difference between CHK and CHK2 is that the effective address specified by CHK2 points to a *pair* of bounds: The lower bound is followed by the upper bound (remember that CHK has a fixed lower bound of zero). A CHK2 instruction can take .B, .W, and .L extensions, which means that the boundary value can be a byte, word, or longword. As in the case of the CHK instruction, a CHK2 instruction causes an exception if the specified value is outside the boundary ranges. Another difference between CHK and CHK2 is that the former tests only a data register, whereas the latter may test a data register or an address register. This is a most sensible extension, since the programmer can now easily test whether a pointer is within the range of permitted addresses.

For example, a CHK2.L $1234,A2 instruction compares the contents of address register A2 with the longwords found in address locations $1234 (i.e., the lower bound) and $1238 (i.e., the upper bound). Although the CHK2 instruction can be employed in a variety of applications, its principal use is in the testing of array subscripts in a high-level language. Suppose a programmer writes the code TIME(J) := 12. If the array subscript J has been incorrectly evaluated, the processor will attempt to access data outside the space allocated to the array TIME. However, by employing a CHK2 instruction to test the array address against the bounds of the array at run-time, the danger of array bound errors can be removed. We provide an example of bounds testing when we introduce the CMP2 instruction.

The relationship between the upper and lower bounds (i.e., U and L) specified by a CHK2 instruction can be illustrated diagrammatically. As you can see, the contents of the specified register may be less than the lower bound, may be greater than the upper bound, or may fall between the bounds.

Note that if the CHK2 instruction is used to test an address register and the size is a word or a byte (yes—a byte!), the bounds are sign extended to 32 bits and the resultant operands are tested against the full 32 bits of the specified address register. However, if the contents of a data register are tested and the operation size is a byte or a word, only the appropriate low-order bits of the data register are tested.

If the upper and lower bounds are the same, the valid range is a single value. For signed comparisons, the arithmetically smaller value should be the lower bound, while for unsigned comparisons, the logically smaller value should be at the lower bound. This statement is not as confusing as it seems. In 4-bit arithmetic, the unsigned bounds might be 2 (lower) and 6 (upper), which correspond to 0010 and 0110, respectively. However, in signed two's complement arithmetic, the bounds might be -2 (lower) and $+4$ (upper), which correspond to 1110 and 0100, respectively.

The CMP2 ⟨ea⟩,Rn instruction is identical to the CHK2 instruction except that it sets condition codes rather than taking an exception when Rn is out of bounds. When the contents of Rn is compared with its lower and upper bounds, the Z and C bits of the condition code register are updated. The Z-bit is set if the contents of Rn are equal to either bound and cleared otherwise. The C-bit is set if the contents of Rn are out of bounds and cleared otherwise. The relationship between the contents of the register being tested and the Z and C bits can be better appreciated from the following diagram.

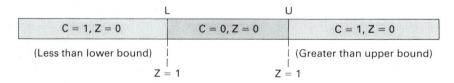

Consider an application of a CMP2 instruction (we have chosen CMP2 rather than CHK2 because a CMP2 does not cause a trap, which we have not yet discussed in detail). A programmer is using a one-dimensional array of bytes, TABLE, and wishes to access the jth element, where j is in data register D0.W. The following fragment of code tests for an array bound error at run time.

```
SIZE        EQU         <size>          The size of the array TABLE in bytes
TABLE       DS.B        SIZE            Save SIZE bytes for the array TABLE
LOWER       DS.L        <lower>         Reserve longword for lower bound of TABLE
UPPER       DS.L        <upper>         Reserve longword for upper bound of TABLE
            :
*                                       Set up array bounds
            LEA         TABLE,A0        A0 points to the start of the array
            MOVE.L      A0,LOWER        Store the lower bound of the array
            LEA         SIZE(A0),A0     Calculate the upper bound of the array
            MOVE.L      A0,UPPER        Store the upper bound of the array
            :
*                                       Perform boundary tests
            LEA         TABLE,A0        A0 points to the start of the array
            LEA         (A0,D0.W),A0    A0 points to the jth element
            CMP2.L      LOWER,A0        Test for out of bounds element
            BCS         ERROR           IF carry set THEN error
            .                                           ELSE continue
            .
ERROR       .                           Deal with error condition
```

The practical difference between a CHK2 and a CMP2 instruction is that a CHK2 instruction uses an operating system call (i.e., a trap) to deal with the out-of-range condition, but a CMP2 instruction simply modifies the condition codes and leaves any recovery action to the programmer.

The EXT Instruction

The 68000 has a sign-extend instruction, EXT, which sign extends an 8-bit byte to a word (i.e., EXT.W Dn) or a 16-bit word to a longword (i.e., EXT.L Dn). The 68020 has a new instruction, EXTB.L Dn, which sign extends a byte to a longword by taking bit 7 of Dn and copying it into bits 8 to 31 of Dn. Without the EXTB.L instruction, you would have to sign extend a byte by:

```
EXT.W Dn        Sign extend byte to a word
EXT.L Dn        Sign extend word to longword
```

New 68020 Instructions

The 68020 provides a very modest set of new instructions. Some of these are related to the use of external coprocessors and are not dealt with here. One instruction pair, call module (CALLM) and return from module (RTM), is used only by operating systems and is not considered further in this chapter. CALLM and RTM are not implemented by the 68030, which implies that these instructions were not a good idea. In fact, they require an external memory management system such as the PMMU if they are to be fully implemented. Another very special pair of instructions are compare and swap (CAS) and compare and swap 2 (CAS2). These two instructions are intended for use in sophisticated applications such as multiprocessor systems and interrupt-driven multitasking systems. Chapter 7 says a little more about these topics.

The two new groups of 68020 instructions of immediate interest to us here are the BCD pack and unpack group and the bit field group. The pack and unpack instructions are unimportant if you are not interested in BCD operations (in which case we suggest that you skip ahead to the much more interesting bit field instructions).

The PACK and UNPK Instructions

The PACK and UNPK instructions are used in conjunction with BCD arithmetic and are intended to simplify the conversion between characters input in coded form (e.g., ASCII 7-bit code) and the internal representation of BCD data. For example, the ASCII code for the character '7' is $37, whereas the BCD representation of the decimal number 7 is 0111. The syntax of the PACK instruction is

```
     PACK      −(Ax),−(Ay),#〈adjustment〉
or   PACK      Dx,Dy,#〈adjustment〉.
```

We can note two things immediately. First, the PACK instruction has *three* operands; second, it can take only two addressing modes (data register direct or register indirect with autodecrementing).

The effect of the PACK instruction is to translate the source data into the destination data by means of the 〈adjustment〉, which is a literal. If we consider the data register form of the instruction (i.e., PACK Dx,Dy, # 〈adjustment〉), the adjustment is added to the value contained in the source register and then bits (11:8) and (3:0) of the result are concatenated and placed in bits (7: 0) of the destination. Note that the literal, 〈adjustment〉, used by the PACK instruction is zero for both ASCII and EBCDIC characters. The effect of a PACK Dx,Dy, # 〈adjustment〉 instruction in RTL is

```
[Temp(0:15)] ← [Dx(0:15)] + 〈adjustment〉
[Dy(0:3)]       ← [Temp(0:3)]; [Dy(4:7)] ← [Temp(8:11)]
```

Let's look at an application of the PACK instruction. Consider the effect of PACK D0,D1,#0, where the source register D0(0:15) contains the ASCII characters for '4' and '2' (i.e., D0 = $3432). The PACK instruction adds the literal (i.e., 0) to D0 to produce a result of $3432. In this case the literal is zero and there is no change. In the next step, the least significant nibbles of this result (i.e., 4 and 2) are extracted and concatenated in the destination register D1 to give D1 = $XX42. The X's indicate that the most significant byte of D1.W is unaffected by this instruction. See figure 2.19.

The preceding example demonstrates how the PACK instruction takes two 8-bit character codes representing 4-bit values and packs them into a single byte so that they can take part in BCD arithmetic operations.

As you might imagine, the UNPK instruction performs the inverse operation. The syntax of UNPK is either UNPK Dx,Dy, # 〈adjustment〉 or UNPK −(Ax), −(Ay), # 〈adjustment〉. In this case, the two BCD digits in the source byte are separated and used to form the least significant nibble of 2 bytes (i.e., they

FIGURE 2.19 Use of the PACK instruction

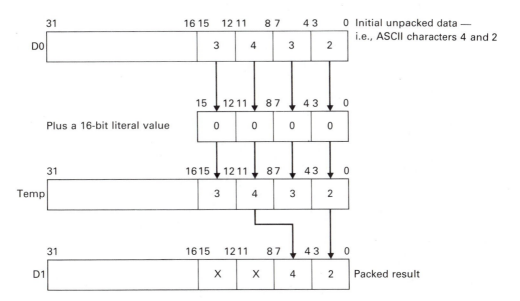

are unpacked). Then the literal specified by ⟨adjustment⟩ is added to give the 16-bit destination operand. The effect of UNPK Dx,Dx, #⟨adjustment⟩ in RTL form is

[Dy(0:3)] ← [Dx(0:3)]; [Dy(8:11)] ← [Dx(4:7)]
[Dy(0:15)] ← [Dy(0:15)] + adjustment

Consider now the action of UNPK D0,D1, #$3030, where D0 = $25. In the first step the contents of D0 are unpacked to give $0205, and in the second step the literal $3030 is added to give $3235 in D1, corresponding to the ASCII characters for '2' and '5' (see figure 2.20).

Bit Field Instructions

The 68020's bit field group of instructions represents, possibly, the most significant enhancement to the 68000's instruction set from the point of view of the programmer writing applications programs.

Conventional microprocessors are byte, word, or longword oriented and operate on data located at a byte or a word boundary. Being restricted to one of these boundaries does not usually cause a problem, since many of the data structures employed by programmers fit within these boundaries quite naturally. For example, ASCII encoded characters (which may be 7-bit or 8-bit values) fit within byte boundaries, just as 64-bit IEEE-format floating-point numbers fit into word or longword boundaries.

Sometimes programmers have to deal with data structures that do not fall within these *natural* boundaries. Suppose a programmer is working with 17-bit data structures. The 68000 programmer must fit each item into two consecutive words (16 bits in one and a single bit in the other) or into a byte and a word. Not only is

FIGURE 2.20   Use of the UNPK instruction

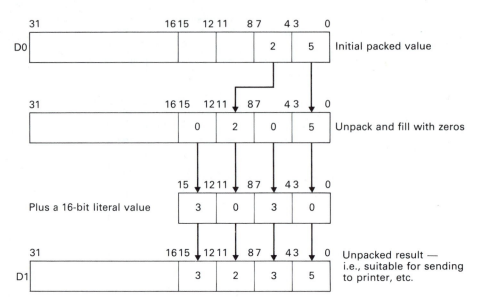

this arrangement inefficient in terms of storage, it is cumbersome in terms of operations on the 17-bit items. The 68020's bit field operations permit the programmer to forget about byte, word, and longword boundaries and to handle data items falling across these boundaries that have been imposed by the hardware.

Since a bit field is an arbitrary group of bits that can fall anywhere within the processor's address space, it is tempting to wonder just what bit field instructions might be designed to do. Because a bit field is an arbitrary data structure, conventional arithmetic operations are meaningless. The answer to our question is that the bit field instructions are designed largely to copy strings of bits between registers and arbitrary locations in memory and to test the bit fields. Once bit fields have been transferred from memory to a register, they can be processed in the normal way by the conventional instructions. We will now briefly describe the bit field instructions provided by the 68020.

BFEXTU   The extract a bit field and deposit it in a data register instruction is the bit field equivalent of the MOVE ⟨source⟩,Dn instruction. If the size of the bit field is less than 32 bits, it is loaded into the low-order bits of the data register and the high-order bits are set to zero.

BFEXTS   The extract a bit field (signed) and deposit it in a register instruction moves the bit field into the data register and sign extends it to 32 bits.

BFINS   The insert the bit field in the data register into memory instruction is the bit field equivalent of a MOVE Dn,⟨destination⟩ instruction.

BFTST   The test bit field instruction tests the specified bit field and sets the CCR accordingly. The N-bit is set if the most significant bit of the bit field is 1 and the Z-bit is set if all the bits of the bit field are 0s.

BFCLR   The bit field clear instruction tests the bit field exactly like BFTST and then clears all the bits of the bit field.

BFSET   The bit field set instruction tests the bit field exactly like BFTST and then sets the bits of the bit field to 1.

BFCHG   The bit field test and change instruction behaves exactly like a BFTST instruction, except that the bits of the bit field are all inverted after the test.

BFFFO   The find first one in bit field instruction is the only instruction that actually performs a calculation on the bits of a bit field (although the bit field is not modified). The bit field at the specified address is read and scanned by the processor. The location of the first 1 bit in the bit field is then loaded into the specified data register. We interpret the "first 1" as the most significant bit of the bit field that is set to the value 1. The location of the first 1, which is loaded into the destination register, is specified as the offset of the bit field itself plus the offset of the first 1. The precise meanings of offset and field width are given later. However, if the offset is 7 and the bit field is equal to 00000011011010, the BFFFO instruction will return the value $7 + 6 = 13$ in the specified data register (because the first one is 6 bits from the leftmost bit). If no 1 is found (i.e., the bit field is all 0s), the value returned is the offset plus the field width. A BFFFO instruction can be used in floating-point arithmetic to locate the most significant bit of a mantissa without going through the slow process of shifting and testing. Equally, it can be used in bit-mapped data structures to scan past strings of zeros.

The assembler syntax of the 68020's bit field instructions is

BFCHG	⟨ea⟩{offset:width}	Bit field change
BFCLR	⟨ea⟩{offset:width}	Bit field clear
BFEXTS	⟨ea⟩{offset:width},Dn	Bit field signed extract
BFEXTU	⟨ea⟩{offset:width},Dn	Bit field unsigned extract
BFFFO	⟨ea⟩{offset:width},Dn	Bit field find first one
BFINS	Dn,⟨ea⟩{offset:width}	Bit field insert
BFSET	⟨ea⟩{offset:width}	Bit field set
BFTST	⟨ea⟩{offset:width}	Bit field test.

Bit field instructions take either three or four operands. For example, the bit field clear instruction has the syntax BFCLR ⟨ea⟩{offset:width}, and the insert bit field instruction has the syntax BFINS Dn,⟨ea⟩{offset:width}. Since a bit field is a user-defined data structure that is not aligned on a byte boundary, three pieces of information are needed to define it: its size (i.e., its width), its location (i.e., a byte address and an offset from that position), and its value.

The base address of a bit field is specified in the conventional way by an effective address. For example, the instruction BFCLR $1000{4:12} refers to a bit field at a base address $1000. The offset (in this case 4) is measured from field bit zero of the base address, and the width of the field is a positive integer in the range 1 to 32 that tells us how many bits there are in the field. Bit fields wider than 32 bits are not supported by the 68020. The instruction BFCLR $1000{4:12} is interpreted as "clear the 12 bits of the bit field whose location is 4 bits from byte $1000." However, there is one rather confusing point with which we must come to terms—the way in which the bits of the bit field are numbered.

Figure 2.21 illustrates the relationship between base byte, base bit, field offset, and field width. The bytes are numbered consecutively $0, 1, \ldots, i - 1, i, i + 1, \ldots, n$, and the individual bits of each byte are numbered 0 to 7, where 0 is the least significant bit (this number is fundamental to the 68000 family). The effective address

FIGURE 2.21   How field data is located in memory

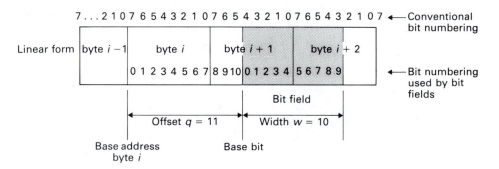

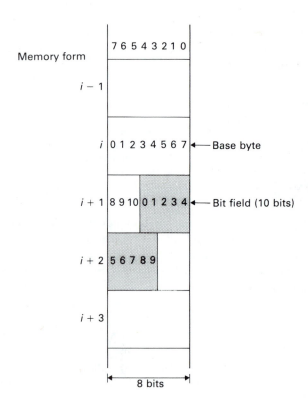

in the instruction points to the base address of the bit field, which, in Figure 2.21, is byte i. The first bit (called the base bit) of the bit field is located q bits away from the base byte. Note that the bit field is located at q bits from the *most significant bit* of the base byte and not q bits from its least significant bit, as you might expect. The bit field itself extends from the base bit to the base bit plus the field width minus 1.

Having stated this fact, we need to look at figure 2.21 a little more closely. Once again it is necessary to stress that the field offset, q, is measured from the most significant bit of the base byte. The bits of the actual bit field itself are numbered 0 to $w - 1$ in the same sense as the field offset (i.e., in the opposite direction of the numbering of the bits of individual bytes).

The effective address (i.e., the base byte) of each of the bit field instruction is specified in the "normal" way by any of the 68020's addressing modes (apart from those employing autoincrementing and autodecrementing). The offset may be specified by either a literal in the range 0 to 31 or by a data register that permits an offset in the range -2^{31} to $2^{31} - 1$. The bit field width may be specified by a literal in the range 1 to 31 (or 0, which specifies 32 bits) or by the contents of a data register modulo 32. Note again that a value of zero for the bit field width is interpreted as a width of 32.

Typical legal bit field instructions are

```
BFCLR       (A0){5:7}
BFCLR       (D2,A3){D6:12}
BFCLR       (d8,PC,A4){30:D2}
BFCLR       $1234{D5:D6}
```

When a bit field offset is specified by the contents of a data register, the range is -2^{31} to $2^{31} - 1$, which means that the offset can be on either side of the base byte (i.e., at a lower address or a higher address).

What then is the purpose of bit field instructions? Bit field operations make it easy to operate on data structures that are not byte oriented. Such structures are associated with graphics and, for example, with disk data structures. Without bit field operations, it would be very tedious to manipulate arbitrary data structures using only byte operations and shifting. For example, consider the operation BFCLR $1000{20:13}. This has the effect of clearing the 13 bits of the specified bit field that lies 20 bits from bit 7 of the base byte $1000, as illustrated in figure 2.22.

FIGURE 2.22 Effect of BFCLR $1000{20:13}

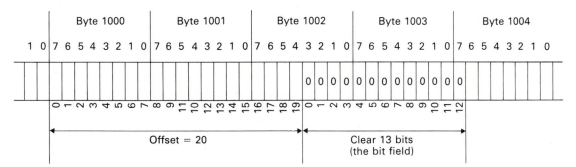

FIGURE 2.23   The free-sector bit map of a disk

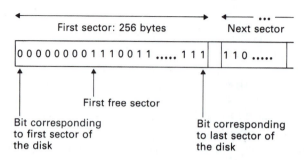

A typical application of bit field instructions might be found in the realm of disk file systems. A disk is made up of a number of tracks, each of which is composed of a series of sectors. When a file is created, new sectors are allocated to it; when a file is deleted, its sectors are released. A simple way of keeping track of sectors is to create a bit map in the first sector of the first track. Suppose a sector contains 256 bytes (i.e., 2048 bits). Each of these bits is associated with a sector, as illustrated in figure 2.23. If a sector is free, the corresponding bit in the bit map is set, and if it is allocated, the corresponding bit is clear.

Now suppose the operating system wishes to create a new file. It must read the bits of the bit map, one by one, until it finds a 1 bit. It then allocates the corresponding sector to the file, clears the bit to mark the sector as allocated, and searches for the next unallocated sector, and so on, until the file has been created. Reading a bit map is not difficult in 68000 code. A byte at a time can be read and the 8 bits of the byte checked in sequence. The 68020's bit field instructions make the task much simpler.

If would be nice if we could turn the entire sector containing the free sector bit map into a single 2048-bit-wide bit field and then use the instruction BFFFO (find first 1 in bit field) to locate the first free sector. However, since the maximum bit field width is 32 bits, we must regard the free sector bit map as a sequence of 64 bit fields of length 32 (since $64 \times 32 = 2048$). Figure 2.24 describes a free-sector map in

FIGURE 2.24   Using bit fields to implement a free-sector bit map

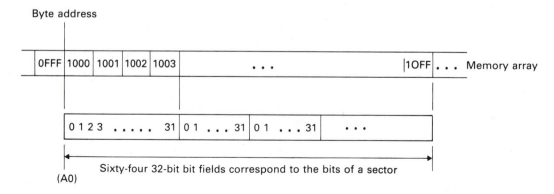

terms of bit fields. A fragment of code to search the sector map is provided next. Remember that the BFFFO instruction returns the location of the first 1 in the bit field or the field width plus the offset if no 1 is found.

```
FIELDS      DS.L          64
*                         Assume that the sector map is made up of 64 longwords
*                         (this assumption means that we can use 32-bit fields)
*
            LEA           FIELDS,A0        A0 points to sector map
            MOVE.W        #63,D7           Up to 64 bit fields to search
LOOP        BFFFO         (A0){0:32},D1    Look for a free sector
            LEA           4(A0),A0         Increment pointer to next bit field
            CMP.B         #32,D1           IF D1 = 32 THEN no free sector found
            BNE           FOUND            IF D1 not 32 THEN free sector found
            DBRA          D7,LOOP          REPEAT UNTIL all 64 fields tested
            BRA           FULL             IF here THEN no free sector located
FOUND       LEA           −4(A0),A0        Wind back the pointer
            BFCLR         (A0){D1:1}       Clear bit to claim sector
  :
FULL        deal with disk full
```

When John Hodson, who was then Motorola's Northern Europe Training Manager, first saw this program, he suggested that it be rewritten to make even better use of the BFFFO instruction.

```
            CLR.L         D0               Initial bit field offset = 0
            LEA           FIELDS,A0        A0 points at sector bitmap
            MOVE.W        #64−1,D7         Up to 64 bit fields to test
LOOP        BFFFO         (A0){D0:32},D0   Look for free sector, IF found Z = 0 and
*                                          bit field offset from (A0) is loaded
*                                          into D0, ELSE Z = 1 and bit field offset
*                                          plus 32 is loaded into D0.
            DBEQ          D7,LOOP          Decrement counter until Z = 0 or all 64
                                           fields searched without success.
            BEQ           FULL             IF Z = 1, no 1's found and disk is full
            BFCLR         (A0){D0:1}       ELSE claim sector. D0 = sector number
```

In John's version of the example, the BFFFO (A0){D0:32},D0 instruction looks for the first 1 in the 32-bit-wide bit field that is offset by the contents of D0 from the base byte pointed at by A0. If a 1 is found, its total offset from the base byte is loaded into D0. If a 1 is not found, the initial offset plus the field width (i.e., 32) is loaded into D0. That is, D0 is the offset pointer, which is incremented by 32 until either a 1 is found or until all sectors have been checked. Note also that the BFFFO instruction sets the Z-bit of the CCR if a 1 is not found in the bit field. We use this fact in the DBEQ D7,LOOP instruction to terminate the loop if a 1 is found.

Another example of the application of bit field instructions can be taken from the world of computer graphics. Consider the bit plane of Figure 2.25, which contains an image made up of individual pixels. A pixel is a picture element that may be on or off to create a dot or no dot. In this example, a rectangular part of the bit plane is to be copied from one place to another. The image is 64 pixels (8 bytes) wide, and a 15-bit field 6 lines deep is to be moved down by 10 lines and to the right by 3 pixels. Assume that the address of the source image is in A0.

FIGURE 2.25 Use of bit field operations in bit-mapped graphics

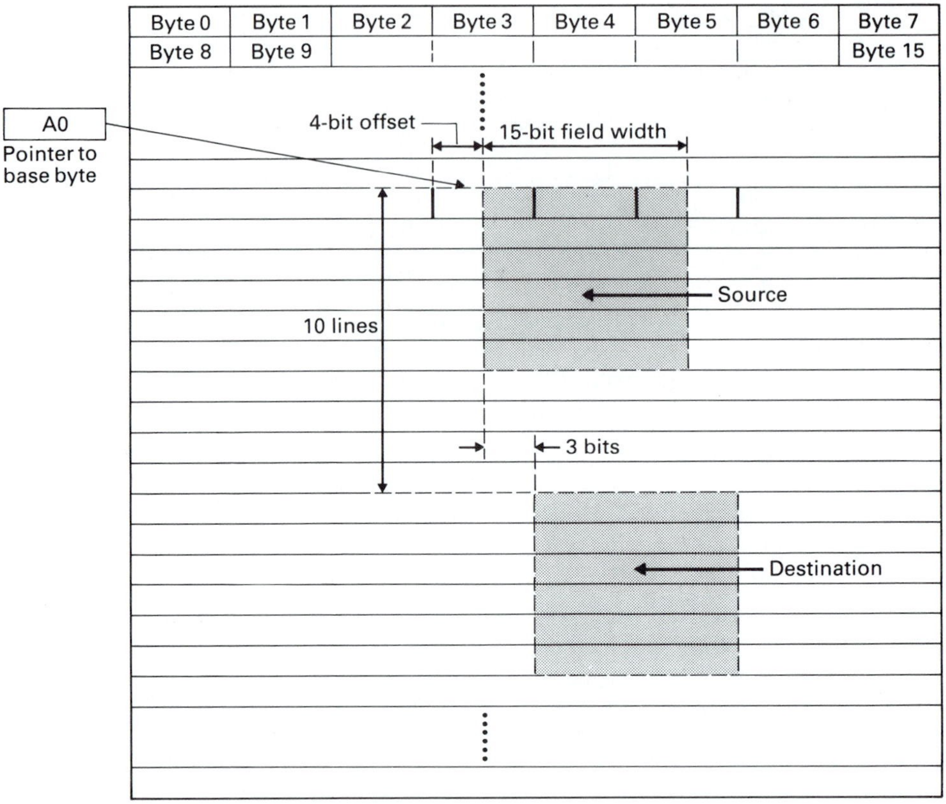

We can carry out the pixel translation by reading a line of the image using the BFEXTU instruction and then copying it to its destination with a BFINS instruction.

```
           MOVE.W      #5,D0              Six lines to move
LOOP       BFEXTU      (A0){4:15},D1      Copy a line of the image to D1
           BFINS       D1,80(A0){7:15}    Copy it to its destination
           LEA         8(A0),A0           Update the pointer by 8 bytes (one line)
           DBRA        D0,LOOP            Repeat until all lines moved
```

A Motorola application report by Bob Reims, AR219, provides an excellent example of the use of bit field instructions. Beims points out that high-level languages such as C often pack several small variables into a single word. For example, a single 16-bit word can hold a 6-bit variable and two 5-bit variables. Packing variables saves memory space at the expense of the time taken to access the individual variables.

Since most microprocessors lack specific instructions to manipulate packed variables, packing and unpacking is rather cumbersome. Consider the following two routines, which retrieve and store, respectively, a packed variable. This variable is 3 bits wide and is packed into bits 7 to 9 of a 16-bit word. We could have used the terms *stored* and *loaded* instead of *packed* and *unpacked*. For example, if the bit field is ABC, the packed word would look like this: XXXXXXABCXXXXXX. We first perform the packing and unpacking in 68000 code and then in 68020 code.

```
*              Read the packed variable from memory and store it in bits 0-2 of D0
*
*
LOAD    MOVE.W    #$0380,D0    Load the mask word 0000001110000000
        AND.W     <ea>,D0      Packed word is masked to bits 7-9 in D0
        LSR.W     #7,D0        Justify D0 to get bits in least significant position
        RTS

*
*              The 3-bit variable to pack is initially in bits 0-2 of D0 and
*              is to be packed in bit 7-9 of the specified memory location.
*              Note that the store operation must not modify any other bits
*              of the packed word (i.e., bits 0-6 and 10-15)
*
STORE   LSL.W     #7,D0        Shift bit field to be stored to bits 7-9
        MOVE.W    #$FC7F,D1    Load mask word 1111110001111111 in D1
        AND.W     <ea>,D1      Get packed word in D1, clear bits 7-9
        OR.W      D0,D1        Insert bit field from D0 in bits 7-9 of D1
        MOVE.W    D1,<ea>      Store packed word in memory
        RTS
```

Since the action of the preceding code might not be immediately clear, consider an example in which the bit pattern 010 is packed into bits 7-9 of the word at the specified effective address and is to be unpacked and loaded into D0.

```
*                            [<ea>] = 0101010101010100 (initial packed string)
LOAD    MOVE.W    #$0380,D0    D0 = 0000001110000000 (mask bits = bits 7 to 9)
        AND.W     <ea>,D0      D0 = 0000000100000000 (mask D0 to bits 7-9)
        LSR.W     #7,D0        D0 = 0000000000000010 (left justify bit field)
        RTS
```

In the next example, the bit field 110 in bits 0-2 of D0 is to be packed into bits 7-9 of the word at the specified effective address.

```
STORE   LSL.W     #7,D0        D0 = 0000001100000000 (move variable to bits 7-9)
        MOVE.W    #$FC7F,D1    D1 = 1111110001111111 (mask to clear bits 7-9)
        AND.W     <ea>,D1          1010101010101010 (data before packing)
*                             D1 = 1010100000101010 (clear bits 7-9)
        OR.W      D0,D1        D1 = 1010101100101010 (insert bits 7-9 from D0)
        MOVE.W    D1,<ea>      Store packed word in memory
        RTS
```

As you can see, there is nothing remarkable about these fragments of code. The code is simply rather long-winded. Now consider the use of the bit field instructions BFEXTU and BFINS. BFEXTU is a bit field extract unsigned instruction that extracts a bit field from the specified effective address, zero extends the result to 32 bits, and loads it into the destination data register. Its assembly language form is

BFEXTU ⟨ea⟩{offset:width},Dn. We can, therefore, recode the preceding load (i.e., pack) operation as the single instruction

    BFEXTU ⟨ea⟩{6,3},D0

Note that the field width is 3 bits and that the offset is 6 because bit field operations number bits from left to right, whereas the bits of a word are numbered from right to left (i.e., the bit field is 6 bits from bit 15 of the effective address). Figure 2.26 shows how the bits are numbered in this example.

Similarly, the BFINS instruction can be used to insert a bit field. Its assembly language form is BFINS Dn,⟨ea⟩{offset:width} and its effect is to take the bit field from the low-order bits of the specified data register and insert them into the bit field at the effective address. In terms of the preceding example, we can recode the store operation as

    BFINS D0,⟨ea⟩{6:3}

Beims provides a second example of bit field applications that manage to combine both the power of the 68020's new addressing modes and its bit field operations in a single instruction. The reader who is unfamiliar with the 68020's indirect addressing modes should read the next section before working through this example. Beims constructs a system in which a set of records is stored in memory and a pointer to the records is pushed onto the stack at an offset FILEPTR below the top of the stack. This situation might arise when a subroutine is called to process the records and the address of the records is passed on the stack. Figure 2.27 illustrates the data structure relevant to this example.

To access the required data, the longword on the stack at address SP + FILEPTR must first be read. The longword address is a pointer to the record structure, and the actual record can be found at some offset from the beginning of the records. Assume that we wish to access record 5, which is at offset REC5. Once record 5 has been located, it is necessary to extract an item within the record. Data register D3.W points to the longword that contains the required bit field. Data register D4 contains the offset from the start of the bit array to the most significant bit of the longword pointed at by D3. In this case we assume that the bit field to be accessed is 24 bits wide.

The bit field can be accessed by the single instruction:

    BFEXTU   ([FILEPTR,SP],REC5,D3.W*4){D4:24},D0

We now look at the 68020's new addressing modes.

FIGURE 2.26   Inserting a bit field

Bit of word	15	14	13	12	11	10	09	08	07	06	05	04	03	02	01	00
Offset	0	1	2	3	4	5	6									
Bit field							0	1	2							

FIGURE 2.27 Using bit field instructions to access a complex data structure: executing a BFEXTU ([FILEPTR,SP],REC5,D3.W∗4){D4:24},D0 instruction

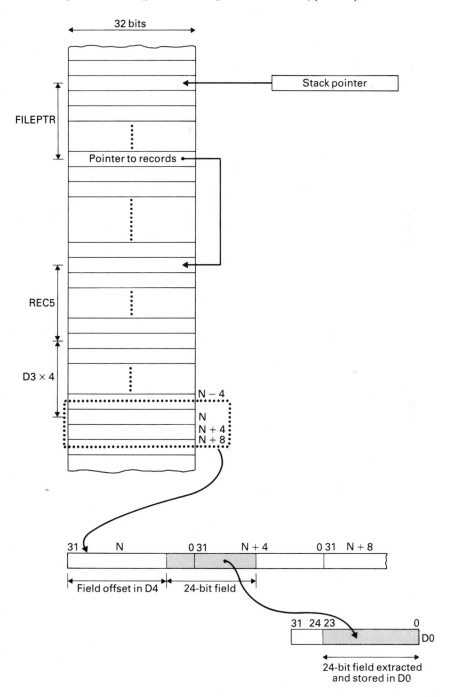

The 68020's New and Extended Addressing Modes

The 68000 has all the basic addressing modes you would expect to find on most 8- or 16-bit microprocessors (with the exception of the 6809's indirect addressing mode). However, the 68020 introduces powerful new addressing modes with several variations, making it much more sophisticated than most other microprocessors. The 68020's new addressing modes are also implemented by the 68030 and the 68040. When we talk about the 68020's new addressing modes in this section, we also mean those of the 68030 and the 68040.

You might almost say that the 68020 represents the high point of micro-processor development—in the sense that future developments will probably be in different directions. For example, it is not likely that we will see new generations of microprocessors like the 68020 but with more and more increasingly complex addressing modes and instruction sets. Instead, we might expect to see micro-processors that have streamlined and simplified addressing modes (e.g., such as the RISC processors), microprocessors that have been designed to execute high-level languages directly, or microprocessors that incorporate system functions such as memory management and memory caches (e.g., such as the 68030).

Extended Addressing Modes

Some of the 68000's addressing modes have been enhanced by the 68020, just as the 68000's operation set has been enhanced. These modifications or enhancements are extensions of some 68000 addressing modes to include 32-bit offsets. For example, the 68000's register indirect with index and program counter indirect with index addressing modes express the effective address as d8(An,Xn) and d8(PC,Xn), respectively. The 68020 permits the constant offset to be either d8, d16, or d32 (i.e., a byte, word, or longword). Note that 68020 literature (and current 68000 literature) describes the syntax of one of the 68000's addressing modes slightly differently. The 68000 effective address d8(An,Xn) is written in 68020 terminology as (d8,An,Xn), although there is no difference in the way in which it is actually evaluated (i.e., ea:=d8+[An]+[Xn]). Note that the 68020 supports several variants on this mode. For example, a 32-bit constant can be used to provide the constant when calculating an effective address (d32,An).

Memory Indirect Addressing

Although a glance at the 68020's instruction manual might lead you to believe that the 68020 has quite a few new addressing modes not found on the 68000, a purist could argue that there is really only one new addressing mode: memory indirect addressing, in which the effective address generated by an instruction points to a memory location that points to the actual operand to be accessed. The reason that the 68020 seems to have a lot of new addressing modes is that memory indirect addressing is implemented with a large number of options. Memory indirect address-ing has been included to make it easier to access arrays and similar data structures.

Consider simple memory indirect addressing (which the 68020 does not directly support; it is synthesized from the 68020's more complex general addressing modes by setting literals to zero or by suppressing them). The effective address is written [⟨address⟩] and is calculated by reading the contents of memory location ⟨address⟩ and using the 32-bit value at that address to access the actual location of the operand. An instruction of the form MOVE D0,[$1234] would have the effect

$$[D0] \leftarrow [M([M(1234)])]$$

This expression can be better understood if it is split into two parts:

```
Temp_ea ← [M(1234)]
[D0]     ← [M(Temp_ea)]
```

Figure 2.28 illustrates memory indirect addressing. One advantage of this addressing mode is clear. It provides the same function as address register indirect, except that instead of having eight address registers, there is an "index register" for each longword of the processor's memory space. By operating on the contents of the effective address in memory, we can skip about an array or similar data structure, just as we could by operating on the contents of an address register.

The 68020 provides two basic memory indirect addressing modes: memory indirect postindexed and memory indirect preindexed. 68020 literature also refers to two other memory indirect addressing modes: program counter memory indirect postindexed and program counter memory indirect preindexed. These are really variations on memory indirect addressing in which we use the program counter to express a relative displacement instead of using an address register, which expresses an absolute displacement. The syntax of these new addressing modes is as follows.

ADDRESSING MODE	ASSEMBLER SYNTAX
Memory indirect postindexed	([bd,An],Xn,od)
Memory indirect preindexed	([bd,An,Xn],od)
PC memory indirect postindexed	([bd,PC],Xn,od)
PC memory indirect preindexed	([bd,PC,Xn],od)

FIGURE 2.28 Indirect addressing

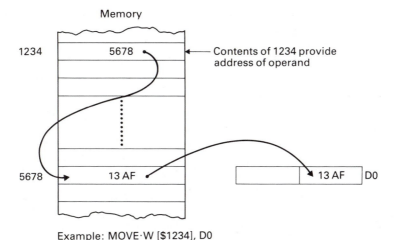

Example: MOVE·W [$1234], D0

Here bd is a 16-bit or 32-bit constant, Xn is an address or data register (the contents of Xn may be scaled by multiplying them by 1, 2, 4, or 8, as shown later), and od is a second 16-bit or 32-bit literal. The full power of these new addressing modes is not immediately apparent. We shall indicate how these modes are used shortly. However, these addressing modes can be used in simpler forms by suppressing the displacements bd and/or od to zero and by omitting An or Xn. When a constant is *suppressed*, it is omitted from the encoding of the instruction rather than simply setting it to zero. Suppressing a constant makes the assembled code more compact. The 68020's numerous indirect addressing modes provide the programmer (or compiler) with a series of options. For example, both the following two instructions are legal examples of memory indirect postindexed addressing:

```
MOVE     D0,([$12345678,A0],D4,$FF000000)
```

and

```
MOVE     D0,([A0])
```

Using the 68020's ability to suppress constants and registers in addressing modes, we can take the full effective address ([d32,An,Rx],d32) and create a large number of options by simply erasing any unnecessary components of the effective address. Following are some of the possible legal options (we use the term *legal* here, since some of the options are not sensible). The meaning of these addressing modes will become clearer when you have read the remainder of this section; at the moment we are simply interested in demonstrating the wide range of options available.

([d32,An,Rx],d32)	Full effective address
([An,Rx])	Rub out displacements
([An],d32)	Rub out inner displacement
([Rx])	Rub out inner and outer displacements and An
(An,Rx)	Rub out displacements and indirection
(Rx)	We can even have data register indirect
[(d32)]	Strange but legal; Offset from 0 = address of address
( )	Even stranger but still legal; address zero
[( )]	As ( ), but an indirect address

Before continuing we demonstrate one simple advantage of memory indirect addressing. A table of longword addresses, each of which corresponds to a subroutine entry point, is pointed at by A0. Suppose that the number in D0 indicates which subroutine is to be called. We can call the appropriate subroutine (in 68000 code) by the following action:

```
LSL.L      #2,D0        Multiply subroutine number in D0 by 4
MOVE.L     (A0,D0),A1   Get subroutine address in A1
JSR        (A1)         Call the subroutine
```

Note that the contents of D0 have to be multiplied by 4 because subroutine addresses are one longword (4 bytes), but the subroutine identifier in D0 is a 1-byte value. By means of the 68020's indirect addressing modes, we can write:

```
JSR        ([A0,D0*4])  Call the subroutine specified by D0
                        That is, [PC] ← [M([A0] + 4 × [D0])]
```

In this example the contents of A0 are added to the contents of D0 multiplied by 4. The resulting effective address is used to access the table. The longword at this address is read and loaded into the program counter to call the subroutine. The 68020's memory indirect addressing mode has allowed us to replace three 68000 instructions by a single 68020 instruction. Moreover, scaling does not modify the contents of the register scaled (see figure 2.29).

Three constants, or literals, are employed by the 68020's new memory indirect addressing modes and are written bd, od, and sc in 68020 literature. The base displacement, bd, corresponds approximately to the 68000's d8 and d16 offsets, except that bd must be either a 16-bit or a 32-bit literal. The base displacement is added to an address register to give a location in memory, which is read to provide a pointer to the actual operand.

The outer displacement, od, is a 16-bit or 32-bit literal and has no 68000 equivalent. The constant od is added to the value read from memory to give the actual address of the operand. The scale factor, sc, is a constant whose value is 1, 2, 4, or 8 and is used to scale the contents of an index register. The scale factor has no explicit 68000 equivalent, although it is used implicitly in certain operations.

Consider the operation MOVE.W (A0)+,D0. The contents of address register A0 are incremented by 2 after it has been used as an index register. Had the instruction been MOVE.L (A0)+,D0, A0 would have been incremented by 4. In other words, the contents of an autoincremented address register are incremented by 1 scaled by 1, 2, or 4.

The need for a scale factor is related to the 68020's ability to support byte, word, and longword operands. Clearly, successive bytes differ by one location, successive words differ by two locations, successive longwords differ by four locations, and successive quadwords differ by eight locations because the 68000 family's memory is byte addressed, irrespective of the type of data being accessed. Suppose that A0 points to an array of elements, Y, and D0 contains an index i. The address of

FIGURE 2.29 Accessing a jump table

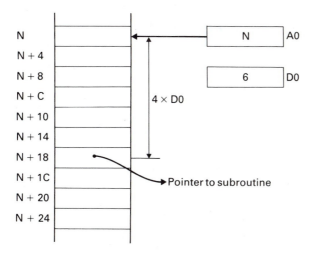

the ith element is given by $[A0] + [D0]*sc$, where sc is 1, 2, 4, or 8 for byte, word, longword, or quadword (i.e., 64-bit) data elements. Unfortunately, since the 68000 has no automatic mechanism for scaling (apart from autoincrementing and autodecrementing), the 68000 programmer must carry out the calculation.

For example, in 68000 code we would write:

```
LSL.L      #s,D0           Scale ith element (s = 0, 1, 2 for sc = 1, 2, 4)
LEA        (A0,D0.L),A1    Calculate address of ith element
```

The 68020 provides explicit scaling in conjunction with memory indirect addressing. The index register, Xn, may be an address or data register and is written Xn*sc in assembler form. Typical effective addresses might be written:

```
MOVE      D0,([A0,D4*1])      Scale factor = 1
MOVE      D0,([A0,D4*4])      Scale factor = 4
```

Figure 2.30 demonstrates how the scale factor relates to the size of data objects. Of course, the built-in scale factors of the 68020 cannot be used with data objects of arbitrary size.

The effect of the base displacement is best illustrated by means of an example. Suppose we have a list of students and each student has a record consisting of six elements. Each of the elements corresponds to the student's results in that subject (see figure 2.31).

There are many ways of organizing the data structure of figure 2.31. One is to choose a seven-element structure consisting of the student's name followed by the six results. An alternative is to create list of pointers, one per student, where each pointer points to the appropriate student's results (see figure 2.31). This example takes the latter approach and demonstrates how the three constants, bd, sc, and od are related.

FIGURE 2.30 Scale factor (only factors of 1, 2, and 4 illustrated)

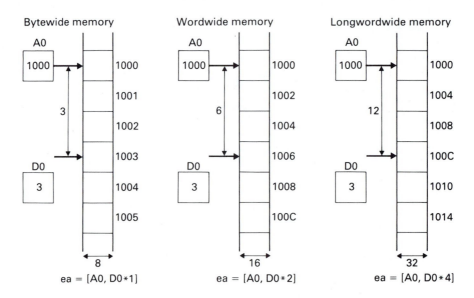

FIGURE 2.31 Memory indirect addressing with preindexing

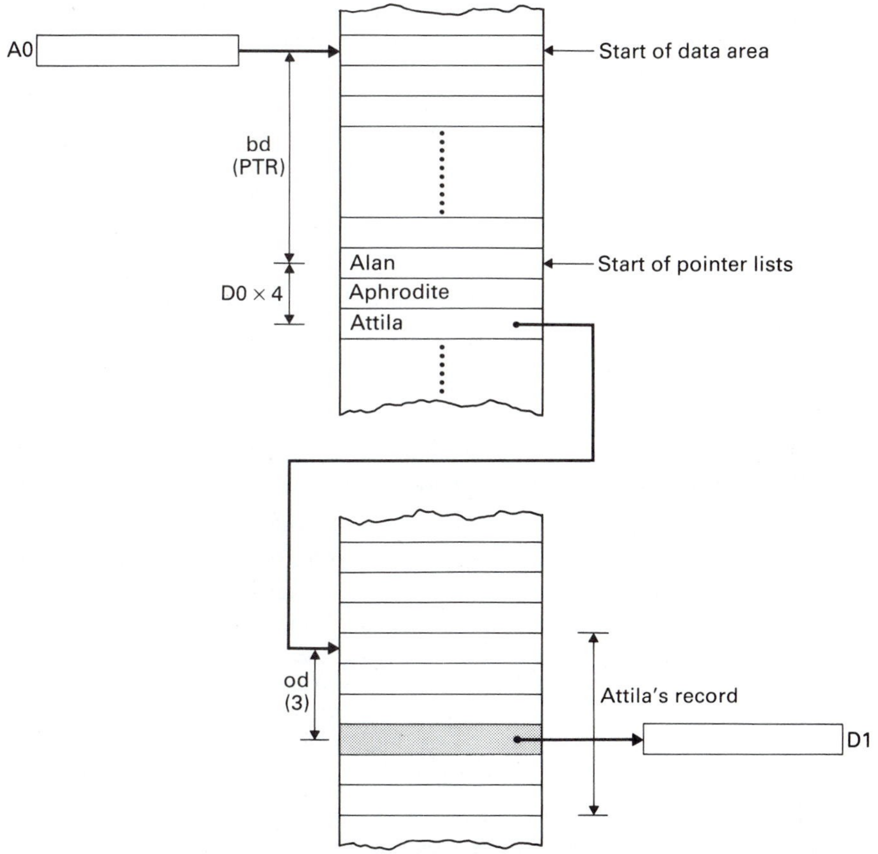

Effect of MOVE·B ([PTR, A0, D0∗4], 3), D1

Register A0 in figure 2.31 points to the base of a region of memory devoted to the students' records. This region may include other items of related data. The base displacement, bd, points to the start of the list of students with respect to the start of the region of data—that is, the first student's entry is at address [A0] + bd. Of course, if A0 had been loaded with the address of the first student's entry, the base displacement could have been set to zero and the addressing mode simplified by omitting bd).

The index of the student selected is in data register D0. Since each entry in the table of pointers is a longword, we have to scale the contents of D0 by 4. The effective address of the pointer to the selected student's record is, therefore, [A0] + bd + 4∗[D0]. The 68020 reads this pointer, which points to the start of the student's record. Suppose we want to know how the student performed in computer science, which is the fourth out of the six results. We need to access the fourth item in the table (i.e., item 3 because the first item is numbered zero). The outer displacement provides us with a facility to do this. When the processor reads the pointer from memory, it adds the outer displacement to it to calculate the actual effective address of the desired operand.

If this example were to be coded for the 68000, the assembly form might look like this:

```
PTR     EQU         ⟨pointer to record offset⟩
        LSL.L       #2,D0                   Multiply the student index by 4
        LEA         PTR(A0,D0.L),A1         Calculate address of pointer to record
        MOVEA.L     (A1),A1                 Read the actual pointer
        ADDA.L      #3,A1                   Calculate address of CS result
        MOVE.B      (A1),D1                 Read the result
```

The same calculation can be carried out by the 68020 using memory indirect addressing with preindexing:

```
MOVE.B      ([PTR,A0,D0.L*4],3),D1
```

Before looking at another example of the 68020's two memory indirect addressing modes, it is worthwhile looking at how they differ. Figure 2.32 illustrates the effect of postindexing and preindexing.

Example of Memory Indirect Addressing It is not easy to provide both simple and realistic applications of memory indirect addressing, since many examples of memory indirect addressing involve complex high-level language data structures and are inappropriate in this text. The following example demonstrates how memory indirect addressing can be used even by the assembly language programmer. This example uses two concepts we have not yet introduced, the TRAPcc instruction and the exception stack frame. In brief, a TRAPcc instruction causes a "TRAP," or call to the operating system, to take place if condition cc is true. For example, TRAPCS causes an operating system call if the carry bit in the CCR is 1. A TRAPcc instruction has three formats, one with no extension, one with an extension word (TRAPcc.W) and one with two extension words (TRAPcc.L). These extensions are literal values and are parameters that can be read by the operating system.

When a TRAPcc instruction is encountered and the specified condition cc is true, a call to the operating system (i.e., the TRAPcc handler) is made, and certain information saved on the stack is pointed at by the stack pointer, A7. Figure 2.33 illustrates the state of the system immediately after a TRAPCS.W #d16 instruction has been executed and the trap taken. You do not have to understand the details of exception handlers; all that you need know for the purpose of this example is that the stack frame pointed at by A7 contains the address of the next instruction after the instruction that caused the exception (i.e., the next instruction after the TRAPCS.W #d16) at location [A7] + 8.

Suppose the TRAPcc handler needs to examine the literal following the TRAPcc. This literal is stored at address N + 2 in figure 2.33. We can use memory indirect addressing to access this literal via the stack pointer, A7:

```
MOVE.W      ([8,A7],−2),D0
```

Eight is added to the contents of the stack pointer to get the address of the "address of the TRAPcc exception." The contents of this location (i.e., [A7] + 8) are

FIGURE 2.32 68020's two basic modes of memory indirect addressing

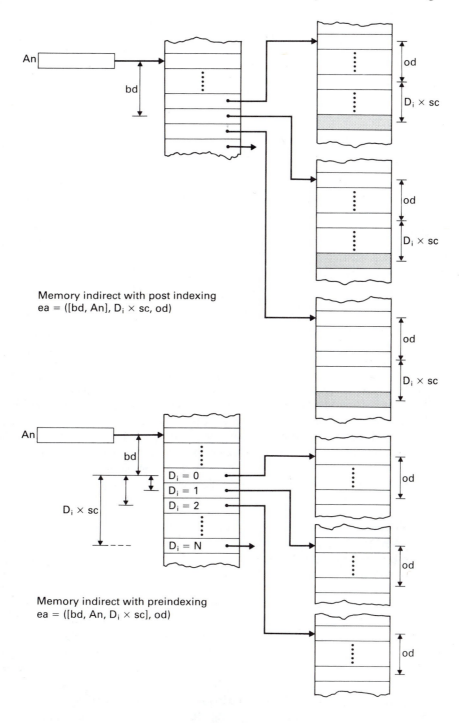

Memory indirect with post indexing
ea = ([bd, An], $D_i \times sc$, od)

Memory indirect with preindexing
ea = ([bd, An, $D_i \times sc$], od)

FIGURE 2.33   Effect of executing a TRAPCS.W #data instruction

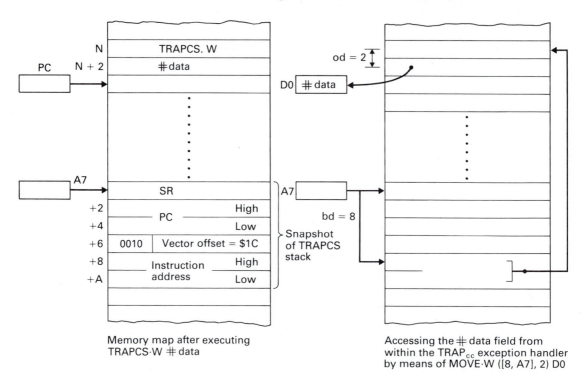

Memory map after executing
TRAPCS·W #data

Accessing the #data field from
within the TRAP$_{cc}$ exception handler
by means of MOVE·W ([8, A7], 2) D0

read to give the address of the TRAPcc. The outer displacement, 2, is added to
this value to give the address of the literal following the TRAPCS.W. This literal is
then loaded into D0. As we have seen, this entire sequence is carried out by a single
68020 instruction. By the way, we assumed that the trap had been called by a
TRAPcc instruction with a word extension. If we did not know this, the exception
handler would have examine the bit pattern of the TRAPcc instruction in order to
determine the length of the operand following it.

2.9   SPEED AND PERFORMANCE OF MICROPROCESSORS

Before we leave the architectures of the 68000 and the 68020, we should comment
on their relative performances. Any engineer choosing a microprocessor for a given
project should select the most appropriate device. We do not have the time and
space to present a full discussion of the relative merits of microprocessors here. We
will just concentrate on *speed*. We might think that there would be some simple
metric whereby we could compare the speed of one microprocessor with another.
No such metric exists, and the comparison of microprocessors is a sensitive and often
acrimonious area (especially when competing devices are compared). There are many

reasons why it is so difficult to compare the speed of two microprocessors. Some of these reasons are as follows:

1. *Clock speed.* Do you compare two chips at equal clock speeds or do you compare them at the top speed specified by their manufacturers? Sometimes it is unfair to use the same speed in comparisons because the clock is used to trigger internal operations, and a microprocessor may have an apparently high clock rate simply because more internal operations may have to be triggered.

2. *Meaningless MIPs.* A classic metric for computers is the MIP (million instructions per second). Although a processor with a high MIPs rating might seem faster than another, the MIP tells us only how fast instructions are executed. It tells us nothing about what the instructions actually do. A microprocessor with a low MIPs rating might be more powerful than one with a high MIPs rating. An example at the end of this section demonstrates the truth of this point.

3. *Memory system.* In today's high-performance systems with microprocessors operating at 20 MHz and above, it is often necessary to slow the processor down by inserting *wait states* whenever it accesses external memory. Therefore, the effect of memory access time should be taken into account when discussing processor speeds. It would be wrong to quote a microprocessor as having some high figure of merit if that figure could not be attained using conventional memory.

4. *Optimum use of registers.* If two processors, one with many registers and one with few registers, are compared, misleading results might be obtained if both sets of registers are used in the same way—that is, care must be taken to use the registers of the register-rich processor in an optimal fashion if its power is to be realized.

5. *Special addressing modes.* Some microprocessors have special addressing modes that can be used to write highly compact and fast code—under certain circumstances. For example, the 68000 can use 16-bit absolute addresses (rather than 32-bit addresses) to access data within the first and last 32K bytes of its address space. It is dangerous to compare the code of two processors when one is running in its general or unrestricted mode and the other is using these special addressing modes (which may be effective only with programs below a certain size).

6. *Relevant benchmarks.* Some benchmarks are misleading because they test irrelevant features of the processor. For example, a benchmark that includes a large amount of input/output activity might tell very little about the performance of the processor (but a lot about the rest of the system).

7. *Use of cache.* Microprocessors such as the 68020 have on-chip instruction caches, and the 68030 has an on-chip data cache as well as an instruction cache. These caches can have a profound effect on the result of benchmarks. Since the cache speeds up the processor by reducing the memory-CPU traffic, the nature of the benchmark becomes critical. For example, a tight loop that permits almost all the instructions and data to reside in the on-chip cache might run twice as fast as code that does not permit the use of the cache (i.e., in-line code with no loops). The use of internal cache negates the effect of wait states in external memory (ie., accessing information from the bus incurs wait states; accessing it from the cache does not).

8. *Effect of pipelines in the processor.* Modern advanced microprocessors (especially RISC types) have several internal units that act on data in parallel. For example, one instruction can be executed at the same time the effective address of another instruction is calculated. The type of instructions in the pipeline and the nature of their effective addresses affects the performance of the processor. For example, the performance of some heavily pipelined processors is reduced by branch instructions, since a branch can negate some of the work done by the pipeline. If the instruction following a branch is loaded into the pipeline, it must be discarded if the branch is taken, because it is no longer the next instruction in sequence.

For these reasons, any published benchmarks have to be carefully interpreted. In one case, manufacturer A demonstrated that its processor was 1.5 times as fast as manufacturer B's processor. Manufacturer B rewrote the benchmarks and demonstrated that the converse case was true—chip B was 1.5 times as fast as chip A. Such stories demonstrate that unless one processor is very much faster than another, it is very difficult to compare two processors.

An interesting example from Motorola report BRE 322/DD illustrates the minefield of the benchmark by comparing the 68000 and the 68020. We will implement the high-level language construct: IF COUNT[[i]] $\langle\rangle$ 0 THEN... first using identical code on both the 68000 and the 68020 and then using 68020 code. Note that D1 = i, D3 = temporary value, A2 = temporary pointer, A5 = base pointer, CLASS = $1C26, and COUNT = $1C40. The code that runs on both the 68000 and 68020 occupies 13 words of assembly language.

Although the same 13 words of code run on both the 68000 and the 68020, these processors behave differently in terms of the number of clock cycles and bus cycles they execute. The following data give the clock cycles required to execute each instruction and the bus cycles used to access external memory for both the 68000 and the 68020. Note that the 68020 requires both fewer clock cycles and bus cycles (the latter because it is assumed that the 68020's code is in cache and it does not have to read instructions from external store).

68000 VERSION		68020 VERSION		CODE	
CLOCK CYCLES	BUS CYCLES	CLOCK CYCLES	BUS CYCLES		
4	1	2	0	MOVE.W	D1,D3
8	1	4	0	LSL.W	#1,D3
12	2	6	0	LEA	0(A5,D3.W),A2
12	3	7	1	MOVE.W	CLASS(A2),D3
8	1	4	0	LSL.W	#1,D3
12	2	6	0	LEA	0(A5,D3.W),A2
12	3	7	1	TST.W	COUNT(A2)
11	2	6	0	BEQ	ELSE
79	15	42	2		

We can write code to perform the same action using some of the 68020's special features. In this case the size of the code is 8 words.

CLOCK CYCLES	BUS CYCLES	CODE	
9	1	MOVE.W	(CLASS,A5,D1.W*2),D3
9	1	TST.W	(COUNT,A5,D3.W*2)
6	0	BEQ	ELSE
24	2		

Using the preceding data for each of the three cases, we can now calculate the time taken to execute the constructs and the corresponding MIPS.

68000		68020 (using 68000 code)		68020 (with 68020 code)	
TIME	MIPs	TIME	MIPs	TIME	MIPs
6.32 μs	1.27	2.52 μs	3.17	1.44 μs	2.08

In this example, the 68020, using its powerful addressing modes, executes code over 400% faster than the 68000. The 68020 using its new addressing modes executes the code almost twice as fast as a 68020 running 68000 code while running at a lower number of MIPs! In other words, it's not just how fast you run code that matters—it's what you do with it.

SUMMARY

In this chapter, we have introduced the 68000's internal architecture, instruction set, and addressing modes. Although the 68000 has some powerful new instructions such as the DBcc, its real power lies in its multiple-length data operations (byte, word, longword), its large and regular array of data and address registers, and its wealth of addressing modes. The addressing modes of the 68000 make it very easy to write position-independent code and to handle the complex data structures associated with today's high level languages.

If we wish to converse with French people in their native language, we must learn French, and that means spending hours learning irregular verbs. Programming in assembly language is very much the same as learning a foreign language: we must learn the instruction set and addressing modes of a microprocessor before we can go on to write programs of any reasonable size. In this chapter we have laid the foundation for chapter 3, in which we examine the writing of assembly language programs, and for chapter 11, in which we present a monitor written in 68000 assembly language.

We have introduced the 68000's internal architecture, its instruction set, and its addressing modes. One of the most important features of the 68000 is its great simplicity. We have discovered that, for example, the 68000 MOVE instruction replaces many of the data transfer instructions peculiar to other microprocessors. Even better, we have found out that the 68000 supports byte, word, and longword operations and that the programmer can use them simply by adding the suffix .B, .W, or .L, respectively, to an instruction. What could be easier? Although we have only hinted at their power here, it is the addressing modes of the 68000 that make it so easy to write complex code in a clear and compact fashion. Moreover, these addressing modes allow us to handle both the complex data structures and the position-independent code required by today's sophisticated microcomputers. We have also looked at the 68020's architectural's highlights: its new instructions and its powerful new indirect addressing modes.

Although we have provided only an overview of the 68000 family's assembly language and instruction set, the reader should now be in a better position to follow the assembly language monitor in chapter 11.

Problems

1. For the following memory map, evaluate the following expressions, where [N] means the contents of memory location N. The purpose of this problem is to illustrate the calculation and the meaning of effective addresses. Assume that all addresses are decimal.

For example, [3] = 4.

a. [7]
b. [[4]]
c. [[[0]]]
d. [2 + 10]
e. [[9] + 2]
f. [[9] + [2]]
g. [[5] + [13] + 2*[14]]
h. [0]*3 + [1]*4
i. [9]*[10]

00	12
01	17
02	7
03	4
04	8
05	4
06	4
07	6
08	0
09	5
10	12
11	7
12	6
13	3
14	2

2. Each of the following expressions conforms to the 68000 assembler. Assume that all addresses are byte values, expressed in decimal form, and that [D0] = 0, [A0] = 4, [A1] = 2, [PC] = 10. Using the memory map of problem 1, explain the action of the following instructions:

    a. LEA   (A0),A3   b. LEA   −2(A0),A3
    c. LEA   12(A0,D0),A3   d. MOVE.B −2(A0,D0),D4
    e. MOVE.B 2,D4   f. MOVE.B 2(PC),D2
    g. MOVE.B 1(PC,A1),D7   h. ADD.B   12,D0
    i. ADD.B   D0,4(A0,A1)

3. What is the difference between the following instructions?

```
ADDI    #⟨data⟩,⟨ea⟩
ADD     #⟨data⟩,⟨ea⟩
```

4. Both ADDQ and MOVEQ are used with *"small"* operands. Apart from the obvious difference (ADDQ adds and MOVEQ moves), what is the difference between these two instructions?

5. What are the differences between a BRA and a JMP instruction?

6. The MOVEM.W (A7)+,D0/A0 instruction can be very dangerous because it contains a trap for the unwary. What is the trap?

7. What is the largest program that can be addressed by processors with the following number of address bits? The largest program is the same as the processor's address space and is measured in bytes.

    a. 12 bits   b. 16 bits
    c. 24 bits   d. 32 bits
    e. 48 bits

8. A 68000-based system is to be used for word processing. Assume that all its memory space is to be populated by read/write memory and that 1M byte is allocated to the operating system, word processor and text buffer space. The remaining memory space is free to hold a text file. If the average English word contains five ISO/ASCII-encoded characters and one page of text is composed of approximately 35 lines of 12 words, how many pages of text can be held in the 68000's memory space at any instant? If the 68000 is replaced by a 68020, how much text can be held in memory?

9. Suppose a 68000 is to be used to process digital images (e.g., in a laser printer). An image is made up of an array of *n* by *n* pixels. Each pixel may have one of 16 grey levels (from all white to all black). If a picture measures 8 in. by 8 in., what is the maximum resolution that can be supported by a 68000 (in terms of pixels per inch)? Assume that 4M bytes of the 68000's address space are required for the operating system and all other software and cannot be used to store image data.

10. What is the effect of the following sequence of 68000 instructions? Explain what each individual instruction does, and then explain the overall effect of the three instructions. If these instructions were to be replaced by more efficient 68020 code, what would that code be?

```
LEA       $1234,A0
MOVE.L    (A0),A0
JMP       (A0)
```

11.   Draw a memory map corresponding to the following sequence of assembler directives.

```
ORG      $001000
DS.B     12
DC.B     'Input Error'
DC.L     1234
DS.L     2
```

12.   All the following 68000 assembly language instructions are incorrect. In each case, what is the error?

a.	MOVE.L	D2,A4	b.	EOR.W	(A2)+,D4
c.	ADDQ.L	#12,D2	d.	MOVEA.B	#4,A3
e.	ADDQ.B	#0,A4	f.	ANDI.B	#FC,D6
g.	LEA	(A3)+,A4	h.	LEA	(A3),D3
i.	MOVEA.L	A4,D7	j.	MOVEP.L	#7,D6
k.	MOVE.B	D2,12(PC)	l.	SWAP	A4
m.	EXG.B	D3,A4	n.	UNLK	D6
o.	ANDI	D4,D5	p.	ASL.L	#9,D3
q.	BRA.B	2741	r.	EOR	(A3)+,D4
s.	DIVU.B	D4,D5	t.	LEA.B	#4,A3
u.	NOT.W	D3,D7	v.	RTS.B	D3
w.	DIVU.W	D3,A4	x.	CLR.L	A2
y.	CMPM.B	(A3)+,(A4)	z.	MOVEQ.B	#$42,D7

13.   The 68000 does not permit the operation CLR A0 (because address register direct is not a legal addressing mode for a CLR instruction). Give two ways of clearing the contents of an address register (each using a single legal instruction from the 68000's instruction set).

14.   What are the advantages and disadvantages of separate address and data registers in the way in which they are implemented in the 68000?

15.   The 68000 implements negative addresses because the contents of address registers are treated as signed two's complement values. What is a *negative* address and how is it used?

16.   The 68000 has addressing modes that are indicated by −(Ai) and (Ai)+ in 68000 assemblers. What are they (i.e., what do they do?) and why is one used only with autoincrementing and one with autodecrementing?

17.   The 68000 has the following condition code bits: C, V, X, N, and Z. Which is the "odd one out" and why?

18.   What does the term *effective address* mean and how is it used?

19.   The NOP (no operation instruction) has a 16-bit format. Suppose you could make it a 32-bit instruction. How would you use the additional 16 bits to increase its functionality— that is, can you think of any way of extending the NOP instruction?

20.   The 68000 has special instructions for use with small literals (e.g., MOVEQ and ADDQ). Some assemblers do not acknowledge these instructions explicitly and automatically make use of them (i.e., the programmer writes ADD #1,D0 and the assembler chooses the code corresponding to ADDQ #1,D0). Is this a good idea, since it permits compact instruction coding without forcing the programmer to learn about the Q options?

21.   Explain how the following assembler directives are used:

a. EQU   b. ORG
c. DC   d. DS

22. What is the meaning of the expression *sign extension* and how is it related to the 68000?

23. Write a sequence of instructions to calculate the parity of the low-order 7 bits of D0. If the parity is even, write 0 into bit 7 of D0 and if it is odd write 1 into bit 7 of D0. For example, if D0(7:0) = X0111011 (where X = don't care), the routine yields the value 10111011.

24. Write suitable 68000 assembly instructions to perform the following operations:

[D1] ← [D0]   and   [D1(0:15)] = 0

25. What is the difference between the MULU and MULS (and DIVU and DIVS) instructions?

26. What is the action performed by a MULU #10,D0 operation? Write the contents of D0.L before and after this instruction is executed (assume that D0 = $12345678 initially).

27. What is the difference between the following instructions?
   a. ADD   b. ADDQ
   c. ADDI   d. ADDA

28. Write a sequence of instructions to reverse the order of the bits of register D0. That is, D0(0) ← D0(31); D0(1) ← D0(30); ... D0(31) ← D0(0).

29. Describe the operation of the stack pointed at by A7. In which direction does it grow as items are pushed onto the stack?

30. Write a program to input a sequence of bytes and store them sequentially in memory, starting at location $00 2000. The sequence is to be terminated by a null byte, $00. Then print the even numbers in this sequence. Assume that an input routine, IN_CHAR at $F0 0000, inputs a byte into D0 and that OUT_CHAR at $F0 0004 outputs a byte in D0.

31. Write a subroutine to sort an array of N 8-bit elements into descending order. On entry to the subroutine, A2 contains the first (i.e., lowest) address of the array and D1 contains the size of the array (i.e., number of bytes).

32. Explain the meaning and significance of each of the following terms:
   a. Position-independent code (PIC)   b. Self-modifying code
   c. Reentrant code

33. A memory-mapped VDT displays the 1024, 8-bit characters starting at $00 F000 on a CRT terminal as 16 lines of 64 characters. The address of the top left-hand character is $00 F000 and the address of the bottom right-hand character is $00 F3FF.

Design a subroutine to display the ASCII character in D0 in the next free position of the display. A cursor made up of a row count and a column count points to the next free position into which a character is to be written. As each character is received, it is placed in the next column to the right of the current column counter.

Certain characters affect the position of the cursor without adding a new character to the display. These characters are as follows:

Carriage return	(ASCII $0D)	Move the cursor to the leftmost position on the current row
Line feed	(ASCII $0A)	Move the cursor to the same column position on the next row
Back space	(ASCII $08)	Move cursor back one space left
Space	(ASCII $20)	Code for space character

When the cursor is positioned on the bottom line of the display, a line feed causes all lines to move up one row (i.e., scroll up). This action creates a new, clear bottom line and causes the previous top line to be lost.

Construct a subroutine to implement this memory-mapped display. On entry to the subroutine, A0 points to the current row position and A1 points to the column position.

34. Design a cross-assembler for the hypothetical UNIFORM1 microprocessor to run on the 68000. Test the cross-assembler by assembling a UNIFORM1 program both by hand and by means of your cross-assembler.

The syntax of UNIFORM1 is

```
OPERATION      ⟨ea⟩
OPERATION      ⟨ea⟩,Ri
```

where OPERATION is the operation to be carried out, ⟨ea⟩ is an effective address, Ri is one of eight registers, and ⟨address⟩ is a 16-bit address. The structure of an instruction in UNIFORM1 is

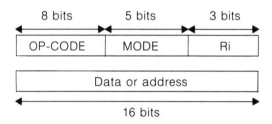

Valid registers are: R0, R1, ..., R7.
Valid addressing modes are

OPERATION	(Rj),Ri	Indexed addressing, modes 0 to 7
OPERATION	#⟨data⟩,Ri	Literal addressing, mode = 8
OPERATION	⟨address⟩,Ri	Absolute addressing, mode = 9
OPERATION	⟨address⟩	Jump, mode = 10
OPERATION	Rj,Ri	Register to register, modes 11 to 18
OPERATION	Ri	Monadic operation on Ri, mode 19
OPERATION	⟨address⟩	Monadic operation on memory, mode 20

The instructions and their op-codes provided by UNIFORM1 are

ADD	$00	CLR	$09
AND	$01	CMP	$0A
ASL	$02	LSL	$0B
ASR	$03	LSR	$0C
BEQ	$04	MOVE	$0D
BNE	$05	NEG	$0E
BCC	$06	OR	$0F
BCS	$07	SUB	$10
BRA	$08		

35. What are the major architectural differences between the 68000 and the 68020? If you were part of the 68020 design team, are there any other changes you would have included?

36. Describe how the 68020 has enhanced (a) the 68000's multiply instructions and (b) the 68000's divide instructions.

37. Why do bit field instructions number bits from the most significant bit of the effective address of the bit field?

38. Why do some of the 68020's new addressing modes incorporate a scale factor?

39. What is the difference between postindexing and preindexing (when applied to the 68020's memory indirect addressing)?

PROGRAM DESIGN

An engineer armed only with a knowledge of a microprocessor's assembly language would probably design rather poor programs. In this chapter we look at five of the ingredients of good program design: (1) top-down design, (2) modularity, (3) structured programming, (4) testability, and (5) recoverability. An appreciation of these topics allows us to design programs that are both large and reliable. All these topics are relevant to the design of both assembly language and high level language programs. Indeed, we would argue that these topics are equally relevant to the design of hardware.

One of the greatest difficulties facing the writer of assembly language programs is knowing where to begin. In the second part of this chapter we introduce the program design language (PDL), which is a form of high-level language that allows us to express a program algorithmically before we begin to code it into assembly language. We illustrate the application of PDL by providing two worked examples.

A few years ago a friend described the newly emerged microprocessor as "the last bastion of the amateur." He meant that although computer technology (both hardware and software) had come a long way by the mid-1970s, leaving little room in the industry for a sloppy or nonprofessional approach, the microprocessor soon changed that situation. Its very low cost brought it to anyone who wanted to investigate its properties. Therefore, people with little or no formal training or experience were using and programming microcomputers. Moreover, the tiny memories available in the 1970s and the lack of development tools forced many people to write programs in assembly language and to develop bad habits such as spaghetti programming. Some were even forced to write programs in hexadecimal machine code form.

Today, the status quo of the 1970s has been resumed and a professional approach to software design for microprocessors is the rule rather than the exception. It is no longer acceptable to throw together assembly language programs. This situation is not because of fashions in programming or sheer snobbishness, but because of the poor results that follow from a blind approach to programming in which the programmer attempts to solve a problem by immediately attempting to code it.

Few people would now argue that there is any intrinsic merit in programming in assembly language. Programmers use it when they do not have a compiler that produces efficient machine code from a high-level language. Also, some programmers are forced to employ assembly language when they require the greatest possible speed or when their high-level language does not permit them to do what they want.

In the final part of the chapter we look at the way in which an assembly language program can be constructed by means of a program design language (PDL). A PDL allows a programmer to take advantage of the program design techniques outlined in this chapter. Two examples of program design using PDL are presented.

3.1 TOP-DOWN DESIGN

Top-down design, or "programming by stepwise refinement," offers a method of handling large and complex programs by decomposing them into more tractable components. Top-down design is an iterative process that seeks to separate the goals of program design from the means of achievement. In other words, we must first decide what we want to do and later think about how to do it. A task is initially expressed in terms of a number of subtasks. Each of these subtasks is, in turn, broken down into further subtasks. Figure 3.1 illustrates this point. At the top of figure 3.1 (the highest level of abstraction) is a statement of the problem to be solved. The solution may be broken down into two separate actions, t_1 and t_2. At this level no details about the implementation of these subtasks need be considered. In turn, subtask t_1 may be divided into three further subtasks, $t_{1,1}$, $t_{1,2}$, and $t_{1,3}$. This process is repeated until each subtask has been fully elaborated and is expressed in terms of the most primitive actions available to the designer. Generally speaking, the lowest level of a task is, according to Shooman, "... small enough to grasp mentally and to code at one sitting in a straightforward and uncomplicated manner."

FIGURE 3.1 Top-down design

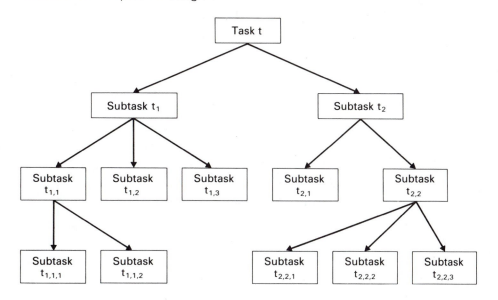

Figure 3.2 shows how this approach can be applied to the design of a disk operating system. The problem has, initially, been split up by a functional decomposition in which each of the subtasks carry out a well-defined function. At the highest level of task definition, the operating system is split into three basic subtasks: a command interpreter, a task scheduler, and a disk file manager (DFM).

Consider the DFM, which can itself be broken down into a number of operations on the data structures stored on the disk, and so on. This approach groups together relevant operations and their associated data at specific levels and attempts to stop irrelevant detail obscuring the action at a particular level. Thus, the operating system's file utilities manipulate only entire files; they are not interested in the actual organization and structure of the files. These functions belong to the DFM. Equally, the DFM is concerned only with operations on individual records; it relegates any operations involving the disk drive itself to lower-level software. An advantage of this approach is that each subtask can be validated independently of its own subtasks. Generally, a program no longer has to be tested as a single entity and its individual components, the subtasks, can be tested independently just like their hard-

FIGURE 3.2   Top-down decomposition

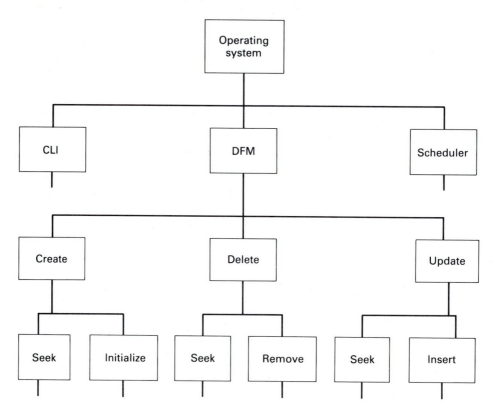

NOTE: For clarity, this diagram is not fully elaborated.

ware equivalents. Note that in figure 3.2 some of the subtasks appear several times; for example, the create, delete, and update subtasks each require the same lower-level subtask (i.e., seek).

One popular approach to program construction is called "top-down design and bottom-up coding." With this approach the problem is decomposed into levels of abstraction (top-down design), but the system is implemented by coding the lowest levels first (bottom-up coding); for example, when a word processor is coded, the elements dealing with the basic input/output of characters are dealt with first, then the elements that format the input/output, and so on.

System Specification

Before a system can be designed, it must be specified. The specification of a system is a statement of the goals that it must achieve. By setting goals, the specifier of the system gives the designer a method of objectively assessing any system he or she creates. Sometimes the specification can be related to an international standard; for example, a personal computer might have a "serial interface" or a "serial interface conforming to the EIA RS232C standard." In the latter case a copy of the specification can be obtained by the buyer to verify the operation of the system.

A tightly specified system is generally more reliable than a loosely specified system because a tight specification covers all possible eventualities (i.e., operating and input conditions). A loosely specified system may fail if certain input conditions have not been catered for; for example, a command in a machine code debugger may be designed to display the contents of memory location Address_1 through location Address_2. A poorly designed program would start by displaying the contents of Address_1, Address_1 + 1, and so on. A tightly specified system would first check that Address_1 was less than Address_2.

In addition to defining design goals, a helpful procedure is to define "nongoals." In other words, as well as defining what a system should do, we should also state what the system does not attempt to achieve. This action does not mean that obvious nongoals need be stated; there is little point in saying that a personal computer is not intended to fry eggs! A nongoal provides the system with explicit limitations that might otherwise be unclear; for example, a designer may state that security is a nongoal of an interface to store data on a floppy disk. This nongoal means that the data on the disk is not encrypted and can be read by any other interface without undergoing a decryption process. However, such a process does not mean that the user cannot suitably encrypt the source data before it is presented to the interface for recording on the disk.

3.2 MODULAR DESIGN

One of the most significant features of modern electronic systems is their modularity. A complex circuit is invariably decomposed into several less complex subsystems, called *modules*. The advantage of such an approach is that the modules can be designed and tested independently of the parent system. A module made by one manufacturer can be replaced by one from another manufacturer, as long as the two modules have the same interface to the rest of the system and are functionally equivalent. In the world of software, modularity is an attempt to treat software like hardware by creating software elements called modules. The principal requirement of a software module is that it is concerned only with a single, logically coherent task. For example, sin x can be considered as a module—it takes an input, x, and returns a result (i.e., the sine of x). It performs a single function and is *logically coherent* because it carries out only those actions necessary to calculate sin x. A module is not just an arbitrary segment of a larger program. An old story is told about the dark ages of programming when a programmer was asked to introduce modularity into his programs. He took a ruler and drew a line after every 75 statements. Those lines were his modules.

A software module is analogous to a hardware element because it has a number of inputs and a number of outputs and can be "plugged into a system." The module processes incoming information to yield one or more outputs. The internal operation of the module is both irrelevant to and hidden from its user, just as the transport properties of electrons in a semiconductor are irrelevant to the programmer of a microprocessor. The advantages of software modules are broadly the same as those of hardware modules. Modules can be tested and verified independently of the parent system and they can be supplied by manufacturers who know little or nothing about the parent system. Figure 3.3 illustrates the concept of a module.

The disk file monitor (DFM), introduced when we were describing top-down design, can be regarded as such a module. It is entered at one point, with details of the action to be carried out passed as parameters to the module. A return is made from the module to its calling point with parameters passed back as appropriate. It is no more reasonable to enter a module at some other point than it is to drill a hole in a floppy disk controller and to attach a wire to the silicon chip. Indeed, the whole point of modular design is to make the design, production, and testing of hardware and software almost identical; for example, the module for sin x takes an input and generates an output, sin x, together with relevant status information. Passing information to the inner working of sin x or getting information from within sin x is quite meaningless to higher level modules.

Module Coupling and Module Strength

A module is sometimes described in terms of two properties: *coupling* and *strength*. Coupling, as its name suggests, indicates how information is shared between one module and other modules. Tightly coupled modules share common data areas. This situation is regarded as undesirable because it makes it harder to isolate the action of one module from the action of other modules. We can liken this to a person with pneumonia and liver disease: If the person dies, how do you know which illness was the cause?

A module exhibiting loose coupling has data that is entirely independent of other modules. Any data it accesses is strictly private to the module and does not

FIGURE 3.3 Module

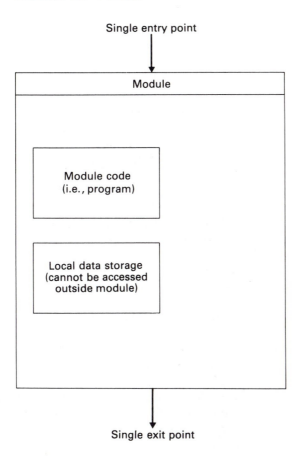

Single entry point

Module

Module code
(i.e., program)

Local data storage
(cannot be accessed
outside module)

Single exit point

interact with other modules. This condition makes it easier to debug systems because the programmer knows that data associated with a module is not modified by other modules. Therefore, errors can be speedily localized.

The strength of a module is a measure of its "modularity." In the story about the programmer who divided a program up into units of 75 lines, the resulting modularity is very weak because the modules are entirely arbitrary and functions are distributed between modules in a random fashion.

A strong module is one that performs a single task; for example, the sin x module discussed previously is strong because it does nothing more than calculate the sine of an input x. A weak module performs more than one task; for example, a module dealing with input from the keyboard, a disk drive, and a serial interface is relatively weak. It is a module in the sense that it performs logically related functions (i.e., input), but it is weak because several functions have been grouped together in one module.

The strength of a module is important, because strong modules are relatively easy to test as they perform only one function. Equally, a strong module from one supplier is not difficult to replace with a functionally equivalent module from another supplier. A weak module carries out several tasks, making it more difficult to test.

Passing Parameters to Modules

A module implemented as a subroutine must transfer information between the subroutine and the program calling it. The only exception occurs when a subroutine is called to trigger some event; for example, a subroutine can be designed to ring a bell or to sound an alarm. Simply calling the subroutine causes the predetermined action to take place. No communication exists between the subroutine and the program calling it.

Consider now the application of subroutines to inputting or outputting data. Obtaining a character from a keyboard or transferring one to a CRT terminal requires the execution of a number of actions and is inherently device dependent. Consequently, input and output transactions are frequently implemented by calling the appropriate input or output subroutine.

Suppose a program invokes the subroutine PUT_CHAR in order to display a single character on the CRT terminal. The character is printed by executing the instruction BSR PUT_CHAR and the whole process takes place automatically, due to the processor's stack mechanism, which takes care of the return address. In this example, the program has to transfer just one item of information to the subroutine, namely, the character to be displayed.

In this case, where the program passes only a single character to the subroutine, one of the eight data registers serves as a handy vehicle to transfer the information from the calling program to the subroutine; for example, if the character to be displayed is in register D7 and register D0 is used to carry the character to the subroutine, the following code may be written:

```
MOVE.B     D7,D0
BSR        PUT_CHAR
```

The preceding method of transferring data between a program and a subroutine by means of one or more registers is popular and is frequently used where the quantity of data to be transferred is very small or the processor is well endowed with registers. The advantage of this method is that it permits both position-independent code and reentrancy. Position independence is guaranteed because no absolute memory location is involved in the transfer of data and reentrancy is possible as long as the reuse of the subroutine saves the registers employed to transfer the data before they are reused.

The only disadvantage in passing information to and from subroutines via registers is that it reduces the number of registers available for use by the programmer. Moreover, the quantity of information that can be transferred is limited by the number of registers. In the case of the 68000, it is theoretically possible to transfer up to 15 longwords of data in this way. This total is made up of eight data registers and seven address registers. Address register A7, the stack pointer, cannot itself be used to transfer data.

Mechanisms for Parameter Passing

At this point it is worthwhile mentioning a number of concepts relevant to parameter passing. We need to distinguish between the ways in which parameters are passed and their implementation. The two basic ways of passing parameters are transfer by value and transfer by reference. In the former, the actual parameter is transferred; whereas in the latter, the address of the parameter is passed between program and subroutine. This distinction is important because it affects the way in which parameters are handled. When passed by value, the subroutine receives a copy of the parameter. Therefore, if the parameter is modified by the subroutine, the "new value" does not affect the "old value" of the parameter elsewhere in the program. In other words, passing a parameter by value causes the parameter to be cloned and the clone to be used by the subroutine. The clone never returns from the subroutine.

When a parameter is passed by address (i.e., by reference), the subroutine receives a pointer to the parameter. In this case, only one copy of the parameter exists, and the subroutine is able to access this unique value because it knows the address of the parameter. If the subroutine modifies the parameter, it modifies the parameter globally and not only within the subroutine.

The mechanism by which information is passed to subroutines generally falls into one of three categories: a register, a memory location, or the stack. We have already seen how a register is used to transfer an actual value. A region of memory can be treated as a mailbox and used by both the calling program and subroutine, with one placing data in the mailbox and the other emptying it. However, the stack mechanism offers the most convenient method of transferring information between a subroutine and its calling program.

Passing Parameters by Reference (i.e., Address)

Suppose a subroutine is written to search a region of memory containing text for the first occurrence of a particular sequence of characters. The sequence we are looking for is stored as a string in another region of memory. In this example, the subroutine requires four pieces of information: the starting and ending addresses of both the region to be searched and the string to be used in the matching process. Figure 3.4 illustrates this problem.

The information required by the subroutine is the four addresses, $00 1000, $00 100D, $00 1100, and $00 1103. Note that we are passing the parameters by reference, because the subroutine receives their addresses; we are not passing the actual parameters (i.e., the text strings) themselves. Eventually, the subroutine returns the value $00 1007. Although it is possible to transfer all these addresses via registers, an alternative technique is to assemble the four parameters into a block somewhere in memory and then pass the address of this block. Figure 3.5 shows how the block is arranged.

The only information required by the subroutine is the address $00 2000, which points to the first item in the block of parameters stored in memory. The following fragment of code shows how the subroutine is called and how the subroutine deals with the information passed to it. In this example, A0 is used to pass the address of the parameter block to the subroutine and A3, A4, A5, and A6 are used by the subroutine to point to the beginning and end of the text and of the string to be matched.

```
                    LEA       $002000,A0     Set up address of parameter block in A0
                    BSR       MATCH          Call string matching subroutine

                       .
                       .
                       .


        MATCH       MOVEM.L   A3–A6, – (SP)  Save address registers on stack
                    MOVEA.L   (A0) + ,A3     Put starting address of text in A3
                    MOVEA.L   (A0) + ,A4     Put ending address of text in A4
                    MOVEA.L   (A0) + ,A5     Put starting address of string in A5
                    MOVEA.L   (A0) + ,A6     Put ending address of string in A6

                       .
                       .                     } Body of subroutine
                       .


                    MOVEM.L   (SP) + ,A3–A6  Restore address registers from stack
                    RTS
```

In the preceding example, A0 is used as a pointer to the parameter block to obtain the four addresses required by the subroutine. We have passed a parameter by

FIGURE 3.4   Memory map of the string-matching problem

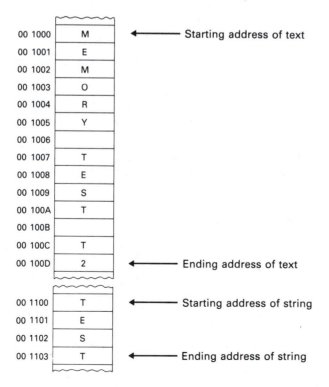

FIGURE 3.5 Passing a block of parameters by their address

00 2000	0000 1000
00 2004	0000 100D
00 2008	0000 1100
00 200C	0000 1103

All that the subroutine needs is the starting address of the block of parameters, that is, $00 2000.

its address (i.e., the address of the parameter block) and have transferred this address in a register (A0). Unfortunately, this method of storing information in a block whose location is fixed in memory cannot always be employed because such a program cannot be used reentrantly. Clearly, if the parameters are stored in a block with a fixed address (i.e., static memory location), any attempt to interrupt the subroutine and to reuse it must result in the new values overwriting the parameters. A much better approach is to use the stack to pass parameters or pointers to parameters. In this case, if the subroutine is interrupted, the new parameters are pushed onto the stack on top of the old parameters. When the interrupting program has used the subroutine, a return from interrupt is made with the stack in the same condition it was in immediately prior to the interrupt. Chapter 6 deals with interrupts in more detail.

Stack and Parameter Passing

The stack is useful not only for storing subroutine return addresses in such a way that subroutines may call further subroutines but also for transferring information to and from a subroutine. All that needs to be done is to push the parameters (or their addresses) on the stack before calling the subroutine. The following program fragment shows how this is done for our string-matching algorithm. In this example, we transfer all parameters by reference.

```
                PEA        TEXT_START        Push text starting address
                PEA        TEXT_END          Push text ending address
                PEA        STRING_START      Push string starting address
                PEA        STRING_END        Push string ending address
                BSR        STRING_MATCH      Call subroutine for matching
                LEA        16(SP),SP         Adjust stack pointer

                 .
                 .
                 .

STRING_MATCH    LEA        4(SP),A0          Put pointer to parameters in A0
                MOVEM.L    A3–A6,–(SP)       Save working registers on stack
                MOVEM.L    (A0)+,A3–A6       Get parameters off stack

                 .
                 .                           }  Body of subroutine
                 .

                MOVEM.L    (SP)+,A3–A6       Restore working registers
                RTS
```

In this example, the instruction PEA (push effective address) pushes the effective address of the operands onto the stack, but instead we could have used MOVE.L #TEXT_START,—(SP). Figure 3.6 illustrates the state of the stack during the execution of the preceding code. The first instruction in the subroutine, LEA 4(SP),A0, loads A0 with the starting address of the last parameter pushed onto the stack. We must add 4 to the stack pointer because the return address is at the top of the stack. The instruction MOVEM.L (A0)+,A3–A6 pulls the four addresses off the stack pointed at by A0 and deposits them in address registers A3 to A6 for use as required. Note that these parameters are left on the stack pointed at by SP after a return from the subroutine is executed.

In the calling program, the instruction LEA 16(SP),SP is executed after a return from the subroutine has been made. This instruction replaces the contents of the stack pointer with the contents of the stack pointer plus 16. Consequently, the stack pointer is restored to the position it was in before the four 32-bit parameters were pushed onto it.

As we have already pointed out, passing parameters on the stack facilitates position-independent code and permits reentrant programming. If a subroutine is interrupted, the stack builds upward and information currently on the stack is not overwritten.

FIGURE 3.6   Register, stack, and memory usage as program is executed

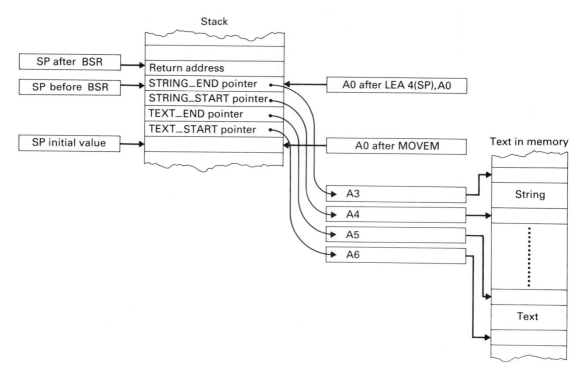

Stack and Local Variables

In addition to the parameters passed between itself and the calling program, most subroutines need a certain amount of *local workspace* for their temporary variables. The word local means that the workspace is private to the subroutine and is never accessed by the calling program or other subroutines. In a few circumstances, it is quite feasible to allocate a region of the system's memory space to a subroutine requiring a work area at the time the program is written. The programmer simply reserves fixed locations for the subroutine's variables in a process called *static allocation*. This process is entirely satisfactory for subroutines that are not going to be used reentrantly or recursively.

If a subroutine is to be made reentrant or is to be used recursively, its local variables must be bound up not only with the subroutine itself but with the occasion of its use. In other words, each time the subroutine is called, a new workspace must be assigned to it. Once again, the stack provides a convenient mechanism for implementing the dynamic allocation of workspace.

Closely associated with dynamic storage techniques for subroutines are the *stack frame* (SF) and the *frame pointer* (FP). The stack frame is a region of temporary storage at the top of the current stack. Figure 3.7 illustrates a stack frame.

Figure 3.7 shows how a stack frame is created merely by moving the stack pointer up by *d* locations. This can be done at the start of a subroutine. Note that the 68000 stack grows toward the low end of memory and therefore the stack pointer is decremented. Reserving 200 bytes of memory is achieved by LEA -200(SP),SP. Once the stack frame has been created, local variables are accessed by any addressing mode that uses A7 as a pointer. Before a return from the subroutine is made, the stack frame must be collapsed by LEA 200(SP),SP.

FIGURE 3.7 Stack frame

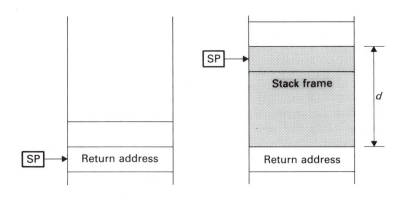

(a) The stack immediately after   (b) The allocation of a stack
      a subroutine call   frame by a subroutine

Link and Unlink Instructions

The simple scheme of figure 3.7 has been mechanized in the 68000 by a complementary pair of instructions, LINK and UNLK. These are relatively complex in terms of their detailed application but are conceptually simple. The great advantage of LINK and UNLK is their ability to let the 68000 manage the stack automatically and to make the memory allocation scheme entirely reentrant.

If the stack frame storage mechanism is to support reentrant programming, a new stack frame must be reserved each time a subroutine is called. As each successive stack frame is of a (possibly) variable size, its length must be preserved somewhere. The 68000 uses an address register for this purpose. In the following description of LINK and UNLK, a 16-bit signed constant, d, represents the size of the stack frame. At the start of a subroutine, the registers that must be saved are first pushed on the stack and then the LINK instruction is executed to create the stack frame belonging to *this* subroutine. The following code achieves the desired effect:

```
MOVEM.L    D0–D7/A3–A6, – (SP)    Save working registers on the stack
LINK       A1, # – 64              Allocate 64 bytes (16 longwords)
                                   of storage on this stack frame
```

In this example, all working registers are saved on the stack, the temporary storage allocation is 64 bytes, and address register A1 is used by LINK. Note that the minus sign is needed because the stack grows toward low memory. The action executed by LINK A1, # – 64 is defined in RTL terms as follows:

```
LINK: [SP]        ← [SP] – 4       Decrement the stack pointer by 4
      [M([SP])]   ← [A1]           Push the contents of address register A1
      [A1]        ← [SP]           Save stack pointer in A1
      [SP]        ← [SP] – 64      Move stack pointer up by 64 locations
```

Note that the old contents of A1 are not destroyed by this action; they are pushed on the stack. Similarly, the old value of the stack pointer is preserved in A1. In other words, the LINK destroys no information and, therefore, it is possible to undo the effect of a LINK at some later time. The state of the stack before a subroutine call, after the call and the MOVEM operation, and after a LINK instruction is given in figure 3.8.

After stage (c) in figure 3.8, the programmer can use the stack frame area as required; for example, LEA (SP),A2 loads the first free address of the frame into address register A2 and then A2 can be used as an offset in all references to the stack frame. Equally, all data references can be made with respect to the stack pointer or to A1. Life becomes interesting when the subroutine calls another subroutine with its own stack frame. Figure 3.9 depicts this situation.

In figure 3.9 at the initial state (a), subroutine 1 is being executed and has its own stack frame, labeled "Stack frame 1." Let us suppose that a second subroutine is invoked and another LINK A1, # – d executed. The second stack frame can be of any size and is not necessarily related to stack frame 1. This situation is illustrated in figure 3.9 at state (b). Register A1 now contains the value of the stack pointer immediately before the creation of stack frame 2; that is, A1 points to the location of the "old A1." The "old A1" is, of course, a pointer to the base of subroutine 1's

FIGURE 3.8 Effect of a LINK instruction on the stack

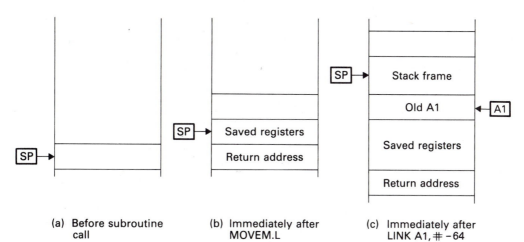

(a) Before subroutine call

(b) Immediately after MOVEM.L

(c) Immediately after LINK A1, # −64

FIGURE 3.9 Effect of a second call on the stack

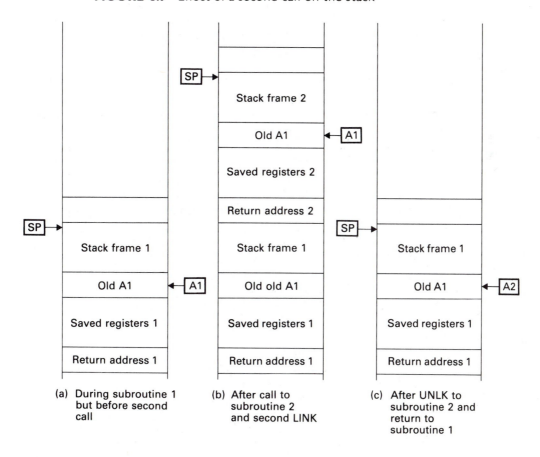

(a) During subroutine 1 but before second call

(b) After call to subroutine 2 and second LINK

(c) After UNLK to subroutine 2 and return to subroutine 1

stack frame, which holds the "old old A1." Because A1 points to the base (i.e., highest address) of the stack frame, all local variables can be accessed by register indirect addressing with displacement, where A1 is the register used to access them.

The next step is to show how an orderly return from subroutine 2 to subroutine 1 can be made. At the end of subroutine 2, the following sequence is executed:

```
UNLK        A1                    Deallocate subroutine 2's stack frame
MOVEM.L     (SP)+,D0–D7/A3–A6     Restore working registers from the stack
RTS                               Return to calling point
```

The RTL definition of UNLK A1 is

```
UNLK:       [SP] ← [A1]
            [A1] ← [M([SP])]
            [SP] ← [SP]+4
```

We can put this definition more clearly by stating that the stack point is first loaded with the contents of address register A1. Remember that A1 contains the value of the stack pointer just before stack frame 2 was created. In this way, stack frame 2 collapses. The next step is to pop the top item off the stack and place it in A1. This process has two effects: it returns both the stack and the contents of A1 to the points where they were located before LINK was executed.

Following the UNLK, the working registers can be pulled off the stack and a return to subroutine 1 made. The execution of subroutine 1 continues from the point at which it left off.

The key to LINK and UNLK is the use of A1 to hold the base of the stack frame and the storage of the *previous* contents of A1 below the base of the stack frame. The LINK and UNLK instructions have been included to support recursive procedures. A recursive procedure is one that calls itself.

The frame pointer is important because it points to the region of the stack allocated to the current procedure. All parameters passed to and from the local workspace can be accessed by means of the frame pointer. Moreover, the frame pointer can be used to access any local variables created by the procedure.

Consider the following example. A subroutine requires three parameters. Two parameters are called by *value* (P and Q) and one is called by *reference* (R). In this example, we shall use the subroutine to calculate $R = (P^2 + Q^2)/(P^2 - Q^2)$. Although it is not always necessary to save working registers on the stack, we do it in the following example because it is good practice. Having to remember that calling a certain subroutine corrupts the values of D6 and A6 is not as good as ensuring that D6 and A6 are not corrupted by the subroutine. We use the LINK instruction to create a two-longword stack frame (which the subroutine can use as temporary storage).

```
MOVE.W      D0,–(SP)     Push value of P on stack
MOVE.W      D1,–(SP)     Push value of Q on stack
PEA         R            Push reference to R on stack
BSR         CALC         Call the subroutine
LEA         8(SP),SP     Clean up stack by removing P, Q, R
:
```

CALC	MOVEM.L	D6/A6, −(SP)	Save working registers on the stack
	LINK	A0, # −8	Establish 2 longword frame for the stack
	MOVE.W	22(A0),D6	Get value of P from stack
	MULU	D6,D6	Calculate P^2
	MOVE.L	D6,−4(A0)	Save P^2 on the stack frame
	MOVE.L	D6,−8(A0)	Save P^2 on the stack frame—twice
	MOVE.W	20(A0),D6	Now get Q from the stack
	MULU	D6,D6	Calculate Q^2
	ADD.L	D6,−4(A0)	Store $P^2 + Q^2$ on the stack frame
	SUB.L	D6,−8(A0)	Store $P^2 − Q^2$ on the stack frame
	MOVE.L	−4(A0),D6	Qet $P^2 + Q^2$
	DIVU	−6(A0),D6	Calculate $(P^2 + Q^2)/(P^2 − Q^2)$
	LEA	16(A0),A6	Get pointer to address of R
	MOVEA.L	(A6),A6	Get actual address of R
	MOVE.W	D6,(A6)	Modify R in the calling routine
	UNLK	A0	Collapse the stack frame
	MOVEM.L	(SP) +,D6/A6	Restore working registers
	RTS		

Figure 3.10 illustrates the structure of the stack corresponding to the problem. Before the subroutine is called, the two MOVE.W ⟨register⟩, −(SP) instructions push the parameters P and Q onto the stack. The following instruction, PEA R, pushes the address of R onto the stack. Note that P and Q each take up a word and R takes up a longword because it is an address rather than the value of R.

After calling the subroutine, the return address is pushed on the stack. We also save the working registers, D6 and A6 which take up a total of two longwords on the stack above the return address.

The LINK A0, # −8 instruction pushes the old value of A0 (the frame pointer) on the stack and moves the stack pointer by 8 bytes. A0 is loaded with the value of the stack pointer immediately before it was moved 8 bytes. That is, A0 is the current frame pointer and is now pointing at the base of the stack frame. We can access all stack values by reference to A0 (we can also use the SP as a reference point). It is better to use A0 (i.e., the FP) than the stack pointer to refer to items in the stack frame, since A0 will be constant for the duration of this procedure, whereas the stack pointer may be modified.

At this stage the work of the subroutine can be carried out and parameters can be accessed from the stack. We get P and Q as values from the stack and then access R indirectly by reading its address from the stack. We access the value of P by MOVE.W 22(A0),D6 because A0 points to the base of the stack frame and we have to step past the old A0 (4 bytes), the saved registers (8 bytes), the return address (4 bytes), and R and Q (4 and 2 bytes) to access P itself. The total offset is 22 bytes. In this case we calculate the value of P^2 and store this longword at address −4(A0) the temporary storage created by the LINK A0, −8 instruction. We do this to demonstrate how the stack frame's space can be used to hold temporary variables. Note that the division instruction, DIVU −66(A0),D6, has an effective address of −6(A0). This corresponds to the low order word of $P^2 − Q^2$ on the stack, since the DIVU instruction takes a 16-bit source.

Accessing R, which is passed by reference, is more complex. The instruction LEA 16(A0),A6 loads the address of the parameter on the stack. However, the

FIGURE 3.10   Example of the use of a LINK instruction

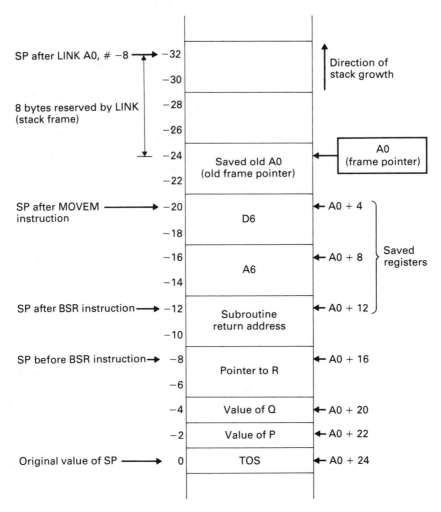

parameter on the stack is the address of R, and we must use the instruction MOVEA.L (A6),A6 to get a pointer to parameter R itself in A6.

After the procedure has done its work, the UNLK A0 instruction collapses the stack frame, discards the current frame pointer and restores the FP to its value immediately before the subroutine was called. Finally, the saved registers are restored from the stack and a return from subroutine is executed. In this example, the stack is in the state it was in before the subroutine call. However, since there are three parameters, taking a total of 8 bytes, left on the stack, we must (should?) clean up the stack by executing an LEA 8(SP),SP.

This example should demonstrate three things. The first is how the LINK and UNLK instructions work. The second is how the stack can be used to pass parameters both by reference and by value. The third is the inherent difficulty in assembly language programming! For example, in order to reference parameter P we had to execute MOVE.W 22(A0),D6, which meant that we had to have a clear picture of the location of P with respect to the frame pointer, A0. The chance of making a mistake in dealing with the stack in programs such as this is very great indeed. Such complex stack manipulation is best left to compilers.

3.3 STRUCTURED PROGRAMMING

Structured programming offers a semiformal method of writing programs and avoids the ad hoc methods of program design in widespread use before the late 1960s. The purpose of structured programming is threefold: It improves programmer productivity, it makes the resulting programs easier to read, and it yields more reliable programs. Essentially, structured programming techniques start from the axiom that all programs can be constructed from three fundamental components: the sequence, a generalized looping mechanism, and a decision mechanism. The rise of structured programming is largely attributed to the overenthusiastic use of the GOTO (i.e., JMP) by programmers in the 1960s. A program with many GOTOs provides a messy flow of control, making it very difficult to understand or to debug.

The *sequence* consists of a linear list of actions that are executed in order, one by one. If the sequence is P1, P2, P3, P1 is executed first, then P2, and then P3. The actions represented by P1, and so on, may be single operations or processes. The *process* is similar to the module described previously and has only one entry point and one exit point. Indeed, this very "process" is the one that is expanded into subtasks during top-down design.

The looping mechanism permits a sequence to be carried out a number of times. In many high-level languages, the looping mechanism takes the form DO WHILE or REPEAT UNTIL. The decision mechanism, which often surfaces as the IF THEN ELSE construct, allows one of two courses of action to be chosen, depending on the value of a test variable. By combining these three elements, any program can be constructed without using the GOTO statement.

Note that the pendulum has swung back a little way, and a few people now consider the total banishment of the GOTO to be a little unwise. Sometimes, in small doses, it can be used to good effect to produce a more elegant program than would otherwise result from keeping rigidly to the philosophy of structured programming.

Because of the importance of the decision and the loop mechanisms in structured programming, we will examine their form in high-level language and show how they can be implemented in assembly language.

Conditional Structure

The ability of a computer to make decisions can be called *conditional behavior*. As well as showing how the 68000 implements conditional behavior, it is a worthwhile exercise to demonstrate how such behavior appears in high-level languages. Consider two entities, L and S. L is a logical expression yielding a single logical value, which may be true or false. S is a statement that causes some action to be carried out. In what follows, the term conditional behavior is called control action, the most primitive form of which is expressed as

```
IF L THEN S
```

The logical expression L is evaluated and, if true, S is carried out. If L is false, S is not carried out and the next action following this control action is executed. For example, consider the expression IF INPUT_1 > INPUT_2 THEN OUTPUT = 4. If, say, INPUT_1 = 5 and INPUT_2 = 3, the logical expression is true and OUTPUT is made equal to 4.

A more useful form of this control action is expressed as

```
IF L THEN S1 ELSE S2
```

Here S1 and S2 are alternative statements. If L is true, then S1 is carried out; otherwise S2 is carried out. There are no circumstances where neither S1 nor S2, or both S1 and S2, may be carried out. Figure 3.11 illustrates the construct in diagrammatic form. For example, now consider the expression IF INPUT_1 > INPUT_2 THEN OUTPUT = 4 ELSE OUTPUT = 7. In this case, if INPUT_1 is greater than INPUT_2, OUTPUT is made equal to 4. Otherwise it is made equal to 7.

The IF L THEN S1 ELSE S2 control action can be extended to a more general form in which one of a number of possible statements, S1, S2, ..., Sn, is executed. As the logical expression L yields only a two-valued result, an expression generating an integer value must be used to effect the choice between S1, S2, ..., Sn. The multiple-choice control action is called a CASE statement in Pascal or a SELECT statement in some versions of BASIC. Here we use the term CASE statement, which is written:

```
CASE I OF
I1: S1
I2: S2
 :  :
In: Sn
END
```

The integer expression I is evaluated and, if it is equal to Ii, statement Si is executed. All statements Sj, where $j \neq i$, are ignored. Note that if I does not yield a value in the range I1 to In, an error may be flagged by the operating system. In some high-level languages, an exception is raised that provides an alternative course of action called an *exception*. Figure 3.12 illustrates the CASE statement.

In addition to the IF and the CASE statements, all high-level languages include a looping mechanism that permits the repeated execution of a statement S. There are four basic variants of the looping mechanism:

1. DO S FOREVER. Here statement S is repeated "forever." Because forever is an awfully long time, S must contain some way of abandoning or exiting the loop. Typically, an IF L THEN EXIT mechanism can be used to leave the loop. EXIT is a label that identifies a statement outside the loop. Strictly speaking,

FIGURE 3.11 IF L THEN S1 ELSE S2 construct

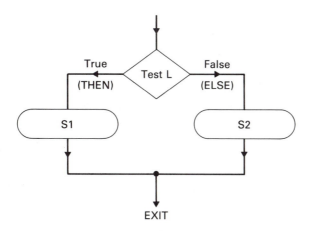

the IF ... THEN EXIT conditional behavior does not conform to the philosophy of structured programming. However, life without it can be very difficult.

2. FOR I:=N1 TO N2 DO S. In this case, the control variable I is given the successive integer values N1, N1 + 1, ..., N2, and statement S is executed once for each of the N2 − N1 + 1 values of I. This variant is the most conventional form of looping mechanism and is found in most high-level languages. Some programmers avoid this construct, because it leads to apparent ambiguity if N1 = N2 or if N1 > N2. In practice, high-level languages define

FIGURE 3.12 CASE construct

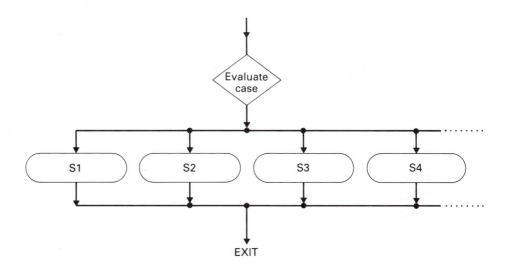

FIGURE 3.13   WHILE construct

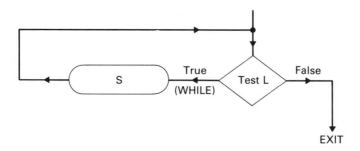

exactly what does happen when N1 > N2—but they do not all do the same thing.

3. WHILE L DO S. Whenever a WHILE construct is executed, the logical expression L is first evaluated and, if it yields a true value, the statement S is carried out. Then the WHILE construct is repeated. If, however, L is false, statement S is not carried out and the WHILE construct is not repeated. Figure 3.13 illustrates the action of a WHILE construct.

4. REPEAT S UNTIL L. The statement S is first carried out and then the logical expression L is evaluated. If L is true, the next statement following REPEAT ... UNTIL L is carried out. If L is false, statement S is repeated. The difference between the control actions of REPEAT ... UNTIL and WHILE ... DO is that the former causes S to be executed at least once and continues until L is true, whereas the latter tests L first and then executes S only if L is true. In other words, the REPEAT ... UNTIL tests the logical expression after executing S, whereas the WHILE ... DO tests the logical expression before executing S. Furthermore, REPEAT ... UNTIL terminates when L is true, but WHILE ... DO terminates when L is false. Figure 3.14 illustrates the REPEAT ... UNTIL statement.

FIGURE 3.14   REPEAT ... UNTIL construct

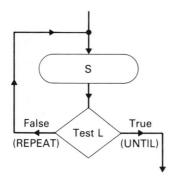

Implementing Conditional Expressions in Assembly Language

All the preceding high-level constructs can readily be translated into assembly language form. In the following examples, the logical value to be tested is in register D0 and is either 1 (true) or 0 (false). The action to be carried out is represented simply by S and consists of any sequence of actions.

```
*              IF L THEN S
*
               TST.B      D0            Test the lower-order byte of D0
               BEQ.S      EXIT          If zero then exit, else continue
               S                        Action S
EXIT
```
****
```
*              IF L THEN S1 ELSE S2
*
               TST.B      D0            Test the lower-order byte of D0
               BEQ.S      ELSE          If zero then S2 (ELSE part)
               S1                       If not zero then S1 (THEN part)
               BRA.S      EXIT          Skip past ELSE part
ELSE           S2                       Action S2 (ELSE part)
               :
EXIT
```
****
```
*              FOR I = N1 TO N2
*
               MOVE.B     #N1,D1        D1 is the loop counter
NEXT           S                        Action S (body of loop)
               ADDQ.B     #1,D1         Increment loop counter
               CMP.B      #N2+1,D1      Test for end of loop
               BNE        NEXT          Repeat until counter = N2 + 1
EXIT
```
****
```
*              WHILE L DO S
*
REPEAT         TST.B      D0            Test the lower-order byte of D0
               BEQ.S      EXIT          If zero then quit loop
               S                        Body of loop (action S)
               BRA        REPEAT        Repeat
EXIT
```
****
```
*              REPEAT S UNTIL L
*
NEXT           S                        Body of loop (action S)
               TST.B      D0            Test the lower-order byte of D0
               BEQ        NEXT          Repeat until true
EXIT
```
****
```
*              FOR I = N    DOWNTO −1     {using the DBRA instruction}
*
               MOVE.W     #N,D1         D1 is the loop counter
NEXT           S                        Action S (body of loop)
               DBRA       D1,NEXT       Decrement D1 and loop if not −1
```

3.4 TESTABILITY

In an ideal world where all stories have a happy ending and toast invariably falls buttered-side upward, design for testability would not be needed. Unfortunately, real-world components sometimes fail, circuits are incorrectly assembled, and the tracks on printed circuit boards break or become shorted together. Therefore, although a microcomputer may have been well designed, when finished it may sit miserably on the bench refusing to do anything.

Manufacturers of microcomputers must test their products. The amount of effort put into testing depends on the application for which the computer is intended and the economics of the situation. A manufacturer churning out ultra low cost microcomputers may perform minimal testing and offer a free replacement policy, as the cheaper alternative may be to replace (say 10 percent of the computers than to fully test all of them. However, the designer of an automatic landing system for a large jet has to deliver a reliable product; simply letting the consumers do their own testing in the field might have unfortunate consequences.

Design for testability is as important for software components as for hardware elements. The designers of electronic hardware approach testability by providing test points at which the value of a signal can be examined as it flows through a system. Because almost all digital systems operate in a closed-loop mode, the test engineer must break the loop in order to separate cause and effect. Typically, this procedure involves either single stepping the processor or injecting known information into, say, the processor's address pins and observing the effect on the rest of the system. The same tests can be applied to software design. Test points become the interfaces between the modules. Open-loop testing is carried out by breaking the links between software modules, just as is done with hardware. Generally, test points in software are implemented by breakpoints. These are software markers that print out a snapshot of the state of the processor whenever they are encountered during the execution of a program.

Two philosophies of program testing exist: top-down and bottom-up. Bottom-up testing involves testing the lowest level components of a system first. For example, the DFM of figure 3.2 might be tested by first verifying the operation of software primitives such as read/write a sector. When the lowest-level components have been tested, the next level up is examined; in this case, "insert module" of the DFM might use the read/write sector primitives that have already been tested.

Bottom-up testing is complete when the highest level of the system has been tested, which is an attractive philosophy because it is relatively easy to implement. A module is tested by constructing the software equivalent of the *fixture* or *jig* used to test hardware in industry. The software fixture provides a testing environment for the module, which cannot stand on its own; for example, a module to read sectors would require software to pass the necessary track and sector addresses to it, examine the returned data, and display the error status at the end of an operation. Note that high level modules may call low level modules during the test, safe in the knowledge that the low level modules have already been tested.

In top-down testing, the highest levels are tested first. This procedure does not require much in the way of a fixture, as the high level modules are tested by giving them the task that the application software is to perform. This philosophy is in line with the top-down design approach and has the advantage that major design errors may be spotted in an early phase. Because high level modules call lower level modules, top-down testing cannot rely on untested lower level modules. In order to deal with this situation, lower level modules are replaced by "stubs." These stubs are dummies that simulate the modules they replace. For example, in testing the DFM, a "read sector" operation must be replaced by a stub that obtains data from an array rather than from the disk drive.

In recent years the term *white-box testing* has become fashionable. If a black box represents the *unknown*, then its converse, the *white* box, represents the known. When we say that a black box is unknown, we mean that we do not know how it operates internally. All we can do is observe its inputs and outputs—not its internal state. The implication of this statement is that a black box must be tested exhaustively with all possible inputs. If we know something about the contents of the black box (i.e., how it operates and how it is put together), we can devise tests that exercise its various functions. We call such a system a *white box*.

For example, a 16-bit multiplier can be treated as a black box and tested by applying to it all possible combinations of inputs (i.e., 2^{32} different values). If we know, however, that each of the two inputs are constrained to lie within certain ranges due to some property of the system using the multiplier, white-box testing can be performed by testing only those inputs that can occur in practice.

3.5 RECOVERABILITY

Recoverability, or exception handling, is the ability of a system to cope with erroneous data and to recover from certain classes of error. Just as students are taught that real gates do not always behave as expected, they should be taught that real software sometimes encounters faulty data. They can then consider the design of software filters in the same way as the design of hardware filters to remove or to attenuate electrical noise.

The next step is to decide what action to take if a software module fails to achieve the intended results because of faulty software or hardware elsewhere in the system. For example, what does a DFM do if one of the procedures it calls fails to execute an operation such as reading a track? Clearly, permitting the system to hang up in an infinite pooling loop is as bad as letting a hardware device hang up because it has not received a handshake from a peripheral. Designers should always consider the possible forms that recovery may take. Several attempts can be made to read the track to distinguish between a soft and a hard error. If the error persists, a graceful recovery may be attempted. A user who has spent 2 hours editing a text file would not be very happy if the system collapsed due to a minor error. A better approach would be to save as much of the text as possible and then report the error.

Exception handling is, to some extent, a controversial subject, as a poorly designed or ill-conceived error-recovery mechanism may be far worse than nothing at all. The 68000 provides several exception-handling mechanisms, some at a software level and some at a hardware level.

3.6   PSEUDOCODE (PDL)

This section presents a possible technique for the production of assembly language programs. The history of computer science, like everything else, is two steps forward and one step backward. As the hardware of computers has advanced, a corresponding progress in software has been achieved. The first computers were programmed in machine code. Progress was later made to assembly language, to early high-level languages such as FORTRAN, and then to block-structured high-level languages (HLLs) such as Pascal or Modula 2. Although a high-level language compiled into machine code by an automatic compiler may not execute as rapidly as an equivalent program written by hand in assembly language, few people would argue against the proposition that the HLL greatly improves programmer productivity and program reliability over assembly language programming.

Unfortunately, microprocessor-based machines have traditionally had small random access memories and poor secondary storage facilities. Consequently, relatively primitive languages such as BASIC have evolved, and these have tended to be interpreted rather than compiled. An interpreted BASIC program is very much slower than either a compiled BASIC or a pure assembly language program. This situation has forced those who require low execution times to take the one step backward and return to assembly language programming. Indeed, some applications such as animated graphics or speech processing make assembly language programming almost mandatory.

One of the reasons for the unpopularity of assembly language (at least in professional circles) is the difficulty of writing, debugging, documenting, and maintaining such programs. The productivity of a programmer writing in, say, Pascal is almost certainly far greater than that of one writing in assembly language.

Even though the early 1980s saw a return (by some) to assembly language programming, no excuse exists for developing assembly language programs in a sloppy or ad hoc form. Today's programmers have a great many tools to help them write programs that were not available to their predecessors. One software tool employed by some assembly language programmers is called *pseudocode* or *program development language* (PDL). A PDL offers a way of writing assembly language programs using both top-down design techniques and structured constructs. Unlike real high-level languages, the PDL is a personalized pseudo HLL; that is, the programmers may design their own PDLs. The result is that the conventions adopted by one programmer may not be those adopted by another, but as long as any given PDL is self-consistent the end product will not be sloppy.

Characteristics of a PDL

A PDL is nothing more than a convenient method of writing down an algorithm before it is coded into assembly language form; for example, a flowchart could be considered to be one type of PDL. Clearly, a program in PDL form is easier to code than to try and code a problem into assembly language without going through any intermediate steps. The features of a PDL are summed up as follows:

1. A PDL represents a practical compromise between a high-level language and an assembly language but lacks the complexity of the former and the "obscurity" of the latter.

2. The purpose of a PDL is to facilitate the production of reliable code in circumstances in which a high-level language is not available or not appropriate.

3. A PDL shares some of the features of HLLs but rejects their overall complexity; for example, it supports good programming techniques including top-down design, modularity, and structured programming. Similarly, it may support primitive data structures.

4. A PDL provides a shorthand notation for the description of algorithms and allows the use of plain English words to substitute for entire expressions. This feature gives it its strong top-down design facilities.

5. A PDL is extensible. The syntax of a PDL can be extended to deal with the task to which it is applied.

Using PDL

The best way to introduce a PDL is by means of an example. Consider a 68000-based system with a software module that is employed by both input and output routines and whose function is to buffer data. When called by the input routine, a character is added to the buffer and when called by the output routine, a character is removed from it. The operational parameters of the subroutine are as follows:

1. Register D0 is to be used for character input and output. The character is an 8-bit value and occupies the lowest-order byte of D0.

2. Register D1 contains the code 0, 1, or 2 on entering the subroutine. Code 0 is interpreted as clear the buffer and reset all pointers. Code 1 is interpreted as place the character in D0 into the buffer. Code 2 is interpreted as remove a character from the buffer and place it in D0. We may assume that a higher-level module ensures that only one of 0, 1, or 2 is passed to the module.

3. The location of the first entry in the buffer is at $01 0000 and the buffer size is 1,024 bytes. Pointers and scratch storage may be placed after the end of the buffer, as long as no more than 32 bytes of storage are used.

4. If the buffer is full, the addition of a new character overwrites the oldest character in the buffer. In this case, bit 31 of D0 is set to indicate overflow and cleared otherwise.

5. If the buffer is empty, the subtraction of a new character results in the contents of the lower byte of D0 being set to zero and its most significant bit set as in parameter 4.

6. Apart from D0, no other registers are modified by a call to this subroutine.

Figure 3.15 shows the memory map corresponding to this problem. The map can be drawn at an early stage as it is relatively straightforward. Obviously, a region of 1,024 bytes ($400) must be reserved for the buffer together with at least two 32-bit pointers. IN_POINTER points to the location of the next free position into which a new character is to be placed and OUT_POINTER points to the location of the next character to be removed from the buffer. At the righthand side of figure 3.15 is the logical arrangement of the circular buffer. This arrangement provides the programmer with a better mental image of how the process is to operate.

The first level of abstraction in PDL is to determine the overall action that the module is to perform. This can be written as follows:

```
Module: Circular_buffer
        Save working registers
        Select one of:
                        Initialize system
                        Input a character
                        Output a character
               Restore working registers
        End Circular_buffer
```

FIGURE 3.15 Circular buffer

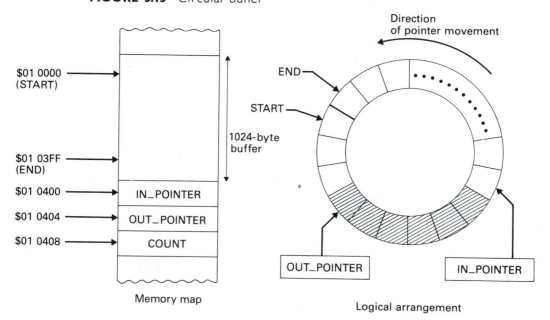

Memory map

Logical arrangement

Note that this PDL description is written in almost plain English: any programmer should be able to follow another programmer's PDL. Try doing that in LISP or FORTH! No indication is given of how any action is to be carried out, and the only control structure is the selection of one of three possible functions. The next step is to elaborate on some of these actions.

```
Module: Circular_buffer
        Save working registers
        IF [D1] = 0 THEN Initialize END_IF
        IF [D1] = 1 THEN Input_character END_IF
        IF [D1] = 2 THEN Output_character END_IF
        Restore working registers
End Circular_buffer

Initialize
        Count:=0
        In_pointer:=Start
        Out_pointer:=Start
End Initialize

Input_character
            Store new character
            Deal with any overflow
End Input_character

Output_character
            IF buffer NOT empty THEN Get character from buffer
                                ELSE Set error flag, return null character
            END_IF
End Output_character
```

At this point, the PDL is getting fairly detailed. Both the module selection and the initialization routines are complete. We still have to work on the input and output routines because of the difficulty in dealing with overflow and underflow in a circular buffer.

Looking at the circular buffer of figure 3.15, it seems reasonable to determine the state of the buffer by means of a variable, COUNT, that indicates the number of characters in the buffer. If COUNT is greater than zero and less than its maximum value, a new character can be added or one removed without any complexity. If COUNT is zero, the buffer is empty and we can add a character but not remove one. If COUNT is equal to its maximum value and therefore the buffer is full, each new character must overwrite the oldest character as specified by the program requirements. This last step is tricky because the next character to be output (the oldest character in the buffer) is overwritten by the latest character. Therefore, the next character to be output will now be the oldest surviving character, and the pointer to the output must be moved to reflect this condition.

Rewriting the entire module in PDL is not necessary, as the first two sections have been resolved into enough detail to allow coding into assembly language. The input and output routines from which the assembly language coding begin are as follows:

Input_character
Store new character at In_pointer
In_pointer:=In_pointer + 1
IF In_pointer > End THEN In_pointer:=Start END_IF
IF Count < Max THEN Count:=Count + 1
ELSE
BEGIN
Set overflow flag
Out_pointer:=Out_pointer + 1
IF Out_pointer > End THEN Out_pointer:=Start
END_IF
END
END_IF
End Input_character

Output_character
IF Count = 0 THEN return null and set underflow flag
ELSE
BEGIN
Count:=Count − 1
Get character pointed at by Out_pointer
Out_pointer:=Out_pointer + 1
IF Out_pointer > End THEN Out_pointer:=Start END_IF
END
END_IF
End Output_character

The program design language has now done its job and the routines can be translated into the appropriate assembly language.

```
CIRC      EQU *          This module implements a circular buffer
          MOVEM.L   A0–A1,−(SP)       Save working registers
          BCLR.L    #31,D0            Clear bit 31 of D0 (no error)
          CMPI.B    #0,D1             Test for initialize request
          BNE.S     CIRC1             IF not 0 THEN next test
          BSR.S     INITIAL           IF 0 THEN perform initialize
          BRA.S     CIRC3             and exit
CIRC1     CMPI.B    #1,D1             Test for input request
          BNE.S     CIRC2             IF not input THEN output
          BSR.S     INPUT             IF 1 THEN INPUT
          BRA.S     CIRC3             and exit
CIRC2     BSR.S     OUTPUT            By default OUTPUT
CIRC3     MOVEM.L   (SP)+,A0–A1       Restore working registers
          RTS                         End CIRCULAR
*
INITIAL   EQU *          This module sets up the circular buffer
          CLR.W     COUNT             Initialize pointers
          MOVE.L    #START,IN_POINTER      Set up In_pointer
          MOVE.L    #START,OUT_POINTER     Set up Out_pointer
          RTS
*
INPUT     EQU *          This module stores a character in the buffer
          MOVEA.L   IN_POINTER,A0     Get pointer to input
          MOVE.B    D0,(A0)+          Store character in buffer, update pointer
          CMPA.L    #END+1,A0         Test for wrap-round
          BNE.S     INPUT1            IF not end THEN skip reposition
          MOVEA.L   #START,A0         Reposition input pointer
```

```
INPUT1      MOVE.L      A0,IN_POINTER         Save updated pointer
            CMPI.W      #MAX,COUNT            Is buffer full?
            BEQ.S       INPUT2                IF full THEN deal with overflow
            ADDQ.W      #1,COUNT              ELSE increment character count
            RTS                                    and return
INPUT2      BSET.L      #31,D0                Set overflow flag
            MOVEA.L     OUT_POINTER,A0        Get output pointer
            LEA         1(A0),A0              Increment Out_pointer
            CMPA.L      #END+1,A0             Test for wrap-round
            BNE.S       INPUT3                IF not wrap-round THEN skip fix
            MOVEA.L     #START,A0             ELSE wrap-round Out_pointer
INPUT3      MOVE.L      A0,OUT_POINTER        Update Out_pointer in memory
            RTS                                    and return
*
OUTPUT      TST.W       COUNT                 Examine state of buffer
            BNE.S       OUTPUT1               IF buffer not empty output character
            CLR.B       D0                    ELSE return null output
            BSET.L      #31,D0                     set underflow flag
            RTS                                    and exit
OUTPUT1     SUBQ.W      #1,COUNT              Decrement COUNT for removal
            MOVEA.L     OUT_POINTER,A0        Point to next character to be output
            MOVE.B      (A0)+,D0              Get character and update pointer
            CMPA.L      #END+1,A0             Test for wrap-round
            BNE.S       OUTPUT2               IF not wrap-round THEN exit
            MOVEA.L     #START,A0             ELSE wrap-round Out_pointer
OUTPUT2     MOVE.L      A0,OUT_POINTER        Restore Out_pointer in memory
            RTS
```

The greatest flaw in the operational characteristics of our circular buffer is the way in which it deals with overflow. We designed the program to overwrite the oldest unread character in the buffer with the newest character. A more usual and simple approach is to reject any new characters once the buffer is full. In order to provide the host program with sufficient warning that the buffer is becoming dangerously full, a circular buffer sometimes returns a 2-bit code that indicates its state; for example, 0,0 = less than one-quarter full, 0,1 = less than half full, 1,0 = less than three-quarters full, and 1,1 = full.

Using PDL to Design a
Command Line Interpreter (CLI)

The design of a command line interpreter (CLI) provides a somewhat more extended example of the application of a PDL to the construction of assembly language programs. A CLI is a program that collects a line of text from a processor's keyboard, parses it to extract a command, and then executes the command. Clearly, the CLI is often found in many text editors, word processors, monitors, and even interpreted languages.

The particular CLI to be designed in this case is intended to form part of a microprocessor system's monitor. Such a monitor is part of many low-cost, single board, educational computers and permits students to enter an assembly language program (sometimes in hexadecimal form only), preset memory locations, and then to execute the program. Frequently, a monitor provides debugging aids, allowing the students to follow the execution of their programs by displaying intermediate results on a display terminal. Chapter 11 considers further the development of a very basic monitor for a 68000 system.

The specifications of the CLI are given as follows:

1. Text is entered into a 64-character (i.e., byte) buffer and is terminated by a carriage return. The longest possible string that we may input is 63 characters plus a carriage return. Note that the carriage return is to be stored in the buffer.

2. In addition to the carriage return terminator, the CLI is to recognize the back space and the abort (ASCII control A) characters. A backspace deletes the last character received. If the line buffer is empty, inputting one or more back-spaces has no effect. An abort character forces an immediate exit from the CLI routine with the carry bit of the CCR set to a logical 1. Otherwise a return from the CLI is made only after a carriage return has been received. In this case, the state of the carry bit of the CCR is determined by the interpreter part of the routine.

3. Should more than 63 alphanumeric characters be received, representing a buffer overflow condition, the message "Buffer full—reenter command" is displayed on a new line, a further newline command issued, and the input procedure initiated as if the routine were being invoked for the first time.

4. Once a command line has been successfully input, it is interpreted. The first word in the input buffer is matched against the entries in a command table. If a successful match is found, the appropriate jump (transfer) address of the command is loaded into address register A0, the carry bit of the CCR cleared, and a return from the CLI made. If a match cannot be found, a return from the CLI is executed with the carry bit of the CCR set. In this case the contents of address register A0 are undefined.

5. The CLI may assume the existence of a subroutine GETCHAR that, when called, returns with an 8-bit ASCII character in the lower byte of data register D0. We assume that GETCHAR takes care of any character echo. The subroutine NEWLINE moves the cursor to the left-hand column of the next line on the display. The subroutine PSTRING displays the text string composed of ASCII characters occupying consecutive byte locations and terminated by a null (zero byte) pointed at by address register A4. These three subroutines have no effect on any register not mentioned previously.

6. No register other than A0 and the CCR must be affected by the CLI following a return from the CLI.

7. The structure of the command table is defined as in figure 3.16.

Each entry has four fields: (1) a byte that indicates the length of the command name in the table, (2) a byte that indicates the minimum number of characters that must be matched (e.g., If the value is three then MEM, MEMO, MEMOR, or MEMORY may be typed to invoke the MEMORY command), (3) the command string itself, whose length is given by entry (1), and (4) the 4-byte transfer address.

FIGURE 3.16 Structure of the command table

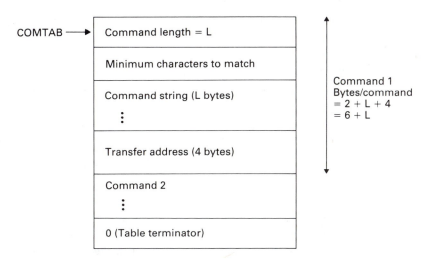

For example, the first two entries in the command table may be as shown in figure 3.17.

Having specified the problem, the next step is to translate it into a sufficiently elaborated PDL that its ultimate translation into 68000 assembly language involves no complex steps; that is, no single PDL statement should require translating into a stream of assembly language requiring embedded branch instructions of any kind.

FIGURE 3.17 Command table

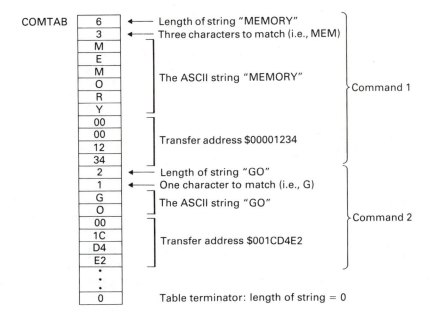

PDL Level 1

The lowest level of PDL is rather trivial and is included here for the sake of completeness and to illustrate the process of top-down design.

Module: CLI

      Initialize variables

      Assemble the input line

      Interpret the line of text

      Exit

End CLI

PDL Level 2

At level 2 we cannot fully elaborate any part of the level 1 design as we do not, for example, yet know what variables must be initialized. We do know that registers must not be corrupted by this program, so we can make a note of this condition. Similarly, we can begin to consider the type of actions to be carried out by the major steps in level 1.

Module: CLI
    Initialize
        Save working registers
        Initialize any variables
        Clear buffer
    Assemble
        REPEAT
            Get character
            Deal with special cases
            Store character
        UNTIL   character = carriage return
    Interpret
        Match command with command table
        IF successful THEN Extract transfer address
                   ELSE Set error indicator
        END_IF
    Exit
        Restore working registers
End CLI

PDL Level 3

At this stage, we can begin to consider the details of the actions defined at level 2. If we are lucky, this is the highest level of PDL necessary; otherwise another iteration (i.e., a higher PDL level) will be necessary. A little common sense tells us that this routine requires few variables other than pointers for the line buffer and for the table with which the match is made. However, there are enough variables to require some form of type declaration block as an aid to understanding the program.

Initialize

      Array of bytes: Comtab
      Array of bytes: Buffer
      Longword: Buffer_pointer
      Longword: Table_pointer
      Longword: Next_entry
      Longword: Transfer_address
      Byte: String_length
      Byte: Chars_to_match
      Byte: Buffer_length = 64
      Equate: Buffer_start = Buffer
      Equate: Buffer_end: = Buffer_start + Buffer_length − 1
      Equate: Buffer_pointer: = Buffer_start
      Equate: Table_address = Comtab
      Equate: Back_space = $08
      Equate: Abort = $01
      Equate: Carriage_return = $0D

      Save working registers on system stack
      Clear_buffer

End initialize

Clear_buffer

      FOR I: = 0 TO Buffer_length − 1
         Buffer[I]: = 0
      END_FOR

End Clear_buffer

Assemble

   REPEAT
     IF Buffer_pointer = Buffer_end − 1
       THEN BEGIN
            Newline
            Display error message
            Newline
            Buffer_pointer: = Buffer_start
            Clear_buffer
            END
     END_IF
     Get character
     IF character = Abort THEN Set Error_flag
                           Exit
     END_IF
     IF character = Back_space
       THEN BEGIN
           IF Buffer_pointer > Buffer_start
              THEN Buffer_pointer: = Buffer_pointer − 1
           END_IF
           END
       ELSE BEGIN
           Store character
           Buffer_pointer: = Buffer_pointer + 1
           END
     END_IF
  UNTIL character = carriage_return

End Assemble

Interpret

```
                Table_pointer: = Table_address
                REPEAT
                    Buffer_pointer: = Buffer_start
                    String_length: = Comtab[Table_pointer]
                    IF  String_length = 0 THEN
                                                    Set Fail
                                                    Exit
                    END_IF
                    Chars_to_match: = Comtab[Table_pointer + 1]
                    Next_entry: = Table_pointer + String_length + 6
                    Table_pointer: = Table_pointer + 2
                    Match
                    IF  Success THEN
                                    BEGIN
                                    Transfer_address: = Comtab[Next_entry − 4]
                                    Exit
                                    END
                                    ELSE Table_pointer : = Next_entry
                    END_IF
                END_REPEAT
        End Interpret
```

Match

```
                Set Fail
                REPEAT
                    IF  Buffer[Buffer_pointer] ≠ Comtab[Table_pointer]
                            THEN  BEGIN
                                    Set Fail
                                    Exit Match
                                    END
                            ELSE  BEGIN
                                    Table_pointer: = Table_pointer + 1
                                    Buffer_pointer: = Buffer_pointer + 1
                                    END
                    END_IF
                    Chars_to_match: = Chars_to_match − 1
                    UNTIL  Chars_to_match = 0
                    Set Success
        End Match
```

Exit

```
                Restore registers from system stack
                Return
        End Exit
```

Having reached this level of detail, we can readily translate the PDL into the 68000 assembly language. As no fixed storage is necessary, apart from the line buffer, it is possible to use the 68000's registers as the pointers and temporary storage.

```
        BUFFER_LENGTH       EQU   64
        BACK_SPACE          EQU   $08
        ABORT               EQU   $01
        CARRIAGE_RETURN     EQU   $0D
        BUFFER_START        EQU   $008000
        BUFFER_END          EQU   BUFFER_START + BUFFER_LENGTH − 1
```

```
COMTAB          EQU    $009000
*
* Registers       A0 = Pointer to transfer address
*                 A1 = Buffer_pointer
*                 A2 = Table_pointer
*                 A3 = Next_entry
*                 A4 = Pointer to error message
*                 D0 = Contains character from console
*                 D1 = String_length
*                 D2 = Chars_to_match
*
***************************************************************************
*
* INITIALIZE saves the working registers and clears the line buffer.
* Note that EXIT restores the registers
*
                ORG    $001000                    Program origin
INITIALIZE      MOVEM.L  A1–A4/D0–D2,–(SP)        Save registers
                BSR    CLEAR                       Clear the buffer
*                                                  Fall through to ASSEMBLE
*
*
***************************************************************************
*
* ASSEMBLE reads a line of text and stores it in the line buffer.
* A back space deletes the previous character and an abort forces
* an exit from the CLI with the C flag set
*
ASSEMBLE        CMPA.L   #BUFFER_END+1,A1         Test for end of line buffer
                BNE.S    ASSEMBLE1               IF not full THEN continue
                BSR    NEWLINE                    ELSE display error
                LEA    MESSAGE,A4                  Point to message
                BSR    PSTRING                    Print it
                BSR    NEWLINE
                BSR.S    CLEAR                     Clear buffer
ASSEMBLE1       BSR    GETCHAR                     Input an ASCII character
                CMPI.B   #ABORT,D0                Test for abort
                BNE.S    ASSEMBLE2               IF not abort THEN skip exit
                ORI    #$01,CCR                    ELSE set error flag
                BRA.S    EXIT                      and leave CLI
ASSEMBLE2       CMPI.B   #BACK_SPACE,D0           Test for back space
                BNE.S    ASSEMBLE3               IF not back space THEN skip
                CMPA.L   #BUFFER_START,A1         Test for buffer empty
                BEQ    ASSEMBLE1                  IF empty THEN get new character
                LEA    –1(A1),A1                   ELSE move back pointer
                BRA    ASSEMBLE1                   and get new character
ASSEMBLE3       MOVE.B   D0,(A1)+                  Store the input
                CMPI.B   #CARRIAGE_RETURN,D0      Test for terminator
                BNE    ASSEMBLE                    IF not terminator repeat
*                                                  ELSE fall through to interpret
***************************************************************************
*
* INTERPRET takes the first word in the line buffer and tries
* to match it with an entry in COMTAB
*
```

```
INTERPRET    LEA       COMTAB,A2            A2 points to command table
INTERPRET1   LEA       BUFFER_START,A1     A1 points to line buffer
             CLR.L     D1                  Clear all 32 bits of D1
             MOVE.B    (A2),D1             Get string length
             BNE.S     INTERPRET2          IF not zero THEN continue
             ORI       #$01,CCR            ELSE set error flag
             BRA.S     EXIT                and return from CLI
INTERPRET2   MOVE.B    1(A2),D2            Get character to match from table
             LEA       6(A2,D1),A3         Get address of next entry
             LEA       2(A2),A2            Point to string in COMTAB
             BSR.S     MATCH               Match it with line buffer
             BCC.S     INTERPRET3          IF C clear THEN success
             LEA       (A3),A2             ELSE point to next entry
             BRA       INTERPRET1          Try again
INTERPRET3   LEA       −4(A3),A0           Get transfer address pointer
             MOVEA.L   (A0),A0             Get transfer address from COMTAB
*                                          Fall through to exit
*
************************************************************************************
* EXIT is the only way out of CLI. Restore registers and exit with
* C clear for success and C set for failure. If C = 0, the
* transfer address is in A0
*
EXIT         MOVEM.L   (SP)+,A1−A4/D0−D2   Restore registers
             RTS                           Return
*
CLEAR        MOVE.W    #BUFFER_LENGTH−1,D0 Clear the input buffer
             LEA       BUFFER_START,A1
CLEAR1       CLR.B     (A1)+
             DBF       D0,CLEAR1
             LEA       BUFFER_START,A1     Reset A1 to start of buffer
             RTS
*
************************************************************************************
*
* MATCH compares two strings for equality. The number of characters
* to match is in D2
* A1 = Buffer_pointer, A2 = Table_pointer
*
MATCH        CMPM.B    (A1)+,(A2)+         Compare two characters
             BNE.S     MATCH1              IF not equal THEN return
             SUBQ.B    #1,D2               ELSE decrement match count
             BNE       MATCH               IF not zero test next pair
             ANDI      #$FE,CCR            ELSE success clear C flag
             RTS                           and return
MATCH1       ORI       #$01,CCR            Set fail flag
             RTS                           and return
```

SUMMARY

In this chapter we have looked at some of the issues raised by the design of assembly language programs. Now that the semiconductor manufacturers have given us such powerful microprocessors that support high-level constructs and modularity, creating assembly language programs in an undisciplined way becomes unreasonable. If programmers are still forced to resort to assembly language programming for reasons of speed or efficiency, they should use every tool available to them to produce reliable programs. Indeed, only by making use of the five components of program design identified in this chapter can we construct large and reliable assembly language programs that can be maintained by other people.

Unlike assembly language itself, program design techniques are not handed down from the manufacturer of the microprocessor. To a certain extent they are an art to be learned by the student. In a text of this length, it is possible to provide only guidelines. Programmers learn their trade only by writing programs. However, as M. L. Shooman says in his book *Software Engineering*, "The lessons in this chapter (i.e., program design tools and techniques) are not difficult to grasp, but many have found them hard to implement. The difficulty lies in the necessity of changing work habits and approaches which are intuitive and comfortable, and replacing them with a different set of disciplined principles." Note the emphasis Shooman places on discipline. The temptation to start coding straight from the problem is sometimes almost irresistible. Only by adopting a top-down approach and by making use of structured programming techniques can the programmer hope to achieve effective results. The program design language introduced in this chapter provides a method of designing assembly language programs while making use of top-down design and structured programming.

Problems

Because this chapter has dealt with the topic of program design, the problems here are longer than those appearing elsewhere in the text. These problems should be regarded as "projects."

1. Write a subroutine to move a block of memory from one location to another. Before entering the subroutine, the following addresses are pushed on the stack (in this order): the starting address of the block to be moved, the ending address of the block to be moved, the starting address of the block's new location. By *starting address*, we mean the lowest address in the block.

Take care to consider all the possible cases (e.g., overlapping blocks). Your subroutine should return a carry bit = 0 if the routine is successful and a carry bit = 1 if there is an error. (What are the possible errors that might be encountered?)

2. Describe the actions carried out by the LINK and UNLK instructions. Explain how they are used.

3. Suppose the LINK and UNLK instructions did not exist. What code would be needed to replace the following sequence?

```
LINK  A5, # -$20
  :
UNLK  A5
```

4. Describe the operation of the PEA instruction and explain how it is used.

5. The IEEE format for a 32-bit floating-point number N is defined as:

S	E	F

where

S = sign bit (0 = positive, 1 = negative mantissa)

E = 8-bit exponent biased by 127

F = 23-bit fractional mantissa, together with an implicit leading 1 (the integer is not stored)

$N = (-1)^S \times 2^{(E-127)} \times (1 \cdot F)$

    a. Write a program to add two 32-bit floating-point numbers.

    b. Write a program to multiply two floating-point numbers.

    c. Write a program to convert a floating-point number to decimal form.

    d. Write a program to convert a decimal number to floating-point form.

6. Text is normally stored in character form with a byte holding each ASCII value. This is relatively inefficient for the storage of plain text in English; for example, the word *the* requires 24 bits of storage. Encoding it as a single symbol reduces the storage needed. Design a program to input text and compress it (using your own algorithm), and a program to reconstitute the compressed text and to print it out.

7. A keyboard encoder detects the closure of a switch at one of 64 cross-points in an 8 by 8 matrix. The output port initially places the value 11111110 on $D_{00}-D_{07}$ and the input port reads $D_{00}-D_{07}$. If no key is pressed, the input reads 11111111. If, say, the leftmost key on row 0 is pressed, 01111111 is read.

    The output port then puts 11111101 on the 8 rows and the input is read. This interrogates switches in the second row. The process is repeated sequentially, with the output going from 11111011 to 01111111 and back to 11111110. In this way, all 8 rows are continually scanned.

    Write a program to control such a keyboard and produce a 6-bit code for each keystroke. The code returned by the program is 0, 0, R2, R1, R0, C2, C1, C0, where R stands for row and C for column. Assume that a procedure OUT_BYTE places the lower byte of D0 on lines D_{00} through D_{07} and that IN_BYTE reads the contents of D_{00} through D_{07} and places the result in the lower-order byte of D0.

THE 68000 FAMILY HARDWARE INTERFACE

A microprocessor cannot function on its own and must be connected to external components to create a microcomputer. We will now examine the nature of the interface between the 68000 and the components needed to turn it into a viable system. We are not concerned with the internal operation of the 68000, but rather with the conditions that must be satisfied if it is to be incorporated in a system. In this chapter, we concentrate on the aspects of the 68000 that are common to all systems using this device. The more sophisticated facilities of the 68000 are glossed over and are dealt with in a later chapter.

All computers spend much of their time reading data from memory or writing it to memory. Consequently, this chapter places great emphasis on the 68000's interface to memory and its read and write cycles. This topic introduces the timing diagram and the protocols required to ensure a reliable exchange of data between the CPU and memory.

We also include an introduction to the 68020's interface. Fortunately for the 68000 systems designer, even though the 68020 has 32-bit address and data buses, it looks very much like the 68000 as far as the rest of the microprocessor system is concerned. Therefore, it is very easy to move from 68000 systems design to 68020 and 68030 systems design. However, we concentrate here on some of the significant differences between the ways in which the 68000 and the 68020 communicate with their memory.

Having looked at the principles governing the 68000's interface, we conclude the chapter by providing a simple example of a 68000 microcomputer system. We do this in order to demonstrate just what pins are essential and what pins are not absolutely required in a basic system.

4.1 68000 INTERFACE

The 68000 has 64 pins that may be arranged in nine groups as shown in figure 4.1. Each group of pins is labeled by the specific function performed by that group. For the purposes of this chapter, the functions of these nine groups are divided into three

FIGURE 4.1   Pinout of the 68000 and logical grouping of pins

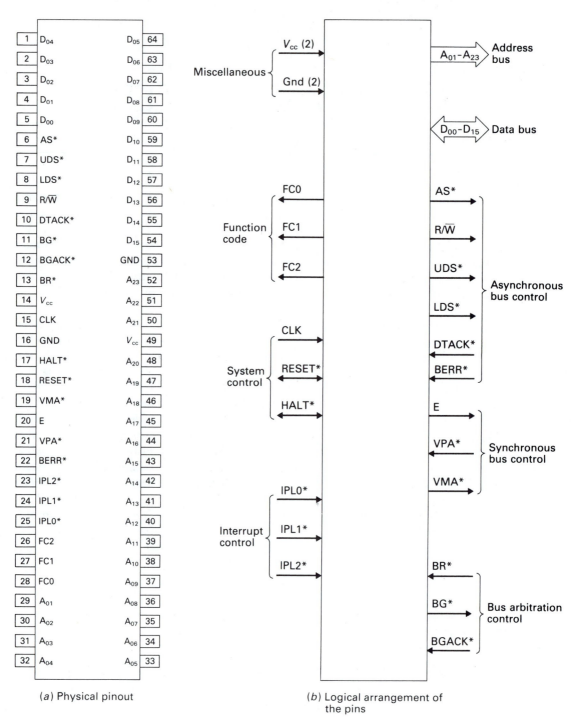

(a) Physical pinout

(b) Logical arrangement of the pins

categories: the system support pins, the memory and peripheral interface pins, and the special-purpose pins not needed in a minimal application of the processor. Table 4.1 shows how the pins may also be classified by the direction of information flow. A pin can act as an input, an output, or a dual-function input/output pin.

The system support pins are those essential to the operation of the processor in every system. These include the power supply, the clock input, the reset input, and similar functions. The memory and peripheral interface group of pins are those that connect the processor to an external memory subsystem. Special-purpose pins provide functions not needed in a basic system. These include interrupt management and bus arbitration facilities. Usually, special-purpose pins can be strapped to ground or V_{cc} (if they are inputs) and forgotten and left open circuit (if they are outputs).

System Support Pins

The first group of pins to be described consists of the system support pins, comprising the power supply, clock, reset, and halt functions. Throughout the chapter, and elsewhere, we have adopted the convention that an asterisk following a name indicates that the signal is active-low. The term *asserted*, when applied to a signal, means that it is placed in its active state. *Negated* means that it is placed in its inactive state. For example, the AS* output is asserted when it is in an electrically low state (i.e., at approximately ground potential).

Power Supply In common with most other digital logic elements found in microprocessor systems, the 68000 requires a single $+5$ V power supply. Two V_{cc} (i.e., $+5$ V) pins and two ground (i.e., 0 V) pins are provided, thereby reducing the voltage drop between the V_{cc} terminals of the chip and the V_{cc} conductors within the chip itself.

Clock The clock input is a single-phase TTL-compatible signal from which the 68000 derives all its internal timing. As the 68000 uses dynamic storage techniques internally, the clock input must never be stopped or its minimum or maximum pulse widths violated. Current versions of the 68000 have maximum clock rates of between 8 and 16.67 MHz. A memory access is called a *bus cycle* and consists of a minimum of four clock cycles. An instruction consists of one or more bus cycles.

RESET* The active-low reset input of the 68000 forces it into a known state on the initial application of power. When a reset input is recognized by the 68000, it loads the supervisor stack pointer, A7, from memory location zero ($00 0000$) and then loads the program counter from address $00 0004$. The detailed sequence of actions taking place during a reset operation is dealt with in the section on exception handling. For correct operation during the power-up sequence, RESET* must be asserted together with the HALT* input for at least 100 ms—an incredibly long time by most microprocessor standards. This delay provides time for the chip's internal back-bias generator to generate the bias voltage required by its MOSFET transistors. At all other times, RESET* and HALT* must be asserted for a minimum of ten clock periods.

TABLE 4.1 Input/output characteristics of the 68000's pins

SIGNAL NAME	MNEMONIC	TYPE	OUTPUT CIRCUIT
Power supply	V_{cc}	—	—
Ground	GND	—	—
Clock	CLK	Input	—
Reset	RESET*	I/O	OD
Halt	HALT*	I/O	OD
Address bus	$A_{01}-A_{23}$	Output	TS
Data bus	$D_{00}-D_{15}$	I/O	TS
Address strobe	AS*	Output	TS
Read/write	R/$\overline{W}$	Output	TS
Upper data strobe	UDS*	Output	TS
Lower data strobe	LDS*	Output	TS
Data transfer acknowledge	DTACK*	Input	—
Bus error	BERR*	Input	—
Enable	E	Output	TP
Valid memory address	VMA*	Output	TS
Valid peripheral address	VPA*	Input	—
Bus request	BR*	Input	—
Bus grant	BG*	Output	TP
Bus grant acknowledge	BGACK*	Input	—
Function code output	FC0, FC1, FC2	Output	TS
Interrupt priority level	IPL0*, IPL1*, IPL2*	Input	—

TS = tristate output	Input = input to the 68000
TP = totem-pole output	Output = output from the 68000
OD = open-drain output	I/O = input or output

RESET* also acts as an output from the 68000 under certain circumstances. Whenever the processor executes the software instruction RESET, it asserts the RESET* pin for 124 clock cycles, which resets all external devices (i.e., peripherals) wired to the system RESET* line, but does not affect the internal operation of the 68000; that is, it allows peripherals to be reset without resetting the 68000 itself.

HALT*   Like the RESET* input, the active-low HALT* input is bidirectional and serves two distinct functions (in addition to that described previously). In normal operation, HALT* is an active-low input to the 68000. When asserted by an external device, HALT* causes the 68000 to stop processing at the end of the current bus cycle and to negate all control signals. Tristate outputs (data and address buses) are floated. However, the function code outputs are not floated when the 68000 is in the halt state.

One application of the HALT* input is to enable the 68000 to execute a single bus cycle at a time. By asserting HALT* just long enough to permit the processor to execute a single bus cycle, the 68000 can be stepped through a program cycle by cycle, which may be used to debug a system. The HALT* input has other applications that are not relevant in simple systems. In most basic 68000 systems, the HALT* pin is, effectively, connected to the RESET* pin.

Sometimes HALT* can act as an output. Whenever the 68000 finds itself in a situation from which it cannot recover (the so-called double bus error, which is dealt with in chapter 6), it halts and asserts HALT* to indicate what has happened.

Another application of the HALT* input in sophisticated systems (to be described later) is in dealing with certain types of memory error. If the system memory fails to respond correctly to a read or write cycle, the HALT* pin can be used in conjunction with the BERR* (bus error) pin to force the 68000 to repeat or to rerun the bus cycle.

Figure 4.2 indicates how the system support pins are connected in a basic 68000 circuit. This diagram is intended to give the reader an idea of the "overhead" required to support the 68000; its operational details are considered later.

Memory and Peripheral Interface

This group of pins takes up 44 of the 68000's 64 pins and is used to read data from memory and to write data to it. As the 68000 treats peripherals exactly like memory components, the same pins are also used by all input/output transactions.

Address Bus   The address bus is provided by A_{01} to A_{23}, permitting 2^{23} 16-bit words to be uniquely addressed. The processor uses the address bus to specify the location of the word it is writing data into or reading data from. From table 4.1 it can be seen that the address bus is driven by tristate outputs, allowing the address bus to be controlled by a device other than the CPU under certain conditions.

The address bus has an auxiliary function and supports vectored interrupts (see chapter 6). Whenever the 68000 is interrupted, address lines A_{01}, A_{02}, and A_{03} indicate the level of the interrupt being serviced. During this so-called interrupt acknowledge phase, address lines A_{04} to A_{23} are set to a logical one level.

Data Bus   The data bus is 16 bits wide and transfers data between the CPU and its memory or peripherals. It is bidirectional, acting as an input during a CPU read cycle and as an output during a CPU write cycle. The data bus has tristate outputs which can be floated to permit other devices to access the bus. When the CPU executes an operation on a word, all 16 data bus lines are active. When it executes an operation on a byte, only D_{00} to D_{07} or D_{08} to D_{15} are active. During an interrupt acknowledge cycle, the interrupting device identifies itself to the CPU by placing an "interrupt vector number" on D_{00} to D_{07}.

The address and data buses operate in conjunction with five control signals: AS*, UDS*, LDS*, R/$\overline{W}$, and DTACK*. These signals are used to sequence the flow of information between the CPU and external memory.

AS*   The address strobe is active-low and, when asserted, indicates that the contents of the address bus are valid.

FIGURE 4.2   Circuit required by the 68000's system support pins

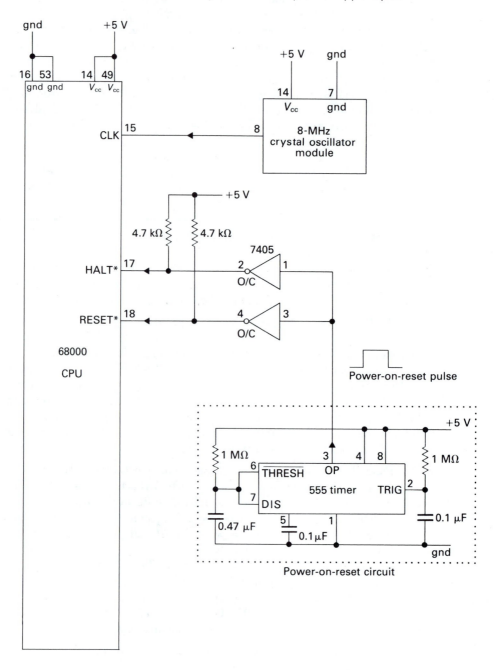

R/W The R/$\overline{W}$ (read/write) signal provided by the 68000 determines the nature of a memory access cycle. Whenever the CPU is reading from memory R/$\overline{W}$ = 1, and whenever it is writing to memory R/$\overline{W}$ = 0. If the CPU is performing an internal operation, R/$\overline{W}$ is always 1; that is, R/$\overline{W}$ is never in a logical zero state unless the CPU is executing a write to a memory location or a peripheral. However, the state of the R/$\overline{W}$ pin is undefined when the 68000 floats its control bus whenever it relinquishes control of the memory bus. Therefore, the designer must arrange for the R/$\overline{W}$ line to be pulled up to V_{cc} when the CPU is not controlling it. One of the most important aspects of microprocessor systems design is the avoidance of an unintentional write to a memory component.

UDS* and LDS* The 68000 accesses memory via a 16-bit wide data bus. However, special provisions have to be made to enable it to access a byte of data instead of a word. When the 68000 accesses a word, both UDS* and LDS* are asserted simultaneously. When it wishes to access a single byte, UDS* is asserted if it is the upper byte (D_{08} to D_{15}) or LDS* if it is the lower byte (D_{00} to D_{07}). Table 4.2 defines the relationship between UDS*, LDS*, R/W, and the data bus. Note that the expression DS* is sometimes used to indicate "UDS* and/or LDS*."

When we introduce the 68020's memory interface later in this chapter, we will find that the 68020 lacks separate upper and lower data strobes. Since the 68020 has a true 32-bit data bus, it requires *four* data strobes to select one or more bytes of a longword. Instead of four data strobes the 68020 has a single data strobe, an A_{00} pin, and two *size control* signals. Designers must use these signals to generate their own byte select signals.

DTACK* The active-low data transfer acknowledge input to the 68000 is a handshake signal generated by the device being accessed, and indicates that the contents of the data bus are valid and that the 68000 may proceed. When the

TABLE 4.2 Control of the data bus by UDS* and LDS*

R/$\overline{W}$	UDS*	LDS*	OPERATION	D_{08}–D_{15}	D_{00}–D_{07}
0	Negated	Negated	No operation	Invalid	Invalid
0	Negated	Asserted	Write lower byte	D_{00}–D_{07}	Data valid
0	Asserted	Negated	Write upper byte	Data valid	D_{08}–D_{15}
0	Asserted	Asserted	Write word	Data valid	Data valid
1	Negated	Negated	No operation	Invalid	Invalid
1	Negated	Asserted	Read lower byte	Invalid	Data valid
1	Asserted	Negated	Read upper byte	Data valid	Invalid
1	Asserted	Asserted	Read word	Data valid	Data valid

NOTE: In a byte write operation, the data on D_{00}–D_{07} is replicated on D_{08}–D_{15} for a lower-order byte access and the data on D_{08}–D_{15} is replicated on D_{00}–D_{07} for a higher-order byte access. This result is due to the current implementation of the 68000 and is not guaranteed in future versions of the 68000. However, this curious "anomaly" once caused me to spend several hours debugging a faulty 68000 system.

processor recognizes that DTACK* has been asserted, it completes the current access and begins the next cycle. If DTACK* is not asserted, the processor generates wait-states (i.e., it idles) until it is or until an error state is declared. DTACK* may be generated by a timer that is triggered by the beginning of a valid memory access. When the timer counts up to a predefined value, it forces the DTACK* input to the 68000 low. This timer must be supplied by the system designer.

The way in which the asynchronous bus control group operates is dealt with in some detail when the 68000 read and write cycles are described. The 68000 has three synchronous bus control pins (E, VPA*, and VMA*) that are not required in all systems and are described later.

Special-Function Pins of the 68000

The 68000's pins in this group perform functions that are not necessarily needed in all applications of the processor. They are included here for the sake of completeness. Later chapters show how they are employed in sophisticated microprocessor systems. The special-function pins of the 68000 fall into four groups: bus error control, which enables the 68000 to recover from certain types of error within the memory system; bus arbitration control, which allows more than one CPU to share the address and data buses; the function code outputs, which define the type of operation being executed by the 68000 and are used to control memory accesses in some systems; and the interrupt control interface, which allows a peripheral to signal its need for attention and permits the CPU to identify the source of the interrupt.

Bus Error Control

An active-low bus error input, BERR*, is used by the microprocessor system to inform the 68000 that something has gone wrong with the bus cycle currently being executed. We may argue that this feature is one of the attributes distinguishing the 68000 from all 8-bit microprocessors and from some 16-bit microprocessors. The provision of a BERR* input permits the 68000 to recover gracefully from events that would spell disaster to other processors.

Sometimes an access is made to a memory location that is either faulty or nonexistent. The latter case may occur when a spurious address is generated due to a software error or it may be that the actual memory available in the system is less than the operating system "thinks." Whenever external logic detects such an anomaly, it asserts BERR*. The precise nature of the action taken by the 68000 on recognizing that BERR* has been asserted is rather complex and is also dependent on the current state of the HALT* input. The behavior of the 68000 under these circumstances is dealt with later. For the moment we can state that the 68000 will either try to repeat (i.e., rerun) the faulty cycle or will generate an exception and inform the operating system of the bus error.

Strictly speaking, we could regard the BERR* input as part of the 68000's memory interface. It does not take part in normal memory access operations and is used only to help the processor deal with faulty memory accesses.

Bus Arbitration Control

When the 68000 has control of the system address and data buses, we call it the bus master. Many microprocessor systems include a mechanism whereby other microprocessors (or DMACs, direct memory access controllers) can also take control of the system bus. Consequently, some arrangement is necessary to inform the current bus master that the bus is needed by another device. The 68000 has three pins dedicated to bus arbitration control: bus request (BR*), bus grant (BG*), and bus grant acknowledge (BGACK*). *Arbitration* is the term used to describe the sequence of events that takes place when a number of potential masters request the bus simultaneously and one of them must be selected as the next bus master. The logic necessary to perform the arbitration is not part of the 68000 and must be designed to suit the user's own application. Some 68000 users employ the bus arbitration control logic to simplify the design of dynamic memory systems. We return to this topic in chapter 7. The three bus arbitration control pins of the 68000 are now briefly described.

BR* BR* is a bus request input and, when asserted, informs the CPU that another device wishes to take control of the system bus. All devices capable of being a bus master may drive the active-low bus request input with open-drain or open-collector outputs. Whenever a device wishes to take control of the bus, it first asserts BR*, signaling its intent to the 68000.

BG* The 68000 asserts its active-low BG* (bus grant) output in response to the assertion of the BR* input, thereby indicating to the potential bus master that the current bus master is going to release control of the bus at the end of the current bus cycle. Once BG* has been asserted, the potential bus master can negate its BR* output.

BGACK* When the potential bus master detects that BG* has been asserted in response to its bus request, it asserts its bus grant acknowledge (BGACK*) output, thereby informing the old bus master that the new bus master is now controlling the bus.

Figure 4.3 provides a timing diagram that illustrates the relationship between the 68000's BR*, BG*, and BGACK* signals.

Function Code Outputs

In principle, a microprocessor simply reads instructions from memory, interprets them, and operates on data either within the processor itself or within the memory system. In practice, the operation of the processor is rather more complex because it may have to interact with external events through the interrupt mechanism. Moreover, the processor accesses different types of information in memory: instructions, data, the stack, and so on. Information about the nature of the operation being executed by the processor is often very important to the system designer; for example, such information can be employed to prevent one user from accessing a region of memory that may "belong" to another user or to the operating system.

This information is called function or status information and is provided by microprocessors (directly or indirectly) in varying amounts. The 68000 has three

FIGURE 4.3 The 68000's bus arbitration signals

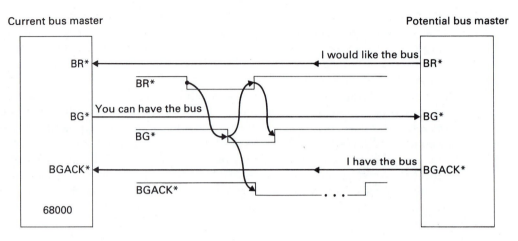

processor status outputs, FC0, FC1, and FC2, that indicate the type of cycle currently being executed. The function code becomes valid approximately half a state earlier than the contents of the address bus. Although the function code from the 68000 is not needed to build a working microcomputer, it can be used to enhance the operation of the system. One way of looking at the function code is to regard it as "an extension of the address bus." The significance of this statement will be made clear in chapter 7 when we discuss memory management.

Table 4.3 shows how FC0, FC1, and FC2 are interpreted. Of the eight states in table 4.3, three are marked "undefined, reserved," which tells us that these states may be reassigned in future versions of the 68000. Function code output FC2 distinguishes between two modes of operation of the 68000: supervisor and user. FC1 and FC0 divide memory space into data space and program space.

TABLE 4.3 Interpreting the 68000's function code output

FUNCTION CODE OUTPUT			PROCESSOR CYCLE TYPE
FC2	FC1	FC0	
0	0	0	(Undefined, reserved)
0	0	1	User data
0	1	0	User program
0	1	1	(Undefined, reserved)
1	0	0	(Undefined, reserved)
1	0	1	Supervisor data
1	1	0	Supervisor program
1	1	1	Interrupt acknowledge (CPU space)

We can see from table 4.3 that, apart from interrupt acknowledge cycles, the 68000 is always in one of two states: user or supervisor. The concept of user and supervisor states does not exist for most 8-bit microprocessors or for some 16-bit devices. User and supervisor states have a meaning only in the world of multitasking systems, where two or more programs (tasks) are running concurrently. The supervisor state is said to be the state of highest privilege, and certain instructions may be executed only in this state. In general, the supervisor state is closely associated with the operating system, and the less privileged user state is associated with user programs running under the operating system. By restricting the privileges available to the user state, individual programs are capable of causing less havoc if they crash.

The supervisor state is in force when the S bit of the processor status word is true. All exception processing (e.g., interrupt handling, bus error, and reset) is performed in the supervisor state, regardless of the state of the processor before the exception occurred. Consequently, the 68000 always powers up in the supervisor state. A change from supervisor to user state can be carried out under program control, but it is impossible to move from the user to supervisor state by any sequence of instructions other than traps and operating system calls. Only by the generation of an exception can a transfer from user to supervisor mode be made. Chapter 6 is devoted to the topic of exception handling by the 68000.

Table 4.3 also shows how it is possible to determine whether the processor is accessing program or data. The region of memory containing data is called *data space*, and the region containing instructions, *program space*. The meaning of the word space in this context is closer to the mathematician's use of the word (e.g., vector space) than to the everyday meaning.

By the way, there is a relationship between function codes and addressing modes. In Chapter 2 we said that program counter relative addressing could be used only with source operands and not destination operands. For example, the instruction MOVE.W Table(PC),D3 is *legal*, whereas MOVE.W D3,Table(PC) is *illegal*. The 68000 accesses program space whenever the program counter is used to calculate an effective address. This rule applies to operands as well as to instructions. Consequently, the effective address Table(PC) yields a function code output corresponding to program space rather than to data space, as you might expect. The 68000 does not permit a program counter relative address to be used as a destination operand, since that would involve writing to program space. Although we have not yet covered interrupts, it is worth pointing out that the initial reset vector (i.e., program entry point) and the initial value of the stack pointer are fetched from program (supervisor) space.

The function code outputs indicate an access to data space whenever operands are read (apart from those using a program counter effective address), all operands are written, and exception vectors are fetched (apart from the reset vector mentioned before).

The function code denoted by FC0 = FC1 = FC2 = 1 is called an *interrupt acknowledge* and indicates that the 68000 is currently responding to an interrupt request by a peripheral. The 68010 and later processors employ the state FC2,FC1, FC0 = 1,1,1 to indicate CPU operations other than interrupt acknowledge cycles (e.g., coprocessor communications). Consequently, this function code output is now more properly referred to as *CPU space*. Since other members of the 68000 family use accesses to CPU space for various purposes, it is necessary to distinguish between different CPU space types in sophisticated systems. The 68010, 68020, and 68030 employ address bits A_{16} to A_{19} indicate the type of CPU space being accessed. Because the 68000 has only one CPU address space (i.e., interrupt acknowledge), it is not normally necessary to decode A_{16}–A_{19} (which are all one) to distinguish an interrupt acknowledge cycle from other CPU space cycles. However, designers of 68000-based systems should decode A_{16}–A_{19} during interrupt acknowledge cycles to enable their systems to be upgraded to the 68020.

One advantage of dividing memory space into program and data spaces is that it becomes possible to prevent a program from corrupting the data space of another program by detecting any access to program space that would corrupt the program. In chapter 2, we discovered that some of the 68000's instructions are privileged and that they can be executed only when the CPU is operating in the supervisor mode.

Figure 4.4 illustrates how the function code outputs might be applied to the control of memory systems by generating suitable enable signals for each block of memory. A region of memory space is devoted to the supervisor (i.e., operating system) and cannot be accessed when the 68000 is operating in the user mode. Note how we are using one of the 68000's hardware facilities (i.e., its function code outputs) to protect a software facility (i.e., the operating system).

To make the example more interesting, we divide the user space into two regions: user program and user data. The user program space can be accessed from both the user program state and from the supervisor data state. Why? Because the operating system might be expected to load the user program into user memory from disk. When the operating system transfers this program, the data is transferred

FIGURE 4.4 Using the 68000's function code outputs

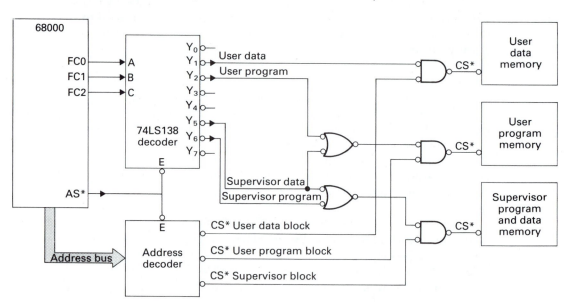

in the supervisor data mode, which means the user program space must be accessible from supervisor data space. Note also that the user data space can be accessed only when the 68000 is in the user data mode. In this case, we argue that user data space is private to the user and even the operating system should not be permitted to access it.

Interrupt Control Interface

Three interrupt control inputs (IPL0*, IPL1*, IPL2*) are used by an external device to indicate to the 68000 that it requires service. These interrupts are encoded into eight levels (0 to 7). Level 0 has the lowest priority and indicates that no interrupt is requested. Level 7 is the highest priority interrupt. The interrupt mask bits, I_2, I_1, and I_0, of the status register determine the level of interrupt that will be serviced.

An interrupt request, indicated by a 3-bit code on IPL0*, IPL1*, IPL2*, is serviced if it has a higher value than that currently indicated by the interrupt mask bits in the status register. A level-7 interrupt is handled rather differently because it is always serviced by the 68000.

Many peripherals capable of generating an interrupt have only a single interrupt request output. Consequently, most 68000-based microcomputer systems must use a priority encoder circuit to convert up to seven levels of interrupt request into a three-bit code that can then be fed into IPL0* to IPL2*.

Chapter 6 describes the logic required to implement fully a 68000 interrupt system. Here we will indicate only the type of logic required to make use of the 68000's interrupt-handling facilities. Figure 4.5 demonstrates how a single 74LS148 priority encoder transforms one of seven interrupt requests on IRQ1*–IRQ7* into a 3-bit code on IPL0*–IPL2*. Since there are seven levels of interrupt request, it is necessary to inform the interrupting device of the current level of interrupt being processed by the 68000. The AND gate connected to FC0–FC2 detects an access to interrupt acknowledge space and uses it to enable the 74LS138 three-line-to-eight-line decoder. During an interrupt acknowledge, the 68000 puts the level of the interrupt on bits A_{01} to A_{03} of the address bus, which is decoded by the 74LS138 into seven levels of interrupt acknowledge. Some designers would put the interrupt logic in a PAL to save space and increase functionality.

Synchronous Bus Control

One important difference between the 68000 and many other microprocessors is that the 68000 is able to carry out asynchronous data transfers between itself and memory or peripherals. In order to understand the nature of asynchronous data transfers, synchronous transfers will be looked at first.

In a synchronous data transfer, the processor provides an address and some form of timing signal. Figure 4.6 demonstrates a simple synchronous data transfer—a CPU read from memory. At point A, a read cycle begins with the falling edge of the clock. At B, the CPU generates an address corresponding to the memory location being accessed. At C, the memory yields its data for the CPU to read. The current cycle ends at D with the falling edge of the clock. The time between C and D is

FIGURE 4.5 External logic required to provide interrupt request and acknowledge signals

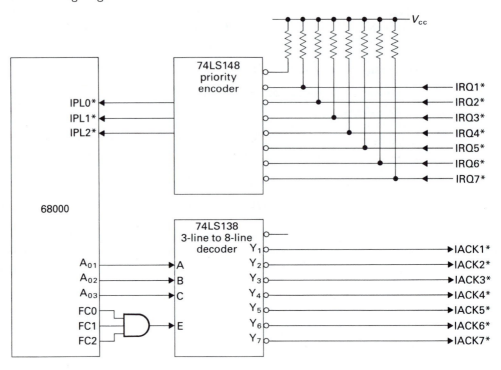

called the data setup time of the CPU and is the time for which the CPU demands that the data be valid before the end of the cycle. In this arrangement the clock must allow enough time for the memory to access its data. If sufficient time is not allowed and the setup time is violated, the data obtained by the CPU may be invalid.

FIGURE 4.6 Synchronous data transfer

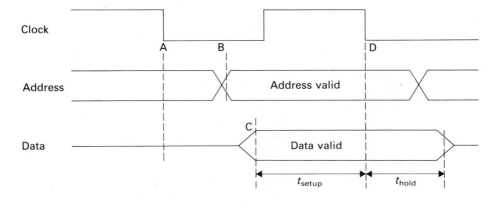

Strictly speaking, the 68000's synchronous bus control group of signals is not needed: All data transfers may take place asynchronously. The synchronous bus control group has been included entirely to simplify the interface between the 68000 and peripherals designed for use with the 6800, 6809 (or 6502) 8-bit synchronous bus microprocessors; that is, this group of signals makes the 68000 look like a 6800 to certain types of peripheral. Section 8.1 on interfacing techniques takes a more detailed look at the synchronous interface. Three control signals are included in this group: VPA* (valid peripheral address), VMA* (valid memory address), and E (enable).

VPA* The active-low valid peripheral address input is used by a device to indicate to the 68000 that a synchronous peripheral is being accessed. When the processor recognizes that VPA* has been asserted, it initiates a synchronous data transfer by means of VMA* and E.

VMA* VMA* is an active-low valid memory address output from the 68000 and indicates to the peripheral being addressed that there is a valid address on the address bus. The assertion of VMA* by the CPU is a response to the assertion of VPA* by an addressed peripheral.

E The enable output from the 68000 is a timing signal required by all 6800 series peripherals and is derived from the 68000's own clock input. One E cycle is equal to ten 68000 clock cycles. The E clock is nonsymmetric: it is low for six clock cycles and high for four.

No defined phase relationship exists between the processor's own clock and the E clock. The E clock runs continuously, independently of the state of the 68000. Later we shall show how the synchronous bus group is used to interface the 68000 to typical 6800 series peripherals.

Note that the synchronous bus control signals VPA*, VMA*, and E are not implemented by the 68020 and later processors. If you are using these advanced microprocessors, you should also be using modern 68000-series peripherals.

Asynchronous Bus Control

An asynchronous data transfer is rather more complex than its synchronous counterpart, as can be seen from the simplified read cycle timing diagram in figure 4.7. The processor generates a valid address at point A and then asserts an address strobe at B. When the memory detects the address strobe, it places data on the data bus, which becomes valid at point C. The memory then informs the processor that it has valid data by asserting a data acknowledge signal at point D. The processor detects that the data is now ready, reads it, and negates its address strobe to indicate that it has read the data (point E). The memory then negates its data acknowledge signal at point F to complete the cycle. The sequence of operations described in figure 4.7 is called a *fully interlocked handshake*.

FIGURE 4.7 Asynchronous data transfer in a CPU read cycle

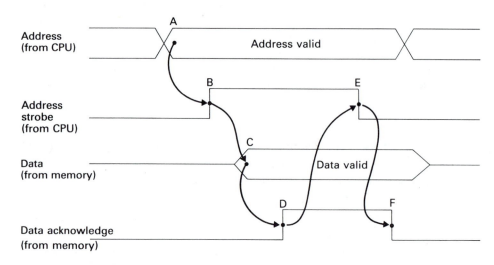

NOTE: Control signals shown are active-high.

The 68000 is not fully asynchronous because its actions are synchronized with its clock input. Although it may extend its memory access until an acknowledgment is received, the operation can be extended only in increments of one clock cycle.

4.2 TIMING DIAGRAM

Having explained the basic purpose of the pins of the 68000, we can now look at some of these functions in more depth. In this section we examine the read and write cycles of the 68000, which represent the most fundamental transactions between a processor and its external environment. Traditionally, the timing diagram has been used to illustrate the detailed operation of a microprocessor or a memory component. A timing diagram shows the relationship between the signals involved in a read/write cycle and time. Although the timing diagram is an educational aid, because it presents visually the relationship between a number of signals, it is principally a design tool. It enables an engineer to match components of different characteristics so that they will work together.

As we have not yet examined memory components and their characteristics, the memory-timing diagrams presented here do not include fine detail. Chapter 5 examines the characteristics of memory components in more depth.

In recent years, the timing diagram has been supplemented by what may best be called a *protocol diagram* or *timing flowchart*. The protocol diagram is an abstraction of the timing diagram, which seeks to remove all detail in order to provide only the most essential information to the reader. The read and write cycles of the 68000 are explained here by means of both protocol flowcharts and timing diagrams.

68000 Read Cycle

This section considers the sequence of events taking place when the 68000 reads data from memory using its address and data buses in conjunction with the asynchronous group of bus control signals. The 68000 can read either a 16-bit word or an 8-bit byte in a single read cycle. Because very little difference exists between these operations, only a word operation is described. The difference between a byte read and a word read is commented on later.

Figure 4.8 gives the protocol flowchart for a 68000 read cycle. Any read cycle involves two parties: the reader and the read. The reader is the 68000 and is represented by the bus master in figure 4.8. A bus master is the active device that is currently controlling the system bus; at any instant there can be only one bus master. A system can support several 68000s in a system, but only one may be the master at a time. Equally, a device other than a CPU may simulate a 68000 to gain control of the bus. The left-hand side of the diagram displays the actions carried out by the master (the 68000). Each block is labeled by the words in its top line. The numbered lines below the header describe the sequence of actions carried out by that block.

The right-hand side of the diagram displays the actions carried out by the slave during the transfer of information. The slave is, of course, the memory being accessed by the master. A slave can also be a peripheral or any device that simulates memory. The protocol diagram is read from top to bottom so that the action "Address the slave" carried out by the master is followed by the slave with the action "Output the data." Note that actions within boxes may or may not take place simultaneously.

Figure 4.8 does not supply precise timing relationships or details of critical events; for example, in the block labeled "Output the data," it is the action of asserting DTACK* that allows the master to continue with the action "Acquire the data." This requirement is not evident from figure 4.8, and therefore the diagram does not tell the whole story.

The essential feature of a 68000 asynchronous read cycle is the interlocked handshaking procedure taking place between the master and the slave. A read cycle starts with the master indicating its intentions by setting up an address and forcing R/$\overline{W}$ high. By asserting AS*, UDS*, and/or LDS*, the CPU is saying, "Here's an address from which I wish to read the data." The slave detects the valid address strobe (AS*) together with the data strobe(s) and starts to access the data. The slave asserts DTACK*, informing the processor that it may proceed. DTACK* is the handshake from the slave to the processor and acknowledges that the slave has (or is about to have) valid data available. The microprocessor systems designer must provide suitable circuits to generate the appropriate delay between the start of a read (or write) cycle and the assertion of DTACK*. If DTACK* is not asserted, the

FIGURE 4.8 Protocol flowchart for a 68000 read cycle

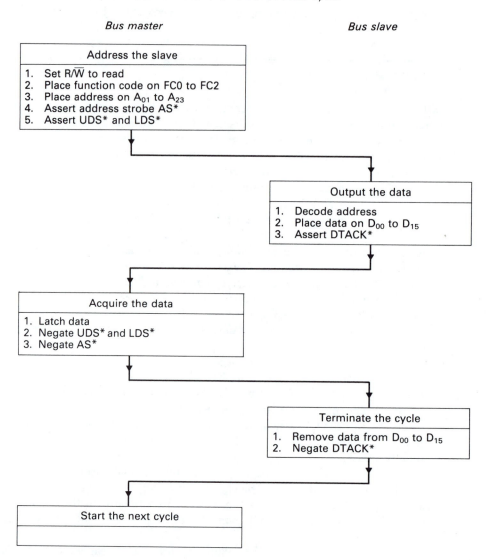

Bus master   *Bus slave*

Address the slave
1. Set R/$\overline{W}$ to read 2. Place function code on FC0 to FC2 3. Place address on A_{01} to A_{23} 4. Assert address strobe AS* 5. Assert UDS* and LDS*

Output the data
1. Decode address 2. Place data on D_{00} to D_{15} 3. Assert DTACK*

Acquire the data
1. Latch data 2. Negate UDS* and LDS* 3. Negate AS*

Terminate the cycle
1. Remove data from D_{00} to D_{15} 2. Negate DTACK*

Start the next cycle

master will, theoretically, wait forever. As we shall see, the 68000 has provision for dealing with the failure of a slave to complete a handshake by asserting DTACK*. When the master recognizes DTACK*, it terminates the cycle by latching the data and negating the address and data strobes, thereby inviting the slave to terminate its actions by removing data from the bus and negating DTACK*.

Before looking at the timing diagram of the 68000, some fundamental concepts of timing diagrams need to be reviewed. Figure 4.9 shows four possible ways of displaying the timing diagram of a humble positive-edge triggered D flip-flop. This device has been chosen because it represents the basic memory element of which

FIGURE 4.9 Timing diagram of a D flip-flop

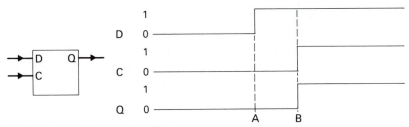

(a) The idealized form of the timing diagram

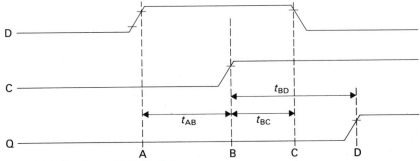

(b) The actual behaviour of a D flip-flop

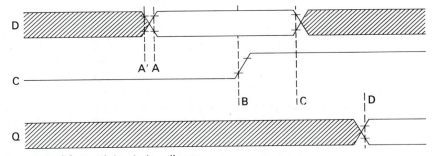

(c) The general form of the timing diagram

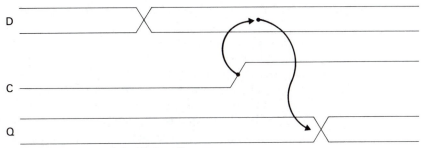

(d) An alternative form of the timing diagram

many memory systems are actually composed and because it latches data just as a microprocessor does. In other words, the microprocessor timing diagram is a scaled-up version of the timing diagram of the D flip-flop.

Figure 4.9a illustrates the idealized form of the timing diagram. All signals are either at a logical 0 (lower) level or a logical 1 (upper) level. The transition between logic levels takes place instantaneously. At point A, the D input rises to a logical 1 level. At point B, the flip-flop is clocked, and the D input is latched and transferred to the Q output.

Figure 4.9b illustrates the actual behavior of a D flip-flop. All transitions between states are represented by sloping lines to show that they are never instantaneous. The sloping line is illustrative and does not indicate the actual rise- or fall-time of the signal. Timing diagrams are almost never drawn to scale.

The gradual transition between logic levels poses the question, "When does a signal actually change state?" The answer is, of course, that when a signal at the input of a logic element passes the device's switching threshold, the device responds to its new logical input. Unfortunately, the switching threshold for logic elements is seldom quoted in their specification. In any case, it varies from device to device. Many semiconductor manufacturers specify the switching characteristics of their devices by referring all timing to the point at which a signal passes a given level.

Thus, the reference level for output voltages is V_{OL} and V_{OH}, representing the guaranteed maximum output voltage in a logical zero state and the guaranteed minimum output in a logical one state, respectively. Similarly, the reference levels for inputs are V_{IL} (the maximum input guaranteed to be recognized as a low level) and V_{IH} (the minimum input guaranteed to be recognized as a high level). For Schottky TTL devices, the values of V_{OL}, V_{OH}, V_{IL}, and V_{IL} are 0.4, 2.7, 0.8, and 2.0 V, respectively. Some manufacturers specify the reference points as 10 and 90 percent of the high-level output of a gate; others choose the midpoint as the reference.

In figure 4.9b the time between the points A and B is labeled t_{AB} and is measured from the point at which the D input has reached its logical 1 level (V_{IH}) and the point at which the clock has reached its logical one level. The value of t_{AB} is called the *data setup time* and represents the minimum time for which data must be stable at the input of the flip-flop before it is clocked. At point C, the D input has left V_{IH} and is returning to a low level. The time between B and C (t_{BC}) is called the *data hold time* and is the minimum time for which the data must be held stable after the flip-flop has been clocked.

As a result of clocking the flip-flop, its Q output changes state at point D. The time t_{BD} is the maximum time taken for the output to become valid following a clock pulse.

So far, we have been concerned with specific changes of state (D changing from a low to a high level). Figure 4.9c gives the more general form of a timing diagram. The D input is represented by two parallel lines at logical zero and logical one levels. We are not interested in the actual value of the D input. Our concern rests with the points at which changes occur. At point A' the levels begin to change state, and at point A they have reached V_{IL} and V_{IH}. Point A is used as the reference point from which the setup time is measured. Prior to point A, the space between the parallel lines representing the D input is shaded and indicates that the data is invalid.

Similarly, after the data hold time has been exceeded (point C), the D input may change once more. Between points A and C the unshaded area represents the period for which the data must be stable.

An alternative way of showing timing diagrams is given in figure 4.9d. Sometimes we wish to emphasize the relationship between signals on a cause and effect basis. The "cause" in figure 4.9d is the rising edge of the clock input. A line from this edge is drawn to the D input, showing that the rising edge of C causes D to be sampled. A line from the point at which D is sampled is drawn to the point at which Q changes state. This shows that the cause, C, results in the effect, Q.

A highly simplified version of a 68000 read cycle is presented in figure 4.10a. It

FIGURE 4.10 A 68000 read cycle

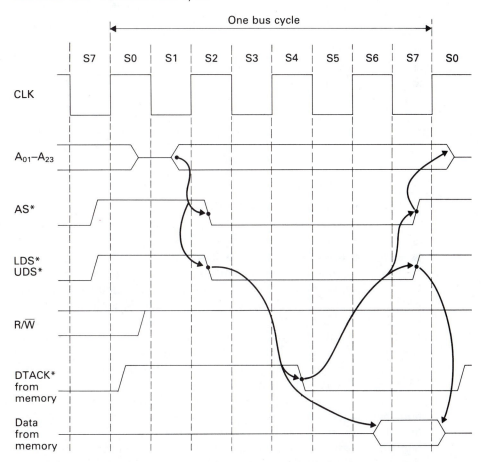

NOTE: The lines shown indicate indirect actions. For example, although the assertion of DTACK* in state S4 leads to the negation of AS* and UDS*/LDS* in state S7, the negation of these signals is actually triggered by the falling edge of the clock at the end of state S6.

(a) Simplified timing diagram

FIGURE 4.10   (*Continued*)

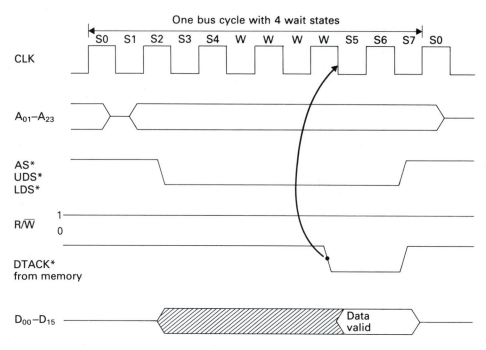

NOTE: Wait states are introduced until DTACK* is asserted.

(*b*) Read cycle with the insertion of wait states

is simplified because we have omitted timing details (i.e., parameters) and have included only those signals of immediate interest (e.g., FC0–FC2 are omitted). Each bus cycle consists of a minimum of four clock cycles and is divided into eight states labeled S0 to S7. All machine cycles start in state S0 with the clock high and end in state S7 with the clock low. As we shall see later, the machine read cycle may be extended indefinitely by the insertion of wait states (each with a duration of one-half clock cycle) between clock states S4 and S5, allowing the 68000 to be operated with a mixture of both fast and slow memory components. Figure 4.10b illustrates the effect of wait states on a read cycle.

Figure 4.10 is designed to show the relationship between the 68000's asynchronous bus signals and between these signals and the states of the clock. During the first state, S0, all signals are negated with the exception of R/$\overline{\text{W}}$, which becomes high (i.e., read) for the remainder of the current machine cycle. In the following description of the 68000, all times given are for the 8-MHz version, unless stated otherwise.

In state S1, the address on A_{01} to A_{23} becomes valid and remains so until state S0 of the following cycle. In state S2 the address strobe, AS*, goes low, indicating that the contents of the address bus are valid. At this point we are tempted to say,

"Why do we need AS*, as the falling edge of S2 can be used to indicate that the address is valid?" The answer to this question lies in the variations between different versions of the 68000. In the 12.5-MHz version, the possibility exists that AS* will not go low until state S3. The relationship between the clock and the 68000's signals is not what matters to the designer: it is the relationship between the signals themselves.

It is important to note that timing diagrams are rarely drawn to scale and that the picture presented can be misleading without an appreciation of the actual timing parameters. For example, we might show a signal such as AS* as changing state in clock state S2, and yet in some extreme circumstances it may actually go low as late as state S3. Consequently, if we were to attempt to use the falling edge of S2 to latch AS*, we might miss it. Details of the 68000 and 68020 read/write cycle timing diagrams are given elsewhere in this text.

In a read cycle, the timing specifications of the upper and lower data strobes (UDS* and LDS*) are the same as AS*. The falling edge of UDS* and/or LDS* initiates the memory access and, at the same time or after a suitable delay, triggers a data transfer acknowledge, DTACK*. Remember that the designer of the microprocessor system is responsible for providing suitable logic to control DTACK*. The delay between a data strobe going low and the falling edge of DTACK* must be sufficient to guarantee that enough time is allowed to access the memory currently being read. If DTACK* does not go low at least 20 ns before the end of state S4, wait states are introduced between S4 and S5 until DTACK* is asserted (see figure 4.10b). In figure 4.10a the assertion of the data strobe causes memory to be accessed and data to appear on the data bus in state S6, although the actual time depends on the access time of the memory being read.

During the final state of the current bus cycle, S7, both AS* and LDS*/UDS* are negated and the data is latched into the 68000 internally. The negation of these strobes causes the memory to stop putting data on the data bus and to return the bus to its high impedance (floating) state. DTACK* must be negated after the strobes have been negated. The address bus is floated (i.e., tristated) in the following S0 state and the read cycle is now complete.

The microcomputer designer needs to know the restrictions placed on his or her design by the timing diagram of a microprocessor. Figure 4.11 provides a more detailed read cycle timing diagram of the 68000. Table 4.4 gives the value of some of the read cycle timing parameters for the 68000L8.

The 68000 clock input is specified by three parameters. Its period, t_{cyc}, must not be less than 125 ns (for an 8-MHz clock) or more than 250 ns (i.e., minimum clock frequency = 4 MHz). The maximum limit of the clock period, 250 ns, is determined by the way in which the 68000 stores data internally as a charge on a capacitor. If the 68000 is not clocked regularly, internal data is lost, leading to unpredictable behavior of the processor.

Limits are also placed on the times for which the clock may be in either a high or a low state. A glance at table 4.4 reveals that the clock input should have an approximately symmetrical waveform with equal high- and low-times.

The address bus is floated within t_{CHADZ} seconds (80 ns maximum) of the start of S0. At no more than t_{CLAV} seconds (70 ns maximum) from the start of S1, the new address is placed on the address bus. The address strobe, AS∗, is asserted no less than t_{AVSL} seconds (30 ns minimum) after the address has stabilized. A key parameter is t_{AVSL}, because if the designer uses it to latch the address, he or she must choose a device with a setup-time less than t_{AVSL}.

FIGURE 4.11 Detailed timing diagram of the 68000 read cycle

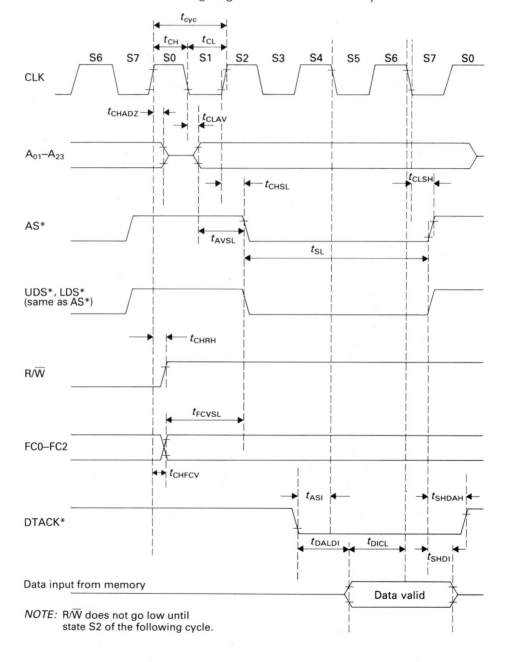

NOTE: R/W̄ does not go low until state S2 of the following cycle.

TABLE 4.4 Basic read cycle timing parameters of the 68000L8 (all values in ns)

PARAMETER NAME	SYMBOL	MINIMUM	MAXIMUM
Clock period	t_{cyc}	125	250
Clock width (low)	t_{CL}	55	125
Clock width (high)	t_{CH}	55	125
Clock high to address bus high impedance	t_{CHADZ}		80
Clock low to address valid	t_{CLAV}		70
Address valid to AS* low	t_{AVSL}	30	
Clock high to AS*, DS* low	t_{CHSL}	0	60
Clock low to AS*, DS* high	t_{CLSH}		70
AS*, DS* width low	t_{SL}	240	
Clock high to R/$\overline{W}$ high	t_{CHRH}	0	70
Clock high to FC valid	t_{CHFCV}		70
FC valid to AS* low	t_{FCVSL}	60	
Asynchronous input DTACK* setup time	t_{ASI}	20	
AS*, DS* high to DTACK* high	t_{SHDAH}	0	245
Data in to clock low setup-time	t_{DICL}	15	
DS* high to data invalid (data hold-time)	t_{SHDI}	0	
DTACK* low to data in setup-time	t_{DALDI}		90

R/$\overline{W}$ is set high at the beginning of a read cycle no more than t_{CHRH} seconds (70 ns maximum) after the start of state S0 and stays high for the remainder of the current cycle. In practice, the designer can therefore forget about R/$\overline{W}$ during a read cycle, as it is high well before the other parameters are valid and remains high until well after they have changed.

The 68000 puts out its function code no more than t_{CHFCV} seconds (70 ns maximum) after the start of state S0 and no sooner than t_{FCVSL} seconds (60 ns minimum) before AS* is asserted. Consequently, the function code behaves like an address and can be latched by AS* at the same time as an address. The function code is available earlier than the address in order to give a memory management unit in a sophisticated system time to decode the type of memory being accessed. This topic is developed in chapter 7.

The key parameter governing the DTACK* handshake from the peripheral is its setup time, t_{ASI} (20 ns minimum), before the falling edge of state S4. If DTACK* is asserted before its minimum setup time, the next state will be S5. If DTACK* does not meet this setup time, the processor introduces wait states after S4, until DTACK* is asserted at least t_{ASI} seconds before the falling edge of the next 68000 clock input.

The data from the memory being accessed is placed on the data bus and must satisfy setup and hold times similar to the input of the D flip-flop described earlier. The data must be valid at least t_{DICL} seconds (15 ns minimum) before the beginning of state S7.

During state S7, both address and data strobes are negated no more than t_{CLSH} seconds (70 ns maximum) after the falling edge of the clock. In order to meet the data hold time of the 68000, the contents of the data bus must be stable for at least t_{SDHI} seconds (0 ns minimum) after the rising edge of the strobes. Here the minimum value of 0 ns means that the data may become invalid concurrently with the rising edge of AS* or LDS*/UDS*.

Memory Timing Diagram

Before we can perform any actual timing calculations on a 68000 read cycle, we have to look at some of the characteristics of memory. In this chapter we are concerned only with how the 68000 executes a memory access. The details of memory system design are left until chapter 5. For our present purposes, the memory device used to illustrate read/write cycles can be regarded as a "generic memory component." Figure 4.12 shows the timing diagram of a 6116P-4 static random access memory component and figure 4.13 shows the pins through which it communicates with a microprocessor.

The 6116P is byte orientated, and each read or write operation must involve an 8-bit byte. Two 6116Ps must be configured in parallel to permit the 68000 to access one 16-bit word at a time. The 6116P has 11 address inputs, labeled A_0 to A_{10}, allowing $2^{11} = 2,048$ locations to be uniquely accessed.

FIGURE 4.12 Read cycle timing diagram of a 6116P-4 static RAM

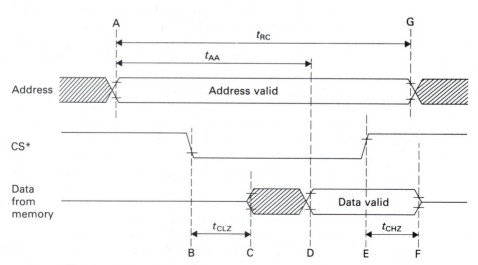

NOTE: R/$\overline{W}$ is high for the duration of the read cycle and OE* is low. The read access is controlled entirely by CS*.

FIGURE 4.13 Pinout of the 6116P-4 static RAM

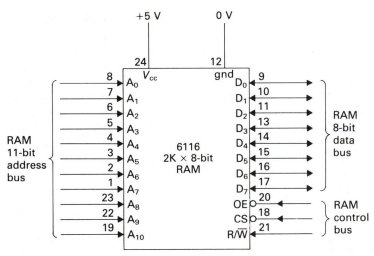

Three inputs, R/$\overline{\text{W}}$, OE$*$, and CS$*$, control the operation of the memory. Read/write (R/$\overline{\text{W}}$) determines the direction of data transfer during a memory access. Output enable (OE$*$) is an active-low control input that, when asserted, turns on the output circuits of the data drivers during a read access. Chip select (CS$*$) is an active-low input that enables the chip during a read or write access.

The 6116P is a static memory component that does not latch its address input. This means that that the current memory cell is accessed as soon as an address is stable at the chip's address terminals. In figure 4.12 the address is stable between points A and G. This time is denoted by t_{RC}, the read cycle time, and has the minimum value of 200 ns (see table 4.5). Consequently, successive accesses to the 6116P must be separated by at least 200 ns. A cycle time of 200 ns is relatively long by the standards of today's high-speed memory components. When we introduce the 68020's memory interface later in this chapter, we will look at the characteristics of a high-performance static RAM.

Neither the R/$\overline{\text{W}}$ line nor OE$*$ appear in figure 4.12. We assume that R/$\overline{\text{W}}$ is high for the entire cycle and that OE$*$ is permanently low. Unless OE$*$ is used as a

TABLE 4.5 Read cycle timing parameters of the 6116P-4

PARAMETER	MNEMONIC	MINIMUM	MAXIMUM
Read cycle time	t_{RC}	200 ns	
Address access time	t_{AA}		200 ns
Chip select to output not floating	t_{CLZ}		15 ns
Chip deselect to output floating	t_{CHZ}	0 ns	50 ns

method of turning on and off the data bus drivers independently of the CS* input, it may be permanently asserted.

At point B, CS* is asserted to execute the read operation. In a 68000-based system, it is normally derived from one of the data strobes (UDS* or LDS*). The action of CS* in a read cycle is to turn on the chip's data bus drivers. Consequently, the data bus comes out of its high impedance state, and data appears on the bus at point C, t_{CLZ} seconds after CS* is asserted. The maximum value of t_{CLZ} is 15 ns, which means that if CS* is asserted shortly after the current address has become valid, the data appearing on the data bus will not be valid. This condition is shown by the shaded region in figure 4.12. Not until point D, t_{AA} seconds after point A, do the contents of the data bus become valid and capable of being read by the computer. The duration between points A and D (i.e., t_{AA}) is the access time of the memory and is quoted as not more than 200 ns.

Chip select is negated at point E, which causes the data bus to be floated at point F. The maximum duration between points E and F is t_{CHZ} and is quoted as 50 ns.

Connecting the 6116P RAM to a 68000 CPU

As an example of how the 68000 read cycle parameters are related to its operation, consider the interface between the 68000 and a typical memory component. Figure 4.14 shows how two 6116P 2K × 8 RAMs can be connected to a 68000 CPU. This circuit will work, although microprocessor systems sometimes isolate memory components from the 68000's address and data buses by means of buffers or data bus drivers. This subject is dealt with in chapter 10, when buses are treated in more depth.

The data bus of the 68000 is connected directly to the data buses of the 6116Ps. RAM1 is connected to D_{00} to D_{07}, and RAM2 to D_{08} to D_{15}. Note that this diagram raises an interesting practical problem. The data bus of all 6116Ps has its pins labeled D_0 to D_7. As RAM1 has its D_0 connected to the 68000's D_{00}, etc., no problem exists. However, RAM2 has its D_0 connected to the 68000's D_{08}, etc., generating a problem of terminology. What do we (the writer, engineer, draftsperson) do when a line called X from component A is connected to a line called Y from component B? Clearly, we have a situation in which the same line has two different names.

The way out of this dilemma is to label lines by their system function and the inputs and outputs of integrated circuits by the names used by the chip manufacturers. Whenever the name used by a chip manufacturer to define a pin differs from the name of a line connected to that pin, the manufacturer's name will appear only within the box representing the chip.

Address lines A_{01} to A_{11} from the 68000 are connected to the address inputs of the two 6116P RAMs. The address inputs of the RAMs are wired in parallel, so that the same location is accessed in each chip simultaneously. Note also that A_{01} of the 68000 is connected to A_0 of the two 6116Ps—illustrating this problem of terminology.

The R/W input of each RAM is connected to the logical AND of AS* and R/W from the 68000. Therefore, the RAM can take part in a write cycle only when

FIGURE 4.14 Connecting the 6116P-6 static RAM to a 68000 CPU

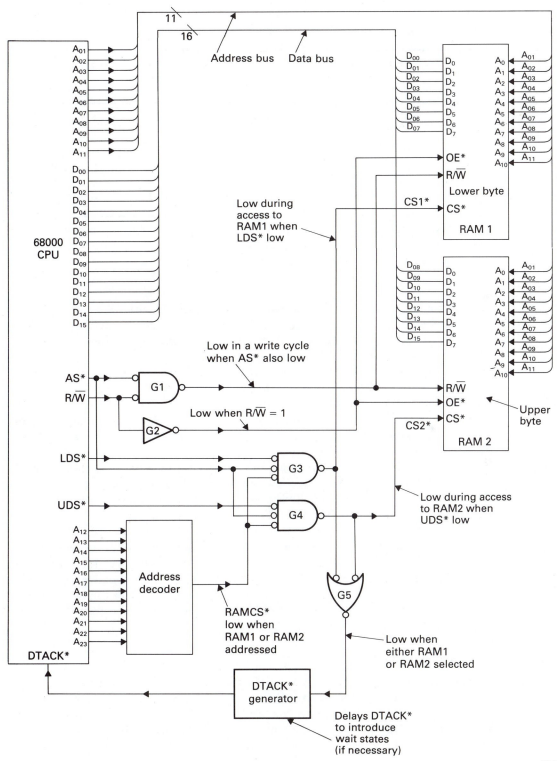

TABLE 4.6 Generating CS1* and CS2* from the 68000's strobes

INPUTS				OUTPUTS		OPERATION
AS*	RAMCS*	UDS*	LDS*	CS2*	CS1*	
1	X	X	X	1	1	No operation
X	1	X	X	1	1	No operation
0	0	0	0	0	0	Word read
0	0	0	1	0	1	Upper byte read
0	0	1	0	1	0	Lower byte read
0	0	1	1	1	1	No operation

NOTE: X = don't care (may be 1 or 0)
1 = true (positive logic)
0 = false (positive logic)

AS* is asserted. Each OE* is connected to the complement of $R/\overline{W}$ from the 68000 so that the RAM drives the data bus only in a CPU read cycle. Only the active-low chip select, CS*, inputs of the two RAMs are treated differently. Before dealing with CS*, a little needs to be said about address decoding.

Address lines A_{01} to A_{11} of the 68000 select one of 2K unique locations within the RAMs. Address lines A_{12} to A_{23} define 2^{12} or 4K possible blocks of 2K (note that 4K blocks of 2K words = 8M words). In order to uniquely assign the 2K words of RAM to one of these 4K possible blocks, address lines A_{12} to A_{23} must take part in a decoding process whereby only one of the 4K possible values spanned by these address lines is used to generate CS*.

The simplest possible address decoder is formed from a 74LS133 thirteen-input NAND gate whose output is active-low only when all address inputs are true. Thus, the RAMCS* output of the NAND is asserted whenever an address in the 2K-word (i.e., 4K-byte) range $FF F000 to $FF FFFF appears on the address bus.

Table 4.6 shows how the three signals, AS*, UDS*, and LDS*, from the CPU are combined with the RAMCS* signal from the address decoder to generate CS1* and CS2*. From table 4.6, it can be seen that CS2* is asserted (i.e., low) whenever AS*, RAMCS*, and UDS* are all asserted. Therefore, a simple negative logic AND gate can be used to derive CS2* from these signals. Similarly, CS1* can be generated in the same way, using LDS* instead of UDS*. Figure 4.14 demonstrates how little logic is needed to perform these functions.

Read Cycle Calculations

Having described the 68000's read cycle, the read cycle of a typical memory component, and a possible connection between the CPU and memory, the next step is to determine whether the CPU-RAM combination violates any timing restrictions. We proceed by drawing the timing diagram of the 68000's read cycle and then overlaying it with that of the memory being accessed.

The principal timing parameter of the RAM is its access time t_{AA}, which must be sufficient to meet the data setup-time of the CPU (i.e., t_{DICL}). Figure 4.15 relates the essential features of the 68000's timing diagram to those of the 6116P RAM. From the falling edge of S0 to the falling edge of S6, three full clock cycles take place, a total time of $3 \times t_{cyc}$. During this time, the contents of the address bus become valid (t_{CLAV}), the memory is accessed (t_{AA}), and the data setup-time is met (t_{DICL}). Thus, the total time for this action is given by $t_{CLAV} + t_{AA} + t_{DICL}$. Putting the two equations together we get

$$3 \times t_{cyc} > t_{CLAV} + t_{AA} + t_{DICL}$$

or

$$t_{AA} < 3 \times t_{cyc} - t_{CLAV} - t_{DICL}$$

or

$$t_{AA} < 3 \times 125 - 70 - 15 \qquad \text{(all values in ns)}$$
$$< 290 \text{ ns}$$

The RAM must have an access time of less than 290 ns to work with the 68000L8 at 8 MHz. As the quoted value of t_{AA} for the 6116P is 200 ns, the access time criterion is satisfied by a reasonable margin. An interesting consideration is what the demands on t_{AA} would have been if a 12.5-MHz version of the 68000 had been used. The value of t_{AA} is now given by

$$t_{AA} < 3 \times 80 - 55 - 10$$
$$< 175 \text{ ns}$$

Therefore, the 6116P RAM cannot be used at 12.5 MHz without the addition of any wait states.

The next criterion to be considered is the value of the data hold-time ($t_{SHDI} = 0$ ns minimum) required by the CPU following the rising edge of AS*. No problem arises here because we can see from figure 4.15 that the address does not change until the start of state S0 in the next cycle, which means that the data from the RAM is valid (nominally) throughout state S7. Following the rising edge of AS*/UDS*/LDS*, the data bus drivers are turned off in the RAM. However, the data bus driver will not be floated instantly and the data hold time of 0 ns will be met.

The control of CS* presents no problem. As CS* is derived from AS*, RAMCS*, and UDS*/LDS*, it is asserted very early in a read cycle, approximately 10 ns (t_1) after the falling edge of AS*. This process turns on the data bus drivers in the RAM early in the cycle, although the data is invalid until after the RAM's access time has been met. At the end of a read cycle, CS* is negated when AS* rises no more than t_{CLSH} (70 ns) after the falling edge of state S6. The data bus is floated no more than $t_{CLSH} + t_2 + t_{CHZ}$ seconds after the start of S7. For a 68000L8 and 6116P-4 combination with $t_2 = 10$ ns, the guaranteed turn-off time is $70 + 10 + 60 = 140$ ns.

FIGURE 4.15 Read cycle timing diagram of a 68000 and 6116P-4 combination

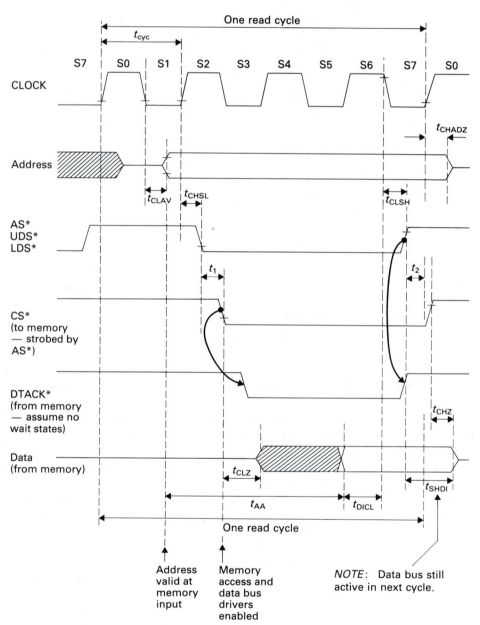

t_1 = address strobe valid to CS* low
t_2 = address strobe invalid to CS* high

As the duration of S7 is nominally 62.5 ns, the data bus may not be floated until up to 77.5 ns into the following S0. Fortunately, the next access does not begin until S2, and so no chance of bus contention occurs; that is, the next access must not try to put data on the data bus until all the data bus drivers have been turned off following the current cycle. Chapter 10, which examines buses, looks at the read/write cycle timing diagram from the point of view of bus contention, the conflict arising when two devices try to drive the same bus simultaneously.

Chapters 5 and 7 also look at the design of memory systems in greater detail. However, at the end of this chapter we will look at the 68020's timing parameters and demonstrate that parameters can sometimes be open to more than one interpretation.

68000 Write Cycle

During a write cycle, the 68000 transmits a byte or word of data to either a memory component or a memory-mapped peripheral. When a byte is written, only 8 bits of data are transferred and the appropriate data strobe asserted. When a word is written, D_{00} to D_{15} transfer the word and both UDS* and LDS* are asserted simultaneously.

Figure 4.16 gives the protocol flowchart for a write cycle, which is very similar to the corresponding read cycle flowchart of figure 4.8. The essential differences are

1. The CPU provides data at the start of a write cycle.
2. The bus slave reads this data.

A simplified timing diagram for a 68000 write cycle is given in figure 4.17. At the start of the cycle, an address is placed on A_{01} to A_{23}, and AS* is asserted, followed by R/$\overline{W}$. Unlike the corresponding read cycle, the data strobes UDS* and LDS* are *not* asserted concurrently with the address strobe. The 68000 does not assert UDS* or LDS* until after the contents of the data bus have stabilized. Consequently, the data strobe can be used by memory to latch data from the CPU. After R/$\overline{W}$ has been asserted, the CPU places data on the data bus. Once the data is valid, one or both data strobes are asserted; that is, DS* is asserted approximately one clock period after AS* has gone low.

If DTACK* is asserted before the falling edge of the S4 clock, the write cycle is terminated without the addition of wait states. Otherwise wait states are introduced until DTACK* is asserted (and meets its setup time) before the falling edge of the processor's clock.

At the end of a write cycle, AS* and DS* are negated simultaneously in response to the earlier assertion of DTACK*.

A more detailed write cycle timing diagram is given in figure 4.18. Table 4.7 defines the parameters in figure 4.18 and gives their values for 8- and 12.5-MHz versions of the 68000. From figure 4.18 it can be seen that the sequence of events at the beginning of a write cycle is

1. Address stable.

2. AS∗ asserted.

3. R/$\overline{\text{W}}$ brought low.

4. Data valid.

5. Data strobe asserted.

Each of these events is separated by a nonzero period of time except for the

FIGURE 4.16 Protocol flowchart for a word write cycle

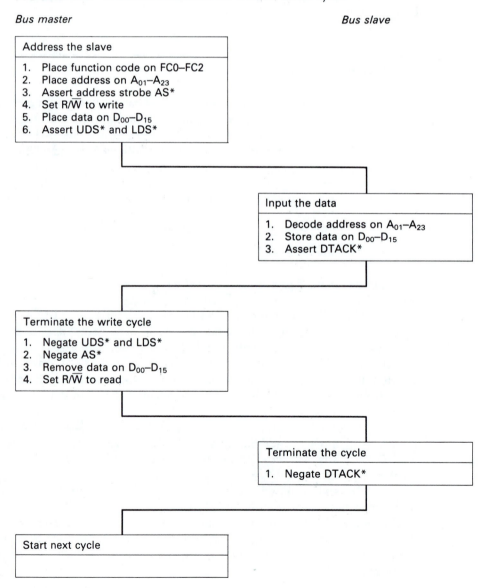

Bus master   *Bus slave*

Address the slave

1. Place function code on FC0–FC2
2. Place address on A_{01}–A_{23}
3. Assert address strobe AS*
4. Set R/$\overline{\text{W}}$ to write
5. Place data on D_{00}–D_{15}
6. Assert UDS* and LDS*

Input the data

1. Decode address on A_{01}–A_{23}
2. Store data on D_{00}–D_{15}
3. Assert DTACK*

Terminate the write cycle

1. Negate UDS* and LDS*
2. Negate AS*
3. Remove data on D_{00}–D_{15}
4. Set R/$\overline{\text{W}}$ to read

Terminate the cycle

1. Negate DTACK*

Start next cycle

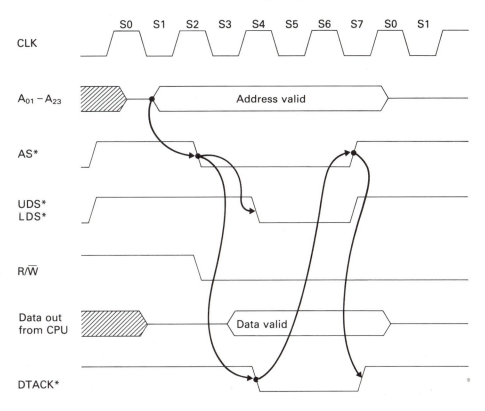

FIGURE 4.17 Simplified write cycle timing diagram for the 68000

NOTE: The data strobe (UDS*/LDS*) is not asserted in a write cycle until one clock cycle after AS*. This allows the memory to use UDS*/LDS* to latch from the CPU.

12.5-MHz 68000, whose address stable to AS* low (t_{AVSL}) is 0 ns minimum. This sequence is well suited to most memory systems, as they require the address to be stable before R/$\overline{W}$ makes its active transition.

At the end of the write cycle, the address and data strobes are negated; R/$\overline{W}$ is set high and the data bus is floated. However, some memory components require that their W* input be negated before their CS* inputs go inactive-high.

Before we consider the interface between a 68000 CPU and a memory component, the timing diagram of a memory component needs to be examined. Figure 4.19 gives the write cycle timing diagram of a 6116 static RAM. Table 4.8 defines the parameters given in figure 4.19. Note that in figure 4.19 the chip's output enable, OE*, is high throughout the write cycle.

The operation of the write cycle is entirely straightforward, and we do not have to worry about complying with any difficult timing requirements (this is not always the case). An address from the CPU is presented to the memory component and its CS* and WE* inputs are asserted. Note that WE* is "write enable"; it is called various names by chip manufacturers: WE*, W*, and R/$\overline{W}$. A write cycle ends

FIGURE 4.18 The 68000 write cycle timing diagram

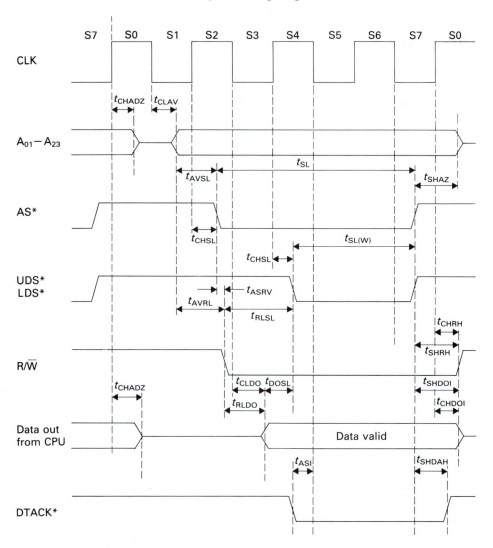

with either CS* or WE* being negated. Many memory components like the 6116 internally combine CS* and WE* so that the rising edge of either CS* or WE* terminates the write access.

In figure 4.19 an address is presented to the chip and CS* and WE* asserted. A write cycle ends by either CS* or WE* being negated. Data from the processor must be valid t_{DW} seconds before the rising edge of CS* (or WE*) and remain valid t_{DH} seconds afterward. Note that an address must be valid at least t_{AS} seconds before WE* is asserted and remain valid t_{WR} seconds after WE* has been negated.

TABLE 4.7 Write cycle timing parameters of the 68000

PARAMETER NAME	SYMBOL	8 MHz		12.5 MHz	
		MIN.	MAX.	MIN.	MAX.
Clock period	t_{cyc}	125	250	80	250
Clock high to data and address bus high impedance	t_{CHADZ}		80		60
Clock low to address valid	t_{CLAV}		70		55
Clock high to AS*, DS* low	t_{CHSL}	0	60	0	55
Address valid to AS* low	t_{AVSL}	30		0	
Clock low to AS*, DS* high	t_{CLSH}		70		50
AS*, DS* high, to address invalid	t_{SHAZ}	30		10	
R/$\overline{W}$ low to DS* low (write)	t_{RLSL}	80		30	
AS* width low	t_{SL}	240		160	
DS* width low (write cycle)	$t_{SL(w)}$	115		80	
Address valid to R/$\overline{W}$ low	t_{AVRL}	20		0	
AS* low to R/$\overline{W}$ valid	t_{ASRV}		20		20
Clock high to R/$\overline{W}$ high	t_{CHRH}	0	70	0	60
AS*, DS* high to R/$\overline{W}$ high	t_{SHRH}	40		10	
Clock low to data out valid	t_{CLDO}		70		55
R/$\overline{W}$ low to data bus low impedance	t_{RLDO}	30		10	
Data out valid to DS* low (write)	t_{DOSL}	30		15	
Data hold from clock high	t_{CHDOI}	0		0	
DS* high to data out invalid	t_{SHDOI}	30		15	
DTACK* setup-time	t_{ASI}	20		20	
AS*, DS* high to DTACK*	t_{SHDAH}	0	245	0	150

TABLE 4.8 Write cycle parameters for figure 4.19

PARAMETER NAME	SYMBOL	MINIMUM	MAXIMUM
Write cycle time	t_{WC}		150
Chip select low to end of write	t_{CW}	90	
Write recovery time	t_{WR}	10	
Address valid to end of write	t_{AW}	120	
Address setup time	t_{AS}	20	
Write pulse width	t_{WP}	90	
Data setup time	t_{DW}	40	
Data hold time	t_{DH}	10	

FIGURE 4.19 Write cycle timing diagram of 6116 static RAM

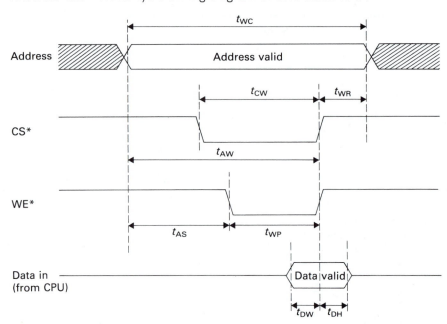

Write Cycle Calculations

The final step in considering the 68000's write cycle is to relate it to the write cycle of a typical memory component. For this exercise we use the circuit of figure 4.14. Figure 4.20 gives the write cycle timing diagram for the circuit of figure 4.14 with the appropriate timing parameters for both the 68000 and the 6116P-4 static RAM.

As in the case of the read cycle, the constraints on the memory component can all be written in terms of the 68000's parameters. These calculations are given in table 4.9. The final column in table 4.9 reports the excess time between the param-

TABLE 4.9 Write cycle parameters of the 6116-20 in terms of the 68000's parameters at 8 MHz

MEMORY PARAMETER	PARAMETER EXPRESSED IN 68000 TERMS	VALUE	REQUIRED	EXCESS
t_{WC}	$t_{AVSL} + t_{SL} + t_{SHAZ}$	$30 + 240 + 30 = 300$	150 min	150
t_{CW}	$t_{SL(w)}$	115	90 min	25
t_{WR}	t_{SHAZ}	30	10 min	20
t_{AW}	$t_{AVSL} + t_{SL}$	$30 + 240 = 270$	120 min	150
t_{AS}	t_{AVRL}	20	20 min	0
t_{WP}	$t_{SL} - t_{ASRV}$	$240 - 20 = 220$	90 min	130
t_{DW}	$t_{DOSL} + t_{SL(w)}$	$30 + 115 = 145$	40 min	105
t_{DH}	t_{SHDOI}	30	10 min	20

FIGURE 4.20 Write cycle timing diagram of a 6116-68000 combination

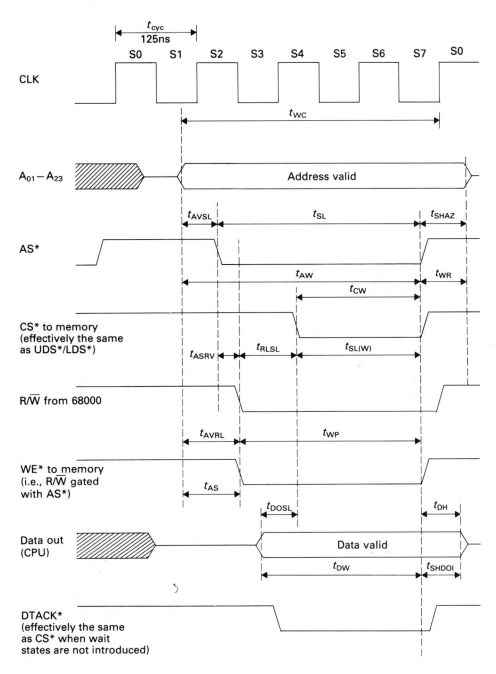

eter required by the 6116 and that provided by the 68000 at 8 MHz. In all cases, the excess value is positive, which indicates that the appropriate requirement is satisfied. In almost all cases the excess is quite large. The one exception is the address setup time of the 6116, t_{AS}, which has a zero margin.

4.3 68020 INTERFACE

In this section we are going to introduce the memory interface of the 68020 and the 68030. Since this section sometimes refers to topics introduced in later chapters, readers may wish to omit this discussion of the 68020 until they have covered the 68000. Since the 68020 and 68030 are relatively similar from an interfacing point of view, we concentrate only on two aspects: their bus timing and its dynamic bus-sizing mechanism. The 68030 has a 68020-compatible interface but includes both a high-speed synchronous interface plus a burst read mechanism (used to fill its cache memory). These aspects of the 68030 are not covered in this chapter.

Just as the 68020's software is largely compatible with the 68000's software, the 68020's asynchronous bus interface is largely compatible with the 68000's asynchronous bus interface. However, there are differences; most are quite subtle, but some are rather more dramatic. The most dramatic change is the 68020's ability to support *dynamic bus sizing*, which permits it to execute bus cycles with operands of any size to memory ports of any size. The 68020 can, for example, perform long-word accesses to byte-wide memory (the system automatically converts a longword access into four consecutive byte accesses). It can also perform word and longword accesses to an operand at a nonword boundary (i.e., at an odd address). The 68000 would generate an address error exception if the programmer attempted to do this. We will return to these aspects of the 68000 and the 68020 in chapter 6.

A more subtle difference between the 68020 and the 68000 concerns the timing of the read/write bus cycles and the provision of extra pins, telling external systems more about what the 68020 is doing.

68020's Asynchronous Bus Interface

Figure 4.21 illustrates the pins that make up the 68020's interface. Apart from the provision of true 32-bit address and data buses, most of the other interface groups are the almost same as the corresponding 68000 groups. Here we are concerned only with the 68020's asynchronous bus interface. This group includes four signals, SIZ0, SIZ1, DSACK0∗, and DSACK1∗, which implement the bus sizing mechanism we mentioned earlier. We will not go into detail at this stage, and all we need to note is that the 68020 uses its bus size outputs, SIZ0 and SIZ1, to tell the memory how much data it would like to transfer in the current bus cycle (i.e., 1, 2, 3, or 4 bytes). This is not a misprint—the 68020 can, indeed, transfer 3 bytes in a bus access. The *two* data transfer acknowledge inputs, DSACK0∗ and DSACK1∗, perform the same function as the 68000's DTACK∗ (i.e., indicating that the bus cycle can be

FIGURE 4.21 The 68020's interface

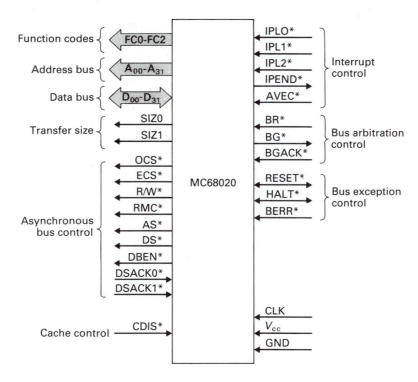

completed) but also tell the processor how wide the memory is and therefore how much data has been transferred.

The following paragraphs describe the 68020's new pins in the asynchronous bus interface group.

OCS∗: Operand Cycle Start   The OCS∗ pin is an output that indicates the beginning of the first external bus cycle for an instruction prefetch or for the first bus cycle of an operand transfer. That is, OCS∗ is asserted to tell you that the 68020 is just beginning a new instruction or operand fetch. Figure 4.22a illustrates the timing of the operand cycle start signal. Although the OCS∗ pin might provide useful information to a logic analyzer, an in-circuit emulator, or to a memory management unit, it plays no important role in general 68020-based microcomputer systems and may, for all practical purposes, be ignored.

ECS∗: External Cycle Start   The ECS∗ pin is an output that is asserted at the beginning of *all* bus cycles, irrespective of type. Note that OCS∗ is asserted for the *first* bus cycle of either an instruction or an operand access (but not for subsequent cycles), whereas ECS∗ is asserted for all bus cycles. Figure 4.22b illustrates the timing of the external cycle start signal. As in the case of OCS∗, the ECS∗ is the earliest indication provided by the 68020 that it is going to perform a bus cycle, and therefore ECS∗ can sometimes be employed to trigger the hardware used to make a memory access.

ECS* is asserted even in a bus cycle that fetches an instruction from the 68020's instruction cache (see chapter 7 for a discussion of cache memory). Such an instruction fetch is aborted by a hit from the cache memory. In this case, the address strobe, AS*, is not asserted. Note that ECS* is asserted very early in a bus cycle and offers the only means of identifying an S0 state (the first clock state in a read or

FIGURE 4.22   OCS* and ECS* timing diagram

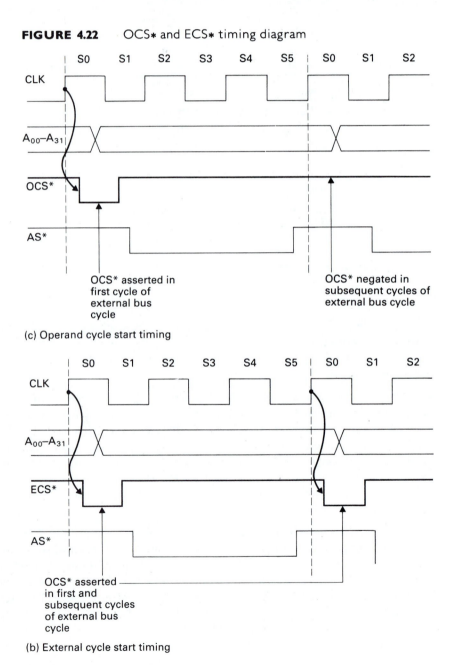

(c) Operand cycle start timing

(b) External cycle start timing

write access). ECS* is useful for a state decoder, but note that AS* must be detected in state S1 to indicate that the bus cycle has not been aborted because of a cache hit.

RMC*: Read-Modify-Write Cycle Both the 68000 and the 68020 can perform read-modify-write cycles (RMW cycles), in which an operand is read from external memory, updated by the processor (i.e., modified), and then written back to the memory. The importance of a RMW cycle is that the bus cycle is considered to be *indivisible* and must be executed from beginning to end without interruption. That is, no other device may take control of the data bus (no matter how temporarily) until the RMW cycle has been executed to completion. Some operations, such as ADDI #$1234,(A0), also read an operand from memory, carry out an operation, and then write the result back to memory. These are not true indivisible read-modify-write cycles, because another device may access the bus in between the read and write phases of the cycle.

The 68000 has just a single instruction that executes a RMW cycle (TAS = test and set), whereas the 68020 has two new instructions (CAS and CAS2 = compare and swap with operand). We will not go into details here; we simply note that the RMW cycle is important in multiprocessor systems and is designed to stop another processor gaining control of the bus at a critical point. Basically, RMW instructions prevent two processors from grabbing the same memory at the same time. We will say a little more about the RMW cycle in chapter 7. Some literature calls a RMW cycle a *locked* cycle.

The 68000 indicates a RMW cycle indirectly by not negating its AS* output between the read and write phases of the indivisible (i.e., locked) bus cycle, as figure 4.23a demonstrates. You cannot tell, in advance, that the 68000 is executing an RMW cycle. Figure 4.23b shows how the 68020's RMC* output indicates to external hardware that the current bus cycle is locked. As you can see, RMC* remains asserted for the duration of the RMW cycle. The RMC* pin is used only in sophisticated multiprocessor systems. No other device (i.e., potential bus master) may take control of the bus while RMC* is asserted. Just think of RMC* as a *do not disturb* signal. Note that the RMC* output is necessary only in multiprocessor environments and not DMA environments, since the 68020 will not issue a bus grant (BG*) output on TAS, CAS, and CAS2 instructions until after the RMW cycle.

DBEN*: Data Bus Enable The DBEN* pin is an output indicating that external data bus buffers should be enabled. DBEN* tells the buffers that there is data on the data bus. During a read access DBEN* is asserted approximately one clock cycle after the beginning of the bus cycle and is negated when the data strobe is negated (figure 4.24). In a write access, DBEN* is asserted at approximately the same time as AS* and held active for the duration of the cycle (figure 4.25).

DBEN* effectively qualifies (i.e., validates) the R/$\overline{\text{W}}$ line and may be employed to ensure that the direction of bus transceivers is selected before their outputs are enabled. Bus transceivers are dealt with when we describe buses in chapter 10. Like the other new control outputs described previously DBEN* is not necessary in many 68020 systems.

The 68020's four new bus control signals (OCS*, ECS*, RMW*, and DBEN*) endow the 68020 with more *functionality* than the 68000. However as we have

FIGURE 4.23   68000 and 68020 RMW cycles

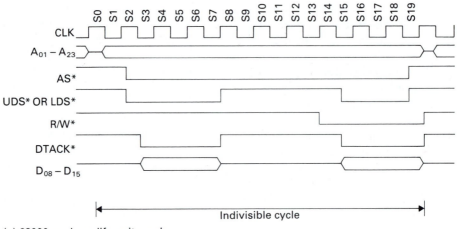

(a) 68000 read-modify-write cycle

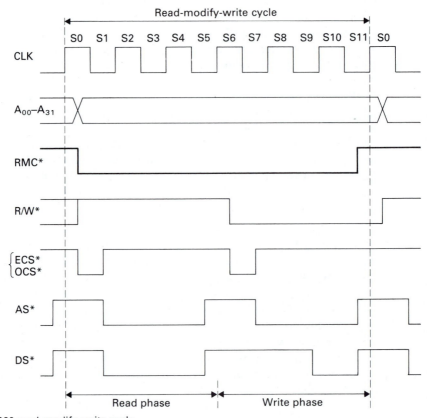

(b) 68020 read-modify-write cycle

FIGURE 4.24 DBEN∗ timing in a read cycle

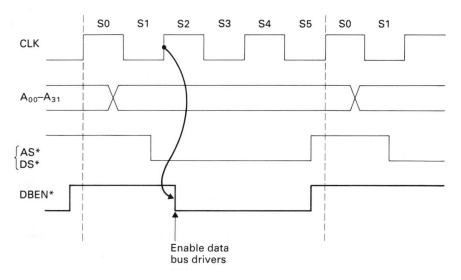

Enable data
bus drivers

FIGURE 4.25 DBEN∗ timing in a write cycle

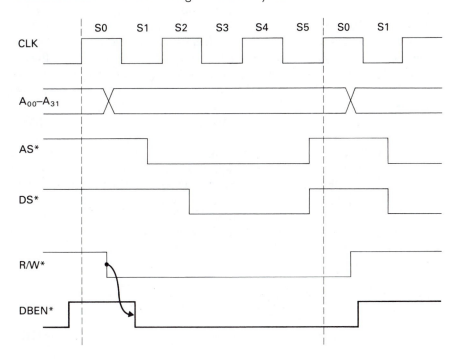

already stressed, these four output pins can be entirely neglected and left as open circuits in the majority of 68020-based systems.

DS∗: Data Strobe   The 68000 has two data strobes, LDS∗ and UDS∗, which indicate an access to a low-order byte or to a high-order byte, respectively, during a bus access. The 68020 has a single pin, DS∗, that acts as a data strobe in the asynchronous handshake. To compensate for the lack of separate upper and lower data strobes, the 68020 has an A_{00} pin (unlike the 68000). We look at how the 68020 accesses bytes, words, and longwords in the next section. The timing of the 68020's DS∗ output is essentially the same as the 68000's UDS∗ and LDS∗ strobes (i.e., asserted at the same time as AS∗ in a read cycle and asserted one clock after AS∗ in a write cycle).

68020 Read Cycle

Although we have already described the 68000's read and write cycle timing diagrams, we are now going to look at the 68020's bus cycle timing. We do this for two reasons. The first is that an appreciation of the way in which a microprocessor communicates with external memory is vital to the systems designer. The second is that we introduce a new concept into our discussion of timing diagrams.

Figure 4.26 describes the full details of the 68020's read cycle for reference purposes. Since we are not interested in all the 68020's pins and are concerned only with the fundamental timing requirements in the read cycle, figure 4.27 presents a much simpler picture of the read cycle timing by omitting several control outputs and by presenting the two data transfer acknowledge inputs, DSACK0∗ and DSACK1∗, as DSACK∗. A quick glance at figure 4.27 might suggest that the 68000 and 68020 read cycles are entirely identical. This is not so, as the 68000 has an eight-state bus cycle, whereas the 68020 has a six-state bus cycle. Even at the same clock rate, the 68020 is faster than the 68000 because of its improved internal design.

The first question we need to ask ourselves is, What is the maximum memory access time that can be tolerated without the introduction of wait states? To keep things simple, we will use static RAM and assume that its access time is measured from the point at which its chip-select input is asserted (rather than when its address input is first valid). Some RAMs measure their access from address valid and some from chip select asserted. If we forget about delays in address decoders and buffers, we can regard the 68020's data strobe, DS∗, as being the same as the memory device's chip select input, CS∗. The memory's access time is, therefore, the period between DS∗ low and the point at which the data must be valid before the falling edge of state S4. The time for which DS∗ is low is defined as t_{SWA} in the following equations.

We can calculate the memory's access time, t_{acc}, from the equation

$$t_{SWA} = t_{acc} + t_{DICL} + t_{CLSH}$$

$$t_{acc} = t_{SWA} - t_{DICL} - t_{CLSH}$$

$$= 120 \text{ min} - 10 \text{ max} - 40 \text{ max} = 70 \text{ ns} \qquad \text{(at 12.5 MHz)}$$

FIGURE 4.26 Full details of the 68020's read cycle

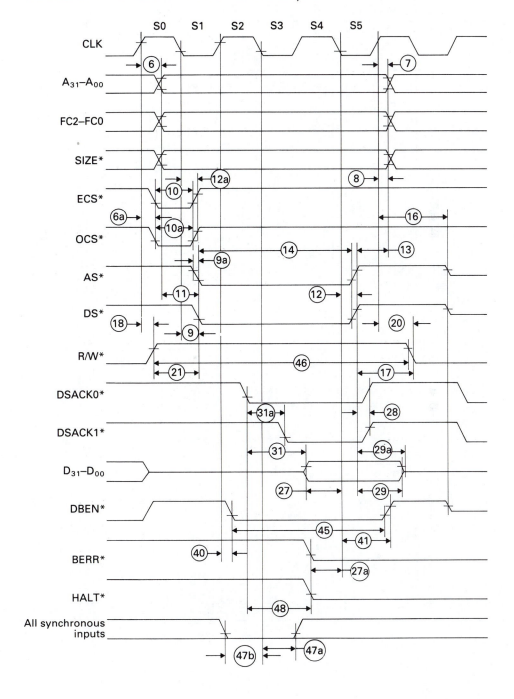

FIGURE 4.26 (Continued)

No.	Characteristics	12.5 MHz Min	12.5 MHz Max	16.67 MHz Min	16.67 MHz Max	20 MHz Min	20 MHz Max	25 MHz Min	25 MHz Max	33.33 MHz Min	33.33 MHz Max	Unit
6	Clock high to address, FC, size, RMC* valid	0	40	0	30	0	25	0	25	0	21	ns
6A	Clock high to ECS*, OCS* asserted	0	30	0	20	0	15	0	12	0	10	ns
7	Clock high to address, data, FC, size, RMC*, high impedance	0	80	0	60	0	50	0	40	0	30	ns
8	Clock high to address, FC, size, RMC*, invalid	0	—	0	—	0	—	0	—	0	—	ns
9	Clock low to AS*, DS* asserted	3	40	3	30	3	25	3	18	3	15	ns
9A	AS* to DS* assertion (read) (skew)	-20	20	-15	15	-10	10	-10	10	-10	10	ns
9B	AS* asserted to DS* asserted (write)	47	—	37	—	32	—	27	—	22	—	ns
10	ECS* width asserted	25	—	20	—	15	—	15	—	10	—	ns
10A	OCS* width asserted	25	—	20	—	15	—	15	—	10	—	ns
10B	ECS*, OCS* width negated	20	—	15	—	10	—	5	—	5	—	ns
11	Address, FC, size, RMC* valid to AS* (and DS* asserted read)	20	—	15	—	10	—	6	—	5	—	ns
12	Clock low to AS*, DS* negated	0	40	0	30	0	25	0	15	0	15	ns
12A	Clock low to ECS*, OCS* negated	0	40	0	30	0	25	0	15	0	15	ns
13	AS*, DS* negated to address, FC, size RMC* invalid	20	—	15	—	10	—	10	—	5	—	ns
14	AS* (and DS* read) width asserted	120	—	100	—	85	—	70	—	50	—	ns
14A	DS* width asserted write	50	—	40	—	38	—	30	—	25	—	ns
15	AS*, DS* width negated	50	—	40	—	38	—	30	—	23	—	ns
15A	DS* negated to AS* asserted	45	—	35	—	30	—	25	—	18	—	ns
16	Clock high to AS*, DS*, R/W*, DBEN* high impedance	—	80	—	60	—	50	—	40	—	30	ns
17	AS*, DS* negated to R/W* invalid	20	—	15	—	10	—	10	—	5	—	ns
18	Clock high to R/W* high	0	40	0	30	0	25	0	20	0	15	ns
20	Clock high to R/W* low	0	40	0	30	0	25	0	20	0	15	ns
21	R/W* high to AS* asserted	20	—	15	—	10	—	5	—	5	—	ns
22	R/W* low to DS* asserted (write)	90	—	75	—	60	—	50	—	35	—	ns
23	Clock high to data out valid	—	40	—	30	—	25	—	25	—	18	ns
25	DS* negated to data out invalid	20	—	15	—	10	—	5	—	5	—	ns
25A	DS* negated to DBEN* negated (write)	20	—	15	—	10	—	5	—	5	—	ns
26	Data out valid to DS* asserted (write)	20	—	15	—	10	—	5	—	5	—	ns
27	Data-in valid to clock low (data setup)	10	—	5	—	5	—	5	—	5	—	ns
27A	Late BERR*, HALT* asserted to clock low setup time	25	—	20	—	15	—	10	—	5	—	ns

No.	Characteristics	12.5 MHz Min	12.5 MHz Max	16.67 MHz Min	16.67 MHz Max	20 MHz Min	20 MHz Max	25 MHz Min	25 MHz Max	33.33 MHz Min	33.33 MHz Max	Unit
28	AS*, DS* negated to DSACKx*, BERR*, HALT*, AVEC* negated	0	110	0	80	0	65	0	50	0	40	ns
29	DS* negated to data-in invalid (data-in hold time)	0	—	0	—	0	—	0	—	0	—	ns
29A	DS* negated to date-in (high impedance)	—	80	—	60	—	50	—	40	—	30	ns
31	DSACKx* asserted to data-in valid	—	60	—	50	—	43	—	32	—	17	ns
31A	DSACKx* asserted to DSACKx* valid (DSACK* asserted skew)	—	20	—	15	—	10	—	10	—	10	ns
32	RESET* input transition time	—	1.5	—	1.5	—	1.5	—	1.5	—	1.5	Clks
33	Clock low to BG* asserted	0	40	0	30	0	25	0	20	0	20	ns
34	Clock low to BG* negated	0	40	0	30	0	25	0	20	0	20	ns
35	BR* asserted to BG* asserted (RMC* not asserted)	1.5	3.5	1.5	3.5	1.5	3.5	1.5	3.5	1.5	3.5	Clks
37	BGACK* asserted to BG* negated	1.5	3.5	1.5	3.5	1.5	3.5	1.5	3.5	1.5	3.5	Clks
37A	BGACK* asserted to BR* negated	0	1.5	0	1.5	0	1.5	0	1.5	0	1.5	Clks
39	BG* width negated	120	—	90	—	75	—	60	—	50	—	ns
39A	BG* width asserted	120	—	90	—	75	—	60	—	50	—	ns
40	Clock high to DBEN* asserted (read)	0	40	0	30	0	25	0	20	0	15	ns
41	Clock low to DBEN* negated (read)	0	40	0	30	0	25	0	20	0	15	ns
42	Clock low to DBEN* asserted (write)	0	40	0	30	0	25	0	20	0	15	ns
43	Clock high to DBEN* negated (write)	0	40	0	30	0	25	0	20	0	15	ns
44	R/W* low to DBEN* asserted (write)	20	—	15	—	10	—	10	—	5	—	ns
45	DBEN* width asserted — Read	80	—	60	—	50	—	40	—	30	—	ns
45	DBEN* width asserted — Write	160	—	120	—	100	—	80	—	60	—	ns
46	R/W* width valid (write or read)	180	—	150	—	125	—	100	—	75	—	ns
47A	Asynchronous input setup time	10	—	5	—	5	—	5	—	5	—	ns
47B	Asynchronous input hold time	20	—	15	—	15	—	10	—	10	—	ns
48	DSACKx* asserted to BERR*, HALT* asserted	—	35	—	30	—	20	—	18	—	15	ns
53	Data out hold from clock high	0	—	0	—	0	—	0	—	0	—	ns
55	R/W* valid to data bus impedance change	40	—	30	—	25	—	20	—	20	—	ns
56	RESET* pulse width (reset instruction)	512	—	512	—	512	—	512	—	512	—	Clks
57	BERR* negated to HALT* negated (rerun)	0	—	0	—	0	—	0	—	0	—	ns
58	BGACK* negated to bus driven	1	—	1	—	1	—	1	—	1	—	Clks
59	BG* negated to bus driven	1	—	1	—	1	—	1	—	1	—	Clks

FIGURE 4.27   Simplified version of the 68020's read cycle

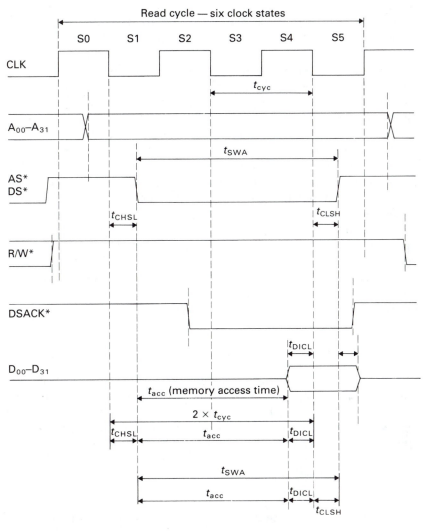

If we look at figure 4.27 again, we can find another way of calculating the read access time. Instead of employing the time for which the data strobe is low, t_{SWA}, as a basic metric, we will consider the time from the start of the S1 state (i.e., falling edge of the S0 clock) to the time at which the data is latched (i.e., the falling edge of the S4 clock). This period is two clock cycles and is made up of the delay to DS* low, the memory's access time and the 68020's data setup time. Therefore, we can write:

$$2t_{cyc} = t_{CHSL} + t_{acc} + t_{DICL}$$

$$t_{acc} = 2t_{cyc} - t_{CHSL} - t_{DICL}$$

$$= 160 - 40 \text{ max} - 10 \text{ max} = 110 \text{ ns} \qquad \text{(at 12.5 MHz)}$$

You can see that we now have two values for the access time, one of 70 ns and one of 110 ns. Taking into account the premium that must be paid for high-speed memory, this difference is not trivial. More to the point, why do we have two different values for t_{acc} when we used Motorola's own data for both calculations?

I once carried out a similar calculation on the blackboard, and then asked the students which of the two values an engineer should take in practice. Their answer was that if we are looking for a worst-case value, we should take the minimum access time, which in this case would be 70 ns. Consider my counterargument. Both these calculations are carried out using worst-case data. Therefore, both calculations yield pessimistic, but guaranteed, results. Consequently, the higher access time of 110 ns is guaranteed, even though a different calculation gives us a more pessimistic result.

Why do some worst-case calculations yield unreasonably pessimistic values? To answer this you have to look at the source of the parameters themselves. Suppose some hypothetical signal goes low for two clock cycles, as described by figure 4.28. One parameter, t_{LOW}, tells us when it goes low with respect to the clock and another parameter, t_{HIGH}, tells us when it goes high. The length of the signal is therefore $2t_{cyc} - t_{LOW} + t_{HIGH}$. Now suppose we are interested in the worst-case minimum time for which the signal is asserted low t_A. This is given by $2t_{cyc} - t_{LOW}$ (max) $+ t_{HIGH}$ (min). Furthermore, we will assume that a clock cycle is 50 ns and that the delays t_{LOW} and t_{HIGH} vary between 20 ns and 40 ns (determined by measurement). The minimum low time for the signal on these assumptions is, therefore,

$$t_A = 100 - 40 + 20 = 80 \text{ ns}$$

There is a flaw in the calculation for t_A. The measurements were probably obtained by sampling devices from the production line. A device with a low value for t_{LOW} would probably have a low value for t_{HIGH}, since the same mechanism (i.e., the device physics) is at work in both cases. In other words, we can fall into a trap when using worst-case parameters, because it might be physically impossible for one parameter to be worst case in one direction and another parameter to be a worst case in the opposite direction.

FIGURE 4.28 Illustration of the problems in interpreting timing parameters

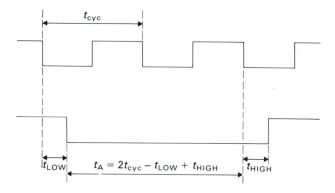

If we return to figure 4.27 and the calculation yielding the worst-case value for the access time of 70 ns, we find that the value was derived from the worst-case low time for the data strobe (i.e., t_{SWA} (min)) and the worst-case low-to-high delay time for DS* following the falling edge of the S4 clock (i.e., t_{CLSH} (max)). It is highly improbable that one mechanism would make the down time for DS* as short as possible while keeping DS* low beyond the falling edge of the S4 clock for as long as possible.

It is interesting to see what happens if we consider using a very high speed version of the 68020 with a clock speed of 33.333 MHz (i.e., a clock cycle time of 30 ns). We will use the expression that gave the most optimistic result and recalculate the maximum access time as

$$t_{acc} = 2t_{cyc} - t_{CHSL} - t_{DICL}$$
$$= 60 - 15 \text{ max} - 5 \text{ max} = 40 \text{ ns}$$

Fast static RAM chips with access times of 40 ns and less are available, but they are very expensive and are usually quite small (in comparison with today's 4M-bit DRAMs). Adding two wait states (i.e., 30 ns) to the 68020 at 33.333 MHz relaxes the access time requirement to a more reasonable $40 + 30 = 70$ ns. Adding four wait states increases the maximum access time to 100 ns. Remember that the 68000 and 68020 require the insertion of an even number of wait states.

68020 Write Cycle

The 68020's write cycle is very much like the corresponding 68000 write cycle: AS* is asserted at the start of the bus cycle and the data strobe; DS* is asserted one clock cycle later. Figure 4.29 provides a simplified version of the 68020 write cycle with only the salient details. We do not intend to spend too much time covering the 68020's write cycle and will just determine the limiting parameters of a typical static RAM in a write cycle.

Figure 4.30 illustrates the four most important parameters governing a static RAM's write cycle. These parameters are the cycle time (effectively the time from the point at which the address is valid to the end of the write cycle, indicated by CS* or W* high), the chip's W* active-low time, and the data setup- and data hold-times. Of course, there are variations from one device to another. We can calculate limiting values for these four parameters from figure 4.29 and figure 4.30 as follows.

Cycle time

$$t_{CHAV} \text{ max} + t_{cycle} - t_{SNAI} = 3t_{cyc}$$
$$t_{cycle} = 3t_{cyc} - t_{CHAV} + t_{SNAI}$$
$$t_{cycle} = 3 \times 80 - 40 + 20 = 220 \text{ ns} \quad \text{(at 12.5 MHz)}$$
$$t_{cycle} = 3 \times 30 - 21 + 5 = 74 \text{ ns} \quad \text{(at 33.333 MHz)}$$

FIGURE 4.29 Simplified version of the 68020's write cycle timing diagram

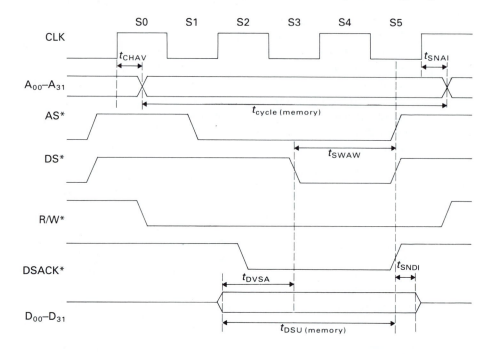

Write Pulse Time (W∗ low) For this calculation we assume that the 68020's R/$\overline{\text{W}}$ output is qualified (i.e., strobed) with its data strobe, DS∗. By qualifying R/$\overline{\text{W}}$ with DS∗ we ensure that W∗ is not active-low unless the 68200 is actively executing a write cycle. Consequently, the W∗ low time is effectively the same as the 68020's

FIGURE 4.30 Write cycle timing diagram of a static RAM

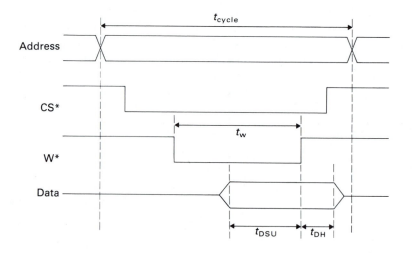

DS* low time. This time is

t_{SWAW} = 50 ns min at 12.5 MHz

t_{SWAW} = 25 ns min at 33.333 MHz

When selecting suitable memory for use with the 68020, the cycle time and W* low time are likely to be the dominating (i.e., limiting) parameters.

Data Setup Time The memory's data setup time, t_{DSU}, in a write cycle is the period between the point at which data from the 68020 is valid to the end of the write cycle when CS* (or W*) goes high.

$t_{DSU} = t_{DVSA} + t_{SWAW}$

t_{DSU} = 20 ns min + 50 ns min = 70 ns min at 12.5 MHz

t_{DSU} = 5 ns min + 25 ns min = 30 ns min at 12.5 MHz

Since many memories have very short data setup times, it is unlikely that any problem with data setup times will be encountered by the 68020 systems designer.

Data Hold Time The memory's data hold time, t_{DH}, in a write cycle is the period of time for which the data must remain valid after the end of the write cycle. This time is given by

$t_{DH} = t_{SNDI}$

t_{DH} = 20 ns min at 12.5 MHz

t_{DH} = 5 ns min at 33.333 MHz

Now that we have looked at the 68020's timing requirements, we are going to look at how it deals with variable-sized memory ports.

Dynamic Bus Sizing

Dynamic bus sizing is the term used to describe the 68020's remarkable ability to support any combination of operand size with any combination of memory port width. In other words, you can fill the 68020's memory space with 8-bit-, 16-bit-, and 32-bit-wide blocks of memory and never have to worry about how the 68020 accesses these blocks. At the start of a bus cycle the 68020 says to the memory, "This is the size of the data I would like to send to you." The memory might then reply, "I can't take it all—just give me a couple of bytes and send the rest later." Dynamic bus sizing is conceptually easy to understand, but its details are rather involved. The reason for this complexity is the large number of combinations of operand size (three) and of memory port widths (three), plus the ability to handle misaligned operands. Here, we just present an overview of dynamic bus sizing and concentrate on its implications for the systems designer.

Before looking at the details of the 68020's dynamic bus sizing mechanism, we will put the topic into its context. Life was simple in the days of the 8-bit microprocessor: The data bus was 8 bits wide and all memory ports were 8 bits wide. A processor simply addressed a memory location and then read from or wrote to that location using all 8 bits of its data bus.

With the advent of 16-bit microprocessors (or, to be more precise, processors with 16-bit data buses, such as the 68000), matters became more complex. You might have noticed the expression *memory port* in the preceding text. We have to introduce the idea of a memory port to distinguish between the width of the processor's logical memory access as viewed by the programmer (e.g., .B, .W, .L) and the width of the physical memory. For example, an 8-bit memory port can participate in an 8-bit memory access in a single bus cycle. If a 16-bit value is to be read from or written to the 8-bit memory port, it is necessary to execute two bus cycles. A 32-bit memory port can take part in 32-bit accesses, and most 32-bit ports are arranged so that they can also take part in 16-bit and 8-bit accesses as well.

If 16-bit microprocessors had only 16-bit memory ports, just as their 8-bit counterparts had only 8-bit memory ports, there would be no problem. But 16-bit microprocessors have to coexist with 8-bit devices (e.g., memory-mapped peripherals) and with byte accesses, as well as with 16-bit memory ports and word accesses—that is, a 16-bit microprocessor still has to access 8-bit data values. As we have seen, the 68000 deals with 8-bit accesses by using address bits A_{01} to A_{23} of its address bus to access a word on a word boundary and then using its data strobes UDS* and LDS* to select one or both bytes of the specified word. Note that the 68000 does not directly support longword accesses. A longword is accessed as two consecutive words.

Even though the 68000 is byte addressable, word and longword values cannot be accessed at odd byte boundaries. This limitation is not imposed by a fundamental restriction of the 68000, but rather by a decision made by the designers of the 68000. For example, if a word were to be accessed at location $1001, it would be necessary first to address the word at location $1000 and access byte $1001 and then to access the word at location $1002 and access byte $1002. Instead of implementing this scheme, the 68000's designers have simply made it illegal to access a word operand at an odd byte boundary. It is, therefore, the responsibility of the programmer to ensure that the rule is not violated. A word that is located at an odd byte address is said to be *misaligned*.

The 68000 supports 16-bit memory ports. We can locate an 8-bit device such as a memory-mapped ACIA or PI/T at an odd byte address and strobe it with LDS* or at an even byte address and strobe it with UDS*. In later chapters we examine such peripherals. Note that the 68000 does not support 8-bit memory ports. You can think about why this is so; we provide the answer later.

Modern 32-bit microprocessors such as the 68020 introduce a new layer of complexity by permitting longword, word, and byte accesses. Figure 4.31 illustrates how a hypothetical 32-bit processor's data bus interface might be organized. Since byte addressing is still necessary, the address bus extends from A_{02} to A_{31} and accesses a longword at a longword boundary. Address bits A_{00} and A_{01} exist only within the hypothetical processor in figure 4.31 and are not required by the memory system, because all accesses are restricted to longword boundaries. Four data strobe outputs, BE0*, BE1*, BE2*, and BE3*, permit the processor to access any or all of

FIGURE 4.31 Dealing with byte and word accesses in a hypothetical 32-bit system

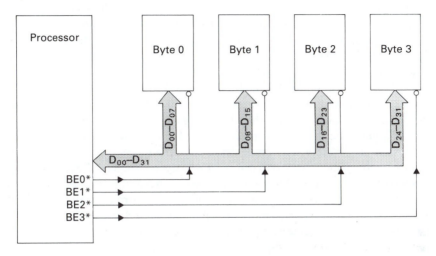

the selected longword's 4 bytes (the mnemonic BEi* = byte enable i is used, although DSi* = data strobe i would have done just as well). The 80386 microprocessor implements this type of arrangement. The 68020 implements a rather complex mechanism for accessing the individual bytes of a longword operand, as we shall soon discover.

Before discussing the details of the 68020's dynamic bus sizing mechanism, we really need to ask ourselves what we wish to know about bus sizing and why. In fact, there are really three distinct groups of people who are interested in the way in which a processor deals with variable-sized operands and variable-sized memory ports: programmers, students of computer architecture and organization, and systems designers.

Programmers are interested only in how the movement of information between the CPU and memory affects the way in which a program is written or with the performance of the program. For example, 68000 programmers must not locate either word or longword instructions and data operands at odd-byte boundaries, as that would probably cause a fatal error. In practice, this means that programmers must take care not to *generate* the following code:

```
            ORG        $1000
DATA1       DS.W       1
DATA2       DS.B       1
DATA3       DS.W       1
            :
            MOVE.W     DATA3,D2
```

Operand DATA1 is located at address $1000, DATA2 is at address $1002, and DATA3 is at address $1003 (since DATA2 requires a single byte). If, later in the program, the word operand DATA3 is accessed by MOVE.W DATA3,D2, the 68000 will attempt to read a word at a misaligned address $1003 and generate an address error exception. The 68020, which permits misaligned operands, is quite willing to accept the preceding code. Misaligned operands cause the processor to run more slowly, since two or more bus cycles are needed to access the operand. Note that we said programmers should not *generate* this code rather than *write* the code, because most assemblers will automatically insert a null byte to avoid a misaligned word address—that is, assemblers automatically protect programmers from some of their errors.

Students of computer architecture and organization are interested in how dynamic bus sizing is implemented internally. They are concerned with the minute details of bus sizing, because they will have to design tomorrow's processors.

Systems designers are interested neither in the effect of bus sizing on performance nor in how it is implemented internally. Systems designers are given the task of connecting the 68020 to real memory components. Systems designers must understand how the 68020's pins are used to create an interface between the CPU and external memory. Of course, systems designers must appreciate how the 68020 implements dynamic bus sizing, but not at the same level as the student of computer architecture and organization. In this book we view the 68020 from the point of view of the systems designer.

The 68020 has a kit of parts (i.e., input and output pins) that it uses to implement dynamic bus sizing. We will introduce dynamic bus sizing by describing this kit, which includes the following components.

Address Bits A_{01} and A_{00} These two address bits are used to identify the byte addressed at the current longword boundary specified by A_{02} to A_{31}. We can regard address bits A_{01}, A_{00} as an offset to the byte of the longword pointed at by A_{02} to A_{31}. As you might expect, the encoding of A_{01} and A_{00} is

A_{01}	A_{00}	OFFSET
0	0	0 bytes
0	1	1 byte
1	0	2 bytes
1	1	3 bytes

Size bits SIZ1 and SIZ0 The 68020's SIZ0 and SIZ1 output pins have no counterpart on the 68000. SIZ0 and SIZ1 inform the memory system how much data the 68020 would like to transfer in the current bus cycle and are encoded as follows.

DATA TO BE TRANSFERRED	SIZ1	SIZ0
Longword	0	0
Three bytes	1	1
Word	1	0
Byte	0	1

Do not be alarmed by the SIZ1,SIZ0 combination 1,1, which indicates that 3 bytes remain to be transferred. If a longword operand is misaligned so that 1 byte lies in one longword and the next 3 bytes lie in the next longword, the 68020 simply accesses the operand by reading 1 byte and then the remaining 3 bytes in the next bus cycle. You must understand that SIZ1 and SIZ0 represent a *request* to transfer a certain amount of data and not a *demand*.

It is very important to appreciate that SIZ1 and SIZ0 signify the 68020's desire to transfer the specified amount of data. Two data transfer acknowledge signals from the port being accessed inform the 68020 how much data can actually be transferred.

Data Acknowledge Strobes DSACK1∗ and DSACK0∗   The 68000 has a single data transfer acknowledge input, DTACK∗, which provides the processor with a handshake from the memory indicating that the current bus cycle may proceed to completion. If a bus cycle starts and DTACK∗ is not received asserted active-low, the 68000 will introduce wait states until either DTACK∗ does go low or BERR∗ is asserted to terminate the cycle with a bus error.

The memory or peripheral currently being accessed uses the *two* data acknowledge strobes to perform two tasks. The first task is to terminate the current bus cycle, which takes place when either or both of DSACK0∗ and DSACK1∗ is asserted. The second task is to tell the 68020 how wide the currently accessed port is. We will briefly look at DSACK0∗ and DSACK1∗ timing before discussing their role in dynamic bus sizing. Figure 4.32 describes the timing restrictions placed on the 68020's data transfer acknowledge inputs. In addition to the timing requirements faced by DTACK∗ in 68000 systems, DSACK0∗ and DSACK1∗ must comply with a skew limitation. That is, if both DSACK0∗ and DSACK1∗ are asserted in the same bus cycle, the maximum time between their assertion (i.e., their skew) must not exceed a specified value. Figure 4.32 tells the systems designer to take care when designing DSACK0∗ and DSACK1∗ circuits.

When the 68020 begins a bus cycle, it waits for DSACK0∗ and DSACK1∗ before terminating the cycle. The information supplied by the two data transfer

FIGURE 4.32   Timing details of data acknowledge strobes DSACK0∗ and DSACK1∗

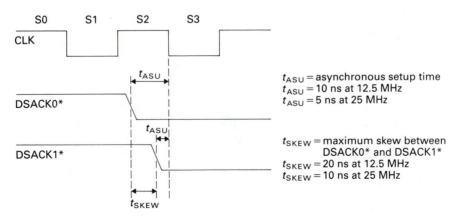

t_{ASU} = asynchronous setup time
t_{ASU} = 10 ns at 12.5 MHz
t_{ASU} = 5 ns at 25 MHz

t_{SKEW} = maximum skew between DSACK0∗ and DSACK1∗
t_{SKEW} = 20 ns at 12.5 MHz
t_{SKEW} = 10 ns at 25 MHz

acknowledge strobes informs the 68020 whether the port is an 8-bit port, a 16-bit port, or a 32-bit port. The encoding of DSACK0∗ and DSACK1∗ is as follows.

DSACK1∗	DSACK0∗	INTERPRETATION
1	1	No data acknowledge—insert wait states.
1	0	Data acknowledge—data bus port size is 8 bits.
0	1	Data acknowledge—data bus port size is 16 bits.
0	0	Data acknowledge—data bus port size is 32 bits.

As we have just said, DSACK0∗ and DSACK1∗ inform the 68020 whether the currently addressed port is 8 bits, 16 bits, or 32 bits wide. In order to understand what this means, consider the 68000. All 68000 instructions and word operands must reside in 16-bit memory ports (although these ports can be implemented as two 8-bit ports side by side). The 68000 can, of course, access the upper or lower byte of a 16-bit port.

Remember the question posed earlier in this section: Why doesn't the 68000 support 8-bit ports even though it can employ 8-bit peripherals? The answer is that these 8-bit peripherals are treated not as true 8-bit ports but as if they were part of a 16-bit memory port. For example, if you memory-map an 8-bit device with two internal locations at address $10 0000, you map the two addresses at $10 0000 and $10 0002. The location $10 0001 does not exist.

The inability of the 68000 to support true 8-bit ports can sometimes be a nuisance when testing 68000 systems, because the monitor in ROM must be stored in two 8-bit chips to create a 16-bit wide memory port. You cannot use a single 8-bit ROM, since the 68000 cannot read an instruction by performing two consecutive byte accesses.

The 68020 permits you to put the monitor in a single 8-bit-wide EPROM, since it can access words and longwords from an 8-bit memory port. For example, suppose that the 68020 wants to read a longword from the 8-bit EPROM. The processor sets its SIZ1,SIZ0 outputs to 0,0 to indicate that it wishes to access a longword. The 8-bit memory port returns DSACK1∗,DSACK0∗ = 1,0 to indicate that only 1 byte can be transferred. The 68020 then executes a bus cycle with SIZ1,SIZ0 = 1,1 to indicate that it wishes to transfer 3 bytes. The process continues until all 4 bytes of the longword operand have been transferred in four consecutive bus cycles.

From the preceding commentary you should now appreciate that the 68020 permits the systems designer to adopt any combination of memory port width without paying a penalty other than a reduction in speed due to the need to execute multiple bus cycles. Dynamic bus sizing allows the systems designer to select the optimum cost:performance ratio for his or her system. Thirty-two-bit memory ports can be used for greatest speed, or 16-bit and 8-bit ports can be used to reduce the cost of the system. For example, if system constants are stored in expensive battery-backed CMOS RAM, it might be much more cost effective to employ a single byte-wide component as an 8-bit memory port than to use four of these chips in parallel to create a fast 32-bit port (which is not really necessary).

The relationship between the 68020's data bus and 32-bit, 16-bit, and 8-bit memory ports is illustrated in figure 4.33. The first thing to note is that there is no misprint. An 8-bit port is indeed connected to data bits D_{24}–D_{31} and not to D_{00}–D_{07}, as you might intuitively expect. Similarly, a 16-bit data port is connected to data lines D_{16}–D_{31} (and not to D_{00}–D_{15}). Don't worry about this arrangement of ports. When the 68020 tells memory it wishes to transfer a longword, it puts the longword on D_{00}–D_{31} and tells the memory about the longword by setting SIZ1, SIZ0 to 0,0. Now suppose that the memory replies that the data transfer is to be a byte. Since the 68000 stores data with the least significant byte of a word or a longword at the *highest* address, that byte will appear on D_{24}–D_{31} of the data bus. Clearly, it would be inefficient to rerun the bus cycle with the data in the *correct place*. Instead, we connect data bits D_{24}–D_{31} to the 8-bit port.

Figure 4.34 demonstrates how the 68020 numbers bytes internally and how it would send a longword to a 32-bit, 16-bit, or 8-bit port, respectively. As you can see

FIGURE 4.33 Relationship between the 68020 and 32-bit, 16-bit, and 8-bit memory ports

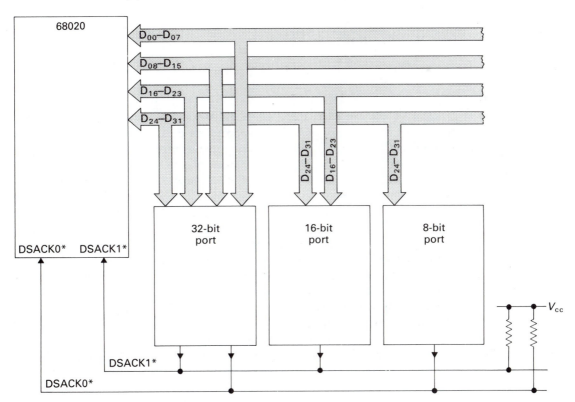

FIGURE 4.34 68020's method for numbering bytes internally and transferring operands between memory ports

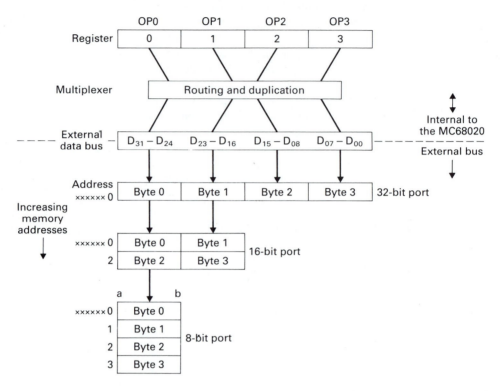

from figure 4.33, we can easily configure these variable-sized ports. All we have to do is to connect the memory port to the appropriate data bus lines and to return the appropriate DSACK1*,DSACK0* signals during a memory port access.

Suppose we wish to design a 16-bit memory port. How do we go about it? From the toolkit of signals made up of SIZ1,SIZ0 and A_{01}, A_{00}, plus DS*, we have to synthesize two byte enable signals—UDS* and LDS* (we selected these names to make them *compatible* with the 68000's byte enable signals). Table 4.10 is taken from the 68020's user's manual and demonstrates the relationship between these signals. Figure 4.35 provides a circuit diagram of a 16-bit memory port and its interface to the 68020 using the data from table 4.10. Note once again that the 16-bit memory port is connected to the 68020's data bus lines $D_{16}–D_{31}$ rather than to $D_{00}–D_{15}$.

The memory interface circuit of figure 4.35 can also be used to implement a *platform board*, which is a small PCB containing a 68020 and the associated UDS*/ LDS* generation circuitry. A platform board allows you to plug a 68020 processor into the socket of a 68000 in a 68000-based system. The platform board contains a 68020 and all the interface logic necessary to make the 68020 look like a 68000 to the rest of the system. In this way, a 68000 in an existing system can be replaced by

TABLE 4.10 Deriving byte enable signals from SIZ1,SIZ0, and A_{01},A_{00} for a 16-bit memory port

	68020 OUTPUTS				DERIVED DATA PROBES	
TRANSFER SIZE	SIZ1	SIZ0	A_{01}	A_{00}	UDS*	LDS*
Byte	0	1	0	0	0	1
			0	1	1	0
			1	0	0	1
			1	1	1	0
Word	1	0	0	0	0	0
			0	1	1	0
			1	0	0	0
			1	1	1	0
Three bytes	1	1	0	0	0	0
			0	1	1	0
			1	0	0	0
			1	1	1	0
Longword	0	0	0	0	0	0
			0	1	1	0
			1	0	0	0
			1	1	1	0

a 68020 to enhance the system's performance. Note, however, that a true 68020 platform board that closely mimics the 68000 would have to include a replacement interface for the 68000's synchronous bus control signals (E, VPA*, VMA*) together with some means of stretching the 68020's bus cycle from six to eight clock cycles.

Designing an interface to an 8-bit memory port is very simple indeed, as figure 4.33 demonstrates. All we have to do is to connect the port's D_{00}–D_{07} data lines to D_{24}–D_{31} from the 68020 and to enable the memory port by DS* from the 68020. When the port is accessed, it returns DSACK1*,DSACK0* = 1,0. For example, if a longword is read from the 8-bit port, the 68020 first accesses it at a longword boundary with SIZ1,SIZ0 = 0,0 and A_{01},A_{00} = 0,0. Since the port responds to all accesses with DSACK1*,DSACK0* set to 1,0, the 68020 must execute three further read cycles: one with SIZ1,SIZ0 = 1,1 and A_{01},A_{00} = 0,1; one with SIZ1,SIZ0 = 1,0 and A_{01},A_{00} = 1,0; and one with SIZ1,SIZ0 = 0,1 and A_{01},A_{00} = 1,1.

Most high-performance 68020 systems will locate much of their memory in high-speed 32-bit memory ports in order to take best advantage of the 68020's 32-bit data bus. As the memory must be individually byte addressable, we need some means of generating individual byte select signals for each of the 4 bytes of a

FIGURE 4.35 The 16-bit memory port

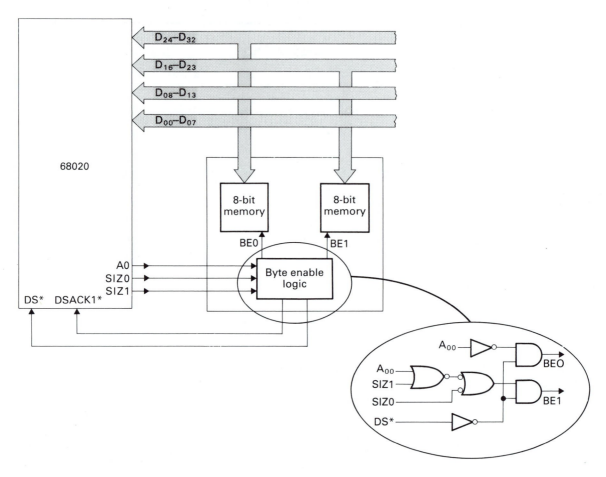

longword. Table 4.11 provides the relationship between SIZ1,SIZ0 and A_{01},A_{00} and the four byte enable signals BE0*, BE1*, BE2*, and BE3*. When BE0* is asserted, the memory block that puts data on address lines D_{24}–D_{31} is enabled. Similarly, BE1* enables bits D_{16}–D_{23}, BE2* enables bits D_{08}–D_{15}, and BE3* enables bits D_{00}–D_{07}. Figure 4.36 provides the circuit of a 32-bit port based on the data in table 4.11. You should, of course, appreciate that the 68020 permits the design of a system with any combination of 8-bit, 16-bit, and 32-bit memory ports. It is the DSACK1* and DSACK0* signals returned from the memory port being accessed that tell the 68020 how to proceed.

Now that we have looked at the 68020's memory interface, we will conclude this chapter with a look at the design of a simple 68000-based system. Chapter 5 will return to the theme of the 68000's memory interface.

FIGURE 4.36 The 32-bit memory port

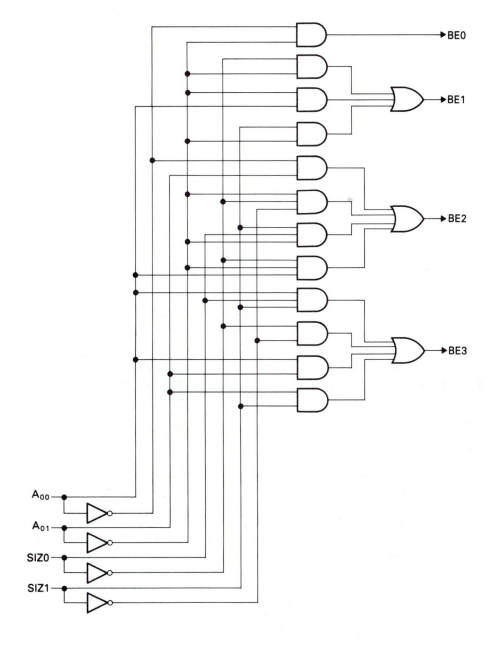

TABLE 4.11 Deriving byte control signals from SIZ1, SIZ2, and A_{01}, A_{02} for a 32-bit memory port

TRANSFER SIZE	68020 OUTPUTS				DERIVED BYTE STROBES			
	SIZ1	SIZ0	A_{01}	A_{00}	BE0*	BE1*	BE2*	BE3*
Byte	0	1	0	0	0	1	1	1
			0	1	1	0	1	1
			1	0	1	1	0	1
			1	1	1	1	1	0
Word	1	0	0	0	0	0	1	1
			0	1	1	0	0	1
			1	0	1	1	0	0
			1	1	1	1	1	0
Three bytes	1	1	0	0	0	0	0	1
			0	1	1	0	0	0
			1	0	1	1	0	0
			1	1	1	1	1	0
Longword	0	0	0	0	0	0	0	0
			0	1	1	0	0	0
			1	0	1	1	0	0
			1	1	1	1	1	0

4.4 MINIMAL CONFIGURATION USING THE 68000

People occasionally ask, "How few chips does it take to build a microcomputer with a 68000 CPU?" In some ways this is an unfair question, because it tries to pin down the 68000 to a largely spurious figure of merit (i.e., a minimum chip-count design). This question takes no account of performance and is based on a rather dubious assumption that low chip count is related to low cost or ease of construction. Having given this warning, we are now going to look at a low chip-count 68000 microcomputer. Our motives are twofold. We want to show which pins of the 68000 are essential to a simple microcomputer and which pins can be "forgotten" in a minimal design. Secondly, sometimes we need to produce a really small system, either as a teaching aid to illustrate the microprocessor or as a stand-alone controller. At this point in the text, the design of a 68000-based system is intended only to demonstrate the preceding points. Many of the details of the design cannot fully be appreciated until other sections of this text have been read.

Although it is possible to design a 68000 microcomputer subject to the constraint of a minimum chip count, this is a rather pointless exercise, as the addition of one or two extra chips results in a vastly increased level of performance. Instead, we will design a system subject to the following constraints:

1. The microcomputer is to be used in a stand-alone mode and requires only a power supply and an external terminal.

2. It is intended to be used as a classroom teaching aid to demonstrate the characteristics of the 68000.

3. It must have a 16K-byte EPROM-based monitor.

4. Its speed (i.e., clock cycle time) is of little or no importance.

5. It must have at least 4K bytes of read/write memory.

6. It must have at least one RS232C serial I/O port and one parallel port.

7. It must be possible to expand the memory and peripheral space of the microcomputer later.

8. Interrupts and multiprocessor capabilities are not needed, but again the possibility of adding them later should exist.

The first step in designing our minimal system is to consider the major components, the ROM, RAM, and peripherals. The ROM is provided by two 8K × 8 components, the RAM by two 2K × 8 devices, and the peripherals by a 6821 peripheral interface adaptor (PIA) and a 6850 asynchronous communications interface adaptor (ACIA). Static RAM is used because it does not require the complex support circuitry needed by dynamic RAM (described in chapter 7). Figure 4.37 shows how the memory components are arranged in the microcomputer module.

The next step is to consider the memory and peripheral support circuitry. Clearly, the 16K bytes of ROM and the 4K bytes of RAM have to be selected out of the 68000's 16M bytes of memory space. The actual location of these devices within this space is largely unimportant, as long as the reset vectors are located at $00 0000. Consequently, the 16K bytes of ROM are situated at $00 0000 to $00 3FFF.

The circuit diagram of the control circuitry of the minimal single board computer is given in figure 4.38. Address decoding is carried out by three integrated circuits: IC1a, IC1b, IC2a, and IC3. These circuits divide the memory space in the region $00 0000 to $01 FFFF into eight blocks of 16K bytes. The first three consecutive blocks at the upper end of the memory space are devoted to ROM, RAM, and peripherals, respectively.

Whenever the Y0∗ or Y1∗ outputs of IC3 go active-low, signifying the selection of ROM or RAM, the output of NAND gate IC2b goes high. This output is complemented by open-collector inverter IC5a to become the processor's DTACK∗ input. Note that no delay is applied to DTACK∗, so we must match the processor's speed to its memory carefully.

The Y2∗ output of IC3 goes active-low whenever a peripheral is addressed in the 16K-byte memory space $00 8000 to $00 BFFF. Y2∗ is buffered by IC5b and IC5c in order to permit the VPA∗ input of the CPU to be driven by an open-collector gate. In this way, other open-collector outputs may drive VPA∗ if they are added latter. Y2∗ is further decoded by IC6 to generate peripheral chip selects for the PIA and ACIA. Note that IC6 is enabled by VMA∗ and LDS∗, which means that

FIGURE 4.37 Block diagram of a minimal 68000-based microcomputer

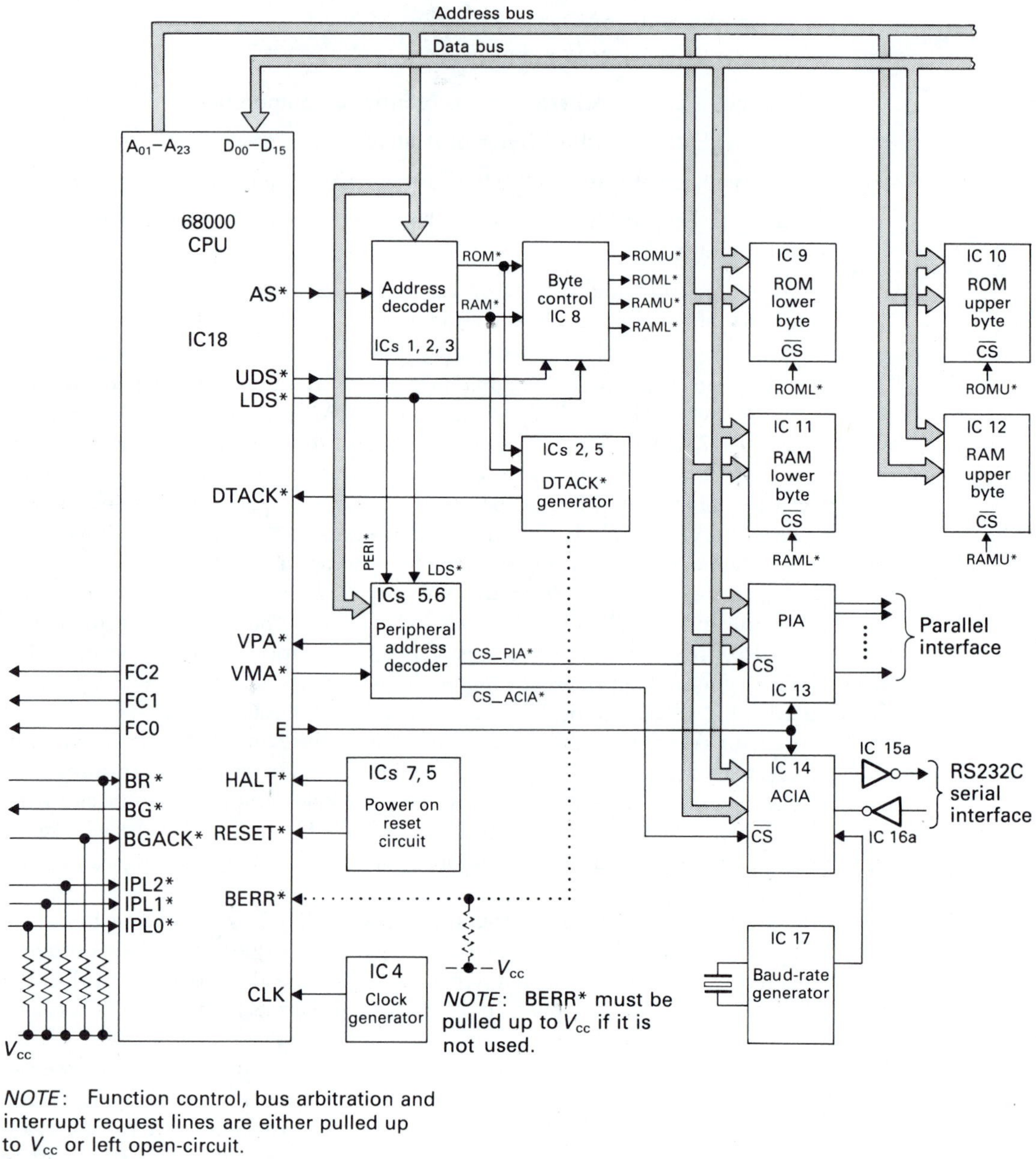

NOTE: BERR* must be pulled up to V_{cc} if it is not used.

NOTE: Function control, bus arbitration and interrupt request lines are either pulled up to V_{cc} or left open-circuit.

FIGURE 4.38 Circuit diagram of a minimal 68000-based microcomputer (control section)

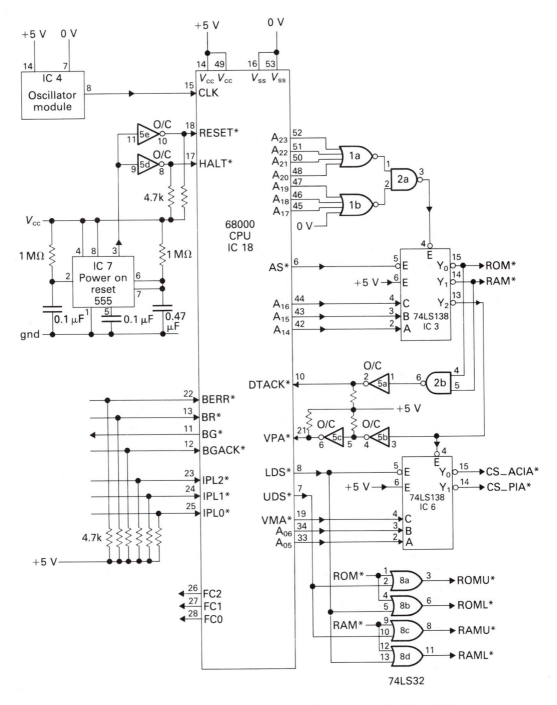

a peripheral is synchronized to a 68000 synchronous cycle operation (triggered by VPA* being asserted) and that the CPU must address a lower byte to select a peripheral.

The power-on-reset circuit forces RESET* and HALT* low when the system is initially switched on. A monolithic DIP clock generator chip supplies the processor with its clock signal.

In this application the interrupt request inputs, IPL0* to IPL2*, are pulled up by resistors to their inactive state. The function code outputs, FC0 to FC2, are not required and are left unconnected. Both the bus request (BR*) and bus grant acknowledge (BGACK*) inputs are pulled up into their inactive-high states by resistors. The bus error input (BERR*) is not used and is also pulled up by a resistor. The 6850 ACIA requires its own clock, which is supplied by baud-rate generator IC17. Its serial inputs are buffered by a line receiver, IC16, and its outputs by a line transmitter, IC15. Chapter 9 describes the 6850 ACIA in greater detail.

In all, this minimal 68000 system contains 18 integrated circuits. It works as it stands and can be expanded to become a more sophisticated system.

Critique of the Minimal Computer

The minimal computer of figures 4.37–4.39 is practical, but only just so. It lacks various features whose inclusion would cost little in terms of the chip count but which would considerably enhance the system. Some of the areas in which the minimal computer could be improved are as follows.

Control of DTACK*

The circuit of figure 4.38 exhibits a poor implementation of the DTACK* input to the 68000. Two problems have not been considered. The first concerns the operational speed of the processor. If the CPU is to run at its maximum rate and moderately fast RAM is used, we need to delay DTACK* only when the slower EPROM-based read-only memory is accessed. Figure 4.40 shows how individual DTACK* delays can be generated, one for RAM accesses (if necessary) and one for EPROM accesses. The second problem concerns the possibility of accesses to unimplemented memory. If a read or write access is made to memory not decoded in figure 4.39, the DTACK* input is not asserted and the processor will hang up indefinitely. In figure 4.40 a watchdog circuit is used to overcome this difficulty. When AS* is asserted, a timer is triggered. The timer is reset by the rising edge of AS*. If DTACK* is not asserted, the timer is "timed-out" and the BERR* input to the 68000 is asserted to indicate a bus error, thereby allowing the processor to detect and to recover from this error.

Control of Interrupts

Although operation of the 68000 or any other processor in an interrupt-driven mode is not necessary, it is worthwhile to provide some form of interrupt facility in a general-purpose digital computer. Figure 4.40 shows how seven levels of interrupt request input can be provided by a 74LS148 priority encoder.

FIGURE 4.39 Circuit diagram of a minimal 68000-based microcomputer (memory and peripherals)

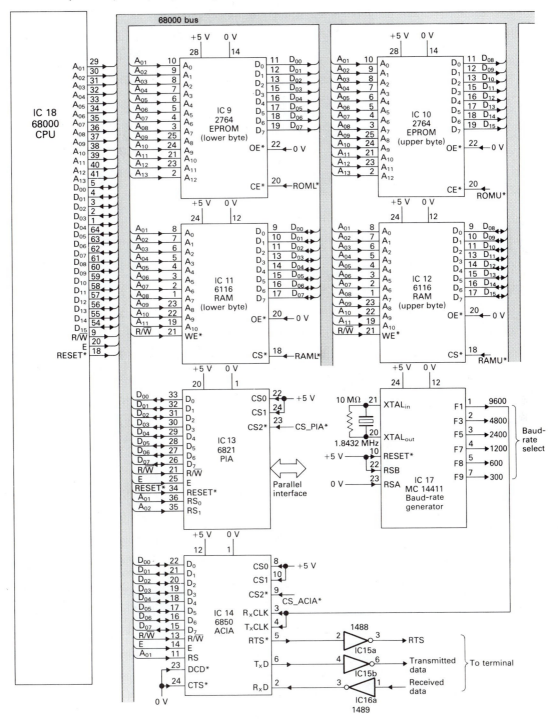

FIGURE 4.40 Turning the minimal microcomputer into a general-purpose single-board computer

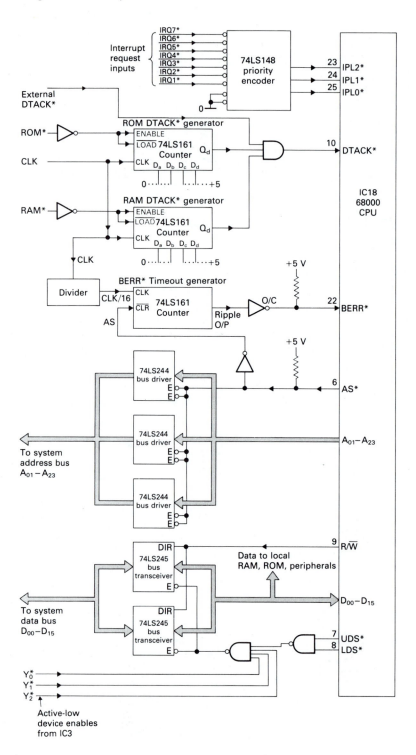

External Bus Interface

If a microprocessor system is to be expanded, it must be able to communicate with external systems via a bus. In a large system with many memory components or peripherals, connecting the 68000's pins directly to a system bus is impossible because the CPU cannot supply the current necessary to drive the distributed capacitance of the bus and all the inputs connected to it. Therefore, special-purpose circuits called *bus drivers* or *buffers* are interposed between the processor and the system bus. In addition to the bus drivers themselves, provision of control circuitry to avoid data bus contention is necessary, which could occur when the CPU reads from memory local to the processor module.

SUMMARY

In this chapter we have introduced the hardware of the 68000 microprocessor from the point of view of the engineer who is going to design a microcomputer around the 68000. Our main focus of attention has been the 68000's data transfer bus, which the microprocessor uses to communicate with its memories and peripherals. An understanding of this bus and of the way in which information flows across the bus is essential to the designer of all microprocessor systems. Consequently, we have also examined the timing diagram of both the 68000 and a typical memory component because the art of microcomputer design consists largely of reconciling the timing diagram of a microprocessor with those of its memory and peripheral devices. We expand on the themes of timing diagrams and buses in chapters 5 and 10, respectively.

We have described some of the 68000's more esoteric pins such as its interrupt control pins and its bus arbitration control pins. Later chapters show how these pins are used in 68000-based microprocessor systems.

The relationship between the 68000, its pins, and a microcomputer was highlighted by the design of a simple single-board microcomputer at the end of this chapter. We do not intend that the reader should be in a position to understand fully the design of this microcomputer at this stage. The microcomputer has been introduced to provide an example of a complete system and to give an idea of the overall complexity of a basic microcomputer based on the 68000. Later chapters examine the various aspects of a microcomputer in greater detail and show how more complex systems can be built.

In addition, we have examined the 68020's interface and described how this processor is connected to an external memory system. We have also used the 68020 as an excuse to look at timing diagrams in more detail and have described its dynamic bus sizing mechanism.

Problems

1. What is the difference between an asynchronous memory access and a synchronous memory access? How do these differences affect the designer of microprocessor systems? If a microprocessor could have only one type of interface, what would be the most suitable overall choice?

2. Can the 68000 be operated in a synchronous mode by strapping its DTACK* input to AS*? If the possibility exists to do this, what advantages and disadvantages would this mode of operation have?

3. Why cannot the 68000 be single-stepped through instructions simply by halting its clock after AS* has been negated at the end of a memory access?

4. What are the advantages and disadvantages of the protocol diagram as a design tool? In other words, what does each diagram reveal and what does it hide?

5. What is the meaning of data setup-time and data hold-time? Can either of these values be zero? Can they be negative? If the answer is yes to either of the last two questions, what does it imply?

6. What are the advantages and disadvantages of the 68000's single R/W* output as opposed to separate active-low read and write strobes, RE* and WR*, employed by some microprocessors?

7. In a 68000 read cycle, AS* and UDS*/LDS* are asserted simultaneously. In a write cycle, UDS*/LDS* is asserted approximately one clock cycle after AS*. What is the reason for this?

8. If a memory access cannot be completed by the slave asserting DTACK*, a watchdog timer on the CPU card asserts BERR* to force the processor out of its memory access. Suppose an engineer wished to know the value of FC0 to FC2, A_{01} to A_{23}, and D_{00} to D_{15} at the time BERR* was asserted? What logic would be necessary for this?

9. What is the maximum read cycle access time that a memory component may have if it is to be employed in an 8-MHz system with no wait states (use the parameters of table 4.4)? If a single wait state is permitted, what is the new maximum access time?

10. For the 60000-6116 combination, the following expressions relate 60000 parameters to 6116 parameters during a write cycle. Show, by means of a timing diagram, that these expressions are valid.

a. $t_{WC} = t_{AVSL} + t_{SL} + t_{SHAZ}$
b. $t_{CW} = t_{SL(w)}$
c. $t_{WR} = t_{SHAZ}$
d. $t_{AW} = t_{AVSL} + t_{SL}$
e. $t_{AS} = t_{AVRL}$
f. $t_{WP} = t_{SL} - t_{ASRV}$
g. $t_{DW} = t_{DOSL} + t_{SL(w)}$
h. $t_{DH} = t_{SHDOI}$

11. Using the equations of question 10, the parameters of the 68000 (table 4.7), and the parameters of the 6116 (table 4.8), investigate the write cycle for a 200-ns 6116 and a 12.5-MHz 68000.

12. Figure 4.41a and b gives the circuit diagram of two typical DTACK* generators. Analyze their operation by providing a suitable timing diagram. The data on the shift register and the counter is given in figure 4.42a and b, respectively.

13. Design a circuit to permit a single bus cycle at a time to be executed. The HALT* line must normally be held in its active-low state and be negated long enough for the 68000 to execute a single bus cycle.

FIGURE 4.41 The DTACK∗ generator

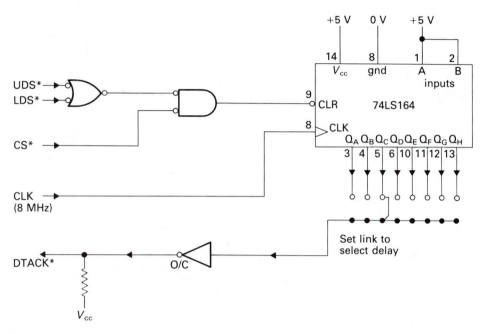

(a) Using a shift register

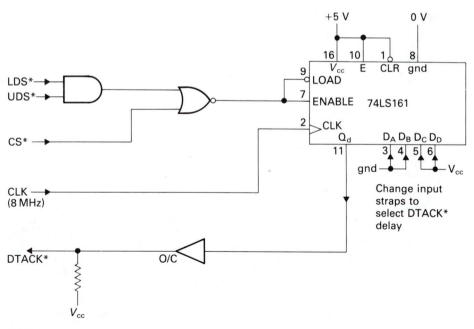

(b) Using a counter

FIGURE 4.42   The 74LS164 shift register and the 74LS161 binary counter

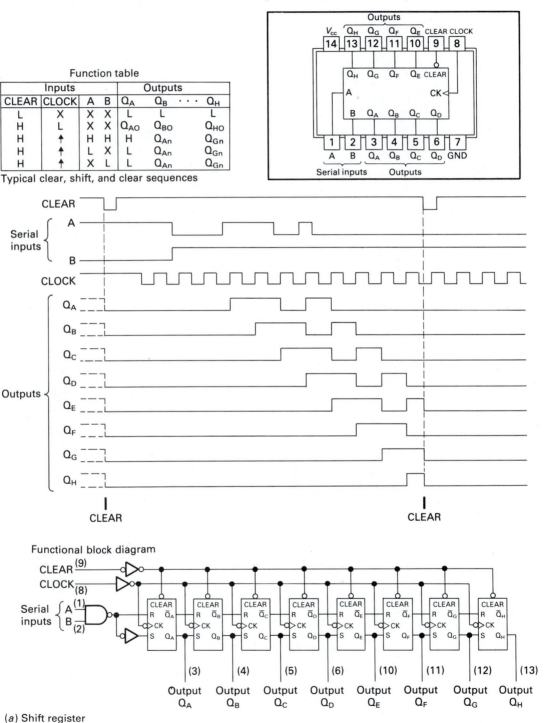

(a) Shift register

FIGURE 4.42 (*Continued*)

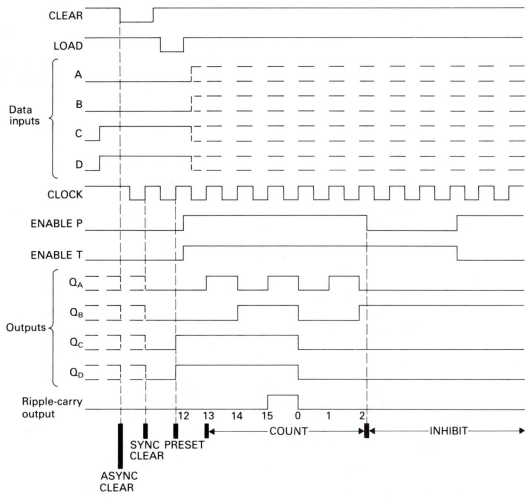

(*b*) Binary counter

14. The quoted physical address space of the 68000 is 16M bytes (i.e., 2^{24} bytes). We could maintain that it is 64M bytes. What argument would we use and would we be correct?

15. Unlike other byte-addressable microprocessors that employ an A_{00} pin as a least significant bit to distinguish between an odd and even byte, the 68000 does not connect A_{00} from an address register (or the program counter) to a pin. How, then, does the 68000 implement byte accesses?

16. What signals does the 68000 use to control asynchronous bus cycles?

17. The 68000, like any other similar digital device, can recognize a signal within tens of nanoseconds. Why, then, must the 68000's RESET* input be asserted for at least 100 ms after the initial application of power?

18. What are the functions of AS*, UDS*, and LDS*?

19. What is a synchronous bus cycle?

20. What is an asynchronous bus cycle?

21. Why do RESET* and HALT* have O/D (open-drain) outputs?

22. The function code outputs FC0–FC2 are active-high outputs, in contrast with the 68000's other control outputs. Do you think that there is a reason for this?

23. The 68000 has 32-bit address registers that hold bits A_{00}–A_{31}. However, the 68000 has only address pins A_{01}–A_{23}. What are the reasons for this? What are the effects of this on both the programmer and the hardware designer? How is it possible to use the fact that there are no A_{24}–A_{31} pins to good advantage?

24. When the 68000 executes a write operation to an odd byte, it asserts LDS* and places data on D_{00}–D_{07}. What happens to D_{08}–D_{15} during this operation and why?

25. Design a decoder that produces five outputs, one for each of the following operations:

Valid memory access to user data space

Valid memory access to user program space

Valid memory access to supervisor data space

Valid memory access to supervisor program space

Valid memory access to interrupt acknowledge space

26. Describe the 68000's read cycle, explaining the actions that take place and the relationship between them.

27. With reference to the 68000's read cycle timing diagram, answer the following questions.
   a. When is AS* asserted with respect to the clock?
   b. When is AS* negated?
   c. What is the relationship between AS*, LDS*, and UDS*?
   d. What is the minimum setup time for DTACK*?
   e. When should DTACK* be negated?
   f. What are the setup and hold times for data?
   g. Describe the R/W timing diagram.

28. What are *misaligned operands* and how do the 68000 and the 68020 deal with them?

29. What is the difference between the way in which the 68000 and the 68020 use the function code outputs?

30. What is the function of the 68020's OCS* and ECS* and what is the difference between them?

31. What is a read-modify-write cycle and what is the difference in the ways in which the 68000 and the 68020 implement it?

32. What are the advantages and disadvantages of the 68020's bus sizing mechanism?

33. Why does the 68020 connect an 8-bit port to data lines D_{24}–D_{31} and not D_{00}–D_{07} ?

34. What role is played by SIZ0 and SIZ1 and by DSACK0* and DSACK1* in the 68020's bus sizing mechanism?

35. If you had to design a 32-bit address and data interface to a modern microprocessor, how would you go about it? What functions would you include (e.g., dynamic bus sizing)?

MEMORIES IN MICROCOMPUTER SYSTEMS

In this chapter we look at the *immediate access* or *random access* memory subsystem that holds programs and data required by the CPU. As memory systems design involves so many different concepts, the chapter is divided into three sections. The first section deals with address decoding strategies and the second looks at circuits needed to interface a block of memory to a microprocessor's address bus. The third section examines the actual memory components themselves and considers their timing diagrams and their interface to the microprocessor's data bus. In order to keep the size of this chapter within reasonable bounds, more advanced memory topics such as dynamic memory, cache memory, and error-correcting memory are dealt with in chapter 7.

5.1 ADDRESS DECODING STRATEGIES

A systems designer has the task of taking a memory component and then mapping it onto the 68000's address space. The "memory component" may be a single device (e.g., a memory-mapped peripheral) or a group of devices that implement a block of memory. Once the designer has performed the mapping, he or she has to design the actual circuit that will perform the mapping. All designers of microcomputers are confronted with the problem of synthesizing the most cost-effective address decoder subject to constraints of economics, reliability, testability, versatility, board area, chip count, power dissipation, and speed. The way in which these factors determine the design of address decoders is discussed in this chapter.

Memory Space

Before we introduce address decoding, a problem of notation must be dealt with. Strictly speaking, the 68000 is a machine that can address one of 8M words of memory with its 23 address lines, designated A_{01} through A_{23}. However, the 68000 permits the access of individual bytes of memory. Two data strobes, UDS* and LDS*, select one or both halves (i.e., bytes) of the memory word addressed by A_{01} through A_{23}. Because it permits byte accesses, the 68000 numbers its memory locations from $00 0000 to $FF FFFF, each pair of locations corresponding to a single word. The problem here is whether to treat the 68000 as having 16M addressable locations of 8 bits or 8M addressable locations of 16 bits.

For the purpose of this chapter, both conventions are used. We need to consider the 68000 as being byte addressable (i.e., 16M bytes) because it operates as a byte-addressable machine from the programmer's point of view. Equally, it appears as a word-addressable machine to the designer of address decoders, because the data control strobes, UDS* and LDS*, take no part in the address decoding process.

Figure 5.1 represents the 16M bytes of uniquely addressable locations accessible by the 68000 as a column, or linear list, from $00 0000 to $FF FFFF. An even address refers to the upper byte of a word, accessed when UDS* is asserted, and an odd address refers to the lower byte, accessed when LDS* is asserted.

The 16M bytes in figure 5.1 constitute an address space that is said to be spanned by the 68000's 23 address lines. The term "spanned" is used because any location in the address space is "reached" or specified by a unique value on A_{01} through A_{23}. The address space of figure 5.1 can be partitioned into blocks, each block containing a number of consecutive memory locations. Figure 5.2 shows the arrangement of three of these blocks and is called a *memory map*. Blocks may correspond to logical entities such as programs, subroutines, or data structures or to actual hardware devices such as ROM, read/write memory, or peripherals.

If we could get a single memory component spanning the entire memory space of the 68000, the problem of address decoding would not arise. Each of the 68000's 23 address lines would be connected to the corresponding address input of this

FIGURE 5.1 The 68000's address space

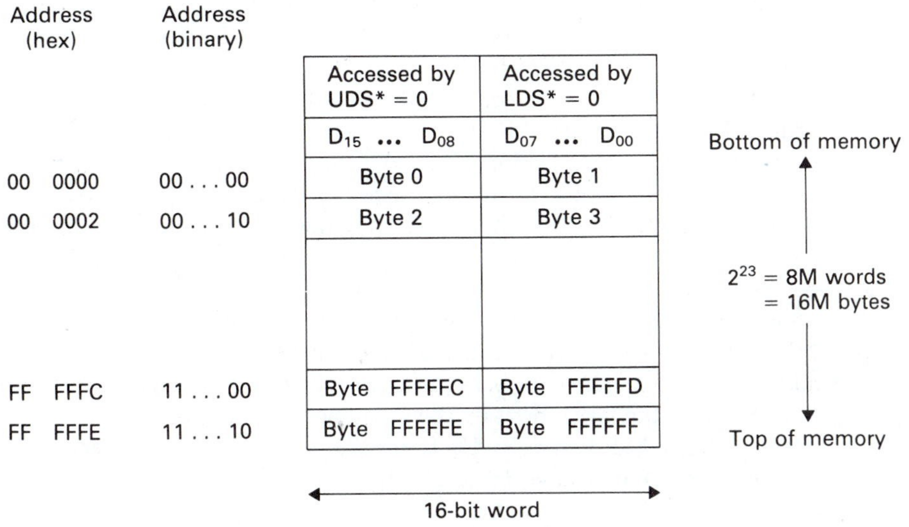

Address (hex)	Address (binary)	Accessed by UDS* = 0	Accessed by LDS* = 0	
		D_{15} ... D_{08}	D_{07} ... D_{00}	Bottom of memory
00 0000	00 . . . 00	Byte 0	Byte 1	
00 0002	00 . . . 10	Byte 2	Byte 3	
				2^{23} = 8M words = 16M bytes
FF FFFC	11 . . . 00	Byte FFFFFC	Byte FFFFFD	
FF FFFE	11 . . . 10	Byte FFFFFE	Byte FFFFFF	Top of memory

16-bit word

FIGURE 5.2 Memory map

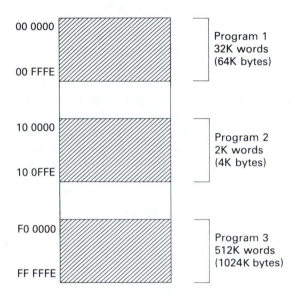

00 0000

00 FFFE

Program 1
32K words
(64K bytes)

10 0000

10 0FFE

Program 2
2K words
(4K bytes)

F0 0000

FF FFFE

Program 3
512K words
(1024K bytes)

16M-byte device. However, not only do microcomputers have memory components with fewer than 16M bytes of internal storage, but they also invariably employ a wide range of devices whose internal storage may vary from 1M locations to just 1 location. It is this broad range of storage capacities that makes the life of the microcomputer designer so difficult.

In this chapter we discuss only the 68000, since moving from the 68000 to the 68020 or 68030 hardly affects the design of address decoders. The 68020's dynamic sizing mechanism has no effect on the design of address decoders, since bus sizing operates at the byte, word, and longword level. As far as address decoders are concerned, the only additional complexity associated with the 68020/30 is the need to decode address lines A_{24} to A_{31}.

Suppose the designer of a 68000-based microcomputer decides that 2K words of ROM and 2K words of RAM are needed for a certain application. Eleven address lines from the system address bus, A_{01} through A_{11}, are connected to the corresponding address inputs of the two memory components, M1 and M2, as shown in figure 5.3. (Note that the 2K × 16-bit memory components are hypothetical examples. A real system would use two 2K × 8-bit devices. Sixteen-bit wide memory components have been "invented" to simplify the system—although there are some 16-bit wide EPROMs.) Whenever a location spanned by A_{01} through A_{11} (i.e., 2^{11} = 2K words or 4K bytes) is addressed in M1, the corresponding location is also addressed in M2. The data buses of both memory components are connected to the data bus from the CPU. Because the data lines of M1 and M2 are connected together with D_i from M1 wired to D_i from M2, the data bus drivers in both memory components need to have tristate outputs, thereby avoiding the situation in which

FIGURE 5.3 Connecting two 2K memory components to an address bus

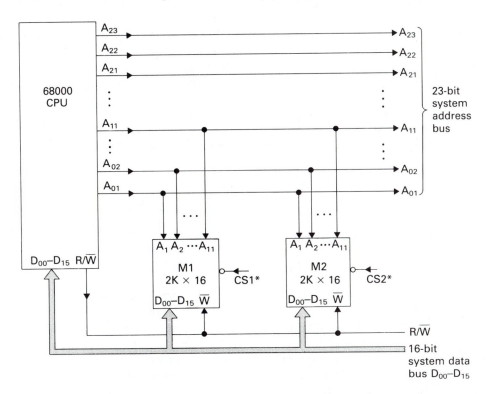

both outputs attempt to drive the bus to different logic levels simultaneously. Chapter 10 deals with this topic in more detail. Each memory component has an active-low chip-select input with which it controls its data bus drivers. Whenever the memory component is enabled by asserting its chip-select input, it is able to take part in a read or a write cycle. Negating the chip-select input of a memory component by putting it in an inactive-high state causes its data bus drivers to be turned off and its data bus outputs to be floated.

Let the chip-select input to M1, CS1*, be made a function of address lines A_{12} to A_{23}, where CS1* = F1(A_{12}, A_{13}, ..., A_{23}). Similarly, let CS2* be made a different function of A_{12} to A_{23}, where CS2* = F2(A_{12}, A_{13}, ..., A_{23}). The art of address decoding is to select functions F1 and F2 so that there is at least one combination of A_{12} to A_{23} that makes CS1* low and CS2* high and at least one combination that makes CS2* low and CS1* high. Under these circumstances, the conflict between M1 and M2 is resolved and the memory map of the system now consists of two disjoint blocks of memory, M1 and M2. This chapter considers three strategies for decoding A_{12} to A_{15} (i.e., choosing F1 and F2). These strategies are full address decoding, partial address decoding, and block address decoding.

Full Address Decoding

A microprocessor is said to have full address decoding when each addressable location within a memory component responds only to a single, unique address on the system address bus. In other words, all the microprocessor's address lines must be used to access each physical memory location, either to specify a given device or to specify an address within it.

Full address decoding can be applied to the problem of distinguishing between two memory components, M1 and M2, by constructing a logic network that uses address lines A_{12} to A_{23} to select either M1 or M2 but not M1 and M2. One of the many possible solutions is:

1. M1 is selected whenever

$$A_{12} A_{13} A_{14} A_{15} A_{16} A_{17}\ A_{18} A_{19} A_{20} A_{21} A_{22} A_{23} = 0\ 0\ 0\ 0\ 0\ 0\ 0\ 0\ 0\ 0\ 0\ 0$$

2. M2 is selected whenever

$$A_{12} A_{13}\ A_{14} A_{15} A_{16} A_{17} A_{18} A_{19} A_{20} A_{21} A_{22} A_{23} = 1\ 0\ 0\ 0\ 0\ 0\ 0\ 0\ 0\ 0\ 0\ 0$$

Figure 5.4 shows the memory map and the address decoder circuit for this solution.

Now let us look at a more complex example of full address decoding. The designer of a 68000 microcomputer has an application requiring 10K words of ROM arranged as a block of 2K words plus a block of 8K words, 2K words of random access read/write memory, 2 words for peripheral 1, and 2 words for peripheral 2. Five memory devices are therefore to be decoded; these will be called ROM1 (2K), RAM (2K), ROM2 (8K), PERI1 (2), and PERI2 (2), respectively. These devices each have an active-low chip-select input and may be located anywhere within the system's memory map, with the sole exception of ROM1, which must respond to addresses in the range $00 0000 to $00 0FFF. Note that we can write the range as $00 0000 to $00 0FFF or $00 0000 to $00 0FFE, depending on whether we choose the last byte or the last word of the address block.

The first step in solving this problem is to construct an address table. Such a table is given in table 5.1, where the vertical columns represent the 23 address lines A_{01} to A_{23} and the rows represent the five memory components to be decoded. A cross, ×, in an address column means that that address line takes part in the selection of a location within the specified component. A 1 or 0 in an address column means that that address line must be 1 or 0 to select that component.

The address lines to be decoded for each memory component must be selected as either 0 or 1. How we perform this procedure is, to a certain extent, unimportant. In table 5.1, we selected the lowest possible address for each device subject to the constraint that an *n*-word block starts at an *n*-word boundary; for example, the 8K ROM starts on an 8K boundary. To achieve full address decoding we need only to decode *every* address line that does not select a location within a device and to ensure that no two devices can be accessed simultaneously.

We can see that when full address decoding is applied to the example in table 5.1, the two peripherals each require 22 address lines to be decoded, as only one

FIGURE 5.4 Resolving the conflict between the memory components by full address decoding

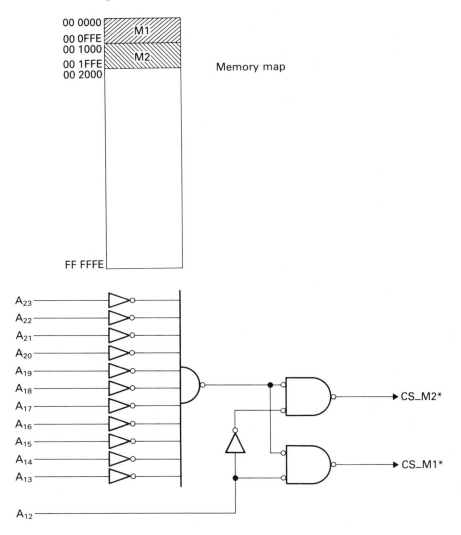

A possible address decoder for the above memory map

address line, A_{01}, selects a location (one of two) within the peripheral. A suitable address decoding arrangement is given in figure 5.5. Note that PERI1 is selected whenever A_{16}–A_{23} and A_{02}–A_{14} are all 0 and A_{15} is a logical 1. Twenty-two address lines need to be decoded and two 74LS133 thirteen-input NAND gates are pressed into service for this purpose. Whenever the outputs of both NAND gates are simultaneously low, the CS* output of the AND gate goes low, selecting PERI1. This circuit highlights one of the paradoxes of microcomputer design. The microcomputer and peripheral are available as two single chips and yet the address decoding circuit

TABLE 5.I Address decoding table illustrating full address decoding

DEVICE				ADDRESS LINE																	
	23	22	21	20	...	15	14	13	12	11	10	09	08	07	06	05	04	03	02	01
ROM1	0	0	0	0	...	0	0	0	0	X	X	X	X	X	X	X	X	X	X	X
RAM	0	0	0	0	...	0	0	0	1	X	X	X	X	X	X	X	X	X	X	X
ROM2	0	0	0	0	...	0	1	X	X	X	X	X	X	X	X	X	X	X	X	X
PERI1	0	0	0	0	...	1	0	0	0	0	0	0	0	0	0	0	0	0	0	X
PERI2	0	0	0	0	...	1	0	0	0	0	0	0	0	0	0	0	0	0	1	X

FIGURE 5.5 Full address decoding network corresponding to table 5.1

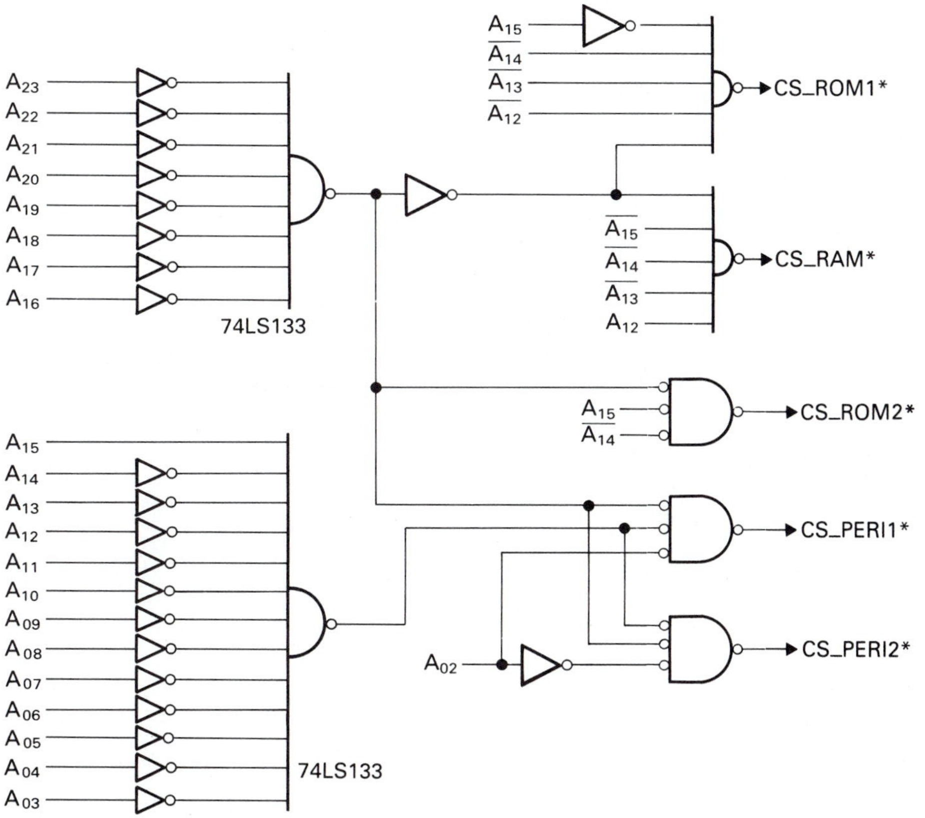

of figure 5.5 requires a total of seven chips (assuming that three hex invertors are used).

Partial Address Decoding

Partial address decoding is so called because all the address lines available for address decoding do not take part in the decoding process. Partial address decoding is the simplest—and, consequently, the most inexpensive—form of address decoding to implement. Figure 5.6 shows how the two 4K-byte blocks of memory depicted in figure 5.3 can be connected to a system address bus in such a way that both blocks of memory are never accessed simultaneously. The potential conflict between M1 and M2 is resolved by connecting CS1* directly to the highest-order address line, A_{23}, and by connecting CS2* to A_{23} via an invertor. In this way, M1 is selected whenever $A_{23} = 0$ and M2 is selected whenever $A_{23} = 1$.

We have now succeeded in distinguishing between M1 and M2 for the cost of a single invertor, but a heavy price has been paid. As M1 is selected by $A_{23} = 0$ and

FIGURE 5.6 Using partial address decoding to distinguish between two memory components

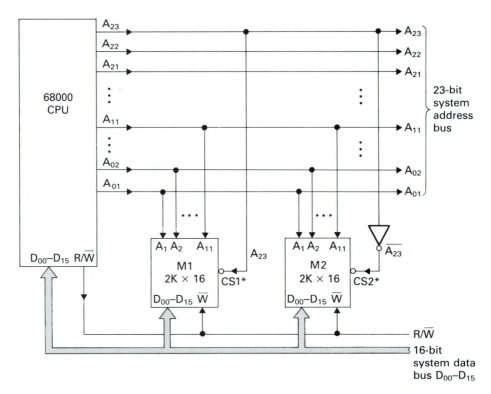

M2 by $A_{23} = 1$, obviously either M1 or M2 must always be selected. Thus, although the address bus can specify 16M different byte addresses, this decoding arrangement allows only 8K different locations to be accessed. Address lines A_{12} to A_{22} take no part whatsoever in the address decoding process and therefore have no effect on the selection of M1 or M2. Figure 5.7 shows the memory map corresponding to this arrangement. We can see that the memory space M1 appears 2,048 (i.e., 2^{11}) times in the lower half of the memory map and M2 is repeated 2,048 times in the upper half.

Partial address decoding was popular in the early days of the 8-bit microprocessor and is still found in small, dedicated systems where low cost is of paramount importance. The penalty paid for employing partial address decoding is that it prevents full use of the microprocessor's available memory space and causes difficulties when expanding the memory system at a later date.

Example of Partial Address Decoding

Consider now a more reasonable example of partial address decoding. We will take the same problem used in the preceding section on full address decoding. An address decoding table for this problem is given in table 5.2.

FIGURE 5.7 Memory map corresponding to figure 5.6

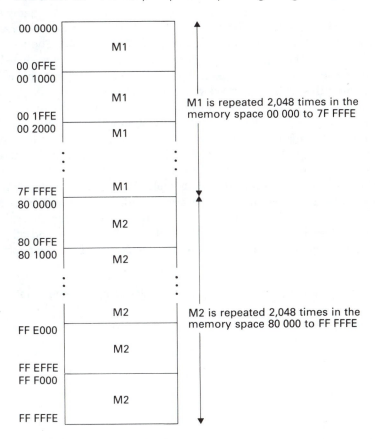

TABLE 5.2 Address table illustrating partial address decoding

DEVICE	ADDRESS LINE																				
	23	22	21	20	...	15	14	13	12	11	10	09	08	07	06	05	04	03	02	01
ROM1	0	0	0							×	×	×	×	×	×	×	×	×	×	×
RAM	0	0	1							×	×	×	×	×	×	×	×	×	×	×
ROM2	0	1				×	×	×	×	×	×	×	×	×	×	×	×	×	×	×
PERI1	1	0																		×
PERI2	1	1																		×

NOTE: An address entry that is neither a 1 nor a 0 is a don't care condition; that is, that address line does not take part in the selection of the device.

The first step is to fill the five rows with ×s for each address line used to select a location within the appropriate memory component; for example, address lines A_{01} through A_{11} select a location within ROM1 ($2^{11} = 2K$), whereas address lines A_{01} through A_{13} select a location within ROM2 ($2^{13} = 8K$). The next step is to select conditions for the higher-order address lines, which distinguish between the five memory components. One of the many possible ways of doing this is illustrated in table 5.2. From this table we can see that no combination of A_{21}, A_{22}, and A_{23} can be used to select two or more devices simultaneously. The reader may wonder why we have chosen A_{21}, A_{22}, and A_{23} to distinguish among these five components. The answer concerns the matter of style. We could have perfectly easily been able to decode, say, A_{14}, A_{15}, and A_{16} to perform the same function. In that case, address lines A_{17} to A_{23} would remain undecoded.

Having drawn the address decoding table, the primary addressing range of each component can be determined. The primary addressing range is calculated by setting all don't care conditions to zero and then reading the minimum and maximum address range taken by the component when the ×s are all 0s and all 1s, respectively. A slight complication arises because the 68000 uses byte addressing, so that

FIGURE 5.8 Implementing the partial address decoding scheme of table 5.2

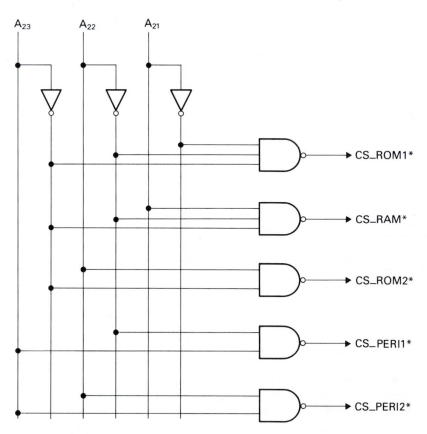

TABLE 5.3 Improved partial address decoding scheme

DEVICE	A_{23}	A_{22}	A_{21}	A_{20}
ROM1	0	0	0	0
RAM	0	0	0	1
ROM2	0	0	1	
PERI1	0	1	0	
PERI2	0	1	1	
SPACE	1			

an imaginary $\times$ representing A_{00} must be placed to the right of A_{01} in each row of table 5.2.

Consider first ROM1. Its primary address range is given by 000(000000000)00000000000[0] to 000(000000000)11111111111[1]. The don't care conditions have been placed in round brackets and A_{00}, in square brackets. This condition corresponds to an address range of \$00 0000 to \$00 0FFF (i.e., 4K bytes). Similarly, PERI2 occupies the primary address range 11(00000000000000000000)0[0] to 11(00000000000000000000)1[1], or from \$C0 0000 to \$C0 0003.

A diagram of one possible implementation of the partial address decoding arrangement is given in figure 5.8. It is highly improbable that any designer would choose the address decoding scheme of table 5.2 or figure 5.8 because the memory map cannot be expanded. No further memory devices can be added as the existing memory devices fill the entire 8M words of memory space. Table 5.3 demonstrates a simple way of building some flexibility into the system.

In this case, the condition $A_{23} = 0$ is made a necessary condition to select all five memory components, which now take up the lower half of the memory space from \$00 0000 to \$7F FFFE. The 4M-word memory space for which $A_{23} = 1$ is now available for future use and is labeled "SPACE" in table 5.3.

In spite of its simplicity, partial address decoding is often shunned because the memory space allocated to a single memory device is repeated many times; for example, PERI1 in table 5.2 occupies one word of memory that is repeated 1M (2^{20}) times. Therefore, if a spurious memory access is made because of an error in a program, harm can possibly be caused by corrupting the memory location that responded to the spurious access. Equally, a limited form of partial address decoding is sometimes found in systems with large numbers of address lines because so many address lines are expensive to decode.

Block Address Decoding

Block address decoding is a compromise between partial address decoding and full address decoding. It avoids the inefficient use of memory space associated with partial address decoding by dividing the memory space into a number of blocks. The blocks are fully address decoded. If necessary, these blocks may be subdivided into smaller blocks.

In a typical application of block address decoding, a microprocessor's memory space may be divided into sixteen blocks of equal size by a single low-cost component. In the days of the 8-bit microprocessor, its 64K-byte memory space could easily be split into 16 manageable blocks, each of 4K bytes. Today, dividing the 68000's 8M words of memory space into 16 blocks yields a relatively massive block size of 512K words. Each of these blocks is four times larger than the entire memory space of a typical 8-bit microprocessor. Even so, splitting the 68000's memory space into 16 blocks of 512K words would enable the system of table 5.2 and figure 5.8 to be built with a single address decoding component and would allow expansion from 5 memory devices to 16 without the addition of extra logic.

In any real microcomputer, a mixture of partial address decoding, full address decoding, and block address decoding may be used to cater for the system's particular needs; for example, a system may apply full address decoding to address lines A_{23} to A_{17}, to select a block of 64K words from the 68000's memory space of 8M words. This 64K is then divided into 16 blocks of 4K by a block address decoder. Some of these 4K blocks can be used to select 4K ROMs. Finally, partial address decoding may be used to locate two peripherals, each of four words, within one of these 4K pages. An example of this type of arrangement is provided later.

5.2 DESIGNING ADDRESS DECODERS

A number of ways exist of implementing the address decoding techniques described in section 5.1. In general, address decoding techniques can be divided into four groups: address decoding using random logic, address decoding using m-line-to-n-line decoders, address decoding using PROMs, and address decoding using programmable logic arrays, programmable gate arrays, or programmable array logic. Each of these techniques has its own advantages and disadvantages, the nature of which depends on the type of system being designed, the scale of its production, and whether or not it needs to be expandable.

Address Decoding with Random Logic

Random logic is the term describing a system constructed from small-scale logic such as AND, OR, NAND, and NOR gates and invertors. When address decoding with random logic is implemented, the chip-select input of a memory component is derived from the appropriate address lines by means of a number of SSI gates as required. The address decoding circuits of figures 5.5 and 5.8 both use random logic.

Address decoding entirely with random logic is found in relatively few systems, because it is rather costly in terms of the number of chips required. Moreover, it is always tailor-made for a specific application and therefore lacks the flexibility inherent in some other forms of address decoding circuit.

The only advantage of random logic address decoding is its speed. As it is tailor-made for any given system, it can use the fastest logic available and can therefore achieve the minimum propagation delay from address valid to chip-select valid. Sometimes, the very low cost of SSI gates also aids the case in favor of random logic address decoding. However, we should appreciate the fact that a rather large number of chips may be required, which increases the cost of designing and testing the microcomputer and reduces the board area available for memory and peripheral components.

Address Decoding with *m*-line-to-*n*-line Decoders

The problem of address decoding can be greatly diminished by means of data decoders that decode an *m*-bit binary code into one of *n* outputs, where $n = 2^m$. These devices effectively carry out the block address decoding described earlier in this chapter. The three most popular decoders in the range of MSI TTL circuits are the 74LS154 four-line-to-sixteen-line decoder, the 74LS138 three-line-to-eight-line decoder, and the 74LS139 dual two-line-to-four-line decoder. Figures 5.9 to 5.11 give the pinouts and truth tables for the 74LS154, 74LS138, and 74LS139, respectively. All three decoders have active-low outputs, making them particularly suitable for address decoding applications, because almost all memory components have active-low chip-select inputs. Here we will discuss only the 74LS138 decoder, as the other two are identical in principal and differ only in detail.

The 74LS138 Three-Line-to-Eight-Line Decoder

The most popular of the *m*-line-to-*n*-line decoders is the 74LS138, which decodes a three-line input into one of eight active-low outputs, as indicated by figure 5.10. In addition to its three-line input, the 74LS138 has three enable inputs, of which two are active-low and one is active-high. Therefore, the 74LS138 is very easy to apply to address decoding circuits. Figure 5.12 demonstrates the versatility of the 74LS138 in five configurations. In each case we can assume that the decoders are being used in conjunction with a 68000 processor and that they are to be strobed by the processor's AS* output. Note that the address decoding ranges chosen in this example are entirely arbitrary and have been selected to illustrate the principles involved rather than to represent any real system.

The difference between the examples in figure 5.12c and d is that the latter employs all the decoder's enable inputs to reduce the size of the eight decoded address blocks as far as possible, thereby yielding a block size of 8K words using only two chips. In figure 5.12c, where the enable inputs are not used to decode address lines, the minimum block size is 128K words. In figure 5.12d the minimum block size is 8K words.

The final example, figure 5.12e, adds a little logic and a five-input NOR gate to a two-74LS138 circuit to achieve a resolution of 512 words. This type of circuit might be used to decode eight blocks of memory space for allocation to memory-mapped peripherals. Note that the AS* strobe in figure 5.12e has been applied to the

FIGURE 5.9   The 74LS154 four-line-to-sixteen-line decoder

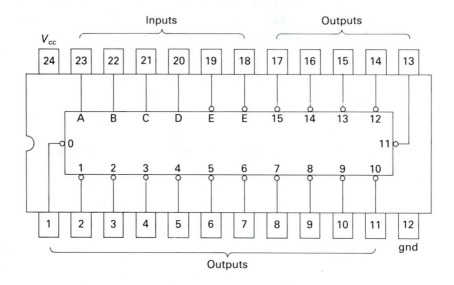

74LS154 truth table

Inputs					Outputs																
E̅1	E̅2	D	C	B	A	0	1	2	3	4	5	6	7	8	9	10	11	12	13	14	15
0	0	0	0	0	0	0	1	1	1	1	1	1	1	1	1	1	1	1	1	1	1
0	0	0	0	0	1	1	0	1	1	1	1	1	1	1	1	1	1	1	1	1	1
0	0	0	0	1	0	1	1	0	1	1	1	1	1	1	1	1	1	1	1	1	1
0	0	0	0	1	1	1	1	1	0	1	1	1	1	1	1	1	1	1	1	1	1
0	0	0	1	0	0	1	1	1	1	0	1	1	1	1	1	1	1	1	1	1	1
0	0	0	1	0	1	1	1	1	1	1	0	1	1	1	1	1	1	1	1	1	1
0	0	0	1	1	0	1	1	1	1	1	1	0	1	1	1	1	1	1	1	1	1
0	0	0	1	1	1	1	1	1	1	1	1	1	0	1	1	1	1	1	1	1	1
0	0	1	0	0	0	1	1	1	1	1	1	1	1	0	1	1	1	1	1	1	1
0	0	1	0	0	1	1	1	1	1	1	1	1	1	1	0	1	1	1	1	1	1
0	0	1	0	1	0	1	1	1	1	1	1	1	1	1	1	0	1	1	1	1	1
0	0	1	0	1	1	1	1	1	1	1	1	1	1	1	1	1	0	1	1	1	1
0	0	1	1	0	0	1	1	1	1	1	1	1	1	1	1	1	1	0	1	1	1
0	0	1	1	0	1	1	1	1	1	1	1	1	1	1	1	1	1	1	0	1	1
0	0	1	1	1	0	1	1	1	1	1	1	1	1	1	1	1	1	1	1	0	1
0	0	1	1	1	1	1	1	1	1	1	1	1	1	1	1	1	1	1	1	1	0
0	1	×	×	×	×	1	1	1	1	1	1	1	1	1	1	1	1	1	1	1	1
1	0	×	×	×	×	1	1	1	1	1	1	1	1	1	1	1	1	1	1	1	1
1	1	×	×	×	×	1	1	1	1	1	1	1	1	1	1	1	1	1	1	1	1

ENABLE∗ input of the second (i.e., lower-order) decoder, rather than to the first decoder. In this way the delay in the decoding circuit has been minimized. You should appreciate that the multiple-level address decoders of figure 5.12c, d, and e may cause timing problems in high-performance systems due to the cumulative address decoding delay.

FIGURE 5.10 The 74LS138 three-line-to-eight-line decoder

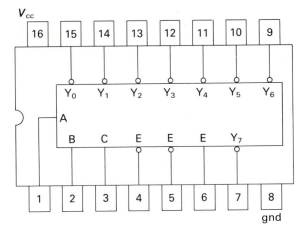

74LS138 truth table

Inputs						Outputs							
$\overline{E1}$ $\overline{E2}$ E3			C	D	A	Y0	Y1	Y2	Y3	Y4	Y5	Y6	Y7
1 1 0			×	×	×	1	1	1	1	1	1	1	1
1 1 1			×	×	×	1	1	1	1	1	1	1	1
1 0 0			×	×	×	1	1	1	1	1	1	1	1
1 0 1			×	×	×	1	1	1	1	1	1	1	1
0 1 0			×	×	×	1	1	1	1	1	1	1	1
0 1 1			×	×	×	1	1	1	1	1	1	1	1
0 0 0			×	×	×	1	1	1	1	1	1	1	1
0 0 1			0	0	0	0	1	1	1	1	1	1	1
0 0 1			0	0	1	1	0	1	1	1	1	1	1
0 0 1			0	1	0	1	1	0	1	1	1	1	1
0 0 1			0	1	1	1	1	1	0	1	1	1	1
0 0 1			1	0	0	1	1	1	1	0	1	1	1
0 0 1			1	0	1	1	1	1	1	1	0	1	1
0 0 1			1	1	0	1	1	1	1	1	1	0	1
0 0 1			1	1	1	1	1	1	1	1	1	1	0

Example of the Application of Block Address Decoders

Consider the following example of an address decoder for a 68000-based system using 74LS138 decoders. The bottom end of memory is populated by 16K words of ROM arranged as four blocks of 4K words. Each block is implemented by two 4K × 8 EPROMs. Up to eight memory-mapped peripherals in the 64-word (128-byte) range $01 0000 to $01 007F are needed and each peripheral occupies 8 words. The 8K words of read/write memory are provided by eight blocks of 1K words in the range $02 0000 to $02 3FFF. Each block is implemented by four 1K × 4 static memory components.

The memory map corresponding to the preceding arrangement is given in figure 5.13, and the address decoding scheme is in table 5.4. In this example full

FIGURE 5.11 The 74LS139 dual two-line-to-four-line decoder

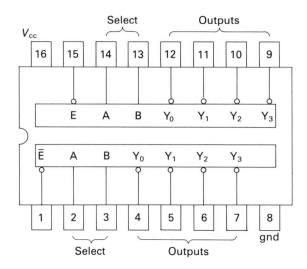

74LS139 truth table

Inputs				Outputs			
Ē	B	A		Y0	Y1	Y2	Y3
1	×	×		1	1	1	1
0	0	0		0	1	1	1
0	0	1		1	0	1	1
0	1	0		1	1	0	1
0	1	1		1	1	1	0

address decoding is employed, so that all address lines must be used in either the selection of a device or of a location within a device.

From table 5.4 we can see that a necessary condition for the selection of all devices is that A_{18} through A_{23} are all logical 0s. This condition strongly suggests that the active-low enable inputs of a decoder or a NOR gate should be used to decode these six high-order address lines. Figure 5.14 shows one possible way of implementing the device-mapping scheme of table 5.4. Preliminary address decoding is performed by ICs 1a and 2. Together, these divide the 256K-word memory space in the region $00 0000 to $07 FFFF into eight 32K-word pages. Note that the devices in the memory map of figure 5.13 are arranged so that ROM is on page 0, peripherals on page 1, and RAM on page 2. Thus, IC1a and IC2 both decode the eight higher-order address lines to provide first-stage decoding for all other devices and also provide for future expansion of the system by employing outputs Y3* to Y7* of IC2.

The address decoding of the ROMs is handled by another three-line-to-eight-line decoder, IC3. The 32K-word page zero selected by IC2 is divided into eight pages of 4K words by IC3, and the lower four pages are used to select the ROMs.

FIGURE 5.12 Some applications of the three-line-to-eight-line decoder

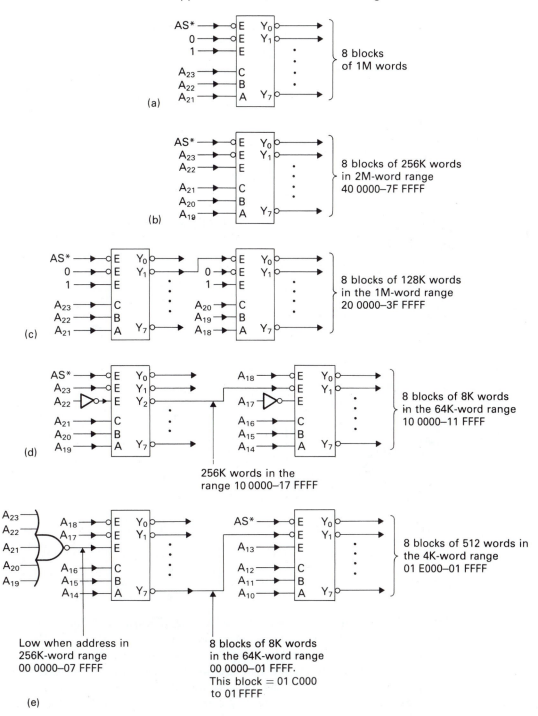

FIGURE 5.13 Memory map of a microcomputer

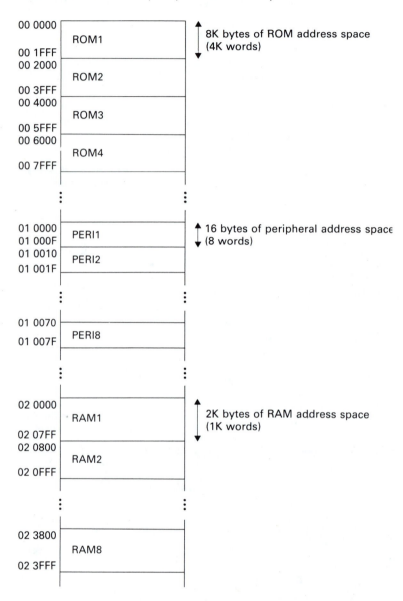

Note that outputs Y4* to Y7* of IC3 are not used and permit future expansion to 32K words of ROM without any additional logic.

As the read/write memory components occupy only 1K words of memory space, a further two address lines, in addition to those decoded by IC2, must take part in the selection of the RAM. Because the two active-low enable inputs of the RAM address decoder, IC6, are already connected to IC2 and the address strobe, a

TABLE 5.4 Address decoding table for figure 5.13

DEVICE	ADDRESS RANGE	23	...	17	16	15	14	13	12	11	10	09	08	07	06	05	04	03	02	01	(00)
												ADDRESS LINE									
ROM1	00 0000–00 1FFF	0	…	0	0	0	0	0	X	X	X	X	X	X	X	X	X	X	X	X	X
ROM2	00 2000–00 3FFF	0	…	0	0	0	0	1	X	X	X	X	X	X	X	X	X	X	X	X	X
ROM3	00 4000–00 5FFF	0	…	0	0	0	1	0	X	X	X	X	X	X	X	X	X	X	X	X	X
ROM4	00 6000–00 7FFF	0	…	0	0	0	1	1	X	X	X	X	X	X	X	X	X	X	X	X	X
PERI1	01 0000–01 000F	0	…	0	1	0	0	0	0	0	0	0	0	0	0	0	0	X	X	X	X
PERI2	01 0010–01 001F	0	…	0	1	0	0	0	0	0	0	0	0	0	0	0	1	X	X	X	X
⋮	⋮																				
PERI8	01 0070–01 007F	0	…	0	1	0	0	0	0	0	0	0	0	0	1	1	1	X	X	X	X
RAM1	02 0000–02 07FF	0	…	1	0	0	0	0	0	0	X	X	X	X	X	X	X	X	X	X	X
RAM2	02 0800–02 0FFF	0	…	1	0	0	0	0	0	1	X	X	X	X	X	X	X	X	X	X	X
⋮	⋮																				
RAM8	02 3800–02 3FFF	0	…	1	0	0	0	1	1	1	X	X	X	X	X	X	X	X	X	X	X

FIGURE 5.14   Implementing the address decoding scheme of table 5.4

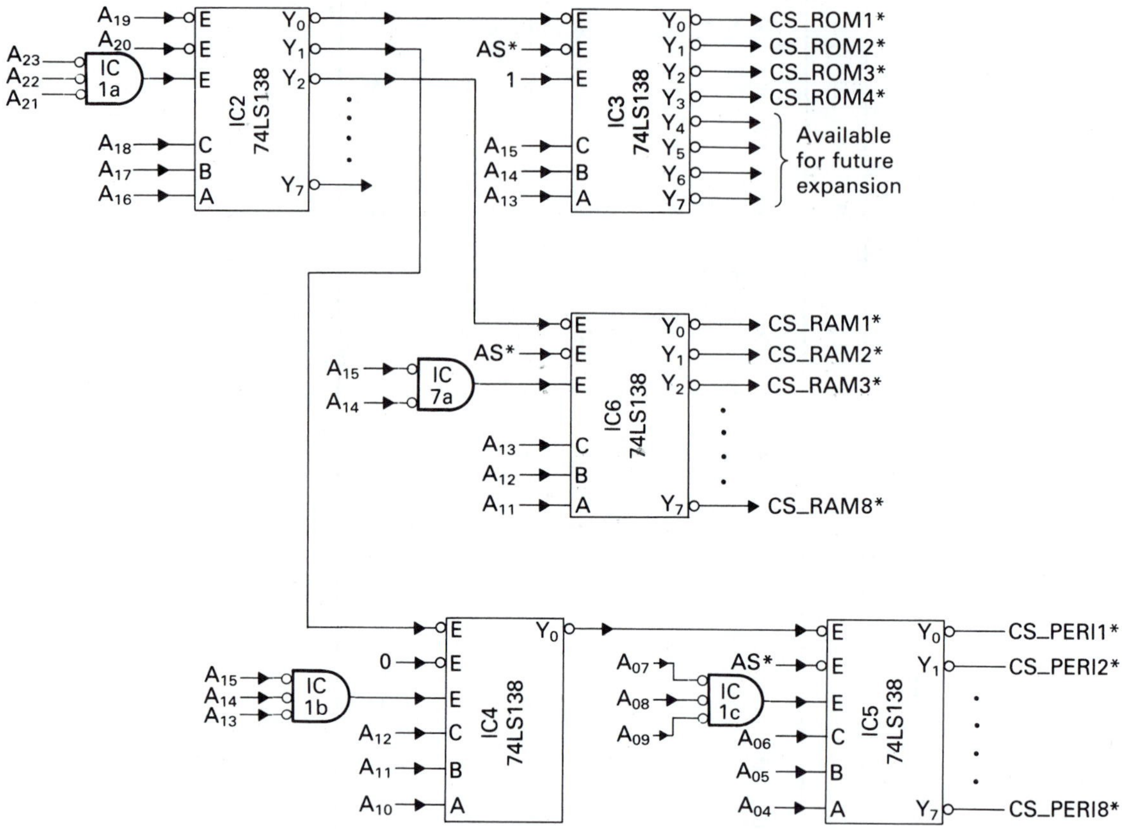

two-input AND gate (NOR gate in positive logic), IC7a, is used to decode A_{15} and A_{14}. Address lines A_{11} to A_{13} select one of eight 1K-word blocks of RAM whenever IC6 is enabled.

The selection of the peripherals is a little more complicated, as a further nine address lines take part in the selection of the peripheral address space (i.e., A_{07} through A_{15}). This procedure would normally indicate that three 74LS138s are necessary to fully decode the peripherals. However, as the peripherals are selected when address lines A_{07} through A_{15} are all zero, two AND gates, IC1b and IC1c, may possibly be used to reduce the number of 74LS138s to two. Figure 5.14 shows how to do this. The process could have been carried further and IC4 eliminated by random logic.

Address Decoding with PROM

As address decoding involves nothing more complex than the generation of one or more chip-select outputs from a number of address inputs, any technique applied to the synthesis of Boolean functions can also perform address decoding. We have used

m-line-to-n-line decoders to do this because they naturally exploit the block structure of memory. Another device most suited to this role is the programmable read-only memory, PROM.

The PROM has m address inputs, p data outputs, and a chip-select input, as illustrated in figure 5.15. This diagram shows the logical arrangement of the PROM and an example of its use as an address decoder. Whenever it is enabled, the m-bit address at its input selects one out of 2^m possible p-bit words and applies it to the p data terminals. Thus, the PROM is nothing more than a look-up table. Instead of designing an address decoder by solving the Boolean equations relating the various

FIGURE 5.15 PROM as an address decoder

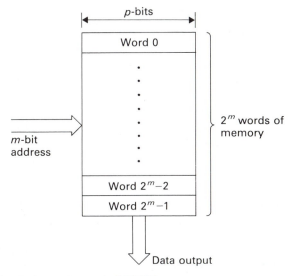

(a) Logical arrangement of PROM

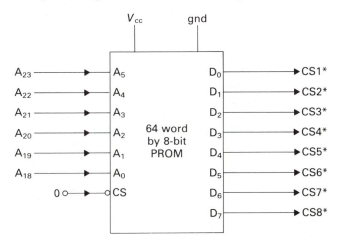

(b) PROM decodes high-order 6 address bits from CPU

device-select signals to the appropriate address lines, or by dividing memory into blocks with an *m*-line-to-*n*-line decoder, it is possible to program the PROM with the truth table directly relating addresses to device-select outputs.

When a PROM is disabled (i.e., deselected), its outputs float (assuming it has tristate drivers). Therefore, the outputs of the PROM need to be pulled up whenever it is disabled in order to force the active-low chip-select inputs into their inactive-high states. Alternatively, the PROM address decoder can be permanently enabled, with its enable (chip-select) input hard-wired to ground.

Unfortunately, one snag arises when PROMs are used as address decoders. As stated previously, a PROM with *m* address inputs stores 2^m words, each of *p* bits. The total capacity of this device is therefore $p \times 2^m$. Clearly, as *m* grows, the total number of bits increases exponentially. The relationship between PROM capacity and the size of the smallest block of memory decoded by it is given in table 5.5. We assume that the data width of the PROM is $p = 8$ bits and that it decodes the *m* higher-order bits in a 68000 system. The column headed $24 - m$ gives the number of undecoded address lines (including A_{00}).

From table 5.5 we can see that a very large PROM is needed to decode the 68000's address lines to select a reasonable block size; for example, if a PROM were to entirely decode a 68000's address bus and the smallest memory component were 2K words, it would be necessary to choose a value of *m* equal to 12. An 8-bit × 4K PROM has a capacity of 32K bits, making it a rather large device to program, especially during development work.

Because microprocessors like the 68000 have enormous address spaces, com-

TABLE 5.5 Relationship between decoded block size and capacity of a PROM

m	2^m (words)	$p \times 2^m$ (bits)	$24 - m$	DECODED BLOCK SIZE $2^{(24-m)}$
3	8	64	21	2M bytes (1M words)
4	16	128	20	1M bytes (512K words)
5	32	256	19	512K bytes (256K words)
6	64	512	18	256K bytes (128K words)
7	128	1,024	17	128K bytes (64K words)
8	256	2,048	16	64K bytes (32K words)
9	512	4,096	15	32K bytes (16K words)
10	1,024	8,192	14	16K bytes (8K words)
11	2,048	16,384	13	8K bytes (4K words)
12	4,096	32,768	12	4K bytes (2K words)
13	8,192	65,536	11	2K bytes (1K words)
14	16,384	131,072	10	1K bytes (512 words)
15	32,768	262,144	9	512 bytes (256 words)
16	65,536	524,288	8	256 bytes (128 words)

m = number of address inputs to the PROM
$p = 8$ = width of PROM's data output bus

pared to their 64K-byte ancestors, some of the basic (i.e., first-stage) address decoding can be left to random logic devices and the fine address decoding is carried out by PROMs or m-line-to-n-line decoders. In the early days of the 8-bit microprocessor, when memories were generally very small, quite a few address lines were decoded. As time passed, memories with much larger capacities appeard and we were able to decode, say, 2K-byte components with a single three-line-to-eight-line decoder.

The situation with modern microprocessors is similar to the early days of the 8-bit processor, as far as address decoding is concerned. Even the 64K-bit (i.e., 8K-byte) static read/write RAM requires ten address lines to be decoded (i.e., A_{14} to A_{23}) when it is used with a 68000. Consequently, most 68000 systems have been forced to resort to multilevel address decoding. Where several types of memory components share the same module (e.g., the single-board microcomputer), we often find it convenient to select a page of the processor's memory space by the simplest random logic decoder available and then to subdivide the page into the memory space of individual components by other means.

Example of the Design of a PROM-Based Address Decoder

Consider the following example. A designer wishes to produce a single-board 68000 system with memory space containing ROM, RAM, and peripherals. The design criteria stress two points: The decoder must be cheap and it must be versatile. As memory components decline in price, the designer hopes to improve the system later, when high-density memories become even cheaper. Initially, the design calls for a minimum address space per ROM/RAM to be 1K words (2K bytes). A further criterion is that all ROM/RAM must be fully address decoded and the total peripheral I/O space limited to 1K words divided between eight peripherals. The basic system is to have the memory map defined by table 5.6.

At first sight, the allocation of memory may seem strange or arbitrary. The three ROMs occupy $00 0000 to $00 2FFF, the RAM from $00 C000 to $00 C7FF,

TABLE 5.6 Memory space allocation for an SBC

DEVICE	ORGANIZATION	MEMORY SPACE		ADDRESS RANGE
		WORDS	BYTES	
ROM1	2K × 8	2K	4K	00 0000–00 0FFF
ROM2	2K × 8	2K	4K	00 1000–00 1FFF
ROM3	2K × 8	2K	4K	00 2000–00 2FFF
RAM1	1K × 4	1K	2K	00 C000–00 C7FF
PERI1	2 × 8	128	256	00 E000–00 E0FF
PERI2	2 × 8	128	256	00 E100–00 E1FF
PERI3	4 × 8	128	256	00 E200–00 E2FF

and the peripherals from $00 E000 to $00 E2FF. We have chosen these ranges to permit later expansion without the problem of read-only memory overflowing into read/write memory space. When the system is fully expanded, the 2K × 8 ROMs can be replaced by 8K × 8 ROMs and then the ROM address space will extend from $00 0000 to $00 BFFF. The next step is to prepare an address table (table 5.7) to help us to determine the best way of arranging the address decoding circuitry.

From table 5.7 we can see that address lines A_{16} through A_{23} perform a page selection, as they are constant and independent of the device selected. These lines can therefore be decoded by a single logic element with eight inputs and a single output. If the peripherals are regarded as a single entity occupying 1K words of memory space, the additional address lines to be decoded in the selection of this block are A_{11} through A_{15}. The discrimination between peripherals is best done by decoding A_{08} to A_{10} with a three-line-to-eight-line decoder.

The decoding of A_{11} through A_{17} is performed by a 32-word × 8-bit PROM. Figure 5.16 provides the basic details of a possible implementation of table 5.7. Address lines A_{16} through A_{23} must all be active-low to enable the 32-word PROM. As is usual in 68000-based systems, the address decoder is enabled by AS∗. When the PROM is disabled by the negation of AS∗, its data outputs are all pulled up by resistors to their inactive-high levels. Whenever the PROM is enabled, address lines A_{11} to A_{15} interrogate one of its 32 locations and yield an 8-bit data value that directly controls the chip-select inputs of the memory devices to be decoded. When the peripheral group is selected by D_3 from the PROM going active-low, address lines A_{08} to A_{10} are further decoded by the three-line-to-eight-line decoder, to select one of the three peripherals.

This circuit displays some of the considerable advantages PROM address decoders have over random logic or *m*-line-to-*n*-line decoders. PROM-based decoders are able to select components having different memory sizes. Here, both 2K-word ROM memory spaces and 1K-word RAM memory spaces are selected with the same PROM. To see how to do this, consider table 5.8, which is both a partial address decoding table (i.e., it shows only address lines A_{11} to A_{15}) and a listing of the contents of the PROM.

The leftmost column of table 5.8 displays the address range decoded by that row (i.e., one of 32) of the table. The second column gives the five address inputs of the PROM, labeled A_0 to A_4, and relates them to the corresponding five address lines of the 68000, labeled A_{11} to A_{15}. The rightmost column provides the data appearing on the PROM's output lines corresponding to the address in the middle column. These data outputs form the four device selects (D_4 to D_7), the peripheral group enable (D_3), and DTACK∗ (D_2).

Consider the selection of ROM1. This device is selected whenever a valid address in the range $00 0000 to $00 0FFF is put out by the processor. Note that this is a 4K-byte (2K-word) range and that the addressing range corresponding to any row of table 5.8 is only 2K bytes (1K words). Therefore, two rows in the table must be dedicated to RAM1. Thus, whenever $A_{15} = 0$, $A_{14} = 0$, $A_{13} = 0$, $A_{12} = 0$, and $A_{11} = 0$, or $A_{15} = 0$, $A_{14} = 0$, $A_{13} = 0$, $A_{12} = 0$, and $A_{11} = 1$, ROM1 is selected. Because ROM1 is selecting whenever $A_{11} = 0$ or $A_{11} = 1$, this address line is a don't care value in the selection of ROM1. If ROM1 occupied 4K words of memory

TABLE 5.7 Address decoding scheme for table 5.6

DEVICE	ADDRESS SPACE		06	07	08	09	10	11	12	13	14	15	16	17	18	19	20	21	22	23
		ADDRESS LINE																		
ROM1	00 0000–00 0FFF		X	X	X	X	X	X	0	0	0	0	0	0	0	0	0	0	0	0
ROM2	00 1000–00 1FFF		X	X	X	X	X	X	1	0	0	0	0	0	0	0	0	0	0	0
ROM3	00 2000–00 2FFF		X	X	X	X	X	X	0	1	0	0	0	0	0	0	0	0	0	0
RAM1	00 C000–00 C7FF		X	X	X	X	X	0	0	0	1	1	0	0	0	0	0	0	0	0
PERI1	00 E000–00 E0FF		X	X	0	0	0	0	0	1	1	1	0	0	0	0	0	0	0	0
PERI2	00 E100–00 E1FF		X	X	1	0	0	0	0	1	1	1	0	0	0	0	0	0	0	0
PERI3	00 E200–00 E2FF		X	X	0	1	0	0	0	1	1	1	0	0	0	0	0	0	0	0

TABLE 5.8 Programming the address decoder PROM

ADDRESS RANGE OF THE 68000	SYSTEM ADDRESS LINES					SYSTEM DEVICE ENABLES					
	A_{15}	A_{14}	A_{13}	A_{12}	A_{11}	ROM1	ROM2	ROM3	RAM1	PERIs	DTACK*
	PROM ADDRESS INPUTS					PROM DATA OUTPUTS					
	A_4	A_3	A_2	A_1	A_0	D_7	D_6	D_5	D_4	D_3	D_2
00 0000–00 07FF	0	0	0	0	0	0	1	1	1	1	0
00 0800–00 0FFF	0	0	0	0	1	0	1	1	1	1	0
00 1000–00 17FF	0	0	0	1	0	1	0	1	1	1	0
00 1800–00 1FFF	0	0	0	1	1	1	0	1	1	1	0
00 2000–00 27FF	0	0	1	0	0	1	1	0	1	1	0
00 2800–00 2FFF	0	0	1	0	1	1	1	0	1	1	0
00 3000–00 37FF	0	0	1	1	0	1	1	1	1	1	1
00 3800–00 3FFF	0	0	1	1	1	1	1	1	1	1	1
00 4000–00 47FF	0	1	0	0	0	1	1	1	1	1	1
00 4800–00 4FFF	0	1	0	0	1	1	1	1	1	1	1
00 5000–00 57FF	0	1	0	1	0	1	1	1	1	1	1

Address											
00 5800–00 5FFF	0	1	0	1	1	1	1	1	1	1	1
00 6000–00 67FF	0	1	1	0	0	1	1	1	1	1	1
00 6800–00 6FFF	0	1	1	0	1	1	1	1	1	1	1
00 7000–00 77FF	0	1	1	1	0	1	1	1	1	1	1
00 7800–00 7FFF	0	1	1	1	1	1	1	1	1	1	1
00 8000–00 87FF	1	0	0	0	0	1	1	1	1	1	1
00 8800–00 8FFF	1	0	0	0	1	1	1	1	1	1	1
00 9000–00 97FF	1	0	0	1	0	1	1	1	1	1	1
00 9800–00 9FFF	1	0	0	1	1	1	1	1	1	1	1
00 A000–00 A7FF	1	0	1	0	0	1	1	1	1	1	1
00 A800–00 AFFF	1	0	1	0	1	1	1	1	1	1	1
00 B000–00 B7FF	1	0	1	1	0	1	1	1	1	1	1
00 B800–00 BFFF	1	0	1	1	1	1	1	1	1	1	1
00 C000–00 C7FF	1	1	0	0	0	1	1	1	0	1	1
00 C800–00 CFFF	1	1	0	0	1	1	1	1	1	1	0
00 D000–00 D7FF	1	1	0	1	0	1	1	1	1	1	1
00 D800–00 DFFF	1	1	0	1	1	1	1	1	1	0	1
00 E000–00 E7FF	1	1	1	0	0	1	1	1	1	1	1
00 E800–00 EFFF	1	1	1	0	1	1	1	1	1	1	1
00 F000–00 F7FF	1	1	1	1	0	1	1	1	1	1	1
00 F800–00 FFFF	1	1	1	1	1	1	1	1	1	1	1

FIGURE 5.16 PROM-based address decoder to implement table 5.8

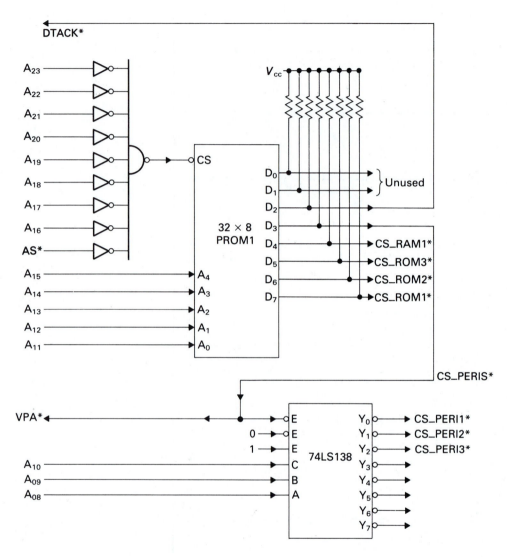

space in the range $00 0000 to $00 1FFF, the first four entries in the D_7 column of table 5.8 would all be 0.

The memory space dedicated to RAM1 in the range $00 C000 to $00 C7FF is 1K words, so only a single 0 appears in the corresponding row of table 5.8. Similarly, the block of peripherals is enabled by D_3 whenever an address in the range $00 E000 to $00 E7FF is placed on the address bus. We can see from figure 5.16 that a 74LS138 provides a third level of address decoding to give eight blocks of 128 words. Only three peripherals are currently implemented, allowing for expansion to eight without any additions or changes to the existing address decoding logic.

Figure 5.17 gives a more detailed and slightly modified version of this address decoder. In this case, the primary address decoding is performed by the 9-bit comparator IC1, an Am29809. Here, only 8 of the 9 bits to be matched are used. When input $A_i = B_i$ for i = 1 through 9 and the device is enabled by $G* = 0$, its $E_{out}*$ pin goes active-low, enabling the secondary decoder, a 32 × 8-bit PROM, IC2. Outputs D_3 to D_7 of the PROM select the memory devices as above. The function of output D_2 requires further explanation. A glance at table 5.8 reveals that D_2 is active-low whenever a zero appears in columns D_4 to D_7. Thus, D_2 is the logical OR of D_4 to D_7 (negative logic). This allows D_2 to be connected to the CPU's DTACK* input after passing through a suitable delay generator, if necessary.

The decoding of memory-mapped peripherals is almost exactly the same as any other memory component. However, as the reader will discover in chapter 8, the 68000 has a special provision for dealing with certain peripherals originally designed for 6800-based systems. These devices are interfaced to the 68000 by means of its synchronous bus and are enabled by the 68000's VMA* output. The peripheral group output, D_3, does not take part in the generation of DTACK* as, for the purposes of this example, we assume that the peripherals are 6800-series devices and operated synchronously.

The peripheral select output of IC2, D_3, enables the peripheral decoder IC3, a three-line-to-eight-line decoder. D_3 is also connected to the 68000's VPA* input, so that a synchronous bus cycle is started whenever a peripheral is addressed. The three-line-to-eight-line decoder is enabled by UDS*, restricting all peripherals to the upper byte of a word, and by VMA*, which synchronizes a peripheral access to the 68000's E clock.

The actual selection of the ROMs themselves is performed by three dual two-line-to-four-line decoders, ICs 4, 5, and 6. For convenience, only two of these are shown in figure 5.17. The lower byte of a ROM is selected when its CS* from the PROM and LDS* from the CPU are low during a read cycle, which corresponds to output Y2* from the two-line-to-four-line decoder. The upper byte of a ROM is selected in the same way, but in this case UDS* is used instead of LDS*. Note that ROMs can be selected only in a read cycle. This measure is a wise precaution and avoids the data bus contention that would occur if the processor attempted to write to a ROM enabled during a memory write access. Indeed, the circuit could be made more sophisticated by using the Y0* outputs of the two-line-to-four-line decoders to detect a write access to a ROM and then forcing BERR* low to indicate a faulty bus cycle.

Selecting the RAM is somewhat easier, as the select signal from the PROM need only be strobed by UDS* or LDS* to select the upper and lower bytes of a word, respectively.

Advantages and Disadvantages of PROM Address Decoders

There are two great advantages of using a PROM as an address decoder: its ability to select blocks of differing size and its remarkable versatility. A PROM that decodes m address lines divides the memory space into 2^m equal blocks. In the preceding example, a 32K-word page was divided into 2^5 blocks of 1K words. Larger blocks

FIGURE 5.17 More complete address decoder to implement table 5.8

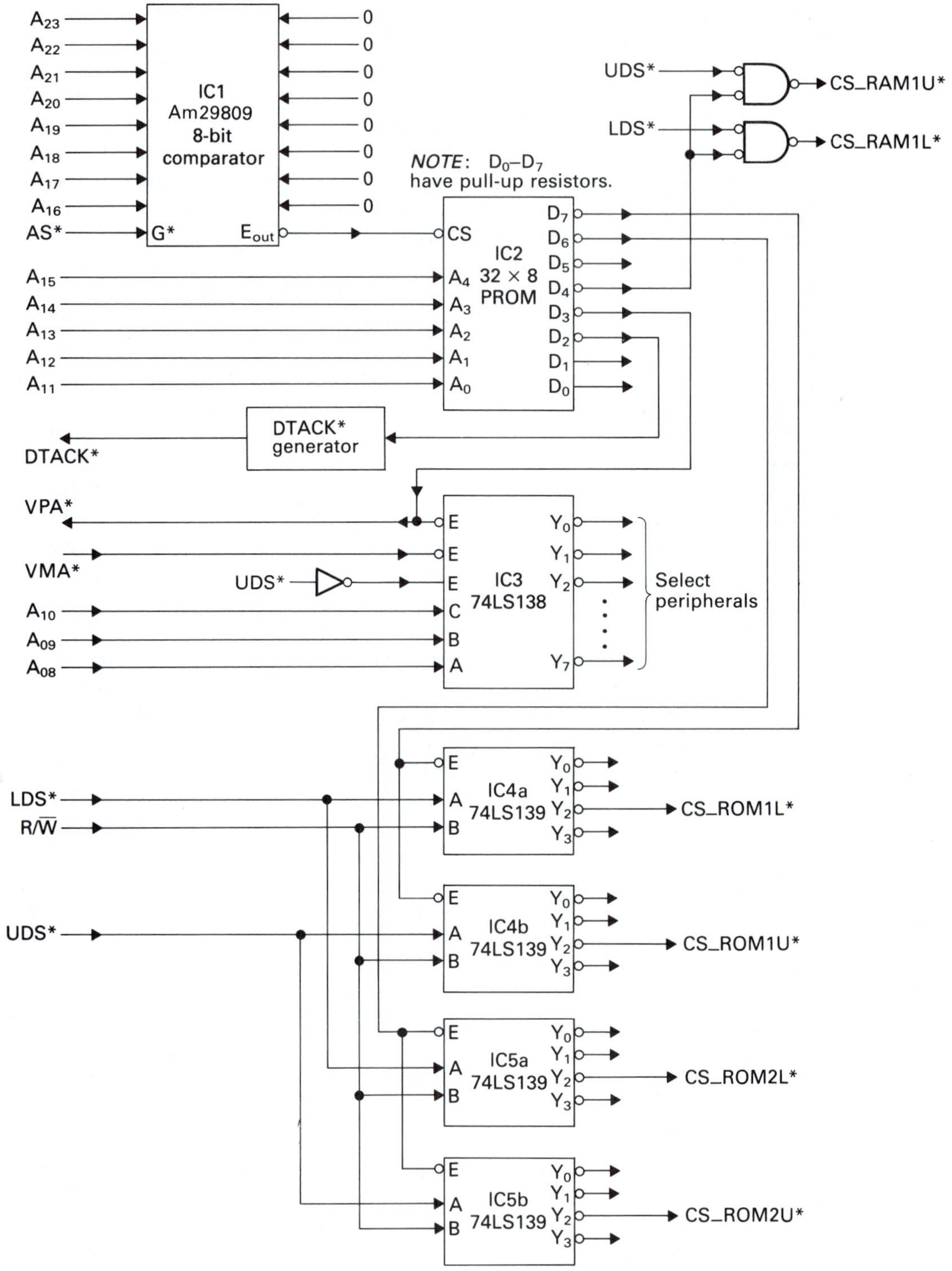

than the minimum size can be decoded simply by increasing the number of active entries (in our case, zeros) in the appropriate data column of the PROM's address/data table. The size of the block of memory decoded by a data output is equal to the minimum block size multiplied by the number of active entries in the appropriate data column. A general expression for the decoded block size of a device is given by

$$B = p \times 2^{(m-s-q)}$$

where

B = decoded block size
p = number of active entries in the appropriate data column
m = number of address lines from the CPU
s = number of address lines in primary (i.e., first-level) address decoding
q = number of address lines decoded by the PROM

As an example of the application of this formula, consider the address decoding of ROM1 in table 5.8. The values for p, m, s, and q are 2, 23, 8, and 5, respectively, giving a value for B of $2 \times 2^{(23-8-5)} = 2 \times 2^{10}$, or 2K words.

Address decoding by PROM is versatile because the selection of devices is determined by the programming of a PROM and not by the physical wiring of a decoder. This procedure makes it possible to configure a new system simply by programming a new PROM. In the example of table 5.8, we may replace, say, ROM1 by a larger version (e.g., a pair of 8K $\times$ 8 ROMs) just by increasing the number of zeros in the D_7 column of the PROM decoder. Therefore 8K words of ROM would require eight 0s, occupying memory space from $00 0000 to $00 3FFF.

The major disadvantage of the PROM has already been stated; that is, an excessively large look-up table is needed to decode more than about eight address lines. A large PROM is not only more expensive than a small PROM but takes longer to program and is more difficult to test.

Address Decoding and Its Impact on Timing

When we looked at the 68000's read and write cycle timing diagrams in chapter 4, we ignored the effects of address decoders and all other associated logic. An address decoder employs the higher-order address lines to synthesize the memory-select signals. Consequently, there must be a delay between the time at which the address is first valid and the time at which the appropriate chip-select is asserted. The actual delay depends on both the speed of the components used to implement the decoding logic and the number of address decoding components in series (i.e., the number of levels or stages in the decoder).

A simple (i.e., single-stage) address decoder in an 8-MHz 68000 system is unlikely to cause timing problems. However, a multistage (multilevel) address decoder using several logic elements in series (e.g., the address decoder of figure 5.17) may incur such a long delay that some action must be taken. For example, you might have to introduce wait states or employ faster memory. Alternatively, you can redesign the address decoder to make it faster. The next topic in this chapter introduces some of the components that make it easier to design both fast and complex address decoding circuits.

Some students read the section on PROM-based address decoders and then implement these decoders using conventional EPROMs. Since many small EPROMs have access times of the order of 300 to 400 ns, the use of such devices in an address decoding circuit will guarantee that the system does not operate correctly.

Although at the beginning of this chapter we said that the 68020 and the 68030 do not affect the design of address decoding circuits, we must qualify this statement. It is true that the address decoding logic of 68020-based and 68000-based microcomputers perform the same functions. But, their implementation may be very different. A fast 68020 or a 68030 provides little enough time for memory to access its data. Therefore, you must ensure that the address decoding logic does not eat into this precious access time. Decoders in 68020 systems use the fastest logic families (LS TTL is frequently too slow) and avoid multistage logic circuits.

Address Decoding with FPGA, PLA, and PAL

Up to the late 1970s, the systems designer had two basic elements with which to construct digital subsystems: the random logic element and the read-only memory. We have already seen how both of these are applied to address decoding circuits. The random logic element gives an optimum design from the point of view of speed, and the ROM-based decoder provides flexibility and compactness at the cost of a slightly slower speed and a restriction on the number of variables that can be handled economically. Today, new families of logic elements have appeared, giving the designer the best of both the random logic world and of the ROM world. These new devices have several names, but they all share one property—they are general-purpose logic elements and are configured by programming.

The simplest and most primitive of the new breed of programmable logic elements is the field programmable gate array, FPGA. The expression *field programmable* means that the array can be programmed or configured by the user in "the field" as opposed to the factory. Once an FPGA has been programmed, it cannot be reprogrammed. Figure 5.18 shows the internal arrangement of the 82S103 FPGA. Sixteen inputs, labeled A_0 to A_{15}, are converted into their true and complementary forms (i.e., A_i and A_i*) within the chip. Each of the resulting 32 terms is fed to the inputs of nine 32-input NAND gates. These inputs are passed through fusible links, which may be made open circuit (i.e., "blown") or closed circuit (i.e., left intact) during the FPGA's programming. Consequently, the output of the NAND gate can be made a function of between 1 and 16 inputs in either their true or inverted forms. Alternatively, we may say that the output of the NAND gate is a function of all 16 variables in either their true, complemented, or don't care form. The outputs of the NAND gates are fed via programmable EOR gates to nine output pins. Therefore, the outputs can be programmed to be active-high or active-low and the gates made to appear as AND or NAND gates, respectively. Finally, the outputs may be floated

FIGURE 5.18 The 82S103 FPGA

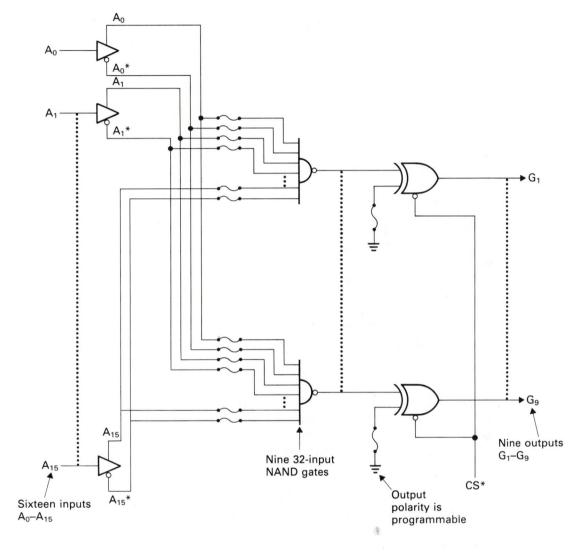

Sixteen inputs
A_0–A_{15}

Nine 32-input
NAND gates

Output
polarity is
programmable

Nine outputs
G_1–G_9

by disabling the chip by means of its active-low CE* input. The 82S102 is identical to the 82S103 but has open-collector rather than tristate outputs.

The 82S103 FPGA is a delight to use. In this one package, 9 outputs can be synthesized from the products of 16 input terms, with each term appearing in a true, false, or don't care form. In a 68000-based microcomputer, the 82S103 decodes up to 16 address lines (i.e., A_{08} to A_{23}), giving a minimum block size of 128 words or 256 bytes. As the 82S103 can decode so many inputs, the possibility often exists of dedicating some of these inputs to 68000 control functions, such as AS*, UDS*, and VMA*, while leaving plenty of inputs for address decoding. The following simple example shows its advantages.

Example of the Application of an FPGA

In the following example one thing should be appreciated. The example has been chosen to illustrate features of the FPGA. The reader should realize that this situation is generally true of any example in any book. However, in a real situation the FPGA will probably not be such a perfect device for the job. Life is never as neat as textbook examples suggest. Figure 5.19 gives the memory space of a 68000-based microcomputer with 8K words of ROM, 8K words of RAM, and three memory-mapped peripherals. The aim of the address decoder designer is to minimize the chip count in the decoder.

We can see from figure 5.19 that the memory components are located consecutively at the bottom of the memory space and the peripherals at the top end of the memory space. Furthermore, the peripherals are allocated a fixed quantity of memory space, subject to the provision of at least 16 words per peripheral. The peripherals are to occupy only odd addresses for which LDS* is active-low.

The first thing to consider in this application is the allocation of control lines to the FPGA. Both UDS* and LDS* take part in the selection of the upper and lower bytes of memory space, and VMA* is necessary to select the peripherals. Furthermore, we can use R/$\overline{\text{W}}$ to make certain that the ROM is selected only during a read cycle. The number of devices to be selected is seven, assuming that 8K × 8 chips are used to implement the ROM and RAM. This leaves two out of the FPGA's nine outputs free. These outputs can be used to provide a DTACK* and a VPA* input to the 68000. Table 5.9 shows how the 82S103 is programmed to implement the addressing scheme of figure 5.19. A dash (—) in a column indicates that the input is a don't care condition.

FIGURE 5.19 Memory map of a 68000-based microcomputer

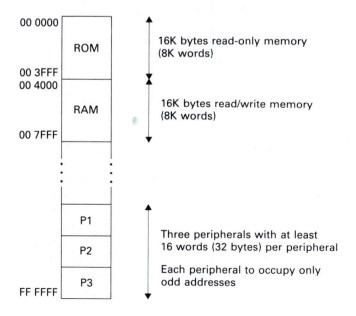

TABLE 5.9 Implementing figure 5.19 with an 82S103 FPGA

DEVICE	CONNECTIONS TO THE FPGA INPUTS FROM A 68000 SYSTEM															
	$R/\overline{W}$	UDS*	LDS*	VMA*	A_{23}	A_{22}	A_{21}	A_{20}	A_{19}	A_{18}	A_{17}	A_{16}	A_{15}	A_{14}	A_{13}	A_{12}
	A_{15}	A_{14}	A_{13}	A_{12}	A_{11}	A_{10}	A_{9}	A_{8}	A_{7}	A_{6}	A_{5}	A_{4}	A_{3}	A_{2}	A_{1}	A_{0}
								FPGA INPUTS								
ROMU	1	0	—	—	0	0	0	0	0	0	0	0	0	0	—	—
ROML	1	—	0	—	0	0	0	0	0	0	0	0	0	0	—	—
RAMU	—	0	—	—	0	0	0	0	0	0	0	0	0	1	—	—
RAML	—	—	0	—	0	0	0	0	0	0	0	0	0	1	—	—
DTACK*	—	—	—	—	0	0	0	0	0	0	0	0	0	—	—	—
PERI1	—	—	0	0	1	1	1	1	1	1	1	1	1	1	0	1
PERI2	—	—	0	0	1	1	1	1	1	1	1	1	1	1	1	0
PERI3	—	—	0	0	1	1	1	1	1	1	1	1	1	1	1	1
VPA*	—	—	0	—	1	1	1	1	1	1	1	1	1	1	—	—

Consider first the selection of the two 8K × 8 ROM components, occupying from $00 0000 to $00 3FFF. The same address lines select both ROMs, as they share identical word address spaces. However, the upper ROM is selected only when UDS* is asserted and the lower ROM when LDS* is asserted. Note that R/W̄ must be high to select the ROMS.

The RAMs are selected by an address in the range $00 4000 to $00 7FFF. As in the case of the ROMs, one is enabled by UDS* and one by LDS*. The state of the R/W̄ line then represents a don't care condition, as the RAM may be written to or read from.

DTACK* is asserted whenever an address in the range $00 0000 to $00 7FFF appears on the address bus. Unfortunately, DTACK* is not synchronized with UDS* or LDS* as the FPGA cannot provide a logical OR capability. In a real system this procedure may have to be done externally.

The peripherals each occupy 2K words of memory because address lines A_{12} to A_{23} are left to be decoded by the FPGA after four of its inputs have been dedicated to control functions. Had fewer control lines been decoded by the FPGA, more address lines could have been decoded and the peripheral block sizes reduced. All peripherals are selected only when LDS* is asserted. Whenever an odd address in the 8K-word range $FF C000 to $FF FFFF appears on the address bus, the VPA* output of the FPGA is asserted, causing the processor to assert VMA* in turn. The assertion of VMA* is a necessary condition for the selection of the peripherals. Note that VPA* is also asserted by an address in the range $FF C000 to $FF C7FF, which is not used in this application. This situation is an irritation but is not particularly dangerous.

FIGURE 5.20 FPGA operated as an address decoder

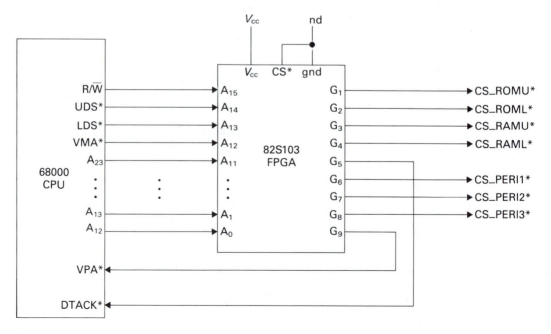

Figure 5.20 shows how the FPGA, programmed according to table 5.9, is connected to a 68000 CPU. Of all the address decoders described so far, the FPGA requires least in the way of support circuitry.

Unlike the PROM, the FPGA can decode 16 address lines without requiring an excessive amount of programming. Unfortunately, the FPGA's outputs do not have a logical OR capability with, say, one output being the logical OR of three other outputs. Other types of programmable logic element now exist that can remedy this situation.

PLA

One of the first of the field programmable logic elements to become widely available was the field programmable logic array, FPLA. Before we look at this device, examination of its near neighbor, the PROM, proves instructive.

Figure 5.21 illustrates the logical structure of a PROM. An n-bit address input is decoded into one of 2^n outputs, and that output is then used to look up an m-bit word in a table of 2^n words. The contents of each word in the table are programm-

FIGURE 5.21 Logical structure of a PROM

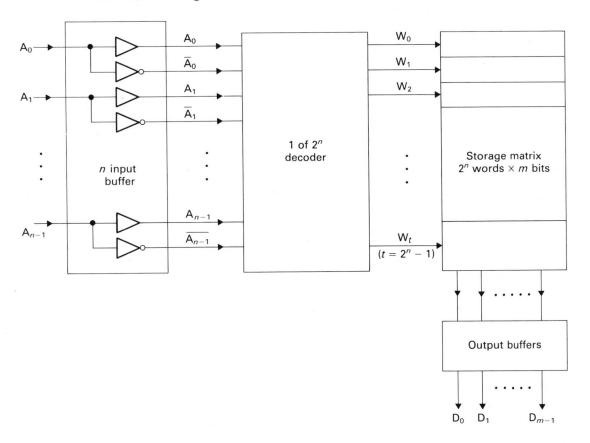

able by the user. It is essential to note that for each of the 2^n possible inputs a corresponding word in the storage matrix always exists.

Figure 5.22 illustrates the way in which a PROM is arranged in terms of its internal gate structure. The n-bit input is decoded into one of 2^n product terms by a fixed array of AND gates. This array is said to be fixed, because it is not programmable by the user. The outputs from the AND gates are fed to OR gates (the storage matrix); for example, in figure 5.22 an input A_2, A_1, $A_0 = 0, 0, 1$ causes the $\overline{A_2} \cdot A_1 \cdot A_0$ product term to be asserted, with the result that any connections between this product line and an OR gate force the output of that gate to be true. In other words, the OR matrix is programmed by the user.

The great disadvantage of the PROM, from the point of view of the systems designer, is its exhaustive storage array. Limited storage is often a hindrance simply

FIGURE 5.22 Structure of a PROM in terms of gate arrays

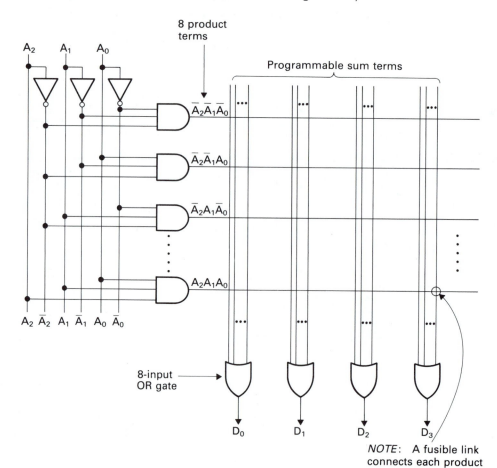

NOTE: A fusible link connects each product term to one of the 8 inputs of the OR gates.

FIGURE 5.23 Structure of the FPLA in terms of gate arrays

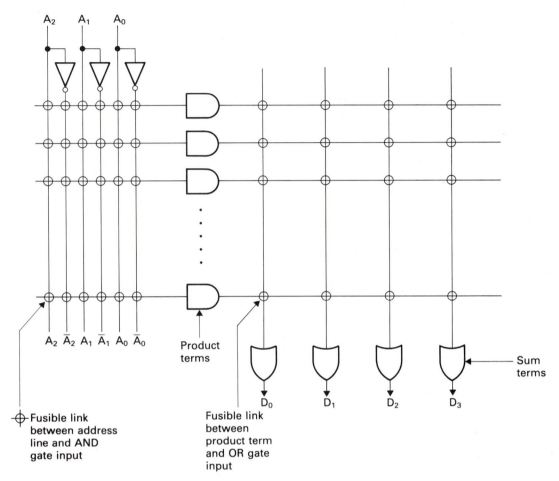

because every possible product term must have its own storage location, whether or not that product term represents a don't care condition. The FPLA is a development of the PROM, which remedies this situation.

Figure 5.23 shows the arrangement of an FPLA in terms of gate arrays. This circuit is almost identical to the PROM, except that the n-line-to-2^n-line decoder has been replaced by a programmable array of AND gates. Now instead of having 2^n AND gates, each with all of the product terms in their true or complement forms, a vastly reduced number of AND gates exists whose inputs may be variables, their complements, or don't care states. A typical FPLA has 48 AND gates for 16 input variables, compared with the 65,536 required by a 16-input PROM. The 82S100 is a typical $16 \times 48 \times 8$ FPLA. The description $16 \times 48 \times 8$ means that it has 16 inputs, 48 storage locations (i.e., 48 product terms), and 8 outputs.

Because the FPLA has a programmable address decoder implemented by the AND gates, product terms can be created that contain between 1 and n variables in exactly the same way as the FPGA described earlier. Indeed, except for the OR matrix, the FPLA would be equivalent to the FPGA.

The principle application of the FPLA is in the synthesis of relatively complex logic systems (often state machines) that would otherwise require many random logic components. Generally speaking, the FPLA is not really appropriate as an address decoder, as simpler devices are often adequate. However, its programmable OR matrix can be helpful in ORing product terms, as the following example demonstrates.

Suppose a $16 \times 48 \times 8$ FPLA is to be applied to the example introduced in figure 5.19 and table 5.9. We can generate product terms for the selection of the devices as follows:

ROMU: $\quad P0 = R/\overline{W} \cdot UDS* \cdot \overline{A_{14}} \cdot \overline{A_{15}} \cdot \overline{A_{16}} \cdot \overline{A_{17}} \cdot \overline{A_{18}} \cdot \overline{A_{19}} \cdot \overline{A_{20}} \cdot \overline{A_{21}} \cdot \overline{A_{22}} \cdot \overline{A_{23}}$

ROML: $\quad P1 = R/\overline{W} \cdot LDS* \cdot \overline{A_{14}} \cdot \overline{A_{15}} \cdot \overline{A_{16}} \cdot \overline{A_{17}} \cdot \overline{A_{18}} \cdot \overline{A_{19}} \cdot \overline{A_{20}} \cdot \overline{A_{21}} \cdot \overline{A_{22}} \cdot \overline{A_{23}}$

RAMU: $\quad P2 = UDS* \cdot A_{14} \cdot \overline{A_{15}} \cdot \overline{A_{16}} \cdot \overline{A_{17}} \cdot \overline{A_{18}} \cdot \overline{A_{19}} \cdot \overline{A_{20}} \cdot \overline{A_{21}} \cdot \overline{A_{22}} \cdot \overline{A_{23}}$

RAML: $\quad P3 = LDS* \cdot A_{14} \cdot \overline{A_{15}} \cdot \overline{A_{16}} \cdot \overline{A_{17}} \cdot \overline{A_{18}} \cdot \overline{A_{19}} \cdot \overline{A_{20}} \cdot \overline{A_{21}} \cdot \overline{A_{22}} \cdot \overline{A_{23}}$

PERI2: $\quad P4 = LDS* \cdot VMA* \cdot \overline{A_{12}} \cdot A_{13} \cdot A_{14} \cdot A_{15} \cdot A_{16} \cdot A_{17} \cdot A_{18} \cdot A_{19} \cdot A_{20} \cdot A_{21} \cdot A_{22} \cdot A_{23}$

PERI3: $\quad P5 = LDS* \cdot VMA* \cdot A_{12} \cdot A_{13} \cdot A_{14} \cdot A_{15} \cdot A_{16} \cdot A_{17} \cdot A_{18} \cdot A_{19} \cdot A_{20} \cdot A_{21} \cdot A_{22} \cdot A_{23}$

VPA*: $\quad P6 = LDS* \cdot A_{14} \cdot A_{15} \cdot A_{16} \cdot A_{17} \cdot A_{18} \cdot A_{19} \cdot A_{20} \cdot A_{21} \cdot A_{22} \cdot A_{23}$

These are the seven product terms to be programmed into the AND gate array. The eight outputs of the FPLA, D_0 to D_7, are formed from the logical ORs of the product terms. Note that this arrangement supports only two peripherals, as the FPLA has eight outputs as opposed to the FPGA's nine. In this example the sum terms are defined as

CSROMU: $\quad D_0 = \overline{P0}$

CSROML: $\quad D_1 = \overline{P1}$

CSRAMU: $\quad D_2 = \overline{P2}$

CSRAML: $\quad D_3 = \overline{P3}$

CSPERI2: $\quad D_4 = \overline{P4}$

CSPERI3: $\quad D_5 = \overline{P5}$

DTACK*: $\quad D_6 = \overline{P0 + P1 + P2 + P3}$

VPA*: $\quad D_7 = \overline{P6}$

The only advantage exhibited by the FPLA over the FPGA here is that DTACK* is the logical OR of the chip-select outputs of the ROM and RAM. Thus, the FPLA does not generate a DTACK* if a write access is made to a ROM, whereas the FPGA implementation does.

PAL

The most recent family of programmable logic elements is the programmable array logic, PAL. The PAL is not to be confused with its more complex neighbor, the PLA, discussed previously. The PAL is an intermediate device falling between the simple gate array and the more complex programmed logic array (PLA). The PLA has both programmable AND and OR arrays, while the PAL has a programmable AND array but a fixed OR array. In short, the PAL is an AND gate array whose outputs are ORed together in a way determined by the manufacturer of the device.

Figure 5.24 illustrates the principle of the PAL with a hypothetical three-input PAL with three outputs. The inputs, A_0 to A_2, generate six product terms, P0 to P5. These product terms are, of course, user programmable and may include an input variable in a true, complement, or don't care form. The product terms are applied to three two-input OR gates to generate the outputs D_0 to D_2. Each output is the

FIGURE 5.24 Principle of the PAL

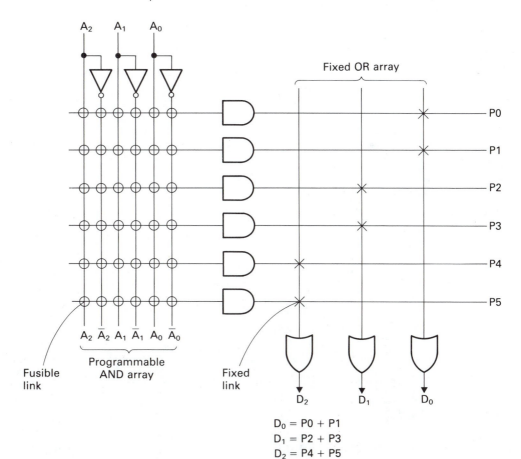

$$D_0 = P0 + P1$$
$$D_1 = P2 + P3$$
$$D_2 = P4 + P5$$

logical OR of two product terms. Thus, $D_0 = P0 + P1$, $D_1 = P2 + P3$, and $D_2 = P4 + P5$. Note that we have chosen three pairs of products. We could have chosen three triplets so that $D_0 = P1 + P2 + P3$, $D_1 = P4 + P5 + P6$, and so on. In other words, the way in which the product terms are ORed together is an arbitrary function of the device and is not programmable by the user.

The structure of the PAL springs from the observation that most designers do not require large numbers of sum terms and the absence of a programmable OR array is of little importance. In fact, throughout this section on address decoders the reader will have observed that few OR expressions have been necessary. As far as address decoding is concerned, OR expressions arise almost exclusively from groups of decoded devices; for example, DTACK* is asserted when any of the memory components decoded by the PAL is accessed. Similarly, VPA* is asserted whenever a peripheral is accessed.

Figure 5.25 gives the details of just some of the PALs now available with active-low outputs. They are designated by the number of inputs, output polarity,

FIGURE 5.25 Some typical PALs

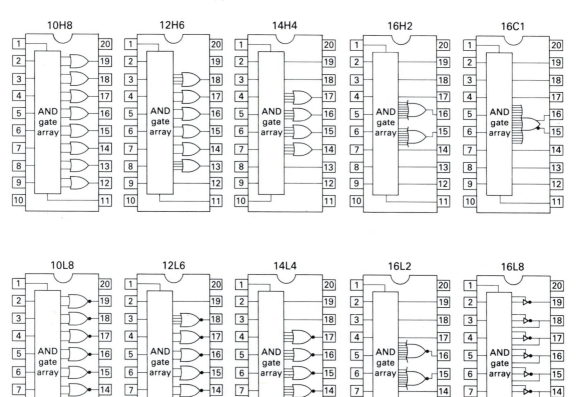

and the number of outputs; for example, the 14L4 has 14 inputs and 4 active-low outputs.

Address Decoding Using a Programmable Address Decoder

You can design address decoders with any of the general-purpose programmable devices we have just described. The need for versatile, high-speed address decoders has led to the design of several special-purpose programmable address decoders. Figure 5.26 describes the 18N8 address decoder, which is available in a 20-pin package.

The 18N8 has ten dedicated inputs and eight product terms (i.e., outputs). Since the *outputs* can be programmed to act as *inputs*, the 18N8 can be configured as $10 + X$ inputs and $8 - X$ outputs, where $X = 0$ to 7. Actually, X could be 8, but there is not a lot of demand for a logic element with no output. A very important aspect of the 18N8 is its low input-to-output propagation delay of less than 6 ns.

If you inspect figure 5.26 you can see that each input/output pin may be programmed to function as a simple output by *enabling* its output buffer. If the buffer is *disabled*, the input/output pin may be programmed as an input by multiplexing the input/output pin onto the AND gate array. Finally, if *both* the output buffer and the multiplexer are enabled, the input/output pin acts as an output, which is also fed back into the array. In this way, we can generate, say, a product term $I_1 \cdot I_2 \cdot I_3 \cdot I_4$ that is connected to an output pin and that is also fed back into the array. Consequently, this product term can be used as an input to other product terms.

Consider a simple example of the 18N8. Suppose we wish to design a 68000 system with DRAM, EPROM, and a block of peripherals. The DRAM is to be implemented with sixteen 1M-bit devices, organized as two arrays each of 1M bytes. One array forms the lower byte and is strobed by LDS* and the other the upper byte strobed by UDS*. The total size of the DRAM is 2M bytes, and the DRAM is to be mapped in the region $80 0000 to $9F FFFF. The ROM is implemented with two 128K-byte chips in the 256K-byte range $00 0000 to $03 FFFF. Input/output space occupies the 64K-byte range $FF 0000 to $FF FFFF. The total number of device selects required by this arrangement is therefore five (i.e., DRAML, DRAMU, ROML, ROMU, and I/O). We can use one of the 18N8's outputs to generate a composite DTACK*, which is asserted whenever a memory component is selected. Accessing I/O space does not cause DTACK* to be asserted, since we assume that each peripheral generates its own DTACK*.

Since an 18N8 has 8 programmable input/outputs and we require 6 of them as dedicated outputs, there is a total of 12 inputs available for address decoding. The next step is to construct an address table for this problem.

DEVICE	RANGE	A_{23}	A_{22}	A_{21}	A_{20}	A_{19}	A_{18}	A_{17}	A_{16}	A_{15}	...	A_{01}	A_{00}
ROM	00 0000–03 FFFF	0	0	0	0	0	0	×	×	×	...	×	×
DRAM	80 0000–9F FFFF	1	0	0	×	×	×	×	×	×	...	×	×
I/O	FF 0000–FF FFFF	1	1	1	1	1	1	1	1	×	...	×	×

FIGURE 5.26   Organization of the 18N8 programmable address decoder

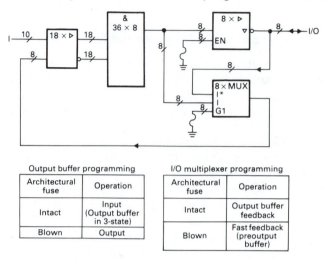

Output buffer programming	
Architectural fuse	Operation
Intact	Input (Output buffer in 3-state)
Blown	Output

I/O multiplexer programming	
Architectural fuse	Operation
Intact	Output buffer feedback
Blown	Fast feedback (preoutput buffer)

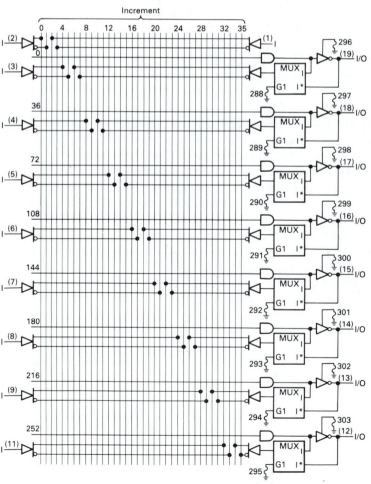

FIGURE 5.27 Using the 18N8 programmable address decoder

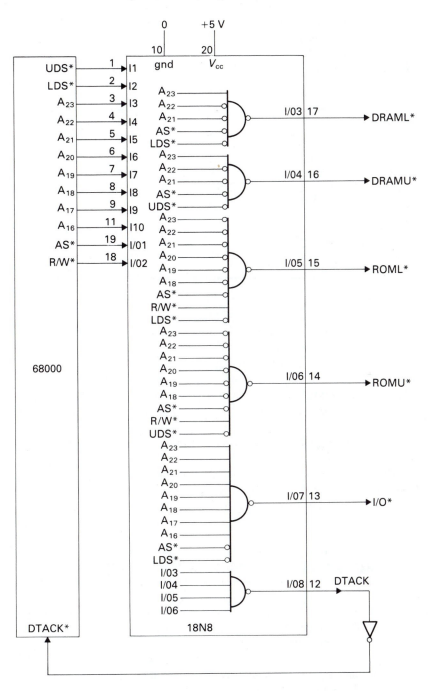

As you can see, we need to employ address lines A_{16}–A_{23} to fully decode each block of memory. Decoding these address lines requires eight inputs, leaving us with four inputs to play with. We can connect two of the remaining inputs to LDS* and UDS* in order to provide upper and lower block selection. The remaining two inputs can be connected to AS* (as a master strobe) and to the 68000's $R/\overline{W}$ output (to ensure that the ROM is selected only in a read cycle). We can, therefore, rewrite the address table to include all device selects, AS*, UDS*/LDS*, and $R/\overline{W}$.

DEVICE	OUTPUT	18N8 INPUTS															
		I/O1 AS*	I/O2 $R/\overline{W}$	I1 UDS*	I2 LDS*	I3 A_{23}	I4 A_{22}	I5 A_{21}	I6 A_{20}	I7 A_{19}	I8 A_{18}	I9 A_{17}	I10 A_{16}	I/O3	I/O4	I/O5	I/O6
ROML*	I/O5	0	1	×	0	0	0	0	0	0	0	×	×				
ROMU*	I/O6	0	1	0	1	0	0	0	0	0	0	×	×				
DRAML*	I/O3	0	×	×	0	1	0	0	×	×	×	×	×				
DRAMU*	I/O4	0	×	0	×	1	0	0	×	×	×	×	×				
I/O*	I/O7	0	×	×	0	1	1	1	1	1	1	1	1				
DTACK	I/O8	×	×	×	×	×	×	×	×	×	×	×	×	1	1	1	1

Note that pins I/O3 to I/O6 appear *twice* in the table. They appear once because these pins provide four memory device–select outputs. They appear again because they are fed back into the array as inputs to generate the composite DTACK. Since the 18N8 cannot generate sum terms, we are forced to generate DTACK by NANDing the four outputs. If the 18N8 is programmed so that DTACK = $\overline{\text{DRAML} \cdot \text{DRAMU} \cdot \text{ROML} \cdot \text{ROMU}}$, the DTACK output is low when none of these devices is selected. If one of the devices is selected, the corresponding output goes low and DTACK goes high. Consequently, the DTACK output of the 18L8 is active high and must be connected to the 68000's DTACK* input via an inverter.

The structure of the address decoder using the 18L8 is given in figure 5.27. Now that we have covered the address decoder, we are going to look at the static RAM in more detail than we did in chapter 4.

5.3 DESIGNING STATIC MEMORY SYSTEMS

In this section we are going to look at the semiconductor components used to store programs and data in a microcomputer. Here the word *semiconductor* is needed to distinguish between the fast semiconductor memories that can store or retrieve information in a time comparable with the cycle time of a processor and the much slower electromechanical memories such as disks or tapes operating at speeds many orders of magnitude lower than semiconductor memories.

Although today's memory components play a passive role in the organization of a computer, as they perform no arithmetic or logical operations on the data they store, they have played a most active role in the development of microprocessor systems. The relationship between the microprocessor and its memory is analogous to that between an automobile and the highways. The active element (processor or automobile) is useless without its resources (memory or highways). In the last decade the greatest progress has been made in the realm of memory technology rather than in microprocessor design. The truth is that microprocessors are much more powerful than they were, but once 1,024-bit memories were the state of the art, while today 16M-bit memories are rolling off the production line. Less than a decade has seen an improvement of four orders of magnitude in the density of memory components. It is doubtful that Intel would call their 8086 microprocessor 1000 times more powerful than their 8080A, or Motorola their 68000 microprocessor 1000 times more powerful than their 6809. Not only has the capacity of memory components increased dramatically in the last decade, but significant improvements have been made in their speed, their ease of use, and their power consumption.

Advances in memory technology are not important merely because they have allowed larger programs to be run on microprocessors; they have paved the way for the new generation of 32-bit microprocessors. In the early days of the 8-bit microprocessor, most programs were written in assembly language. Although this approach is suitable for very small programs, or for tightly coded programs with limited memory, or for optimized code, it is not suited to most of today's applications. Modern programs are often very large and the techniques used to create assembly language programs are no longer appropriate.

The current approach to program design is to rely heavily on high-level languages, choosing, wherever possible, the best language for the job. This method requires large memories to hold the source program, compiler, operating system, and other software tools.

Although low-cost memory has made large microprocessor systems possible, it is still feasible to design more modest systems with relatively little memory. By developing software on a large system and transferring the resulting object code (i.e., machine code) to a small system, we are able to produce dedicated computers with the minimum memory needed to carry out their intended functions. Small memory makes economic sense in embedded systems.

Static Random Access Memory Characteristics

In cases where a microprocessor system operates as a general-purpose digital computer, the bulk of the immediate access memory is likely to be read/write random access memory, because a wide range of different programs are run on the computer. When a microprocessor is dedicated to a specific application such as a chemical process controller, the majority of the immediate access memory is more likely to be implemented as read-only memory, because the application-oriented program is never altered. The designer of any microcomputer must decide how the read/write

RAM is to be implemented. Should it be implemented with static or with dynamic RAM?

Static read/write RAM is often the designer's first choice, because it is so much easier to use than dynamic read/write RAM. Unlike dynamic memories, static memories do not require any action to be taken to periodically refresh their contents; nor do they require the address multiplexing circuitry peculiar to dynamic memory.

The basic circuit diagram of a typical NMOS static-storage cell is given in figure 5.28. The most significant feature of this cell is that six transistors are required to store a single bit of information. As we shall see later, dynamic memory cells store their data as an electrical charge on the internal capacitance of a single transistor, and therefore require fewer transistors per cell than static memories. Because more components per cell exist in a static memory, a dynamic memory of a given chip (i.e., silicon die) size can always store more data (approximately four times as much) than a corresponding static memory chip of the same size.

From the preceding remarks, we should observe that the designer of a memory system has the choice between low-cost, high-density dynamic RAM with its more complex control circuitry and the more expensive static RAM that is easier to use. In any given situation the designer must weigh up all the relative merits of both systems.

Memory Configuration

Over the years the density of memory chips has increased and the price per bit declined. As each new memory chip appears, it has the effect of depressing the price of its more humble antecedents. Table 5.10 gives the details of some of the static memory components currently available. Some represent the next generation of chips

FIGURE 5.28   Static RAM memory cell

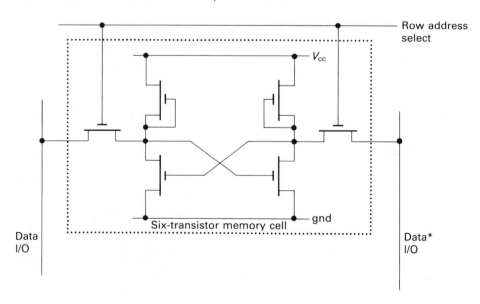

TABLE 5.10 Characteristics of some static RAM chips

STATIC RAM	TOTAL CAPACITY	ORGANIZATION	POWER (mW)	ACCESS TIME (ns)	PINS
TMS 4044–45	4,096	4,096 × 1	495	450	18
MK4118	8,192	1,024 × 8	400	250	24
TMS 4016–25	16,384	2,048 × 8	495	250	24
HM 6167H	16,384	16,384 × 1	400	45	20
MCM6268–20	16,384	4,096 × 1	550	20	20
HM 6264P–15	65,536	8,192 × 8	200	100	28
MB84256–10	262,144	32,768 × 8	350	100	28
HM6208	262,144	65,536 × 4	300	35	24
MCM6206	262,144	32,768 × 1	600	35	28
TC55100PL–70	1,048,576	1,31,072 × 8	350	70	32

and are currently designed only for very high performance equipment. These chips are tomorrow's mainstream devices.

The trend in memory component design is to create either 1-bit wide memories or 8-bit ("bytewide") memories (some 4-bit and 16-bit wide devices are also available). People who design memory modules do not lose much sleep when deciding whether to choose 1- or 8-bit wide chips for their product. A simple rule of thumb is: If it is possible to use 1-bit wide components then do so. For example, suppose that a microcomputer is equipped with 2K words by 16 bits of RAM and that 16K-bit chips are to be used for economic reasons. The designer has no reasonable choice other than the arrangement of figure 5.29. The system address bus is connected to the eleven address inputs of both chips in parallel, so that when an 8-bit location is addressed in one chip, the corresponding location is addressed in the other. Bits D_0 to D_7 from the low-order chip are connected to bits D_{00} to D_{07} of the data bus, and bits D_0 to D_7 of the high-order chip are connected to bits D_{08} to D_{15}. The chip-select and read/write inputs of both memories are connected together. In the following examples we will forget about byte addressing for convenience.

Now consider the design of a 16K word by 16-bit memory. A designer is able to use either sixteen 16K by 1 chips or sixteen 2K by 8 chips. Although the outcome is nominally the same, figures 5.30 and 5.31 show the results of using 16K × 1 and 2K × 8 chips, respectively. In figure 5.30 all the memory components' address lines are connected in common to the system address bus. Each memory component contributes one data line, which is connected to the appropriate line of the system data bus.

Things are rather different in figure 5.31. The chips are arranged as eight pairs, one member of the pair being connected to data lines D_{00} to D_{07} of the system data bus and the other connected to data lines D_{08} to D_{15}. All 11 address lines from each of the RAMs are connected to the system address bus. The fact must immediately be obvious that only 11 address lines are taking part in the selection of a memory location instead of the 14 required by a 16K-word array. The additional address lines

FIGURE 5.29 2K × 16-bit memory organization with 2K × 8 chips

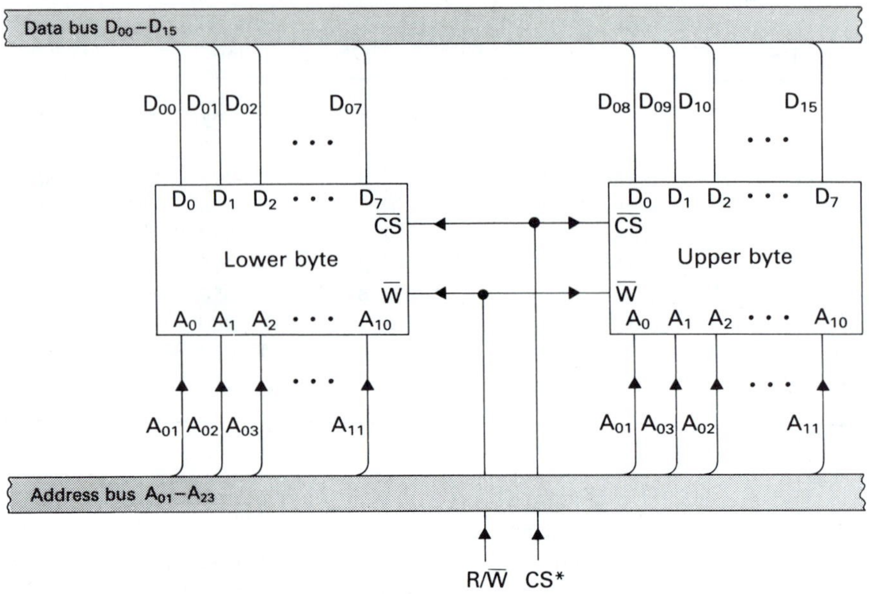

A_{12}, A_{13}, and A_{14} are decoded into one of eight lines, each of which selects one of the eight pairs of RAMs; that is, the 16K memory space has been partitioned into eight blocks of 2K, and a decoder is needed to distinguish between the blocks.

Clearly, the arrangement of figure 5.31 is inferior to that of figure 5.30 because extra logic is required without providing any added benefit whatsoever. Another two reasons indicate why 16K × 1 chips beat 2K × 8 chips. The 2K × 8 chip is available in a 24-pin DIP package with a nominal area (or footprint) of 1.2 × 0.6 = 0.72 in². As the 16K × 1 chip has three more address lines but seven less data lines, it can be packaged in a 20-pin DIP chip with a footprint of only 1.0 × 0.3 = 0.3 in². Finally, the arrangement of figure 5.30 means that each data line from the system bus is loaded by just one data input/output connection to a RAM chip. However, in figure 5.31 each data line from the system bus is connected in parallel to the corresponding data line of each of the eight pairs of 2K × 8 RAM. This arrangement represents an eightfold increase in data bus loading, reducing the noise immunity of the data bus.

CMOS RAM

Just as houses can be built of straw, wood, or brick, semiconductor devices can be manufactured by a number of different technologies. The three most important processes are bipolar, NMOS, and CMOS. Bipolar technology has been around longest,

FIGURE 5.30 Memory organization 1—16K words × 16 with 16K × 1 chips

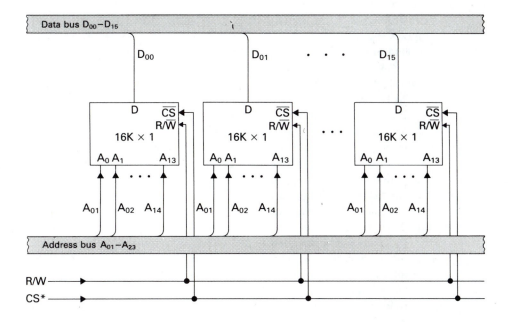

FIGURE 5.31 Memory organization 2—16K words × 16 with 2K × 8 chips

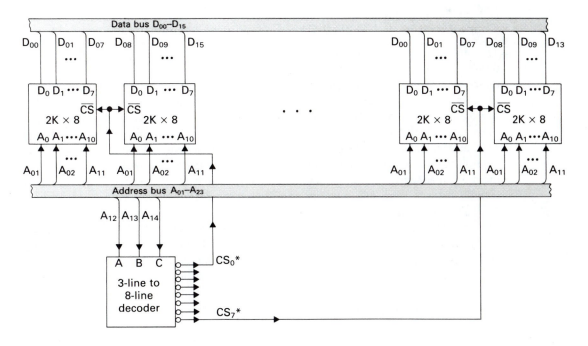

and is seen most frequently in TTL (transistor-transistor logic) devices. The advantage of bipolar logic is its great speed. Unfortunately, bipolar devices cannot yet be constructed with the same density as NMOS and CMOS components. In any case, bipolar logic is power-hungry and consumes large quantities of energy, making the cooling of equipment a limiting factor in systems design.

NMOS technology has replaced the earlier PMOS technology and is used to fabricate the majority of microprocessors and memory devices. Although NMOS is not as fast as bipolar logic, it consumes much less power. A close relative of NMOS is CMOS logic, which has the highly desirable property of consuming very little power. Not long ago the engineer was faced with a nasty dilemma: choose bipolar logic for high speed but at the expense of being forced to implement everything in small- or medium-scale integration; choose CMOS logic for low-power operation but suffer very slow switching speeds, a density less than NMOS, and a hefty price tag; or choose NMOS as a compromise.

Semiconductor manufacturers have tried to combine the best features of all three technologies by improving the speed and packing density of CMOS while bringing down its price. Before continuing we should explain that CMOS devices are not quite as frugal with power as some people believe. A CMOS logic element consumes an appreciable amount of power only when it changes state, so that the power consumption rises with the rate at which a system is clocked. At a sufficiently high clock rate, CMOS systems can consume more power than equivalent NMOS systems. However, when idle, a CMOS component consumes an amazingly tiny amount of power. Because of this factor CMOS memories may be operated from small batteries when the prime source of power (i.e., the public electricity supply) is interrupted.

Characteristics of a Typical CMOS Memory Component

The pinout and internal arrangement of a typical 8K × 8-bit static CMOS RAM, the 6264LP, is given in figure 5.32. In a world plagued by a general lack of standards, a reasonable measure of agreement has been made on the pinout of memory components. Not only are similar memory components from different manufacturers plug compatible, but read-only memory and read/write memory can be interchanged. Thus, the 6264LP read/write memory is pinout compatible with 8K × 8 EPROMs such as the 2764. This fact is important because the same PCB can be used for read-only or read/write memories. Moreover, engineers can write programs, test them in read/write memory, and then, only when they are working correctly, commit them to read-only memory and replace the RAM with ROM.

The principal features of the 6264LP and similar static CMOS read/write memories are their low power consumption when active (typically 200 mW) and their tiny power consumption when in a standby mode (typically 0.1 mW). The standby mode comes into operation when the V_{cc} supply voltage is reduced from its normal 5 V to no less than 2.0 V. This allows computers to be designed with nonvolatile memory by powering CMOS chips from small batteries when the system is not connected to the line supply.

FIGURE 5.32 Pinout and internal arrangement of the 6264LP-I0. (Reprinted by permission of Hitachi Limited)

■ BLOCK DIAGRAM

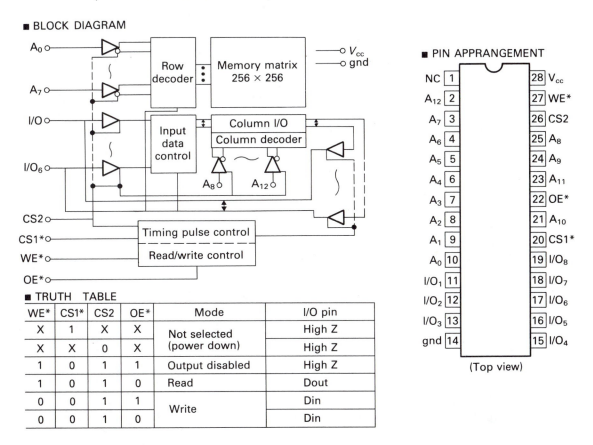

■ TRUTH TABLE

WE*	CS1*	CS2	OE*	Mode	I/O pin
X	1	X	X	Not selected (power down)	High Z
X	X	0	X		High Z
1	0	1	1	Output disabled	High Z
1	0	1	0	Read	Dout
0	0	1	1	Write	Din
0	0	1	0		Din

■ PIN APPRANGEMENT

NC	1	28	V$_{cc}$
A$_{12}$	2	27	WE*
A$_7$	3	26	CS2
A$_6$	4	25	A$_8$
A$_5$	5	24	A$_9$
A$_4$	6	23	A$_{11}$
A$_3$	7	22	OE*
A$_2$	8	21	A$_{10}$
A$_1$	9	20	CS1*
A$_0$	10	19	I/O$_8$
I/O$_1$	11	18	I/O$_7$
I/O$_2$	12	17	I/O$_6$
I/O$_3$	13	16	I/O$_5$
gnd	14	15	I/O$_4$

(Top view)

6264 Read Cycle

A slightly simplified version of the 6264LP-10 timing diagram is given in figure 5.33. The suffix 10 denotes that this device has a 100-ns access time. In order to read data from the chip, the appropriate address must be applied to the memory, CS1* brought low, CS2 high, and OE* low. Data becomes valid after the longest of t_{AA}, t_{CO1}, or t_{CO2} has been satisfied. To the designer this condition means that the address, CS1*, and CS2 must be asserted as close together as possible if the minimum access time of the 6264LP-10 is to be achieved. If, for example, the address becomes valid but CS1* and CS2 are asserted 70 ns later, then the data does not become valid until 70 ns + t_{CO1} = 170 ns later.

The output enable line, OE*, causes the data bus lines to assume a low impedance state no sooner than 5 ns (t_{OLZ}) after OE* is asserted and no later than 50 ns. These figures are of interest to the systems designer who is going to use OE* to control data bus contention. The 6264LP-10 may be operated with OE* permanently

FIGURE 5.33 Simplified read cycle timing diagram of an 6264LP-10 RAM

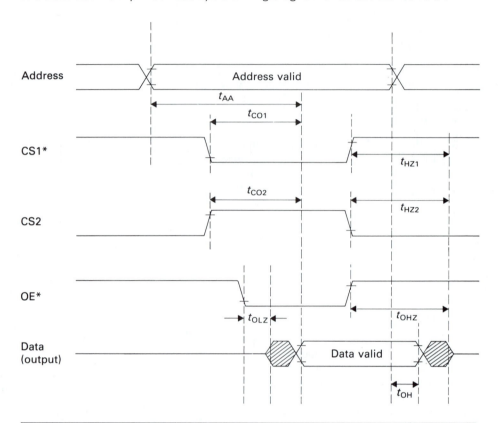

Symbol	Parameter	Value for 6264P-10
t_{AA}	Address access time	100 ns max.
t_{CO1}	Chip select to output	100 ns max.
t_{CO2}	Chip select to output	100 ns max.
t_{HZ1}	Chip select to output float	35 ns max.
t_{HZ2}	Chip select to output float	35 ns max.
t_{OLZ}	Output enable to output low Z	5 – 50 ns
t_{OHZ}	Output disable to output float	35 ns max.
t_{OH}	Output data hold	10 ns min.

grounded, in which case the chip is controlled solely by CS1*, CS2, and WE*. If OE* can be grounded and forgotten, why have the manufacturers provided it? The short answer is that the 6264LP-10 comes in a 28-pin package and only 25 pins are strictly necessary to implement an 8K × 8 read/write RAM. Of the other three pins, one is NC (not connected = not used), one is a second chip-select input, and the remaining pin has been given the function of turning the data bus drivers on and off during a read cycle. The long answer concerns bus contention in microprocessor systems.

A memory's OE* pin can be used explicitly to turn off its data bus drivers. The bus drivers are, of course, turned off when CS* is negated, and, therefore, OE* is not vital to the operation of a memory component (which is just as well, since many memories lack an OE* pin). However, using OE* to turn off the data bus drivers reduces the danger of bus contention in systems in which CS* might remain active-low beyond the end of a read cycle.

By the way, this particular memory device specifies the same maximum value for data bus floating from CS* high ($t_{HZ1} = 35$ ns) as for data bus floating from OE* high (t_{OHZ}). Little is to be gained by using OE* to force the data bus buffers into a high impedance state at the end of a read cycle. Many memories have very much lower values for "data bus floating from OE* high" than they do for "data bus floating from CS* high."

Bus Contention and Static Memories

The microprocessor systems designer has to worry as much about what happens at the ends of a read or write access as he or she does about its middle; that is, you cannot look at a *single* bus cycle without considering what happens *before* and *after* it. We will now look at the effect of two consecutive read cycles and then a read cycle followed by a write cycle.

Figure 5.34 illustrates a hypothetical but realistic situation. Two memory components, M1 and M2, are connected to a system's address and data bus. During memory read cycle 1, memory M1 is selected, and during memory read cycle 2, memory M2 is selected.

The timing diagram shows the behavior of the data outputs from M1 and M2, which are labeled data 1 and data 2, respectively. Suppose that M1 has data bus drivers with relatively long turn-off times. Therefore, the data bus from M1 is in a low-impedance state well into cycle 2. Now suppose that M2 has data bus drivers with relatively fast turn-on times, so that the data bus from M2 goes into a low impedance state very early in cycle 2. As the data buses from both memories are connected to the same system data bus, a period follows when two devices are simultaneously trying to drive it. This situation is shown by the shaded portion in figure 5.34 and is potentially harmful to the system.

Another form of data bus contention is related to a memory write cycle. Suppose a microprocessor begins a write cycle and puts out a valid address early in the cycle, as illustrated in figure 5.35. If the address is valid at time t_{AV}, the memory will be selected t_{CS} seconds later, where t_{CS} is the delay involved in decoding the address. The time at which the data bus from the memory assumes a low impedance state is given by $t_{AV} + t_{CS} + t_{OE}$, where t_{OE} is the time delay between chip select going low and the data bus low impedance state. In other words, the memory is behaving as if the current cycle were a read cycle.

All read/write memory components automatically disable their data bus drivers when their WE* input is forced low. Consequently, any data bus contention cannot take place later than $t_{AV} + t_{WD} + t_{WE}$, where t_{WD} is the delay between address valid and WE* going low and t_{WE} is the time required to turn off the memory's data output drivers when WE* goes high.

FIGURE 5.34 Relationship between output enable and bus contention in a read cycle

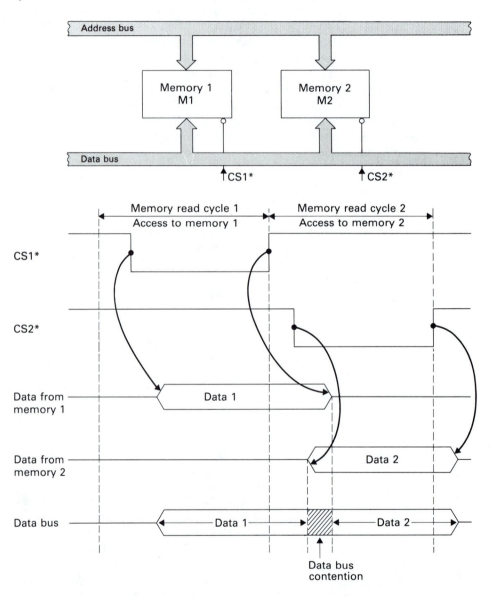

Data bus contention occurs in the period between the times when the data bus buffers of the memory are turned on by chip select (CS*) going low and then off by WE* going low. This period is given by $(t_{AV} + t_{WD} + t_{WE}) - (t_{AV} + t_{CS} + t_{OE}) = t_{WD} + t_{WE} - t_{CS} - t_{OE}$. If this value is zero or negative, there is no problem, but if it is positive, some action must be taken to avoid this type of contention.

FIGURE 5.35 Relationship between output enable and bus contention in a write cycle

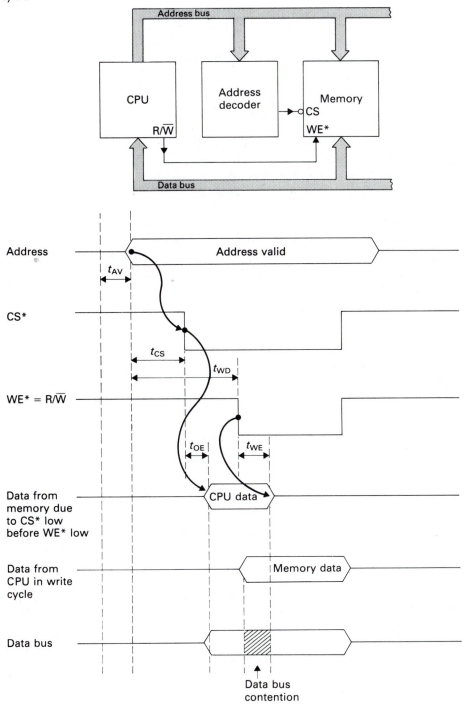

Returning to the 6264LP-10 read cycle timing diagram of figure 5.33, another critical period during the read cycle comes at its end. The data bus buffers are turned off by one of CS1*, CS2, or OE* becoming inactive. The data bus begins to float no later than 35 ns after the chip is deselected. Should the contents of the address bus change before the data bus is floating, the 6264LP-10 is guaranteed to hold the contents of the data bus for at least t_{OH} seconds (10 ns) after the address has changed.

The 6264 Write Cycle

Figure 5.36 gives a simplified version of an 6264LP-10 write cycle timing diagram for which OE* is low throughout the entire cycle. A valid write cycle takes place when CS1* is low, CS2 is high, and WE* is low. The write cycle begins when the last of these three signals is asserted and ends when the first of them is negated. The only setup requirement for these three control signals is that the contents of the address bus be valid for at least t_{AS} (i.e., 0 ns) before they are asserted. Clearly, letting the address settle after CS1*, CS2, or WE* is asserted may result in an erroneous write to a random address location.

In order to sustain a write cycle, both CS1* and CS2 must be asserted for $t_{CW} = 80$ ns (t_{CW} is the smaller of t_{CW1} and t_{CW2}) and WE* for at least $t_{WP} = 60$ ns.

FIGURE 5.36 Simplified version of the 6264LP-10 write timing diagram

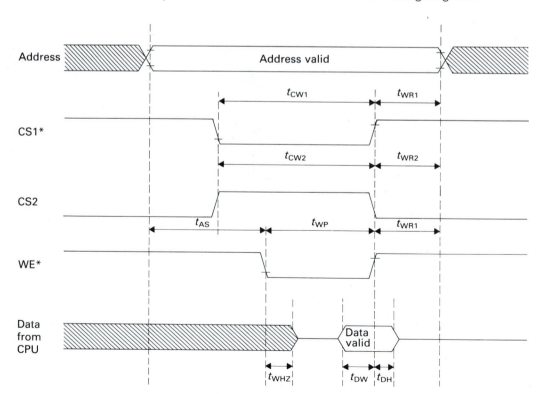

After the end of a write cycle, the contents of the address bus must not change for a period called the *write recovery time*. This time is t_{WR_1}; it is 5 ns minimum in the case of a write cycle being ended by CS1∗ or by WE∗ being negated or $t_{WR_2} = 15$ ns minimum in the case of a write cycle terminated by the negation of CS2.

As OE∗ is asserted for the duration of the write cycle, the data bus driver may be in a low-impedance state for up to $t_{WHZ} = 35$ ns maximum following the falling edge of WE∗. We have already seen that data bus contention must not be allowed during this time. The CPU must place data on the memory's data lines at least $t_{DW} = 40$ ns before the termination of a write cycle and maintain it for at least $t_{DH} = 0$ ns after the end of the cycle.

An Example of a Modern High-Density 128K × 8 CMOS RAM

Now that we have described the characteristics of a relatively small 8K × 8-bit RAM, we will look at a much larger 128K × 8-bit RAM, the 628128. Since, in principle, the 628128 behaves much like its smaller predecessors, there is little point in plowing through all the details of this device.

We have provided some of the relevant details of the 628128 in figure 5.37. Instead of presenting the chip's pinout, its interface circuit details, and its timing information in three separate figures, we have combined all the important information in one diagram. The 68000 microprocessor is on the left-hand side and the 628128 RAM is on the right-hand side. We have also included the logic required to generate the RAM's chip select and WE∗ (write enable) inputs.

Between the CPU and the RAM, we have included the appropriate signal waveforms together with their relevant timing parameters. All parameters are those appropriate to the RAM. Note that in order to deal with both the read and write timing diagrams without ambiguity, we have replicated the data bus and the R/$\overline{\text{W}}$ line.

CMOS Memory and Battery Backup

Like many other CMOS memory components, the 6264LP-10 is able to operate in a special power-down or data-retention mode, in which the supply voltage, V_{cc}, is reduced to no less than 2.0 V. Under these conditions the memory must be deselected with either CS1∗ at no less than 0.2 V below V_{cc} or CS2 at no more than 0.2 V above ground. When powered down, the 6264LP-10 consumes less than 50 μA through its V_{cc} pin. Such a low current can readily be supplied by a small on-board battery. Figure 5.38 gives the low V_{cc} data-retention waveform for the case where the memory is deselected by CS1∗ going high. We must conform with this sequence if data is not to be lost.

In order to bring the 6264LP-10 into a power-down mode safely, CS1∗ must first rise to at least 2.2 V while V_{cc} is at its nominal 5.0 V level. This action has the effect of deselecting the chip. The V_{cc} supply may then be reduced to its data-retention value of 2.0 V minimum. Note that in figure 5.38, as V_{cc} falls, CS1∗ tracks it, and that CS1∗ must not drop below $V_{cc} - 0.2$ V during this process. When

FIGURE 5.37 Details of the 628128's 128K by 8-bit CMOS RAM

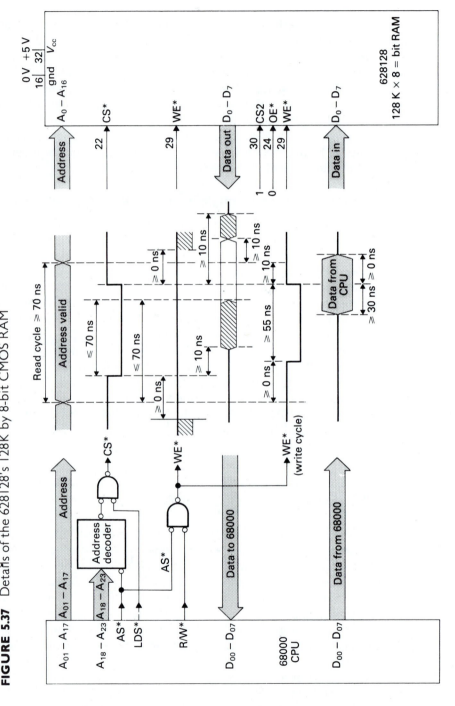

FIGURE 5.38 Low V_{cc} data-retention waveform controlled by CS1*

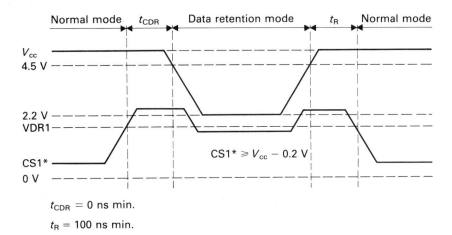

t_{CDR} = 0 ns min.

t_R = 100 ns min.

V_{cc} = 2.0 V, the memory is in its power-down mode, consuming no more than 250 μW. To bring the memory out of its power-down mode, V_{cc} must be returned to its normal value of 5 V. During this transition, CS1* must track V_{cc} up to at least 2.2 V. Once V_{cc} has settled, the chip can be accessed after a period equal to the read cycle time (100 ns).

In theory, the power-down mode of this and other CMOS memories is quite unremarkable. In practice, the procedure can be a bit of a nightmare, because three very practical problems arise in the design of battery-backed-up or "nonvolatile" CMOS read/write memories: the switchover from normal V_{cc} to its standby value, the control of chip select or write enable during this time, and the control of the memory's other inputs (i.e., address lines) while the system is in standby mode. These problems will be dealt with later in this section. First we must look at the power supply requirements of CMOS memories.

The designer of battery-backed-up CMOS memory is faced with the choice of the battery that will actually supply the standby power. This choice can be difficult, because several types of battery are available and many considerations are involved in the design of a standby power supply. Apart from cost and physical size, the four most important parameters affecting the selection of a battery are its type, its capacity (measured in ampere-hours), its temperature range, and its self-discharge current.

There are two basic types of battery: primary and secondary. A primary battery delivers its current, becomes exhausted, and is thrown away. A secondary battery is rechargeable and can be topped up whenever necessary. Popular primary cells are carbon-zinc, alkaline, silver oxide, mercury, or lithium-iodine. Secondary cells are typically lead-acid or nickel-cadmium (Ni-Cd). Most battery backup systems employ secondary cells rather than nonrechargeable primary cells. One reason for this is due to a characteristic of all cells, called self-discharge, which means that cells gradually discharge even though no current is being drawn from their terminals. Therefore, these cells have a finite shelflife, beyond which they cannot be relied upon. If battery-backed-up memory is to be very reliable and to operate without a frequent change of batteries, secondary cells must be used. However, the lithium-iodine primary cell is an exception to this rule and has a negligible self-discharge current, remaining active for at least 10 years. Moreover, the lithium-iodine battery is less affected by temperature than other types and can operate over the range from -54 to $74°C$.

Storage cells are normally the designer's first choice for battery-backed-up memory arrays, as they can be recharged from the line power supply while the system is running. Of the various types of storage cell, the nickel-cadmium cell is generally preferred, because it is of moderately low cost, it does not contain a spillable liquid (as do some lead-acid cells), it is available in small sizes suitable for direct mounting on printed circuit boards, and its output voltage is constant as it is discharged. As a single nickel-cadmium cell has an output of 1.2 V, two or three in series need to be connected to obtain the 2.4 or 3.6 V needed to furnish the standby voltage.

Unfortunately, nickel-cadmium cells have a high self-discharge current of approximately 1 percent per day at $15°C$ to as much as 8 percent per day at $50°C$. Therefore, unless the cell is fully charged when a power failure occurs, its useful life may be severely limited. Furthermore, the self-discharge current limits the time that the system can be operated before the battery must be recharged.

This simplest type of battery backup circuit is given in figure 5.39, where a single diode, D1, is placed between the system power supply and the CMOS power supply. When the anode of the diode is at 5 V, it conducts, providing a CMOS V_{cc} and a charging current to the nickel-cadmium battery. The resistor R_c, in series with the battery, limits the charging current. A recommendation is given that nickel-cadmium batteries be charged at a current of $C/10$, where C is the capacity of the cell in ampere-hours. If a typical battery has a capacity of 100 mA-h, the charging current should ideally be 10 mA. The charging resistor is given by $10(V_{cc} - V_{bat})/C$. Two 1.2-V cells in series with 100 mA-h capacity require a current-limiting resistor of $10(5 - 2.4)0.1 = 260 \ \Omega$.

If the main supply in figure 5.39 falls, diode D1 becomes reverse biased and ceases to conduct. Now the CMOS V_{cc} is supplied by the battery. Unfortunately, the

FIGURE 5.39 Single-diode battery isolation

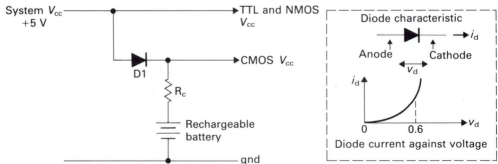

simple scheme of figure 5.39 presents a potential difficulty. In normal operation the TTL and NMOS V_{cc} supply is 5 V. The CMOS supply is 5 V less the voltage drop across the diode, which is approximately 0.6 V for a silicon diode. Therefore, the CMOS V_{cc} supply sits at 4.4 V. Since the CMOS memories are driven by NMOS or TTL signals, it is possible for an input to exceed the CMOS V_{cc} value (i.e., 4.4 V). This condition may harm the CMOS device, because bipolar junctions within the chip can be forward biased, causing very high values of I_{cc} to be drawn.

Figure 5.40 provides a possible solution to this problem. A second diode is placed in series with the V_{cc} supply to TTL and NMOS devices. Now the TTL V_{cc} and CMOS V_{cc} track each other, stopping the TTL V_{cc} from rising appreciably above the CMOS V_{cc}. This circuit requires a main voltage supply of 5.6 V to allow for the 0.6-V drop across D2. Unfortunately, many microprocessor systems provide only a 5-V supply to the individual modules of the system. The circuit of figure 5.40 could be used if the diodes were germanium types rather than silicon. A germanium diode has a forward voltage drop of only 0.2 V, so that a system supply of 5 V would be suitable. Alas, a germanium diode also has an appreciable leakage current when it is reverse biased. The period for which a battery can back up the main supply is therefore reduced, as some current will flow from the battery through the diode to ground.

An alternative arrangement is given in figure 5.41, where a PNP transistor, T2, supplies the CMOS V_{cc} in normal operation. When T2 is conducting, the voltage across it is 0.2 V, which means that the CMOS V_{cc} closely matches the TTL V_{cc}. If the main supply falls, transistor T1 is turned off, turning off T2 and permitting the battery to supply the CMOS V_{cc}. All the circuits described in figures 5.39 to 5.41 automatically solve the CS1∗: V_{cc} tracking problem if CS1∗ is pulled up to V_{cc} (standby) during the power-down/standby/power-up modes.

Another area to which the designer of battery-backed-up CMOS memories must pay careful attention is the state of the inputs to the CMOS device when it is powered down. Figure 5.42 shows a typical CMOS input stage. Components D1, D2, and R form a protective network designed to eliminate the CMOS circuit's

FIGURE 5.40 Dual-diode battery isolation

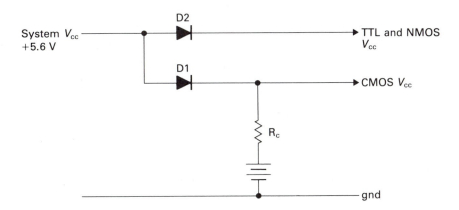

FIGURE 5.41 Isolation of the battery by a transistor circuit

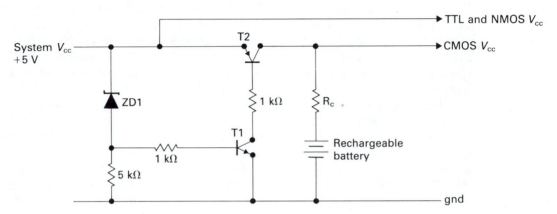

susceptibility to static electricity, and do not affect the circuit under normal operation. The key to CMOS's low power consumption can be found in the P-type and N-type metal oxide transistors in series (T1 and T2, respectively, in figure 5.42). Because only one transistor in the pair is in the on-state (i.e., conducting) at a time, no direct current path exists between V_{cc} and ground, apart from a tiny leakage current through the off-transistor.

If the input to a CMOS gate is ever allowed to float, a possibility exists that it may settle at a level midway between V_{cc} and ground, which has the effect of turning on both the P-channel transistor and the N-channel transistor simultaneously. Under these circumstances a direct path goes between V_{cc} and ground and an appreciable current can flow through both output transistors in series. Although this current is not large by the standards of microcomputer systems, it can be a magnitude or two greater than the V_{cc} power-down current taken by the memory component.

FIGURE 5.42 Input circuit of a CMOS gate

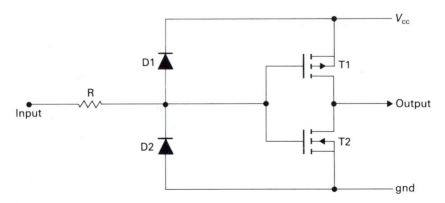

Floating inputs to CMOS devices can be eliminated by careful attention to the interface between the memory chip and the rest of the system. The inputs to the CMOS device must either be pulled up to the CMOS V_{cc} or down to ground. Some authorities suggest that CMOS inputs should have pull-up resistors to V_{cc} or pull-down resistors to ground, to stop any inputs floating while the system is powered down. However, many low-power Schottky devices used to drive CMOS memories have low impedance outputs (unlike that of standard TTL) when their V_{cc} is grounded, as it is during the power-down mode. Consequently, if LSTTL gates drive CMOS inputs, pull-up resistors must not be used, because current will flow through them and the LSTTL driver from V_{cc} to ground. Equally, pull-down resistors are not needed, as the LSTTL outputs automatically pull the CMOS input down to ground level.

If a guarantee can be made that the low-power Schottky logic driving the CMOS chip has a low impedance output when its $V_{cc} = 0$ V, then it is an excellent driver for all the CMOS inputs that may safely be in a low state during the power-down mode. Of course, the active-low chip-select input cannot be driven by LSTTL. A recommendation is also made that the write enable input to the CMOS memory be pulled up during the power-down mode. This condition is not essential if the active-low chip-select input is in a logical one state, but it is a wise precaution.

The possible circuit diagram of an 8K × 16-bit memory using two 6264LP-10s is given in figure 5.43. Address and data lines are buffered by four LSTTL buffers operating from the TTL V_{cc} supply. The CS1* input of the two memories is derived from (CS* + UDS*) and (CS* + LDS*), as in any other 68000-based system. Normally, OR gates would be used to generate these functions, but since no OR gate with open-collector outputs is available, NAND gates with inverted inputs are used to achieve the same effect. Similarly, the write enable inputs of the memories are driven by a NAND gate with an open-collector output. All TTL gates are driven from the TTL V_{cc} supply.

When the system is powered down, address and data inputs are clamped at a low level by the LSTTL bus drivers. The outputs of the open-collector gates are pulled up to V_{cc} CMOS by the resistors labeled R_p, making it impossible for data within the memory to be corrupted. When power is reapplied to the system, the write enable and CS1* inputs automatically track V_{cc} CMOS as it rises toward V_{cc} TTL.

During the power-down mode, the current supplied by the battery drives the memory components together with leakage currents flowing through the decoupling capacitor C, the reverse-biased diode, and the outputs of the open-collector gates. The current taken by the memories is typically 1 to 50 μA maximum. By using a tantalum low-leakage capacitor and a low-leakage silicon diode, these components should cause little drain from the battery.

The leakage current into the open-collector gates is somewhat higher: A maximum value of 100 μA is quoted for LSTTL gates. Therefore, the current supplied by the battery can be up to 3 × 100 μA + 2 × 50 μA, or 400 μA maximum. A PCB-mounting Ni-Cd with a capacity of 100 mA-h should be able to supply this current for 250 hours, or over 10 days.

Figure 5.43 shows the active-high CS2 input of the memories connected to V_{cc}

FIGURE 5.43 Example of a CMOS memory module backed up by battery

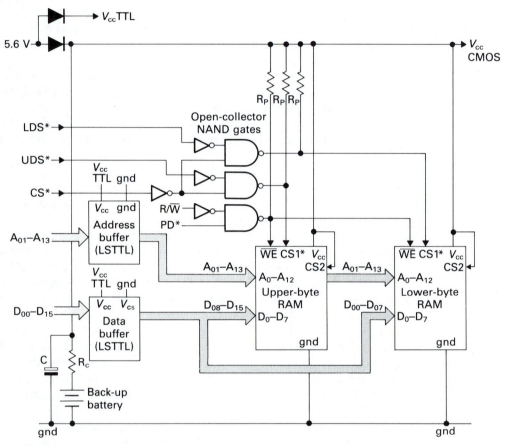

NOTE: PD* = power_down*
(low when V_{cc} below nominal value)

CMOS. This input can also be used to force a power-down mode when V_{cc} TTL falls. In figure 5.44 a CMOS operational amplifier controls CS2. The operational amplifier is a CMOS type and is powered from V_{cc} CMOS. The noninverting input of the op-amp is given by V_{cc} TTL$[R2/(R1 + R2)]$ and the inverting input by V_{cc} CMOS$[R4/(R3 + R4)]$. The resistors are selected to make the noninverting input more positive than the inverting input. In normal operation V_{cc} TTL is equal to V_{cc} CMOS, and the op-amp's output is driven high to enable CS2. When V_{cc} TTL falls, the output of the op-amp drops to ground—0.1 V—and disables the memory components. It should be noted that the ICL7611 CMOS op-amp has an output slew rate of only 0.016 V/μs when operated at 10 μA. Thus the output will take approximately 300 μs before CS2 drops to its lower level.

FIGURE 5.44 Controlling the 6264LP-10's CS2 input by a CMOS op-amp

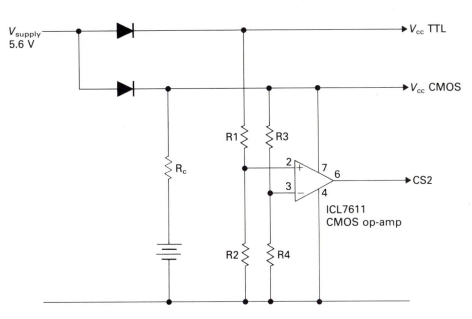

EPROM

The erasable and programmable read-only memory, EPROM is a nonvolatile memory component that can be programmed and reprogrammed by the user with relatively low-cost equipment. Unlike most static and dynamic memory components, EPROM is largely available in bytewide form. Some wordwide EPROMs are now in use. Typical EPROMs vary from 2K × 8 to 256K × 8 bits.

The function of an EPROM is to store programs and data that are never, or only infrequently, modified. An alternative to the EPROM is the mask programmed ROM. These ROMs are cheaper than EPROMs in large-volume production but cannot be reprogrammed and are not used unless the scale of production is sufficient to absorb the initial cost of setting up the mask.

EPROMs are found mainly in four applications: in embedded systems where they hold firmware, in personal computers where they hold the operating system and/or interpretors for HLLs, in bootstrap loaders in general-purpose digital systems, and in the development of microprocessor systems. In an embedded system, programs are held in EPROM, as backing store is not appropriate in most cases. General-purpose systems must have at least sufficient EPROM to hold the bootstrap loader that reads the operating system from disk. Some manufacturers put much of the operating system in EPROM to increase the speed of the system and to reduce the demands on the system read/write memory.

The EPROM is useful in developing microprocessor systems because a program can be developed on one computer, stored in EPROM, and then plugged into the system under development.

Characteristics of EPROM

An EPROM memory cell consists of a single NMOS field-effect transistor and is illustrated in figure 5.45. A current flows between the V_{ss} and V_{dd} terminals through a positive channel. By applying a charge to a gate electrode, the current flowing in the channel can be turned on or off. A special feature of the EPROM is the "floating gate," which is insulated from any conductor by means of a thin layer of silicon dioxide—an almost perfect insulator. By placing or not placing a charge on the floating gate, the transistor can be turned on or off and hence a 1 or a 0 can be stored in the memory cell.

As the floating gate is entirely insulated, we are left with the problem of getting a charge on it. The solution is to place a second gate close to it but insulated from it. By applying a charge to this second and noninsulated gate, some electrons flow between the noninsulated and the floating gates. The voltage necessary to achieve this action is typically 12–25 V. In general, EPROMs are not programmed while they are in their "normal" environment. They are plugged into a special-purpose EPROM programmer because they require a non-TTL voltage during programming and a write cycle is very much longer than a typical read cycle. Although circuits can be designed that program EPROMs in the equipment where they are used, such an approach is most rare.

Another reason why EPROMs are programmed in special-purpose equipment is due to their method of erasure. Once an EPROM has been programmed, its stored data can be erased only by removing the charge trapped on the floating gates. The charge is removed by exposing the surface of the chip to ultraviolet (UV) light. Consequently, the EPROM chip must be mounted behind a transparent window, removed from inside its equipment, and placed under a UV lamp. The window above

FIGURE 5.45 Structure of an EPROM memory cell

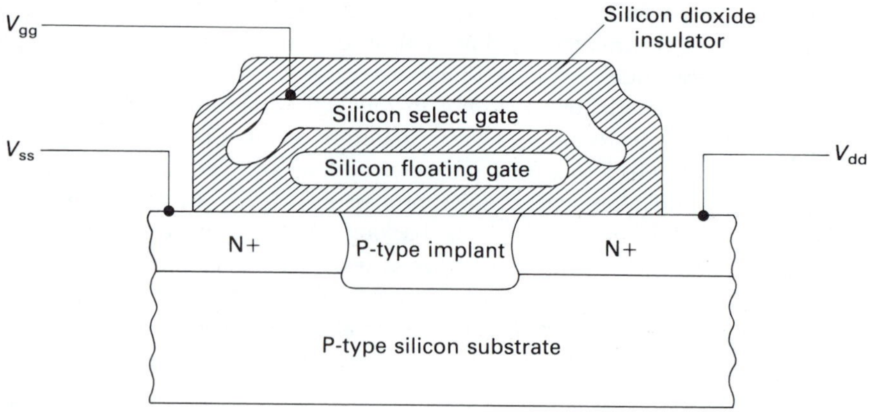

the EPROM chip is made of quartz, because glass is opaque to UV. As the chip must be removed to erase it, the need to place it in a special programmer creates no further hardship.

Although microprocessor systems have been plagued by a lack of standardization, one area of limited success is in EPROM pinout. Two widely used series of EPROMs exist, the 25-series and the 27-series. The 27-series is frequently preferred because its EPROMs are "compatible" with the pinout of typical byte-wide static RAMs; that is, equipment can be designed so that RAM and EPROM chips are interchangeable with little effort.

Figure 5.46 gives the pinout of some of the 27-series EPROMs. Note that the pinouts are arranged so that equipment can be designed to accommodate several different sizes of EPROM by simply modifying jumpers on a PCB; for example, a 24-pin 2732 (4K × 8) EPROM will fit in a 28-pin socket wired for a 2764 (8K × 8) EPROM with minimal effort. The 2732 is plugged in so that its pin 12 (ground) is in the pin 14 position in the 28-pin socket. Pin 26 on a 2764 is marked NC (not connected) in figure 5.46 and is connected to $+5$ V to become the 2732's V_{cc} supply.

Using an EPROM is simplicity itself. Figure 5.47 gives the timing diagram of a 27256 EPROM and figure 5.48 shows how two 27256s are connected to a 68000 CPU. An address is applied to the 27256's 15 address inputs, CS* is asserted, and OE* is forced active-low when R/$\overline{W}$ is high. We could strap OE* to ground, as the data output buffers are also enabled by CS*. Some CPUs use OE* to avoid data bus contention. The only two considerations in the design of EPROM memories are the access time calculation and the danger of data bus contention.

In a 68000 system, access time is taken care of by providing the appropriate DTACK* delay. Most EPROMs are slower than static RAM and have access times of 200–450 ns from address valid. Unfortunately, EPROMs have relatively long values of OE* high (or CS* high) to data bus floating. A typical value is 80–150 ns, which means that the EPROM is driving the data bus well into the next cycle. Care should be taken to avoid any other device driving the data bus until this time has elapsed.

Wordwide EPROM

Eight-bit EPROMs made it very easy to design 8-bit microprocessor systems with a minimum chip count, since the width of the EPROM matches that of the processor's data bus. The introduction of modern 16- and 32-bit microprocessors made it impossible to put a monitor or bootstrap loader in a single EPROM. Some processors like the 68008 have a 16-bit *internal* architecture but use an 8-bit data bus and can therefore be interfaced to a bytewide EPROM. The 68020 has a 32-bit data bus, but it is still able to use bytewide EPROMs because of its dynamic bus-sizing mechanism, which we described in chapter 4.

However, microprocessors with 16-bit data buses and no bus-sizing mechanism are forced to use two bytewide EPROMs, side by side, to create a 16-bit memory. Designing systems with pairs of EPROMs presents no insurmountable difficulty, but programming them can be moderately irritating. You have to create a binary file of

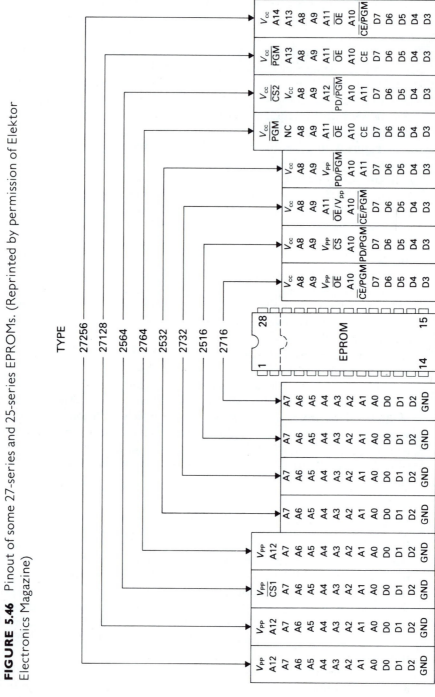

FIGURE 5.46 Pinout of some 27-series and 25-series EPROMs. (Reprinted by permission of Elektor Electronics Magazine)

FIGURE 5.47 Read cycle timing diagram of an EPROM

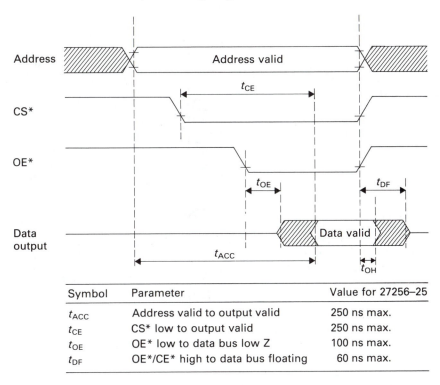

Symbol	Parameter	Value for 27256–25
t_{ACC}	Address valid to output valid	250 ns max.
t_{CE}	CS* low to output valid	250 ns max.
t_{OE}	OE* low to data bus low Z	100 ns max.
t_{DF}	OE*/CE* high to data bus floating	60 ns max.

the program to be burnt into the EPROMs and split the file into two parts—one containing even bytes and the other, odd bytes. Then you have to program two 8-bit EPROMs, one with even bytes and one with odd bytes. Finally, you have to plug the two EPROMs into the target system (taking care not to switch them over). Some of the microprocessor development systems (and EPROM programmers) designed for 16/32-bit processors now provide all the facilities needed to divide a program between two (or even four) EPROMs.

   Some semiconductor manufacturers have produced a range of 16-bit, or *word-wide*, EPROMs to simplify the design of 16-bit systems. A typical CMOS wordwide EPROM, the 27C220, has a capacity of 2M bits and is internally arranged as 128K × 16 bits. Figure 5.49 illustrates the pinout of a 27C220, which is available with an access time of 150 ns.

   An obvious question to ask is, How can a 16-bit-wide EPROM be used in a 68000 system that is byte-oriented and that employs two data strobes, one to access the upper byte and one to access the lower byte? Figure 5.50 demonstrates the structure of an interface between the 27C220 and a 68000. As you can see, it is indeed impossible to read a single byte from the EPROM. All accesses are wordwide (even if the 68000 reads just a single byte), and the EPROM is enabled by the assertion of *either* UDS* or LDS*. If the 68000 reads the byte on D_{00} to D_{07} by asserting LDS*, it does not matter that the upper byte is also placed on D_{08} to D_{15}

FIGURE 5.48   Connecting a 27256 EPROM to a 68000 CPU

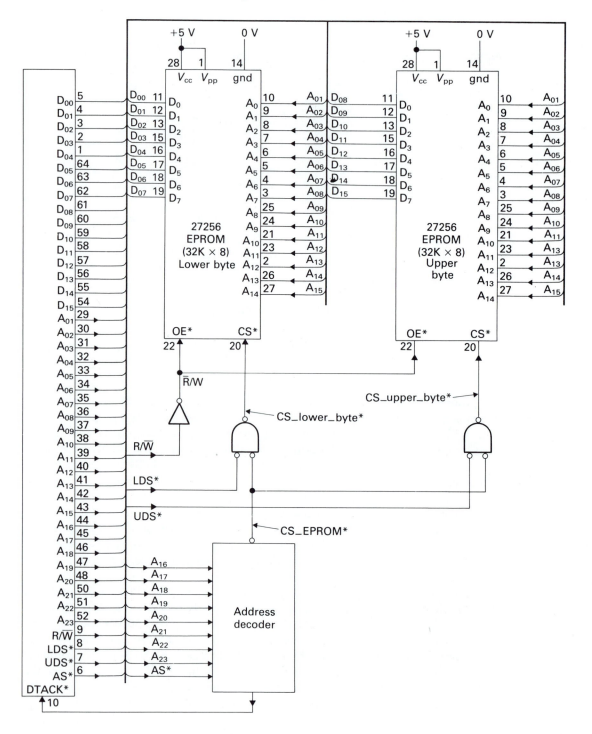

FIGURE 5.49 The 27C220 128K × 16-bit wordwide EPROM

		Pin names	
$A_0 - A_{18}$	Addresses
CE*	Chip enable
OE*	Output enable
$O_0 - O_{15}$	Outputs
PGM*	Program
N.C.	No internal connect
D.U.	Don't use

27C220

8M	4M	1M			1M	4M	8M
A_{18}	V_{pp}	V_{pp}	V_{pp} ☐1 ⌒ 40☐ V_{cc}	V_{cc}	V_{cc}	V_{cc}
CE*	CE*	CE*	CE* ☐2 39☐ PGM*	PGM*	A_{17}	A_{17}
O_{15}	O_{15}	O_{15}	O_{15} ☐3 38☐ A_{16}	NC	A_{16}	A_{16}
O_{14}	O_{14}	O_{14}	O_{14} ☐4 37☐ A_{15}	A_{15}	A_{15}	A_{15}
O_{13}	O_{13}	O_{13}	O_{13} ☐5 36☐ A_{14}	A_{14}	A_{14}	A_{14}
O_{12}	O_{12}	O_{12}	O_{12} ☐6 35☐ A_{13}	A_{13}	A_{13}	A_{13}
O_{11}	O_{11}	O_{11}	O_{11} ☐7 34☐ A_{12}	A_{12}	A_{12}	A_{12}
O_{10}	O_{10}	O_{10}	O_{10} ☐8 33☐ A_{11}	A_{11}	A_{11}	A_{11}
O_9	O_9	O_9	O_9 ☐9 32☐ A_{10}	A_{10}	A_{10}	A_{10}
O_8	O_8	O_8	O_8 ☐10 31☐ A_9	A_9	A_9	A_9
gnd	gnd	gnd	gnd ☐11 30☐ gnd	gnd	gnd	gnd
O_7	O_7	O_7	O_7 ☐12 29☐ A_8	A_8	A_8	A_8
O_6	O_6	O_6	O_6 ☐13 28☐ A_7	A_7	A_7	A_7
O_5	O_5	O_5	O_5 ☐14 27☐ A_6	A_6	A_6	A_6
O_4	O_4	O_4	O_4 ☐15 26☐ A_5	A_5	A_5	A_5
O_3	O_3	O_3	O_3 ☐16 25☐ A_4	A_4	A_4	A_4
O_2	O_2	O_2	O_2 ☐17 24☐ A_3	A_3	A_3	A_3
O_1	O_1	O_1	O_1 ☐18 23☐ A_2	A_2	A_2	A_2
O_0	O_0	O_0	O_0 ☐19 22☐ A_1	A_1	A_1	A_1
OE*/V_{pp}	OE*	OE*	OE* ☐20 21☐ A_0	A_0	A_0	A_0

(which is ignored by the 68000 during byte accesses). Similarly, reading the upper byte on D_{08} to D_{15} also puts the low-order byte on D_{00} to D_{07} (which is ignored by the 68000).

Programming EPROMs

As we have already stated, EPROMs are not programmed in their target systems but are first removed and erased under a UV light source (which takes about 20 minutes). Then they must be programmed in an EPROM programmer, which is usually a general-purpose commercial EPROM programmer. After discussing how the conventional EPROM is programmed, we shall shortly introduce two special classes of EPROM device that are well suited to programming inside the target system (the flash EEPROM and the EEPROM).

Consider the way in which a typical modern EPROM is programmed (we will use the 27C220's data). We employ the terms *typical* and *modern*, since the way in which an EPROM is programmed varies from device to device. Moreover, fast programming algorithms have been developed to keep the time taken to program a modern high-density chip within reasonable bounds. Indeed, it is the sheer range of programming algorithms that has ensured the success of the commercial EPROM programmer.

Figure 5.51 describes the interface between the 27C220 and the EPROM programmer and provides the timing diagram for a write cycle. The 27C220 EPROM is programmed by first applying the appropriate data to its data pins and the address of the location to be written to its address pins. Its V_{pp} pin is raised to 12.75 V, a low level is applied to its CE* pin, a TTL high level is applied to its OE* pin, and a low

FIGURE 5.50 Using the wordwise EPROM

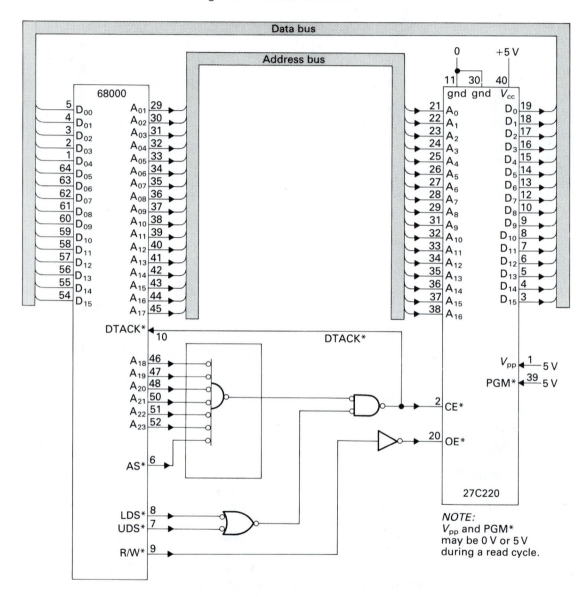

level is applied to its PGM* pin (the program pin). Note that this EPROM is programmed with the voltage at its V_{pp} pin raised from the normal TTL-compatible 5-V level to 6.25 V.

As you can see from figure 5.51, the 27C220's write cycle looks like that of a static RAM. However, the EPROM's write cycle timing parameters are far longer than those of a static RAM. These parameters illustrate another reason why EPROMs are not normally programmed in their target machines. You should also

FIGURE 5.51 The EPROM write interface and timing diagram

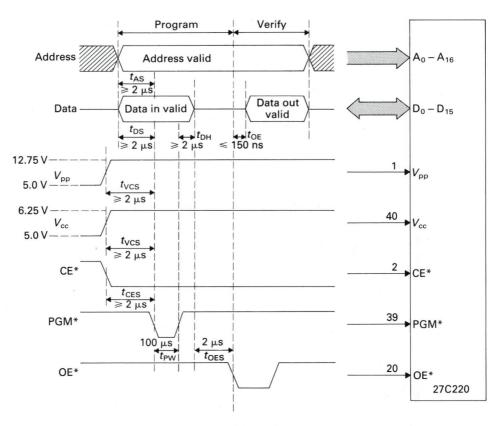

appreciate that the voltage applied to the V_{pp} pin during programming varies from device to device. First-generation devices often required a V_{pp} of 21 V.

Note that the very nature of an EPROM (i.e., its isolated floating gate separated by a very thin insulator) makes it sensitive to excessive voltages at its V_{pp} pin; that is, you can easily damage an EPROM permanently by exceeding its stated maximum V_{pp} voltage. Manufacturers recommend that the V_{cc} pins of EPROMs should be decoupled by a 4.7-μF capacitor (for every eight devices) and that a 0.1-μF ceramic capacitor should be connected between each V_{cc} pin and ground. These capacitors are needed to prevent the V_{cc} voltage from drooping when the short— but high—current pulse is taken during programming (when the programming is actually done on the card).

If you look at the timing diagram of figure 5.51 again, you will see that it also includes a read cycle after the write cycle. The read cycle is necessary to verify that the data has really been stored.

Yesterday's EPROMs had relatively small capacities and could have their entire contents modified (after first erasing them, of course) simply by writing the appropriate data to each location in turn. If you were to take the same approach with today's large EPROMs, it would take forever to program a high-capacity chip (since each

FIGURE 5.52   A typical EPROM programming algorithm

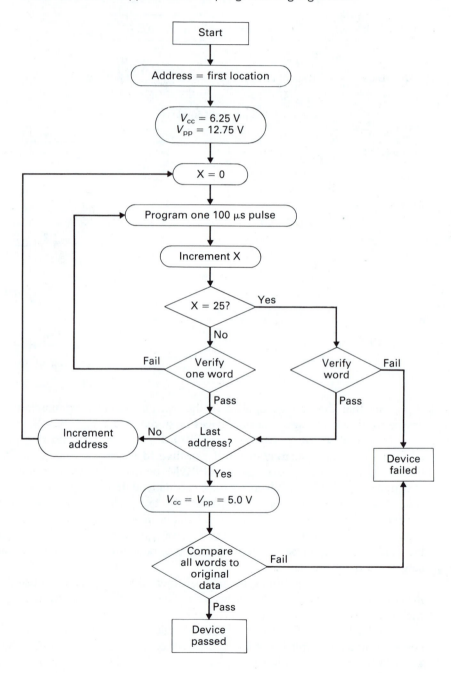

write cycle would have to be long enough to ensure correct programming in the worst case). Semiconductor manufacturers have developed algorithms that permit the programming of their EPROMs to be speeded up by a factor of 100 or so. The basis of the algorithm is the use of a short programming pulse, which is repeated until the data has been correctly written into a given location.

Figure 5.52 illustrates a typical programming algorithm. The write pulse has a duration of 100 μs, and, after each write cycle, the data is read back. If the data has been correctly stored, the next location is programmed. If the data has not been stored, another attempt at writing is made. Note that programming attempts are cumulative, as each write pulse injects more charge on to the floating gate. In this example, the maximum number of attempts is set to 25. If the data has not been correctly stored after 25 pulses, the device is deemed faulty.

Flash EEPROM

The EPROM, which has been in common use since the 1970s, suffers from two limitations: It has to be taken out of its target system to be erased and it requires non-TTL voltages to program it. As you would expect, semiconductor manufacturers have undertaken much development work to overcome their limitations. One programmable memory device closely related to the EPROM is the flash EEPROM (the acronym EEPROM is used to indicate *electrically erasable* PROM). A flash EEPROM can be erased electrically without the need for a UV source.

Figure 5.53 illustrates the structure of a conventional EPROM memory cell together with the corresponding cell of a flash EEPROM. You might be forgiven for asking what the difference between these two devices is. The difference lies in the thickness of the insulating layer (silicon oxynitride) between the floating gate and the surface of the MOS transistor. A conventional EPROM has an insulating layer that is about 300 Å thick, whereas a flash EEPROM's insulating layer is only 100 Å thick. Note that 1 Å = 1 × 10^{-9} m.

FIGURE 5.53 Structure of EPROM and flash EEPROM memory cells

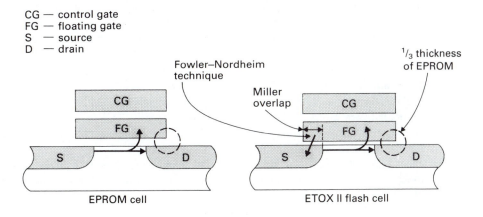

CG — control gate
FG — floating gate
S — source
D — drain

When a conventional EPROM is programmed, the charge that carries the data is transferred to the floating gate by an avalanche effect, which causes electrons to burst through the oxynitride insulating layer. These electrons are sometimes called *hot electrons* because of their high levels of kinetic energy (i.e., speed). The charge on the floating gate is removed during erasure by means of UV light, which gives the electrons enough energy to cross the insulating layer.

A flash EEPROM is programmed in exactly the same way as an EPROM (i.e., by hot electrons). However, the insulating layer in a flash EEPROM is so thin that an entirely new mechanism can be used to transport electrons across it when the chip is erased. This mechanism is known as *Fowler-Nordheim tunneling* and is a quantum mechanical effect. When a high voltage in the range of 12 to 20 V is applied across the insulating layer, electrons on the floating gate are able to tunnel through the layer, even though they do not have enough energy to cross the barrier. Indeed, an electron on one side of the barrier (i.e., the insulating layer) *disappears* and *reappears* on the other side of the barrier. Quantum mechanical effects often defy common sense. Erasing a flash EEPROM takes about 1 s.

Flash EEPROMs have a separate V_{pp} pin to supply the high voltage required for programming, exactly like conventional EPROMs, although the high voltage at the V_{pp} pin can be present at all times. It is not necessary to turn V_{pp} off when the device is not being programmed. If the V_{pp} pin is connected to ground (or even left open), the flash memory behaves exactly like a conventional EPROM (i.e., you can only read from it).

Due to the structure of flash EEPROMs, it is impossible to erase individual cells (i.e., bits). Usually, you can erase either the entire chip or just a part of the chip. The memory space of a flash EEPROM is divided into *sectors*, or *blocks*, with a typical capacity of 1024 bytes. Whether you can perform only a full chip erase or a sector erase as well as a chip erase depends on the design of the chip.

Thus, the flash EEPROM runs off a 5-V and a 12-V supply, is electrically erasable and reprogrammable, and is just a little more expensive than the UV-erasable EPROM. Unfortunately, it cannot be programmed, erased, and reprogrammed without limit. Repeated write and erase cycles eventually damage the very thin insulating layer. Some first-generation flash EEPROMs are guaranteed to perform only 100 erase/write cycles, although devices are now available with lifetimes of at least 10,000 cycles. These are *guaranteed* figures; a more realistic value is probably an order of magnitude greater. That does not mean, of course, that you should design a system that attempts to perform more than the guaranteed number of write cycles.

In the early 1990s flash EEPROM has found its niche as a method of storing large amounts of firmware that has to be updated occasionally (in situ). Flash EEPROMs are available with the same density as EPROMs. However, flash EEPROMs are more expensive than conventional EPROMs, because their die size is 20 percent larger than an EPROM with a corresponding capacity. Flash EEPROMs are cheaper than *pure* EEPROMs (which are discussed later).

The behavior of a flash EEPROM in a read cycle is exactly like that of an EPROM. Figure 5.54 provides the pinout and read-cycle timing details of a typical 1-M-bit flash EPROM.

FIGURE 5.54 Pinout and read cycle timing of IM-bit flash EEPROM

Pin names

$A_0 - A_9$	Column address input
$A_{10} - A_{16}$	Row address input
CE*	Chip enable
OE*	Output enable
WE*	Write enable
I/O_{0-7}	Data input (write) output (read)
N.C.	No internal connection
V_{pp}	Write/erase input voltage

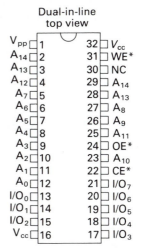

Dual-in-line
top view

Symbol	Parameter	E48F010 −200		Unit
		Min.	Max.	
t_{RC}	Read cycle time	200		ns
t_{AA}	Address to data		200	ns
t_{CE}	CE* to data		200	ns
t_{CE}	OE* to data		75	ns
t_{DF}	OE*/CE* to data float		50	ns
t_{OH}	Output hold time	0		ns

Read timing

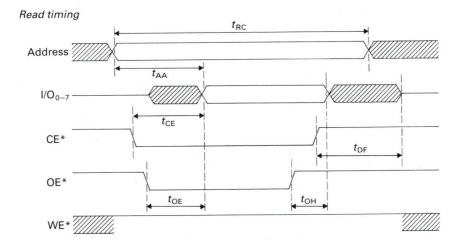

Flash EEPROMs are programmed in much the same way as the EPROMs we described earlier. They even employ programming algorithms similar to those of EPROMs to speed up programming. Their interface to the system is like that of a static RAM, and they have a conventional WE* (active-low write-enable) input. Figure 5.55 illustrates the write cycle timing diagram of a flash EEPROM. Note that the write cycle time, t_{WC}, is very long—over 100 μs. Fortunately, such a long cycle time does not place an undue burden on the systems designer. The flash EEPROM has an on-chip timer and associated control circuits that automatically ensure the appropriate signal delays without the use of external hardware. In other words, you

FIGURE 5.55   Write cycle timing of a flash EEPROM

| Symbol | Parameter | 48F010 −200 | | Unit |
		Min.	Max.	
t_{PS}	V_{pp} setup time	2		μs
t_{VPH}	V_{pp} hold time	250		μs
t_{CS}	CE* setup time	0		ns
t_{CH}	CE* hold time	0		ns
t_{OES}	OE* setup time	10		ns
t_{OEH}	OE* hold time	10		ns
t_{AS}	Address setup time	20		ns
t_{AH}	Address hold time	100		ns
t_{DS}	Data setup time	50		ns
t_{DH}	Data hold time	0		ns
t_{WP}	WE* pulse width	100		ns
t_{WC}	Write cycle time	100	150	μs
t_{AR}	Write recovery time		1.5	ms

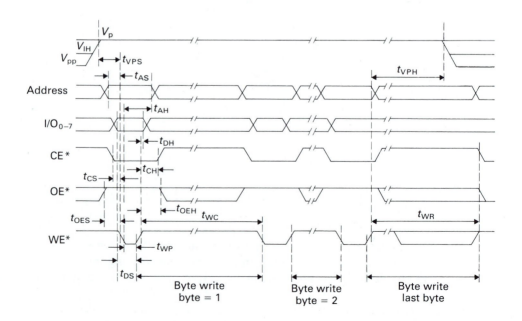

treat the flash EEPROM like a static RAM and begin a write cycle; the device itself then completes the cycle.

When a byte is written to a flash EEPROM, it is internally latched and the host processor is freed to continue other operations while the memory completes its internal operations. A byte is written to the device as if the EEPROM were a static RAM, and the write operation is repeated, typically, ten times before the data is read back and verified. Although flash EEPROMs require no special external hardware to support them, they do require careful attention to the way in which they are accessed (read, written to, and erased). Some flash EEPROMs can be programmed a byte at a time, whereas others require an entire sector (e.g., 1,024 bytes) to be written in one operation.

Like the EPROM, all the bytes of a flash EEPROM are set to $FF when it is erased. However, it should be noted that the *erase interface* (software and hardware) of flash EEPROMs varies from manufacturer to manufacturer. For example, all the bytes of a SEEQ 47F010 1M-bit flash EEPROM are erased simultaneously, whereas the SEEQ 48F010 can be programmed either to perform a *chip erase* or a *sector erase* on 1,024 bytes. An Intel 28F010, like the SEEQ 47010, has only a chip-erase mode. Some flash EEPROMs permit the programming voltage, V_{pp}, to remain at $+12$ V at all times, but others require that V_{pp} be reduced to $+5$ V to carry out certain operations. The clear message here is that flash EEPROMs have not yet reached the same level of standardization as EPROMs.

Consider the erase mode of a SEEQ 48F010. A sector erase is initiated by writing a special code byte to any location in the sector to be erased. The code byte is $FF. Think about it. Since all cells of the EEPROM contain $FF when they are erased, there is never any reason to write $FF to a cell. Consequently, writing the *dummy code* $FF to a memory cell is used to trigger a sector erase. However, the sector erase does not take place until a delay of about 200 μs has elapsed. If another write operation takes place within this 200-μs guard band, the sector erase is aborted. An abort mechanism is necessary to avoid an accidental sector erase, should the value $FF appear in the stream of bytes being written to the flash EEPROM. The sector-erase algorithm can be written in pseudocode as

```
Module Sector_erase
      Set Vpp = +12 V
      FOR I = 1 TO 10
            Wait 2 μs
            Write $FF to any location in the sector
            Wait 500 ms
      END_FOR
      Wait 250 ms
      Verify erasure
End Sector_erase
```

The *entire* SEEQ 48F010 can be erased by first writing any data value to any address with V_{pp} set to $+5$ V or less (this called a *chip-erase select operation*), setting V_{pp} to $+12$ V, waiting a delay, writing $FF to any location, and waiting a further delay. This process is repeated 24 times. The 48F010's chip-erase algorithm can be expressed in pseudocode form as

```
Module Chip_erase
        FOR I = 1 TO 24
                Set V_pp = +5 V
                Write $FF to any address
                Set V_pp = +12 V
                Wait for 2 μs
                Write $FF to any address
                Wait for 500 ms
        END_FOR
        Wait for 250 ms
        Verify all bytes = $FF
End Chip_erase
```

An Intel 28F010 128K × 8-bit flash EEPROM operates in a rather different way to the 48F010. The 28F010 employs a *command register* to initiate the various operations that can be carried out by the chip. The command register is accessed by executing a write operation to any location within the device. You program the 28F010 to perform a particular operation by first writing the appropriate command code to the command register and then performing the intended operation. Some of the command codes are illustrated next.

OPERATION	COMMAND CODE
Read memory	$00
Read device code	$90
Set up erase	$20
Erase verify	$A0
Set up program	$40
Program verify	$C0
Reset	$FF

The code $00 is written to the 28F010 to put it in its normal read mode. Note that once $00 has been written to the 28F010, the flash EEPROM is available for read cycles until it receives another command. The 28F010 automatically loads the read command into its control register during the V_{pp} power-up phase.

The *read device code* command is carried out by writing $90 to the command register and then reading the contents of locations $00 and $01. Location $00 returns the manufacturer's code and $01 returns the device code. This mechanism makes it possible to identify a Flash EEPROM automatically. The SEEQ 48F010 has a similar (but less convenient) identification mechanism called a *silicon signature*, which is activated by elevating the voltage at the A_9 pin to 12 V.

The 28F010 is fully erased by first writing $20 to its command register to trigger a *set up erase* command. This command is followed by a second write of $20. As in the case of the 48F010, the 28F010 uses a double write mechanism to reduce the probability of an accidental erase operation. Of course, an erase operation is not possible unless V_{pp} is raised to its high level (i.e., 12 V).

The *set up program* operation is used to prepare the 28F010 for programming. After $40 has been written into the command register, the next write operation stores the data on the data bus in the location on the address bus. Once a byte has been written, the *program verify* command is used to read back the data. That is, the programmer loads $C0 in the command register and then reads the byte just programmed. When the byte is read back during a program verify operation, it is the last byte programmed; the contents of the address bus are ignored during a program verify. Note that you must perform a verify operation after you write a byte—the verify command is not optional!

The reset command, which is initiated by writing $FF to the command register, aborts any write or erase operations. You cannot read the memory again until you have issued a read command.

We can express the 28F010's programming algorithm in pseudocode form as

```
Module 28F010_program
        Set V_pp to +12 V
        REPEAT
                Write_attempts = 0
                REPEAT
                        Write_attempts := Write_attempts + 1
                        Write set up command {write $40}
                        Write the data to appropriate address
                        Wait 10 μs
                        Write program verify command {$C0}
                        Wait 6 μs
                        Read data from device
                UNTIL Data valid OR Write_attempts = 25
                IF Write_attempts = 25 THEN EXIT to ERROR
                Select next address
        UNTIL device programmed
        Set V_pp to +5 V
End 28F010_program
```

Using Flash EEPROMs

It is not difficult to use a flash EEPROM, as long as you pay careful attention to the specific software required to perform read and write operations. Figure 5.56 illustrates a possible interface between a SEEQ 48F010 128K-byte flash EEPROM and a 68000 microprocessor. The 48F010 supports a sector-erase and a sector-write mechanism (i.e., you cannot write a single byte). The similar 47F010 flash EEPROM supports the writing of a single byte but not a sector erase (you must erase the entire chip).

As you can see from figure 5.56, the 48F010 looks, to the processor, very much like any other EPROM. In this example, we have employed a switchable V_{pp} supply (taken from SEEQ Tech Brief 29). There are three reasons why we might wish to make V_{pp} switchable. The first reason is that some flash EEPROMs require V_{pp} to be reduced to V_{cc} for certain operations. The second reason is that turning off V_{pp} makes it impossible to erase a flash EPROM accidentally. Finally, the specification sheets of some flash EEPROMs state that V_{pp} should not be set to 12 V until after V_{cc} has become established.

The V_{pp} supply of figure 5.56 is derived from a +12.5-V source and fed to the flash EEPROM's V_{pp} pin via transistor switch T_1. When the control signal, V_{pp}ON, is

FIGURE 5.56   Circuit of interface between a 68000 and a flash EEPROM

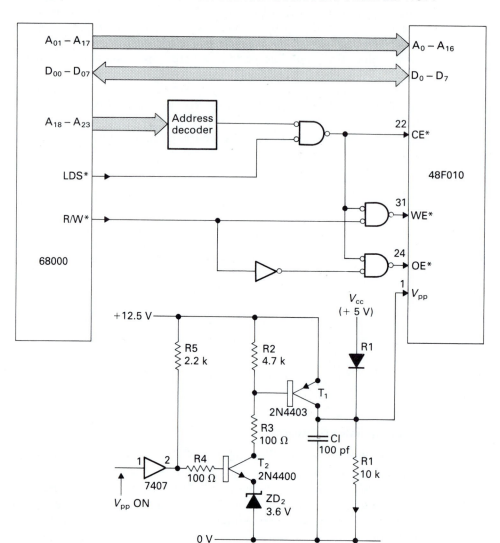

high, transistor T_2 is turned on and a current flows through the chain R_2, R_3, T_2, and ZD_2. The function of the 3.6-V zener diode between T_2's emitter and ground is to ensure that T_2 cannot be turned on until its base rises to over approximately 4 V. When T_2 is turned on, the current flowing through R_2 forward-biases the base-emitter junction of T_1, which causes T_1 to turn on and provide the flash EEPROM with V_{pp} at 12.5 V (less T_1's collector-emitter saturation voltage).

The function of capacitor C_1 is to smooth the rise and fall of V_{pp}. Resistor R_1 provides a discharge path for C_1 when V_{pp} is turned off.

When V_{pp}ON is forced low, both T_1 and T_2 are turned off and the 12.5-V supply is removed from the V_{pp} terminal. Since the V_{pp} terminal is connected to 5 V via diode D_1, which becomes forward biased when V_{pp} drops below 5 V, the voltage at the V_{pp} terminal does not fall below about 4.4 V.

We can erase a sector of the 48F010 and then rewrite it by means of the following algorithm, expressed in pseudocode. As you can see, the sector is erased 24 times and then verified. The sector is written 7 times and then verified. If a faulty byte is found during verification it is rewritten 5 times. If, after six cycles of verification, all bytes have not been correctly written, the device is assumed to be faulty.

```
Module 48F010_sector_erase_write_algorithm

        Set Vpp to 12 V
        Wait 2 µs
        FOR I = 1 TO 24 {perform 24 sector erases}
            Write $FF to any address in sector
            Wait 500 ms
        END_FOR
        Wait 250 ms
        Success = true
        FOR I = 0 to 1023 {Verify sector erase}
            IF Chip(I) ≠ $FF THEN Success = false
        END_FOR
        IF Success = false THEN EXIT_fail

        Set Vpp to 12 V
        Wait 2 µs
        FOR I = 0 TO 7 {perform seven sector writes}
            FOR J = 0 TO 1023 {write a sector}
                Write byteⱼ to addressⱼ
                Wait 75 µs
            END_FOR
        END_FOR
        Wait 1.5 ms

        FOR I = 1 TO 6 {perform six sector verifies}
            Verify = true
            FOR J = 0 TO 1023 {read and verify a sector}
                IF byteⱼ not correctly written THEN
                                        Verify = false
                                        FOR K = 1 TO 5
                                            Rewrite byteⱼ
                                            Wait 75 µs
                                        END_FOR
                                        Wait 1.5 ms
            END_FOR
            IF Verify = true THEN EXIT_success
        END_FOR
        EXIT_fail
End 48F010_sector_erase_write_algorithm
```

Note that the delays in this algorithm are normally provided in software by the writer of the algorithm.

EEPROM

The electrically erasable and reprogrammable ROM is similar to its cousin, the flash EEPROM. In fact, the major difference between the EEPROM and the flash EEPROM is that the flash EEPROM uses Fowler-Nordheim tunneling to erase data and hot electron injection to write data, whereas pure EEPROMs use the tunneling mechanism both to write and erase data. Table 5.11 illustrates the difference between the three programmable devices we have discussed in this section.

EEPROMs (or E²PROMs as they are sometimes called) are more expensive than flash EEPROMs and generally have smaller capacities. The size of the largest state-of-the-art flash memory is usually four times that of the corresponding EEPROM. However, modern EEPROMs do run from single 5-V supplies and are rather more versatile than flash EEPROMs. Like the flash memory, they are *read-mostly* devices, with a lifetime of 10,000 erase/write cycles. Modern EEPROMs have fast access times (as low as 35 ns) but still have long write cycle times (10 ms).

We will now discuss the operation of a second-generation EEPROM, the 1024K-bit 28C010. Here, the term *second generation* is used to indicate a high-density, single 5-V, high-functionality device (in contrast with first-generation EEPROMs). Since EEPROMs use a tunneling mechanism to write data, a write cycle takes longer than one involving hot electron injection. Some EEPROMs require that their control signals be active for the duration of the write cycle, thereby increasing the complexity of the interface. The 28C010 has an on-chip timer that takes care of all timing during write and erase operations.

A read cycle to a 28C010 is exactly like a read cycle to any other EPROM. An address and data is applied to the chip; it is enabled (CE* low); its output enable (OE*) is asserted low, and its write enable (WE*) is negated high. The 28C010-120 has an access time of only 120 ns from address valid.

The 28C010 automatically performs a byte-erase operation before a new byte is written. Consequently, the 28C010 looks exactly like a static RAM as far as the programmer is concerned (since you do not have to erase data before writing it). A byte is written by providing the chip with an address and data, asserting CE* and WE*, and negating OE*. Since the 28C010 latches its inputs during a write cycle,

TABLE 5.11 EPROM differences

DEVICE	EPROM	FLASH EEPROM	EEPROM
Normalized cell size	1.0	1.0–1.2	3.0
Programming mechanism	Hot electron injection	Hot electron injection	Tunneling
Erase mechanism	UV light	Tunneling	Tunneling
Erase time	20 min	1 s	5 ms
Minimum erase	Entire chip	Entire chip (or sector)	Byte
Write time (per cell)	$< 100 \ \mu s$	$< 100 \ \mu s$	5 ms
Read access time	200 ns	200 ns	35 ns

the processor interface must provide a stable address on the falling edge of the WE∗ or CE∗ inputs (whichever goes low last) and valid data on the rising edge of CE∗ or WE∗ (whichever goes high first).

If you attempt to perform a read access to the 28C010 after a write cycle has begun but has not been completed, the EEPROM will return the complement of bit d_7 of the last byte written (irrespective of the location being read). This behavior means that you can confirm that a byte has been written by reading it back (i.e., data polling) until the value of d_7 returned is the same as that stored. Figure 5.57 illustrates the read and write cycle timing diagrams of the 28C010, together with an interface to the 68000.

In order to speed up average write times, the 28C010 has a 128-byte page buffer. The programmer can choose a page (selected by the state of address lines A_7–A_{16}) and then write to any of the 128 bytes on this page. If the programmer loads bytes into the page buffer with a delay of no longer than 200 μs between consecutive bytes, the 28C010 saves them in its page buffer and does not store them in its array. Once the page buffer has been loaded (with from 1 to 128 bytes) and a delay of over 200 μs has elapsed, the contents of the buffer are copied into the array. The effect of the page buffer is to speed up the average write time of the EEPROM.

The 28C010 demonstrates one of the interesting developments in memory technology—the intelligent memory. By *intelligent* we mean the ability of a memory to perform operations other than simple read and write cycles. For example, it is possible to force the 28C010 into a read-only mode to prevent accidental or malicious reprogramming (or even erasure during a faulty power-up or power-down sequence). The intelligent mode is activated (as in the case of the flash EEPROM) by means of a sequence of operations that are highly unlikely to occur in practice. Several bytes must be loaded into certain locations with less than a 200-μs delay between consecutive loads, if the command is to be recognized. For example, the 28C010 is put into its read-only mode (and taken out of the read-only mode) by executing the following sequences:

ACTIVATE READ-ONLY MODE	DEACTIVATE READ-ONLY MODE
1. Write $AA to address $5555.	1. Write $AA to $5555.
2. Write $55 to address $2AAA.	2. Write $55 to $2AAA.
3. Write $A0 to address $5555.	3. Write $80 to $5555.
	4. Write $AA to $5555.
	5. Write $55 to $2AAA.
	6. Write $20 to $5555.

Other operations controlled by writing special sequences to the EEPROM are chip erase (i.e., erase the entire contents of the chip) and disable auto-erase (i.e., the automatic erase that takes place before each byte is written). These facilities can speed up the average operation of the chip by reducing the write cycle time.

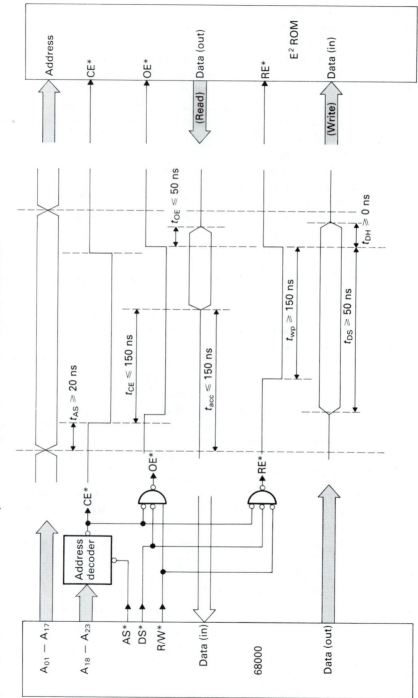

FIGURE 5.57 Read and write cycles of the 28C010 EEPROM and its interface

The use of special data sequences to force a memory to carry out certain operations is an interesting trend. There is nothing to stop a designer from building any arbitrary level of intelligence into a memory. For example, you could design a memory that would look for the occurrence of a string within its contents. At the same time, the user must be careful not to generate any of these control sequences by accident. We might argue that the probability of writing three specific bytes to three specific addresses occurring randomly is 1 in $(2^8 \times 2^{16})^3 = 1$ in 2^{72}. This value is unimaginably small. However, we might still worry about being in an aircraft controlled by such a memory. Imagine that, due to congestion, air-traffic control asked a pilot to carry out a long series of maneuvers and that this series triggered a memory's auto-erase function.

Although flash EEPROMs need a high voltage to induce the tunneling required to erase data stored on the floating gate, some manufacturers supply devices that operate from a single 5-V supply. The required high voltage is generated on-chip by an oscillator and a charge pump.

SUMMARY

A situation in which a memory component could simply be bolted onto a microprocessor to create a microcomputer (apart from I/O devices) would be very convenient. We have demonstrated that interfacing memory components to a microprocessor is, in fact, no trivial task. The designer has to construct an address decoder to map the memory space of address components onto the address space of the microprocessor. An address decoder can be designed using the address decoding techniques and devices introduced at the start of this chapter. The actual circuit employed must take account of the type of memory map required by the processor and whether or not flexibility has to be built into the system.

We have described the characteristics of the static memory component and in chapter 7 look at its more complex counterpart, the dynamic RAM. Anyone wishing to connect memory components to a microprocessor must be aware of the two important design considerations discussed here. The first is the relationship between the timing characteristics of the memory component and the timing requirements of the microprocessor. The second is the danger of data bus contention that arises when the memory and the microprocessor both attempt to drive the data bus at the same time.

A memory component of particular interest described in this chapter is the CMOS static RAM, which is able to enter a power-down mode in which it retains data while consuming virtually no power. This ability to store data when the microcomputer is powered down can be used to hold system data when the microcomputer is switched off.

Recent innovations like the EEPROM and the flash EEPROM were also described. These types of nonvolatile memory device will continue to evolve and will have an ever-increasing impact on the design of microprocessor systems (particularly on diskless systems).

Problems

1. Design address decoding networks to satisfy the following memory maps:

   a. RAM1   00 0000–00 FFFF   b. ROM1   00 0000–00 3FFF
   RAM2   01 0000–01 FFFF   ROM2   00 4000–00 7FFF
   I/O_1   E0 0000–E0 001F   ROM3   00 8000–00 BFFF
   I/O_2   E0 0020–E0 003F   RAM   04 0000–07 FFFF
                                                   I/O   08 0000–08 00FF

   Compare and contrast the various ways of implementing these address decoders.

2. A manufacturer designs a single-board computer with eight pairs of bytewide EPROMs to hold system firmware. The designer decides to cater for three EPROM sizes: 4K × 8, 8K × 8, and 16K × 8. Design an address decoder that will allow the size of each EPROM to be user selectable by means of jumpers on the PCB.

3. A 68000 system is to have up to eight 256K word pages of read/write memory, up to eight 8K word pages of EPROM, and up to eight 128 word pages of I/O space. The designer wishes to make the 68000 memory map *user definable* under software control. Therefore, the address from the CPU must be compared with addresses set up by the user in order to generate the necessary device-select signals. Design an arrangement that will do this. Note that the address decoding network itself is to be permanently mapped at the address $FF FFE0 and that, following a reset, the first page of ROM is mapped at $00 0000–$00 3FFF.

4. A region of up to 16K bytes of the 68000's memory space, whose lowest address is $80 0000, is to be devoted solely to the 68000's supervisor stack. Design an address decoder to implement this arrangement (using 8K × 8 static RAMs). You are to make maximum use of the 68000's function code outputs in this address decoder.

5. What is meant by the terms partial address decoding and full address decoding?

6. What are the advantages and the disadvantages of the PROM as an address decoder in comparison with decoders constructed from conventional TTL and MSI logic elements?

7. Why are *bytewide* memory components frequently preferred by the designers of small memory systems and *bitwide* memory components preferred by the designers of large memory systems?

8. How is an EPROM able to store data in the absence of electrical power?

9. Why are address decoders not located within the microprocessor itself?

10. What is battery-backed memory? What are its advantages and disadvantages in comparison with members of the EPROM family?

11. The 68000 does not have either I/O memory space or special I/O instructions. Discuss the advantages and disadvantages of memory-mapped I/O in comparison with I/O using dedicated I/O memory space.

12. What are the criteria by which address decoders are judged? That is, what factors does an engineer take into account when selecting a particular address decoder for use in a microprocessor system?

13. What does *primary address range* mean when it is applied to a system using partial address decoding?

14. A 68000 system uses two 512K-bit EPROMs to create a 16-bit-wide block of memory that is selected when $A_{23} = 1$, $A_{22} = 1$, $A_{21} = 0$, $A_{20} = 1$, and $A_{19} = 0$. What is the primary address range of this block of EPROM?

15. How can a memory's OE∗ input be used to reduce the dangers of bus contention?

16. A microcomputer designer decides to implement the simplest possible partial address decoder. Three 1M-byte blocks of memory are arranged so that the first block is selected whenever $A_{23} = 1$, the second when $A_{22} = 1$, and the third when $A_{21} = 1$. This decoder is so simple that it is nonexistent. The designer simply connects the appropriate address line (i.e., A_{21} or A_{22} or A_{23}) to the memory block's CS∗ input via an inverter. Would this arrangement work? What, if any, are the dangers inherent with this system?

17. What are the fundamental differences among EPROM, flash EEPROM, and E²PROM?

EXCEPTION HANDLING AND THE 68000 FAMILY

In this chapter we examine two closely related topics: interrupts and exceptions. Most readers will already be familiar with the interrupt, which is a specific instance of the more general exception. As its name suggests, an exception is an event that alters the normal execution of a program. We are going to describe how the 68000 implements exceptions in general and interrupts in particular. As we shall see, exception handling is the one topic that most intimately combines the software and hardware aspects of a microprocessor.

An *interrupt* is a message to the CPU from an external device seeking attention. Such devices are frequently I/O peripherals of the type to be described in later chapters (e.g., the serial and parallel interface). Part of this chapter shows how the 68000 family implements its versatile interrupt-handling mechanism and how it uses interrupts. We look at the interrupt from both hardware and software standpoints. Other exceptions to be described range from bus errors caused when the processor fails to complete a memory access to software errors caused by, for example, an attempt to execute an illegal instruction. We complete this chapter by briefly examining the relationship between exceptions and real-time systems.

Since any discussion of exceptions covers several distinct but interrelated topics, we provide a short overview of exception handling to set the scene before we look at exceptions in detail.

6.1 OVERVIEW OF 68000 FAMILY EXCEPTION-HANDLING FACILITIES

Readers might notice an element of repetition in this chapter. This repetition is intentional, since, in many ways, it is more difficult to write about the 68000's exception-handling facilities than any of the 68000's other attributes. The difficulty stems from the way in which all aspects of exception handling are interrelated—you cannot describe one aspect without referring to the others. Moreover, exceptions bridge three important components of a microcomputer: the hardware, the software, and the operating system.

Microprocessors belonging to the 68000 family operate in one of two states: user or supervisor. A microcomputer runs its operating system and its utilities in the supervisor mode and runs its user programs in the user mode. One of the main differences between these two modes is that each mode has its own stack pointer (the SSP in the supervisor mode and the USP in the user mode). Consequently, if a user program corrupts its stack, it does not cause the processor to crash, since the operating system SSP is protected from the user. That is, the user program *itself* crashes, but the operating system remains unaffected.

Interrupts and exceptions are really calls to the operating system and share many of the characteristics of the subroutine. We shall use the term *operating system* throughout this chapter to include both real operating systems and system service routines in embedded systems without an operating system as such. These operating system calls may be made *explicitly* by the programmer because he or she wishes to employ an operating system facility (e.g., output), or they may be made *automatically* by the 68000 in response to certain types of software or hardware error (e.g., an attempt to execute a nonvalid instruction code or an attempt to access an address at which no memory exists). Some writers divide exceptions into two classes: *Internal* exceptions are those generated by the execution of instructions and *external* exceptions are those caused by actions taken by hardware outside the 68000 chip.

Each type of exception has its own *exception handler* that is responsible for dealing with the recovery from the event that caused the exception. Exception handlers are normally part of the operating system and are written by the systems programmer (rather than the applications programmer). Unlike the subroutine, an exception does not require an explicit address, since the sequence of actions taking place when an exception is triggered is *predetermined* by the designers of the microprocessor. Let us explain what we mean by this. If we call a subroutine to perform an input operation, we might write BSR INPUT, but if we call an exception handler to deal with arithmetic overflow, we might write TRAPV (i.e., trap on overflow). As you can see, the TRAPV requires no address. The TRAPV instruction causes the 68000 to look for the *address* of the interrupt handler in a special table called an *exception vector table*.

The systems programmer must put the address of the actual TRAPV exception handler at the appropriate place in the exception vector table. Similarly, systems programmers have to write the actual exception handlers themselves. We emphasize the term *systems* programmer because exception handlers are the province of the designer of the operating system. Note that the term *exception processing* is used to describe the 68000's response to an exception, and *exception handling* is used to describe what the user does in response to an exception.

When a user program is running and it encounters an exception, it forces the 68000 into its supervisor mode, and the operating system deals with the exception. Of course, a supervisor mode program remains in the supervisor state when it encounters an exception. One important application of exceptions is in implementing input/output operations. For example, the user programmer may force a software exception, called a TRAP, when he or she wishes to perform I/O. By ensuring that all input and output transactions are carried out by the operating system, we make it possible to control the way in which users access I/O devices. For example, programmers are forced to perform I/O in a consistent fashion and should not be able to access devices directly.

As we have already stated, once an exception is accepted by the 68000, it is dealt with by the appropriate exception handler routine. After the exception has been dealt with, a return from exception instruction, RTE (rather like an RTS instruction), restores the processor to the state it was in before the exception. As in the case of subroutines, exceptions can be nested to an arbitrary level. However, interrupts are prioritized and an interrupt with a high priority may not be interrupted by one with a lower priority.

Although chapter 7 covers memory management and virtual memory systems, we must at least mention these topics, since they are intimately connected with exception processing. All we need say here is that virtual memory permits *logical addresses* generated by programs to be mapped onto the *physical address* space of the real memory components. If a logical address is generated and the operand being accessed is currently on disk, the operating system intervenes and moves a block of data from disk to random access memory. Virtual memory systems mean that programmers do not have to worry about where data is to go in memory and even whether the data is in random-access memory or on a high-speed hard disk. Virtual memory goes hand in hand with the 68000 family's user and supervisor modes and exception-handling facilities. Memory management and exception processing are related because the memory management hardware generates a special type of interrupt called a *bus error* to initiate the transfer of data between RAM and disk.

The 68000 supports exception handling and virtual memory to a certain extent, but it suffers from some limitations. The 68010 is an enhanced version of the 68000 with almost exactly the same instruction set but with an ability to support true virtual memory systems. From the point of view of exception handling, the 68020 has all the features of the 68010 plus a few new enhancements. In this chapter we will describe the 68000 and indicate any special exception-handling facilities implemented by the 68010, the 68020, and the 68030 wherever necessary.

Exception handling involves a number of interrelated topics, the following provides a guide to the topics covered in this chapter. We begin by describing some of the characteristics of the interrupt, since most readers will be familiar with this topic. We then provide an overview of how the 68000 family goes about processing interrupts and look at the 68000's user and supervisor modes. Following this, we describe specific 68000 exceptions and describe in greater detail how they are processed. In the middle of the chapter we look at the relationship between hardware and external exceptions, such as interrupts and bus errors. Finally, we look at the relationship between real-time systems and exceptions.

6.2 INTERRUPTS AND EXCEPTIONS

A computer executes the instructions of a program sequentially unless a jump or conditional branch modifies their order or unless a subroutine is called. In such cases, any deviation from the sequential execution of instructions is determined by the programmer. Deviations caused by conditional branches or subroutines can be said to be *synchronous*, because they occur at predetermined points in the program. Under certain circumstances this arrangement is very inefficient. Suppose a microprocessor is reading data from a keyboard at an average rate of 250 characters per minute, corresponding to approximately 4 characters per second. In a typical 68000 system, the processor reads the status of a memory-mapped peripheral to determine whether or not a key has been pressed. If no key has been pressed, a branch is made back to the instruction that reads the status of the peripheral, and the cycle continues until a key is pressed. The following program shows how this is done.

```
KEY_STATUS    EQU      $F00001           Location of input status register
KEY_VALUE     EQU      KEY_STATUS+2      Location of input data register
              LEA      KEY_STATUS,A0     A0 points to key_status
              LEA      KEY_VALUE,A1      A1 points to key_value
TEST_LOOP     BTST     #0,(A0)           REPEAT
              BEQ      TEST_LOOP             Test least significant bit of status
              MOVE.B   (A1),D1           UNTIL least significant bit = 0; Read the data
```

The two instructions BTST #0,(A0) and BEQ TEST_LOOP constitute a *pollling loop*, which is executed until the least significant bit of the status byte is true, signifying that the data from the keyboard is valid.

These two instructions take 20 clock cycles to execute, requiring 2 μs with a 10-MHz clock. Thus, for each key pressed, the polling loop is executed approximately 100,000 times. Quite clearly, this use of a microprocessor is grossly inefficient. In some applications of the microprocessor, the time wasted in executing a polling loop is of little significance, because the computer has nothing better to do with its time. When the user is sitting at the keyboard of a personal computer thinking about the next word to enter, it is of no consequence that the CPU is patiently asking the keyboard if it has a new character every 2 μs or so.

More sophisticated applications of computers cannot afford to let the CPU waste its time executing a polling loop. These computers have many tasks to perform and can always find something useful to do. A queue of programs may be waiting to be run, or some peripheral may need continual attention while another program is being run, or there may be a background task and a foreground task. A technique for dealing more effectively with input/output transactions has been implemented on all microprocessors and is called an *interrupt-handling mechanism*. An *interrupt request line*, IRQ, is connected between the peripheral and the CPU. Whenever the peripheral is ready to take part in an input/output operation, it asserts the IRQ line and invites the CPU to deal with the transaction. The CPU is free to carry out other tasks between interrupt requests from the peripheral.

An interrupt is clearly an *asynchronous* event, because the processor cannot know at which instant a peripheral such as a keyboard will generate an interrupt. In other words, the interrupt-generating activity (i.e., keyboard) bears no particular timing relationship to the activity the computer is carrying out between interrupts. When an interrupt occurs, the computer first decides whether to deal with it (i.e., to service it) or whether to ignore it for the time being. If the computer is doing something that must be completed, it ignores interrupts. The time between the CPU receiving an interrupt request and the time at which it responds is called the *interrupt latency*. Should the computer decide to respond to the interrupt, it must carry out the following sequence of actions:

1. The computer completes its current instruction. All instructions are *indivisible*, which means they must be executed to completion before the 68000 responds to an interrupt. In theory, it is perfectly possible to design a microprocessor that could be interrupted in the middle of an instruction, with the remainder of the instruction being completed after the interrupt has been processed.

2. The contents of the program counter are saved in a safe place, in order to allow the program to continue from the point at which it was interrupted. The program counter is almost invariably saved on the stack so that interrupts can, themselves, be interrupted without losing their return addresses.

3. The state of the processor (e.g., the 68000's status word) is saved on the stack as well as the PC. Clearly, it would be unwise to allow the interrupt service routine to modify, say, the value of the carry flag, so that an interrupt occurring before a BCC instruction would affect the operation of the BCC after the interrupt had been serviced. In general, the servicing of an interrupt should have no effect whatsoever on the execution of the interrupted program. This statement applies to hardware interrupts rather than exceptions whose origin lies in the software.

4. A jump is then made to the location of the interrupt-handling routine, which is executed like any other program. After this routine has been executed, a return from interrupt is made, the program counter is restored, and the system status word is returned to its preinterrupt value.

The interrupt is *transparent* to the interrupted program, since the processor is returned to the state in which it was before the interrupt took place. Before we examine the way in which the 68000 deals with interrupts, it is worth mentioning some of the key concepts used in any discussion of interrupts and exceptions.

Nonmaskable Interrupts

An interrupt request is so called because it is a *request* and, therefore, carries the implication that it may be denied or deferred. Whenever an interrupt request is deferred, it is said to be *masked*. Sometimes it is necessary for the computer to respond to an interrupt no matter what it is doing. Most microprocessors have a special interrupt request input, called a *nonmaskable* interrupt request (NMI). This interrupt cannot be deferred and must always be serviced. A nonmaskable interrupt is normally reserved for events such as a loss of power. In this case, a low-voltage detector generates a nonmaskable interrupt as soon as the power begins to decay. The NMI handler routine forces the processor to deal with the interrupt and to perform an orderly shutdown of the system before the power drops below a critical level and the computer fails completely. The 68000 has a single, level-seven, nonmaskable interrupt request called IRQ7*.

Prioritized Interrupts

An environment in which more than one device is able to issue an interrupt request requires a mechanism to distinguish between an important interrupt and a less important one. For example, if a disk drive controller generates an interrupt because it has some data ready to be read by the processor, the interrupt must be serviced before the data is lost and replaced by new data from the disk drive. On the other hand, an interrupt generated by a keyboard interface probably has from 200 ms to several seconds before it must be serviced. Therefore, an interrupt from a keyboard can be forced to wait if interrupts from devices requiring immediate attention are pending.

For the preceding reasons, microprocessors often support *prioritized* interrupts. Each interrupt has a predefined priority, and a new interrupt with a priority lower than or equal to the current one cannot interrupt the processor until the current interrupt has been dealt with. Equally, an interrupt with a higher priority can interrupt the current interrupt. The 68000 provides seven levels of interrupt priority.

Vectored Interrupts

A *vectored interrupt* is one in which the device requesting the interrupt automatically identifies itself to the processor. Typical 8-bit microprocessors lack a vectored interrupt facility and have only a single interrupt request input (IRQ*). When IRQ* is asserted, the processor recognizes an interrupt but not its source. The processor's interrupt-handling routine must examine, in turn, each of the peripherals that may have initiated the interrupt. To do this, the interrupt-handling routine interrogates an interrupt status bit associated with each of the peripherals.

More sophisticated processors have an interrupt acknowledge output line (IACK) that is connected to all peripherals. Whenever the CPU has accepted an interrupt and is about to service it, the CPU asserts its interrupt acknowledge output. An interrupt acknowledge from the CPU informs the peripheral that its interrupt is about to be serviced. The peripheral then generates an identification number and puts it on the data bus, allowing the processor to calculate the address of the interrupt-handling routine appropriate to the peripheral. This type of interrupt is a vectored interrupt. The 68000 provides the designer with both vectored and nonvectored interrupt facilities.

Now that we have introduced the hardware interrupt, the next step is to examine the more general form of interrupt, the exception.

Exceptions

Exception is a word that has trickled down from the world of the mainframe computer. Like an interrupt, an exception is a deviation from the normal sequence of actions carried out by a computer. Interrupts are asynchronous exceptions, initiated by hardware. Some processors, such as the 68000, support other types of exceptions originating from errors detected by the system hardware. For example, an exception can be generated when the processor tries to read data from memory and something goes wrong, such as an attempt to read from nonexistent memory or the detection of a memory error.

In addition to exceptions raised by external hardware, there are exceptions initiated by software. Some are related to errors, such as an attempt to execute an illegal operation code. Others are actually generated by the programmer. For example, the TRAP instruction acts like a hardware interrupt in the sense that it invokes an interrupt-handling procedure. We will see later that a TRAP (as well as other exceptions) can be used to implement special instructions not normally part of the processor's instruction set. Figure 6.1 illustrates the exceptions implemented by the 68000 family. The characteristics and applications of these exceptions are the subject of this chapter.

The 68000 deals with both hardware and software exceptions in a consistent and *logical* fashion. The exception-handling facilities of the 68000 are among the main factors lifting it out of the world of the 8-bit processor. We will now describe what happens when the 68000 processes an exception.

Brief Overview of 68000-Family Exception Processing

Once the 68000 has accepted an interrupt or exception, the processor treats it rather like a subroutine by saving the program counter on the stack. All exceptions save the status register as well as the PC, and some exceptions save a considerably larger amount of information on the stack (indeed, the actual amount of information saved on the stack also depends on the whether the processor is a 68000, a 68020, etc.).

Each exception is associated with a unique longword location in the *exception vector table*. The processor identifies the type of exception, reads the longword in the vector table, and loads it into the program counter. In other words, the 68000 executes an indirect jump to the exception-handling routine via the exception vector table. Once the exception-handling routine has been completed, an RTE (return from exception) instruction retrieves the information saved on the stack, and the 68000 continues processing normally. Interrupts are treated in a slightly different way. When the 68000 is about to respond to an interrupt, it asks the interrupting device for an identification number and then uses this number to select the desired address in the exception vector table.

The way in which the 68000 processes exceptions can be represented by the following pseudocode. Some of the details are discussed more fully later in this chapter.

```
Module Process_exception
   BEGIN
      [TemporaryRegister] ← [SR]    {save status register in temporary location in CPU}
```

```
              S ← 1                              {force supervisor mode}
              T ← 0                              {turn off trace mode}
              Get VectorNumber                   {calculate exception type number}
              Address ← VectorNumber × 4         {exception vector is at four × vector number}
              Handler ← [M(Address)]             {read table to get address of exception handler}
              [SSP] ← [SSP] − 4                  {predecrement stack pointer}
              [M([SSP])] ← [PC]                  {push program counter}
              [SSP] ← [SSP] − 2                  {predecrement stack pointer}
              [M([SSP])] ← [TemporaryRegister]   {push status register}
              [PC] ← Handler                     {call exception handler}
          END
            ⋮                                    {the exception handler processes the exception}
              ProcessException                   {the handler is provided by the user}
            ⋮
          BEGIN                                  {begin the return from exception sequence—RTE}
              [SR] ← [M([SSP])]                  {restore old SR}
              [SSP] ← [SSP] + 2
              [PC] ← [M([SSP])]                  {restore old PC}
              [SSP] ← [SSP] + 4
          END
      End Process_exception
```

Before we begin to look at the fine details of the 68000 family's exception-processing mechanism, we will demonstrate how you might deal with interrupts caused by a typical serial peripheral (the 6850 ACIA is to be described in chapter 9). Three items are required: code that sets up (i.e., configures) the peripheral, an interrupt handler that deals with each interrupt, and a vector that tells the 68000 where to find the interrupt handler. The following fragments of code illustrate these items. However, you should note that this example is purely illustrative—it simply demonstrates the type of code you need to implement interrupt-driven input.

```
*
*   Set up the ACIA to provide receiver interrupts (see chapter 9)
*
Setup      MOVE.B    #$03,ACIAC     Reset the ACIA
           MOVE.B    #$91,ACIAC     Set ACIA for receiver interrupt, 8 bits, no parity
           RTS
    ⋮
*   ACIA interrupt handler
*
ACIA_Int   MOVEA.L   Pointer,A0     Pick up pointer to input buffer
           MOVE.B    ACIAC,D0       Read the ACIA's status register
           BTST.B    #7,D0          Test bit 7 of the status (IRQ bit)
           BEQ       ACIA_Ret       If zero than no interrupt from the ACIA
           BTST.B    #0,D0          Test bit 0 of status for input ready
           BEQ       ACIA_Ret       If zero then no data available
           MOVE.B    ACIAD,(A0)+    Read ACIA data and store it in the buffer
           MOVE.L    A0,Pointer     Save the pointer
ACIA_Ret   RTE                      Return from exception—all causes
    ⋮
           ORG       $064           Location of autovector level 1
           DC.L      ACIA_Int       Put vector to interrupt handler in the vector table
    ⋮
```

FIGURE 6.1 The 68000 family's exceptions

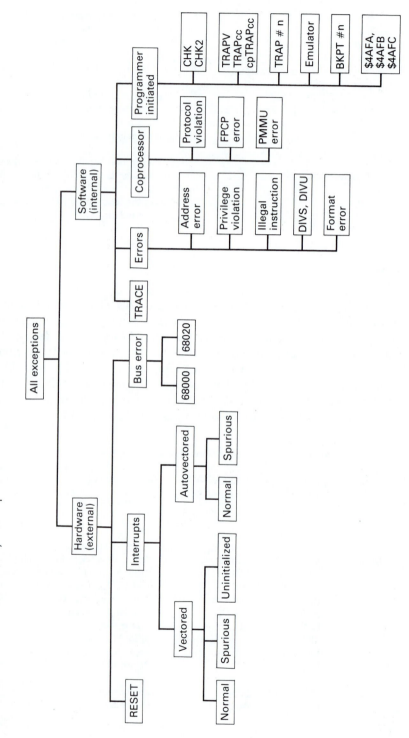

6.3 PRIVILEGED STATES, VIRTUAL MACHINES, AND THE 68000

We are now going to describe the 68000's user and supervisor states and the importance of these states before we look at the fine details of exception handling in Section 6.4.

The way in which the 68000 processes exceptions is intimately bound up with the notion of privileged states associated with that processor. At any instant, the 68000 is in one of two states: user or supervisor. By forcing the individual user programs to operate only in the user state and by dedicating the supervisor state to the operating system, it is possible to provide users with a degree of protection against one program corrupting another. The relationship between privileged states and exception processing is quite simple. An exception always forces the 68000 into the supervisor state. This means that individual user programs have no direct control over exception processing and interrupt handling.

Privileged States

The supervisor state is the higher state of privilege and is in force whenever the S-bit of the status register (i.e., bit 13) is set to 1. Some of the 68000's instructions are said to be *privileged* and can be executed *only* when the 68000 is operating in its privileged supervisor state. The user state is the lower state of privilege and privileged instructions cannot be executed in this state.

Each of the two states has its own stack pointer, so that the 68000 has two A7 registers. The user-mode A7 is called the user stack pointer, USP, and the supervisor-mode A7 is called the supervisor stack pointer, SSP. Note that the SSP cannot be accessed from the user state, whereas the USP can be accessed in the supervisor state by means of the MOVE USP,An and MOVE An,USP instructions.

Following a hard reset, the S-bit is set and the 68000 begins processing in the supervisor state. The supervisor state register map of all members of the 68000 family includes at least two registers that cannot be accessed from the user state (i.e., the SSP and the status byte of the register). By the way, no confusion over the two stack pointers normally arises. Most of the time we do not need to write USP or SSP explicitly, because A7 or SP will invariably suffice. There is no confusion because the user program sees only one stack pointer and the supervisor (operating system) also sees only one stack pointer. The user is *unaware* of the supervisor state, just as the airline passenger is unaware of air traffic control.

Figure 6.2 illustrates the supervisor state registers of the 68000 family. All the registers added to newer members of the 68000 family lie within the supervisor state rather than the user state. That is, enhancements to newer members of the 68000 family are largely concerned with operating systems support. Note that the status register, SR, appears *twice* in figure 6.2, although there is really only one status register. Figure 6.2 has been drawn in this way because the 68020 implements two

FIGURE 6.2   Supervisor state registers of the 68000 family

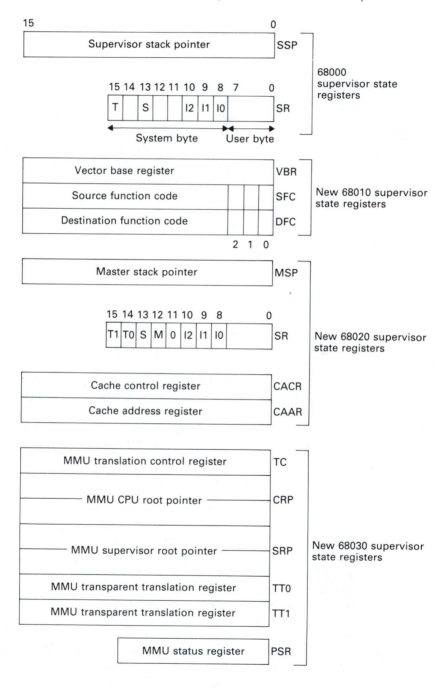

new status bits (undefined in the 68000's status register). We will now describe the 68000's status register bits (the status bits of the 68020 and other models are covered later).

S (Supervisor State) When set, S indicates that the 68000 is in its supervisor state. When clear, it indicates that the 68000 is in its user state.

T (Trace Bit) When clear, the 68000 operates normally. When $T = 1$, the 68000 generates a trace exception after the execution of each instruction. We will deal with the trace exception, which offers a method of debugging programs, later.

I2, I1, I0 (Interrupt Mask Bits) The value of I2, I1, I0 (from 0 to 7) indicates the level of the current interrupt mask. An interrupt will not be serviced unless it is of a higher level than that reflected by the interrupt mask. A level-seven interrupt is nonmaskable and will be accepted even if I2, I1, I0 = 1, 1, 1.

Not all registers in the supervisor register space are concerned with exception handling. To a fair approximation, we can state that the 68010's new registers are devoted to exception handling (and providing a virtual machine facility), the 68020's new registers are devoted to the control of its cache, and the 68030's new registers are devoted to its memory management unit, MMU. These aspects of the 68020 and 68030 are dealt with in chapter 7.

All exception processing is carried out in the supervisor state, because an exception forces a change from user to supervisor state. Indeed, the only way of entering the supervisor state from the user state is by means of an exception. Figure 6.3 shows how a transfer is made between the 68000's two states. An exception causes the S-bit in the 68000's status register to be set and the supervisor stack pointer to be selected at the start of the exception, with the effect that the return address is saved on the supervisor stack and not on the user stack.

How do we know which state the 68000 is in? In chapter 4 we introduced the 68000's three function code outputs, which indicate the type of memory access the 68000 is making. When the processor is in the supervisor state, the function code

FIGURE 6.3 State diagram of user and supervisor state transitions

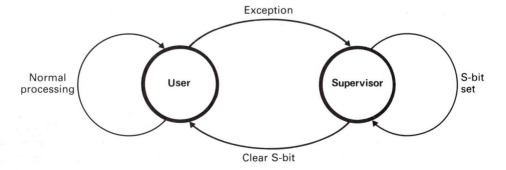

FIGURE 6.4 Using the function codes to protect supervisor memory space

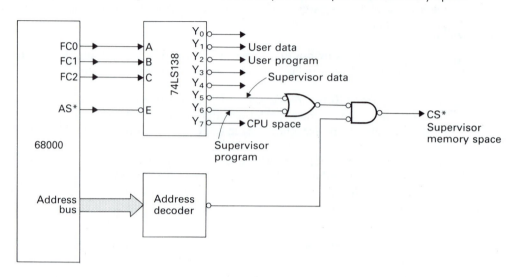

output from the processor (FC2,FC1,FC0) is 1,0,1 if the supervisor is executing a memory access involving data or 1,1,0 if it is accessing an instruction from memory.

It is possible to employ the 68000's function code outputs in address decoding circuits and thereby reserve address space for either supervisor or user applications (or even program or data space). Figure 6.4 demonstrates how it is possible to dedicate a region of address space to supervisor applications and to protect it from illegal access by user programs. A reminder of how the 68000's function codes are interpreted is provided below.

FC2	FC1	FC0	MEMORY ACCESS TYPE
0	0	0	Undefined—reserved
0	0	1	User data
0	1	0	User program
0	1	1	Undefined—reserved
1	0	0	Undefined—reserved
1	0	1	Supervisor data
1	1	0	Supervisor program
1	1	1	IACK space (CPU space)

Note that processors later than the 68000 refer to the function code FC2,FC1,FC0 = 1,1,1 as *CPU space* (rather than *interrupt acknowledge space*). This change of terminology is necessary because the 68010, 68020, and 68030 microprocessors use the function code output 1,1,1 to indicate *breakpoint acknowledge* and *coprocessor cycles* as well as interrupt acknowledge cycles. When a member of the 68000 family executes a CPU space cycle, it uses bits 16–19 of the address bus to define the type of the space cycle, as figure 6.5 illustrates.

The change from supervisor to user state is made by clearing the S-bit of the status register and is carried out by the operating system when it wishes to run a user program. Four instructions are available for this operation: RTE, MOVE.W ⟨ea⟩,SR, ANDI.W #$XXXX,SR, and EORI.W #$XXXX,SR. The "#$XXXX" represents a 16-bit literal value in hexadecimal form. Note that instructions that modify the *status* byte of the SR also modify its *condition code* byte.

The RTE (return from exception) instruction terminates an exception-handling routine and restores the value of the program counter and the old status register stored on the stack before the current exception was processed. Consequently, if the 68000 was in the user state before the current exception forced it into the supervisor state, an RTE restores the processor to its old (i.e., user) state.

A MOVE.W ⟨ea⟩,SR loads the status register with a new value that will force the system into the user state if the S-bit is clear. Similarly, the Boolean operations, AND immediate and EOR immediate, may affect the state of the S-bit. For example, we can clear the S-bit by executing an ANDI.W #$DFFF,SR instruction.

In the user state, the programmer must not attempt to execute certain privileged instructions. For example, the STOP and RESET instructions are not available to the user programmer. Why? Because the RESET instruction forces the 68000's RESET* output low and resets any peripherals connected to this pin. Suppose a peripheral is being operated by the *first* user. If a *second* user causes RESET* to be

FIGURE 6.5　Interpreting the 68000 family's CPU space

The breakpoint acknowledge cycle (68010, 68020, 68030)

31												19				16	15	14	13	12									4	3	2	1	0
0	0	0	0	0	0	0	0	0	0	0	0	0	0	0	0	0	0	0	0	0	0	0	0	0	0	0	0	BKPT #		0	0

The MMU access level control cycle (68030)

31												19				16	15	14	13	12						7	6			0
0	0	0	0	0	0	0	0	0	0	0	0	0	0	0	1	0	0	0	0	0	0	0	0	0	0	MMU register				

The coprocessor communication cycle (68020, 68030)

31												19				16	15	14	13	12							5	4		0
0	0	0	0	0	0	0	0	0	0	0	0	0	0	1	0	CPID			0	0	0	0	0	0	0	0	CP register			

The interrupt acknowledge cycle (68000, 68010, 68020, 68030)

31												19				16	15	14	13	12							5	4	3	2	1	0
1	1	1	1	1	1	1	1	1	1	1	1	1	1	1	1	1	1	1	1	1	1	1	1	1	1	1	1	Level					

pulsed low, the first user program will be adversely affected. Similarly, a STOP instruction has the effect of halting the processor until certain conditions are met. Clearly, the user state programmer must not be allowed to bring the entire system to a standstill.

The whole philosophy behind user/supervisor states is to protect the operating system (or any other type of supervisor program) and other user programs from accidental or malicious corruption by any user program. You should not think that being in the user state somehow limits what you can do. The user state simply controls the way in which certain system resources are accessed. A user can always ask the operating system to perform actions that it cannot carry out directly, although the operating system should deny those requests that are harmful to the system as a whole.

Instructions capable of modifying the S-bit in the upper byte of the status register (i.e., RTE, MOVE.W ⟨ea⟩,SR, ANDI.W #$XXXX,SR, ORI.W #$XXXX,SR, and EORI.W #$XXXX,SR) are not permitted in the *user* mode. Note that no instruction that performs *useful* computation is barred from the user mode, and therefore the user programmer does not lose any of the 68000's power. Only certain operations that affect the operating mode are privileged.

Suppose the programmer is either willful or ignorant and tries to set the S-bit by executing an ORI.W #$2000,SR instruction. If the programmer succeeds in setting the S-bit, he or she will be able to execute privileged instructions. Obviously, there is nothing to prevent the programmer from writing this instruction and running the program containing it.

When the program is run, the illegal operation, ORI.W #$2000,SR, causes a privilege violation exception because the instruction may not be executed in the user state. The resulting exception causes a change of state from user to supervisor. That is, by *attempting* to enter the supervisor state, the user has been *forced* into the supervisor state. "Ah!" you say, "this seems rather like punishing a criminal by giving a reward!" After all, the punishment for trying to enter the supervisor state is to be forced into it. A change of state has indeed been made, but the programmer has now *lost control of the processor*. The effect of attempting to execute an ORI.W #$2000,SR is to raise a *privilege violation exception*, forcing a jump to the routine dealing with this type of exception. We will describe the way in which exception states are entered and processed shortly.

Once the exception-handling routine dealing with the privilege violation has been entered, the user no longer controls the processor. The operating system has now taken over. In other words, in attempting to enter the supervisor state through the front door, the user has fallen through a hole in the floor and is now trapped. The 68000 is in the supervisor state, but it is executing the privilege violation exception handler and not the user program. It is highly probable that the exception-handling routine will deal with the privilege violation by terminating the user's program.

Before we describe the 68000's exceptions, we will explain how the 68010 and later members of the 68000 family have extended their architecture to provide more supervisor state facilities. Readers not interested in these processors may turn to section 6.3.

Architectural Enhancements of the 68010, 68020, and 68030

For the purpose of this discussion, the supervisor model of the 68010 consists of the five registers depicted in figure 6.6. Here we will concentrate on the 68010, as all the enhancements described by figure 6.6 are included in the 68020 and 68030.

The 68010's new registers are its vector base register, VBR, and the alternate function code registers, SFC and DFC. The alternate function code registers allow the systems programmer to force a particular value on the FC0–FC2 outputs during a memory access; we will return to this point later. The VBR may be used to relocate the 68010's exception vector table containing the pointers to each of the exception handlers anywhere within the processor's memory space (discussed later). The 68020 and 68030 also have a third stack pointer called the *master stack pointer*, MSP, in addition to the 68000's USP and SSP. The 68020 literature renames the supervisor stack pointer the ISP (interrupt stack pointer) or A7' (A7 prime). The MSP is sometimes referred to as A7" (A7 double prime). As we shall see, the ISP can be dedicated to interrupt handling, and the MSP can be used for all other exception handling.

These new registers require a new instruction to access them. The 68010 employs a move to or from the control register instruction, MOVEC, to access these

FIGURE 6.6 Enhancements of the 68000's supervisor mode registers

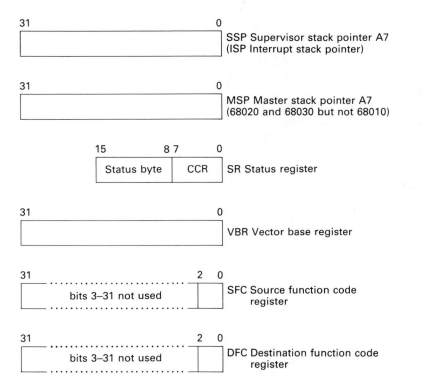

three registers (VBR, SFC, DFC). MOVEC is a privileged instruction with the assembly language form

```
MOVEC      Rc,Rn
```

or

```
MOVEC      Rn,Rc
```

where Rn is a general register (A0–A7 or D0–D7) and Rc is a control register (i.e., VBR, SFC, or DFC). Note that MOVEC takes a longword operand (even through only 3 bits of the SFC and DFC registers are defined). When, for example, a MOVEC SFC,D0 is executed, bits 3 to 31 of D0 are padded with zeros. Since MOVEC permits only register-to-register operations, data must be first loaded into a data or address register before it can be loaded into a control register.

We can relocate the 68010's exception vector table from it's default location $00 0000 to $00 8000 by executing the following instructions:

```
MOVE.L     #$8000,D0
MOVEC      D0,VBR
```

The alternate function code registers, SFC and DFC, are used in conjunction with the new 68010 instruction MOVES (move to or from address space). Before continuing, we have to make it clear that the alternate function code registers have a meaning only in systems with memory management units (or hardware similar to that of figure 6.4) that distinguish between *user address space* and *supervisor address space*. The 68000 employs its function code outputs on FC0 to FC2 to indicate the type of address space being accessed (user or supervisor) to a MMU (memory management unit). If the type of address space accessed does not match the type of address space indicated by FC0 to FC2, the MMU issues a bus error exception by asserting BERR∗. This arrangement of processor and MMU suffers from a subtle flaw. When the 68000 is operating in the *supervisor* mode, it cannot make a memory access to *user* address space, because all its accesses are to supervisor address space (by definition; that's what being in the supervisor state means). What we need is a method of fooling the MMU into thinking that the microprocessor is operating in the user mode when it is, in fact, operating in the supervisor mode. Why should we wish to do this? Such a facility permits the operating system to transfer data to user data space. Equally, it allows the operating system to perform diagnostic tests on user address space.

The 68010 can access any address space when it is in the supervisor mode by means of the privileged instruction MOVES. This has the following assembly language forms:

```
MOVES Rn,⟨ea⟩
```

and

```
MOVES ⟨ea⟩,Rn
```

Rn is a general register (i.e., A0–A7 or D0–D7) and ⟨ea⟩ is an effective address. Note that MOVES permits byte, word, and longword operands. The address space used by the MOVES instruction is determined by the source function code register, SFC, if the source operand is in memory. Similarly, the address space is determined by the destination function code register, DFC, if the MOVES instruction specifies a destination operand. For example, suppose the operating system wishes to read the contents of location $4 0000, which lies in user address space. The following sequence of operations will perform this task:

```
MOVE.L      # %001,D0      Load D0 with the user data function code 0,0,1
MOVEC.L     D0,SFC         Copy the user space code into the SFC register
MOVES.W     $40000,D1      Read data from the user data space
```

The Virtual Machine

We are now going to introduce the idea of the *virtual machine* made possible by the 68010. Virtual machines are required only in rather specialized applications. We intend to introduce the virtual machine by first describing an anomaly in the 68000's instruction set that is corrected in the 68010 and later members of the 68000 family.

The 68000's user mode instructions are entirely compatible with those of the 68010, the 68020, and the 68030, but with one exception. The 68010 implements the 68000's *move from status register instruction* (i.e., MOVE SR,⟨ea⟩) in a different way than the 68000. The 68000's MOVE SR,⟨ea⟩ instruction is not privileged, but the 68010's MOVE SR,⟨ea⟩ instruction is privileged. So, the 68010 programmer cannot read the status register in the user state, and the user is not allowed to know the state of the machine. However, a new 68010 instruction, MOVE CCR,⟨ea⟩, has been implemented to enable the user programmer to read the condition code register. Just why this seemingly curious change has been made leads us to the concept of a virtual machine.

Sometimes we have to run two operating systems (or *applications environments*) on the same machine *concurrently*. Such a situation might arise when users require such very diverse facilities (e.g., business and scientific) that no one operating system is sufficient. When two operating systems are run concurrently, each operating system provides a virtual environment for the programs running under it. That is, each user program sees its operating system as the real (i.e., virtual) operating system of the machine itself. In fact, the operating systems are themselves user tasks running under the actual operating system of the machine.

A similar situation exists when a programmer is developing an operating system. The operating system being developed cannot run in the user mode, because it is an operating system and requires access to the privileged operations associated with the supervisor mode. Equally, it cannot run in the supervisor mode, because it is a user task running under the real (i.e., the actual) operating system. In the following discussion we will call the operating system being developed the *virtual operating system* to avoid confusion with the real operating system.

The solution to our dilemma is to run the virtual operating system in the user mode and to make it appear as if it were really running in the supervisor mode. Whenever the virtual operating system attempts to access a supervisor mode facility, an exception is generated and the *actual* operating system emulates *in software* the requested facility.

The following example indicates how the virtual operating system appears to run on a real machine. Suppose the virtual operating system attempts to read its status register by means of a MOVE SR,D0 instruction. Since the virtual operating system is actually running under the user mode, a privilege violation exception is generated by the attempt to read the status register. When this happens, the actual (i.e., real) operating system supplies an appropriate value for the SR to the virtual operating system.

Although the 68000 implements some of the facilities required by a virtual machine, a major flaw exists in the 68000's structure. A 68000 program running in the user mode can access the status word by means of a MOVE SR,⟨ea⟩ operation. That is, the virtual operating system can access the real operating system's status register.

A true virtual machine should not allow a user task (e.g., the virtual operating system) to see the status register that reflects the status of the real machine. Consequently, the 68010 makes the 68000's MOVE SR,⟨ea⟩ instruction a privileged operation, so that any attempt by a program in the user mode to access the status register results in a privilege violation exception. As we have said the 68010 introduces a new instruction, MOVE.W CCR,⟨ea⟩, to enable user tasks to access the condition code register byte of the status word. Clearly, it is perfectly reasonable for a user task to access the condition code register. The 68010's new MOVE CCR,⟨ea⟩ instruction is not privileged and has the effect of copying the CCR into the lower byte of the word specified by the given effective address. The upper byte of this word is filled with zeros.

Unfortunately, making the move from SR a privileged instruction and introducing a new move from CCR instruction does cause an incompatibility problem between the 68000 and the 68010. A 68010 program with a MOVE SR,⟨ea⟩ instruction will cause a privilege violation exception when run on a 68010. Equally, a 68010 program with a MOVE CCR,⟨ea⟩ instruction will cause an illegal instruction exception when run on a 68000. Either the programs must be modified and reassembled before they are transferred between the 68000 and 68010 or both privilege violation and illegal instruction exception handlers must be written to deal with the compatibility problem automatically. The 68010's virtual machine enhancement is carried over to the 68020 and 68030. We are now going to describe the 68000's exceptions in more detail.

6.4   EXCEPTIONS AND THE 68000 FAMILY

A wealth of exception types are supported by the 68000 family. Some exceptions are associated with *external hardware events* such as interrupts, and others are associated with *internally generated events* such as privilege violations. Provision has been made for new types of exceptions in future versions of the 68000.

Exception Types

We are now going to describe the exception types currently implemented by the 68000 family. We first introduce the 68000's exceptions (note that the 68010's exception types are the same as those of the 68000) and then discuss the 68020's and the 68030's new exceptions. The way in which these exceptions are implemented is described later.

Reset   An externally generated reset is caused by forcing both the 68000's RESET* and HALT* pins low for at least ten clock pulses (or for longer than 100 ms on power-up); this reset is used to place the 68000 in a known state at start-up or following a totally irrecoverable system collapse. A reset loads the program counter and the supervisor stack from memory and sets up the status register. The reset is a unique exception, because there is no *return from exception* following a reset.

Bus Error   A bus error is an externally generated exception, initiated by hardware driving the 68000's BERR* pin active-low. It is a *catchall* exception, because the systems designer may use it in many different ways, and it is provided to enable the processor to deal with hardware faults in the system. A typical use of the BERR* input is to indicate either a faulty memory access or an access to a nonexistent memory. The bus error is also used in systems with memory management or virtual memory. There are major differences in the way in which the 68000 and the 68010 (and later processors) implement bus error exception processing.

Interrupt   The 68000 implements a conventional, but powerful, hardware interrupt mechanism. A peripheral uses the 68000's three active-low binary encoded interrupt request inputs, IPL0*–IPL2*, to signal one of seven levels of interrupt. To obtain maximum benefit from the interrupt request inputs, you have to employ an eight-line-to-three-line priority encoder to convert one of seven interrupt request inputs from peripherals into a 3-bit code for IPL0*–IPL2*. The eighth code represents no interrupt request. The designers of the 68000 have made the interrupt-handling facilities of the 68000 unusually flexible by including provision for prioritized and vectored interrupts.

Address Error   An address error exception occurs when the 68000 attempts to access a 16-bit word or a 32-bit longword at an odd address. If you think about it, attempting to read a word at an odd address would require *two* accesses to memory—one to access the *odd* byte of an operand and the other to access the *even* byte at the next address. Address error exceptions are generated when the programmer does something foolish. Consider the following fragment of code:

```
LEA      $7000,A0    Load A0 with $0000 7000
MOVE.B   (A0)+,D0    Load D0 with the byte pointed at by A0 and increment A0 by 1
MOVE.W   (A0)+,D0    Load D0 with the word pointed at by A0 and increment A0 by 2
```

The third instruction results in an address error exception, because the previous operation, MOVE.B (A0)+,D0, causes the value in A0 to be incremented by one from $7000 to $7001. Therefore, when the processor attempts to execute MOVE.W (A0)+,D0, it finds it is trying to access a *word* at an *odd address*.

In many ways, an address error is closer to an exception generated by an event originating in the hardware than by one originating in the software. The bus cycle leading to the address error is aborted, as the processor cannot complete the operation. The 68000 generates an address error if an attempt is made to read a word (or a longword) instruction or an operand at an odd address. Since the 68020's dynamic bus sizing mechanism permits operands to cross word boundaries, address errors are not generated when the 68020 reads a *misaligned* operand. However, an address error exception is generated if the 68020 attempts to read an instruction at an odd address.

Illegal Instruction   In the "good old days" of the 8-bit microprocessor, it was fun finding out what effect unimplemented op-codes had on the processor. For example, if the value $A5 did not correspond to a valid op-code, an enthusiast would try and execute it and then see what happened. This situation was possible because the control unit (i.e., instruction interpreter) of most 8-bit microprocessors was implemented by random logic. Such control units will interpret the bit pattern of any instruction as a sequence of operations (some of which have no meaningful effect and some of which might perform a useful operation).

To reduce the number of gates in the control unit of the CPU, some semiconductor manufacturers have not attempted to deal with illegal op-codes. After all (you might erroneously argue), if users try to execute unimplemented op-codes, they deserve everything they get. In keeping with the 68000's approach to programming, an illegal instruction exception is generated whenever an operation code is read that does not correspond to the bit pattern of one of the 68000's legal instructions.

Divide by Zero   If a number is divided by zero, the result is meaningless and often indicates that something has gone seriously wrong with the program attempting to carry out the division. For this reason, the 68000's designers decided to make any attempt to divide a number by zero an exception-generating event. Good programmers should write their programs so that they never try to divide a number by zero; therefore the divide-by-zero exception should not arise. It is intended as a fail-safe device to avoid the meaningless result that would occur if a number were divided by zero.

CHK Instruction   The check register against bounds instruction, CHK, has the assembly language form CHK $\langle$ea$\rangle$,Dn and has the effect of comparing the contents of the specified data register with the operand at the effective address. If the lower-order word in the register Dn is negative or is greater than the upper bound at the effective address, an exception is generated. For example, when the instruction CHK D1,D0 is executed, an exception is generated if

$$[D0(0:15)] < 0 \quad \text{or} \quad [D0(0:15)] > [D1(0:15)]$$

Oddly enough, the CHK instruction works only with 16-bit words and therefore cannot be used with an address register as an effective address. The CHK exception has been included to help compiler writers for languages such as Pascal that have facilities for the automatic checking of array indexes against their bounds. The 68020 provides two useful extensions to the 68000's CHK exception, which we described in chapter 2.

Privilege Violation   If the processor is in the user state (i.e., the S-bit of the status register is clear) and attempts to execute a privileged instruction, a privilege violation exception occurs. In addition to any logical operation that attempts to modify the state of the status register (e.g., ANDI #data,SR), the following three instructions cannot be executed in the user state: STOP, RESET, and MOVE ⟨ea⟩,SR.

Trace   A popular method of debugging a program is to operate in a *trace mode*, in which the contents of all registers are printed out after each instruction has been executed. The 68000 has a built-in trace facility. If the T-bit of the status register is set (under software control), a trace exception is generated after each instruction has been executed. The exception-handling routine called by the trace exception can be constructed to offer programmers any facilities they need. As we shall discover, the 68020 extends the 68000's trace function by permitting tracing either of every instruction, or only those instructions that modify the flow of instructions, such as branch or return from subroutine.

Line 1010 Emulator   Operation codes, whose four most significant bits (bits 12–15) are 1010 or 1111, are unimplemented in the 68000 and therefore represent illegal instructions. However, the 68000 generates a special exception for op-codes whose most significant nibble is 1010 (also called line A, or line ten). One of the purposes of this exception is to *emulate* instructions that the 68000 lacks in software. Suppose a version of the 68000 is designed that includes string manipulation operations as well as the normal 68000 instruction set. Clearly, it is impossible to run code intended for the string processor on a normal 68000. But by using 1010 as the four most significant bits of the new floating-point instructions, an exception is generated each time the 68000 encounters one of these instructions. The line 1010 exception handler can then be used to allow the 68000 to emulate its more sophisticated brother.

Line 1111 Emulator   The line 1111 (or line F) emulator behaves in almost exactly the same way as the line 1010 emulator, except that it has a different exception vector number and can, therefore, call a different exception-handling routine. This emulator trap is intended to allow the 68000 to emulate, for example, a floating-point coprocessor. The 68020 and 68030 use bit patterns that begin with 1111 as a means of communicating with 68000-series coprocessors (see chapter 7). If one of these processors encounters such an instruction, it attempts to access the appropriate coprocessor. If, however, a bus error terminates the resulting access, the system initiates a normal line 1111 exception sequence. Line 1111 exceptions are intended for use with 68000-series coprocessors.

Uninitialized Interrupt Vector   The 68000 supports vectored interrupts so that an interrupting device can identify itself and allow the 68000 to execute the appropriate interrupt-handling routine without having to poll each device in turn.

Before a device can identify itself, the programmer must first configure it by loading its interrupt vector register with the appropriate value. If a 68000-series peripheral is unconfigured and yet generates an interrupt, the 68000 responds by raising an *uninitialized interrupt* vector exception. 68000-series peripherals are designed to supply the uninitialized interrupt vector number $0F during an IACK cycle if they have not been initialized by software. Their interrupt vector registers are automatically loaded with $0F following a reset operation.

Spurious Interupt   If the 68000 receives an interrupt request from external hardware and sends an interrupt acknowledge, but no device responds, the CPU generates a spurious interrupt exception. The spurious interrupt exception prevents the 68000 from hanging up should an interrupt request be received on IPL0*–IPL2* and no peripheral respond to the ensuing interrupt acknowledge. To implement the spurious interrupt exception, external hardware is required to assert BERR* following the nonappearance of either DTACK* or VPA* a reasonable time after an interrupt acknowledge has been detected.

TRAP (Software Interrupt)   The 68000 provides 16 instructions of the form TRAP #I, where I = 0, 1, ..., 15. These instructions are available to the user programmer. When a TRAP instruction is executed, an exception is generated and 1 of 16 exception-handling routines is called. Thus, TRAP #0 causes TRAP exception-handling routine 0 to be called and so on. The TRAP #I instruction is very useful. Suppose we write a program that is to run on all 68000 systems. The greatest problem in designing *portable* programs is in implementing input or output transactions. One 68000 system may deal with input in a very different way than every other 68000 system. However, if everybody agrees that, for example, TRAP #0 means input a byte and TRAP #1 means output a byte, then the software becomes truly portable. All that remains to be done is for an exception handler to be written for each 68000 system to actually implement the input or output as necessary. The TRAP exception is dealt with in more detail in section 6.5.

TRAPV Instruction   When the trap on overflow instruction, TRAPV, is executed, an exception occurs if the overflow bit, V, of the condition code register is set. Note that an exception caused by dividing a number by zero occurs *automatically*, whereas TRAPV is an instruction equivalent to: "IF V = 1 THEN exception ELSE continue." The 68020 extends the TRAPV instruction to a general form TRAPcc, where cc is any of the 68020's conditions.

Double Bus Fault   A double bus fault is not really an exception in its own right but is a situation in which two exceptions occur in close proximity. Suppose a 68000 system experiences a bus error (or an address error) exception and the processor begins exception processing by saving the program counter on the stack. Now suppose that a second bus error occurs during the stacking of the PC. The 68000 has nowhere to go. It cannot continue normally because of the original exception and it cannot enter exception processing because of the second exception. In this case, a double bus fault is said to have occurred, and the 68000 halts. Further execution is stopped and the HALT* pin is asserted active-low. Only a hard reset will restart the 68000 following a double bus fault.

New 68020 and 68030 Exceptions

Apart from exceptions associated with the 68000's new instructions (CHK2 and TRAPcc), the 68020's and the 68030's new exceptions are concerned with the coprocessor interface, the format error, and the 68030's memory management unit. We will provide a brief description of these exceptions. Readers not interested in these processors may skip this section.

TRAPcc The TRAPcc exception causes an exception if condition cc is true when the instruction is executed. The condition specified by cc represents 1 of the 68000's 16 conditions, which are the same as the branch (Bcc) conditions. The exception routine called by the TRAPcc instruction is located at the same address as that called by a TRAPV instruction. A TRAPcc instruction has three possible formats:

```
TRAPcc
TRAPcc.W      #⟨d16⟩
TRAPcc.L      #⟨d32⟩
```

A TRAPcc can take no extension, a word extension, or a longword extension. These optional extensions have no effect on the execution of the TRAPcc exception itself and can be used to pass a parameter to the TRAPcc exception-handling routine. Since the TRAPV and TRAPcc exceptions share the same exception vector, it is up to the writer of the exception handlers for these instructions to provide the appropriate course of action.

CHK2 The CHK2 instruction is an extension of the 68000's existing CHK instruction. CHK2 has the assembly language form CHK2 ⟨ea⟩,Rn and can take byte, word, and longword operands. The value in Rn is compared with the lower and upper bounds at the address specified by ⟨ea⟩. An exception is called if

$$Rn < \text{lower bound} \quad \text{or} \quad Rn > \text{upper bound}$$

Note that the CHK and the CHK2 instructions share the same exception vector. That is, they share the same exception-handling routine.

Format Error Exception The various members of the 68000 family save different amounts of information on an exception stack frame, depending on the nature of the exception being processed. For example, the 68000 has two possible stack frames and the 68020 has seven. Each of these stack frames has a *format* that defines its size and the type of information stored in it. When a return from exception is made by an RTE instruction, information from the stack frame must be restored to the processor. The processor (i.e., 68020 or 68030) determines the type of stack frame by reading the format number on the stack frame (which is held in bits 12–15 of the word that contains the vector offset). A format error exception takes place when the processor encounters an RTE instruction and the information saved in the stack frame does not match that specified by the frame's format number.

cpTRAPcc Exceptions The cpTRAPcc exception causes a trap if the selected condition code of the coprocessor is true. Chapter 7 describes the coprocessor. All we need say here is that you can (in a 68020 program) force an exception on the coprocessor's condition code register (which is not the same as the 68020's CCR).

Privilege Violation Exception   In addition to the 68000's privileged instructions, the 68010, 68020, and 68030 generate privilege violation exceptions for the following privileged instructions:

SUPERVISOR MODE INSTRUCTIONS	COPROCESSOR INSTRUCTIONS	MMU INSTRUCTIONS
MOVEC	cpRESTORE	PFLUSH
MOVES	cpSAVE	PLOAD
		PMOVE
		PTEST

Exception Vectors

Having described the various types of exception supported by the 68000, we next explain how the processor is able to determine the location of the corresponding exception-handling routine. Each exception is associated with a *vector*, which is the 32-bit absolute address of the appropriate exception-handling routine. All the 68000's exception vectors are stored in a table of 256 longwords (i.e., 512 words), extending from address $00 0000 to $00 03FF.

A list of all the exception vectors is given in table 6.1, and figure 6.7 shows the physical location of the 256 vectors in memory. The left-hand column of table 6.1

FIGURE 6.7   Memory map of the 68000's vector table

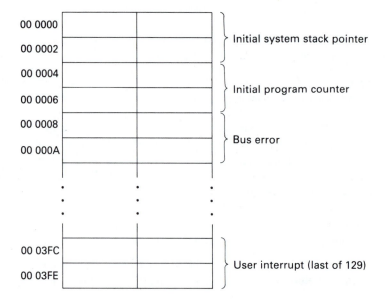

TABLE 6.1 The 68000's exception vector table

VECTOR NUMBER	VECTOR (HEX)	ADDRESS SPACE	EXCEPTION TYPE
0	000	SP	Reset—initial supervisor stack pointer
—	004	SP	Reset—initial program counter value
2	008	SD	Bus error
3	00C	SD	Address error
4	010	SD	Illegal instruction
5	014	SD	Divide by zero
6	018	SD	CHK instruction
7	01C	SD	TRAPV instruction
8	020	SD	Privilege violation
9	024	SD	Trace
10	028	SD	Line 1010 emulator
11	02C	SD	Line 1111 emulator
12	030	SD	(Unassigned—reserved)
13	034	SD	(Unassigned—reserved)
14	038	SD	(Unassigned—reserved)
15	03C	SD	Uninitialized interrupt vector
16	040	SD	(Unassigned—reserved)
⋮	⋮	⋮	⋮
23	05C	SD	(Unassigned—reserved)
24	060	SD	Spurious interrupt
25	064	SD	Level 1 interrupt autovector
26	068	SD	Level 2 interrupt autovector
27	06C	SD	Level 3 interrupt autovector
28	070	SD	Level 4 interrupt autovector
29	074	SD	Level 5 interrupt autovector
30	078	SD	Level 6 interrupt autovector
31	07C	SD	Level 7 interrupt autovector
32	080	SD	TRAP #0 vector
33	084	SD	TRAP #1 vector
⋮	⋮	⋮	⋮
47	0BC	SD	TRAP #15 vector
48	0C0	SD	(Unassigned—reserved)
⋮	⋮	⋮	⋮
63	0FC	SD	(Unassigned—reserved)
64	100	SD	User interrupt vector
⋮	⋮	⋮	⋮
255	3FC	SD	User interrupt vector

gives the *vector number* of each entry in the table. The vector number is a value that, when multiplied by 4, gives the address, or offset, of an exception vector. For example, the vector number corresponding to a privilege violation is 8, and the appropriate exception vector is found at memory location $8 \times 4 = 32 = \$20$. Therefore, whenever a privilege violation occurs, the CPU reads the longword at location $20 and loads it into its program counter.

Although we said that two words of memory space are devoted to each 32-bit exception vector, the *reset exception* (vector number zero) is a special case. The 32-bit longword at address $00 0000 is not the address of the reset-handling routine but the initial value of the supervisor stack pointer. The actual reset exception vector is at address $00 0004. Thus, the reset exception vector requires *four* words of memory in the exception vector table instead of the usual two. The 68000's designers have been very clever here. The first operation performed by the 68000 following a reset is to load the system stack pointer. Loading the supervisor stack pointer is important, because, until a stack is defined, the 68000 cannot deal with any other type of exception. Once the stack pointer has been set up, the reset exception vector is loaded into the program counter and processing continues normally. The reset exception vector is, of course, the initial (or cold-start) entry point into the operating system or, in the case of some embedded systems, the applications program.

Yet another difference exists between the reset vector and all other exception vectors. The reset exception vector and supervisor stack pointer initial value both lie in *supervisor program space*, denoted by SP in table 6.1. Thus, when the 68000 accesses these vectors, it puts out a function code of 1,1,0 on FC2,FC1,FC0, respectively. All other exception vectors lie in *supervisor data space* (SD), and the function code 1,0,1 is put out on FC2,FC1,FC0 when one of these is accessed.,

Certain vectors, numbers 12–14, 16–23, and 48–63, have been reserved for possible future enhancements of the 68000. Some of these have been assigned to the 68010, 68020, or 68030 processors. Table 6.2 illustrates the 68010's, 68020's and 68030's new additions to the vector table. We are now going to look at how the location of the exception vector table affects the hardware engineer.

TABLE 6.2 Additions to the 68000's exception vector table

VECTOR NUMBER	VECTOR (HEX)	ADDRESS SPACE	EXCEPTION TYPE	CPU
6	018	SD	CHK2 instruction	020/300
7	01C	SD	TRAPcc, cpTRAPcc instructions	020/030
13	034	SD	Coprocessor protocol violation	020/030
14	038	SD	Format error	020/030
48	0C0	SD	FPCP Bcc or Scc incorrect	020/030
49	0C4	SD	FPCP inexact result	020/030
50	0C8	SD	FPCP divide by zero	020/030
51	0CC	SD	FPCP underflow	020/030
52	0D0	SD	FPCP operand error	020/030
53	0D4	SD	FPCP overflow	020/030
54	0D8	SD	FPCP signaling no number (i.e., NaN)	020/030
56	0E0	SD	PMMU configuration	020/030
57	0E4	SD	PMMU illegal operation	020/030
58	0E8	SD	PMMU access level violation	020/030

Implementing the Exception Vector Table

All 68000 systems must maintain an exception vector table in memory. Although the complete table occupies 256 longwords (1,024 bytes), it is not strictly necessary to fill it entirely with exception vectors. For example, if the system does not implement vectored interrupts, the memory space from $00 0100 to $00 03FF does not need to be populated with user interrupt vectors. Unless forced to do otherwise, we would always reserve the memory space $00 0000–$00 03FF for the exception vector table, even if we were not using the whole of the table. Furthermore, we would probably preset all unused vectors to point to the spurious exception handler. This action is wholly consistent with the philosophy of always providing a recovery mechanism for events that may possibly happen and that would cause the system to crash if not adequately catered for.

Even the humble 8-bit microprocessor has its own rather small exception vector table, corresponding to the very limited exception-handling facilities of most 8-bit devices. This table is invariably maintained in the same read-only memory that holds the processor's operating system or bootstrap monitor. Putting exception vectors in a read-only memory is good because the table is always there immediately after power-up. On the other hand, it is bad because it is inflexible. Once a table is in ROM, the vectors cannot be modified to suit changing conditions. You can get around this problem by providing a fixed vector in ROM that points to a second vector in read/write memory. The vector in read/write memory can be modified to point to the appropriate exception-handling routine. This approach increases the time taken to respond to an exception.

Because the 68000 is so much more sophisticated than 8-bit devices and a *dynamic*, or *flexible*, response is sometimes required for the treatment of exception-handling routines, the exception vector table is frequently held in read/write rather than in read-only memory. The operating system, held either in ROM or loaded from disk, sets up the exception vector table early in the initialization process following a reset. Unfortunately, there is a big blot on the horizon in the shape of the *reset vector*. The two things that *must* be in read-only memory are the reset vector and the system monitor or bootstrap loader. Clearly, when the system is powered up and the RESET* input asserted, the reset exception vector and supervisor stack pointer, loaded from $00 0004 and $00 0000, respectively, must be in read-only memory.

At first sight, you might think that it is necessary to place the *whole* exception vector table in ROM, since it is not possible to get a 4-word ROM just for the reset vector and a (512 − 4)-word read/write memory for the rest of the table. Hardware designers have solved the problem by locating the exception vector table in read/write memory and overlaying this with ROM whenever an access in the range $00 0000 to $00 0007 is made. There are several ways of dealing with this problem, and we shall provide a simple solution here.

Figure 6.8 describes a possible memory map for a system in which the exception vector table in *read/write memory* is overlaid with ROM whenever the reset vectors are accessed. The 4K bytes of memory in the range $00 0000 to $00 0FFF are implemented by read/write memory. As we shall see, the region of RAM at $00 0000 to $00 0007 is not accessible by the processor. Read/write memory

FIGURE 6.8 Overlaying the read/write exception vector table with ROM. A read access to addresses in the range $00 0000 to $00 0007 automatically retrieves the corresponding data in the range $00 1000 to $00 1007

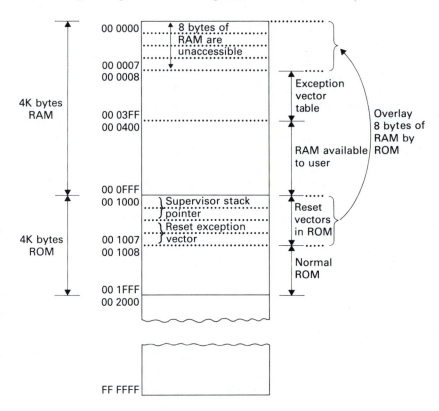

extending from $00 0008 to $00 03FF holds the exception vector table, which is loaded with exception vectors, as dictated by the operating system. The remaining read/write memory from $00 0400 to $00 0FFF is not restricted in use and is freely available to the user or the operating system.

The next 4K bytes of memory space, from $00 1000 to $00 1FFF, are populated by *read-only memory*. The first 8 bytes of this ROM, from $00 1000 to $00 1007, contain the reset vectors. All we have to do is to design the hardware of the 68000 system so that a read access to the reset vectors automatically fetches them from the ROM rather than the read/write memory. That is, whenever the 68000 reads from either $00 0000 or from $00 0004, it actually accesses locations $00 1000 or $00 1004, respectively. This arrangement gives us the best of both worlds: The reset vectors are put in ROM and all other exception vectors are put in read/write memory.

One possible way of remapping the 68000's reset vectors is demonstrated by the circuit of figure 6.9. The read/write and read-only memory elements are supplied by conventional 2K × 8 chips, and their circuitry is entirely straightforward. Read/

FIGURE 6.9 Implementing the overlaid exception vector table of figure 6.8

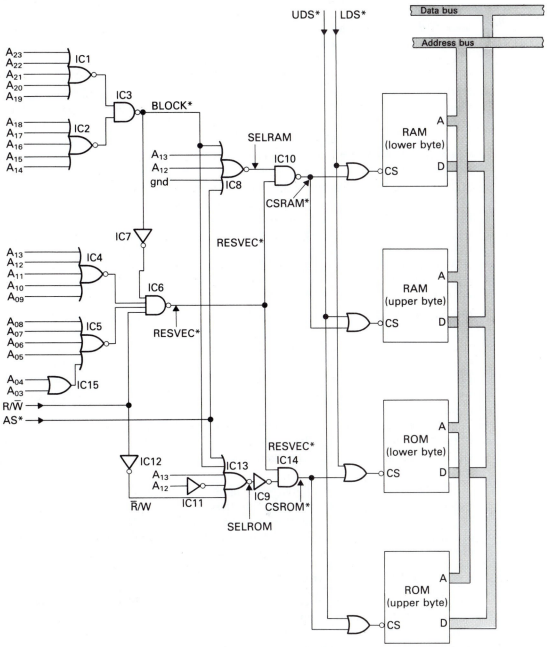

write memory is selected when CSRAM* is active-low and read-only memory is selected when CSROM* is active-low. The part of the circuit of interest to us here is that concerned with the generation of CSRAM* and CSROM*.

Address lines A_{01} to A_{11} take part in selecting locations within the memory components, so the 12 remaining address lines, A_{12} to A_{23}, must take part in the address decoding process if the memory is to be fully decoded. Gates IC1, IC2, and IC3 generate an active-low output (labeled BLOCK*) when A_{14} to A_{23} are all low. The five-input NOR gate, IC8, produces an active-high output, SELRAM, whenever BLOCK* is low, both A_{12} and A_{13} are low, and AS* is active-low. If we were not concerned with remapping the reset vectors, the complement of SELRAM could be used to select the RAM components in the 4K-byte address range $00 0000–$00 0FFF.

The output, SELROM, of the five-input NOR gate, IC13, goes active-high when the 4K-byte read-only memory space $00 1000–$00 1FFF is read by the 68000. Our aim is to detect a read to the reset exception space (i.e., $00 0000–$00 0007) and then *disable* the read/write memory and *enable* the ROM.

ICs 4, 5, 6, 7 and 15 (in conjunction with the memory block select signal BLOCK*) generate an active-low *reset-vector* signal, RESVEC*, whenever the 68000 reads from address $00 0000–$00 0007. The SELRAM signal is NANDed with RESVEC* in IC10 to give the active-low signal, CSRAM*, that enables the read/write memory. During a normal access to RAM, RESVEC* is high. If SELRAM goes high, CSRAM* goes low, selecting the read/write memory. Should an access be made to the reset vectors, RESVEC* goes low, forcing CSRAM* high and thereby deselecting the read/write memory; that is, the read-write memory is disabled whenever the reset vector is fetched by the 68000.

The read-only memory is selected by ICs 13, 9, and 14. The ROM select signal, SELROM, from the IC13 is inverted by IC9 and ANDed with RESVEC* in IC14 to give the active-low ROM-enable signal, CSROM*. During a normal access to ROM in the region $00 1000 to $00 1FFF, both RESVEC* and SELROM are high, forcing CSROM* low.

If a read-access to the reset vectors is made, RESVEC* goes active-low, forcing CSROM* low and enabling the read-only memory. Note that the first 8 bytes of the ROM can be accessed either from addresses $00 0000 to $00 0007 or from $00 1000 to $00 1007.

Later, when we look at the reset input in more detail, we will discuss other ways of dealing with the remapping of the reset vectors. We are now going to describe how the 68010 and later members of the 68000 family deal with the problem of a fixed exception vector table. Readers not interested in these processors may skip ahead to the section on exception processing.

Exception Vectors and the 68010, 68020, and 68030

As we have just seen, whenever the 68000 responds to an exception, it reads an exception vector from the appropriate location in the exception vector table in the region $00 0000 to $00 03FF. The 68010 and later members of the 68000 family calculate the address of an exception vector in exactly the same way as the 68000 but then add the address of the exception vector to the contents of the 32-bit vector base register, VBR, to provide the actual address of the exception vector. Figure 6.10

FIGURE 6.10 Using the VBR to remap the exception vector table

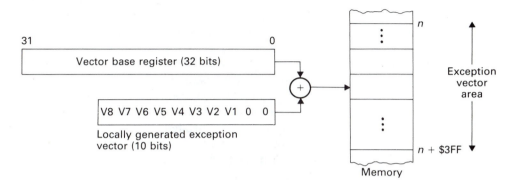

illustrates how the exception vector table can be remapped. As figure 6.10 demonstrates, the VBR permits the exception vector table to be relocated anywhere within the processor's memory space. The VBR is cleared following a hardware reset, so that a 68010 or later initially behaves just like a 68000.

The 68010's VBR supports *multiple* exception vector tables. Simply reloading the VBR selects a new exception vector table anywhere in the 68010's address space. Such a facility might be useful in certain classes of multitasking systems, in which each task maintains its own copy of the exception vector table. For example, the various tasks may treat, say, a divide-by-zero exception in different ways. If each task shared the same exception vector table, the exception handler would have to determine which task generated the current divide-by-zero exception and then call the corresponding procedure. By changing the exception vector table each time a new task is run, exceptions are automatically vectored to the handler appropriate to the task now running.

The VBR also removes the dilemma posed by the *desire* to locate read/write memory in the region $00 0000 to $00 03FF and the *need* to put the two longwords containing the supervisor stack pointer and initial program counter at $00 0000 and $00 0007 in ROM. We can locate ROM in, for example, the range $00 0000–$00 FFFF and include a predefined exception vector table in the range $00 0000–$00 03FF. This initial exception vector table is in ROM and cannot be modified. Once the system is up and running, the exception vector table can be relocated in any suitable block of read/write memory merely by modifying the contents of the VBR. Remember that the VBR is loaded with zero following a hardware reset and, therefore, the 68010 automatically emulates the 68000's mode of exception vector generation until the operating system explicitly loads an offset into the VBR.

Exception Processing

We are now going to look at what happens when the 68000 responds to an exception. Since the 68010, 68020, and later processors handle certain classes of exception in a slightly different way than the 68000, we will consider these cases after we have described the 68000's exception-processing mechanism.

The 68000 responds to an exception in four phases. In phase 1, the processor makes a temporary internal copy of its status register and modifies the current status register ready for exception processing. This process involves setting the S-bit and clearing the T-bit (i.e., TRACE bit). The S-bit is set because all exception processing takes place in the supervisor mode. The T-bit is cleared because the trace mode *must* be disabled during exception processing. Remember that the trace mode forces an exception after the execution of each instruction. If the T-bit were set, an instruction would trigger a trace exception that would, in turn, cause a trace exception after the first instruction of the trace-handling routine had been executed. In this way, an infinite series of exceptions would be generated.

Two specific types of exception have a further effect on the contents of the status byte of the SR. After a reset, the interrupt mask bits are automatically set to indicate an interrupt priority of level 7. That is, all interrupts below level 7 are initially disabled. The status byte is also modified by an interrupt. An interrupt causes the interrupt mask bits to be set to the same level as the interrupt currently being processed. As we shall see, the CPU responds only to interrupts with a priority greater than that reflected by the interrupt mask bits.

In phase 2, the vector number corresponding to the exception being processed is determined. Apart from interrupts, the vector number is generated internally by the 68000 according to the exception type. If the exception is an interrupt, the interrupting device places the vector number on data lines D_{00} to D_{07} of the processor data bus during the interrupt acknowledge cycle, signified by a function code (FC2,FC1,FC0) of 1,1,1. Under certain circumstances, to be described when we deal with interrupts, an external interrupt can generate a vector number internally in the 68000, in which case the interrupting device does not supply a vector number. Once the processor has determined the vector number, it multiplies it by 4 to calculate the location of the exception-processing routine within the exception vector table.

In phase 3, the current *CPU context* is saved on the stack pointed at by the *supervisor stack pointer*, A7. The CPU context is the information required by the CPU to return to normal processing after an exception. A reset does not, of course, cause anything to be saved on the stack, as the state of the system is undefined prior to a reset. Phase 3 of the exception processing is complicated by the fact that the 68000 divides exceptions into *two* categories and saves different amounts of information according to the nature of the exception. The information saved by the 68000 is called the *most volatile portion of the current processor context* and is saved in a data structure called an *exception stack-frame*.

Figure 6.11 shows the structure of the 68000's two types of exception stack frame. The 68000's exceptions are classified into three groups. We will return to this point shortly. The information saved during group 1 or group 2 exceptions (figure 6.11a) is only the program counter (two words) and the system status register, temporarily saved during phase 1. The PC and the SR are the minimum information required by the processor to restore itself to the state it was in prior to the exception.

The 68000's exceptions are divided into three groups, according to their priority and characteristics, and are categorized in table 6.3. Basically, a Group 0 exception originates from hardware errors (the address error has all the characteristics of a

FIGURE 6.11 Stack frames

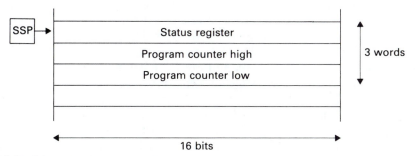

(a) Stack frame for Group 1 and Group 2 exceptions

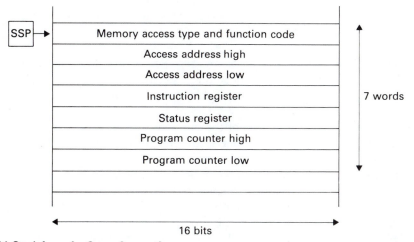

(b) Stack frame for Group 0 exceptions

TABLE 6.3 68000 Exception grouping according to type and priority

GROUP	EXCEPTION TYPE	TIME AT WHICH PROCESSING BEGINS
0	Reset Bus error Address error	Exception processing begins within two clock cycles.
1	Trace Interrupt Illegal op-code Privilege	Exception processing begins before the next instruction.
2	TRAP TRAPV CHK Divide by zero	Exception processing is started by nomral instruction execution.

bus error but is generated internally by the 68000) and often indicates that something has gone seriously wrong with the system. Because of this, the information saved in the stack frame corresponding to a group 0 exception is more detailed than that for groups 1 and 2. Figure 6.11b shows the stack frame for group 0 exceptions (except for a reset, which does not save information on the stack).

Two additional items saved in the stack frame by a group 0 exception are a copy of the first word of the instruction being processed at the time of the exception and the 32-bit address that was being accessed by the aborted memory access cycle. The third new item saved at the top of the stack is a 5-bit code (in bits 4:0 of the top word on the exception frame) giving the function code displayed on FC2,FC1,FC0 at the time the exception occurred, together with an indication of whether the processor was executing a read or a write cycle (R/W bit 4) and whether it was processing an instruction or not (I/N bit 3). For example, if the top of stack is $0012, corresponding to bits 4:0 = 10010, the faulted bus cycle is interpreted as: a read cycle (bit 4 = 1), an instruction-processing cycle (bit 3 = 0), and a user data access (bits 2:0 = 010).

The information saved on a group 0 exception stack frame is largely diagnostic and may be used by the operating system when dealing with the cause of the exception. We shall see later that this statement does not apply to the 68010, 68020, and so on.

By the way, the value of the program counter saved on a group 0 stack frame is the address of the first word of the instruction that leads to the bus fault plus a value between 2 and 10; that is, the program counter value is *indeterminate* and does not point at the next instruction following the exception (as it does in the case of group 1 and 2 exceptions). This uncertainty arises because a bus error can happen at any point during the execution of a long instruction and the 68000 does not store enough internal information to deal correctly with a bus error. For example, a MOVE.L $1234,$3334 instruction might generate a bus error during the instruction fetch, operand fetch, or operand store phases. Although the 68000 cannot itself return from a group 0 exception, it is possible to write a bus error exception handler to use the information on the stack to create a new stack frame from which a return can be made. This procedure is not recommended. If you wish to implement a return from a group exception you should use a 68010 or a later processor.

The fourth, and final, phase of the exception processing sequence consists of a single operation—the loading of the program counter with the 32-bit address pointed at by the exception vector. Once this has been done, the processor continues executing instructions normally. These instructions are, of course, the exception-handling routine.

When an exception-handling routine has been run to completion, the return from exception instruction, RTE, is executed to restore the processor to the state it was in prior to the exception. RTE is a privileged instruction and has the effect of restoring the status register and program counter from the values saved on the system stack. The contents of the program counter and status register just prior to the execution of the RTE are lost.

It is important to stress that an RTE instruction cannot be used after a group 0 exception to execute a return (although, of course, it works perfectly well for all group 1 and group 2 exceptions). There are two reasons why you cannot use an RTE following a bus error exception. The first is that the RTE pulls the program counter and status register off the stack. Since the group 0 stack frame has a different structure than group 1 and 2 frames, an RTE just would not work. The programmer could always try to modify the group 0 stack frame and make it look like a group 1 stack frame before executing the RTE. The second reason is the one we mentioned earlier: The value of the program counter saved on the stack frame after a bus error is not reliable. Some programmers have managed to return from a bus error by means of a clever trick. Since the group 0 stack frame contains the value of the instruction at the time of the bus error, you can read the program counter from the stack frame and then *search* for the instruction that caused the bus error.

Figure 6.12 graphically summarizes the way in which the 68000 processes exceptions. We shall now look at how the newer members of the 68000 family implement exception processing.

Exception Processing and the 68010, 68020, and 68030 Processors

Broadly speaking, the 68010, 68020, and 68030 handle exceptions in the same way as the 68000. These newer microprocessors build on the exception-handling ability of the 68000. It would be reasonable to say that as the applications of microprocessors have grown increasingly sophisticated, the 68000 family has been extended to match the requirements of these new applications. This statement is probably more true when applied to exception handling than any other aspect of microprocessor architecture. Consider the way in which the 68000 family has progressed, as shown in table 6.4.

TABLE 6.4   Summary of 68000 family exception-handling capabilities

TRADITIONAL 8-BIT MICROPROCESSORS	68000 MICROPROCESSOR	68010 MICROPROCESSOR	68020/30 MICROPROCESSORS
Simple interrupt handling with fixed vectors to interrupt handlers and one or two interrupt request inputs. Very limited software exception-handling capabilities.	Seven levels of prioritized and vectored interrupts. Extensive range of software exceptions. User and supervisor modes permit protected operating system.	Same as 68000 but with the ability to recover from a bus error. This makes it possible to implement virtual memory systems. Addition of a vector base register improves its multitasking performance. Making MOVE from SR a privileged instruction transforms the 68010 into a true virtual machine.	Same as 68010 but with two supervisor state stack pointers. One of these can be devoted to interrupts and the other to task control blocks in a multitasking environment. This enhancement makes the 68020 ideal for environments supporting both multitasking and extensive interrupt handling.

FIGURE 6.12 The 68000's exception-processing sequence

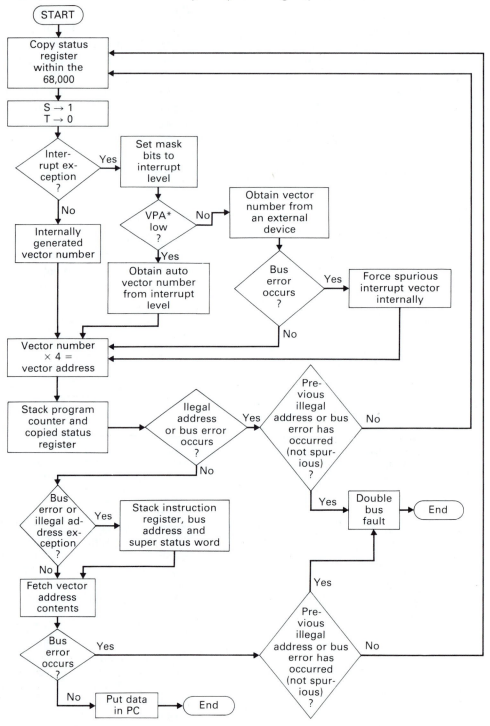

We have already seen that the 68010 improves on the 68000's exception-handling mechanism, because it is able to remap the exception vector table by means of its VBR. The addition of a VBR is not the only change in the way the 68010, 68020, and 68030 process exceptions. All these processors save more information on the exception stack frame than the 68000. Indeed, several different stack frames are necessary to facilitate a return from exception under all circumstances. The 68010 (like the 68000) has just two stack frames, whereas the 68020/030 has six.

Another major extension to the 68000 family's exception processing mechanism implemented by the 68020 and 68030 is the inclusion of an additional supervisor state stack pointer, the *master stack pointer*, MSP. At any instant, one of three stack pointers may be active in a 68020/30 system. Following a reset, the 68020/30's M-bit (bit 12 of its status register) is cleared and the *interrupt stack pointer*, ISP, is selected. The 68020's ISP is the default supervisor state stack pointer and corresponds to the 68000's SSP. However, when the 68020 is in the supervisor mode, its M-bit may be set to select the master stack pointer, MSP. Consequently, the systems mode programmer has a choice of two stack pointers. The relationship between the S- and M-bits and the stack pointers can be summarized as follows:

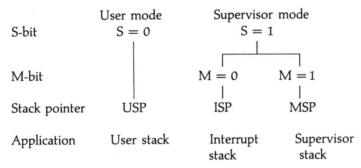

	User mode	Supervisor mode
S-bit	S = 0	S = 1
M-bit		M = 0 M = 1
Stack pointer	USP	ISP MSP
Application	User stack	Interrupt stack Supervisor stack

Although the provision of two supervisor stack pointers seems a little excessive, there is an excellent reason. In a multitasking system several user tasks are run concurrently by switching from task to task so rapidly that the processor appears to be executing the tasks simultaneously. When the processor switches from one task to another, the information required to rerun the task just switched off must be saved. This information is a copy of the processor's registers, its CCR, and its stack pointer (i.e., the USP) and is saved in a structure called a *task control block*, or TCB. The TCB is saved on the supervisor mode stack. At the end of this chapter we look at multitasking in greater detail.

Now suppose that an interrupt occurs during the time a given task is being processed. The interrupt's stack frame is stored on the task's (supervisor state) stack. If interrupts are heavily used, a task might require an enormous TCB just to store interrupt stack frames. Moreover, each task must have an equally large TCB to store interrupt stack frames generated while the task is active. A much more sensible solution is to implement two entirely separate supervisor state stacks. One stack is dedicated to multitasking (and general-purpose supervisor state applications) and the other is dedicated to interrupt processing. That is, interrupts store information on an entirely separate stack than that used to handle all other exceptions. In this way, the systems designer does not have to worry about saving interrupt information within a task's TCB.

The 68020 uses its MSP to maintain a general-purpose stack and its ISP to maintain a stack dedicated to interrupt handling. When M = 0 (following a reset), the 68020 behaves exactly like the 68000/68010 and uses a single supervisor mode stack pointer, the ISP, for all exception processing. When the M-bit of the status register is set to 1 (by the operating system), both the 68020's supervisor state stacks are activated. By switching the master stack pointer each time a new task is activated, a separate master stack pointer value can be assigned to each task in a multitasking environment. Once the M-bit has been set, interrupts use the interrupt stack pointer and not the master stack pointer.

When the 68020/30 processes an *interrupt* (but not any other exception), it tests the status of the M-bit after the processor context has been saved on the currently active supervisor stack. If the M-bit is clear, exception processing continues normally (since there is only one supervisor state stack that is pointed at by the ISP).

If the M-bit in the status register is set, the processor clears it and creates *throwaway* exception stack frame on top of the interrupt stack. Note that there are now *two* interrupt stack frames, one on each of the supervisor state stacks. This second throwaway stack frame on the interrupt stack contains the same program counter value and vector offset as the frame created on top of the master stack. However, the stack frame on top of the interrupt stack has a format number 1 instead of 0 or 9 (stack formats are described later).

The copy of the status word on the throwaway frame is the same as the version in the frame on the master stack, except that the S-bit is set and the M-bit is cleared in the version placed on the interrupt stack. The version of the status word on the master stack may have S = 0 or S = 1, depending on whether the processor was in the user or supervisor state before the interrupt. Interrupt processing then takes place in the normal way, except that the ISP is the active stack pointer.

As we have just said, at the end of the interrupt-processing sequence, the processor's S-bit is set and its M-bit is cleared. When a return from exception is executed, the processor reads the status register from the throwaway frame on the interrupt stack, increments the active stack pointer by 8, and begins RTE processing again. This repetition may seem strange, but remember that there are two stack frames (one on the ISP stack and one on the MSP stack).

Stack Frames of the 68010, 68020, and 68030

We already know that the 68000 has two types of stack frame, one for both group 1 and group 2 exceptions and a more complex stack frame for group 0 exceptions (i.e., bus errors and address errors). The 68020 and 68030 employ a total of six stack frames, according to the type of exception being processed. No simple "68000-type" stack frame is implemented, because all stack frames include at least four words. In fact, the first four words of all six stack frames are identical and comprise (starting at the top of the stack): status register, program counter, stack frame format, and vector offset.

The *format* of the stack frame is stored in bits 12–15 of the word at address

[SP] + 6. Bits 0–11 of the same word contains the *vector offset*, which is the exception number multiplied by 4. That is, the vector offset is the index into the exception vector table and permits the exception handler to determine the nature of the exception that invoked it. The format is read by the processor itself when it executes a return from exception. (Otherwise, how would the processor know how much information to restore when it encounters an RTE?)

Figures 6.13 to 6.18 describe the 68020's six stack frames. Note that the *name* of the stack frame is determined by the binary value of the frame's format. For example, if the format code is 1010, the stack frame is called *stack frame ten*.

Stack frame 0 (figure 6.13) has a four-word structure and is used by the following exceptions:

Interrupts
Format error
TRAP #N
Illegal instruction
A-line and F-line instructions
Privilege violation
Coprocessor preinstruction

Stack frame 1 (figure 6.14) is a *throwaway* frame used during interrupt processing when a change from master state to interrupt state is made. When the M-bit of the SR is set, the exception stack frame is saved on the stack pointed at by the master stack pointer. If the exception is an interrupt, a second copy of stack frame

FIGURE 6.13 Stack frame 0

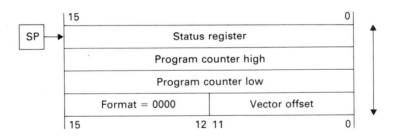

FIGURE 6.14 Stack frame I

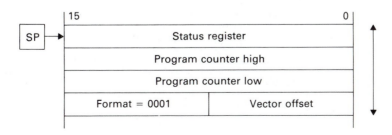

FIGURE 6.15 Stack frame 2

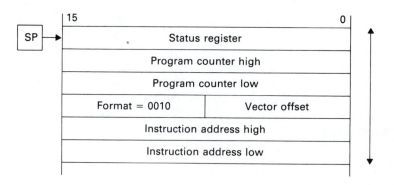

(i.e., a type 1 stack frame) is pushed onto the stack pointed at by the ISP and the M-bit is cleared.

Stack frame 2 (figure 6.15) has a six-word structure and is used by the following exceptions:

CHK, CHK2
ccTRAPcc, TRAPcc, TRAPV
Trace
Divide by zero
MMU configuration
Coprocessor postinstruction

Stack frame 9 (figure 6.16) stores ten words and is used by the following exceptions:

Coprocessor midinstruction
Main-detected protocol violation
Interrupt detected during coprocessor instruction

FIGURE 6.16 Stack frame 9

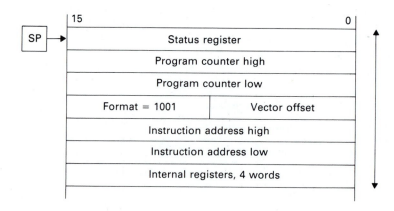

FIGURE 6.17 Stack frame 10

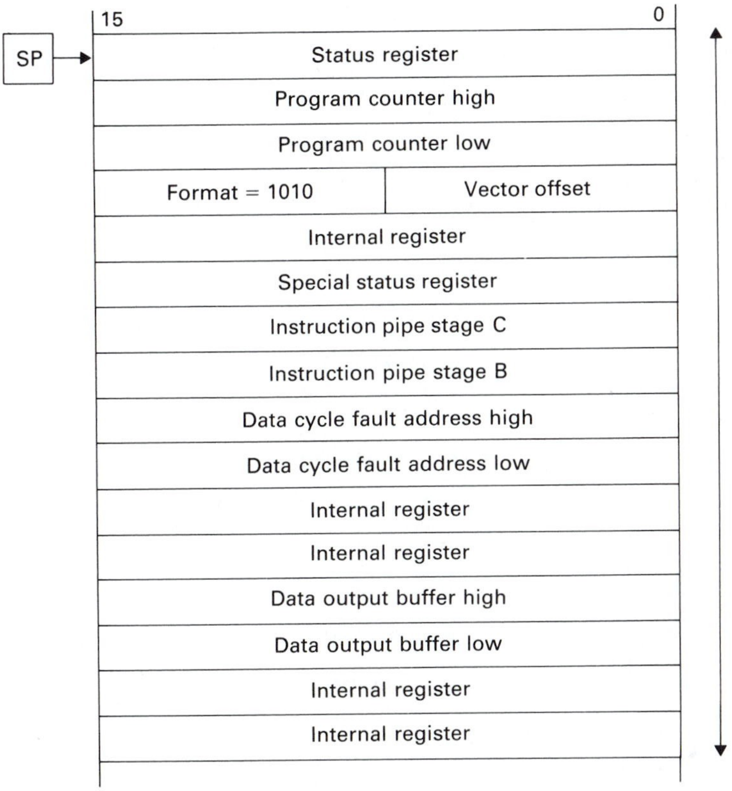

Stack frame 10 (figure 6.17) stores 16 words on the stack and is used when an address error or bus error exception occurs at an instruction boundary. This format is called a short bus cycle fault format, since it is relatively easy to return from an exception at an instruction boundary.

Stack frame 11 (figure 6.18) is the longest stack frame and is composed of 46 words. It is used when an address error or a bus error exception occurs during the execution of the faulted instruction.

Multiple Exceptions and the 68020/30

Since two or even more exceptions can occur at the same time, the processor must prioritize competing exceptions. The 68020/30 provides a fixed prioritization for exceptions, as listed in table 6.5. Note that exceptions that include the prefix "Cp" are exceptions related to the 68020's coprocessor interface. When multiple simultaneous exceptions occur, the processor deals with the highest priority exception and then processes the exception with the next highest priority, and so on. High-priority

FIGURE 6.18 Stack frame 11

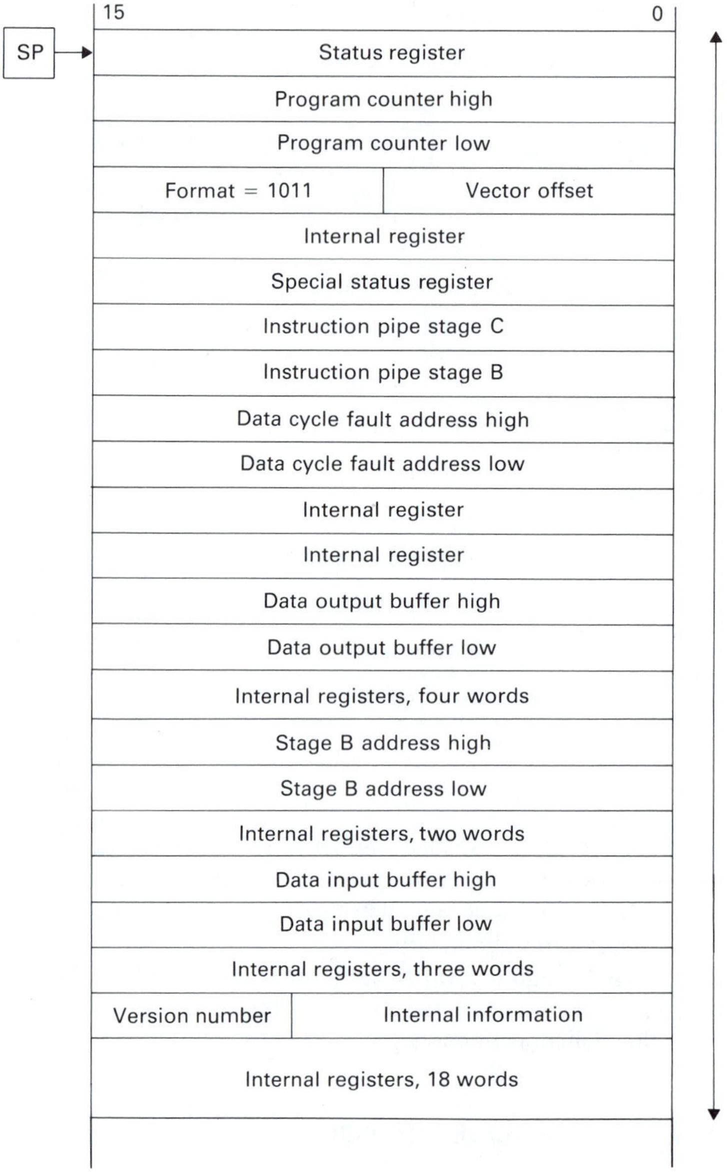

15		0
SP →	Status register	
	Program counter high	
	Program counter low	
Format = 1011		Vector offset
	Internal register	
	Special status register	
	Instruction pipe stage C	
	Instruction pipe stage B	
	Data cycle fault address high	
	Data cycle fault address low	
	Internal register	
	Internal register	
	Data output buffer high	
	Data output buffer low	
	Internal registers, four words	
	Stage B address high	
	Stage B address low	
	Internal registers, two words	
	Data input buffer high	
	Data input buffer low	
	Internal registers, three words	
Version number		Internal information
	Internal registers, 18 words	

exceptions such as bus and address errors are processed immediately—even if another exception is currently being processed.

In the next section we are going to leave the theory of exception processing and look at the hardware required to implement external exception processing.

TABLE 6.5 Exception priority groups for the 68020/30

GROUP	EXCEPTION	CHARACTERISTICS
0.0	Reset	Aborts all processing and does not save the old machine context.
1.0	Address error	Suspends processing and saves
1.1	Bus error	internal machine context.
2.0	BKPT #n, CHK, CHK2, Cp midinstruction, Cp protocol violation, CpTRAPcc, divide by zero, RTE, TRAP #n, TRAPV, MMU configuration	Exception processing is part of the instruction execution.
3.0	Illegal instruction, line A unimplemented line F, privilege violation, Cp preinstruction	Exception processing begins before the instruction is executed.
4.0	Cp postinstruction	Exception processing begins when the
4.1	Trace	current instruction or previous
4.2	Interrupt	exception processing is completed.

6.5 HARDWARE-INITIATED EXCEPTIONS

Three types of exception are initiated by events taking place outside the 68000 and are communicated to the CPU via its input pins: the reset, the bus error, and the interrupt. Each of these three exceptions has a direct effect on the hardware design of a 68000-based microcomputer. We now examine each of these exception types in more detail.

Reset

A reset is a special type of exception, because it takes place under only two circumstances: a power-up or a total and irrevocable system collapse. For this reason, the reset exception has the highest priority and is processed before any other exception that is either pending or being processed. Following the detection of a reset, by the RESET* pin being asserted for the appropriate duration, the 68000 sets the S-bit, clears the T-bit, and sets the interrupt mask level to seven (i.e., SR = $2700). The 68000 then loads the supervisor stack pointer with the longword at memory location $00 0000 and loads the program counter with the longword at memory location

$00 0004. Once this has been done, the 68000 begins to execute its start-up routine. Figure 6.19 illustrates the 68000's reset sequence.

Although the 68000's stack pointer is set up during the reset, the user stack pointer is not. Systems designers must take care not to switch from supervisor state to user state and then forget to preset the user stack pointer. The privileged instruction MOVE An,USP can be employed to set up the USP while the computer is still in the supervisor state. Equally, the user stack pointer can be set up in the user state.

The 68020's reset exception is handled in a similar fashion, but with the following differences: The vector base register, VBR, is reset to zero, *both* trace bits are cleared, the enable and freeze bits of the on-chip cache are cleared, and the contents of the on-board cache are invalidated. That is, following a reset the 68020 and 68030 behave as if they did not have cache memories. In addition, the 68030 invalidates all enties in its data cache and clears the enable bit in the translation control register of

FIGURE 6.19 The 68000's reset sequence

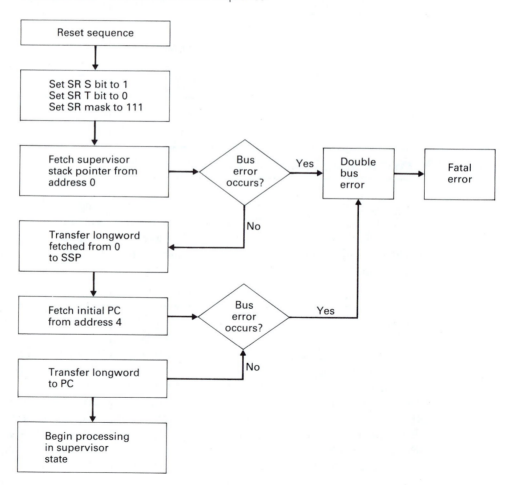

its memory management unit. In short, the 68020 *flushes* its cache and the 68030 bypasses its memory management unit.

In section 6.3 we stated that the exception vector table is frequently located in read/write memory and that the circuit of figure 6.9 can be used to overlay the first 8 bytes of the read/write memory with ROM containing the reset vectors. The only other circuitry associated with the 68000's reset mechanism is that connected to the RESET* pin itself.

The designers of the 68000, in an attempt to minimize the number of pins, have made RESET* a bidirectional input/output, thereby complicating the design of the reset circuitry. In normal operation, the RESET* pin is an input. Almost all microprocessor systems connect every device that can be reset to the RESET* pin. Consequently, all devices are reset along with the 68000 at power-up or following a manual reset. The 68000 is also capable of executing a software RESET instruction, which forces its RESET* pin active-low, resetting all devices connected to it. This facility has been provided to permit a system reset under software control that does not affect the processor itself. The RESET* pin of the 68000 cannot, therefore, be driven by gates with active pull-up circuits. RESET* must be driven by open-collector or open-drain outputs.

Another aspect of a hardware-initiated reset to be noted is that both RESET* and HALT* must be asserted *simultaneously*. It should be remembered that the HALT* pin is also bidirectional and must be driven by an open-collector or open-drain output. If RESET* and HALT* are asserted together following a system crash, they must be held low for at least ten clock cycles to ensure satisfactory operation. However, at power-up they must be held low for at least 100 ms after the V_{cc} supply to the 68000 has become established. The 68020 requires that its RESET* be held low for 520 clock cycles for a satisfactory reset. However, the 68020/030 does not require that HALT* be asserted along with RESET*.

A possible arrangement of a reset circuit for the 68000 is given in figure 6.20. IC1, a 555 timer, generates an active-high pulse at its output terminal shortly after the initial application of power. The timer is configured to operate in an astable mode, generating a single pulse whenever it is triggered. The time constant of the output pulse (i.e., the duration of the reset pulse) is determined by resistor R_2 and capacitor C_2. R_1 and C_1 trigger the circuit on the application of power. The output is buffered and inverted by IC2a, to become the system active-low, power-on-reset pulse (POR*) which can be used by the rest of the system as appropriate.

The output of the timer is also connected to one terminal of a two-input OR gate, IC3. If either input goes high, the output goes high, forcing both HALT* and RESET* low via inverting buffers IC2b and IC2c, respectively. Both buffers have open-collector outputs that are pulled up toward V_{cc} by 4.7-kΩ resistors.

A manual reset facility is provided by an RS bistable constructed from two cross-coupled NAND gates, IC4a and IC4b. The RS bistable debounces the switch to avoid multiple reset pulses. In normal operation, the push button is in the NC (normally closed) position and the output of IC4b is low. When the button is pushed into its NO (normally open) position, the output of IC4b rises, generating a reset pulse. The duration of this pulse is determined by the time for which the button is depressed and is likely to be many orders of magnitude longer than the ten-clock-

FIGURE 6.20 Control of the 68000's RESET* input

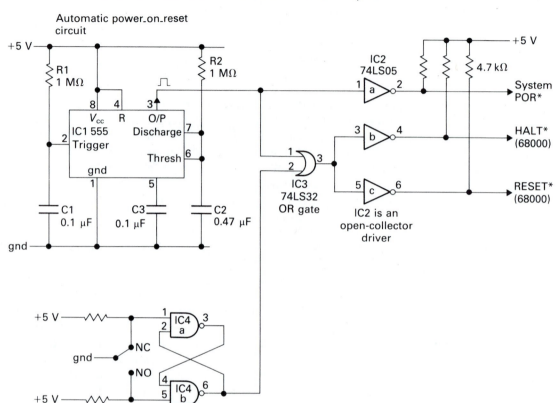

Manual reset circuit

pulse minimum required by the 68000. Once more, we stress that a manual reset should not be used until all other forms of recovery have failed.

Alternative Ways of Remapping the Reset Vectors

Earlier in this section, we looked at how the reset vectors could be relocated in ROM anywhere within the 68000's address space and then remapped to the region $00 0000 to $00 0007 during the reset vector fetch. We are now going to look at two other ways of performing this vector-remapping operation.

The reset vector-remapping scheme we described in figure 6.9 requires that address lines A_{03}–A_{23} be decoded, a rather complex operation requiring at least two or more ICs. It is possible to remap the reset exception vector in a rather more simple fashion. Consider the circuit in figure 6.21. Instead of detecting the address of the reset vector, we can use the fact that the *first operations* the 68000 does after a reset are to read the stack pointer and the reset vector from the exception table.

When RESET* is active-low, the 74LS164 shift register is forced into a reset state and all its outputs are forced low. The Q_D output is used as the MAP* signal

FIGURE 6.21 Using a shift register to remap the reset vectors

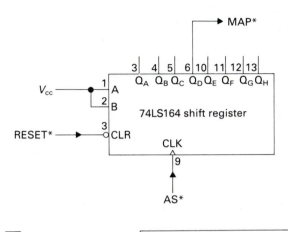

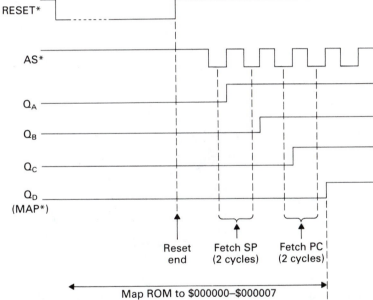

to switch between the RAM and the ROM during the reset vector fetch. The 68000 begins its reset exception processing when RESET* goes inactive-high, which also releases the shift register. Each bus cycle of the 68000 (caused by AS* being pulsed) clocks the shift register. After four bus cycles, the Q_D output of the shift register goes high.

Another way of dealing with the reset exception vector problem is to provide *shadow* ROM. Consider figure 6.22, in which 64K-byte blocks of both ROM and RAM are located in the region $00 FFFF. Of course, only one block of memory is selected at any instant. The simple RS bistable, FF1, is reset at the same time the 68000 system is reset by the POR* signal and the ROM is selected. Therefore, the

FIGURE 6.22 Using read-write memory to shadow ROM

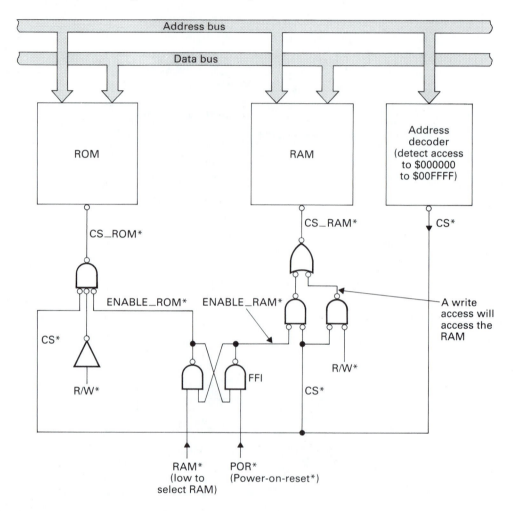

68000 powers up, with the ROM (containing the whole exception table) mapped from $00 0000 to $00 FFFF.

When a read access is executed, data is read from the ROM. However, when a write access is made, data is written into the RAM at the same address, because the RAM-select logic ensures that the RAM is always selected by a write access. Suppose the programmer executes the following sequence:

```
COPY    MOVEA.L   #$00000000,A0    A0 points to the start of the exception table
        MOVE.W    #$7FFF,D0        There are 32K words in the table to move
COPY1   MOVE.W    (A0),(A0)+       Move a word from the ROM to the RAM
        DBRA      D0,COPY1         Repeat until all the table is copied to RAM
        MOVE.B    #1,RAM_SEL       Now select the RAM for reading
```

This strange-looking code reads a word from an address and then writes it into the *same* address. The data is read from the ROM and stored in the RAM at the same address. When all the data has been transferred, the RAM is selected by setting the RS bistable. Now, all accesses, both read and write, are made to the shadow RAM. The ROM is locked out until the system is once more reset.

What, then, are the advantages of this arrangement of shadow RAM? First, the associated control logic is relatively simple. Second, both the monitor and the exception table are in RAM, and each can be modified dynamically. This feature might be very useful during the development of a monitor (a good monitor can be stored in ROM while the new monitor is developed in RAM). Finally, it permits the use of low-speed ROM and high-speed RAM. Suppose the monitor/operating system is large and is located in relatively slow ROM. One effect of this might be to introduce wait states if the 68000 has a high-speed clock. By copying the monitor into faster RAM, wait states can be avoided. But note that this would require a wait state generator to be activated when the ROM is selected and defeated when the shadow RAM is selected. The disadvantage of shadow RAM is that is wastes memory.

Bus Error

A bus error is an exception raised in response to a failure by the system to complete a bus cycle. There are many possible failure modes, and the details of each depend on the type of hardware used to implement the system. Therefore, the detection of a bus error has been left to the systems designer rather than to the 68000 chip itself. All the 68000 provides is an active-low input, BERR*, which, when asserted, generates a bus error exception.

Figure 6.23 gives the timing requirements that the BERR* input must satisfy. In order to be recognized during the current bus cycle, BERR* must fulfill one of two conditions. BERR* must be asserted at least t_{ASI} seconds (the asynchronous input setup time) before the falling edge of state S4, or it must be asserted at least t_{BELDAL} (BERR* low to DTACK* low) seconds before the falling edge of DTACK*. It is necessary to maintain BERR* active-low until t_{SHBEH} seconds (AS* high to BERR* high) after the address and data strobes have become inactive. The minimum value of t_{SHBEH} is 0 ns, implying that BERR* may be negated concurrently with AS* or DS*. It is important to realize that if BERR* meets the timing requirement t_{ASI}, it will be processed in the current bus cycle irrespective of the state of DTACK*.

There are a number of reasons why BERR* may be asserted in a system. Typical applications of BERR* are as follows:

1. *Illegal Memory Access.* If the processor tries to access memory at an address not populated by memory, BERR* may be asserted. Equally, BERR* may be asserted if an attempt is made to write to a memory address that is read-only. A decision as to whether to assert BERR* in these cases is a design decision. It is not mandatory. All 8-bit microprocessors are quite happy to access non-existent memory or to write to ROM. The philosophy of 68000 systems design is to trap events that may lead to unforeseen circumstances. If the processor tries to write to ROM, the operating system can intervene because of the exception raised by BERR*.

FIGURE 6.23   Timing diagram of the 68000's bus error input (BERR*)

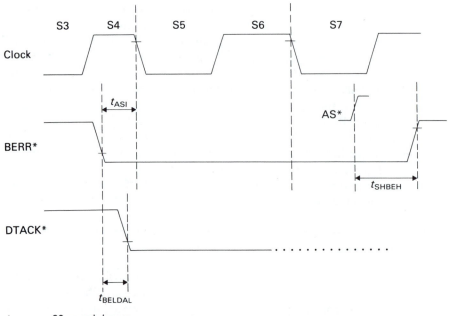

t_{ASI} = 20 ns minimum
t_{BELDAL} = 20 ns minimum
t_{SHBEH} = 0 ns minimum

2.   *Faulty Memory Access*. If error-detecting memory is employed, a read access to a memory location at which an error is detected can be used to assert BERR*. In this way the processor will never try to process data that is in error due to a fault in the memory.

3.   *Failure to Assert VPA*. If the processor accesses a *synchronous bus* device and VPA* is not asserted after some time-out period, BERR* must be asserted to stop the system from hanging up and waiting for VPA* forever.

4.   *Memory Privilege Violation*. When the 68000 is used in a system with some form of memory management, BERR* may be asserted to indicate that the current memory access is violating a privilege. A privilege violation may be caused by an access by one user to another user's program space or by a user to supervisor space. In a system with virtual memory, a memory privilege violation may result from a page-fault, indicating that the data being accessed is not currently in read/write memory. Chapter 7 deals with memory management.

Bus Error Sequence

When BERR* is asserted by external logic and satisfies its setup timing requirements, the processor negates AS* in state S7. As long as BERR* remains asserted, the data and address buses are both floated. When the external logic negates BERR*, the processor begins a normal exception-processing sequence for a group 0 exception. Figure 6.11b shows that additional information is pushed on the system stack by the 68000 to facilitate recovery from the bus error. Once all phases of the exception-processing sequence have been completed, the 68000 begins to deal with the problem of the bus error in the BERR* exception-handling routine.

It must be emphasized that the treatment of the hardware problem that led to the bus error takes place at a software level within the operating system. For example, if a user program generates a bus error, the exception-handling routine may abort the user's program and provide him or her with diagnostic information to help deal with the problem that caused the exception. The information stored on the stack by a bus error exception (or an address error) is to be regarded as diagnostic information only and should not be used to institute a return from exception, as we have already stated. In other words, the 68000 does not support a direct return from a group 0 exception. The 68010, 68020, and 68030 processors can execute a return from bus error exception.

Before we look at how the 68010 and later processors can recover from a bus error exception, we will discuss the ability of all members of the 68000 family to rerun (i.e., repeat) a bus cycle without performing exception processing.

Rerunning the Bus Cycle

It is possible to deal with a bus error in a way that does not involve an exception. If, during a memory access, the external hardware detects a memory error and asserts both BERR* and HALT* simultaneously, the processor attempts to rerun the current bus cycle.

Figure 6.24 demonstrates a rerun cycle. A bus fault is detected in the read cycle and both BERR* and HALT* are asserted simultaneously. As long as the HALT* signal remains asserted, the address and data buses are floated and no external activity takes place. When HALT* is negated by the external logic, the processor will rerun the previous bus cycle using the same address, the same function codes, the same data (for a write operation), and the same control signals. For correct operation, the BERR* signal must be negated at least one clock cycle before HALT* is negated.

A possible implementation of bus error control in a sophisticated 68000-based system might detect a bus error and assert BERR* and HALT* simultaneously. The rising edge of AS* can be used to release BERR* and then HALT* at least a clock cycle later. This guarantees a rerun of the bus cycle. Of course, if the error is a *hard* error (i.e., is persistent), rerunning the bus cycle will achieve little and external logic will once again detect the error. A reasonable strategy would be to permit, say, a single rerun and, on the next cycle, assert BERR* alone, forcing a conventional bus error exception.

We will now look at the type of logic required to rerun a bus cycle. Suppose we have a memory system that returns an active-low error signal when a parity error is detected. We wish to use this error signal to automatically initiate a rerun of the bus cycle (Figure 6.24). All we have to do is to assert BERR* and HALT* simultaneously as soon as the parity error is detected.

The 68000 responds to BERR* asserted by negating its address strobe. We can

FIGURE 6.24 Rerunning the bus cycle

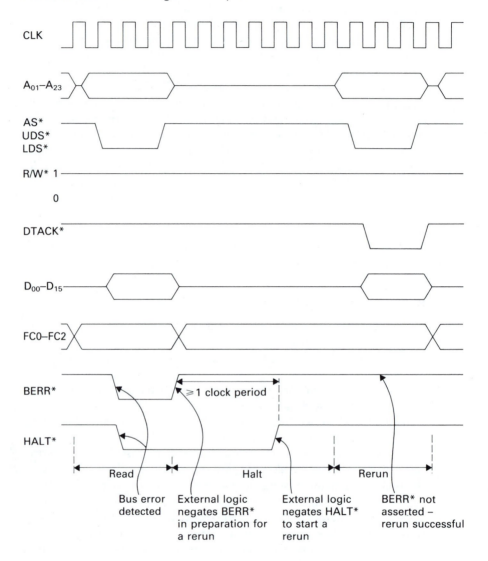

detect the rising edge of the address strobe and use it to negate bus error. After waiting for at least a cycle, we can release HALT* to permit the rerun.

It doesn't take a lot of logic to perform the sequence of operations necessary to rerun a bus cycle, as you can see in figure 6.25. Flip-flop FF1 is held in its reset state until the 68000's address strobe goes low. If, during the bus cycle, the memory returns an error signal while the address strobe is still low, the Q_1* output of flip-flop FF1 goes low to provide the 68000 with a bus error input. The Q_1* signal is also passed through an OR gate (negative logic) to provide a simultaneous HALT* input to the 68000.

FIGURE 6.25 Bus cycle rerun logic

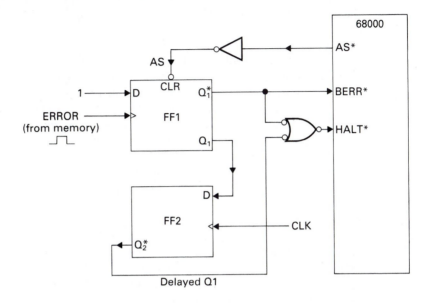

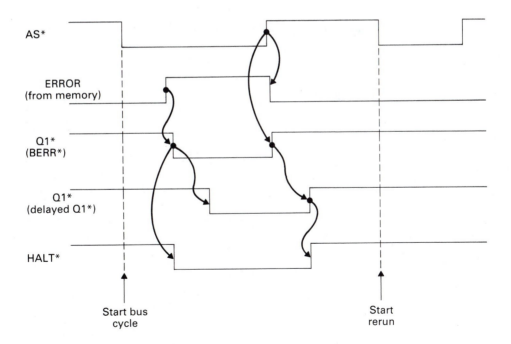

Since the HALT∗ signal must be negated at least a clock cycle after BERR∗, we can use another D flip-flop, FF2, to provide a suitable delay from which we can generate a HALT∗ signal. The HALT∗ input to the 68000 is generated by ORing the output of both the undelayed Q∗ (i.e., Q_1∗) and the delayed Q∗ (i.e., Q_2∗). This means that HALT∗ goes low at the same time as BERR∗ but goes inactive-high a clock cycle later.

Unfortunately, we have forgotten a vital fact in the design of the logic for rerunning a bus cycle following a parity error. If the error turns out not to be a soft error but is, in fact, a hard error, rerunning the bus cycle will not achieve a lot. Indeed, the same bus cycle will be rerun an infinite number of times. What we really want to do is to try rerunning the bus cycle once and, if that also fails, to assert BERR∗ alone to force a normal bus error exception.

You could go to a great deal of effort to design a circuit that permits just one attempt at rerunning a bus cycle. Fortunately, there is a very simple modification we can make to the rerun circuit of figure 6.25. If you think about it, what we want to do is to make the circuit ignore a second parity error that immediately follows the first one (figure 6.26).

A second delay generator, based on a counter circuit, produces a signal that is delayed considerably beyond the HALT∗ signal. This delay must be longer than the time-out period of the bus error generator. We can call this a *guard signal*. The guard signal is fed back to the input of the rerun circuit and is used to block the error signal from memory when GUARD∗ is asserted.

Suppose that a parity error occurs, and the rerun generator is triggered. After HALT∗ has been released, the bus cycle will be rerun (figure 6.27). Unfortunately, in

FIGURE 6.26 Logic necessary to ensure that a bus cycle is rerun only once

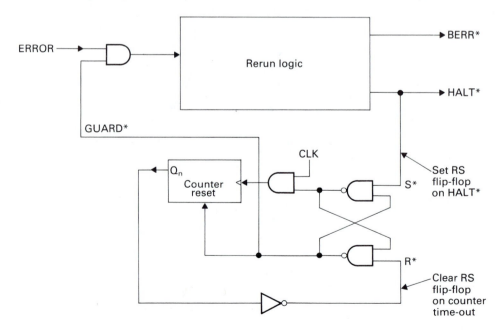

FIGURE 6.27 Timing diagram for figure 6.26

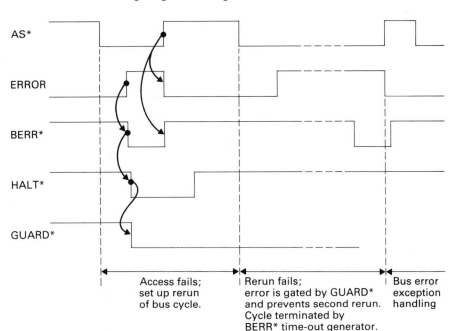

| | Access fails; set up rerun of bus cycle. | Rerun fails; error is gated by GUARD* and prevents second rerun. Cycle terminated by BERR* time-out generator. | Bus error exception handling |

this case the rerun also leads to an error signal from the memory because the error is persistent. However, this second error cannot trigger the rerun generator because the guard signal is blocking any error input.

The system will now hang until the guard signal becomes inactive. If this guard period is long enough, the signal from a bus error generator of the type we described earlier can be used to assert BERR* alone and trigger a normal bus exception.

68010/20/30 and the Bus Error

Probably the most significant advance over the 68000's exception-processing mechanism is the ability of the 68010 and later processors to recover from bus cycles terminated by a bus error exception. Whenever the 68000's BERR* input is asserted (and HALT* is not asserted), a bus error exception is forced, and the appropriate bus error exception handler is executed in software. If the bus error is due to a *page-fault* in a system with memory management, the instruction that caused the fault must be rerun once the operating system has loaded appropriate page in read/write memory from disk.

Unfortunately, as we already know, the 68000 does not store sufficient information on the stack to permit the faulted memory access to be rerun. Processors later than the 68000 save more information on the stack following a bus error than the 68000 and can use this additional information to return from a bus error. In effect,

FIGURE 6.28 The 68010 exception frame

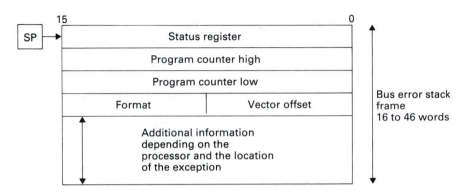

the 68000 *aborts* the instruction that caused a bus error, whereas the 68010 and later *suspend* the instruction. The *generic* bus error exception stack frame employed by the 68010 and later is described in figure 6.28.

The 68010 and later save the most volatile portion of the current processor status in a variable-length stack frame. The most significant 4 bits of the word saved immediately before the low-order word of the program counter provide the format of the current stack frame. The three format codes of interest are

1000 = 68010 bus error format (29 words)

1010 = short 68020 bus error format (16 words)

1011 = long 68020 bus error format (46 words)

The RTE (return from exception) instruction uses the format code to determine the nature of the current stack frame and thereby permit an appropriate return. The information saved on the stack when a bus error exception is processed provides everything the processor needs to continue an instruction. Notice the word *continue* in the previous sentence. The 68010 does not rerun or restart an instruction interrupted by a bus error exception. Instead, it saves sufficient information in the stack frame to continue the instruction from the point at which it was interrupted. After a bus error (or an address error) exception has been processed, an RTE can be used to complete the interrupted instruction. If the faulted bus cycle was a read-modify-write cycle, the entire cycle is rerun, whether the fault occurred during the read or the write operation.

It is possible to return from a 68010 (and 68020 and later) bus error by means of two mechanisms. One is to use the RTE instruction, as you might expect. Alternatively, you can use all the information on the stack frame to *emulate* the faulted bus cycle. In fact, emulating a bus cycle is the only way that you can return from an address error exception. If you were to attempt ro recover with an RTE, you would simply try to execute the same operation that caused the original exception. Doing this would result in an infinite sequence of exceptions and returns.

Interrupts

As we have seen, an interrupt is a request for service generated by an external peripheral. In keeping with the 68000's general versatility, it offers two schemes for dealing with interrupts. One scheme is intended for modern peripherals specially designed for 16-bit processors, whereas the other is more suited to earlier 8-bit 6800-series peripherals.

An external device signals its need for attention by placing a 3-bit code on the 68000's interrupt request inputs, IPL0*, IPL1*, IPL2*. The code corresponds to the priority of the interrupt and is numbered 0 to 7. A level 7 code indicates the highest priority, level 1 is the lowest priority, and level 0 indicates the default state of no interrupt request. Although it is perfectly possible to design peripherals with three interrupt request output lines on which they put a three-bit interrupt priority code it is easier to have a single interrupt request output and to design external hardware to convert its priority into a suitable 3-bit code for the 68000.

The 68000 internally debounces the interrupt level applied to IPL0*–IPL2*, and it is not necessary to employ any external synchronization circuitry. The only restriction of IPL0*–IPL2* is that the interrupt request be maintained until it is acknowledged by the 68000.

Figure 6.29 shows a typical scheme for handling interrupt requests in a 68000

FIGURE 6.29 Interrupt request encoding

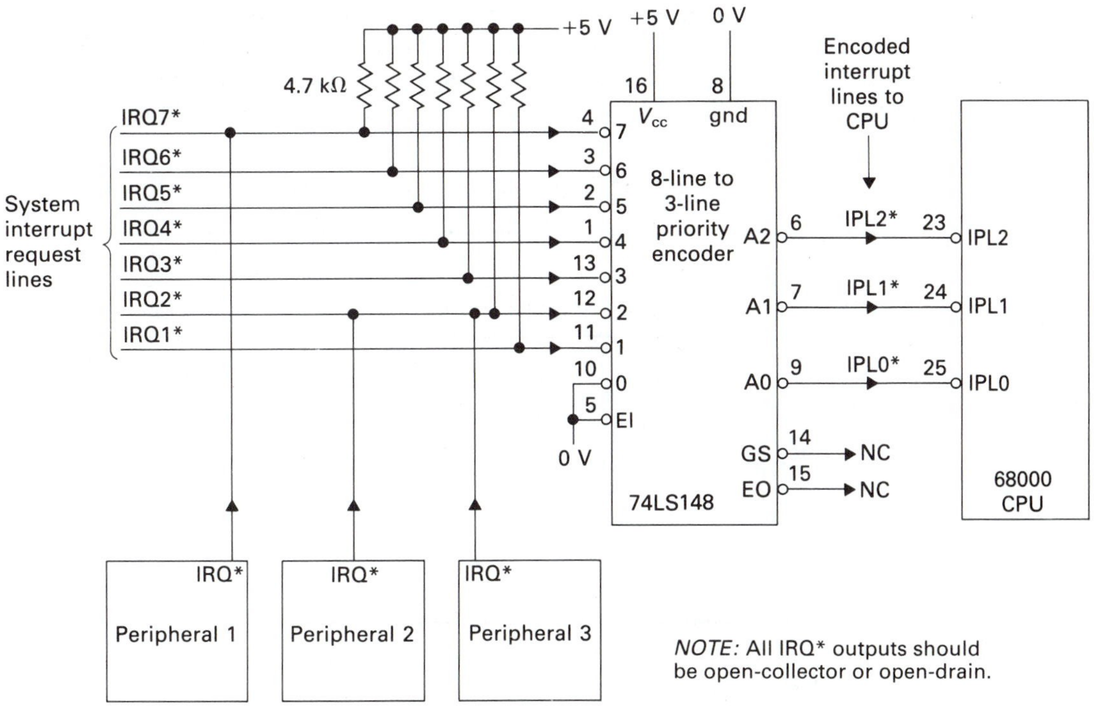

NOTE: All IRQ* outputs should be open-collector or open-drain.

TABLE 6.6 Truth table for the 74LS148 eight-line-to-three-line priority encoder

	INPUTS								OUTPUTS				
EI*	0	1	2	3	4	5	6	7	A2	A1	A0	GS*	EO*
1	×	×	×	×	×	×	×	×	1	1	1	1	1
0	1	1	1	1	1	1	1	1	1	1	1	1	0
0	×	×	×	×	×	×	×	0	0	0	0	0	1
0	×	×	×	×	×	×	0	1	0	0	1	0	1
0	×	×	×	×	×	0	1	1	0	1	0	0	1
0	×	×	×	×	0	1	1	1	0	1	1	0	1
0	×	×	×	0	1	1	1	1	1	0	0	0	1
0	×	×	0	1	1	1	1	1	1	0	1	0	1
0	×	0	1	1	1	1	1	1	1	1	0	0	1
0	0	1	1	1	1	1	1	1	1	1	1	0	1

system. A 74LS148 eight-line-to-three-line priority encoder is all that is needed to translate one of the seven levels of interrupt request into a 3-bit code. Table 6.6 gives the truth table for this device. Input EI* is an active-low enable input, used in conjunction with outputs GS* and EO* to expand the 74LS148 in systems with more than seven levels of priority.

Because the enable input and expanding outputs are not needed in this application of the 74LS148, table 6.6 has been redrawn in table 6.7 with inputs 1 to 7 renamed IRQ1* to IRQ7*, respectively, and outputs A0 to A2 renamed IPL0* to IPL2*. It must be appreciated that all inputs and all outputs are active-low, so that an output value 0, 0, 0 denotes an interrupt request of level 7, whereas an output 1, 1, 1

TABLE 6.7 The truth table for a 74LS148 configured as in figure 6.29

	INPUTS							OUTPUTS		
LEVEL	IRQ1*	IRQ2*	IRQ3*	IRQ4*	IRQ5*	IRQ6*	IRQ7*	IPL2*	IPL1*	IPL0*
7	×	×	×	×	×	×	0	0	0	0
6	×	×	×	×	×	0	1	0	0	1
5	×	×	×	×	0	1	1	0	1	0
4	×	×	×	0	1	1	1	0	1	1
3	×	×	0	1	1	1	1	1	0	0
2	×	0	1	1	1	1	1	1	0	1
1	0	1	1	1	1	1	1	1	1	0
0	1	1	1	1	1	1	1	1	1	1

NOTE: 0 = low level signal ($<V_{IL}$)
           1 = high level signal ($>V_{IH}$)
           × = don't care

denotes a level 0 interrupt request (i.e., no interrupt). But note that the interrupt mask bits of the status register are active-high, so that a level 5 interrupt mask is represented by 101, whereas a level 5 interrupt request on IPL0*–IPL2* is represented by 010.

Inspecting table 6.7 reveals that a logical zero on interrupt request input i forces interrupt request inputs 1 to i − 1 into don't care states. That is, if interrupt IRQi* is asserted, the state of interrupt request inputs IRQ1* to IRQ(i − 1)* has no effect on the output code IPL0* to IPL2*. It is this property on which the microprocessor systems designer relies. Devices with high-priority interrupts are connected to the higher-order inputs. Should two or more levels of interrupt occur simultaneously, only the higher value is reflected in the output code to the 68000's IPL0*–IPL2* pins.

Figure 6.29 demonstrates that the 74LS148 does not restrict the system to only seven devices capable of generating interrupt requests. More than one device can be wired to a given level of interrupt request, as illustrated by peripherals 2 and 3. If either peripheral 2 or 3 (or both) asserts its interrupt request output (IRQ*), a level 2 interrupt is signaled to the 68000, provided that levels 3 to 7 are all inactive. The mechanism to distinguish between an interrupt from peripheral 2 and one from peripheral 3 will be discussed later. Chapter 10, on computer buses, introduces a mechanism called *daisy-changing* that enables several devices to share the same level of interrupt priority and yet permits only one of them to respond to an IACK cycle.

Processing the Interrupt

All interrupts to the 68000 are latched internally and made pending. Group 0 exceptions (reset, bus error, address error) take precedence over an interrupt in group 1. Therefore, if a group 0 exception occurs, it is serviced before the interrupt. A trace exception in group 1 takes precedence over the interrupt, so that if an interrupt request occurs during the execution of an instruction while the T-bit is asserted, the trace exception has priority and is serviced first. Assuming that none of the preceding exceptions have been raised, the 68000 compares the level of the interrupt request with the value recorded in the interrupt mask bits of the processor status word.

If the priority of the pending interrupt is lower than or equal to the current processor priority denoted by the interrupt mask, the interrupt request remains pending and the next instruction in sequence is executed. Interrupt level 7 is treated slightly differently, as it is always processed regardless of the value of the interrupt mask bits. In other words, a level 7 interrupt always interrupts a level 7 interrupt if one is currently being processed. Any other level of interrupt can be interrupted only by a higher level of priority. Note that a level 7 interrupt is edge-sensitive and is interrupted only by a high-to-low transition on IRQ7*.

Once the processor has made a decision to process an interrupt, it begins an exception-processing sequence, as described earlier. The only deviation from the normal sequence of events dictated by a group 1 or group 2 exception is that the interrupt mask bits of the processor status word are updated *before* the exception processing continues. The level of the interrupt request being serviced is copied into the current processor status. This means that the interrupt cannot be interrupted unless the new interrupt has a higher priority. An example should make the effect and implications clearer.

Suppose that the current (i.e., preinterrupt) interrupt mask is level 3. If a level 5 interrupt occurs, it is processed and the interrupt mask is set to level 5. If, during the processing of this interrupt, a level 4 interrupt is requested, it is made pending, even though it has a higher priority than the original interrupt mask. When the level 5 interrupt has been processed, a return from exception is made and the former processor status word is restored. The old interrupt mask was 3, so the pending level 4 interrupt is then serviced.

Unlike other exceptions, an interrupt may obtain its vector number externally from the device that made the interrupt request. As stated earlier, there are two ways of identifying the source of the interrupt, one *vectored* and one *autovectored*. A vectored interrupt is discussed first.

Vectored Interrupt

After the processor has completed the last instruction before recognizing the interrupt and has stacked the low-order word of the program counter, it executes an interrupt acknowledge cycle (IACK cycle). During an IACK cycle, the 68000 obtains the vector number from the interrupting device, with which it will later determine the appropriate exception vector. Figure 6.30 is a protocol flowchart showing the sequence of events taking place during an IACK cycle. It should be obvious from Figure 6.30 that an IACK cycle is nothing special—it is just a modified read cycle. Because the 68000 puts out the special function code 1,1,1 on FC2,FC1,FC0, during an IACK cycle, the interrupting device is able to detect the interrupt acknowledge cycle. At the same time, the level of the interrupt is put out on address lines A_{01}–A_{03}. The IACK cycle should not decode memory addresses on A_{04}–A_{23} and memory components should be disabled when FC2,FC1,FC0 = 1,1,1.

The device that generated the interrupt at the specified level then provides a vector number on D_{00}–D_{07} and asserts DTACK*, as in any normal read cycle. The remainder of the IACK cycle is identical to a read cycle. Figure 6.31 provides the timing diagram of an IACK cycle. However, if the IACK cycle is not terminated by the assertion of DTACK*, BERR* must be asserted by external hardware to force a *spurious interrupt* exception. Note that the IACK cycle falls between the stacking of the low-order word of the program counter and the stacking of the high-order word.

After the peripheral has provided a vector number on D_{00}–D_{07}, the processor multiplies that number by 4 to obtain the address of the entry point to the exception-processing routine from the exception vector table. Although a device can provide an 8-bit vector number, giving 256 possible values, space is reserved in the exception vector table for only 192 unique vectors. These 192 vectors are more than adequate for the vast majority of applications. But note that a peripheral can put out vector numbers 0–63 during an IACK cycle, as there is nothing to stop these numbers being programmed into the peripheral and the processor does not guard against this situation. In other words, if a peripheral is programmed to respond to an IACK cycle with, for example, a vector number 5, then an interrupt from this device would cause an exception corresponding to vector number 5—the value also appro-

FIGURE 6.30 Interrupt acknowledge sequence

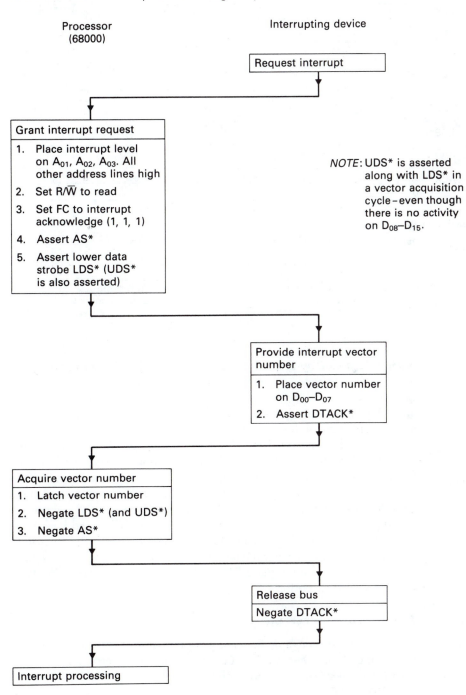

FIGURE 6.31 Interrupt acknowledge and the IACK cycle

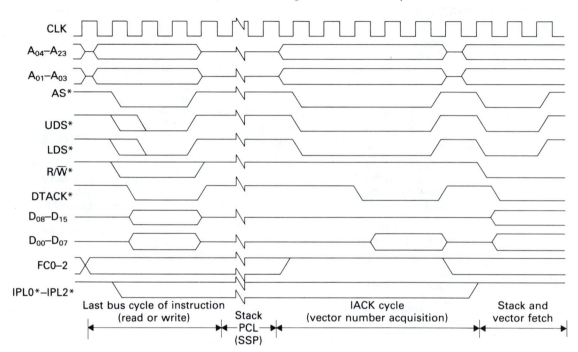

priate to a divide-by-zero exception. Although this feature might be useful at times, it seems an oversight to allow interrupt vector numbers to overlap with other types of exceptions.

A possible arrangement of hardware needed to implement a vectored interrupt scheme is given in figure 6.32. A peripheral asserts its interrupt request output, IRQ5*, which is encoded by IC3 to provide the 68000 with a level 5 interrupt request. When the processor acknowledges this request, it places 1, 1, 1 on the function code output, which is decoded by the three-line-to-eight-line decoder IC1. The interrupt acknowledge output (IACK*) from IC1 enables a second three-line-to-eight-line decoder, IC2, that decodes address lines A_{01}–A_{03} into seven levels of interrupt acknowledge. In this case, IACK5* from IC2 is fed back to the peripheral, which then responds by placing its vector number onto the low-order byte of the system data bus. If the peripheral has not been programmed to supply an interrupt vector number, it should place $0F on the data bus, corresponding to an *uninitialized interrupt* vector exception.

Autovectored Interrupt

As we have just seen, a device that generates an interrupt request must be capable of identifying itself when the 68000 carries out an interrupt acknowledge sequence. This requirement presents no problem for modern 68000-based peripherals such as the 68230 PI/T.

FIGURE 6.32 Implementing the vectored interrupt

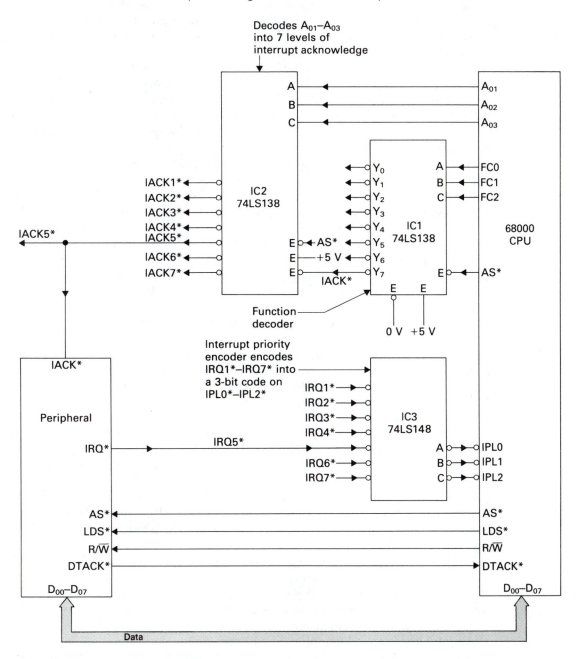

Unfortunately, older peripherals originally designed for 8-bit processors do not have interrupt acknowledge facilities and are unable to respond with the appropriate vector number on $D_{00}-D_{07}$ during an IACK cycle. The systems designer could overcome this problem by designing a subsystem that supplied the appropriate vector as if it came from the interrupting peripheral. Such an approach is valid but a little messy. Who wants a single-chip peripheral that needs a handful of components just to provide a vector number in an IACK cycle?

An alternative scheme is available for peripherals that cannot provide their own vector number. An IACK cycle, like any other memory access, is allowed to continue on to state S4 by the assertion of DTACK*. If, however, DTACK* is not asserted but VPA* is asserted, the 68000 carries out an autovectored interrupt.

Valid peripheral address, VPA*, is an input belonging to the 68000's synchronous data bus control group that was introduced briefly in chapter 4. We will meet it again in chapter 8 on input/output techniques. When asserted, VPA* informs the 68000 that the present memory access cycle is to be synchronous and is to *look like* a 6800-series memory access cycle. If the current bus cycle is an IACK cycle, the 68000 executes a *spurious read cycle*; that is, an IACK cycle is executed, but the interrupting device does not place a vector number on $D_{00}-D_{07}$. Nor does the 68000 read the contents of the data bus. Instead, the 68000 generates the appropriate vector number internally.

The 68000 reserves vector numbers 25–31 (decimal) for its autovector operation (see table 6.1). Each of these autovectors is associated with an interrupt on IRQ1* to IRQ7*. For example, if IRQ2* is asserted, followed by VPA*, during the IACK cycle, vector number 26 is generated by the 68000, and the interrupt-handling routine address is read from memory location $000068.

Should several autovectored interrupt requesters assert the same interrupt request line simultaneously, the 68000 will not be able to distinguish between them. The appropriate autovectored interrupt-handling routine must poll each of the possible requesters in turn—that is, the interrupt status register of each peripheral must be read to determine the source of the interrupt.

The timing diagram of an autovector sequence is given in Figure 6.33; it is almost identical to the vectored IACK sequence of Figure 6.31, except that VPA* is asserted shortly after the interrupter has detected an IACK cycle from FC0–FC2. Because VPA* has been asserted, wait states are introduced into the current read cycle in order to synchronize the cycle with VMA*. Note that this is a dummy read cycle, since nothing is read (the autovector is generated internally and no device places data on $D_{00}-D_{07}$ during the cycle). An IACK cycle differs from a read cycle in two ways: FC0–FC2 = 1,1,1 and VMA* is asserted.

The hardware necessary to implement an autovectored interrupt is minimal. Figure 6.34 shows a possible arrangement involving a typical 6800-series peripheral that requests an interrupt in the normal way by asserting its IRQ* output. The interrupt is prioritized by IC3, and an acknowledge signal is generated by ICs 1 and 2 exactly as in the corresponding autovectored scheme of figure 6.27. The interrupting device cannot, of course, respond to an IACK* signal from IC2. Instead, the appropriate interrupt acknowledge signal from the 68000 is combined with the interrupt request output from the peripheral in an AND gate (negative logic AND). Only

FIGURE 6.33 Timing diagram of an autovectored interrupt

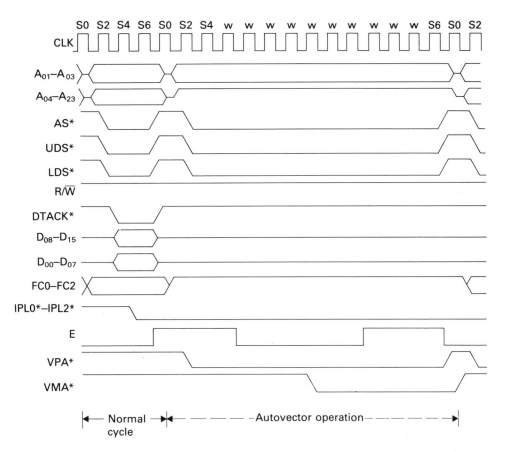

when the peripheral has asserted its IRQ* and the correct level of IACK* has been generated does the output of the AND gate go low to assert VPA* and force an autovectored interrupt.

Use of IRQ7* to Generate Manual Interrupts

Occasionally (sometimes too often), a 68000 system hangs up, and it is necessary to intervene manually. Pressing the reset button is not always a good idea, since the CPU context is lost. An alternative is to use a manual interrupt button connected to IRQ7* via a debouncing circuit to avoid multiple interrupts. IRQ7* is normally chosen because it cannot be masked. When IRQ7* is asserted by pushing the button, the IACK7* must be detected and VPA* must be asserted to generate an autovectored interrupt.

I once encountered a system that asserted IRQ7* to generate a manual interrupt but did not provide any IACK response to complete the cycle; that is, IRQ7*

FIGURE 6.34 Hardware needed to implement autovectored interrupts

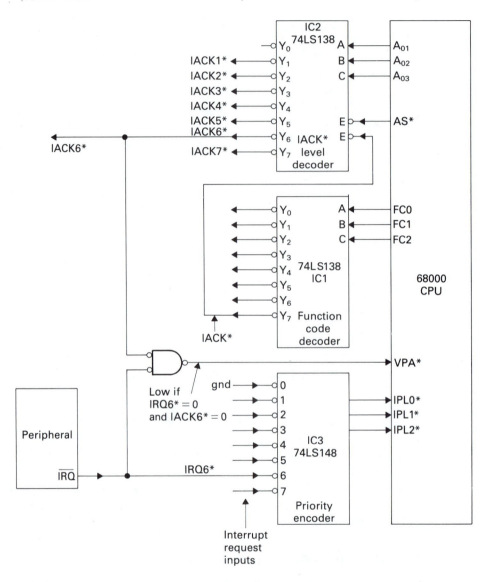

was asserted by hitting a button, and I could find no logic anywhere on the board that responded to the interrupt acknowledge. It took a little time to work out what was happening. If interrupt request is not acknowledged by either the assertion of DTACK* (vectored) or VPA* (autovectored), the cycle is terminated by the assertion of BERR* after a suitable time-out. Under these circumstances, the processor does not generate a bus error exception. Instead, the processor generates a spurious interrupt exception. The manufacturers of this system had used the spurious exception handler routine to deal with manual interrupts.

68020 and Interrupts

Apart from its ability to use an interrupt stack pointer solely for interrupts and a master stack pointer for all other exceptions, the 68020 processes interrupts very much like the 68000. However, there are two minor differences. Since the 68020 does not implement synchronous bus cycles, it has no pins corresponding to the 68000's E, VMA*, and VPA* pins. The lack of a VPA* pin means that the 68020 must find some other way to deal with autovectored interrupts.

When an interrupting device requests an autovectored interrupt, it detects the IACK cycle and responds by asserting the 68020's AVEC* (autovector) input. That is, the 68020 deals with autovectored interrupts in exactly the same way as the 68000. The only difference is that the 68000's VPA* pin has been renamed AVEC*.

The other enhancement of the 68020 is the addition of an interrupt-pending output (IPEND*). IPEND* is asserted by the 68020 to indicate to external hardware that an interrupt request has been recognized and is at a higher priority than that reflected by the interrupt mask. In other words, as its name suggests, IPEND* indicates that an interrupt is pending. IPEND* is negated during the S0 state of the interrupt acknowledge cycle. Most systems will not use the IPEND* output.

Breakpoint Instructions

The 68000 generates an illegal instruction exception if ever it attempts to execute an op-code that is not part of the 68000's instruction set. However, as we have seen, the designers of the 68000 have reserved a special illegal instruction with the assembly language form ILLEGAL and with the op-code $4AFC. The ILLEGAL instruction can be used to generate an illegal instruction exception for test purposes (or for any other purpose desired by the programmer). The special bit pattern $4AFC will always be an illegal op-code in all future enhancements of the 68000.

External hardware such as a logic analyzer can easily detect a 68000 illegal instruction exception by monitoring the address from which the exception vector is read during the exception processing phase. This address is $00 0010. It is much harder for external logic to detect an illegal instruction exception generated by a 68010 and later. The reason for this inability to detect an illegal instruction exception is that the external hardware does not know the address of the exception vector, because the 68010's vector base register, VBR, can remap the exception vector table anywhere within the processor's memory space.

In addition to the illegal op-code $4AFC, the 68010 defines eight other reserved illegal instructions with the bit patterns $4848 to $484F. These instructions are called *breakpoint illegal instructions* and are treated in a special way. A 68010 breakpoint instruction has the assembler form BKPT #⟨data⟩, where ⟨data⟩ is a value from 0 to 7.

In practice, a breakpoint illegal instruction is often inserted into memory by a microprocessor development system (as opposed to appearing as an instruction in a program). When a breakpoint illegal instruction is encountered by the 68010, the CPU begins an illegal instruction exception in the normal fashion. However, the 68010 executes a special bus cycle, called a *breakpoint cycle*, before the normal stacking operations are carried out. A breakpoint cycle is a dummy read cycle in which

FC0 to FC2 are all set high (i.e., a CPU space cycle) and the address lines are all set low (see figure 6.5). The 68010 does not execute a read or a write operation during the breakpoint cycle and continues with normal exception handling, irrespective of whether the cycle is terminated by a DTACK*, BERR*, or a VPA*.

The purpose of the breakpoint cycle is to provide a trigger for external hardware. For example, the breakpoint cycle can be used in system testing. The breakpoint is detected by looking for 1, 1, 1 on FC0–FC2 and 0s on A_{16}–A_{19}. When the breakpoint is executed, the hardware detects it and triggers, for instance, a signature analyzer. A second breakpoint can be used to stop the signature analyzer.

Oddly enough, the 68020 and 68030 implement the breakpoint exception in a different way to the 68010. We say "oddly" because, in general, the 68010's enhancements are carried over to the 68020 and 68030 without modification. The breakpoint is, however, an exception to this rule. When a 68020 or 68030 executes a breakpoint instruction, it performs a breakpoint acknowledge cycle by setting A_{00}, A_{01}, and A_{05}–A_{31} to zero, placing the breakpoint code on A_{02}–A_{04} and executing a read cycle (so far this is the same as the 68010 sequence). The breakpoint code on A_{02}–A_{04} is the data field of the BKPT #⟨data⟩ instruction.

There are two ways of completing this read cycle: One is to assert BERR* and the other is to assert DSACK*. If BERR* is asserted, the illegal instruction exception is taken and the breakpoint exception handler is executed.

FIGURE 6.35 Protocol flowchart for a breakpoint cycle

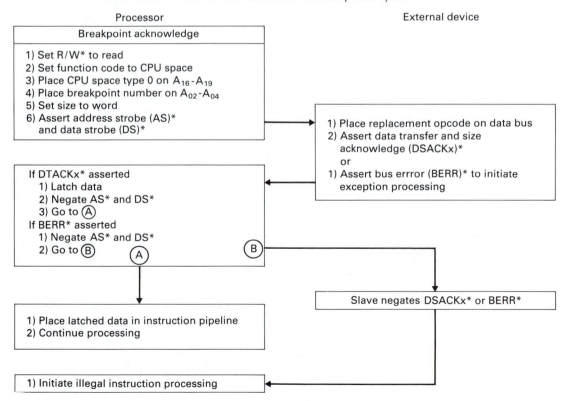

If, however, the read cycle is terminated by the assertion of DSACK*, the instruction word currently on the data bus is read by the 68020/30. This instruction word is supplied by external hardware and is used to *replace* the illegal instruction that is currently in the 68020/30's pipe (i.e., the sequence of instructions prefetched from memory waiting to be executed). All other instructions in the pipe are unaltered, and no stacking or vector fetching takes place. The inserted instruction is executed immediately after the completion of the breakpoint cycle. Figure 6.35 provides the protocol flowchart of a breakpoint cycle terminated by DSACK*, and figure 6.36 provides the associated timing diagram. Figure 6.37 demonstrates the hardware required to detect breakpoints.

FIGURE 6.36 Timing diagram of a breakpoint cycle

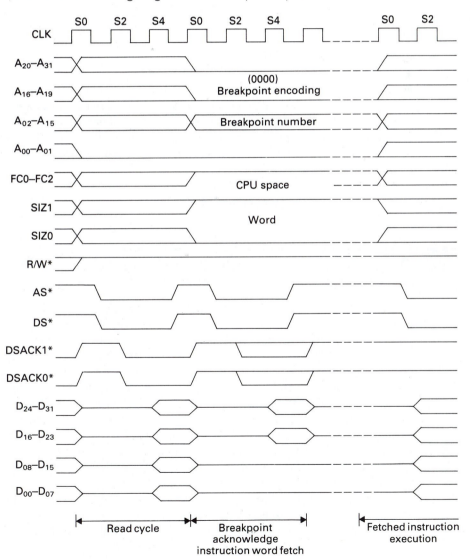

FIGURE 6.37 Possible logic used to detect breakpoint cycles

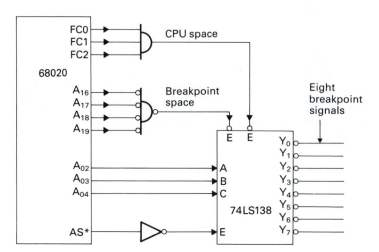

The 68020/30's breakpoint mechanism permits a hardware system to *remove* an instruction from memory and to replace it by a breakpoint code. The target system executes its code normally until the breakpoint is detected. The microprocessor development system can reinsert the code it removed to permit the program to be executed past the breakpoint.

Now that we have looked at the 68000's hardware exceptions, we are going to look at software exception in more detail.

6.6 SOFTWARE-INITIATED EXCEPTIONS

A software-initiated exception occurs as the result of an attempt to execute certain types of instruction—except the address error, which can really be classified as a hardware-initiated interrupt. Software-initiated interrupts fall into two categories: those executed deliberately by the programmer and those caused by certain types of software errors.

Software errors that lead to exceptions include the illegal op-code; the privilege violation; the TRAPV instruction; and the divide-by-zero error. These are all exceptions that are normally caused by something going wrong. Therefore, the operating system needs to intervene and sort things out. The nature of this invervention depends heavily on the structure of the operating system. Often, in a multiprogramming environment, the individual task that raised the exception will be aborted, leaving all other tasks unaffected.

Consider the illegal op-code exception. This exception occurs when the 68000 attempts to execute an operation code that does not form part of the 68000's instruction set. The only way that this can happen unintentionally is when something has gone seriously wrong. For example, an op-code might have been corrupted in memory by a memory error, or a jump might have been made to a region containing nonvalid 68000 code. The latter event frequently results from wrongly computed GOTOs.

Clearly, once an illegal op-code exception has occurred, it is futile to continue trying to execute further instructions, as they have no real meaning. By generating an illegal op-code exception, the operating system can inform users of the problem and invite them to do something about it.

Software exceptions deliberately initiated by the programmer are the trace, the trap, and the emulator. We will deal with each of these in turn.

Trace Exceptions

The trace exception mode is in force whenever the T-bit of the status word is set. After each instruction has been executed, a trace exception is automatically generated, permitting the user to monitor the execution of a program. Since the T-bit has to be set in the user mode to trace a user program and yet it is impossible to set the T-bit in the user mode (the SR can be accessed only by privileged instructions), we have a bit of a problem. The solution is to call a supervisor mode function (e.g., by means of a TRAP instruction) that sets the T-bit of the status register on the supervisor stack before executing an RTE. All the TRAP handler has to do is:

```
ORI.W #$8000,(SP)      Set trace bit of SR on stack
RTE                    Return from exception
```

The simplest trace facility would allow the user to dump the contents of all registers on the CRT terminal after the execution of each instruction. Unfortunately, such a simple use of tracing leads to the production of vast amounts of utterly useless information. For example, if the 68000 were executing an operation to clear an array by executing a CLR.L (A4)+ instruction 64K times, the human operator would not wish to see the contents of all registers displayed after each CLR.L (A4) + had been executed.

A better approach is to display only the information needed. Before the trace mode is invoked, the user informs the operating system of the conditions under which the results of a trace exception are to be displayed. Some of the events that can be used to trigger the display of registers during a trace exception are the following:

1. The execution of a predefined number of instructions. For example, the contents of registers may be displayed after, say, 50 instructions have been executed.

2. The execution of an intruction at a given address. This is equivalent to a breakpoint.

3. The execution of an instruction falling within a given range of addresses, or the access of an operand falling within the same range.

4. An event the same as event 3, but the contents of the register are displayed only when an address generated by the 68000 falls outside the predetermined range.

5. The execution of a particular instruction. For example, the contents of the registers may be displayed following the execution of a TAS instruction.

6. Any memory access that modifies the contents of a memory location—that is, any write access.

Several of the preceding conditions may be combined to create a composite event. For example, the contents of registers may be displayed whenever the 68000 executes write accesses to the region of memory space between $3A 0000–$3A 00FF.

Note that the STOP #⟨data⟩ instruction does not perform its function when it is traced. The STOP instruction causes the 68000 to load the 16-bit literal into its status register and then to stop further processing until an interrupt (or reset) occurs. However, if the T-bit is set, an exception is forced after the status register has been loaded with the literal following the STOP instruction. Upon a return from the trace handler routine, execution continues with the instruction following the STOP, and the processor never enters the stopped condition.

68020/30 Trace Mode

The 68020/30 implements a modest improvement in the 68000's trace mode by including a simple trace filter. The 68020/30's status register has two trace bits, T_1 and T_0, which are bits 15 and 14 of the status register, respectively. The effect of these bits is as follows:

TRACE BITS		TRACE FUNCTION
T_1	T_0	
0	0	No tracing
0	1	Trace on change of program flow
1	0	Trace on any instruction
1	1	Undefined state—reserved

The new mode provided by the 68020/30 is activated when $T_1, T_0 = 0, 1$. A trace exception is generated only when a change of flow takes place, caused by the execution of, for example, a BRA, JMP, TRAP, and return instruction. The 68020/30 considers a change of flow to take place when the status register is modified.

Example: Application of Trace Mode Exceptions

Consider the design of a skeleton generic trace exception handler. The component parts of the trace handler are

1. A longword at the trace exception vector location pointing to the actual trace handler itself.

2. A mechanism for switching on the trace facility. The switching must be carried out in the supervisor state, since any attempt to modify the status register while in the user mode would result in a privilege violation.

3. A mechanism for returning to (or activating) the user program once the trace mode has been turned on. A JMP instruction cannot be used as that would, itself, cause a trace exception once the T-bit of the status register has been set.

4. A trace handler that deals with the trace exception. In this case we will assume that the trace handler simply dumps all registers on the screen.

5. A subroutine, Print_regs, that can be called by the trace handler to perform the actual printing of the registers. We assume that the subroutine takes the registers off the stack.

6. A mechanism that will permit the tracing to continue or to be suspended. We will assume that after each instruction has been executed, the system waits for a character from the keyboard. If the character is a T, the next instruction is printed. If it is not, execution continues without further tracing.

7. A subroutine to input a character from the keyboard and deposit it in D0.B.

Following are the fragments of code required to implement the trace exception handler. The code fragment labeled "Go" is responsible for running the code being traced. Before Go is executed, the supervisor stack pointer must be set up with the status register at the top of stack and the program counter immediately below that.

```
            ORG      $00000024           Location of trace vector
            DC.L     TraceH              The trace handler vector
            :
Go          ORI.W    #$8000,(SP)         Set the trace bit on the stack
            RTE                          Now run the (user) program
            :
*           The trace  handler
TraceH      MOVEM.L  D0–D7/A0–A7,–(SP)   Save all registers on stack (A7 = SSP = dummy register)
            MOVE.L   USP,A0              Grab the user's stack pointer and put it on
            MOVE.L   A0,60(SP)           the stack (overwriting the saved SSP).
            JSR      Print_regs          Display all registers on stack
            JSR      Get_char            See if we want to continue tracing
            CMP.B    #'T',D0             Is the character a T?
            BEQ      Continue            IF it is THEN continue
            ANDI.W   #$7FFF,64(SP)           ELSE turn off trace mode
Continue    MOVEM.L  (A7)+,D0–D7,A0–A6   Restore registers (except A7 which is the USP)
            LEA      4(SP),SP            Move past dummy A7 left on the stack
            RTE                          Return from exception
```

A simple way of running the user program is to deposit the entry point to the program and its initial status register contents on the stack pointed at by the supervisor stack pointer. In this example, we set the trace bit on the stack by means of the ORI.W #$8000,(SP) instruction. Executing an RTE loads the program counter and the status register with the values from the stack. Consequently, the RTE will activate the user program with its T-bit in the status register set.

When the 68000 executes an instruction in the "target" program, a trace exception will be forced, since the T-bit is set. At the start of the trace exception processing, the value of the program counter and the status register will be pushed onto the supervisor stack. The program counter points to the instruction after the instruction that caused the trace exception (i.e., the next instruction), and the status register corresponds to the contents of the status register immediately prior to beginning exception processing.

The trace handler, TraceH, pushes all the 68000's registers onto the supervisor stack by means of a MOVEM.L D0–D7/A0–A7,(SP) instruction. Saving these registers means that we can now use the 68000's registers without worrying about corrupting their preexception values. You will appreciate, of course, that the value of A7 saved on the stack is supervisor stack pointer. We can replace it (on the stack) with the value of the user stack pointer. The trace handler first loads the USP into A0 and then overwrites the old A7 on the stack with the USP. Thus, when the register display routine is called, it will print D0–D7, A0–A6, and the USP.

After the display routine has been called, a character is input into D0. If it is a T, trace exception continues normally. If it is not, we access the status register pushed on the stack by the trace exception and clear its T-bit.

In the next step, the copies of the registers saved on the stack at the time of the exception are reloaded into the 68000's registers, except for A7. Since A7 is the supervisor stack pointer, we do not have to load it. In any case, the value of A7 on the supervisor stack frame is the USP. Having reloaded all registers except A7, it is necessary to tidy up the stack with a LEA 4(SP),SP instruction to "step past" the USP on the stack.

At this stage the stack is in the same state it was in at the start of exception handling. Executing an RTE restores the program counter and the status register to the values they had immediately before the trace exception. The next instruction in the target program is now executed. If the trace bit is still set, a trace exception will be raised and the entire sequence will be repeated. If the trace bit is clear (because we cleared it during the last trace handling), the 68000 will continue normally.

Emulator Mode Exceptions

The emulator mode exceptions provide the systems designer with tools to develop software for new hardware before that hardware has been fully realized. Suppose a company is working on a coprocessor to generate the sine of a 16-bit fractional operand. For commercial reasons, it may be necessary to develop software for this hardware long before the coprocessor is in actual production.

By inserting an emulator op-code at the point in a program at which the sine is to be calculated by the hardware, the software can be tested as if the coprocessor were actually present. When the emulator op-code is encountered, a jump is made to the appropriate emulator-handling routine. In this routine, the sine is calculated by conventional techniques.

Line A Exception

A line A trap has the op-code form $AXXX, where the 12 bits represented by the Xs are user definable. Line A traps are vectored to the location pointed at by the contents of location $28 and are generally employed to synthesize new instructions.

Note that when a line A exception occurs (or any other group 1 or group 2 exception), the exception handler can read the instruction that caused the exception (i.e., the $AXXX) and uses the XXX-field to interpret the instruction.

From the line A exception handler, you can access the actual line A instruction by means of the following code.

```
LEA        2(SP),A0     A0 points at the address of the instruction following the exception
MOVEA.L    (A0),A0      Read this address
MOVE.W     −2(A0),D0    Now read the instruction that caused the exception (the $AXXX)
```

Figure 6.38 illustrates the relationship between the line A trap and the preceding fragment of code. The first instruction, LEA 2(SP),A0, sets A0 to point to the saved program counter on the supervisor stack (the PC is under the status register, which is at the top of the stack). The second instruction, MOVEA.L (A0),A0 loads A0 with the saved program counter. The third instruction, MOVE.W −2(A0),A0 copies the A-line instruction that caused the exception into D0. Note that this instruction uses an offset of −2 because the program counter saved on the exception stack frame points to the instruction after the instruction that caused the trap. Of course, it is even easier to access the line A instruction that caused the trap if you are using a 68020. The instruction MOVE.W ([2,SP],−2),D0 will load the A line op-code into D0.

Line F Exception

In principle, the line F emulator trap that has the format $FXXX behaves exactly like the corresponding line A trap. The only difference is that the line F trap locates the vector to its handler at address $2C. However, the line F instruction is intended to be used to implement coprocessor instructions (in an environment that includes a 68020 or a 68030). When the 68000 encounters a line F trap, it begins exception processing in the normal fashion. However, when the 68020/30 encounters a line F instruction, it assumes that it is a coprocessor instruction and attempts to speak to the coprocessor by executing a bus cycle. If the instruction is a valid coprocessor instruction and a coprocessor is installed in the system, the coprocessor will complete the cycle normally (by returning a data acknowledge strobe). If there is no coprocessor installed, a time-out will occur and a bus error will be generated in the normal way. In this case, the bus error will cause the 68020 to take the line F exception. The line F exception is, therefore, reserved either for coprocessor communication or for the software that will emulate the coprocessor. Note that if the 68020 encounters a line F exception when in the user mode, a privilege violation is made.

TRAP Exception

The trap is almost certainly the most useful user-initiated software exception available to the programmer. Indeed, it is one of the more powerful functions provided by the 68000. To be honest, there are no real differences between traps and emulator exceptions. They differ only in their applications. Sixteen traps, TRAP #0 to

FIGURE 6.38 Accessing a line A instruction

(a) Line A instruction at N triggers exception

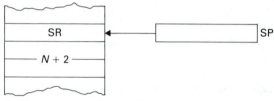

(b) The exception pushes SR and N + 2 on the stack

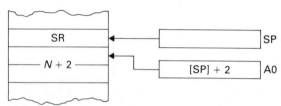

(c) Executing LEA 2(SP), A0 in the exception handler
loads A0 with the address of the return
address on the stack

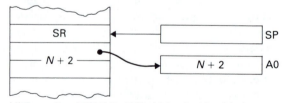

(d) Executing MOVEAL (A0), A0 loads A0 with the
return address on the stack

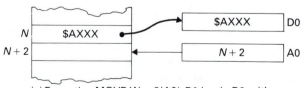

(e) Executing MOVE.W −2(A0), D0 loads D0 with
the A-line op-code

TRAP #15, are associated with exception vector numbers 32 to 47 decimal, respectively.

Just as emulator exceptions are used to provide functions in software that will later be implemented in hardware, the trap exceptions create new operations, or *extracodes*, not provided directly by the 68000 itself. However, the real purpose of the trap is to separate the details of certain housekeeping functions from the user or application-level program.

Consider input/output transactions. These involve real hardware devices, and the precise nature of an input operation on system A may be very different from that on system B, even though both systems put the input to the same use. System A may operate a 6850 ACIA in an interrupt-driven mode to obtain data, whereas system B may use an Intel 8055 parallel port in a polled mode to carry out the same function. Clearly, the device drivers (i.e., the software that controls the ports) in these systems differ greatly in their structures.

Applications programmers do not wish to consider the fine details of I/O transactions when writing their programs. One solution is to use a jump table and to thread all I/O through this table. Figure 6.39 illustrates this approach. It can be seen from figure 6.39 that the applications programmer deals with all device-dependent transactions by indirect jumps through a jump table. For example, suppose that all console input at the applications level is carried out by the instruction BSR GETCHAR. At the address GETCHAR in the jump table, the programmers insert a link (i.e., JMP INPUT) to the actual routine used in their own systems.

This approach to the problem of device-dependency is perfectly respectable. Unfortunately, it suffers from the limitation that the applications program must be tailored to fit on to the target system. We can do this by providing the program with a suitable jump table. An alternative approach, requiring no modification whatsoever to the applications software, is provided by the TRAP exception. The TRAP instruction leads to truly system-independent software.

FIGURE 6.39 The jump table

```
              ORG      $001000            ⎤  Jump Table
GETCHAR       JMP      INPUT              ⎥
OUTCHAR       JMP      OUTPUT             ⎥  Each procedure is vectored to
GETSECTOR     JMP      DISK_IN            ⎥  the actual procedure that carries
PUTSECTOR     JMP      DISK OUT           ⎦  out the appropriate task.


              •
              •
              BSR      GETCHAR    input char      ⎤
              •                                   ⎥  Application program
              •                                   ⎥  (address of subroutines
              •                                   ⎥  not system dependent)
              BSR      PUTSECTOR  write sector     ⎥
              •                                   ⎦
              •
```

When a trap is encountered, the appropriate vector number is generated and the exception vector table is interrogated to obtain the address of the trap-handling routine. Note that the exception vector table fulfils the same role as the jump table in figure 6.39. The difference is that the jump table forms part of the applications program, whereas the exception vector table is part of the 68000's operating system.

An example of a trap handler found on the Motorola ECB computer is known as the TRAP #14 handler. The TRAP #14 exception provides the user with a method of accessing functions within the ECB's monitor software without the user

TABLE 6.8 The functions provided by the TRAP #14 handler on the EBC

FUNCTION VALUE	FUNCTION NAME	FUNCTION DESCRIPTION
255	—	Reserved functions—end-of-table indicator
254	—	Reserved function—used to link tables
253	LINKIT	Append user table to TRAP #14 table
252	FIXDAOD	Append string to buffer
251	FIXBUF	Initialize A5 and A6 to BUFFER
250	FIXDATA	Initialize A6 to BUFFER and append string to BUFFER
249	FIXDCRLF	Move "CR", "LF", string to buffer
248	OUTCH	Output single character to port 1
247	INCHE	Input single character from port 1
246	—	Reserved function
245	—	Reserved function
244	CHRPRNT	Output single character to port 3
243	OUTPUT	Output string to port 1
242	OUTPUT21	Output string to port 2
241	PORTIN1	Input string from port 1
240	PORTIN20	Input string from port 2
239	TAPEOUT	Output string to port 4
238	TAPEIN	Input string from port 4
237	PRCRLF	Output string to port 3
236	HEX2DEC	Convert hex values to ASCII-encoded decimal
235	GETHEX	Convert ASCII character to hex
234	PUTHEX	Convert 1 hex digit to ASCII
233	PNT2HX	Convert 2 hex digits to ASCII
232	PNT4HX	Convert 4 hex digits to ASCII
231	PNT6HX	Convert 6 hex digits to ASCII
230	PNT8HX	Convert 8 hex digits to ASCII
229	START	Restart TUTOR; perform initialization
228	TUTOR	Go to TUTOR; print prompt
227	OUT1CR	Output string plus CR, LF, to port 1
226	GETNUMA	Convert ASCII-encoded hex to hex
225	GETNUMD	Convert ASCII-encoded decimal to hex
224	PORTIN1N	Input string from port 1; no automatic line feed
223–128	—	Reserved
127–0	—	User-defined functions

having to know their addresses. The versatility of a trap exception can be increased by passing parameters from the user program to the trap handler. The TRAP #14 handler of TUTOR (the monitor on the Motorola ECB) provides for up to 255 different functions to be associated with TRAP #14. Before the trap is invoked, the programmer must load the required function code into the least significant byte of D7. For example, to transmit a single ASCII character to port 1, the following calling sequence is used.

```
OUTCH      EQU 248                 Equate the trap function to name of activity
           ⋮
           MOVE.B #OUTCH,D7        Load trap function in D7
           TRAP #14                Invoke TRAP #14 handler
```

Table 6.8 gives a list of the functions provided by the TRAP #14 exception handler of the TUTOR monitor on the ECB.

6.7 INTERRUPTS AND REAL-TIME PROCESSING

There are those who believe that hardware is hardware, software is software, and never the twain shall meet. Although this maxim may be applied to certain areas of this book (for example, address decoding), it cannot be applied to hardware-initiated interrupts. An interrupt is a request for service from some device requiring attention. The request has its origin in hardware, but the response (the servicing of the interrupt) is at the software level. It is, therefore, difficult to deal with one aspect without at least some consideration of the other. We have already examined how a device physically signals an interrupt request and how the 68000 begins executing an interrupt-handling routine. Now we shall briefly look at the impact of the interrupt mechanism on a processor's software.

At the start of this chapter, we said that interrupts were closely associated with input/output transactions. Without an interrupt mechanism, an input or output device must be polled in order to determine whether or not it is busy. During the polling, the CPU is performing no useful calculations. By permitting a peripheral to indicate its readiness for input or output by asserting an interrupt request line, it becomes possible to free the processor to do other work while the peripheral is busy. Implicit in this statement is the assumption that the processor has something else to do while the peripheral is busy. This leads us to the concept of multitasking (the executing of a number of programs, or *tasks*, apparently simultaneously) and the operating system (the mechanism that controls the execution of the tasks).

Multitasking

Multitasking (or multiprogramming) is a method of squeezing greater performance out of a processor (i.e., a CPU) by chopping the programs up into tiny slices and executing slices of different programs one after the other, rather than by executing

FIGURE 6.40 An example of the applications of multitasking

Task A	VDT1	CPU	DISK	CPU	VDT1	Time ⟶

Task B	VDT2	CPU	DISK	VDT2	CPU	Time ⟶

FIGURE 6.41 Scheduling the two tasks of figure 6.40

Resource	Activity					
	Slot 1	Slot 2	Slot 3	Slot 4	Slot 5	Slot 6
VDT1	Task A				Task A	
VDT2	Task B				Task B	
Disk			Task A	Task B		
CPU		Task A	Task B	Task A		Task B

⟶ Time

each program to completion before starting the next. The concept of multitasking should not be confused with multiprocessing, which is concerned with the subdivision of a task between several processors.

Figure 6.40 illustrates two tasks (or *processes*), A and B. Each of these tasks requires several different resources (i.e., input/output via a VDT, disk access, CPU time) during its execution. One way of executing the tasks would be "end to end," with task A running to completion before task B begins. Such serial execution is clearly inefficient, since for much of the time the processor is not actively involved with either task. Figure 6.41 shows how the system can be made more efficient by scheduling the activities carried out by tasks A and B in such a way as to make best use of the resources. For example, in time slot 3, task A is accessing the disk while task B is using the CPU.

If we examine the idealized picture presented by figure 6.41, it is immediately apparent that two components are needed to implement a multitasking system: a scheduler that allocates activities to tasks and a mechanism that switches between tasks. The first is called the *operating system* and the second, the *interrupt mechanism*.

Real-Time Operating System

It is difficult to define a real-time system precisely, as *real-time* means different things to different people. Possibly the simplest definition is that a real-time system responds to a change in its circumstances within a meaningful period. For example, if a number of users are connected to a multitasking computer, its operation can be called real-time if it responds to the users almost as if each of them had sole access to the machine. Therefore, a maximum response time of no more than 10 s must be guaranteed. Similarly, a real-time system controlling a chemical plant must respond to changes in the chemical reactions fast enough to control them. Here the maximum guaranteed response time may be of the order of milliseconds.

Real-time and multitasking systems are closely related but are not identical. The former optimizes the response time to events while trying to use resources efficiently. The latter optimizes resource utilization while trying to provide a reasonable response time. If this is confusing, consider the postal system. Here we have an example of a real-time process. It offers a (nominally) guaranteed response time (i.e., speed of delivery) and attempts to use its resources well (i.e., pickup, sorting, and delivery take place simultaneously). Suppose the postal service were made purely multitasking at the expense of its real-time facilities. In that case, the attempt to optimize resources might lead to the following argument. Transport costs can be kept down by carrying the largest load with the least vehicles. Therefore, all vehicles wait on the East Coast until they are full and then travel to the West Coast, and so on. This would increase efficiency (i.e., reduce costs) but at the expense of degrading response time.

Real-Time Kernel

Operating systems, real-time or otherwise, can be very complex pieces of software with sizes greater than 10M bytes. Here, we are concerned only with the heart (kernel or nucleus) of a real-time operating system, its *scheduler*. The kernel of a real-time operating system has three functions.

1. The kernel deals with interrupt requests. More precisely, it is a first-level interrupt handler that determines how a request should be treated. These requests include timed interrupts that switch between tasks after their allocated time has been exhausted, interrupts from peripherals seeking attention, and software interrupts or exceptions originating from the task currently running.

2. The kernel provides a dispatcher or scheduler that determines the sequence in which tasks are executed.

3. The kernel provides an interprocess (intertask) communication mechanism. Tasks often need to exchange information or to access the same data structures. The kernel provides a mechanism to do this. We will not discuss this topic further, other than to say that a message is often left in a *mailbox* by the originator and then *collected* by the task to which it was addressed.

A task to be executed by a processor may be in one of three states: running, ready, or blocked. A task is *running* when its instructions are currently being executed by the processor. A task is *ready* if it is able to enter a running state when its turn comes. It is *blocked*, or dormant, if it cannot enter the running state when its turn comes. Such a task is waiting for some event to occur (such as a peripheral becoming

FIGURE 6.42 State diagram of a real-time system

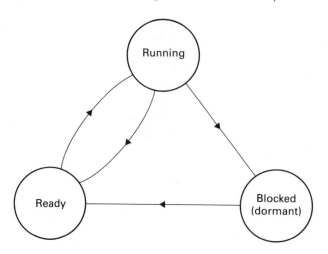

free) before it can continue. Figure 6.42 gives the state diagram for a task in a real-time system.

The difference between a running task and a waiting or blocked task lies in the task's volatile portion. The volatile portion of a task is the information needed to execute the instructions of that task. This information includes the identity of the next instruction to be executed, the processor status word, and the contents of any registers being used by the task.

FIGURE 6.43 Task control block

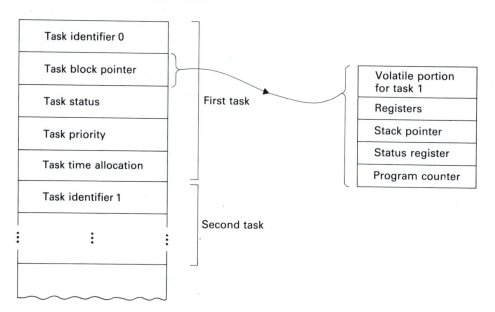

When a task is actually running, the CPU's registers, PC, PSW, data, and address registers, define the task's volatile environment. But when a task is waiting or dormant, this information must be stored elsewhere. All real-time kernels maintain a data structure called a task control block (TCB), which stores a pointer to the volatile portion of each task. The TCB is also called a run queue, task table, task list, and task status table.

Figure 6.43 provides a hypothetical example of a task control block. Each task is associated with an identifier, which may be a name or simply its number in the task control block. The task block pointer, TBP, contains a vector that points to the location of the task's volatile portion. Note that it is not necessary to store the actual volatile portion of a task in the TCB itself.

The *task status* entry defines the status of the task and marks it as running, ready, or blocked. A task is activated (marked as ready to run) or suspended (blocked) by modifying its task status word in the TCB. The *task priority* indicates the task's level of priority, and the *task time allocation* is a measure of how many time slots are devoted to the task every time it runs.

Interrupt Handling and Tasks

The mechanism employed to switch tasks is called a real-time clock (RTC). The RTC generates the periodic interrupts used by the kernel to locate the next runnable task and to run it. But how do we deal with interrupts originating from sources other than the RTC?

There are two ways of dealing with general interrupts and exceptions. One way is to regard them as being outside the scope of the real-time task scheduling system and to service them as and when they occur (subject to any constraints of priority). An alternative and much more flexible approach is to integrate them into the real-time task structure and to regard them as tasks just like any user task. We will adopt this latter approach in our examination of a real-time kernel.

Figure 6.44 describes a possible arrangement of an interrupt handler in a real-time system. When either an interrupt or an exception occurs, the appropriate interrupt-handling routine is executed by means of the 68000's exception vector table. However, this interrupt-handling routine does not service the interrupt request itself. It simply locates the appropriate interrupt-handling routine in the TCB and changes its status from blocked to runnable. The next time that this task is encountered by the scheduler, it may be run. We say "may be" because, for example, a TRAPV (trap on overflow) exception may have a very low priority and may not be dealt with until all the more urgent tasks have been run. Such an arrangement is called a *first-level* interrupt handler. The strategy of figure 6.44 can be modified by permitting the first-level interrupt handler to take *preemptive* action; that is, the interrupt handler not only marks its own task as runnable but suspends the currently running task, as if there had been a real-time clock interrupt.

The real-time interrupt is physically implemented by connecting the output of a pulse generator to one of the 68000's IRQ* inputs. A relatively high priority interrupt (e.g., IRQ5* or IRQ6*) can be reserved for this function. The highest-priority

FIGURE 6.44 Interrupt-handling in a real-time kernel

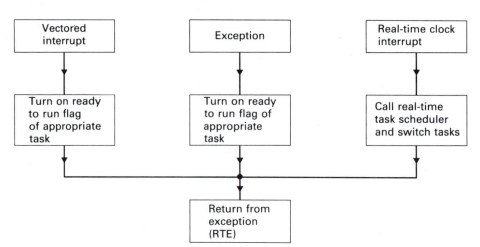

interrupt, IRQ7*, is frequently dedicated to the abort (i.e., panic) button and is used by the engineer to restart a system when it has crashed. However, a good argument can be made for using IRQ7* to switch tasks. Clearly, if we use IRQ6* and the interrupt mask gets set to level 7 (for whatever reason), a task-switching interrupt at level 6 will never be recognized and the system could hang up. Whenever an RTC interrupt is detected, its first-level handler (i.e., the vector table in the 68000) invokes the scheduler part of the real-time kernel.

There are two major approaches to task scheduling, a fixed-priority (or round-robin scheme) and a priority scheme. The round-robin scheme runs tasks in order of their appearance in the TCB. When the task with the highest number (i.e., bottom of the TCB) has been run (or found to be blocked), the next task to be run is the task with the lowest number. In a prioritized scheme, entries in the TCB are examined sequentially, but only a task whose priority is equal to the current highest priority may be run.

All interrupts and exceptions from sources other than the real-time clock simply mark the associated task as runnable. This action changes the task's status from *blocked* to *runnable* and takes only a few microseconds.

Designing a Real-Time Kernel for the 68000

To conclude this chapter on exception handling, we look at the skeleton design of a simple real-time kernel for a 68000-based microcomputer. Tasks running in the user mode have eight levels of priority, from 0 to 7. Priority 7 is the highest, and no task with a priority P_j may run if a task with a priority P_i (where $i > j$) is runnable. Each task has a time-slice allocation and may run for that period before a new task is run.

We can assume that the timed interrupt is generated by a hardware timer which pulses IRQ7* low every 20 ms. All other interrupts and exceptions are dealt with as in figure 6.44.

The highest level of abstraction in dealing with task switching is illustrated in Figure 6.45. A timed interrupt causes the time_to_run allocation of the current task to be decremented. When this reaches zero, the task table is searched for the next runnable task and that task is run. The time_to_run counter is reset to its maximum value before the next task is run. Any interrupt or exception other than TRAP #15 causes the appropriate task to be activated. A TRAP #15 instruction allows user programs to access the kernel and to carry out certain actions. Note that a TRAP #15 exception is not asynchronous—it is executed under program control in the user task currently being run. Figure 6.45 is an elaboration of that of figure 6.44. Here, TRAP #15 has been reserved to provide user access to the real-time system.

A suitable task control block structure for this problem is defined in figure 6.46. A separate task control block is dedicated to each task. The TCBs are arranged as a circular linked list, with each individual TCB pointing to the next one in the chain. The last entry points to the first. Each TCB description occupies 88 bytes: a long-word pointing to the next TCB, a 2-byte task number, an 8-byte name, a 2-byte task status word (TSW), a 70-byte task volatile portion and 2 reserved bytes. Note that in this arrangement, a task's volatile environment is part of the TCB itself. The task volatile portion is a copy of all the working registers belonging to the task at the moment it was interrupted (A0–A7 and D0–D7), plus its program counter and status

FIGURE 6.45 Real-time kernel

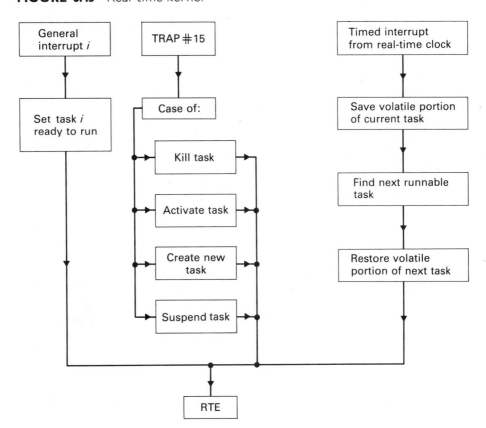

FIGURE 6.46 The task control block arranged as a linked list

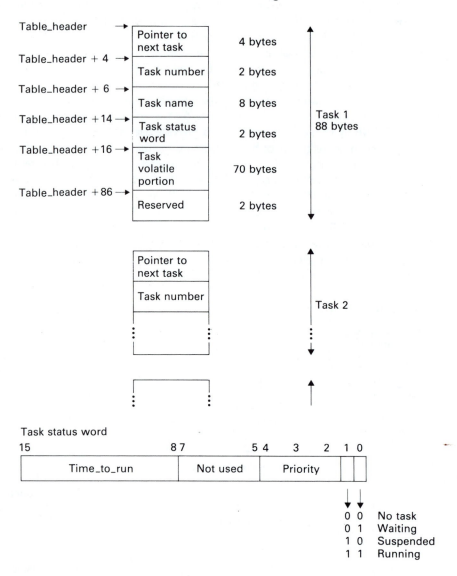

register. The stored value of A7 is, of course, the user stack pointer (USP) and not the supervisor stack pointer. The format of the TSW is also given in figure 6.46. The TSW contains a time_to_run field, a priority field (0 to 7), plus a 2-bit activity field.

The first-level pseudocode program to implement the real-time kernel of figure 6.45 is as follows:

```
Module: Interrupt_i
        Activate interrupt_i
        RTE
End module
```

```
Module: TRAP_#15
        CASE I OF
                    1: Kill_task
                    2: Activate_task
                    3: Create_new task
                    4: Suspend_task
        RTE
End module

Module: Timed_interrupt
        Decrement time_to_run
        IF time_to_run = 0 THEN
                            BEGIN
                            Move working registers to TCB
                            Save USP in TCB
                            Transfer saved PSW from stack to TCB
                            Transfer saved PC from stack to TCB
                            Find next active task in table
                            Load new USP from TCB
                            Restore new working registers from TCB
                            Transfer new PSW, PC from TCB to stack
                            Reset time_to_run
                            END
        END_IF
        RTE
End module.
```

We do not intend to deal with multitasking in great detail, and so the level 1 PDL is only partially elaborated, producing the following level 2 PDL:

```
Module Interrupt_i:
            Calculate address of associated task
            Get TSW_address for this task
            Clear bit 1 of TSW at TSW_address
            Set bit 0 of TSW at TSW_address
            RTE
End module

Suspend_task:
            Calculate address of associated task
            Get TSW_address for this task
            Clear bit 0 of TSW at TSW_address
            Set bit 1 of TSW at TSW_address
            RTE
End suspend_task

Module Timed_interrupt:
                    Global variable: Current_pointer
                                   : Current_priority
                                   : Time_to_run
                    Time_to_run :=  Time_to_run − 1
                    IF  Time_to_run = 0 THEN
                                        BEGIN
                                        Time_to_run := Max_time
                                        Newtask
                                        END
                    END_IF
                    RTE
End module
```

```
Newtask: {This swaps "task volatile environments"}
          Push A0–A6, D0–D7 on system stack
          Task_volatile_pointer:= Current_pointer + 16
          Copy A0–A6, D0–D7 from stack to [Task_volatile_pointer]
          Copy PC, SR from stack to [Task_volatile_pointer + 64]
          Copy USP to [Task_volatile_pointer + 60]
          Mark current task as waiting
          Next_task {Find the next runnable task}
          Task_volatile_pointer:= Current_pointer + 16
          Transfer A0–A6, D0–D7 from TCB to stack
          Transfer USP from TCB to USP
          Transfer PC, SR from TCB to supervisor stack
          Transfer A0–A6, D0–D7 from stack to registers
End Newtask

Next_task: {This locates the next runnable task in the TCB}
          Temp_pointer := Current_pointer
          Temp_priority := Current_priority
          Next_pointer := Current_pointer
          Next_priority := Current_priority
          REPEAT
              IF  Next_priority ⩾ Temp_priority AND  Next_TSW(0:1) = waiting
                    THEN
                              BEGIN
                              Temp_priority := Next_priority
                              Temp_pointer := Next_pointer
                              END
                    END_IF
                    Next_pointer      := [Next_pointer]
                    Next_TSW          := [Next_pointer + 14]
                    Next_priority     := Next_TSW(2:4)
              UNTIL  Next_pointer  = Current_pointer
              Current_pointer     := Temp_pointer
              Current_priority    := Temp_priority
End Next_task
```

Having outlined the PDL required to implement part of the RTL task switching kernel, the next step is to convert the PDL to 68000 assembly language form (asssuming the absence of a suitable high-level language). The only fragments of code provided here are NEWTASK, which switches tasks, and NEXT_TASK, which locates the next runnable task in the list of TCBs. These subroutines are for illustrative purposes only and are too basic for use in a real multitasking system.

```
*************************************************************************************
*
*  Newtask switches tasks by saving the volatile portion of the current task in its
*  TCB, transferring the volatile portion of the next task to run to the supervisor
*  stack and then copying all registers on the stack to the 68000's registers. The
*  new task is then run by copying the new PC and the SR from the TCB to the supervisor
*  stack and then executing an RTE. Newtask runs in the supervisor mode and the
*  supervisor stack is active. Note that all tasks are assumed runnable—bits 1, 0 of
*  the TSW are not used in this example.
*
*  NOTE: All registers (including A7 = SSP) are saved on the
*  supervisor stack at the beginning of the exception processing
```

```
*   routine. The saved value of the SSP is later overwritten with the
*   USP. Doing this, rather than leaving a "space" for the USP on
*   the stack, simplifies the coding.
*
*   A2 = Current_ptr    (points to the TCB of the current task)
*   A1 = Temp_ptr       (temporary pointer to a TCB during the search)
*   A0 = Next_ptr       (pointer to TCB of next task in chain of TCBs)
*   D2 = Current_prty   (the priority of the current task)
*   D1 = Temp_prty      (the priority of a task during the search)
*   D0 = Next_prty      (the priority of the next task in the chain of TCBs)
*
NEWTASK  MOVEM.L   A0–A7/D0–D7,–(SP)   Save all registers on supervisor stack
         MOVE.L    CURRENT_PTR,A2      Pick up the pointer to this (the current
         LEA       16(A2),A2           task) and point at its volatile portion.
*
         MOVE.W    #34,D0
NEW_1    MOVE.W    (SP)+,(A2)+         Transfer all registers on the stack to the current
         DBRA      D0,NEW_1           TCB to save the task (D0–D7, A0–A6, dummy USP, PC, SR)
         MOVE.L    USP,A1             Get the current stack pointer and store
         MOVE.L    A1,–10(A2)         it in A7 position in TCB (i.e., replace SSP by USP)
*
         MOVEA.L   CURRENT_PTR,A2     Restore A2 to top of current TCB before
         BSR.S     NEXT_TASK          finding the next task.
         MOVEA.L   A2,CURRENT_P       Save pointer to new current task
         LEA       86(A2),A2          A2 points at the bottom of the volatile
*                                     portion of the new TCB.
         MOVE.W    #34,D0             Copy all registers in the new volatile
NEW_2    MOVE.W    –(A2),–(SP)        portion to the supervisor stack.
         DBRA      D0,NEW_2
         MOVEA.L   60(A2),A2          Move the USP from the new TCB to the
         MOVE.L    A2,USP             user stack pointer.
*
         MOVEM.L   (SP)+,A0–A6/D0–D7  Move registers on stack to 68000's actual registers
         LEA       4(SP),SP           (except A7, SR, PC). Skip past the A7 position on
         RTE                          the stack by loading PC, SR from the stack into
*                                     the PC and SR.
****************************************************************************************
*
*   Next_task locates the next runnable task
*   A2 = Current_pointer
*   A1 = Temp_pointer
*   A0 = Next_pointer
*   D2 = Current_priority
*   D1 = Temp_priority
*   D0 = Next_priority
*   A2 imports Current_pointer and exports new Current_pointer
*
NEXT_TASK  MOVEM.L  A0–A1/D0–D2,–(SP)  Save working registers
           LEA      (A2),A1            Preset ptrs: Temp_pointer:=Current_pointer
           LEA      (A2),A0                         Next_ptr:=Current_ptr
           MOVE.W   14(A2),D2                       D2:=Current TSW
           ANDI.W   #$001C,D2          Mask D2 (TSW) to priority bits TSW(2:4)
           MOVE.W   D2,D1              D1:=Temp_priority
           MOVE.W   D2,D0              D0:=Next_priority
*
```

```
NEXT_1    CMP.W     D1,D0            REPEAT IF Next_priority < Temp_priority
          BMI.S     NEXT_2                  THEN locate next TCB in list
          MOVE.W    D0,D1                   ELSE Temp_priority:=Next_priority
          LEA       (A0),A1                      Temp_pointer:=Next_pointer
NEXT_2    MOVE.L    (A0),A0          Locate next TCB: Next_ptr:= [Next_ptr]
          MOVE.W    14(A0),D0        Get TSW of next task in list
          ANDI.W    #$001C,D0        Mask D0 (new TSW) to priority bits
          CMPA.L    A0,A2            UNTIL Current_ptr = Temp_ptr
          BNE       NEXT_1
          LEA       (A1),A2          A2 = new current task = temp task
          MOVEM.L   (SP)+,A0–A1/D0–D2  Restore working registers
          RTS                        Return with A2 pointing at new TCB
```

SUMMARY

In this chapter we have discovered that the 68000 supports one of the most comprehensive exception-handling mechanisms found on any microprocessor. Although it is, of course, perfectly possible to design and to program a microprocessor without recourse to either software or hardware exceptions (i.e., interrupts), the exception-handling capability of the 68000 greatly facilitates the design of real-time systems.

The 68000 provides prioritized and vectored interrupts, enabling many peripherals to interact with the CPU in real-time with a much greater efficiency than that of earlier 8-bit microprocessors. Just as importantly is the fact that the 68000 implements its prioritized and vectored interrupt structure with a minimum of additional components.

A special feature of the 68000's exception-handling mechanism is its ability to recover from system faults that would probably spell disaster in other microprocessor systems. By careful control of the 68000's BERR* input, the designer can create a microcomputer that is able to recover from a wide range of faulty bus cycles.

The 68000's software exceptions provide both protection against certain classes of software error (e.g., the illegal op-code) and a means of accessing the operating system via the TRAP. The importance of the TRAP, which links user programs to operating system utilities, cannot be stressed enough. Programs that use TRAPs to perform input/output operations can be made portable and largely system independent.

The 68000's dual operating mode is a feature that appeals most to the designer of secure real-time and multitasking systems. All user tasks are carried out in the user mode, whereas all exception handling takes place in the more privileged supervisor mode. Consequently, user tasks are forced to access system resources in a highly controlled and, therefore, reliable fashion.

Problems

1. Write 68000 assembly language instructions to perform the following operations:
   a. Set the trace bit.
   b. Set the interrupt level to 5.

c. Put the 68000 in the user mode.

d. Clear the trace bit, set the interrupt level to 6 and set user mode.

2. What is the effect of the following operations on the status of the 68000?

a. MOVE.W #$0000,SR   b. MOVE.W #$2700,SR

3. What is the difference between an uninitialized interrupt and a spurious interrupt exception?

4. You are thinking of designing a special version of the 68000 for use in high-speed word processing. A new instruction is to have the form MATCH ⟨source buffer⟩, ⟨target buffer⟩. Its action is to match the character string starting at address ⟨source buffer⟩ with the character string starting at ⟨target buffer⟩. Both strings are terminated by a carriage return. If the source string does not occur within the target string, the carry bit of the CCR is cleared. If the source string occurs within the target string, the carry bit of the CCR is set and the address of the start of the first occurrence of the source string within the target string is pushed on the stack. In order to test the new processor before the "first silicon," you decide to use the 68000's line 1010 emulator trap. Show how you would do this. Remember that the instruction will have the form:

⟨16-bit opcode⟩,⟨32-bit source address⟩⟨32-bit target address⟩.

5. Why is the 68000's supervisor stack pointer automatically loaded from address $00 0000 during the 68000's reset operation? Why is the 68000's user stack pointer not loaded during a reset operation?

6. The 68000 has a TRAPV instruction. Suppose you required a TRAPZ instruction (i.e., trap on zero result). How would you go about implementing this instruction?

7. The 68000 has an instruction ILLEGAL (with the bit pattern 0100 1010 1111 1100). When encountered by the 68000, the CPU carries out the operation:

[SSP] ← [SSP] − 4
[M([SSP])] ← [PC]
[SSP] ← [SSP] − 2
[M([SSP])] ← [SR]
[PC] ← [M(16)]

Explain the action of this sequence of RTL (register transfer language) operations in plain English. Why do you think that such an instruction was implemented by the designers of the 68000?

8. Why is the reset different from all other 68000 exceptions?

9. Why does the location of the 68000's reset vector's cause the system designer so many problems? Why are the 68010, 68020, and 68030 much better in this respect?

10. What is the exception with the highest priority?

11. Why is the 68000's BERR∗ exception so important?

12. Describe the 68000's BERR∗ exception sequence. Explain why the 68000's BERR∗ sequence is limited in comparison with the 68010 and later 68XXX family processors.

13. What is a double bus fault and why is it described as fatal?

14. What is the difference between a vectored and an autovectored interrupt? Under what circumstances are autovectored interrupts employed?

15. The 68000's exception vectors in the exception vector table lie in supervisor data space, with the exception of the two reset vectors. Why do the initial value of the supervisor stack pointer and the initial value of the program counter lie in supervisor program space?

16. The 68000 has a user and a supervisor mode. When the 68000 is in the user mode, it is impossible to execute privileged instructions. If you were a systems hacker, how would you attempt to get around this restriction? If you were a systems designer, what would you do to attempt to make your system hacker-proof?

17. What does context switching mean and how can it be implemented (making best use of the 68000's features)?

18. Why does the 68000 have a supervisor mode (in contrast to many 8-bit microprocessors)? What instructions are privileged and how do they differ from non-privileged instructions?

19. What is the difference between the RTR and RTD instructions?

20. What is the effect of the STOP $\#\langle d16\rangle$ instruction and how do you think it might be used?

21. During an IACK cycle, a peripheral may supply one of 256 possible vector numbers. This includes numbers 0 to 63, which do not apply to vectored interrupts. Design a hardware filter that would prevent any peripheral supplying a number in the range 0–63 during an IACK cycle.

22. Write a trace exception-handling routine that will display the contents of registers whenever an instruction is executed that falls between two addresses, Address_low and Address_high. Hint: Where can you find the instruction that actually causes the trace exception?

23. A large corporation makes computer systems to control sections of rail track. The designers of these systems are not permitted to use interrupts in their designs. All I/O must be polled. Why do you think that this decision has been taken? Hint: Railway control is a high-security application of computers.

24. Suppose you have to use a non-68000-series peripheral, which signals an interrupt by asserting a single active-low IRQ* output. Due to timing considerations, the peripheral must make use of the 68000's vectored interrupt facilities. Design the necessary logic interface to make this peripheral look like a 68000-series component. The logic must respond to an interrupt acknowledge from the 68000 "in the usual way." Attention must also be paid to the way in which the interrupt vector number is initially loaded into the interface. And don't forget the response to uninitialized interrupts.

25. Consider a 68000 system with n peripherals capable of generating a vectored interrupt. Suppose the ith peripheral generates a level I_i interrupt request, which requires t_i seconds to service, and the mean time between interrupts is f_i seconds. How would you investigate the likely behavior of such a system?

26. A print spooler prints one or more files as background jobs while the processor is busy executing a foreground job. Design a basic print spooler that will print a file. Assume the existence of GETCHAR, which reads a character from the disk drive, and PUTCHAR, which sends a character to the printer. The spooler operates in conjunction with a real-time clock, which periodically generates a level 5 interrupt. Clearly state any other assumptions you use in solving this problem.

27. Most 68000 interrupt mechanisms are prioritized, as described earlier in this chapter. Design an interrupt handler with seven inputs (IRQ1*–IRQ7*) and three outputs (IPL0*–IPL2*) that implements a round-robin scheme. In this arrangement, the most recently serviced interrupt level becomes the lowest level of priority and the highest level of priority is the next level in numeric sequence.

28. The 68000 interrupt structure requires several external packages, if its full facilities are to be used. Suppose that only four interrupt levels are needed (IRQ4*–IRQ7*). Design a minimum component circuit that is able to support four levels of vectored interrupt. Hint: Think PROM.

29. A 68000-based microcomputer has the requirement that an interrupt be generated at least every T seconds. Design a circuit to generate an interrupt on IRQ1* every T seconds provided that interrupts IRQ2*–IRQ7* have not been asserted during the previous T seconds.

30. Design a circuit that would assert both BERR* and HALT* for a rerun bus cycle whenever the signal MEMORY_ERROR* is asserted. The circuit should provide for three successive reruns and then, if not successful, generate a BERR* alone.

31. Investigate the rate at which tasks should be switched in a multitasking system. Hint: What is the overhead required to switch tasks? The answer to this question will require you to state any assumptions you have made concerning task-switching.

32. How does the 68010 improve the 68000's exception-handling facilities?

33. What is the difference between the way in which the 68000 and the 68020 handle trace mode exceptions?

34. How does the 68010's VBR enhance the 68000 family's exception-processing facilities (consider both hardware and software)?

35. Suppose you require more than 192 interrupt vectors. Can you locate some of the additional vectors in the exception vector table at locations marked "unimplemented, reserved"?

36. During an IACK cycle, what happens if BERR* is asserted instead of DTACK*?

37. A manual interrupt can be implemented by connecting IRQ7* to a pushbutton. Each time the button is depressed, IRQ7* is pulled low. Why is it necessary to employ a debounced switch?

38. What are the four phases the 68000 carries out during exception processing?

39. What are the differences between TRAPs, illegal instruction exceptions, and line A and line F exceptions?

40. How can a systems designer use the 68000's function code outputs to enhance the design of a microcomputer?

41. Describe how the 68010 and later members of the 68000 family have enhanced the interrupt acknowledge space for which FC0,FC1,FC2 = 1,1,1. Do these actions have any retrospective effects on the designers of 68000 systems?

42. Describe the breakpoint exception and explain how its implementation differs in the 68010 and the 68020/30 microprocessors.

43. What are the functions of the 68010's SFC and DFC registers and how are they used?

44. Why has the 68000's MOVE SR,⟨ea⟩ instruction been made privileged by the 68010?

45. Since the 68020 has a data bus sizing mechanism and can access word and longword values at even boundaries, does an address error exception have any real meaning in 68020 systems?

46. How does the 68020 enhance the 68000's CHK exception?

47. What are the format and the vector offset stored in word 4 of a 68020 stack frame and how are they used (either by the 68000 itself or by the programmer)?

48. What would happen if a 68020 processor attempted to access a nonexistent floating-point coprocessor using valid coprocessor access instruction?

49. What are the differences between the 68020 and 68030's ISP and MSP? Explain the meaning of a throwaway stack and describe how it is employed.

THE 68000 FAMILY IN LARGER SYSTEMS

In this chapter we look at some of the topics concerning the designer of larger systems. We employ the expression *larger systems* to imply microcomputers that have relatively big memories and use components or techniques not strictly necessary in some of the more basic microcomputers. Readers not interested in these topics may omit this section on a first reading of the text. The topics to be discussed here are the application of the 68000's bus arbitration interface (i.e., the BR*, BG*, and BGACK* lines) to multiprocessor systems, the principles of dynamic memory systems design, error-detecting and error-correcting memory, memory management techniques, cache memory, the 68020's coprocessor interface, and the 68040.

Dynamic memory is described in this chapter rather than chapter 5 because it is so much more complex than static memory and requires an entire subsystem to support it. Error-detecting and error-correcting memory goes hand in hand with dynamic memory, because the (soft) failure rate of dynamic memory components frequently makes it necessary to take action to deal with these errors.

Memory management is introduced because it helps a processor to implement sophisticated multitasking systems and ensures that the memory space belonging to one task is protected from illegal access by other tasks. Moreover, the way in which the 68000 family implements memory management is intimately bound up with its user/supervisor operating modes and with the bus error exception. Cache memory is included because the 68020, 68030, and the 68040 all implement internal caches to improve their throughput. An overview of the powerful 68040 microprocessor is provided because, although it is downward compatible with the 68000/20/30 at the software level, it includes several radical enhancements to the 68000 family.

7.1 BUS ARBITRATION CONTROL AND MULTIMASTER SYSTEMS

Up to now we have considered systems with a *single bus master*—that is, microcomputers in which only one device is capable of initiating the transfer of information between itself and a memory component or peripheral. In high-performance systems two or more devices may act as bus masters. Some of these bus masters may be intelligent peripherals and some may be CPUs.

An intelligent peripheral can be designed to move data directly to or from memory without the active intervention of a CPU. Such action is called *direct memory access* and is described further in chapter 8.

In a multiprocessor system, two or more CPUs operate simultaneously on different parts of a problem. For example, one CPU might be reading data from an interface and processing the results while a second processor is digesting the data previously read by the interface. In general, these multiprocessor systems are arranged so that each processor has some—but not all—the resources it needs. For example, a module on a card may contain a 68000 CPU, I/O devices, and several megabytes of read/write memory.

Figure 7.1 illustrates a multiprocessor system with three modules: a slave module and two bus masters, each containing a 68000 CPU. If the system contained only one bus master, the communication between it and the memory module would be exactly as described in chapter 4. That is, the 68000 generates an address, asserts AS∗, sets R/$\overline{W}$ to indicate the data direction, and then reads from or writes to the data bus. The data transfer is acknowledged by the memory module, which asserts DTACK∗ to complete the bus cycle.

In a multimaster system, more than one processor may wish to use the system bus at any time. Consequently, a mechanism called *bus arbitration* is required to decide which processor is entitled to access the bus if a conflict arises. Chapter 10 looks at the way the VMEbus and the NuBus deal with arbitration. Here, we are

FIGURE 7.1　Multiprocessor system

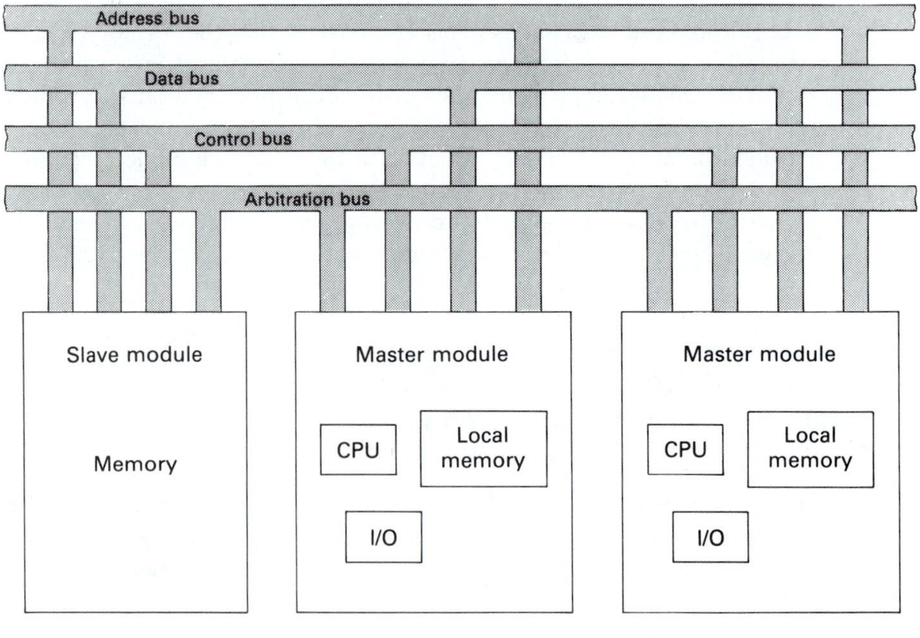

going to examine the facilities provided by the 68000 to help it to operate in a multimaster environment. By the way, you should appreciate that the organization of the system in figure 7.1 differs from that of simple systems with a single processor. The individual processors in figure 7.1 spend much of their time accessing local memory on their own module. When they wish to access memory or a peripheral on another module, they must explicitly request access to the system bus. They cannot just go ahead and put data on the bus as if they owned it.

The 68000 must use at least two signals if it is to implement the multiprocessor system of figure 7.1. One signal is a "would you please relinquish the bus so that I can use it" output from the *potential* bus master and the other is a "yes you can have the bus" output from the *current* bus master. The first signal is called *bus request* and the response is called *bus grant*. Whenever a 68000 (or a DMA controller, etc.) generates the address of a memory location that falls within the address space of another module, a bus request signal is asserted to request the bus from the current bus master. Because the 68000 (i.e., the would-be master) is an asynchronous device, it can wait for the bus to become free.

The 68000 has three signals (BR*, BG*, and BGACK*) enabling it to give up the bus in an orderly fashion in response to a request from a potential bus master. Figure 7.2 provides a protocol flowchart for the 68000's bus arbitration. Incidentally, the 68020 and 68030 perform arbitration in exactly the same way as the 68000 (but not the 68040). Figure 7.3 illustrates the 68000's three bus arbitration pins, two of which are inputs and one of which is an output. If a 68000 is not used in a multi-master system, both the 68000 arbitration control inputs (BR* and BGACK*) can be pulled up to V_{cc}, and all three bus arbitration control pins can be forgotten.

Assume that a certain 68000 is currently the bus master and is using the system bus. When another device wishes to become a bus master, it asserts the 68000's BR* (bus request) input. The 68000 *must* respond to this request. A bus request cannot be masked or prioritized by the 68000 like an interrupt. In chapter 10 we will see that *external* logic is necessary if we wish to implement a prioritized bus request.

After BR* has been asserted and internally latched by the 68000, the CPU asserts its BG* (bus grant) output, which indicates to the requesting device that the current bus master will give up the bus at the *end* of the present bus cycle. BG* is usually asserted as soon as possible by the 68000. However, if the 68000 has already made a decision to execute the bus cycle but has not yet asserted AS* to start this cycle, the 68000 delays the assertion of BG* until AS* has been asserted. If the 68000 ever asserted BG* before AS*, the requesting device might wrongly assume that it had control of the bus.

When the requesting device detects the assertion of BG*, it knows that it is going to get control of the bus and waits until AS*, DTACK*, and BGACK* (bus grant acknowledge) have all been negated before asserting its own BGACK* output. This situation indicates that the current bus master is not accessing the bus (AS* negated), the current bus slave is not accessing the bus (DTACK* negated), and that no other potential master is about to use the bus (BGACK* negated). The requesting device may release (i.e., negate) BR* after the BG* handshake from the current bus master has been detected.

Once BGACK* has been asserted by the new bus master, the old bus master

FIGURE 7.2 Protocol flowchart for bus arbitration

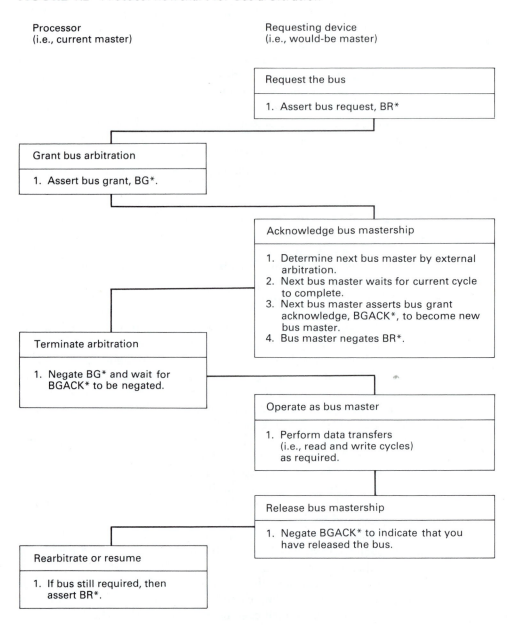

(or any of the potential bus masters) cannot access the bus as long as BGACK∗ is asserted. The old bus master then negates its BG∗ output. Note that, at this stage, only BGACK∗ is being asserted by the new bus master and, therefore, BR∗ can be asserted by another potential bus master. The timing diagram of a bus arbitration sequence is given in figure 7.4.

FIGURE 7.3 Bus arbitration control signals and the 68000

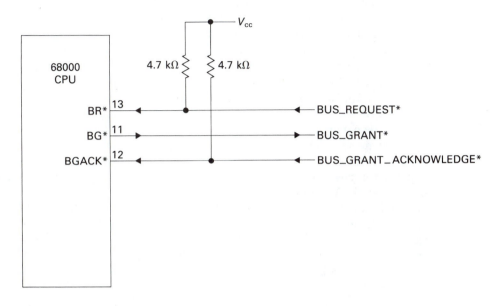

FIGURE 7.4 Timing diagram of a bus arbitration sequence

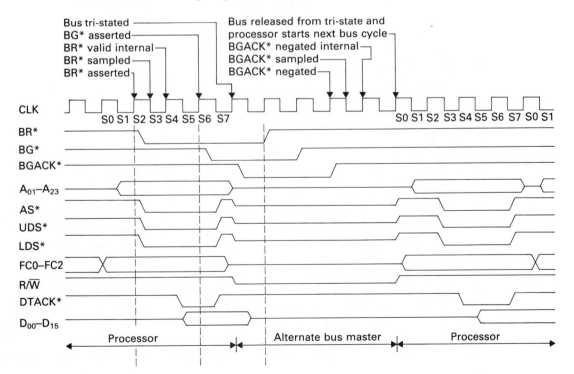

When BGACK∗ is asserted, the 68000 three-states its address bus, function code outputs, data bus, and AS∗, R/$\overline{\text{W}}$, LDS∗/UDS∗, and VMA∗ control outputs; that is, for all practical purposes the 68000 is taken off the system bus. Since all these pins are floated when the 68000 has given up the bus, the designer should take care to avoid a situation in which (even for an instant) no device is driving the bus. Pull-up resistors can be used to force all control inputs into an inactive-high state when they are floated.

If the 68000 is used in a simple arrangement with only one other potential bus master, little or no extra bus arbitration control logic is required. Later in this chapter, we will demonstrate how bus arbitration can be used to control the refresh of dynamic memory. However, whenever a system implements several potential bus masters, some arbitration logic is required to deal with the near simultaneous requests for bus mastership from several devices. As indicated earlier, chapter 10 shows how the VMEbus deals with this.

You can implement a single-line bus arbitration system in simple 68000 systems. If an alternate bus master wishes to force a 68000 off the bus, all the alternate bus master needs do is force the 68000's BGACK∗ input low (i.e., without any prior BR∗, BG∗ handshake). Asserting BGACK∗ takes the 68000 off the bus and three-states its address and data bus drivers. If you do use this mode, take great care not to blast a 68000 in midcycle. A recommended way of using this mode (if you really must) is to wait for the 68000 to negate AS∗, wait a further two clock cycles, and then assert BGACK∗.

Before leaving the topic of arbitration, we should discuss the 68000's test and set (TAS) instruction, which has been explicitly designed to support multiprocessor systems.

Test and Set Instruction

Members of the 68000 family all support an unusual instruction called *test and set*, or TAS. The assembly language form of TAS is TAS ⟨ea⟩ and its effect is:

[ea] − 0; [ea(7)] ← 1

that is, the contents of the specified effective address are tested (i.e., the N and Z bits of the CCR are set accordingly), and the most significant bit of the operand is set to 1. Apart from one detail, the effect of TAS ⟨ea⟩ is equivalent to TST.B ⟨ea⟩ followed by ORI.B #$10000000,⟨ea⟩. This one detail is that a TAS instruction is *indivisible* and must be executed to completion. By indivisible we mean that no other potential bus master can arbitrate for the bus and receive it until the TAS has been completed. When a TAS instruction tests the operand, it performs a *read* cycle, and when it sets bit 7 of the operand, it performs a *write* cycle. The consecutive read and write cycles of a TAS instruction are performed in a unique way by the 68000.

Figure 7.5 provides the timing diagram of the 68000's read-modify-write (RMW) cycle used to execute a TAS instruction. Unlike all other bus cycles, a 68000 RMW cycle requires a minimum of 20 clock states to read the operand, test it, and set its bit 7. The point of greatest interest in figure 7.5 is the behavior of the address

FIGURE 7.5 The 68000's indivisible RMW cycle

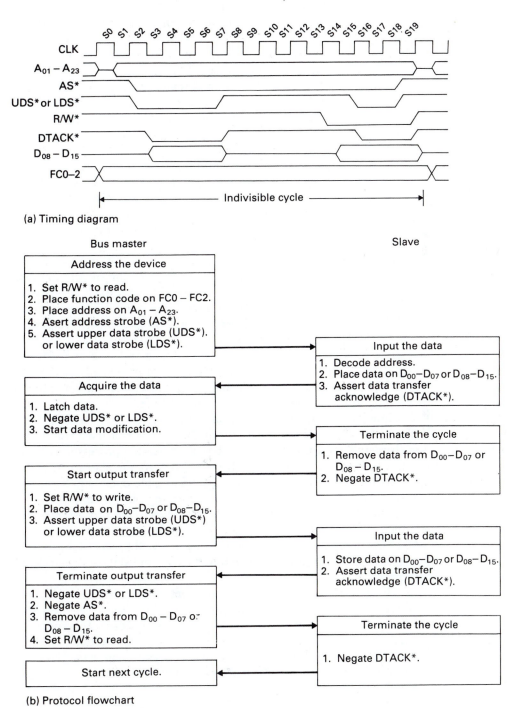

(a) Timing diagram

(b) Protocol flowchart

strobe, AS∗. The address strobe is asserted in bus state S1 and remains asserted until bus state S19. From our earlier discussion of arbitration, it is clear that 68000-based systems can use the state of AS∗ to test whether the processor is accessing the bus or not. As long as AS∗ is asserted, the bus is *marked as occupied*. If a TAS were implemented as two separate instructions, AS∗ would be negated between the *operand test* and the *operand modify* phases of the instructions.

So, why is a TAS instruction so important? In a system with a single processor, the TAS instruction has, indeed, little role to play. Consider now a multiprocessor system with two 68000's using bus arbitration to access common memory. We will call one processor A and the other B. Suppose that processor A wishes to access the disk drive. It can read a status location within the disk drive's controller to determine whether or not the disk drive is busy (in use). Assume that bit 7 of the status register is set to indicate a *busy* condition and clear to indicate a *free* condition. If the drive says that it is free, processor A can set the drive's busy status bit to indicate owner-ship. Then it can go ahead and access the drive. Should processor B attempt to access the drive, it will find that the drive is busy; it must then wait for this device to become available (when processor A releases it).

This scheme for allocating the disk drive might seem flawless—the first pro-cessor to ask for it is granted access. Now consider the situation in which the disk drive is free and both processor A and processor B ask for it at effectively the same time. Such a situation is possible in an asynchronous multiprocessor system in which the various processors operate independently of each other. Processor A tests the status of the disk drive and finds that it is free. Processor B also wants to access the disk drive so it arbitrates for the bus, gets it, and tests the disk drive's status. Pro-cessor B will find that the disk drive is free, since processor A has tested the status bit but has not yet set it. Consequently, both processors think that they have access to the disk drive, and the system is guaranteed to fail.

Now consider how this procedure can be carried out using a TAS instruction. Both processors A and B compete for the bus and one of them will be granted access (see chapter 10 for further details of buses). When the processor that receives access to the bus executes a TAS instruction, it first tests the status of the disk drive and then sets the status bit—all in one single indivisible operation. If the disk drive was busy (i.e., bit 7 of its status byte was set), carrying out the set phase of the operation has no effect. If, however, the disk drive was free, the processor detects that state and then sets bit 7 of the status byte to claim the drive. It should be clear now why TAS is so important—no other processor can test the same status byte between the test phase of the instruction and the set (i.e., marking as claimed) phase of the instruction.

The 68020 and 68030 also implement RMW cycles and TAS instructions. However, unlike the 68000 these processors do release (i.e., negate) the address strobe between the read and the write cycles. Instead of keeping AS∗ asserted throughout a bus cycle, the 68020/30 asserts its RMC∗ output in state S0 and keeps it asserted until state S11 to indicate that it is executing a read-modify-write cycle (note that the 68020 uses a 12-state read-modify-write cycle). As you might expect, either the read or the write (or both) phases of the RMW cycle can be extended by the addition of wait states. The 68020/30 also implement two other instructions with indivisible RMW cycles (CAS and CAS2). These instructions are beyond the scope of this book, and all that we need say is that they are used to maintain a linked-list data structure in a multiprocessor environment (as they are indivisible, a processor may insert or delete an item from the list without another processor intervening).

7.2 DESIGNING DYNAMIC MEMORY SYSTEMS

In this section we are going to look at how dynamic memory operates, the problems inherent in the design of a dynamic memory system, and the way in which dynamic memory components can be interfaced to a 68000-based microcomputer. Apart from the increased speed of the 68020 and 68030, these newer processors are interfaced to DRAM exactly like the 68000.

Dynamic memory is a form of low-cost, random access, semiconductor memory that is usually associated with memory arrays larger than about 64K to 256K bytes. Smaller memory arrays are frequently implemented as static RAM, although there are reasons why some designers may choose to select static RAM for large memories. My favorite apocryphal comment on dynamic memory is found in an article by L. T. Hauck in *Byte* (July 1978). "What's the difference between static RAM and dynamic RAM? Static RAM works and dynamic RAM doesn't." Perhaps the answer should have been, "Static memory works on its own—dynamic memory has to be made to work for you".

Dynamic read/write RAM (DRAM for short) is available in a number of different formats, like its static counterpart. In 1990, the preferred dynamic memory was organized as 1M × 1 bits, although 256K × 1-bit chips were also used. Older 16K and 64K parts are obsolete but are still sold to support existing systems. During 1990 the 4M × 1 and 1M × 4 parts were finding their way into the newer and more sophisticated applications of microprocessors. In the early 1990s, 16M × 1 DRAMs represent the state of the art. In this section we will analyze the operation of a 4M-bit DRAM arranged as 1M × 4 bits.

Dynamic memory stores information as an electrical charge on a capacitor forming the interelectrode capacitance of a metal oxide field-effect transistor. The capacitor is not perfect but is *leaky*, and the charge held on it is gradually lost. Consequently, some mechanism is needed to restore the charge on the capacitor periodically before it leaks away. This mechanism is called *refreshing*; it has to be performed at least once every 16 ms for a 4M-bit chip. Note that smaller DRAMs (e.g., 64K chips) have typical refresh periods of 2 ms. Some versions of 4M-bit DRAMs have a remarkably long refresh period of 128 ms (e.g., 51L4400).

Semiconductor manufacturers have argued, quite rightly, that putting high-density memory chips in physically large packages is irrational because the object of producing compact memory modules is thus defeated. A 1M × 1 RAM requires 20 address lines ($2^{20} = 1M$), so a dynamic RAM component might be expected to have at least 20 address pins, 2 power supply pins, 1 chip-select pin, 1 R/$\overline{W}$ pin, and 1 data pin, or at least 25 pins in all, which would require a 28-pin package, taking up a nominal 1.4 × .6 = .84 in.² of board space.

The majority of dynamic memories save address pins by using a multiplexed address bus, so that a 20-bit address (for a 1M chip) is fed in as two separate 10-bit values, reducing the address bus requirement to 10 pins but requiring two strobes or clocks to latch the address. The RAS* (row address strobe) latches the 10-bit row address, and then the CAS* (column address strobe) latches the 10-bit column address. The address multiplexing and the control of RAS* and CAS* strobes are done off-chip with logic supplied by the user. Consequently, a 1M dynamic RAM can now fit into a 18-pin DIP package, taking up a board space of approximately $.9 \times .3 = .27$ in.^{2} The size of DRAM memories has been further reduced by putting, for example, nine surface-mount DRAM chips on a single small PCB carrier to create a complete memory module (e.g., 1M × 9 bits). Nine bits are frequently used to provide an 8-bit data byte plus a single parity bit.

Figure 7.6 gives the internal arrangement of a 1M × 4-bit dynamic memory and figure 7.7 shows its pinout. This device is available in either a .350-in.-wide J-lead small outline package or a .1-in. zig-zag in-line package (ZIP), as opposed to the more conventional DIP package of earlier and smaller DRAMs. Modern packaging permits the design of very high density memories. Note that the data is stored in four 1,024- by 1,024-bit arrays, each of 1,048,576 bits. We cannot devote space here to delving into the internal operation of the dynamic memory, as its

FIGURE 7.6 Internal arrangement of a typical 1M × 4 DRAM

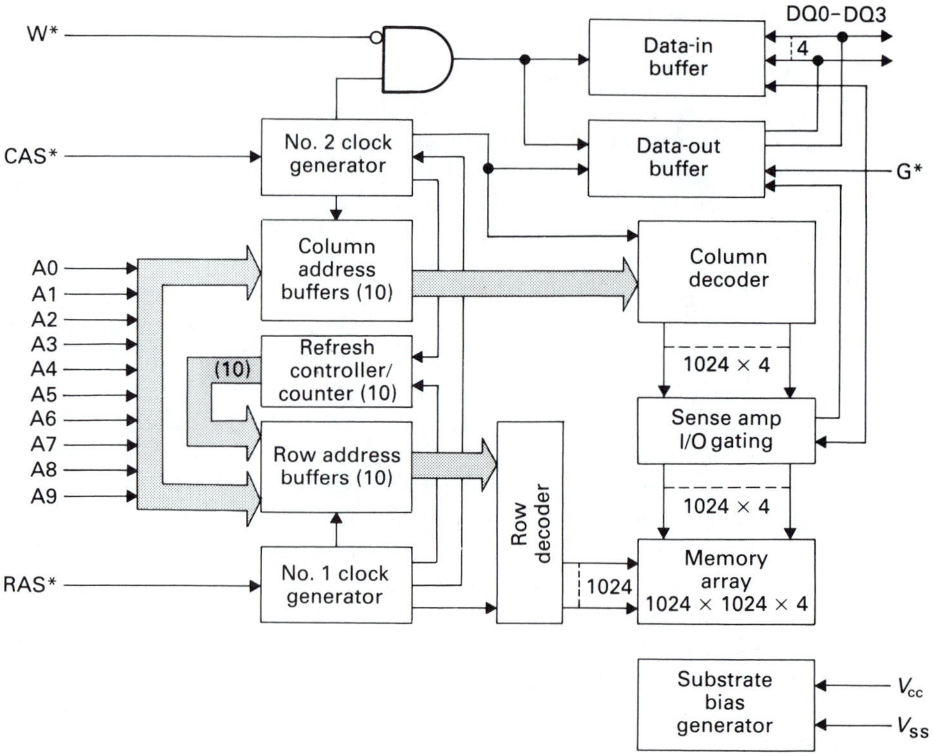

FIGURE 7.7 Pinout of a typical 1M × 4 DRAM

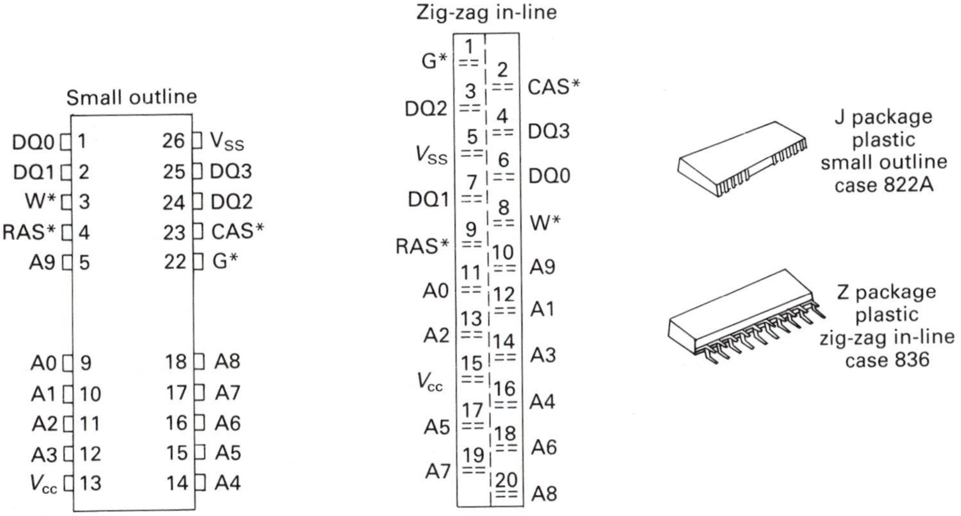

Small outline

DQ0	1	26	V_{SS}
DQ1	2	25	DQ3
W*	3	24	DQ2
RAS*	4	23	CAS*
A9	5	22	G*
A0	9	18	A8
A1	10	17	A7
A2	11	16	A6
A3	12	15	A5
V_{cc}	13	14	A4

Zig-zag in-line

J package
plastic
small outline
case 822A

Z package
plastic
zig-zag in-line
case 836

circuitry is so complex, but it is worthwhile making a few comments about the nature of DRAM. Early dynamic RAMs required three power supplies of +12, +5, and −5 V. The +12 V was necessary to achieve clock pulses of adequate amplitude within the chip and the −5 V provided the substrate bias. Fortunately for the systems designer, current 16K and larger chips operate from the system V_{cc} supply alone. Dynamic memories still need a negative V_{bb} supply, but they now derive it on-chip from an internal V_{bb} generator.

Yet another difficulty associated with dynamic memory is called the *alpha-particle problem*. The capacitance on which each bit of data is stored is exceedingly tiny (both electrically and physically). An alpha particle (i.e., helium ion) passing through a memory cell can cause sufficient ionization to corrupt the stored data. The effect of an alpha particle is to create a so-called soft error, because the cell has not been permanently damaged but has lost its stored data. The alpha-particle contamination comes largely from the encapsulating material. Semiconductor manufacturers have attempted to minimize the problem by careful quality control of the material used to encapsulate the chip. Even with careful quality control, reducing the soft-error rate to zero is impossible.

One approach to soft-error control is to build special memory arrays that can detect—or even detect and correct—soft errors. As long as soft errors are relatively infrequent, this approach yields a very large mean time between undetected soft errors. Error detection and correction are not yet performed inside the memory components and must be provided by the memory systems designer. A typical single error-correcting and double-error-detecting 8-bit memory requires the storage of an additional 5 check bits per byte (i.e., each byte is stored as a 13-bit word). The extra 5 bits are a function of the 8 data bits and are used to correct any single-bit error in the stored word. We return to the topic of error correction in section 7.3.

Dynamic RAM Timing Diagram

Now we are going to examine the timing diagram of a typical DRAM and its specification sheets. The purpose of this exercise is to enable engineers to design memory modules using DRAM chips.

Few things in the known universe are more terrifying than the timing diagram of a dynamic RAM. Not only does this diagram (in fact there are several diagrams) look hopelessly complex, it is peppered with masses of parameters (the 514400 has about 60 parameters). The best way of approaching the dynamic RAM timing diagram is to strip it of all but its basic features. Once this simplified model has been digested, fine details can be added later.

Dynamic Memory Read Cycle

Figure 7.8 gives an outline of the basic dynamic memory timing diagram during a read cycle. In order to put this diagram into context, figure 7.9 shows the arrangement of a 1M-word by 16-bit memory based on sixteen 1M × 1 chips (we could have used four 1M × 4 chips). The 10 address inputs (labeled A_0–A_9) of each of the 16 memory components are connected to the 10 outputs of the address multiplexer,

FIGURE 7.8   Basic read cycle timing diagram of a dynamic RAM

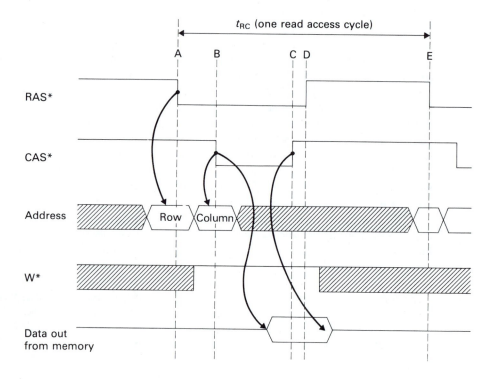

FIGURE 7.9 Arrangement of a 1M-word by 16-bit dynamic RAM module

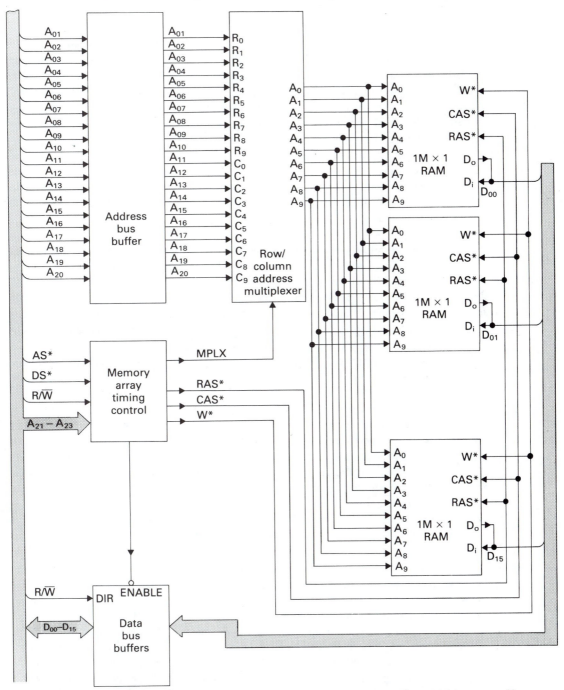

NOTE: A minimal DRAM module contains three elements: the DRAM array itself, an address multiplexer, and a memory array-timing control circuit that controls the multiplexer and generates RAS* and CAS* from the CPU's own timing signals. The timing control circuit also generates the DTACK* handshake. Byte/word control is not included here. Byte/word control (and the 68020's dynamic bus sizing) is implemented by negating RAS* or CAS* to the byte you are not accessing.

MPLX. The inputs to the address multiplexer are A_{01}–A_{10} (the row address) and A_{11}–A_{20} (the column address) from the 68000. Assume that when MPLX is low the *row* address is selected; when it is high, the *column* address is selected. Remember that the 23 address lines from the 68000, A_{01}–A_{23}, select one of 2^{23} word addresses and the two data strobes, LDS* and UDS*, select the lower or upper (or both lower and upper) bytes of the word addressed by A_{01}–A_{23}. Consequently, A_{01} from the 68000 is connected to A_0 at the memory.

DRAMs organized as n bits × 1 often have separate data-in (D_i) and data-out (D_0) pins (see figure 7.9). You can strap these together and treat the pair as a single bidirectional data line. Alternatively, they can be used to provide entirely separate data-in and data-out buses to reduce some of the problems caused by bus contention in high-speed systems.

Some 1M × 4-bit DRAMs do not have the luxury of separate data-in and data-out pins and employ an output enable pin, G*, to switch on the data bus drivers in a read cycle. The G* input behaves exactly like an EPROM's output enable (OE*) pin and is asserted in a read cycle to enable the device's output buffers. The 4M × 1 DRAM's do not have a G* input, since they have separate data input and data output pins. Smaller DRAMs (e.g., 256K or less) do not have a G* input. For the purpose of our description of the operation of DRAMs, we can forget about the G* pin, as it would normally be connected to R/$\overline{W}$ from the 68000 via an inverter.

Four signals in figure 7.9, MPLX, RAS*, CAS*, and W*, control the operation of the memory system. The timing control module must furnish these signals from the available system control signals. In other words, the design of the timing control module will vary from one microprocessor system to another, as each processor has its own unique control signals. Note that figure 7.9 is simplified in two ways. We have not provided the byte/word control required by the 68000, and no facilities for refreshing the dynamic memory are yet available.

We will illustrate the dynamic memory timing diagram with a typical device, the 514400-10, a 100-ns component. A read cycle in figure 7.8 lasts from A to E and has a minimum duration of t_{RC}, the *read cycle time*. The minimum value for t_{RC} is given as 180 ns. Note that the dynamic memory, unlike the static memory, with its equal access and cycle times, has a cycle time much greater (180 ns) than its access time (100 ns). The designer of a dynamic memory system cannot, therefore, begin the next access as soon as the current one has been completed. This restriction arises because the dynamic memory performs an internal operation, known as *precharging*, between accesses.

The first step in a read cycle is to provide the chip with the lower-order bits of the CPU address on its ten address inputs, A_0 to A_9. Then, at point A, the row address strobe, RAS*, is brought active-low to strobe (i.e., latch) the row address into the chip's internal latches. Once the row address had been latched, the low-order address from the processor is redundant and is not needed for the rest of the cycle. Contrast this with the static RAM, which requires that the address be stable for the entire read or write cycle.

The ten higher-order address bits from the CPU are then applied to the address inputs of the memory, and the column address strobe (CAS*) is brought active-low at point B to latch the column address. Now the entire 20-bit address has been acquired by the memory and the contents of the system address bus can change, since the DRAM has captured the address.

Once CAS* has gone low, the addressed memory cell responds by placing data on its data-output terminal, allowing the CPU to read it. At the end of a read cycle, CAS* returns inactive-high and the data bus drivers are turned off, floating the data bus. RAS* and CAS* may both go high together or in any order. It does not matter which goes high first as long as all other timing requirements are satisfied. To make our explanation of the DRAM more tractable, we will break it up into its component parts, beginning with a discussion of the role of the address pins.

Address Timing

Details of the DRAM's address timing requirements are given in figure 7.10, which is just an enlargement of the address bus timing in figure 7.8. In fact, the timing requirements of a DRAM's address bus are effectively the same as those of a typical

FIGURE 7.10 Details of the address timing of a dynamic RAM

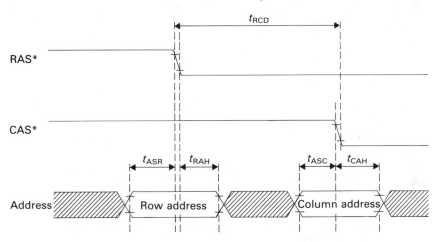

MNEMONIC	SIGNAL NAME	VALUE (ns)
t_{RCD}	Row to column strobe lead time	25–75
t_{ASR}	Row address setup time	0 min.
t_{RAH}	Row address hold time	15 min.
t_{ASC}	Column address setup time	0 min.
t_{CAH}	Column address hold time	20 min.

NOTE: The row address must be valid from t_{ASR} seconds before the falling edge of the row address strobe and t_{RAH} seconds after it. Similarly, the column address must be valid t_{ASC} seconds before and t_{CAH} seconds after the falling edge of the column address strobe. The minimum time between the falling edge of RAS* and the falling edge of CAS* is t_{RCD} and is made up of the row address hold time, the multiplexer switching time, and the column address setup time.

latch. The row address must be stable for a minimum of t_{ASR} seconds (i.e., row address setup time) before the falling edge of the RAS* strobe. As the minimum value of t_{ASR} is quoted as 0 ns, the row address has a zero setup time and does not have to be valid prior to the falling edge of RAS*. In the worst case, it must be valid coincident with the falling edge of RAS*. Once RAS* is low, the row address must be stable for t_{RAH}, the row address hold time, before it can change. The hold time is 15 ns minimum, and it restricts the time before which the column address may be multiplexed onto the chip's address pins.

Once the row address hold time has been satisfied and the column address multiplexed onto the memory's address pins, CAS* may go low. The column address setup time, t_{ASC}, is quoted as 0 ns minimum; that is, CAS* may go low at the same time that the column address becomes valid. After CAS* has gone active-low, the column address must be stable for a further t_{CAH} seconds, the column address hold time, before it may change. Once t_{CAH} (20-ns minimum) has been satisfied, the address bus plays no further role in the current access.

An important parameter in figure 7.10 is t_{RCD}, the row-to-column strobe lead time. For the 514400-10 the minimum value of t_{RCD} is quoted as 25 ns, and the maximum value is 75 ns. It must be appreciated that the limiting values of t_{RCD} are not fundamental parameters of the memory—they are derived from other parameters. The minimum value of t_{RCD} is determined by the row address hold time plus the time taken for the address from the multiplexer to settle.

The maximum value of t_{RCD} is a *pseudomaximum*. This value is not a maximum determined by the device but a maximum that, if exceeded operationally, extends the access time of the memory. We will return to this point later.

Data Timing

Having latched an address in the chip by asserting RAS* and CAS* in turn, data appear at the chip's data pin, as depicted in figure 7.11 (we assume that R/$\overline{W}$ is high for the duration of the read cycle). Only RAS*, CAS*, G*, and the data signals are included in figure 7.11 for clarity. We assume that the address setup and hold times and all relevant parameters have been satisfied.

The data at the data pin is valid no later than t_{RAC}, the access time from row address strobe, following the falling edge of RAS*. This time is, of course, the quoted access time of the chip and is 100 ns for a 514400-10. However, in the world of the dynamic RAM, all is not so simple. The row access time is achieved only if other conditions are met, as we shall see.

The column address strobe has two functions: It latches the column address that interrogates the appropriate column of the memory array, and it turns on the data output buffers. For these reasons, data is not available for at least t_{CAC}, the access time from CAS* low, after the falling edge of CAS*. The maximum value of t_{CAC} is 25 ns. In other words, reading the data is a two-part process: accessing the memory cell and placing the data on the chip's data-out pin. The following two examples should make this distinction clear.

Suppose that CAS* goes low at the minimum time after the falling edge of RAS* (i.e., $t_{RCD} = 25$ ns); data will appear at the data-out pin no later than $t_{RCD} + t_{CAC} = 25$ ns + 25 ns = 50 ns later. At this time, the data is not guaranteed to

FIGURE 7.11 Data access timing of a DRAM in a read cycle

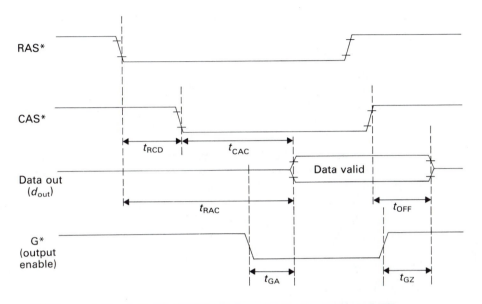

MNEMONIC	SIGNAL NAME	VALUE (ns)
t_{RCD}	Row-to-column strobe lead time	25–75
t_{CAC}	Access time from column address strobe	25 max.
t_{RAC}	Access time from row address strobe	100 max.
t_{OFF}	Output buffer turn off time	0–20
t_{GA}	Output buffer turn on time	20 max.
t_{GZ}	Output buffer turn off time	20 min.

NOTE: In a read cycle, data become valid not more than t_{CAC} seconds after the falling edge of CAS* and not more than t_{RAC} seconds after the falling edge of RAS*. At the end of a cycle, the data bus buffer is turned off no later than t_{OFF} seconds after the rising edge of the first of RAS* or CAS*. Since the 514400-10 has an output enable pin, G*, data bus drivers can be turned on and off in a read cycle by asserting and negating G*, respectively. If G* is asserted throughout the read cycle, the data bus buffers are turned on and off by the assertion and negation of CAS*, just as with any other DRAM.

be valid, as the minimum value of t_{RAC} (i.e., 100 ns) has not been met. However, once t_{RAC} has been satisfied, the data will be valid.

Now suppose that the falling edge of CAS* is delayed beyond the maximum quoted value of t_{RCD}. Say that CAS* is asserted 90 ns after RAS*. The data will not be valid until $t_{RCD} + t_{CAC} = 90$ ns $+ 25$ ns $= 115$ ns later. This value exceeds t_{RAC} by 15 ns.

We should now be able to see why the maximum value of t_{RCD} given in the data sheets of dynamic RAMs is a pseudomaximum. This value is not a maximum determined by the memory; instead, it is a limit that, if exceeded operationally, throws away access time. There is little point in buying a 100-ns chip and then degrading its access time to 150 ns by a careless design that exceeds t_{RCD} (max).

The relationship between $t_{RCD}(max)$, t_{RAC} and t_{CAC} is

$$t_{RCD}(max) = t_{RAC} - t_{CAC}$$

When CAS* goes high at the end of a read cycle, the data bus drivers are turned off and the bus floats t_{OFF} seconds later (t_{OFF} = output buffer turn-off delay). The maximum value of t_{OFF} is 20 ns. RAS* does not play any part in the ending of a read (or write) cycle. RAS* may be negated before or after CAS*, as long as its timing requirements are met.

Remember that DRAMs with an output enable pin, G*, (e.g., 1M × 4-bit chips) require G* to be asserted to turn on the data bus buffers in a read cycle. Data is placed on the data bus no later than t_{GA} seconds after G* goes active low, and the data bus is floated no later than t_{GZ} seconds after G* goes inactive high. You can also permanently tie G* to ground and use CAS* to turn on and off the data bus drivers.

W* Timing

The read cycle timing diagram of the DRAM's W* input is given in figure 7.12. As you can see, this is a very simple diagram; it reveals that W* must be high at least t_{RCS} seconds before the falling edge of CAS* and must remain high until at least t_{RCH}

FIGURE 7.12 Read cycle timing diagram of the W* input of a dynamic RAM

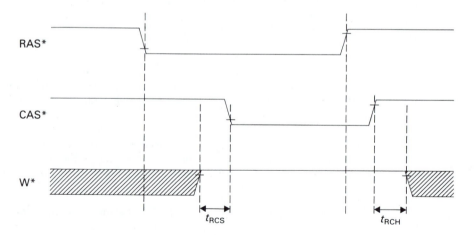

MNEMONIC	NAME	VALUE (ns)
t_{RCS}	Read command setup time	0 min.
t_{RCH}	Read command hold time	0 min.

seconds after the rising edge of CAS*. Both t_{RCS} and t_{RCH} are quoted as 0 ns minimum, which means that W* must be high for a read cycle the entire time that CAS* is low.

RAS* and CAS* Timing

The final part of the read cycle timing diagram is given in figure 7.13 and concerns the timing requirements of the row and column address strobes, RAS* and CAS*. We should appreciate that the RAS* and CAS* clocks are responsible for controlling several internal operations within the chip as well as the more mundane tasks of latching addresses and controlling three-state buffers. Although figure 7.13 looks rather complex with its eight timing parameters, it is entirely straightforward, and there are no parameters that make it difficult to design a DRAM controller. Basically, figure 7.13 illustrates the maximum and minimum times for which RAS* and CAS* must be high and low and the relationship between RAS* and CAS*.

A fundamental parameter of figure 7.13 is t_{RC}, the read cycle time—the minimum time that must elapse between successive memory cycles. This time is quoted as 180 ns for the 514400-10, which has a 100-ns read access time. The

FIGURE 7.13 Timing diagram of the RAS* and CAS* strobes

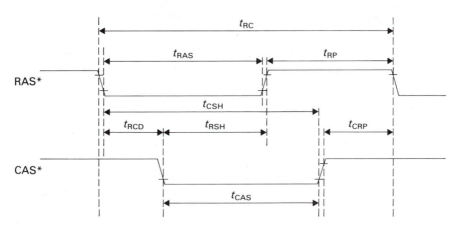

MNEMONIC	NAME	VALUE (ns)
t_{RC}	Random access cycle time	180 min.
t_{RAS}	Row address strobe pulse width	100–10,000
t_{RP}	Row address strobe precharge time	70 min.
t_{CSH}	CAS* hold time	100 min.
t_{RCD}	Row to column strobe lead time	25–75
t_{RSH}	RAS* hold time	25 min.
t_{CAS}	Column address strobe pulse width	25–10,000
t_{CRP}	Column to row strobe precharge time	10 min.

corollary of these figures is that the cycle time must be taken into account when designing memory systems. For example, if a microprocessor has a 150-ns cycle time, this dynamic RAM could not be relied upon, even if its 100-ns read access time were more than adequate. Interestingly, the value of 180 ns for t_{RC} is the minimum value necessary for reliable operation over the device's full temperature range of 0°C to 70°C. If the ambient temperature were guaranteed always to be lower than 70°C, the value of t_{RC} would be improved as the device slows with increasing temperature.

The RAS* clock must be asserted for at least t_{RAS} seconds (the row address strobe pulse width) during each read access. This has a minimum value of 100 ns and a maximum value of 10,000 ns. The maximum value is related to the need to refresh the device and creates no problems, as it is many times longer than a micro-processor's read cycle. The only danger in a 68000 system would arise if DTACK* were not asserted in a read cycle. The processor would hang up with RAS* held low and the DRAM's data would eventually be lost. You can avoid this situation by asserting BERR* to force a bus error after a suitable time-out.

After RAS* has been negated, it must remain high for at least t_{RP} seconds, the row address strobe precharge time. Precharge time is a characteristic of dynamic memories and relates to an operation internal to the chip. The minimum value of t_{RP} is 70 ns and no maximum value is specified, subject to the constraint that the memory needs periodic refreshing. The final constraint on the timing of RAS* is its hold time with respect to CAS*, t_{RSH}. RAS* must remain low for at least t_{RSH} seconds after CAS* has been asserted. The RAS* hold time is quoted as a minimum of 25 ns.

The column address strobe timing requirements are analogous to those of the row address strobe. CAS* must be asserted for no less than t_{CAS} seconds (25 ns), it must be negated for at least t_{CRP} seconds (10 ns) before the falling edge of the next RAS* clock, and it must be asserted for at least t_{CSH} seconds (100 ns) measured from the falling edge of the current RAS* clock.

The full read cycle timing diagram of a 514400-10 dynamic memory is given in figure 7.14 so that all the points discussed so far may be related to each other.

Dynamic Memory Write Cycle

The write cycle timing diagram of a dynamic RAM is more complex than the corre-sponding read cycle diagram, because stringent requirements are placed on both its W* and data inputs. Having already worked through the read cycle timing diagram, we do not need to go through the same material again. Figure 7.15 gives the full timing diagram of a 514400-10 1M × 4 dynamic RAM during a write cycle. Note that this type of write cycle is sometimes called an *early write cycle*, because the DRAM's W* input is asserted before CAS* goes low (i.e., early in a write cycle). We describe the early write cycle here since it most closely matches the 68000's write cycle, since the 68000 asserts its R/$\overline{W}$ output early in a write cycle. DRAMs often implement an alternative write cycle in which the W* input is asserted *after* CAS* goes low.

FIGURE 7.14 Full timing diagram of a DRAM read cycle

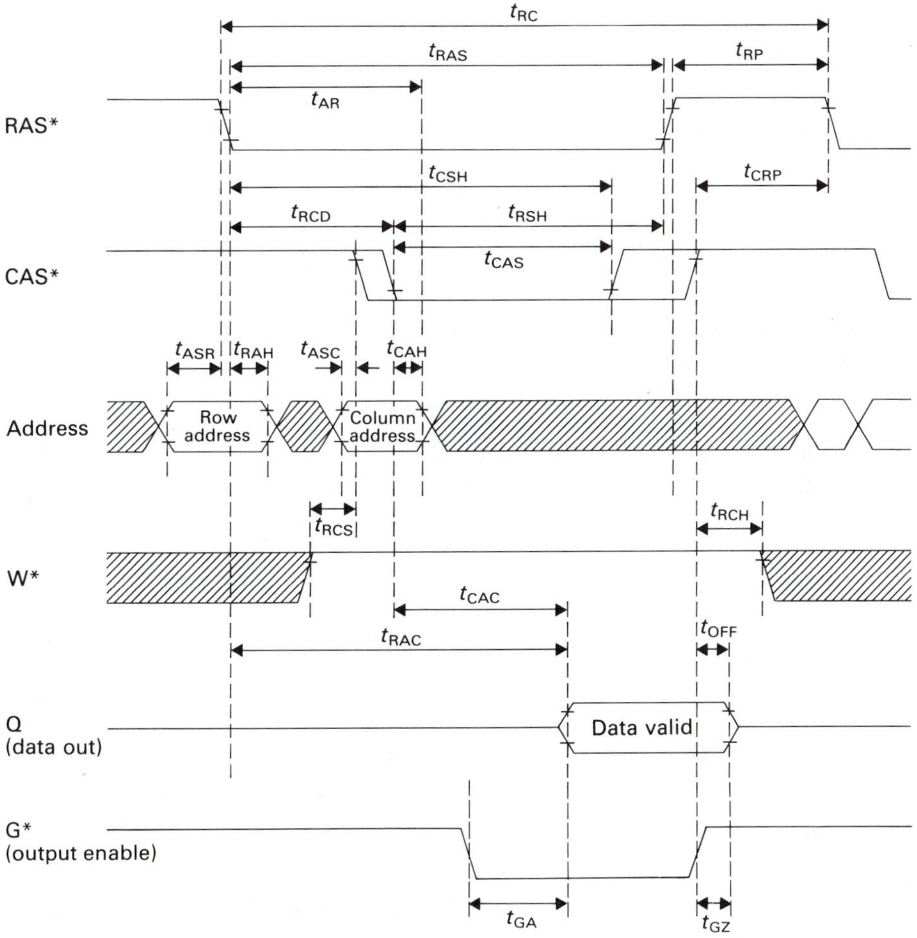

NOTE: G* is available only on certain DRAMs.

Figure 7.16 is a simplified version of the full write cycle timing diagram of figure 7.15. Figure 7.16 includes only parameters that differ between the DRAM's read and write cycle. We can immediately see that all the timing requirements of the RAS*, CAS*, and address inputs are identical in both read and write cycles.

Write Timing

Consider first the requirements of the W* input, which has to satisfy five conditions. It is latched by the falling edge of the CAS* clock and has a setup time of t_{WCS} seconds (remember that we are describing the early write cycle in which W* is asserted before CAS*). The minimum value of t_{WCS} is 0 ns, implying that W* can be asserted concurrently with the falling edge of CAS*. Once asserted, it has a

FIGURE 7.15 Full timing diagram of a DRAM in a write cycle

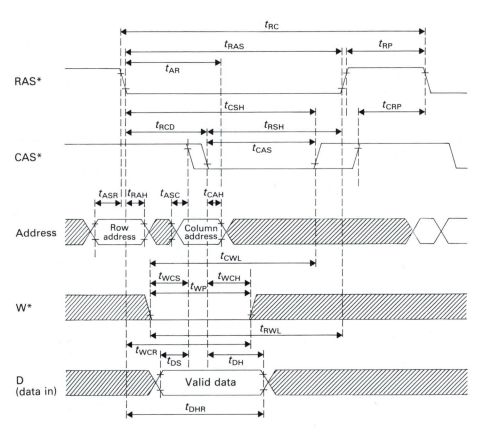

minimum down time of t_{WP} seconds (write pulse width = 20 ns) and must not be negated until at least t_{WCH} seconds (write pulse hold time = 20 ns) after the falling edge of CAS*.

In addition to these parameters, W* must be asserted at least t_{RWL} (write command to row strobe lead time) before the rising edge of RAS* and at least t_{CWL} (write command to column strobe lead time) seconds before the rising edge of CAS*. These times are both quoted as 25 ns. At this point, we can be forgiven for thinking that the dynamic RAM is a hideously complex device and that we would rather stick to static RAM. There is an old British saying, "Look after the pennies and the pounds take care of themselves." So it is with dynamic RAM. Look after the RAS* and CAS* clocks and the W* input will take care of itself—almost. To illustrate this point we will construct a simple example in which the write pulse is made equal to the CAS* clock. This example is purely illustrative and is not intended as a practical circuit.

Figure 7.17 illustrates the situation in which CAS* is made equal to W*. That is, W* goes low for the same period as CAS* in a write cycle. The RAS* pulse (i.e., the time for which RAS* is active low) is given the minimum possible value of

FIGURE 7.16 Details of the write cycle timing diagram of a DRAM

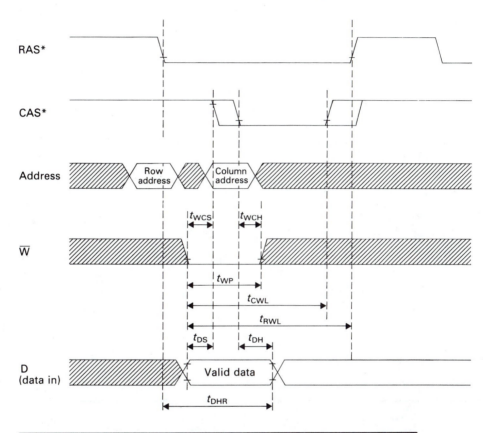

MNEMONIC	NAME	VALUE (ns)
t_{WCS}	Write command setup time	0 min.
t_{WCH}	Write command hold time	20 min.
t_{WP}	Write command pulse width	20 min.
t_{CWL}	Write command to column strobe lead time	25 min.
t_{RWL}	Write command to row strobe lead time	25 min.
t_{DS}	Data setup time	0 min.
t_{DH}	Data hold time	20 min.
t_{DHR}	Data hold time from RAS* low	75 min.

NOTE: Only parameters directly related to the write cycle have been included here. Note that the critical event in a write cycle is the falling edge of CAS* which latches the W* and the data input to the DRAM.

FIGURE 7.17 Dealing with the W* timing in a dynamic RAM

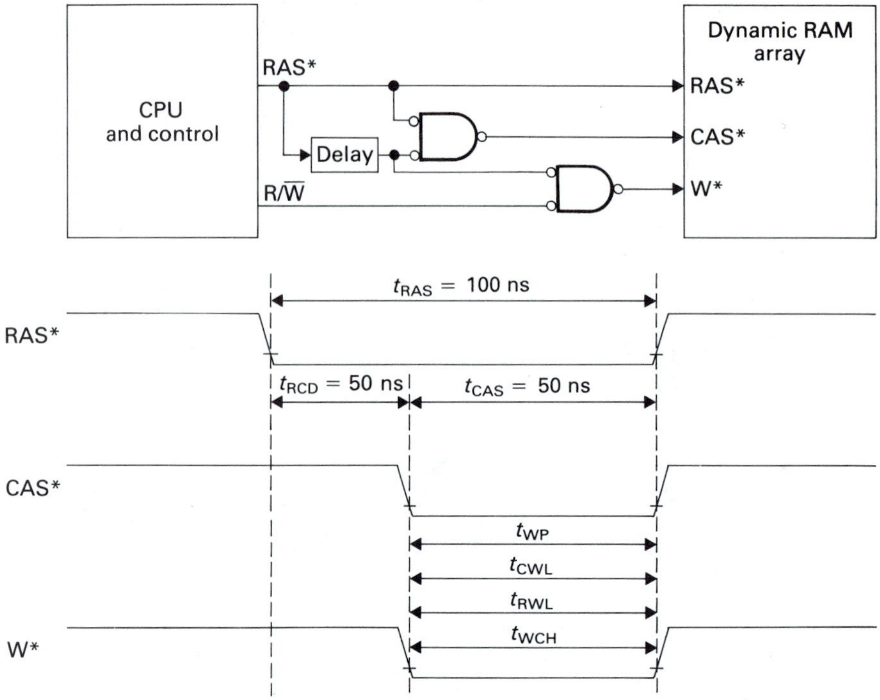

PARAMETER	NAME	MINIMUM VALUE	ACTUAL VALUE	MARGIN
t_{WCS}	Write command setup time	0 ns	0 ns	0 ns
t_{CWL}	Write command to column strobe lead time	25 ns	50 ns	+25 ns
t_{WP}	Write command pulse width	20 ns	50 ns	+30 ns
t_{RWL}	Write command to row strobe lead time	25 ns	50 ns	+25 ns
t_{WCH}	Write command hold time	20 ns	50 ns	+30 ns

t_{RAS} = 100 ns. CAS* is derived from RAS* by delaying its falling edge by 50 ns from RAS*, to yield a value for t_{RCD} of 50 ns and for t_{CAS} of 50 ns (the minimum down time of CAS* is 25 ns). As stated previously, the DRAM's W* input is obtained by gating the R/W̄ output of the 68000 with CAS*. Following the timing diagram of figure 7.17 are the five parameters associated with W* during a write cycle. Also given are their *minimum* requirements, together with the *actual* values achieved by this circuit.

In each case, the right-hand column provides the margin by which the requirement is satisfied. A negative value would indicate a failure to meet a requirement. Since no entry in this column is negative, it can be concluded that this circuit satisfies all constraints on W∗. Note that the margin on the write command setup time, t_{WCS}, is given as 0 ns. A margin as low as this would require careful attention to fine detail in a real circuit. In fact, because the CAS∗ signal in figure 7.17 is delayed with respect to W∗ by a gate, the margin on t_{WCS} is met adequately.

Data Timing in a Write Cycle

Data is written into the memory on the falling edge of the CAS∗ clock. The requirements for data-input timing are entirely straightforward and involve only three parameters (see figure 7.16). The data to be written into the memory must be valid for t_{DS} seconds (the data setup time) before the falling edge of CAS∗ and must be maintained for t_{DH} seconds (the data hold time) following the falling edge of CAS∗. The data setup time is 0 ns (minimum), and the data hold time 20 ns (minimum).

In general, the data timing requirements are not critical during a write cycle. By saying that the parameters are *not critical*, we mean that the DRAM's timing requirements are not difficult to satisfy. However, as the data from the processor is latched into the memory by the falling edge of the CAS∗ clock, which often occurs early in a write cycle, it is necessary for the processor to supply its data output very early in a write cycle. Otherwise CAS∗ must be delayed until data from the processor is available. Figure 7.18 illustrates the combined timing diagram of a 1M × 4 bit DRAM and a 68000 at 8 MHz in a write cycle. DTACK∗ is not included in this figure, as it is assumed that DTACK∗ goes low sufficiently early to permit operation without wait states. The memory systems designer must ensure that the DRAM's data setup, t_{DS}, and data hold, t_{H}, parameters are met by the 68000.

Special DRAM Operating Modes

Dynamic RAMs operate in a conventional read or write access mode in which a single location (be it 1 or 4 bits wide) is accessed. In addition to this standard mode, many modern DRAMs support one or more special access modes. These new modes are made possible because of the structure of the DRAM array and the way in which its address input is multiplexed between rows and columns. Typical special access modes are *page*, *nibble*, and *static column*. More importantly, these modes have been implemented to get around some of the limitations of the DRAM brought about by its need for precharging between accesses. You should appreciate that not all DRAMs support these modes and, in general, a given DRAM might support only one of them.

The *page mode* permits a fast access to any of the column locations of a given row. Suppose that a 1M-bit DRAM is accessed in a normal read or write cycle. The 10-bit row address is first applied and RAS∗ is brought low to latch it and to select one of 1,024 rows. The column address is then applied, CAS∗ is asserted, and a location is accessed. The page mode permits successive accesses to the same row, simply by pulsing CAS∗ and latching a new column address on each falling edge of the CAS∗ strobe. Figure 7.19 illustrates the read cycle timing of a typical page mode access.

FIGURE 7.18 Write cycle timing diagram of a DRAM-68000 combination

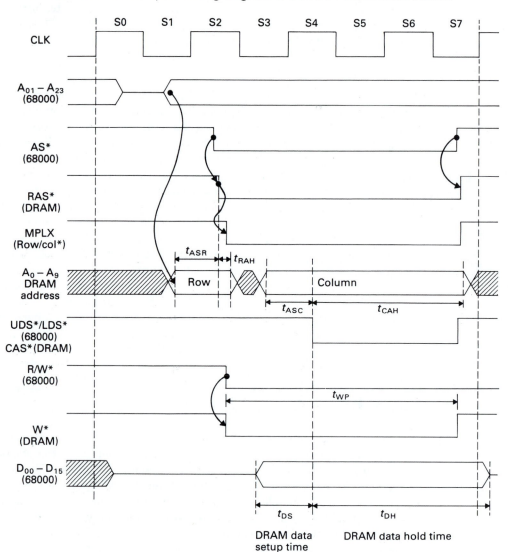

The period between successive page mode column access cycles is t_{PC}, which is 60 ns for a 514400-10. The page mode permits bursts of accesses to a single row address, with each column access separated by only 60 ns. The page mode cycle time compares favorably with this device's normal access time of 100 ns and cycle time of 180 ns. The page mode is particularly useful for transferring bursts of data from consecutive locations (e.g., during block transfers or in raster-scan graphics).

Certain DRAMs support a *nibble mode* that has some of the features of a page mode but is even faster. A nibble mode access begins exactly like a normal DRAM access, with the capture of the row address followed by the column address. Then

FIGURE 7.19 Timing diagram of a page mode access

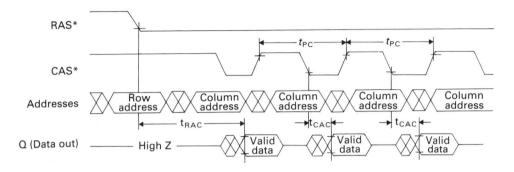

MNEMONIC	NAME	VALUE (ns)
t_{PC}	Fast page mode cycle time	60 min.
t_{RAC}	Access time from RAS* low	100 max.
t_{CAC}	Access time from CAS* low	25 max.

the CAS* strobe can be *cycled* (i.e., strobed), and up to four successive locations can be read from (or written to) without providing new column addresses.

Figure 7.20 illustrates the timing of the nibble mode access. Note that, unlike the page mode, the nibble mode latches just a single column address at the start of the burst. The next three accesses (made by cycling CAS*) take place in the sequence 00, 01, 10, 11, 00, 10, etc. The DRAM itself automatically generates the sequential addresses internally. For example, if we access location $0 1234 and then cycle CAS*, we will access locations $0 1234, $0 1235, $0 1236, and $0 1237 (in that order). The first cycle of a nibble mode takes as long as any other read or write cycle. Subsequent cycles can be performed in as little as 40 ns (for a DRAM with a nominal access time of 100 ns).

The *static column mode* looks rather like the page mode, except that the CAS* strobe is not cycled after the first column access. Figure 7.21 provides a static column mode timing diagram in which a location in a given row is accessed and then successive column locations are accessed simply by providing a new column address. As you can see, this mode is called *static column* because the DRAM is behaving exactly like a static RAM as far as column addressing is concerned. Static column mode DRAMs are used in high-speed applications, where the absence of a CAS* strobe improves system performance by reducing switching noise.

Another special access mode supported many DRAMs is the *read-write cycle*, in which a write cycle immediately follows a read cycle to the same address. A read-write cycle eliminates the need to latch the row and column addresses twice. After the read cycle has been completed, data is applied to the DRAM's data pin and W* is brought low to latch the data. Figure 7.22 shows the read-write cycle. Note that although a DRAM may support a read-write cycle, the cycle cannot be used unless

FIGURE 7.20 Timing diagram of a nibble mode access

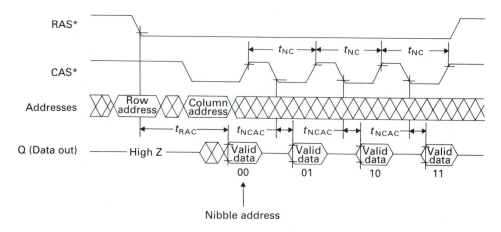

MNEMONIC	NAME	VALUE (ns)
t_{NC}	Nibble mode cycle time	40 min.
t_{RAC}	Access time from RAS* low	100 max.
t_{NCAC}	Nibble mode access from CAS* low	25 max.

FIGURE 7.21 Timing diagram of a static column mode access

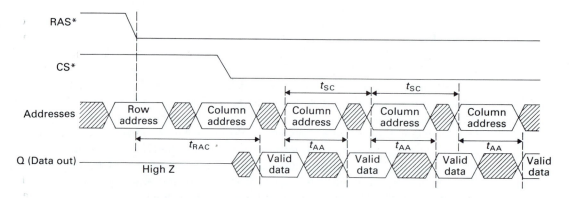

MNEMONIC	NAME	VALUE (ns)
t_{SC}	Static column mode cycle time	50 min.
t_{RAC}	Access time from RAS* low	100 max.
t_{AA}	Access time from address valid	50 max.

FIGURE 7.22 Timing diagram of a read-write cycle

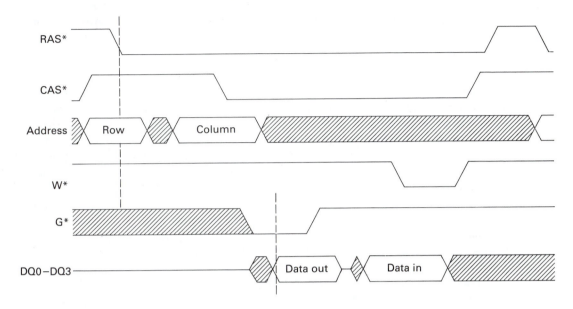

the host microprocessor has a suitable interface that permits a write cycle immediately to follow a read cycle.

Dynamic Memory Refresh

Having dealt with the DRAM's read and write cycles, the next step is to look at how the DRAM is refreshed. Because of the way in which a dynamic memory cell operates, with an automatic write-back of information following a destructive readout, all that is needed to refresh a memory cell is to read its contents periodically (at least once every 16 ms in the case of an 514400). Even better news is that when a particular row of the memory array is accessed, all columns in that row are refreshed, so that only 1,024 refresh operations need be carried out every 16 ms. If refresh operations are distributed evenly in time, a row must be refreshed every 15.6 μs.

DRAMs with low capacities usually have shorter refresh periods than DRAMs with high capacities. For example, DRAMs with capacities of 64K, 256K, and 1M have maximum refresh periods of 2 ms, 4 ms, and 8 ms, respectively. Note that bitwide DRAMs are usually internally organized as several arrays. For example, a 256K chip might be organized as four arrays of 256 rows by 256 columns. Consequently, a 256K chip has 256 row refreshes to perform in 4 ms, rather than the 512 refreshes you might expect.

Refreshing can be carried out either by hardware or by software. If the software can be guaranteed to perform at least 1,024 read or write cycles to the row addresses specified by the chip's A_0 to A_9 address inputs, then no additional hardware is required. Of course, guaranteeing 1,024 refresh cycles every 16 ms can be rather difficult. What happens if the reset button is pushed, or if ... ? We are now going to look at some of the ways in which a DRAM can be refreshed by hardware means.

RAS*–Only Refreshing

The vast majority of dynamic memories rely on hardware-refreshing techniques. Modern DRAMs such as the 514400 usually employ one of three refreshing techniques: RAS*-only refresh, CAS* before RAS* refreshing, and hidden refresh. We look at the RAS*-only refresh mode first (figure 7.23). For the duration of the refresh cycle, CAS* is held inactive-high, and the data-in and W* inputs are don't care conditions. RAS* behaves as it does in a normal memory access and the row address for the refresh is latched exactly in the same way as any other access. Thus, in order to execute a single refresh cycle, the row refresh address is applied to A_0–A_9 and RAS* is brought low for one cycle, while CAS* is held high. The user must supply a row-refresh counter that is incremented after each refresh cycle and the logic necessary to multiplex the refresh address onto the DRAM's address pins.

This refreshing procedure couldn't be simpler, but every silver lining marks out a nasty black cloud. In this case, the simple refresh operation is, indeed, the silver lining. Fitting the refresh cycle into the processor's normal sequence of operations is the black cloud. Somehow a logic subsystem has to take a decision to perform one or more memory refresh cycles and then to interleave them with normal processor

FIGURE 7.23   RAS*-only refresh timing

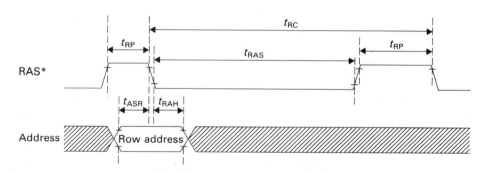

MNENOMIC	NAME	VALUE (ns)
t_{RP}	Row address strobe precharge time	70 min.
t_{RC}	Random access read cycle time	180 min.
t_{RAS}	Row address strobe pulse width	100–10,000
t_{ASR}	Row address setup time	0 min.
t_{RAH}	Row address hold time	15 min.

NOTE: In a RAS*-only refresh cycle, the refresh address is latched by the falling edge of RAS*. CAS* and W* are high throughout the refresh.

memory accesses. Some systems perform burst refreshing and refresh all row addresses in one operation. Others distribute refresh cycles among normal memory accesses (e.g., 1,024 rows in 16 ms = 15.6 μs/row). How refreshing is done depends on the nature of the system. It is very difficult to design a DRAM refresh system without reference to the host processor (in contrast to the design of a static RAM array). We will shortly look at the design of a refresh controller for a 68000 system.

CAS* before RAS* Refreshing

CAS before RAS* refreshing offers a refreshing mode that was not available with earlier generations of DRAMs. As its name suggests, a CAS* before RAS* refresh requires that CAS* make an active-low transition before RAS* goes low. By including a row-refresh address generator on-chip, this mode removes the need to provide an external row-refresh address, greatly simplifying the design of the refresh circuitry. Since normal accesses begin with the sequence RAS* low followed by CAS* low, the reverse sequence can be identified by the chip as a refresh cycle.

A CAS* before RAS* refresh cycle is illustrated in figure 7.24, from which it can be seen that CAS* makes its negative transition t_{CSR} seconds (10 ns minimum) before RAS*. Once RAS* has gone low, CAS* may return inactive-high any time later than t_{CHR} seconds (20 ns minimum). During a CAS* before RAS* refresh cycle, the state of the address bus, the W* pin, and the data pins are all don't care values,

FIGURE 7.24 The CAS* before RAS* refresh operation

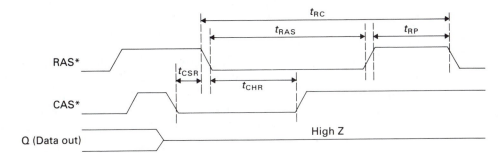

MNEMONIC	NAME	VALUE (ns)
t_{RC}	Read/write cycle time	180 min.
t_{RAS}	RAS* pulse width	100 min.
t_{RP}	RAS* precharge time	70 min.
t_{CSR}	CAS* setup time	10 min.
t_{CHR}	CAS* hold time	20 min.

NOTE: In a CAS* before RAS* refresh, the row-refresh address is generated internally by the DRAM and the input at its address pins are don't care values. If CAS* goes low at least t_{CSR} seconds before RAS* goes low, a CAS* before RAS* refresh is executed. CAS* may be held low while RAS* is pulsed to execute a series of refreshes.

and the data-out pin floats throughout the refresh cycle. The duration of a CAS*
before RAS* refresh cycle is t_{RC} (180 ns minimum for a 514400-10), which is, of
course, the same duration as a normal read or write cycle.

Hidden Refresh Cycle

Not only does the CAS* before RAS* refresh cycle avoid the need for a user-
supplied row-refresh address generator, it can easily be combined with a conventional read or write access in an operation called *hidden refresh*. Hidden refresh allows
refresh cycles to take place while valid data is maintained at the output pin. Holding
CAS* active-low at the end of a read or write cycle, while RAS* is brought inactive-
high for t_{RP} seconds and then low, triggers the refresh cycle. As you can see from
figure 7.25, the hidden refresh cycle is nothing more than a CAS* before RAS* cycle
started while the current access is in progress.

Pin 1 Refreshing

Another form of refreshing is called *pin 1 refreshing*, because it uses pin 1 on certain
dynamic RAMs to perform the refresh cycle. Although it is associated with 64K
DRAMs, we will describe pin 1 refreshing here, since 64K DRAMs are still found in
existing equipment. Figure 7.26 shows the operation of a typical pin 1 refresh mode.
RAS* is held high while the REFRESH* input on pin 1 is forced active-low. The
timing requirements are self-explanatory. The only noteworthy points are t_{FRI} (the
RAS* inactive time during a refresh cycle) and t_{FRD} (REFRESH* to RAS* delay
time). These are relatively long, 370 ns and 320 ns, respectively, and might cause
timing problems in some systems if a pin 1 refresh cycle is to be interleaved with
normal processor cycles.

  As t_{FRI} and t_{FRD} are relatively long compared to the refresh pulse period, t_{FP},
with its 60 ns minimum, it is advantageous to operate the device in a pulsed pin 1
refresh mode in which REFRESH* is pulsed while RAS* is high. A timing diagram
for this is given in figure 7.27. A single row-refresh cycle has a minimum duration of
$t_{FC} = 270$ ns, which is equal to the chip's normal cycle time. In a pulsed refresh

FIGURE 7.25 Hidden refresh cycle

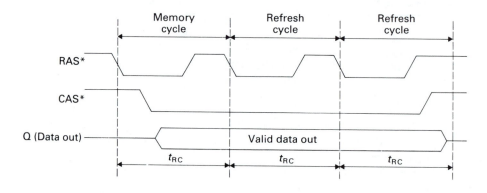

FIGURE 7.26 Pin I refresh mode—single cycle

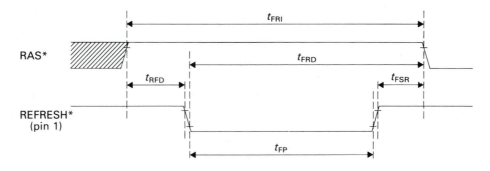

MNEMONIC	NAME	VALUE (ns)
t_{FRI}	RAS* inactive time during refresh	370 min.
t_{RFD}	RAS* to REFRESH* delay	− 10 min.
t_{FRD}	REFRESH* to RAS* delay time	320 min.
t_{FSR}	REFRESH* to RAS* setup time	− 30 min.
t_{FP}	REFRESH* pulse period	60–2000

FIGURE 7.27 Pin I refresh mode—multiple cycles

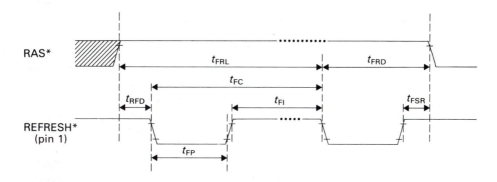

MNEMONIC	NAME	VALUE (ns)
t_{FRL}	RAS* to REFRESH* lead time	370 min.
t_{FRD}	REFRESH* to RAS* delay time	320 min.
t_{FC}	Refresh cycle time	270 min.
t_{RFD}	RAS* to REFRESH* delay	− 10 min.
t_{FI}	REFRESH* inactive time	60 min.
t_{FSR}	REFRESH* to RAS* setup time	− 30 min.
t_{FP}	REFRESH* pulse period	60–2000

mode, a batch of rows is refreshed in one burst rather than by executing 128 separate pin 1 refresh cycles.

Dynamic RAM Controller

Now that we have examined the operation of the DRAM, the next step is to show how it is actually used in microcomputers. A block diagram of the functional parts of a dynamic memory system is given in figure 7.28. The memory is organized as 1M 16-bit words, with independent byte control. During normal (i.e., nonrefresh) operation, the address decoder provides an active-low memory select output, MSEL*, whenever a valid address within the memory array is generated by the 68000. A negative edge on MSEL* triggers the timing generator, which synthesizes all the control signals required by the DRAM array and the address multiplexer. The address multiplexer control signal, MUX, from the timing controller selects the row or column address from the system address bus. A modern DRAM controller would almost certainly employ either CAS* before RAS* refresh or hidden refresh, eliminating the need for an additional refresh address generator.

One problem associated with 16-bit systems is the need to carry out *byte* operations on a word. Although two completely independent memory systems could be designed, one for the lower byte and one for the upper byte, such an approach would be hopelessly inefficient. In order to access data in a dynamic memory chip, both RAS* and CAS* must be asserted. Suppose we design a 16-bit memory array arranged as two 8-bit bytes and provide logic to furnish independent RAS* and/or CAS* strobes to the upper and lower bytes of a word. By negating either RAS* or CAS* to one-half word, that byte is disabled and takes no part in the memory access. Therefore, we can use the RAS* or CAS* strobe to perform byte selection. We assert RAS*/CAS* to both bytes of a word to access the word and assert RAS*/CAS* to half the word to perform a byte operation.

However, because refresh operations are applied to all bits of a word, RAS* should not be gated with UDS* or LDS* if you are using RAS*-only refreshing; that is, the use of the CAS* strobe to perform byte selection is preferable to RAS*. In any case, RAS* should not be gated with UDS*/LDS*, because the 68000's data strobe is asserted late in a 68000 write cycle. Byte selection in figure 7.28 is performed by dividing the CAS* inputs to the four DRAMs into two groups, CASL* and CASU*. CASL* is formed by gating CAS* with LDS*, and CASU* occurs by gating CAS* with UDS*.

The complex part of the dynamic memory system is its timing and control circuit, which must perform normal read/write memory cycles and memory refreshes, as well as handling processor bus handshakes. The way in which memory refreshes are slotted into the operation of the processor is a difficult design decision, and many factors have to be taken into account. For example, should the refreshing be interleaved with normal memory accessing, or should it be done separately by stopping the processor and then carrying out a burst of refresh cycles? The decision is difficult because many factors have to be considered, some of which are economic and some,

FIGURE 7.28 Structure of a dynamic memory module

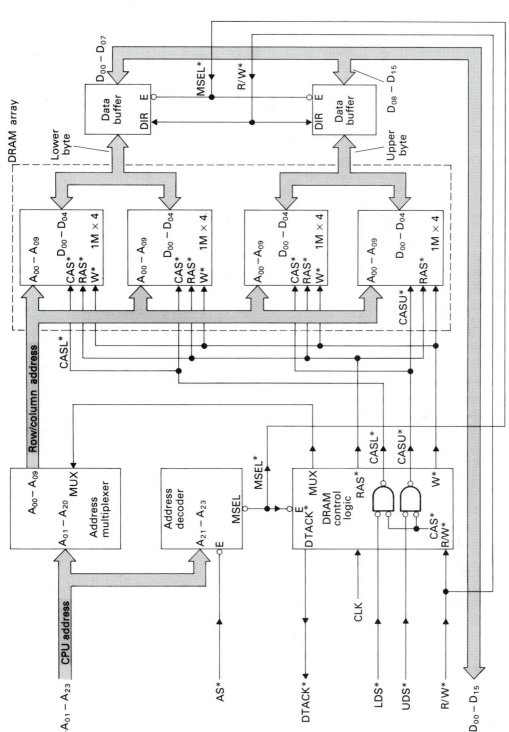

NOTE: When MSEL∗ (memory select) goes low, the 68000 memory control signals (AS∗, LDS∗/UDS∗ and R/$\overline{W}$) are used by the timing and control circuit to generate the multiplexer control signals and the DRAM row and column strobes. CASL∗ strobes the lower byte on D_{00}–D_{07}, and CASU∗ strobes the upper byte on D_{08}–D_{15}. The 68000's bus arbitration control lines can be used to halt the processor while the refresh operation takes place.

technical. In general, interleaving refresh cycles, a process called hidden or transparent refresh, has the advantage that the processor is not slowed down. However, interleaving refresh cycles requires faster memory because a refresh and a normal memory access have to be completed within the same processor cycle. Equally, burst mode refresh does not call for faster memory, but the processor is halted during the refresh process.

DRAM Control and the ECB

We will now look at one of the many possible ways of implementing the control circuits for a DRAM module connected to a 68000 microprocessor. The example is provided by Motorola's 68000 single-board educational computer, the ECB. The 16K × 1 DRAM components on this board are effectively obsolete, but the basic principles are valid for DRAMs of any size. This example has been selected because so many universities and colleges employ the ECB to support their teaching. The principal difference between this circuit and one using state-of-the-art chips is that a modern circuit would probably use CAS* before RAS* refreshing rather than RAS*-only refreshing.

Figure 7.29 provides a block diagram of the ECB. This circuit is essentially the same as that of figure 7.28, except that a row-refresh counter is used to supply the refresh address and a second multiplexer is required to multiplex the refresh address onto the DRAM's address inputs.

Figure 7.30 gives the basic circuit of the timing generator used to control read and write cycles. The actual arrangement of the dynamic memories and the row/column/refresh address multiplexers is not included here, as it is entirely straightforward. The read cycle timing diagram for this circuit is provided in figure 7.31.

The component at the heart of the controller is a 74LS175 4-bit shift register, clocked at twice the rate of the 68000. The ECB's 68000 runs at only 4 MHz, so the DRAM controller is clocked at 8 MHz. The serial input to the shift register is permanently tied to a logical 1 level. When the array is accessed, the output of the address decoder, RAMEN, goes high to request an access. As long as the dynamic memory block is not being accessed (i.e., both LDS* and UDS* high or RAMEN low), the output (CLR*) of the two-input AND gate is low and the shift register is held in its clear state with $Q_a = Q_b = Q_c = Q_d = 0$.

Whenever the processor addresses the array, RAMEN goes active-high, enabling one input to the AND gate. As soon as LDS* or UDS* is asserted, the second input to the AND gate rises and its output, CLR*, becomes high. The shift register is now enabled and a logical 1 is shifted along on each rising edge of the 8-MHz clock.

On the first rising edge of the 8-MHz clock, the Q_a output rises to a high level, since the serial input to the shifter is tied to V_{cc}. The Q_a output from the shift register is NORed with REFRAS from the refresh circuitry (to generate RAS* pulses during refresh cycles) to provide the memory array's RAS* input: All DRAM chips are clocked by RAS* simultaneously.

On the next rising edge of the 8-MHz clock, Q_b goes high and Q_b* is low. Q_b* acts as the row/column multiplex control signal and also gates the R/$\overline{W}$ signal

FIGURE 7.29 Organization of the ECB's DRAM

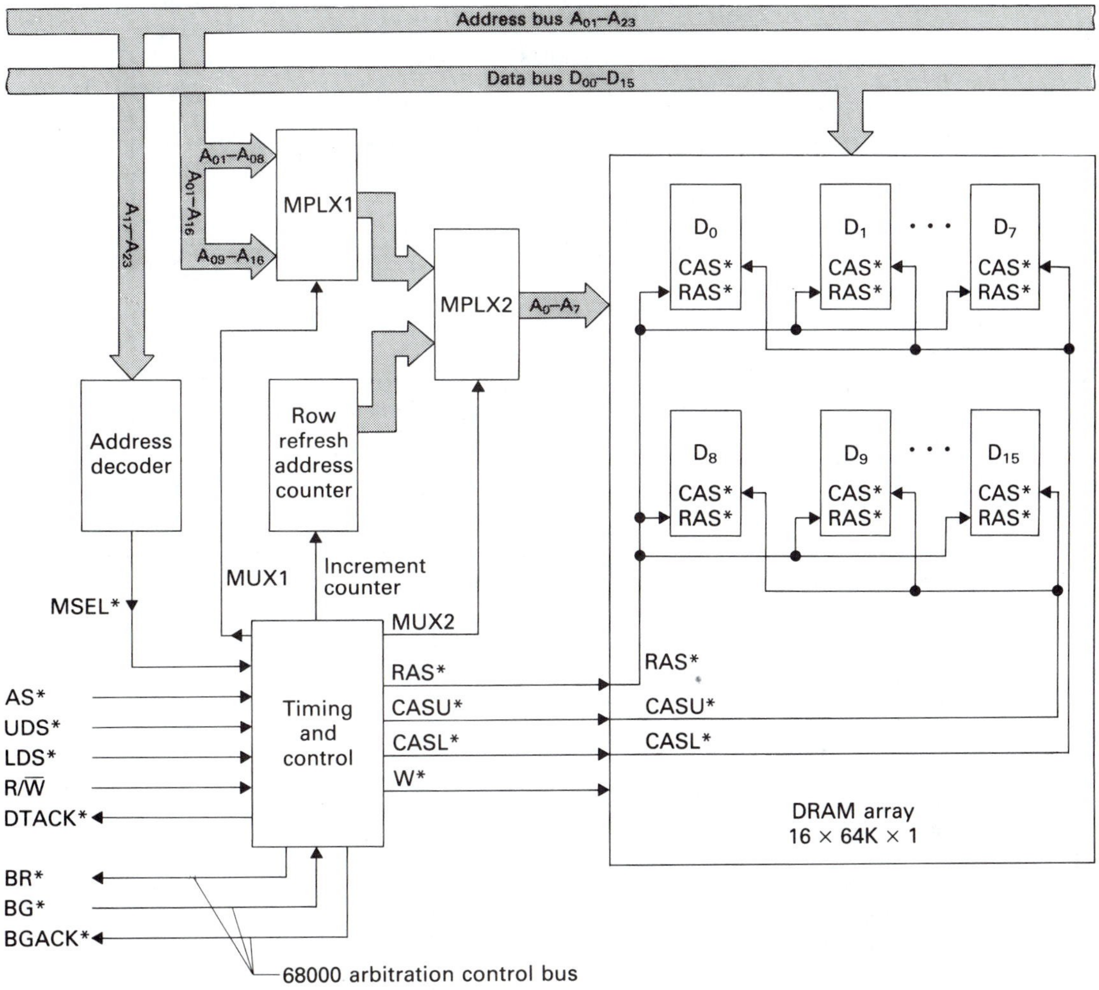

from the 68000. The next rising edge of the 8-MHz clock sets Q_c and resets Q_c* to generate CAS*. A further clock pulse later, Q_d* falls to provide a DTACK*. The falling edge of DTACK* is recognized by the 68000, and the memory access continues with state S5. During state S7, AS* is negated (together with UDS* and LDS*) and the shift register is cleared, forcing RAS*, CAS*, MUX, W*, and DTACK* high simultaneously.

DRAM Refresh on the ECB

The simplified circuit diagram of the dynamic memory refresh generator on the 68000 ECB is given in figure 7.32, and its timing diagram is shown in figure 7.33. A refresh clock, RFCLK, operating at 7.54 kHz signals the need for a burst of refresh

FIGURE 7.30 Dynamic memory control and the 68000

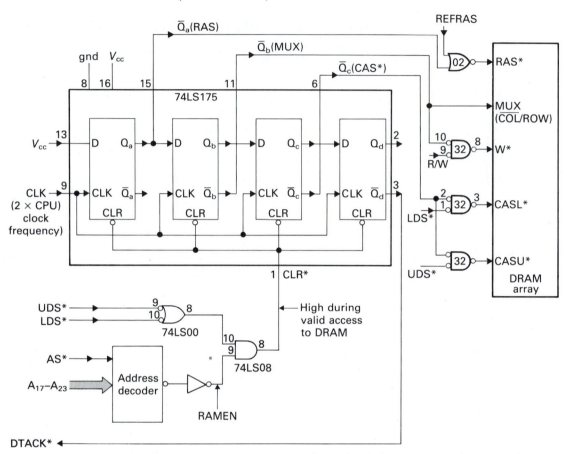

NOTE: A valid access to the DRAM array forces the shift register out of its reset (clear) state. Each clock pulse causes a logical one to be shifted down the register to generate the RAS*, MPLX, and CAS* signals in sequence.

cycles every $1/(7.54 \times 10^3) = 0.133$ ms. This design does not carry out all refreshes in one burst—it performs eight cycles every 0.133 ms, completing all 128 row refreshes in $0.133 \times 16 = 2.128$ ms. By distributing the refresh operation over 16 bursts of eight cycles, the processor is not held up for any appreciable length of time.

The refresh control circuitry employs the 68000's bus arbitration signals, BR*, BG*, and BGACK*, described earlier in this chapter. Further details of bus arbitration are provided in chapter 10 when we look at the VMEbus.

At power-up, POR* (power-on-reset from the processor control circuitry) goes low, clearing D flip-flop FF1 and setting FF2. Any well-designed circuit should be similarly initialized and placed in a *safe state*. In this state, Q_1* (i.e., BR*) is negated (i.e., high) and Q_2* (i.e., NORM/REF) is low, signifying normal operation. When the refresh clock, a simple RC oscillator, generates a rising edge, FF1 is set and BR* is asserted. The 68000 detects the bus request and asserts its bus grant output, BG∗, in response. AND gate G1 detects the condition BG∗ = 0, AS∗ = 1, DTACK∗ = 1 (i.e., BUSFREE) that occurs when the 68000 has relinquished the bus and forces input D_2 of FF2 low. Note that at this time the other two inputs to NOR gate G3 are both low—one because we will assume HALT∗ is negated and the other because $Q_2∗$ (i.e., $\overline{\text{NORM/REF}}$) is low after FF2 has been preset.

When D_2 is forced low by the rising edge of BUSFREE, the Q_2 output of FF2 is cleared on the falling edge of the 1-MHz clock. Q_2 is connected to the 68000's bus grant acknowledge input (BGACK∗) and, while low, stops the processor from regaining control of the bus. At the same time, it forces the output of OR gate G5 low,

FIGURE 7.31 Read cycle timing diagram for figure 7.30

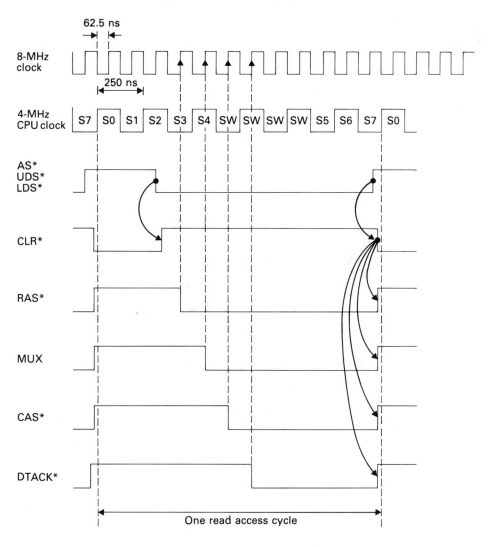

FIGURE 7.32 Dynamic memory refresh and the ECB

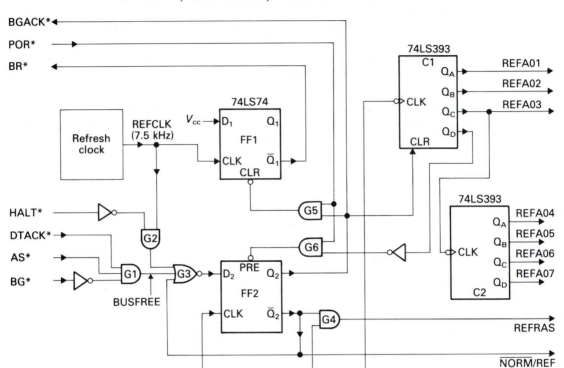

NOTE: C1 and C2 are counters that provide the DRAM array with a 7-bit refresh address REFA01 to REFA07.

clearing FF1 and negating BR∗. Thus, FF1 has done its job in this burst of refresh cycles and is once more in its initial state.

While FF2 is in a cleared state during a refresh operation, its Q_2* output remains high. Q_2* is the $\overline{\text{NORM/REF}}$ signal controlling the address multiplexer to the RAM array. When high, $\overline{\text{NORM/REF}}$ selects the address from the refresh counter (ICs C1 and C2); the refresh address multiplexer is not shown here. Q_2* is also fed back to the D_2 input of FF2 via NOR gate G3, so that once Q_2* is high, the flip-flop is held in this state and no longer depends on the state of BG∗ from the CPU, since BG∗ is automatically cleared following the negation of BR∗. FF2 is now locked up with Q_2* high and can only be released by the assertion of its PRE∗ (preset) input.

The final role played by Q_2* is to gate the 1-MHz clock in AND gate G4, the output of which is the pulsed RAS∗ needed in the refresh cycle. Because Q_2, when low, allows counter C1 to operate, 3 bits of the refresh address appear on REFA_{01} to

FIGURE 7.33 Timing diagram for memory refresh on the ECB

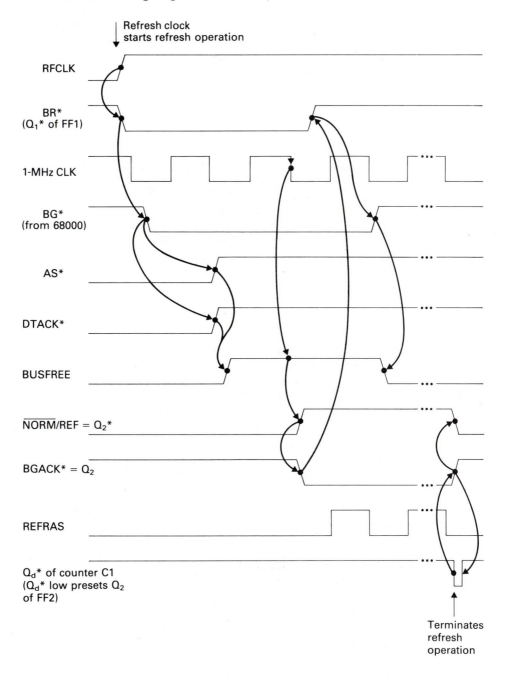

REFA$_{03}$, which forms part of the dynamic RAM's row-refresh address. This counter is clocked at the refresh rate—1 MHz. A second-stage counter, C2, is clocked by C1 after eight refresh cycles and provides the remaining four row-refresh addresses, REFA$_{04}$ through REFA$_{07}$.

After the 3-bit counter C1 has produced eight pulses, its Q$_D$ output rises and disables AND gate G6, which presets FF2, causing Q$_2$ (i.e., BGACK*) to be negated. The processor is freed by releasing BGACK*. At the same time, Q$_2$* (i.e., NORM/REF) goes low, disabling AND gate G4 and removing the refresh clock (REFRAS*). The system is now in its normal state, with BR*, BG* and BGACK* all negated. The only change since the start of the refresh burst is that counter C2 has been advanced by 1, so that the next time the refresh clock generates a pulse, the following eight row addresses will be refreshed.

Other Considerations in Dynamic Memory Design

Although we have largely concentrated on the DRAM from the point of view of its timing diagram, that is not the whole of the story. Dynamic RAM is associated with at least two other problems. The current taken by a DRAM is very bursty and the current taken by the V_{cc} pin can rise at a rate of 50 mA/ns when the RAS* input is asserted. This corresponds to a rate of change of 50 million amps per second. Such an immense rate of change of current can cause the V_{cc} voltage at the terminal of the chip to fall to a point at which erratic operation may occur.

The power supply problem is solved by a combination of attention to the circuit layout and to decoupling. The power lines to each DRAM chip are made as wide as possible to reduce their impedance and a high-quality (r.f.) 0.1-μF capacitor is connected between ground and V_{cc} at each chip—or at least at every other chip. This capacitor provides the current surge required by the DRAM whenever RAS* goes low. It is also necessary to ensure that all gates driving the DRAM's pins (address and data, etc.) are able to supply the current required by the DRAM array.

At the beginning of this section, we stated that DRAMs generate an internal bias voltage. It takes about 200 μs after power-up to generate the required bias. Following this, eight row strobes must be generated to correctly initialize the DRAM. Indeed, if the DRAM is not accessed for a period longer than 16 ms (this should not usually happen unless the designer provides a *sleep mode*, in which virtually all activity in the system stops), it is necessary to awaken the chip with eight row accesses. In short, the system designer should be careful not to access DRAM too soon after applying power to the circuit.

7.3 ERROR DETECTION AND CORRECTION IN MEMORIES

In this section we examine one of the major problems associated with large computer memory systems using DRAMs, that of read errors. A read error is said to occur when the data read back from a memory cell differs from that originally written to the cell. There are two classes of error: the hard error and the soft error. A *hard error* is repeatable and always happens under the same circumstances. Typically, 1 bit of a particular word may be permanently stuck at a logical 1 (or 0) level. Hard errors are due to faults in the memory system and are removed by repairing the damage—that is, by replacing the faulty chip.

A *soft error* is a form of transient error and is not normally repeatable. A certain memory cell may, for example, corrupt its data contents once in 100 years. Such an error is not caused by a fault in the ordinary sense of the word, but the error can cause as much trouble as any hard error. It is almost impossible to prevent occasional soft errors from occurring in a large memory array, particularly if dynamic memory components are employed. However, their effects can be greatly reduced by the following techniques, which attempt to detect errors or even to detect and correct them.

Dynamic memory cells are not as reliable as their static counterparts, in which positive feedback is used to hold a transistor in an on- or off-state. A dynamic memory cell suffers from soft errors due to both ionizing background radiation and pattern sensitivity. The latter problem occurs when a cell is sensitive to particular data patterns stored in adjacent cells. These errors are not frequent, but if the probability of error in a single cell is finite, the mean time between failure (MTBF) in a memory system declines as the size of the array is increased. For example, the MTBF of a memory array composed of 1,000 chips with a failure of 0.001 percent per 1,000 hours is approximately 1 year. Of course, a single soft error in such a large memory array once a year on average is not always of any importance. Occasionally, however, even such a low error rate cannot be tolerated. Medical, aviation, and nuclear power applications are areas where soft errors may prove to have hard consequences.

Error detection (and correction) involves nothing more than encoding data before it is stored in such a way that any change in the data will reveal itself. The advent of low-cost, large dynamic memory systems has led to renewed interest in codes for the detection or detection and correction of errors in memory arrays. Error detection and correction is not a new subject; its growth can be traced to 1948, when C. E. Shannon proved that it is theoretically possible to transmit information over a noisy channel without error as long as the channel capacity is not exceeded. Shannon's work was important because of the then-growing need for reliable data communication over long distances.

Designers can (economically speaking) now apply error detection and correction technology to memory arrays in order to reduce the probability of undetected soft errors. Hard errors are detected by test software; if a chip is found to be faulty, it is replaced. The coding of messages for error detection/correction has become an area of great sophistication, requiring the engineer to have a very strong mathematical background. Fortunately for the computer engineer, the basic principle behind error detection/correction is rather elementary, and the most popular form of error correction used with memory arrays is also one of the oldest and simplest.

Before we continue, we must provide two definitions. An error-detecting code is able to determine that a message has been corrupted and is not valid. Further details about the nature of the error cannot be provided. An error-correcting code both detects and corrects certain types of errors in a message. For example, a typical error-correcting code can correct a single error in an m-bit message and detect—but not correct—a 2-bit error. Note that we use the term *message* because the language and terminology of error-detecting/correcting codes comes largely from the world of telecommunications. In what follows, we use *word* to correspond to *message* in the preceding text. A *source word* applies to the information to be encoded and a *code word* applies to the information after encoding. At the moment, the source word is not associated with a particular bit length. That is, it does not imply a 16-bit word as used throughout the rest of this text.

Forward error correction (FEC) is a technique whereby the source word is encoded to yield the code word, and an error in the code word can be corrected automatically (just by using the code word). For example, we could encode the number 123 as ONE_TWO_THREE. Even if this sequence were to be struck by three errors to give, say, OQE_TWO_THRWK, we could probably correct it (because we know that all valid code words fall in the range ONE, TWO, ..., NINE). This error-correcting technique contrasts with an *automatic retransmission request* (ARQ), which is associated with data links. When operating under the ARQ mode, the receiver requests the retransmission of any message it determines to be in error. Error-correcting memories use forward error-correction techniques, as they cannot request the restorage of faulty data.

The basis of all forms of error correction and detection is redundancy. For example, we have already indicated that a human reader detects a spelling error in a book because English is a highly redundant language. Although there are $26 \times 26 = 676$ possible combinations of two letters, from AA to ZZ, there are only about 30 legal combinations of two letters. By knowing the valid combinations, an error can be detected if a given code differs from all possible valid codes.

The simplest possible binary error-detecting code is the single-bit parity code. A single bit called a *parity bit* is appended to the source word to create the code word. If the code uses even parity, the parity bit is chosen to force the total number of 1s in the code word (including the parity bit itself) to be even. If the code uses odd parity, the parity bit is chosen to force the total number of 1s in the code word to be odd. For example, the 7-bit source word 100 1101 would be encoded as 0100 1101 (even parity with the parity bit in the most significant bit position) and as 1100 1101 (odd parity with the parity bit in the most significant bit position). Consider the example of table 7.1, in which a 2-bit source word is encoded into a 3-bit code word with even parity.

Note that the code word has 3 bits, allowing $2^3 = 8$ possible combinations, but only 4 of these combinations are legal. A full list of possible code words is given in table 7.2, together with an indication of their validity.

On the left-hand side of table 7.2, the sequence of code words is presented in natural binary order, whereas on the right-hand side, it is presented in the Gray-code order. In a Gray-code sequence only 1 bit changes between successive code words. When the Gray code is examined, it can be clearly seen that there are no two adjacent valid code words. Thus, if 1 bit of a valid code word is changed by an error, the resulting code word is invalid and the error can be detected. Unfortunately, if 2 bits are changed, the resulting code word is valid and the error cannot be detected.

TABLE 7.1 The single-bit even parity code

SOURCE WORD	CODE WORD
00	000
01	011
10	101
11	110

↑
Parity bit

Figure 7.34 illustrates the 3-bit even parity code in three-dimensional space. Note, once again, that no two valid code words are adjacent. By adding sufficient redundancy to a code word, we can construct an error-detecting and error-correcting code. Figure 7.35 shows how an error-detecting and error-correcting code can be constructed with a trivial source word of 1 bit and a code word of 3 bits. Although there are $2^3 = 8$ possible code words, there are only two valid codes: 000 and 111. Each valid code is separated by a minimum of 3 bit changes from its valid neighbor. Suppose a single error occurs and 1 bit of the code word is changed. From figure 7.35 we can see that a single error yields a code word one unit from the correct value and two units from the other possible value. Thus, if 100, 010, or 001 is received, the correct code word is assumed to be 000 and the corresponding source word is assumed to be 1. The philosophy behind this code is that single-bit errors are infrequent and, therefore, double-bit errors are very rare, so that when an invalid code word is received, the closest valid code word can be taken as the corrected data value.

These results can be generalized to any number of bits. In what follows, the number of bits in the source word is m and the number of bits in the code word is n.

TABLE 7.2 The 3-bit even parity code

NATURAL BINARY SEQUENCE		GRAY CODE	
CODE WORD	VALIDITY	CODE WORD	VALIDITY
000	Valid	000	Valid
001	Invalid	001	Invalid
010	Invalid	011	Valid
011	Valid	010	Invalid
100	Invalid	110	Valid
101	Valid	111	Invalid
110	Valid	101	Valid
111	Invalid	100	Invalid

FIGURE 7.34 Eight 3-bit even-parity code words

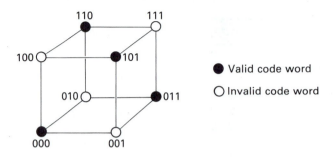

The number of redundant bits is r, where $r = n - m$. These r redundant bits are frequently called *check bits*.

We are now going to calculate the relationship between r, m, and n in an error-correcting code. The total number of *possible* code words is 2^n, and the total number of *valid* code words is 2^m. Consequently, there are $2^n - 2^m$ *error states*.

$$\text{Total error states} = 2^n - 2^m$$
$$= 2^{(m+r)} - 2^m \qquad \text{because } r = n - m$$
$$= 2^m(2^r - 1)$$

In other words, there are $2^r - 1$ error states per valid state. In the previous example of a 3-bit code with $n = 3$ and $m = 1$, there are $2^2 - 1 = 3$ error states per valid code word. Figure 7.35 demonstrates the validity of this result.

An m-bit source word yields 2^m valid code words, providing $n2^m$ possible single-error code words. The factor n appears because each n-bit valid code word can have n single-bit errors, since an error is possible in any one of the n bit positions. In order to correct a single error, there must be at least as many possible error states as words with single-bit errors. If this were not true, two or more single-bit errors

FIGURE 7.35 Three-bit error-correcting code word

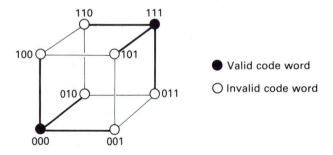

TABLE 7.3   The relationship between source and code wordlength

SOURCE WORD BITS m	CODE WORD BITS n	REDUNDANT BITS r
4	7	3
8	12	4
12	17	5
16	21	5
20	25	5
24	29	5
28	34	6
32	38	6
$\vdots$	$\vdots$	$\vdots$
57	63	6
$\vdots$	$\vdots$	$\vdots$
119	126	7

would share the same error state and it would not be possible to work backward from the error state to the valid code word. Therefore,

$$2^m(2^r - 1) > n2^m$$

$$2^r - 1 > n$$

$$2^r - 1 > m + r$$

Table 7.3 gives the relationship between m and n for $m = 4$ to 119. In each case, the value of r is chosen as the minimum integer satisfying the preceding equation.

Table 7.3 demonstrates few additional bits are required to provide a single-bit error-correction capability to relatively large source wordlengths. Unfortunately, single-bit error-correcting codes become less attractive as the source wordlength drops below about 16 bits.

Hamming Codes

One of the earliest and still the most popular type of single-bit error-correcting code is the Hamming code. An n,m Hamming code has an n-bit code wordlength and an m-bit source wordlength. This class of codes is popular because it is relatively easy to implement with MSI logic elements. In fact, several complete single-chip Hamming encoder/decoder chips are now widely available. The Hamming code adds parity bits to a source word in such a way that the recalculation of the parity bits, after the word has been stored, not only indicates the presence of an error (if any) but also points to its location. In order to appreciate the operation of a Hamming code, consider the 7, 4 Hamming code. The information bits in this code are written I_i and the redundant, or check bits, C_i. The code word can therefore be represented by

7	6	5	4	3	2	1
I_4	I_3	I_2	C_3	I_1	C_2	C_1

Note that the bit positions are numbered 1 through 7 (rather than 0 through 6) and that the check bits, C_1, C_2, C_3, are placed in binary sequence (i.e., 1, 2, 4, 8, ...) in positions 1, 2, and 4. The parity equations for the check bits are derived from table 7.4.

Below each of the bit positions of the code word in table 7.4 is the binary value of the position. For example, bit position 6, representing I_3, is located above binary code $110_2 = 6$. The parity equations are derived by reading across the columns of the binary code lines and including each bit position with a logical 1 in the parity equation; that is,

MSB bit $C_3 \oplus I_2 \oplus I_3 \oplus I_4 = 0$

Middle bit $C_2 \oplus I_1 \oplus I_3 \oplus I_4 = 0$

LSB bit $C_1 \oplus I_1 \oplus I_2 \oplus I_4 = 0$

We can use these three equations to calculate the check bits from the information bits. When the data is retrieved, the check bits can be recalculated and compared with the stored values. If they are the same, we assume no error. If they are different, we assume that an error has occurred. Note that this arrangement detects errors in both the stored information and in the check bits.

An example should make things clearer. Suppose the source word is I_4, I_3, I_2, $I_1 = 1, 0, 1, 1$. We can substitute the values of I_1 to I_4 in the preceding equations to calculate the check bits.

$C_3 \oplus 1 \oplus 0 \oplus 1 = 0;$ therefore, $C_3 = 0$

$C_2 \oplus 1 \oplus 0 \oplus 1 = 0;$ therefore, $C_2 = 0$

$C_1 \oplus 1 \oplus 1 \oplus 1 = 1;$ therefore, $C_1 = 1$

TABLE 7.4   The parity equations for a 7, 3 Hamming code

CODE BIT NUMBER	7	6	5	4	3	2	1	
CODE BIT	I_4	I_3	I_2	C_3	I_1	C_2	C_1	
	1	1	1	1	0	0	0	← MSB of 3-bit number
	1	1	0	0	1	1	0	
	1	0	1	0	1	0	1	← LSB of 3-bit number

The 7-bit code word I_4, I_3, I_2, C_3, I_1, C_2, C_1 is given by 1, 0, 1, 0, 1, 0, 1. Suppose this code word is stored in memory and later read back as 1, 1, 1, 0, 1, 0, 1, with an error in bit position 6. The next step is to recalculate the parity equations using the stored check bits:

MSB bit	$C_3 \oplus I_2 \oplus I_3 \oplus I_4 = 0 \oplus 1 \oplus 1 \oplus 1 = 1$	
Middle bit	$C_2 \oplus I_1 \oplus I_3 \oplus I_4 = 0 \oplus 1 \oplus 1 \oplus 1 = 1$	
LSB bit	$C_1 \oplus I_1 \oplus I_2 \oplus I_4 = 1 \oplus 1 \oplus 1 \oplus 1 = 0$	

The parity equations are no longer all equal to zero. Their value is 110_2, which is the binary value of 6. This points to the position of the bit in error, namely, the bit 6 position. The simplest way of mechanizing a Hamming decoder is to take the stored information bits and use them to recalculate the check bits. An exclusive OR is performed between the stored and recalculated check bits. If the result is zero, the stored data is assumed to be error-free. Otherwise, the result indicates the bit position of the stored word in error.

Consider a second example with information bits 1001. The corresponding code bits are

$$C_3 \oplus 0 \oplus 0 \oplus 1 = 0; \quad \text{therefore, } C_3 = 1$$

$$C_2 \oplus 1 \oplus 0 \oplus 1 = 0; \quad \text{therefore, } C_2 = 0$$

$$C_1 \oplus 1 \oplus 0 \oplus 1 = 0; \quad \text{therefore, } C_1 = 0$$

The codeword is 100 1100. Suppose that bit 3 is corrupted in storage and the code word read from memory is 100 1000. We can recalculate the code bits as

$$C_3 \oplus 0 \oplus 0 \oplus 1 = 0; \quad \text{therefore, } C_3 = 1$$

$$C_2 \oplus 0 \oplus 0 \oplus 1 = 0; \quad \text{therefore, } C_2 = 1$$

$$C_1 \oplus 0 \oplus 0 \oplus 1 = 0; \quad \text{therefore, } C_1 = 1$$

If we perform an exclusive OR between the stored and recalculated check bits, we get $(1, 0, 0) \oplus (1, 1, 1) = (0, 1, 1) = 3_2$. This is, of course, the location of the bit in error. To correct this error, all we need to do is invert the erroneous bit.

Figure 7.36 illustrates the arrangement needed to implement an error-correcting memory (ECM). The m-bit data input is used to calculate an r-bit check word, which is stored along with the data. When the memory is read, the n-bit code word determines the position of any error. This information can then be used to correct a faulty bit.

A more detailed implementation of a 7,4 Hamming code single-bit error-correcting memory is described in figure 7.37. A parity tree formed by three pairs of cascaded EOR gates generates the check bits. Note that the parity circuits perform the same function as the equations described earlier. On reading back the data, the 3-bit position of the error is determined from the stored data and check bits by the

FIGURE 7.36 Conceptual arrangement of error-correcting memory

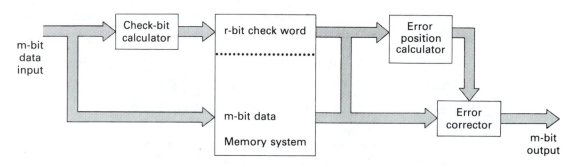

second group of EOR gates. Finally, the 3-bit error position code is decoded into one of eight lines. The least significant output, Y0∗, is active-low and, when asserted, implies that no error has been detected. Four of the other outputs from the decoder feed EOR gates that invert the erroneous data bit to correct it.

FIGURE 7.37 Possible arrangement of a 7, 4 Hamming code ECM

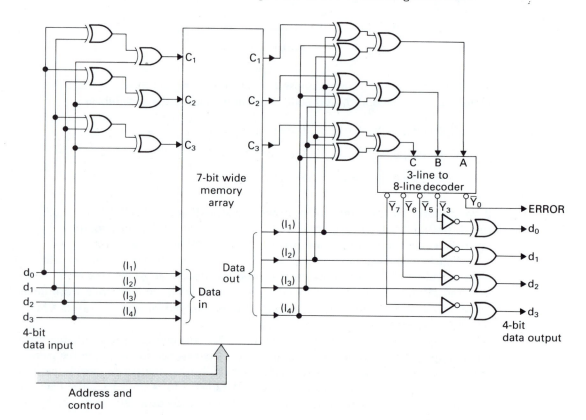

Practical Error-Detection/Correction Systems for Microprocessors

Table 7.3 tells us that only five check bits are required to implement a 16-bit, single-bit error-correcting memory for a microprocessor. Unfortunately, table 7.3 does not tell the whole story. Two particular issues are raised by practical arrangements of error-correcting memories. The first is that the basic Hamming code is not normally used, since its performance can be considerably improved by the addition of another check bit. The second is the problem arising when a 16-bit microprocessor attempts to carry out operations on 8-bit bytes.

Modified Hamming Code

The standard Hamming code described earlier provides single-bit error detection and correction. If a word is corrupted by a double-bit error, the code fails either to correct or, more importantly, to detect it. The reason for this failure to cope with multiple errors can be seen from the parity equations for a 7,4 Hamming code, repeated here:

$$C_3 = I_2 \oplus I_3 \oplus I_4$$
$$C_2 = I_1 \oplus I_3 \oplus I_4$$
$$C_1 = I_1 \oplus I_2 \oplus I_4$$

Suppose I_3 is corrupted in storage. This information bit error affects the value of check bits C_2 and C_3, because it appears only in the equations for these check bits. If both I_1 and I_2 are corrupted, only check bits C_2 and C_3 are affected, because the double-bit error in the calculation of C_1 does not affect the result. However, C_2 and C_3 are also affected by an error in I_3. Therefore, a double error in I_1 and I_2 appears as a single error in I_3.

In order to implement a single-bit error-correcting, double-bit error-detecting code, we need to employ five check bits, arranged so that each bit of the source word is protected by three of the check bits. Table 7.5 gives the modified Hamming adopted by Texas Instruments in their 74LS637 8-bit parallel error detection and correction circuit.

An $\times$ in table 7.5 indicates that the bit of the source word takes place in the generation of the check bit on the same row. For example, $C_0 = d_7 \oplus d_6 \oplus d_4 \oplus d_3$. Note that each data bit is involved in the calculation of exactly three check bits (i.e., there are three $\times$s in each of the columns).

During a write cycle to memory, the 8-bit data word is used to generate a 5-bit check word, which is stored along with the data. On reading back the data and check word, the five check bits are recalculated and, if they are the same as the retrieved check bits, the data is assumed to be error-free.

If a single error occurs in the stored data bits d_0 to d_7, exactly three of the

TABLE 7.5 Modified Hamming code for $m = 8, r = 5$

CHECK BIT	EIGHT-BIT SOURCE WORD							
	d_7	d_6	d_5	d_4	d_3	d_2	d_1	d_0
C_0	×	×		×	×			
C_1	×		×	×		×	×	
C_2		×	×		×	×		×
C_3	×	×	×				×	×
C_4				×	×	×	×	×

recalculated bits differ from the retrieved check bits. If one of the check bits is corrupted, only one error is detected when the check bits are read back.

Any 2-bit error alters an even number of check bits. This fact can be used to interrupt the processor and inform it that an uncorrectable error has been detected. In 68000-based systems, the detection of a multiple error may be used to assert BERR∗ in order to abort the current memory access. Alternatively, BERR∗ can be used in conjunction with HALT∗ to rerun the bus cycle. Three or more simultaneous errors cannot be handled by this code; the error detection circuitry will "see" the three errors as a single correctable error, an uncorrectable error, or no error at all.

As in the case of the basic Hamming code, error correction is achieved by determining the location of the bit in error and then inverting it. Table 7.6 gives the

TABLE 7.6 Locating a single error with a modified Hamming code

ERROR LOCATION		SYNDROME ERROR CODE				
		C_0	C_1	C_2	C_3	C_4
Data bit	d_7	0	0	1	0	1
Data bit	d_6	0	1	0	0	1
Data bit	d_5	1	0	0	0	1
Data bit	d_4	0	0	1	1	0
Data bit	d_3	0	1	0	1	0
Data bit	d_2	1	0	0	1	0
Data bit	d_1	1	0	1	0	0
Data bit	d_0	1	1	0	0	0
Check bit	C_0	0	1	1	1	1
Check bit	C_1	1	0	1	1	1
Check bit	C_2	1	1	0	1	1
Check bit	C_3	1	1	1	1	0
Check bit	C_4	1	1	1	1	0
No error		1	1	1	1	1

error location table corresponding to a single-bit error. The syndrome error code is the EXCLUSIVE OR of the regenerated check word and the retrieved check word.

Problem of Byte Operations on a 16-Bit Word

The most irritating problem associated with codes for error detection/correction is due to the modern 16-bit processor's ability to operate on a whole 16-bit word or just 1 byte of it. Suppose a 16-bit data word is used with a 22,16 modified Hamming code. Whenever a new word is written to memory, the appropriate six check bits are calculated and stored. On reading back the 22-bit code word, automatic single-bit error correction can be performed by the memory array or a multiple error can be signaled, should the need arise.

Imagine that the processor wishes to read 1 byte of a word. Clearly, the whole word must be read to carry out the error-correction process, as check bits are distributed throughout the word. The situation is much worse if we wish to write to 1 byte of a word—that is, 1 byte of a 16-bit word is to be updated while the other byte remains unaffected. Changing half of a 16-bit word is simply not possible. We first need to read the 22-bit codeword, perform an error check on it, and then extract the byte to be retained. Then the new byte from the processor is appended to the retrieved (i.e., retained) byte and the appropriate six check bits are generated for the whole 16-bit word. Therefore, although the error correction circuitry itself is relatively simple, the error-correcting memory system becomes quite complex. Even worse, system throughput is reduced because two memory accesses are necessary for each CPU access: one to read the old word from memory and another to restore half the old word plus the "new" byte.

An alternative, but hardly elegant, arrangement is to design the 16-bit error-correcting memory as two entirely independent 8-bit memories. Not only does this duplicate the error-correcting hardware, but a total of 12 check bits per 16-bit word is required if two 6-bit modified Hamming codes are used. Although this arrangement almost doubles the cost of the main memory, it is still cheaper than relying on the better error performance of static read/write memory if the memory array is sufficiently large.

Figure 7.38 gives the block diagram of an 8-bit ECM using a modified 13,8 Hamming code; figure 7.39 gives its detailed implementation. To construct a 16-bit version, two identical 8-bit ECMs must be used, with one enabled by LDS* and one by UDS*, as shown in figure 7.40. The detailed implementation of figure 7.39 reduces the parts count by replacing EOR parity tree generators with PROM look-up tables. In a read cycle, the five check bits, C_1 through C_5, are generated by applying the eight data inputs to the address inputs of a 256 × 8 PROM. The data outputs of the PROM directly provide the check bits.

On read-back, a second, identical 256 × 8 PROM recalculates the check bits from the retrieved data. The recalculated and stored check bits are fed to five EOR gates to compute the syndrome for the word currently being accessed. This syndrome is applied to another 32-word by 8-bit PROM, which looks up the position of

FIGURE 7.38 Eight-bit error-correcting memory

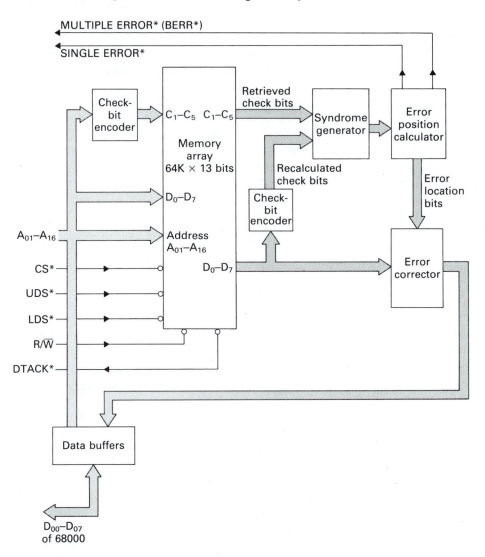

any single-bit error and applies it to the *error location* output. This is fed to a three-line-to-eight-line decoder, which controls the 8-bit programmable inverter. If there is no error, the three-line-to-eight-line decoder is disabled and no data bit is inverted. If the syndrome detects more than one error (i.e., an even number of check bits are corrupted), the multiple error flag from the 32×8 PROM is asserted.

In any practical implementation of 68000-based ECM, the designer is faced with the problem of what to do when one or more errors are detected. You may think that a single error can be forgotten, since it is automatically corrected. In principle, this statement is true. Each time the faulty byte is read, its error is corrected. However, no margin is left for a second and, therefore, uncorrectable error. A better strategy is to detect each single-bit error, inform the processor, and let it write back the word in error. Correcting the error, followed by a write-back, will ensure that the error is removed, as both the data and check bits will then be error-free.

FIGURE 7.39 Implementing 8-bit ECM

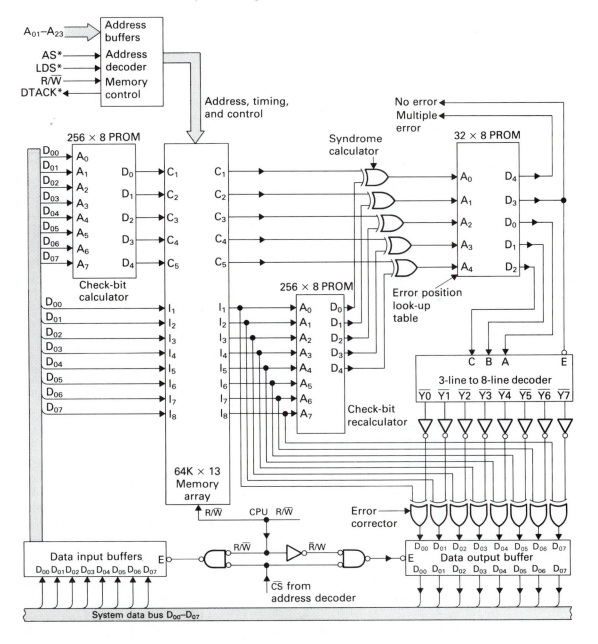

FIGURE 7.40 Sixteen-bit byte-accessible ECM

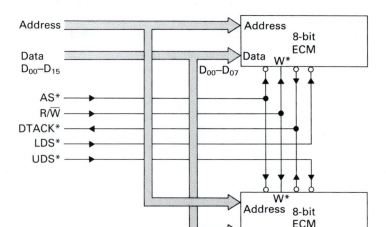

If two or more errors occur, the processor must be informed. This action can be taken by any conventional technique. The processor may be interrupted, or RESET* or BERR* may be asserted to abort the cycle. The latter action seems most reasonable, as a reset is too drastic and an interrupt will not be serviced until the faulty data has been read and, possibly, the harm done.

A useful addition to the ECM we have just described is some form of error-logging circuitry, which may be implemented in software by the processor itself keeping track of each error; it can also be realized by an auxiliary circuit on the ECM module that records the address and data associated with each error. Such an approach makes it possible to determine the state of the memory module's health. If the frequency of soft errors rises above its expected level, an operator can be informed and memory chips can be replaced after the processor has determined their location from the error log.

7.4 MEMORY MANAGEMENT AND MICROPROCESSORS

Memory management is the term applied to any technique that takes an address generated by the CPU and employs it to calculate the actual address in memory being accessed by the processor. The concept of memory management often seems obscure, because there appears to be no reason why an address at the microprocessor's address pins should be tampered with in order to access the appropriate memory location. This section explains why some microprocessor systems implement memory management and how it is achieved. In particular, we look at the 68451 memory management unit and the 68851 paged memory management unit.

Another reason for the air of mystery surrounding memory management is its application only to sophisticated microprocessor systems. Many introductory texts on microprocessors ignore memory management because it is entirely unnecessary in a lot of small-scale and medium-scale microcomputers. In such machines the address generated by the processor is indeed the same address as the location of the data in memory. For example, if you execute the instruction MOVE.W D2,$1234, the contents of D2 are copied into the memory location whose address is 1234_{16}. A computer with memory management may, for example, execute the same instruction but actually access the location $1FBC34_{16}$ in physical memory.

The key to understanding the role of memory management is an appreciation of the meaning of the terms logical address space (LAS) and physical address space (PAS). The term *virtual address space* is also frequently used to describe local address space. In the simplest terms, the *logical address space* of a microprocessor is the address space made up by all the addresses the microprocessor can place on its address bus. That is, it is the address space made up of all the addresses that can be generated by the CPU. Thus, the logical address space of an 8-bit microprocessor with a 16-bit address bus is $2^{16} = 64K$ bytes. The 68000 has a 16M-byte logical address space and the 68020/30 has a massive 4G-byte logical address space.

The size of the logical address space does not depend on the addressing mode used to specify an operand. Nor does it depend on whether a program is written in a high-level language, assembly language, or machine code. The 68000 instruction MOVE.B D4,TEMPERATURE permits the program to specify the logical address of TEMPERATURE as any one of 16M bytes. No matter what technique is used, the 68000 cannot specify a logical address outside the 16M-byte range 0 to $2^{24} - 1$, simply because the number of bits in its program counter is limited to 24 (by the address pins of the 68000 in this case). Of course, in a real system programmers may not be able to choose *any* logical address for their programs and data, since the programmer may select a location at which no actual memory component is located. Strictly speaking, we could say that the 68000 has a logical address space of 2^{32} bytes, because the program counter is a 32-bit register, even though the 68000 has only a 24-bit external address bus.

Physical address space is the address space spanned by all the actual address locations in the processor's memory system. This physical memory is the memory that is in no sense abstract and costs real dollars and cents to implement. In other words, the system's main memory makes up the physical address space. Although the quantity of a computer's logical address space is limited by the number of bits used to specify an address, the quantity of physical address space is frequently limited only by its cost.

We can now see why a microprocessor's logical and physical address spaces may have different sizes. What is much more curious is why a microprocessor might, for example, translate the *logical* address $0000 1234 into the physical address $86 1234.

Five of the fundamental objectives of memory management systems are as follows:

1. To control systems in which the amount of physical address space exceeds that of the logical address space (e.g., an 8-bit microprocessor with a 64K-byte logical address space and 1M bytes of RAM).

2. To control systems in which the logical address space exceeds the physical address space (e.g., a 32-bit microprocessor with a 4G-byte logical address space and 4M bytes of RAM).

3. To protect memory, which includes schemes that prevent one user from accessing the memory space of another user.

4. To ensure efficient memory usage, in which best use can be made of the existing physical address space.

5. To free the programmer from any considerations of where his or her programs and data are to be located in memory. That is, the programmer can use any address he or she wishes, but the memory management system will map the logical address onto an available physical address.

Any real memory management unit may not attempt to achieve all these goals (the first two are mutually exclusive). Note that the second goal (i.e., logical address space greater than physical address space) is especially important to designers of 68000-based systems. In fact, when memory management is applied to this problem, it is frequently referred to as *virtual memory technology*. Virtual memory is almost synonymous with *logical* memory.

Let us first take the case where the physical memory space is greater than the logical memory space. Not very long ago, 8-bit microprocessors had small physical memories due to the relatively high cost of memory components. In those days the programs executed on such microprocessor systems were rather small. Believe it or not, at one time integer BASIC was regarded as something special. By the early 1980s, many general-purpose 8-bit microcomputers were equipped with a full complement of 64K bytes of RAM because of the introduction of low-cost 16K and 64K DRAMs. Such a large memory made bigger programs possible and speeded up operations such as text processing by permitting a larger chunk of the text to reside in immediate access memory at any given instant. Unfortunately, the new low-cost memory chips could not readily be used to create 128K or larger memories without some sleight of hand. Obviously, a CPU with a 16-bit address bus can access a maximum of 64K bytes of logical address space but cannot specify a unique location in 128K bytes (2^{17} bytes) of physical memory without ambiguity.

Memory management techniques solved this problem by *bank switching*. The physical memory is arranged as a number of separate banks of 64K bytes and the processor is allowed access only to one bank of 64K bytes of physical memory space at any time. In other words, the 64K-byte logical address space of the processor is mapped onto a 64K region of the physical address space. For example, the 16-bit logical address $1234 could be mapped onto physical addresses $0 1234, $1 1234, $2 1234, etc. Figure 7.41 illustrates this technique of memory mapping and figure 7.42 shows how it is implemented at the conceptual level by passing the logical address from the processor through an address translation unit (ATU). How this operation is carried out is dealt with later.

Now consider the more interesting problem from the point of view of the designer of systems with 24-bit (or larger) address buses. In such cases, the processor will probably have a larger logical address space than its physical address space.

FIGURE 7.41 Mapping logical address space onto physical address space

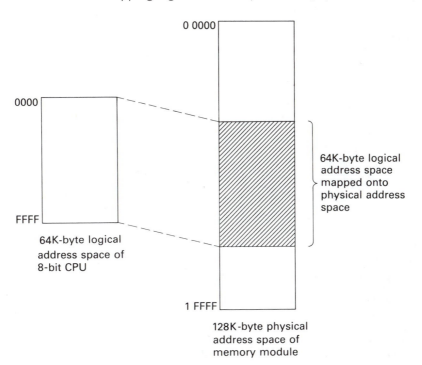

64K-byte logical
address space
mapped onto
physical address
space

64K-byte logical
address space of
8-bit CPU

128K-byte physical
address space of
memory module

After all, not all 68000 systems have a full complement of 16M bytes of random access memory, but there are many 68000 systems with 4M bytes of RAM that need to run programs with a logical address space of more than 4M bytes.

This problem of the available physical address space being smaller than the processor's logical address space is caused by economics and has always plagued the mainframe industry. In the late 1950s, mainframes were available with large logical address spaces (even by today's standards), but they were restricted to tiny 2K or so blocks of RAM. A group of computer scientists at Manchester University in the United Kingdom proposed a memory management technique, now known as virtual memory, to deal with this situation. The logical (or virtual) address space is mapped onto the available physical address space, as shown in figure 7.43. As long as the processor accesses data in the logical space that is mapped onto the existing physical

FIGURE 7.42 Using address translation to achieve memory mapping

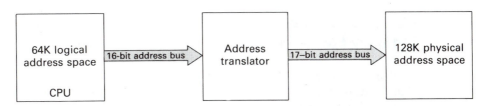

FIGURE 7.43 Mapping part of a large logical address space onto a smaller physical address space

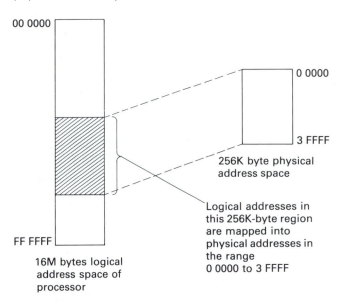

00 0000

0 0000

3 FFFF

256K byte physical address space

Logical addresses in this 256K-byte region are mapped into physical addresses in the range 0 0000 to 3 FFFF

FF FFFF

16M bytes logical address space of processor

address space, all is well. We have been doing this all the way through this text, because we have assumed that an address from the 68000 is passed directly (i.e., unchanged) to the system's address bus. However when the processor generates the logical address of an operand that cannot be mapped onto the available physical address space, we have a problem.

The solution adopted at Manchester University is delightfully simple. Whenever the processor generates a logical address for which there is no corresponding physical address, the operating system stops the current program, fetches a block of data containing the desired operand from its disk store, places this block in physical memory (overwriting old data), and tells the memory management unit that a new relationship exists between logical and physical address space. In other words, the program or data is held on disk and only the parts of the program currently needed are transferred to the physical RAM. The memory management unit keeps track of the relationship between the logical address generated by the processor and that of the data currently in physical memory. This entire process is very complex in its details and requires *harmonization* of the processor architecture, the memory management unit, and the operating system. People dream of simple virtual memory systems and have nightmares about real ones.

Memory Mapping

Although this text is concerned mainly with 16/32-bit microprocessor systems with their large logical memory spaces, this section looks at the problems of microprocessors with smaller logical memory spaces than physical memory spaces. Such microprocessors are normally 8-bit devices. This topic is included here for the sake of completeness, as memory mapping and virtual memory are really inverse operations. In any case, 8-bit microprocessor systems will still be around for some time. Readers not interested in mapping small logical memory spaces may skip ahead to virtual memory.

The most primitive form of memory-mapping system is illustrated in figure 7.44, in which the physical address space is provided by a fixed bank of 32K bytes of memory plus two to eight switchable banks of 32K bytes of memory. The fixed block of physical memory is arranged so that it is selected whenever A_{15} from the processor is a logical zero. Therefore, this region of physical memory is permanently mapped onto the 32K bytes of logical memory in the range $0000 to $7FFF. That is, logical addresses from $0000 to $7FFF have identical physical addresses.

Whenever A_{15} from the processor is high, the three-line-to-eight-line decoder is enabled and one of its outputs goes low to select a bank of 32K bytes of memory. The actual bank of memory selected depends on the state of the 3-bit code stored in the latch. Figure 7.45 shows the memory map corresponding to figure 7.44. The processor's 32K bytes of logical memory space in the range $8000 to $FFFF can be translated to the memory space of any of the eight 32K banks of physical address space. Note that the processor can always access 64K bytes of both logical and physical memory space at any instant. The process of bank switching provides the

FIGURE 7.44 Memory mapping by bank switching

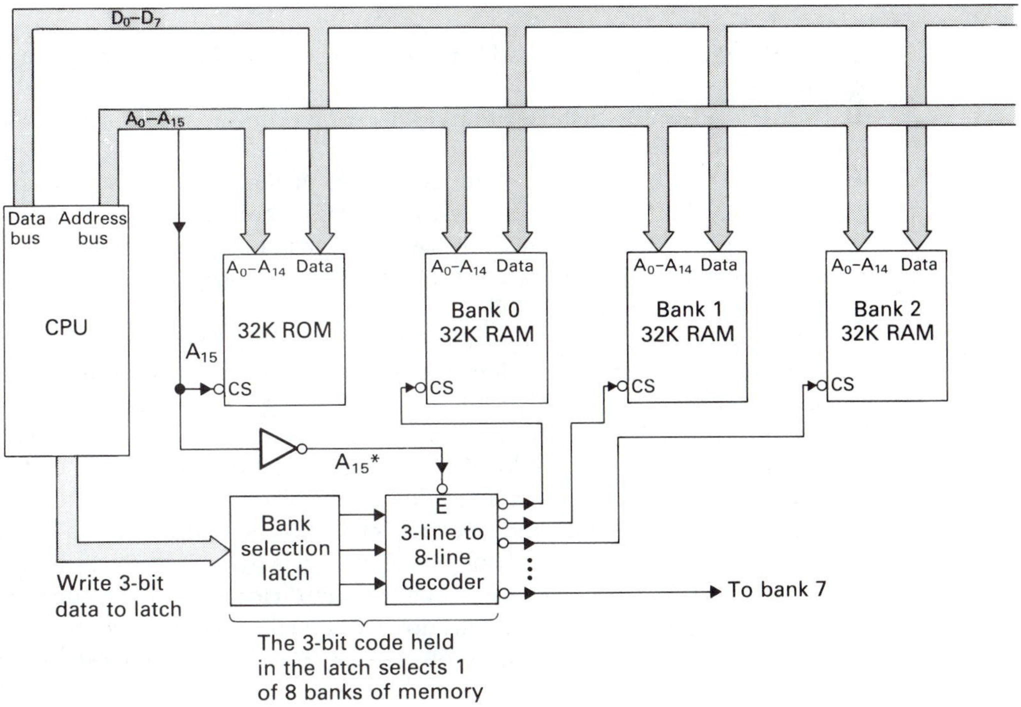

FIGURE 7.45   Memory map corresponding to figure 7.44

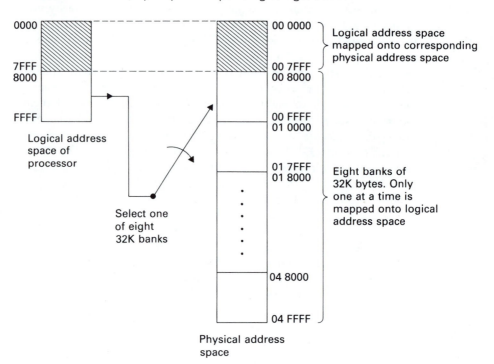

CPU with a window through which it can see one of the eight possible banks of switched memory.

However, bank switching or any other form of memory management is useless without the software necessary to control it. Consider a program running in the lower half of the address space ($0000–$7FFF) in the system described by figures 7.44 and 7.45. Both logical and physical addresses are equivalent and the physical memory space is never switched out (i.e., made inaccessible to the processor), so no special programming problems exist. Now suppose a program in one of the eight switchable banks is to be executed.

A jump to the desired program cannot be executed unless that bank is currently selected. Therefore, before the jump is executed, the calling program in the fixed memory must select the appropriate block by loading its value in the bank-selection latch. Note that once the new bank is selected, the old bank is switched out and cannot be accessed without reloading the bank-selection latch. Thus, if a program running in the fixed memory is accessing a data table in one bank and calls a sub-routine in another bank, then the data table cannot be accessed from the subroutine.

If a program is running in bank A and a jump is to be made to bank B, we need to first select bank B and then carry out the jump. However, if the code in bank A modifies the contents of the bank-selection latch, bank B will be selected immediately and the JMP instruction to B will not be executed, as it is in the locked-out bank A.

Therefore, the next instruction to be executed will be the instruction in bank B with the same logical address as the JMP instruction in bank A.

The only way to effect a jump from one bank to another is to jump first to a location within the fixed memory (where LAS = PAS), switch banks, and then jump to the desired location in the new bank. Therefore, programmers must write their programs with this object always in mind. Consequently, bank switching places a considerable burden on the programmer. We shall soon see that other types of memory management lessen this burden, some to the point at which memory management becomes totally invisible to the user-programmer.

Bank switching is not difficult to implement and is very cheap. It is useful for systems requiring large data tables that can be switched in as required. It can be used equally well to implement systems where the operating system software and other utilities (interpreters, editors, word processors) are held in read-only memory. This facility speeds up the system, as the utilities are switched in rather than loaded from disk. One of the disadvantages of bank switching is the *granularity* of the blocks switched in and out. Clearly, switching tiny blocks of memory would be hopelessly inefficient; memory cards populated with large numbers of low-density memory components would be required.

Indexed Mapping

A much better approach to bank switching is called *indexed mapping*. Instead of performing the bank switching by loading a special latch, the identity of the bank to be selected forms part of the logical address itself. For example, a CPU with a 16-bit address bus may specify a logical address as $XYYY, where X is the 4-bit bank-selection address and $YYY is the 12-bit address within a bank. Therefore, the banks can be switched rapidly and a jump from one bank to another becomes possible, as the address part of the JUMP automatically selects the new bank. In what follows, the term *page* is used rather than *bank* to be consistent with other authors writing on this topic. The words *block*, *bank*, *page*, and *chunk* are interchangeable and all describe the same thing—a unit of memory space. These units are usually of *fixed* size (the actual size depends on the particular application).

Figure 7.46 shows how the p most significant bits from the CPU's logical address interrogate a table of 2^p locations in a *mapping table* to determine the current physical memory block. Each of these locations contains a q-bit value ($q > p$), providing the higher-order q bits of the physical address. For example, an 8-bit microprocessor with $p = 4$ provides 16 pages of 4K bytes. If $q = 8$, the physical memory size is $2^q \times 4K = 256 \times 4K = 1M$ bytes. Once again, we must stress that although there are 256 pages of physical memory, only 16 of these can be accessed at any time, as there are only 16 entries in the mapping table.

The contents of the mapping table permit 16 windows to be opened onto the physical memory space. This table is loaded and controlled by the operating system, so that the processor is always *viewing* the 64K bytes of physical memory currently of most interest. Figure 7.47 shows how the logical address space is mapped onto the physical address space. In this example, $p = 4$ and $q = 8$, to give 4K pages with

FIGURE 7.46 Address translation by indexed mapping

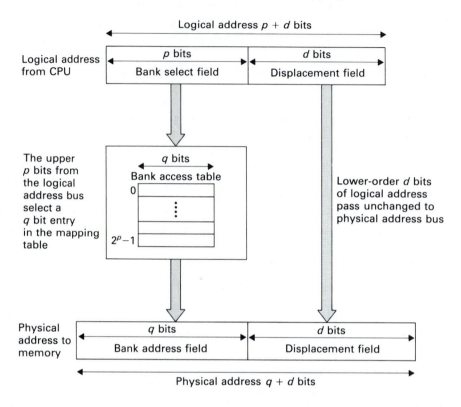

an 8-bit processor. For each of the sixteen 4K pages of logical address space, the page table or mapping registers (middle column of figure 7.47) contain the corresponding page of the physical address space. For example, a logical address $0XXX is mapped onto the physical address $0 0XXX. Similarly, the logical address $4XXX is mapped onto the physical address $1 1XXX

Although the translation of logical addresses into physical addresses is automatically carried out by hardware, the management of the index registers (i.e., translation table) is performed by software—invariably by the operating system itself. There is, however, one exception to this rule. At switch-on or following a system reset, part of the processor's logical address space is mapped onto a fixed region of the physical address space. We must perform this fixed mapping because the state of the address translation table is undefined at switch-on. For example, the logical address range $F000 to $FFFF in a 6809-based microprocessor system may be mapped onto, say, the physical address range $0 0000 through $0 0FFF following the initial application of power. The 6809 stores its reset vector at $FFFE and $FFFF, and the physical address block should contain the initialization routine in ROM to load the operating system from disk and to set up the address-mapping table.

FIGURE 7.47 Logical-to-physical address translation

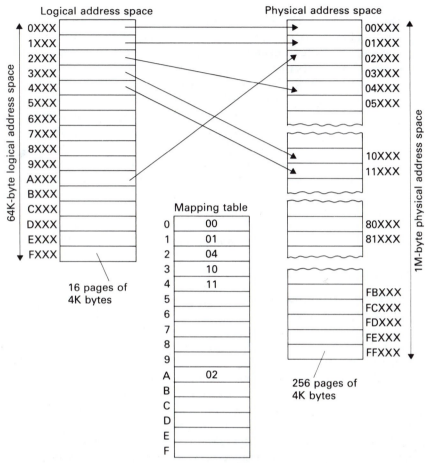

NOTE: The 4 most significant bits of the physical address interrogate the mapping table to provide the 8 most significant bits of the physical address. For example, the logical address $A 123 is mapped into the physical address $0 2123.

The way in which the mapping table is used is relatively simple. If the physical memory is considered to be a treasure house of resources (programs, data structures, text, etc.), the operating system opens windows onto these resources as they are required by any program currently being executed. The only limitation is that the processor cannot directly address more than 64K bytes without the operating system switching in new pages (and, of course, switching out old pages).

Now that we have looked at systems designed to *widen* the address space of 8-bit microprocessors, we are going to consider how memory management can be applied to the world of the 68000.

Virtual Memory

Virtual memory systems serve two purposes: They map logical addresses onto physical address space and they allocate physical memory to tasks running in logical address space. Virtual memory techniques are found in systems where the physical memory space is less than the logical memory space and in multitasking systems.

It would be foolish to pretend that justice can be done to the topic of virtual memory in a text of this size. Although it would be reasonable to expect readers to be able to design a 68000-based microprocessor system after reading this book, it would be unreasonable to expect them to design a 68000 system with virtual memory. Quite simply, virtual memory systems are not one-person efforts. They are designed by teams of designers and programmers and require many hours to produce, because the management of virtual memory is not only rather complex but is also found almost exclusively in systems with multiuser or multitasking operating systems. Therefore, only an overview of virtual memory and microprocessors is given, together with a brief description of the 68451 memory management unit (MMU) and the 68851 paged memory management unit (PMMU).

Memory Management and Tasks

In chapter 6 we introduced the idea of multitasking systems that execute a number of *tasks* or *processes* concurrently by periodically switching between tasks. Clearly, multi-tasking is viable only if several tasks reside in main memory at the same time. If this restriction were not so, the time required to transfer an *old* task to disk and to swap in a *new* task would be prohibitive.

Figure 7.48 demonstrates how memory management is applied to multitasking. Two tasks, A and B, are in physical memory at the same time. The left-hand side of figure 7.48 shows how the tasks are arranged in logical address space. Each task has its *own* logical memory space (e.g., program and stack) but accesses *shared* resources lying in physical memory space. Programmers are entirely free to choose their own addresses for the various components of their tasks. Consequently, task A and task B can each access the same data structure in physical memory, even though they use different logical addresses. That is, each task is aware only of its own copy of data it shares with another task.

A memory management unit (MMU) maps the logical addresses chosen by the programmer onto the physical memory space. As we shall see, any logical address generated by the CPU is automatically mapped to the appropriate physical address by the MMU. Note that the operating system is responsible for setting up the logical-to-physical mapping tables, which is a very complex process and is well beyond the scope of this text. Basically, whenever a new task is created, the operating system is informed of the task's memory requirements. The operating system then searches the available physical memory space for free memory blocks and allocates these to the task. You can imagine that, after a time, the physical memory space may become very fragmented, with the various physical blocks belonging to each task interwoven in a complex pattern. A good operating system attempts to perform

FIGURE 7.48 Mapping logical address space onto physical address space in a multitasking environment

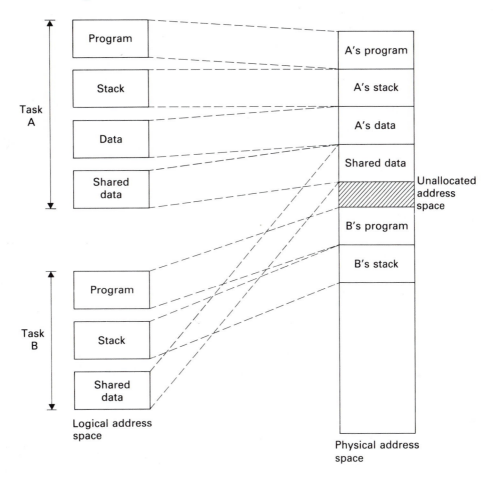

memory allocation efficiently and should not permit large numbers of unused blocks of physical memory. The way with which this fragmentation is dealt depends on both the type of memory mapping implemented and on the operating system.

A powerful feature of memory mapping is that each logical memory block can be associated with various *permissions*. For example, memory can be made read-only, write-only, accessible only by the operating system or by a given task, or shared between a group of tasks. Without this facility, there would be nothing to stop one task from corrupting the memory space belonging to another task. In this introduction we refer to memory *blocks* rather than *pages*, because we shall soon see that there are two fundamental ways of implementing memory management. One uses fixed-sized blocks of memory called *pages* and the other uses variable-sized blocks of memory called *segments*.

Address Translation

In a virtual memory system, where the logical address space may be greater than the available physical address space, the processor might generate a logical address for which no actual physical memory exists. Therefore, the MMU must perform two distinct functions. The first is to map logical addresses onto the available physical memory and the second is to deal with the situation arising when the physical address space "runs out" (i.e., the logical-to-physical address mapping cannot be performed because the data is not available in the random-access memory). Figure 7.49 shows how the first objective is achieved for paged memory systems.

Although memory mapping and virtual memory are inverse operations, the arrangement described by figure 7.49 is remarkably similar to the memory-mapping

FIGURE 7.49 Page table and virtual memory

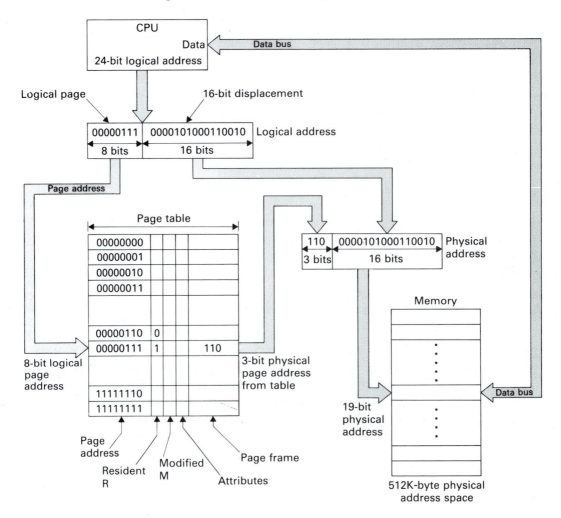

system of figure 7.46. In figure 7.49 (which describes a hypothetical system) the 24-bit logical address from the 68000 processor is split into a 16-*bit displacement*, which is passed directly to the physical memory, and an 8-*bit page address*. The displacement accesses one of 2^{16} locations within a 64K-byte page. The page address specifies the page (one of $2^8 = 256$ pages) currently accessed by the processor. The unit of memory that can hold a page is called a *page-frame*. The distinction between *page* and *page-frame* is between a quantity of data (i.e., the page) and the physical memory in which it is stored (i.e., the page-frame).

The page-table (compare it with the index-mapping table) contains 256 entries, one for each logical page. For example, in figure 7.49 the 8-bit logical page address from the processor is 0000 0111$_2$. In each entry of the page table is a 3-bit page-frame address that provides the 3 most significant bits of the physical address. In this case, the page frame is 110. Notice how the logical address has been condensed from $8 + 16$ bits to $3 + 16$ bits. Therefore, the logical address 00000111 0000101000110010 is mapped onto the physical address 110 0000101000110010.

Although there are 256 possible entries in the page-frame table (one for each *logical* page), the *physical* page-frame address is only 3 bits, limiting the number of physical pages to eight. Consequently, a unique physical page-frame in random access memory cannot be associated with each of the possible logical page numbers. Each page address has a single-bit R-field labeled *resident* associated with it. If the R-bit is set, that page-frame is currently in physical memory. If the R-bit is clear, the corresponding page-frame is not in the physical memory and the contents of the page-frame field are meaningless.

Whenever a logical address is generated and the R-bit associated with the current logical page is found to be clear, an event called a *page-fault* occurs. It is at this point the fun begins. Once a memory access is started that attempts to access a logical address whose page is not in memory because the R-bit was found to be clear, the current instruction must be *suspended*, since it cannot be completed. The BERR* input of a 68000 or a 68010/20/30 is asserted to indicate a bus fault (but remember that in chapter 6 we said it is not easy to restart the 68000 after a bus error exception).

Now the operating system must intervene to deal with the situation. Although the information accessed was not in the random-access physical memory, it will be located in the disk store, which is normally a hard disk drive. The operating system retrieves the page containing the desired memory location from disk, loads it in the physical memory, and updates the page-table accordingly. The suspended instruction can then be executed.

This procedure is not as simple as it looks. When the operating system fetches a new page from disk, it must overwrite a page of the random access physical memory. Remember that one of the purposes of virtual memory is to permit relatively small physical memories to simulate large memories. If we are going to replace *old* pages by *new* ones, we require a strategy to decide which old pages are to go. The classic paging policy is called the *least recently used* (LRU) algorithm. The page that has not been accessed for the longest period of time is overwritten by the new page (i.e., if you haven't accessed this page recently, you are not likely to access it in the near figure).

The LRU algorithm has been found to work well in practice. Unfortunately, the operating system must know when each page is accessed if this algorithm is to work, which somewhat complicates the hardware (each page has to be "date-stamped" after use). Another problem with which the operating system has to deal is the divergence between the data stored in RAM and the data held on disk. If the page fetched from disk contains only program information, it will not be modified in RAM and therefore overwriting it causes no problems. If, however, the page is a data-table or some other data structure, it may be written to while it is in RAM. In this case, it cannot just be overwritten by the new page.

In figure 7.49 we can see that each entry in the table has an M- (modified) bit. Whenever that page is accessed by a write operation, the M-bit is set. When this page is to be overwritten, the operating system checks the M-bit and, if it is set, the operating system first rewrites this page to the disk store before fetching the new page.

Finally, when the new page has been loaded, the address translation table updated, the M-bit cleared, and the R-bit set (to indicate that the page is valid), the processor can rerun the instruction that was suspended. As we discovered in chapter 6, the 68000 does not store enough information following the assertion of BERR∗ (caused by a page-fault) to rerun the suspended instruction. However, the 68010 was introduced a little after the 68000 to deal with this situation. It is nominally identical to the 68000 but has features to allow it to recover fully from a bus error. The 68010's stack frame saved after a bus error exception is increased from 7 words (68000) to 26 words. The more recent 68020 and 68030 also implement an ability to rerun a faulted bus cycle. The reader is encouraged to refer to chapter 6 for further information about the bus error exception.

Clearly, the effort involved every time a page-fault occurs is rather large. As long as page-faults are relatively infrequent, the system works well because of a phenomenon called *locality of reference*. Most data is clustered so that once pages are brought from disk, the majority of memory accesses will be found within these pages. When the data is not well ordered or when there are many unrelated tasks, the processor ends up by spending nearly all its time swapping pages in and out, and the system effectively grinds to a halt. This situation is called *thrashing*.

MC68451 Memory Management Unit

We are now going to describe two 68000-series memory management units. The first such unit is the older MC68451, intended for use with the 68000/68010, and the second is the more sophisticated MC68851, intended for use in 68020/68030 systems. Indeed, the core of the 68851 is located on the 68030's chip. The MC68451 MMU (referred to in this section as the MMU) is intended primarily to manage memory resources in multitasking systems. It also provides virtual memory support. Any virtual memory system using the MMU should also have at least the 68010 CPU, as the 68000 is not really capable of resuming processing after a bus error.

Like many of today's special-purpose peripherals, the 68451 is exceedingly complex, and the space devoted to it here can do little other than to give an insight into its operation. Having digested this section, the reader should then be in a position to tackle its data manual.

The MMU has 32 on-chip address translation registers, permitting up to 32 blocks of logical memory to be mapped onto physical memory. Unlike the schemes described so far in this chapter, these pages are of a user-definable size and extend from a minimum of 256 bytes to a maximum of 16M bytes, in powers of 2. These blocks are referred to as *segments* and the 68451 performs *segmented memory management*, as opposed to *paged memory management*. (The 68851 PMMU that we describe in the next section is a paged MMU.)

Note that MMU's segment sizes may be mixed, so that a single MMU can support pages of 256 bytes and 2M bytes simultaneously. Page sizes varying by powers of 2 are required to implement the *binary buddy algorithm*, which is used to allocate memory space to tasks. The MMU operates either in a stand-alone mode, or up to eight MMUs can be operated in parallel to provide a maximum of $8 \times 32 = 256$ segments. Some feel that this number is too low for today's sophisticated multitasking systems. Here, only single-MMU systems are described. The 68851 PMMU locates its address translation tables in external memory and can therefore support an arbitrarily complex logical-to-physical mapping scheme.

Address Mapping and the 68451 MMU

A simplified block diagram of a 68451 MMU is given in figure 7.50. Figure 7.51 shows how the MMU performs memory management by mapping logical segments onto physical segments. Note that several tasks can share the same physical segment. The MMU maintains a table of 32 72-bit *descriptor registers*. Descriptor registers provide all the information needed to map a logical address within a segment onto its physical address. Each descriptor is composed of the six fields shown in figure 7.50. Three fields are 16 bits wide and perform the actual logical-to-physical address translation, and three are 8 bits wide and are devoted to the control of the mapping process.

A major difference between the 68451 MMU and the type of memory-mapping scheme described earlier (see figure 7.49) lies in the selection of the segment descriptors. In figure 7.49, the page table contains an entry for each possible logical page. That is, the higher-order bits of the logical address interrogate the appropriate entry in the page table and access the required page descriptor. The 68451 employs an *associative* addressing technique to interrogate its table of 32 descriptors. The higher-order logical address bits from the processor plus the function code are fed to each of the 32 descriptors *simultaneously*. Any descriptor that matches (i.e., is associated with) the current logical address takes part in the address translation process. If no descriptor indicates a match, the MMU signals a page-fault by asserting its FAULT* output, which is connected to the 68000's BERR* input.

An associative memory technique is necessary to match the logical segment address from the processor with the available descriptors, because the MMU's segments (page-frames) can be as small as 256 bytes. If an MMU had a descriptor for each of the possible segments, the mapping table would contain not 32, but 65,536 entries. Such a large number of descriptors is necessary, since the MMU employs

FIGURE 7.50   Simplified view of part of the 68451 MMU address translation mechanism

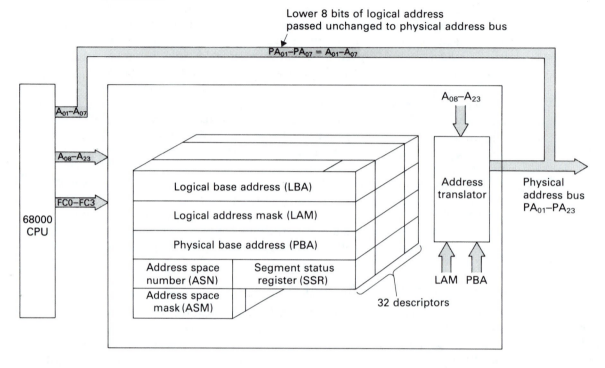

logical address lines A_{08}–A_{23} from the 68000 to select an entry in the table. An associative memory requires that each location in the table store a *tag* as well as its contents. This tag is the 16 highest-order address bits. When a valid logical address is applied to the MMU, it is sent to all 32 descriptors in parallel. Each of the descriptors simultaneously matches its tags with the incoming address to generate a *hit* or a *miss* signal.

In order to understand how the MMU's address translation process operates, we need to define the function of its three 16-bit descriptor registers. These registers are the logical base address (LBR), the logical address mask (LAM), and the physical base address (PBA). Figure 7.52 relates the LBR, PBA, and the LAM to the address translation process.

Logical Base Address (LBA)   The contents of the LBA provide the most significant bits of the logical segment address and therefore define the start of the segment in logical address space. In order to provide a variable segment size, we can mask out some of the bits in the LBA and reduce the effective size of the LBA register from 16 bits to between 1 and 15 bits. This action is carried out by the logical address mask, LAM. For example, if one of the 32 LBAs has the format 1010 11XX XXXX XXXX (where the Xs represent masked bits), any logical address from the 68000 whose most significant 6 bits (i.e., A_{18} to A_{23}) are 10 1011 will be translated into the corresponding physical address, as described shortly.

FIGURE 7.51 Example of segmented memory management

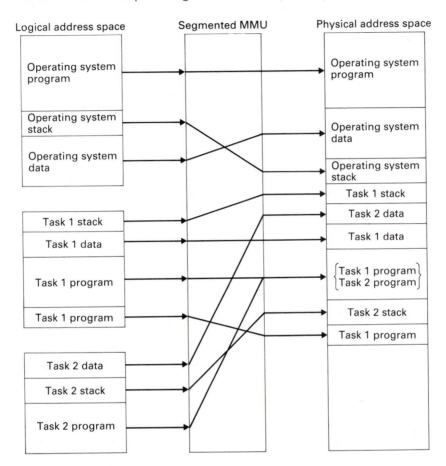

Logical Address Mask (LAM) The LAM is a 16-bit mask that defines the bit positions in the LBA to be used in the definition of the size of a logical segment. A logical 1 in the LAM defines the corresponding bit position in the LBA as part of the segment logical address. A logical 0 in the LAM means that the corresponding bit position in the LBA does not take part in the segment logical address. An example should make the relationship between LBA and LAM clear. Suppose that the size of a logical segment starting at address $04 0000 is 4K bytes and the LAM contains the value $FFF0. The contents of the LBA and LAM are

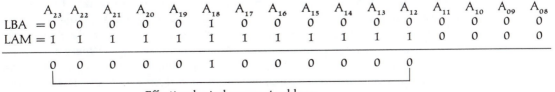

	A_{23}	A_{22}	A_{21}	A_{20}	A_{19}	A_{18}	A_{17}	A_{16}	A_{15}	A_{14}	A_{13}	A_{12}	A_{11}	A_{10}	A_{09}	A_{08}
LBA = 0	0	0	0	0	1	0	0	0	0	0	0	0	0	0	0	
LAM = 1	1	1	1	1	1	1	1	1	1	1	1	0	0	0	0	

0 0 0 0 0 1 0 0 0 0 0 0

Effective logical segment address

FIGURE 7.52   Relationship among the LBA, PBA, LAM, and address translation

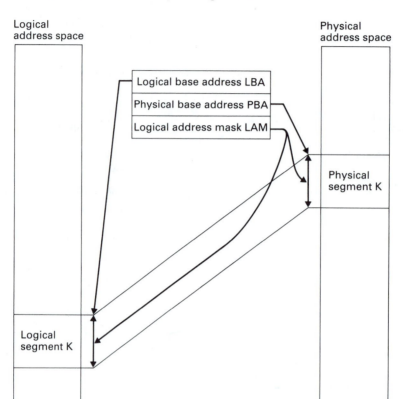

NOTE: The LBA points to the start of the logical segment. The PBA points to the start of the corresponding physical segment. The LAM defines the size of the logical segment. Note that both the logical and physical address spaces must be the same size.

As you can see, the LAM converts the 16-bit logical base address field into a 12-bit address field. Only logical address bits A_{12} to A_{23} from the CPU take part in matching this logical address segment. Bits A_{01} to A_{11} select a location within the 4K-byte segment and are passed directly to the physical address bus. The LBA register of this descriptor responds to any logical address in the range $04\ 0000 to $04\ 0FFF.

    Physical Base Address (PBA)   The PBA is a 16-bit address that is used in conjunction with the logical address from the 68000 and the LAM to calculate the desired physical address. Just as the LBA provides the first address (i.e., lowest) in a logical segment, the PBA provides the first address in a physical segment. The logical address bits corresponding to 0s in the LAM are passed directly through the MMU to become physical address bits (together with A_{01} to A_{07}). Where a bit of the LAM is a logical 1, the corresponding bit of the PBA register is passed to the physical address bus. In other words, the PBA of a descriptor supplies the higher-order bits of its physical address.

Continuing the preceding example, suppose that the contents of the PBA register in the segment descriptor are 0001 1000 0011 0000 (i.e., $1830). If a logical address $04 0123 is generated by the processor, what is the corresponding physical address? This logical address falls in the range determined by the LBA and LAM, so the descriptor responds by calculating the physical address. The LAM register also masks the PBA so that the 12 most significant bits of the PBA form the corresponding physical address bits (i.e., 0001 1000 0011). The remaining 4 bits of the 16-bit higher-order logical address (i.e., 0001) are passed unchanged to the physical address bus. Finally, logical address bits A_{01} to A_{07} do not take part in the mapping process and are transferred directly to the physical address bus. Therefore, the final physical address is

0001 1000 0011 0001 0010 0011,   or $18 3123

At this point the reader may feel a little unhappy because the logical address range and the physical address range of the MMU are *both* 16M bytes. What happened to the problem of the physical address space being less than the logical address space? The answer is that the 68451 MMU is a general-purpose device and can perform memory management, if called upon, over a physical memory space of 16M bytes. However, the use of the MMU may populate this physical address space with, say, only 256K bytes of memory. It is then up to the programmer (in practice, the operating system) to make sure that the segment descriptors map the logical address space of the programs and their data onto this 256K bytes of physical address space.

Address Space Matching

Up to now we have implied that the 68451 MMU automatically carries out its logical-to-physical address translation process using nothing more than the logical address from the CPU and information stored in the descriptor registers. In fact, an important step in the translation process has been omitted.

Basically speaking, each logical address segment is associated with an address space type *by the programmer*. We have already encountered address spaces when we discussed the 68000's function codes (e.g., user program space, supervisor data space, CPU space, etc.) If the processor is not currently addressing the correct type of address space, the MMU does not permit an address translation even though the current logical segment is defined by one of the 32 descriptors.

Whenever the 68000 CPU accesses memory, it puts out a function code on FC0 to FC2 to indicate the type of access it is carrying out. Table 4.3 illustrates the relationship between the function code and type of bus cycle in progress. We can regard these function codes as pointing to different types of address space. For example, when FC2,FC1,FC0 = 0,0,1, the processor is accessing *user data space*, and when FC2,FC1,FC0 = 1,1,0 it is accessing *supervisor program space*. In a way, the function code is an extension of the address bus. Just as A_{23} divides memory space into an upper and a lower 8M-byte page, the function code divides memory space into a 16M-byte supervisor data space, a 16M-byte supervisor instruction space, etc. The 68000 and 68451 combination can therefore be said to have an effective address space of 4 × 16M bytes = 64M bytes, because the MMU recognizes 16 types of address space (not 8 as you might expect).

Although the 68000 has three function code outputs, the MMU has four function code inputs, FC0 to FC3. Three of these come directly from the 68000 CPU and the fourth (FC3) is usually derived from BGACK* via an inverter. By using BGACK* as a pseudofunction code input to the MMU, we are able to define a set of eight address spaces associated with DMA operations or with other processors.

The MMU maintains an *address space table* (AST), as illustrated in figure 7.53. For each of the 16 possible combinations of FC0 to FC3, an 8-bit entry in the AST is assigned a cycle address space number (CASN) by the programmer (or operating system). The CASN usually corresponds to a task number in a multitasking system. For example, when the processor accesses user data space, the appropriate entry in the address space table is accessed and the cycle address space number (i.e., task number) read—this number might be, say, 0000 0011 ($03). Similarly, when the processor accesses user program space, the task number might be 0000 0010 ($02). Remember that the 16 entries (CASNs) in the AST are loaded into the MMU by the programmer and are not determined by the MMU itself. Thus, for every memory access, a cycle address space number is determined by the MMU, depending on the type of address space being accessed. We will shortly describe what the MMU does

FIGURE 7.53 Address space matching

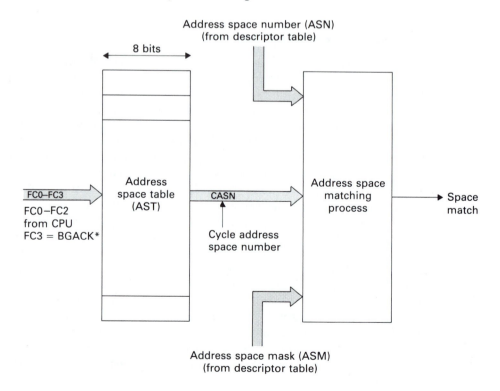

with the CASN, but first we must introduce the segment descriptor's *address space number field*.

Each of the 32 segment descriptors in figure 7.50 contains an 8-bit address space number field (ASN), which associates that segment with a particular ASN and therefore with a particular task. A task is said to comprise all the logical segments with the same ASN. Note that each segment descriptor also has an address space mask (ASM) field. The ASM is used to determine which bits of the segment's ASN should be masked with the value from the address space table produced during the current cycle. The purpose of this is to permit address spaces to be shared between several tasks.

Consider the sequence of events taking place in a CPU memory access. The function code is used to obtain a cycle address space number (CASN) from the MMU's address space table (AST). The CASN is used together with the address space number (ASN) and the address space mask (ASM) from the descriptor to generate a *space match* signal. If space match is not asserted, no logical segment matches the current access. For example, suppose that the 68000 generates the function code 110 (i.e., supervisor program space) and that the MMU's FC3 input is 0. Location 0110 in the AST will be interrogated and the CASN will be read. We will assume that the CASN corresponding to this space is 0011 1010. Assume also that the LBA of one of the 32 logical descriptors matches the current address from the CPU. If this descriptor has, say, an address space number 1100 1010 and an address space mask of 0000 1111, the descriptor's ASN will be masked to the 4 bits 1010 and compared with the masked CASN (i.e., 1010). In this case, these two values match and the address translation can take place. If the ASN in the descriptor does not correspond to the CASN from the AST, no match occurs and the MMU asserts its FAULT* output.

To appreciate the purpose of the address space table and the ASN, we must consider the multitasking environment. Suppose task A is being executed. Let task A be a user task and let the entries in the AST corresponding to user data and user program space both be $01. Assume that the ASMs of all segment descriptors are set to 11111111 ($FF), so that all bits of the ASN from the descriptor must match those from the AST. A descriptor whose ASN is $01 will be used in the address translation process, assuming the processor is currently addressing its logical address range.

At any time, only one task number can be associated with each of the 16 different types of address space. Therefore, the MMU may, theoretically, support up to 16 tasks simultaneously. However, as the various combinations of FC0 to FC2 from the CPU define only five address spaces implemented by the 68000 (and one is only for interrupt acknowledgment), the address spaces of greatest interest to the systems designer are user and supervisor address space and program and data address space. Consequently, in a practical system, an MMU might support two tasks: the supervisor task (i.e., operating system) and the user task. The operating system is able to switch rapidly between several user tasks simply by modifying the CASN values in the user address spaces of the AST.

For example, if user task A has an ASN of $01 and user task B an ASN of $02, just loading $01 or $02 in the FC3,FC2,FC1,FC0 = 0,0,0,1 and 0,0,1,0 entries of the AST selects the appropriate task. Moreover, because only those segment descriptors whose own address space numbers (masked by the ASM) match CASN from the AST in the current bus cycle, several user tasks may share the same logical address space. We are now going to describe the segment status register (SSR) which is used by each of the 32 logical segments to describe the nature of the segment.

Segment Status Register (SSR)

Each descriptor has an 8-bit segment status register that provides additional information about the nature of the segment, although only 6 bits of the SSR have defined meanings. Brief descriptions of these bits are given next. In all but one case, the term *reset* means a reset to segment 0 of the master MMU. The meaning of this will be made clear when we discuss initializing the MMU.

U (Used) The U-bit is set if the segment has been accessed since it was defined. It is cleared by a reset or by writing a 0 into this bit. The operating system may read the U-bit to determine whether the segment is currently active.

I (Interrupt) The I-bit is set (or cleared) under program control and, when set, forces an interrupt whenever the segment is accessed. This bit can be used as an aid to debugging. It is also cleared by a reset.

IP (Interrupt Pending) The IP-bit is set if the I-bit is set when the segment is accessed. This bit indicates that the associated segment was the source of the MMU's interrupt request. It is cleared under software control, by a reset, or when the segment's E-bit is clear.

M (Modified) The M-bit is set by the MMU if the segment has been written to since it was defined. If a segment is to be swapped out of physical memory, the old segment must be saved if the M-bit is set. If the M-bit is clear, the segment can be overwritten. The M-bit is cleared under software control or by a reset.

WP (Write-Protect) The WP-bit is set by the operating system to write-protect the segment. Once the WP-bit has been set, any attempt to write to the logical address space spanned by the segment will cause a write violation and the assertion of the MMU's FAULT* output. The WP-bit is cleared under software control or by a reset.

E (Enable) The E-bit, when set under software control, enables the segment to take part in the address translation process. In other words, setting E = 1 activates the segment and setting E = 0 turns off the segment. The function of the E-bit is to remove any segment descriptors from the pool of 32 until they are actively engaged in the address translation process. The E-bit is cleared by software or by an unsuccessful load descriptor register operation. It is *not* cleared by a reset.

In addition to the 32 descriptors and the 16-entry address space table, the MMU has a set of six 8-bit registers and a temporary descriptor. The registers perform various functions, from providing interrupt vectors to storing global status information. The temporary descriptor has the same structure as other descriptors and can be loaded with data from the system bus. This data is then transferred to any of the 32 descriptors selected by the programmer.

Operating the 6845I MMU

As stated earlier, the precise operational details of the MMU are rather complex, and readers must refer to its data sheet if they want to use this component. Here only the details of most interest are given. In particular, the application of the 68451 in multiple-MMU systems is not considered.

Figure 7.54 shows how the MMU is connected between a 68000 processor and the system's physical address bus. Note that the MMU itself is address mapped onto the 68000's PAB (physical address bus) and not the logical address bus, because its active-low chip select input is derived from PA_{01} to PA_{23}. Due to the limitations of chip packaging, the MMU cannot support a separate data bus (it already has 64 pins), and therefore its data input, D_{00} to D_{15}, from the CPU is multiplexed onto its address outputs, PAD_0 to PAD_{15}. This arrangement requires support circuitry to multiplex the address and to latch the data from D_{00} to D_{15}. Two outputs from the MMU, ED* and HAD*, control the data and address buses.

FIGURE 7.54 Interfacing the MMU to a 68000

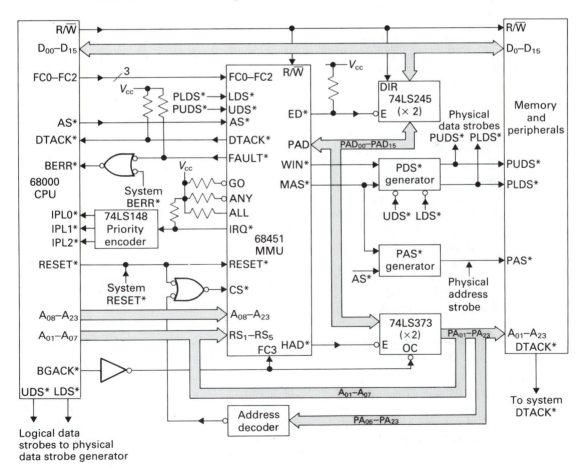

The hold address (HAD∗) output controls the external latches on the address bus. HAD∗ is asserted by the MMU to hold the physical address stable in the latches. The enable data (ED∗) output controls the data transceivers between the 68000 data bus and the MMU. When ED∗ is asserted, the data transceivers are enabled and drive the data bus. The MMU asserts ED∗ only during a read/write access to it, or during an IACK∗ cycle.

The MMU communicates with a 68000 CPU using the conventional AS∗, UDS∗, LDS∗, and DTACK∗ signals. The mapped address strobe (MAS∗) output from the MMU is used to generate mapped data strobes for the system memory. Although AS∗ is asserted early in a read cycle by the CPU, MAS∗ cannot be asserted by the MMU until the address translation process has taken place. Adding an MMU to a system creates a rather severe penalty in terms of the memory cycle time. A memory cycle is extended from 8 cycles to 12 cycles.

The MMU has three control signals that enable it to communicate with other MMUs. These are global operation (GO∗), any (ANY∗), and all (ALL). In single MMU applications, these signals should be pulled up to V_{cc} by a resistor. One MMU must always be designated the master MMU by arranging its CPU interface so that *both* RESET∗ and CS∗ are asserted *simultaneously*. Nonmaster MMUs are reset in the normal way by asserting their RESET∗ inputs alone. Even in single-MMU systems, RESET∗ and CS∗ must be asserted simultaneously.

When a master MMU is reset, segment zero is set up as follows: 0 is loaded into its logical address mask (LAM), its ASN is loaded with 0, its ASM is loaded with $FF, and its E-bit is enabled. The 16 entries in the AST are cleared to 0, and all segments other than 0 are disabled by clearing their E-bits. Because of these actions, all logical addresses from the 68000 are initially passed unchanged to the physical address bus following a reset. This reset sequence gives the operating system an opportunity to set up the MMU by loading the descriptors. Each descriptor is loaded by copying the information into the temporary descriptor and then transferring the information to the appropriate descriptor.

Switching from one task to another is called *context switching* and is performed by the operating system running in the supervisor state. The operating system changes the first two entries in the AST (AST1 = user data space, AST2 = user program space) to the ASN of the new task to be run. The new values of the program counter and status register (obtained from the new task's TCB) are pushed onto the supervisor stack. When an RTE is executed, the new task runs. Context switching is the same as *task switching* described in chapter 6, except that the task number in the AST is also changed each time a new task is run.

68851 Paged Memory Management Unit

The 68451 segmented memory management unit is widely regarded as a brave attempt to introduce a sophisticated memory management mechanism in 68000-based systems. Unfortunately, it lacks the power and flexibility demanded by today's high-performance 68020- and 68030-based microcomputers. The two principal limitations of the 68451 are its speed (it adds four clock states to a 68000's memory access) and its limited number of segments, which makes it difficult to use in systems with complex memory maps.

The organization of the 68451 MMU restricts designers to segments with *exponential sizes* (e.g., 1, 2, 4, 8, 16, ... times the minimum segment size). Consequently, the systems programmer is forced to use the binary buddy (or similar) algorithm to organize the physical memory efficiently. In other words, the programmer cannot readily adopt one of the many algorithms used to manage memories with fixed size pages. On the other hand, the 68451 is relatively simple to use (at least in comparison with the 68851 PMMU) and its variable-sized segments make it easy to cater for systems with relatively few tasks (e.g., an operating system, a single user, and a couple of background activities).

The 68851 paged memory management unit (PMMU) implements a conventional, paged, memory-mapping scheme with a fixed (but user-selectable) page size and is designed to operate in high-speed systems with an address translation penalty of about 35 ns. The PMMU's operating principles are very simple: It takes a logical address from the processor and uses its most significant bits to look up a *page descriptor* containing the address of the corresponding physical page. Unlike the *segmented* 68451 MMU, all the 68851 PMMU's address translations involve a fixed-size page. The page descriptor also contains information about the physical page (e.g., access rights).

However, the sheer wealth of options provided by the 68851 make it appear to be a very complex device. Part of the PMMU's complexity arises from the support it gives to the 68020's CALLM and RETM instructions. In this section we attempt to provide an overview of the PMMU and to describe its basic principles plus a few of its options.

The 68851 is implemented as a coprocessor (see section 7.6) and is therefore well suited to applications in 68020-systems (you would not use it with a 68000). Figure 7.55a illustrates how the pins of the 68851 can be arranged into functional groups and figure 7.55b shows how the 68851 is interfaced to a 68020 at a block-diagram level. Like the 68451 MMU, the 68851 PMMU sits between the processor's logical address bus and the memory's physical address bus. Unlike the 68451, the 68851 does not require additional address multiplexers or data latches, since it is housed in a pin grid array and has sufficient pins for logical address and data buses plus a physical address bus. The 68030 includes a cut-down version of the 68851 on-chip, whereas the 68040 has two internal MMUs, both of which operate in a similar fashion to the 68030's MMU (but are simplified versions of 68030's MMU).

Although the PMMU's page size is fixed in the sense that all pages have the same size, the programmer can select the actual size of the page (from 256 to 32K bytes) in his or her application. As the number of pages supported by the PMMU is, effectively, unlimited, this device does not suffer from many of the limitations of the 68451 MMU. The actual maximum number of pages is given by 4G byte/page size.

Two important differences exist between the way in which the PMMU implements address translation and the simple address translation mechanism using the page-table look-up technique we described earlier. The first is that the PMMU does not keep its page-table on-chip (remember we said that the PMMU supports an *effectively* unlimited number of pages). There is no way in which current technology

FIGURE 7.55a   The PMMU's functional signal groups

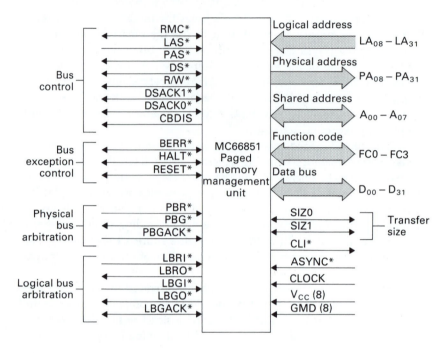

FIGURE 7.55b   Block diagram of interface of a 68851 to a 68020

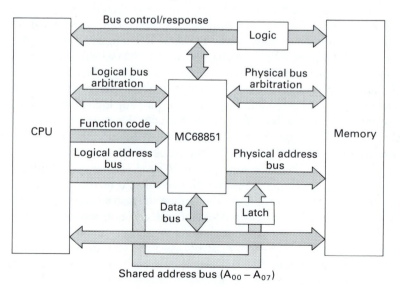

can provide a gigantic on-chip page-table. Instead, the PMMU keeps some *frequently used* page-table entries on-chip in a 64-entry cache and hopes that it will not be necessary to fetch new page-table entries from memory too often.

The second difference between the PMMU and a simple page-table address translator is that the PMMU supports multilevel page-tables. We will soon see what this means in practice.

Suppose a programmer elects to use the PMMU's smallest page size (i.e., 256 bytes) in a 68020-68851 combination. Since eight address lines, A_{00} to A_{07}, select a location within a page, then $32 - 8 = 24$ memory lines must select a page descriptor in the page-table; that is, the page-table has 2^{24} entries because logical address lines A_{08} to A_{31} index into the table (which is at least *three* times as large as the 68000's entire address space). Remember that an entry in the page-table (i.e., a *page descriptor*) must consist of at least a 24-bit physical page address plus the associated assess rights and status information). As the PMMU maintains this page-table in external memory, it must read memory to get each translation entry, considerably extending the processor's access time. If this were the whole truth, then using the PMMU would add an intolerable burden to the processor's average access time.

As we have said, the PMMU maintains a table of the last 64 logical-to-physical translations on-chip in an *address translation cache* (ATC). By the way, the ATC is frequently called a *translation look-aside buffer* (TLB) by other semiconductor manufacturers. Although just 64 out of, say, 2^{24} possible entries seems rather insignificant, it is highly probable that the next memory access made by the processor will use one of the 64 cached page descriptors in the ATC, since memory accesses are frequently highly correlated (i.e., some pages are repeatedly accessed). In the next section we will look at the characteristics of cache memory in greater detail.

Address Translation

The precise mechanism by which the PMMU carries out its logical-to-physical address translation is rather complex, so we will just provide an overview here. This element of complexity largely springs from the way in which the PMMU implements a mapping table with pages as small as 256 bytes without requiring the 2^{24} page-table entries we mentioned previously. One way of looking at the process used by the PMMU to access its logical-to-physical address descriptors in its address translation table is to regard the CPU's logical address in the same way we approach a postal address. A postal address has a series of fields: state, town, and street. These fields help guide the letter from its source to its destination. In the same way, the PMMU takes a logical address from the processor and uses its fields (we say what we mean by fields later) to locate the appropriate entry in the logical-to-physical address translation table. Figure 7.56 illustrates a logical address with multiple fields (in this example only two are shown).

The translation process that takes us from a logical to a physical address is called a *table walk* (figure 7.57). The start of the path to the required entry in the address translation table is the *root pointer*, which is a register in the PMMU. Three root pointers are implemented by the 68851: the supervisor root pointer (SRP), the CPU root pointer (CRP), and the DMA root pointer (DRP). The actual root pointer

FIGURE 7.56 The logical address with multiple fields

31		0
Logical address Field 1	Logical address Field 2	Page address

NOTE: This example provides only two levels of address translation, indexed by field 1 and field 2. The PMMU supports one through four logical address fields.

FIGURE 7.57 Using a logical address to perform a table walk

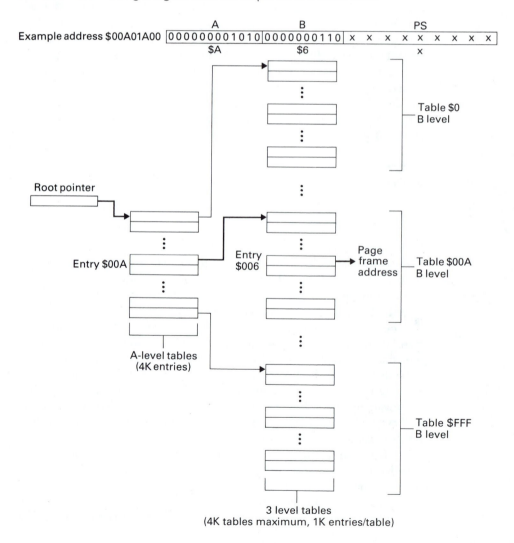

3 level tables
(4K tables maximum, 1K entries/table)

selected for an address translation depends on the type of memory space being accessed (e.g., user or supervisor). These root pointers provide three different sets of address translation tables and therefore permit the separation of user, supervisor, and DMA address spaces. By providing three root pointers, the PMMU can manage three separate tasks simultaneously. Incidentally, the 68030 has two root pointers (SRP and CRP); the DMA root pointer is not implemented because the MMU is on-chip. Therefore DMA on the logical side of the address bus is not possible.

The logical address from the processor is divided into from one to four fields (the number and size of the fields actually used by the PMMU are user-programmable). Imagine that we have chosen two fields (figure 7.57). The most significant field, A, indexes into the first-level address translation table in memory. Each entry in this table, called a *table descriptor*, points to a second-level table. The *least significant* field, B, of the logical address indexes into the second-level table to locate the address of the actual page descriptor for this logical-to-physical translation. Notice that the PMMU employs two basic types of descriptor: *page* descriptor and *table* descriptor.

As you can see from figure 7.57, a logical-to-physical address translation requires *multiple* memory accesses. Moreover, each access might require the fetching of a 64-bit descriptor, which, in this example, makes four 32-bit accesses in all. Remember that the 68551 PMMU stores up to 64 descriptors in its on-chip cache; therefore, the PMMU will only occasionally have to access the external descriptor tables in memory.

PMMU's Protection Mechanism

The PMMU supports a sophisticated memory protection mechanism. In a very modest operating system, you might decide to classify all pages into four types: user, supervisor, read-only, and read-write pages. The PMMU provides a much more versatile protection mechanism (whose facilities will be used fully only in the most powerful of operating systems). To understand how the PMMU operates, you must appreciate the structure of the table descriptors that point to the next level down in the hierarchical address translation tables and the page descriptors that point to the actual physical pages.

The programmer (i.e., operating system designer) may select one of two types of table (and page) descriptor. Figure 7.58a describes the structure of a long table descriptor (called *long* because it is 8 rather than 4 bytes long) that points to the next entry down in the hierarchical address translation tables. As you can see from figure 7.58a, the long table descriptor provides both a translation address (bits 4–31 of the descriptor provide bits PA_{04}–PA_{31} of the physical table address) and a protection mechanism with bits 32 to 63.

Figure 7.58b describes the short table descriptor, which provides less information than a long table-descriptor. You may wonder why the PMMU provides the luxury of two page table descriptors. The answer is simple. A glance at figure 7.57 demonstrates that the page table structure is *hierarchical*. If you use a long descriptor in a table structure, its protection bits apply to all lower levels.

The meanings of bits 32 to 47 in a long table descriptor are as follows.

FIGURE 7.58 Format of a table descriptor

63	62					48	47	46	45	44	43	42	41	40	39	38	37	36	35	34	33	32
L/U			Limit				RAL			WAL			SG	S	0	0	0	0	U	WP		DT

			Table address (PA31 – PA04)									Unused		
31												4	3	2	1	0

(a) Format of a long table descriptor

31									4	3	2	1	0
			Table address (PA31 – PA04)						U	WP		DT

(b) Format of a short table descriptor

DT (Descriptor Type) The two descriptor-type bits tell the PMMU whether another level in the table exists and whether that level uses a short or a long descriptor format. In other words, the DT bits tell the PMMU what type of descriptor to expect in the next level down the table walk.

WP (Write-Protect) The write-protect bit indicates that pages pointed at by this descriptor may not be written to. If you use a long table descriptor at a high level in the address translation tree and $W = 1$, all subsequent levels in the trees and their associated page descriptors will be write-protected.

U (Used) The used bit is initially cleared to zero by the operating system when the descriptor table is set up. The PMMU automatically sets this bit to 1 the first time that the descriptor is accessed. Consequently, you can use this bit in virtual memory systems when deciding whether to swap the page out.

S (Supervisor) When the supervisor bit is set, pages pointed at by this descriptor can be accessed only from the supervisor mode.

SG (Shared Globally) When set to 1, the shared globally bit indicates that the page descriptor may be shared. That is, if $SG = 1$, then all tasks within the system may access the physical page. SG tells the PMMU that only a single descriptor for this page need be held in the ATC.

WAL (Write Access Level) The write access level indicates the minimum privilege level allowed for a pages located via this descriptor. Three bits define eight levels of privilege (see later).

RAL (Read Access Level) The three read access level bits perform the read function corresponding to the WAL bits.

Limit The 15 limit bits provide a lower or upper bound on index values for the next level in the translation table. The limit can range from 0 to 2^{15} in powers of 2. That is, the limit field restricts the size of the next table down. For example, one of the logical address fields may have 7 bits and therefore support a table with 128 entries. However, in a real system you might never have more than, for example, 20 page descriptors at this level. By setting the limit to 5, you can restrict the table to 32 entries (rather than 128).

L/U (Lower/Upper) The lower/upper bit determines whether the *limit* field refers to a lower bound or to an upper bound. If L/U = 0, the limit field contains the unsigned upper limit of the index and all table indices for the next level must be less than or equal to the value contained in the limit field. If L/U = 1, the limit field contains the unsigned lower limit of the index and all table indices must be greater than or equal to the value in the limit field. In either case, if the actual index is outside the maximum/minimum, a limit violation will occur.

Most of the bits in the page-table descriptor are self-evident, but we need to say more about the WAL and RAL bits. The two access-level fields, WAL and RAL, are used in 68020-based systems that provide a hierarchical protection mechanism in conjunction with the call module instruction CALLM. As we have said, this complex mechanism has been dropped from the 68030 and later processors. Basically, a CALLM instruction allows a task operating at one level of privilege to request the use of a module at a higher level of privilege. You can think of this mechanism as extending the 68020's supervisor/user operating mode into a system with nine levels (i.e., a supervisor mode and eight levels of user mode). Privilege level 0 is the highest and 7 is the lowest.

WAL and RAL are used in conjunction with the PMMU's CAL (current access level) register. Suppose that CAL contains 4 and a page is accessed by a read cycle. This task at a privilege level of 4 may not access pages with a higher level of privilege. If the page being accessed has a read access level of 4 to 7, it may be read. If it has an access level of 0 to 3, it cannot be read, and a page-fault is raised.

The end result of a table walk is the page descriptor that is going to be used to perform the actual logical-to-physical address translation. As in the case of the table descriptors, there are two page descriptor formats: a long page descriptor (figure 7.59a) and a short descriptor (figure 7.59b). Remember that the only real difference between table and page descriptors is that table descriptors point to the next level in the translation tables and page descriptors point to the actual physical page. However, a page descriptor has several control bits not found in a table-descriptor. These bits are as follows.

FIGURE 7.59 Format of a page descriptor

63			48	47	46	45	44	43	42	41	40	39	38	37	36	35	34	33	32
Unused				RAL			WAL			SG	S	G	CI		L	M	U	WP	DT

31					8	7	6	5	4	3	2	1	0
Page address (PA31 – PA08)								Unused				

(a) Format of a long page descriptor

31			8	7	6	5	4	3	2	1	0
Page address (PA31 – PA08)			G	CI		L	M	U	WP		DT

b) Format of a short page descriptor

M (Modified)   The modified bit indicates whether the corresponding physical page has been written to by a bus master. The M-bit must be set to zero when the descriptor is first set up by the operating system, since the PMMU may set the M-bit but not clear it. Note that the U-bit (used bit) is set if a table or a page descriptor is accessed, whereas the M-bit is set if the page itself is written to.

L (Lock)   The lock bit indicates that the corresponding page descriptor should be made exempt from the PMMU's page-replacement algorithm. When L = 1, the physical page cannot be replaced by the PMMU. That is, you can use the L-bit to keep page descriptors in the ATC. Suppose your system regularly accesses 100 page descriptors. Clearly, only 64 of them may be stored in the ATC at any instant. You might wish to lock some of the descriptors if their associated pages must be accessed without performing a table walk. For example, descriptors corresponding to a critical section of operating system code (the scheduler) can be locked to keep them in the ATC.

CI (Cache Inhibit)   The cache inhibit bit indicates whether or not the corresponding page is cachable. If CI = 1, the CLI* (cache load inhibit) bit of the PMMU is asserted to inform caches that this access should not be cached. Physical I/O devices or memory shared by another microprocessor must not be cached.

G (Gate)   The gate bit is used in conjunction with the 68020's CALLM (call module) instruction. Remember that these instructions are designed to support the eight-level prioritization mechanism that has been abandoned in the 68030 and later processors.

In addition to the long and short table descriptors, the PMMU supports *invalid* descriptors and *indirect* descriptors. The DT (descriptor-type) field of an invalid descriptor is zero and indicates that the corresponding page is not currently in physical memory but is on disk. Note that an invalid descriptor may appear at any location in the table walk (i.e., at any level) except for the root. When the PMMU encounters an invalid descriptor, it raises a bus error and permits the processor to load the missing pages.

An indirect descriptor is what its name suggests. The prize at the end of a table walk is a page descriptor. However, if an indirect table descriptor is used to replace a page descriptor, it is possible to point to a page descriptor in another table. Figure 7.60 describes the use of an indirect descriptor. Indirect descriptors can be used to share pages between different tasks, although this is not the only way to share pages.

Programmer's Model of the PMMU

Due to its coprocessor interface (described in section 7.6), the 68851 appears to extend the 68020's architecture by adding a new set of registers and instructions. Figure 7.61 illustrates these registers, which, of course, belong only to the 68020's supervisor mode architecture. The user programmer has no "knowledge" of the PMMU.

FIGURE 7.60 Example of the use of an indirect table descriptor

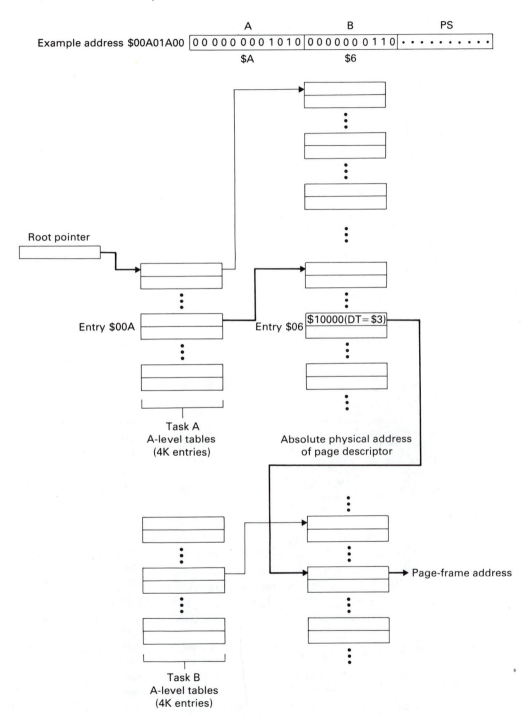

FIGURE 7.61 The PMMU's programmer-visible register set

```
63                                                    32
┌─────────────────────────────────────────────────────┐
│                   CPU root                            │
│                   pointer                             │
└─────────────────────────────────────────────────────┘
31                                                     0
┌─────────────────────────────────────────────────────┐
│                   DMA root                            │
│                   pointer                             │        Address
└─────────────────────────────────────────────────────┘        translation
                                                                control
┌─────────────────────────────────────────────────────┐        registers
│                Supervisor root                        │
│                   pointer                             │
└─────────────────────────────────────────────────────┘

┌─────────────────────────────────────────────────────┐
│                Translation control                    │
└─────────────────────────────────────────────────────┘

15                   8
┌─────────────────────────────┐                                Status
│  PMMU cache status (PCSR)    │                               information
└─────────────────────────────┘                                registers

┌─────────────────────────────┐
│     PMMU status (PSR)        │
└─────────────────────────────┘

7            0
┌────────────┐    ┌────────────┐    ┌────────────┐            Protection
│    CAL     │    │    VAL     │    │    SCC     │            mechanism
└────────────┘    └────────────┘    └────────────┘            control
┌─────────────────────────────┐                                registers
│     Access control (ACR)     │
└─────────────────────────────┘
```

BAD0	BAC0	
BAD1	BAC1	
BAD2	BAC2	
BAD3	BAC3	Breakpoint
BAD4	BAC4	control
BAD5	BAC5	registers
BAD6	BAC6	
BAD7	BAC7	

The PMMU's architecture includes three sets of registers: 3 root pointer registers, 16 breakpoint registers, and several control and status registers. We will describe the functions carried out by these groups of registers in turn.

The three 64-bit root pointers are used to start the table walk for access in supervisor, user, and DMA address spaces, as we described earlier. Figure 7.62 illustrates the structure of a root pointer, which is similar to a table descriptor but lacks the RAL, WAL, S, U, and WP fields. As we will see, it is not necessary to use three root pointers. For many applications of the PMMU, one root pointer will suffice.

Breakpoint Registers The eight pairs of breakpoint registers, BAD0 to BAD7 and BAC0 to BAC7, provide a breakpoint acknowledge facility for the 68020. One pair of registers is associated with each of the 68020's eight breakpoint instruc-

FIGURE 7.62 Format of a root pointer

63		48	47	46	45	44	43	42	41	40	39	38	37	36	35	34	33	32
Limit			0	0	0	0	0	0	SG	0	0	0	0	0	0	0	DT	
Page address (PA31 – PA04)															Unused			
31															4	3	2	1	0

tions. These registers are used only during software analysis and system debugging and are not required for normal operation of the PMMU. When the 68020 executes a breakpoint instruction (BKPT #n), it carries out a breakpoint acknowledge bus cycle by reading from a predetermined address in CPU address space. The PMMU detects this access and responds either by providing a replacement op-code and terminating the cycle normally or by terminating the bus cycle with an illegal instruction exception. The breakpoint registers permit the programmer to force an illegal instruction or to provide a replacement instruction op-code.

Eight breakpoint acknowledge data registers, BAD0 to BAD7, hold replacement op-codes, and eight breakpoint acknowledge control registers, BAC0 to BAC7, determine how many times the op-code replacement should be made (1 to 255) before forcing an illegal instruction exception. For example, if you load BADi with $WXYZ, then the op-code $WXYZ will be loaded into the 68020 whenever the instruction BKPT #i is executed. Bits 0 to 7 of each breakpoint acknowledge register determine the breakpoint skip count, and bit 15 is the breakpoint enable control, BPE. If the BPE bit is clear, the breakpoint acknowledgement is disabled and the breakpoint is terminated by the 68851 asserting BERR*.

If the BPE bit is set, one of the breakpoint acknowledge data registers supplies an op-code in response to a breakpoint instruction, and DSACK0*/DSACK1* is asserted to terminate the bus cycle. However, each time a breakpoint is acknowledged, the breakpoint skip count in the corresponding breakpoint acknowledge control register is decremented. When the skip count reaches zero, the PMMU asserts BERR* to terminate the breakpoint acknowledge cycle and the 68020 begins exception processing for an illegal instruction.

You can use the breakpoint mechanism to monitor the flow of instructions. Suppose you place a breakpoint at the start of a subroutine and put the op-code that was at the breakpoint address in a breakpoint acknowledge data register. By means of the corresponding breakpoint control register, you can either let the breakpoint result in an illegal instruction exception (if the breakpoint skip count is zero) or you can permit the PMMU to respond to a breakpoint by supplying the replaced op-code.

Translation Control Register The PMMU's control and status registers determine the operating mode. Probably the most important PMMU control register is its translation control register, TC, that permits the programmer to define the way in which the PMMU performs its address translation. More specifically, it defines the structure of the address translation tables. Figure 7.63 describes the format of the TC. Basically, the translation control register enables the operating systems designer to

FIGURE 7.63 Format of the translation control register, TC

31	30	29	28	27	26	25	24 23	20 19	16 15	12 11	8 7	4 3	0
E	0	0	0	0	0	SRE	FCL	PS	IS	TIA	TIB	TIC	TID

choose an embarrassingly large number of possible address translation mechanisms by defining the size and number of address translation tables.

The most significant bit of the TC is simply an enable bit. When clear, it effectively removes the PMMU from the system, and all logical addresses are passed unchanged to the PMMU's physical address bus. When E is set, the PMMU performs address translation.

The supervisor root pointer enable bit, SRE, is set to force all supervisor mode accesses to use the supervisor root pointer, SRP, and all user mode accesses to use the CPU root pointer, CRP. If SRE is clear, all accesses use the CRP. That is, the SRE bit lets the programmer choose between two root pointers (depending on the state of the S-bit in the 68020) or a single root pointer.

The function code lookup bit, FCL, is set to enable *function code indexing*. That is, when FCL = 1, the PMMU takes the three function code bits (FC0 to FC2) from the processor and uses them to index into an 8-entry table. This table is the first-level descriptor table; it allows you to have completely separate address translation tables for each function code. If FCL = 0, the root pointer points to the first-level address translation table, which is common to all function codes.

The 4-bit page size field, PS, selects the page size as 2^{PS}, where PS is in the range 8 to 15, giving page sizes 2^8 (256 bytes) to 2^{15} (32K bytes). Values of PS in the range 0 to 7 are illegal.

The 4-bit initial shift field, IS, provides a mask for the IS most significant bits of the logical address. That is, the IS field tells the PMMU to ignore the first IS bits of all logical addresses and to begin translation at address bit $31 - IS$. So, if we wish to use a 24-bit logical address (i.e., make the 68020 look like a 68000), setting IS = 8 has the desired effect.

Four 4-bit fields, TIA, TIB, TIC, and TID, describe the lowest four levels of address translation tables. The first field, TIA, must be nonzero and indicates the number of bits in the logical address that are to be used as an index into this table. For example, if TIA = 6, the first 6 bits of the logical address point to one of $2^6 = 64$ entries (i.e., next-level table descriptors).

Fields TIB, TIC, and TID each provide information about the next-lower translation table, respectively. If any of these fields is zero, address translation tables end at the level above and the levels below must be set to zero. Consider an example with two fields: TIA = 7, TIB = 10, TIC = 0, TID = 0. The first-level table, A, contains 2^7 descriptors. Each of these 128 descriptors points to a second-level B table (i.e., there are 128 second-level tables) containing $2^{10} = 1,024$ entries. The total number of descriptors in the system is, therefore, $128 \times 1024 = 128K$ (this number can be reduced by employing the limit field of the descriptors to prune the trees). Table levels C and D do not exist.

FIGURE 7.64 Translation control register

31						25	24	23	20	19	16	15	12	11	8	7	4	3	0
E	0	0	0	0	0	SRE	FCL	PS		IS		TIA		TIB		TIC		TID	
1	0	0	0	0	0	0	0	1100		1000		0101		0111		0000		0000	

The sum of the six fields (figure 7.63) is made up of IS, TIA, TIB, TIC, TID, and PS and must be equal to 32. Why? Because collectively these translation control register control bits divide up the 32-bit logical address into several fields. IS indicates the number of bits that are not translated. TIA, TIB, TIC, and TID define the number of bits in each of the hierarchical address translation tables. Finally, PS indicates how many bits are to be passed from the logical to the physical address bus untranslated (since these bits access a location within a page).

Once more, it must be stressed that the PMMU is an elaborate all-singing, all-dancing device and that many applications just do not require its sophistication. Imagine that you are designing a modest operating system to run on a computer with no more than 16M bytes of memory and your pages are to be 4K bytes. Furthermore, you do not require multiple address translation tables according to function code or to supervisor or user accesses. Finally, you are happy to accept two levels of address translation (a 32-entry first-level table and a 128-entry second-level table). The translation control register would therefore have the form shown in figure 7.64.

Status Registers The PMMU has two status registers, PCSR and PSR. The PMMU's cache status register, PCSR, is used to indicate the status of the address translation cache (nothing to do with cache memory). The PMMU status register, PSR, contains status information about a descriptor that can be read by the host processor to determine the validity of the descriptor.

Only 5 bits of the PMMU's 16-bit cache status register, PCSR, are defined (figure 7.65). The TA (task alias) field contains the current internal task alias (we will mention the task alias again later). The F (flush) field is set when the PMMU flushes entries in the ATC as a result of a write to the CRP. As you can appreciate, changing the CPU root pointer invalidates many descriptors in the ATC, since they belong to trees pointed at by a previous root pointer. The LW (lock warning) field is set when all entries in the ATC have been locked. Remember that you lock a cached descriptor to prevent it being swapped out when the ATC is full. If you lock all entries in the ATC, the PMMU may grind to a halt (as it will have to perform a table walk for all noncached entries).

FIGURE 7.65 Structure of the PMMU's cache status register

15	14	13	12	11	10	9	8	7	6	5	4	3	2	1	0
F	LW	0	0	0	0	0	0	0	0	0	0	0		TA	

At this stage we will introduce the 68851's root pointer mechanism and its task-aliasing mechanism. However, these facilities are managed entirely by the PMMU and the programmer is unaware of them. The PMMU's invisible (at least to the programmer) root pointer table holds the last eight CPU root pointers (CRPs). This root pointer table is rather like an ATC but for root pointers. Whenever you change a CPU root pointer to run (or rerun) a task, the old CRP is held in the root pointer table. Now, since you explicitly load and reload the CPU root pointer, why does the PMMU need to record old CRPs? It is not really the old CRPs that the PMMU is interested in saving—it is the descriptors in the ATC that were associated with these old CRPs.

Each descriptor in the ATC has a field containing a value in the range 0 to 7 that records its associated CPU root pointer, or *task alias*. Whenever an access is made to the ATC, the task alias of the descriptor is compared with that of the current CPU root pointer. These two values must match before the descriptor can be used.

Now suppose the CPU root pointer is reloaded because of a task change. Obviously, all *old* descriptors in the ATC are redundant because they relate to an earlier task. The PMMU knows that they are invalid because their task alias field does not match that of the current CRP. However, suppose the CPU root pointer is reloaded with a recent value in order to run a task again. Any descriptor in the ATC whose task alias matches the root pointer can be reused. That is, the task alias mechanism exists to make the PMMU more efficient by not throwing away old ATC entries.

Access Control Register The four access control registers (AC, CAL, VAL, and SCC) described in figure 7.66 are used in conjunction with the eight-level privilege mechanism we mentioned earlier. As we said, most memory management

FIGURE 7.66 Structure of the PMMU's access control registers

15	14	13	12	11	10	9	8	7	6	5	4	3	2	1	0	
0	0	0	0	0	0	0	0	MC	0	ALC		0	0	MDS		Access control register

7	6	5	4	3	2	1	0	
Access level			0	0	0	0	0	Current access level

7	6	5	4	3	2	1	0	
Access level			0	0	0	0	0	Validate access level

7	6	5	4	3	2	1	0	
								Stack change control register

systems will not use these registers, primarily because they operate (partially) in conjunction with the special 68020 instructions that are no longer implemented on the 68030 and 68040. The 16-bit access control register, ACR, determines how the PMMU implements its access control mechanism. If the MC (module control) bit of the access control register is clear, the PMMU does not support module call and return instructions, and the access level mechanism is not implemented. If the MC bit is set, the access control register's access level control (ALC) and module descriptor size (MDS) fields are enabled. We do not discuss the PMMU's access control mechanism here (further details can be found in the PMMU users' manual).

We are now going to look at the PMMU's status register, PSR, which is described in figure 7.67. The bits of the PSR are set or cleared by the PMMU when a descriptor is fetched by a PTEST instruction. The meaning of most of the PSR's bits is self-evident. For example, the B-bit (bus error) indicates, when set, that a bus error will be caused if the descriptor is fetched from memory. Bits 0 to 2 of the PSR indicate how many levels of translation table were searched during the table walk to find the descriptor. Note that a value zero in the limit field indicates either that the descriptor was in the ATC or that the search was terminated early.

The PSR reflects the 68851's status, but in a rather different way than the status register of the 68000 etc. The PSR is not automatically updated after each operation (i.e., each address translation). Instead, it is updated only when the programmer executes the explicit PTEST (test logical address) instruction, which is available in two forms: PTESTW and PTESTR. One form is used with *write* accesses and the other with *read* accesses.

The function of the two PTEST instructions is to perform an address translation and then update the status register. Perhaps the PTEST is rather like the 68000's CMP instruction—both these instructions carry out an operation and set status flags but leave everything else unchanged. Basically, the PTEST instruction permits you to access a page and then examine the effects of this access by reading the bits of the status register. In other words, you are saying to the PMMU, "If I carried out this address translation by accessing this page, what would happen?" An important

FIGURE 7.67 Structure of the PMMU's status register (PSR)

15	14	13	12	11	10	9	8	7	6	5	4	3	2 1 0
B	L	S	A	W	I	M	G	C	T	0	0	0	Table levels

B	= bus error
L	= limit violation
S	= supervisor only
A	= access level violation
W	= write-protected page
I	= invalid
M	= modified
G	= gate
C	= globally shared
T	= transparent translation (68030 only)
Levels	= number of table levels

application of the PTEST instruction is in the validation of vectors passed to a procedure. By performing a PTEST on a vector, you can determine whether or not it leads to a legal access.

The syntax of the PTEST instruction is rather complex: PTESTR ⟨fc⟩,⟨ea⟩,#level[,An]. The ⟨fc⟩ field permits you to specify the type of address space in which the descriptor lies (i.e., 0 to 7). The ⟨ea⟩ field provides the effective address of the page you are testing. The #level field defines the search level (i.e., the maximum number of levels to be searched in the hierarchical table's address translation). Finally, the optional [,An] field can be used to receive the physical address of the descriptor. As an example, consider the instruction PTESTR #1,8(SP),#$7,A0. The instruction performs a read test on the page at the effective address 8(SP) (i.e., the effective is at [SP] + 8, which is 8 bytes below the top of the 68000's stack). The test is performed to a page accessed with the function code 1 (i.e., user data space), and the maximum number of search levels that may be performed is 7 (i.e., all levels in the tree must be searched, since 7 is the maximum search depth supported by the PMMU; we leave it to you to work our why it is 7). Finally, the address of the resulting descriptor is placed in A0. Once this instruction has been executed, you can test the bits of the status register, PSR to see what happens.

PMMU's New Instructions

Before we describe some of the MMU's instructions, we will remind you that it has a coprocessor interface; therefore, its registers and instructions appear, to the programmer, as an extension of the 68020's instruction set. In addition to an enhanced register set, the supervisor mode programmer also has several new instructions to access the PMMU. All these instructions are, of course, privileged. Some of the new instructions are as follows.

PMOVE This instruction is used to transfer data to and from the PMMU's internal registers. The amount of data moved is determined by the size of the corresponding PMMU register. For example, PMOVE (A0),CRP will copy the 8-byte descriptor pointed at by A0 into the PMMU's CPU root pointer.

PLOAD Two load instructions, PLOADR ⟨function code⟩,⟨ea⟩ and PLOADW ⟨function code⟩,⟨ea⟩, are used to load an entry into the address translation cache (which is normally done automatically by means of a table walk when you access a page whose descriptor is not already cached). The ⟨ea⟩ field provides the effective address of the logical page to be translated, and the ⟨function code⟩ field defines the type of address space in which the page lies. You can do this to save time during certain critical operations. For example, suppose you know that you are soon going to access a certain region of memory. By loading its descriptor(s) into the ATC, you can remove the overhead of the table walk. You must load one descriptor for each of the pages you wish to install. Note that the PMMU performs a table walk to determine whether the descriptor you are loading already exists in the ACT. If it does, the old descriptor is flushed and the new one added.

PFLUSH   The PMMU flush instruction is used to invalidate the address translation cache and remove old entries. For example, if you are to change the logical-to-physical mapping tables (perhaps because of a task change), you will need to scrub the ATC of old descriptors. Sometimes you can use PFLUSH in a preemptive fashion. If certain tasks or procedures are not needed again, you can remove them from the ATC to make room for more valuable descriptors. Note that the PFLUSH instruction has several formats, depending on whether you wish to perform a complete or a partial flush.

PVALID   The PVALID (validate a pointer) instruction is used in conjunction with the PMMU's mechanism that provides eight levels of privilege. Executing a PVALID VAL,⟨ea⟩ instruction forces the PMMU to check the upper address bits of the destination against the contents of the validate access level register, VAL. If the effective address has a higher privilege level than that reflected by VAL, the PMMU takes an access-level violation exception. Otherwise, execution continues normally.

PBcc   The   PMMU   implements   coprocessor   branch   instructions.   A PBcc ⟨label⟩ instruction causes a branch to the target destination on the condition cc. The condition cc is one of 16 conditions, depending on the state of the B-, L-, S-, A-, W-, I-, G-, and C-bits in the PSR. Similar PMMU conditional instructions are PDBcc, PScc, and PTRAPcc.

PSAVE   The PMMU's PSAVE and the corresponding PRESTORE instructions are used to save and restore the PMMU's internal user-invisible state on the stack. This instruction is very obscure and is required only by processors with multiple operating systems. The only time the PMMU's state must be saved is when one O/S takes over from another.

68030's PMMU

The 68030 includes much of the hardware of a 68851 PMMU on-chip. However, since it was not possible to include all the PMMU's silicon on a 68030 die, the 030's PMMU is a rather simplified version of the 68851. Since the 68030 does not support the 68020's module call/return instructions, the 68030's MMU does not implement the 68851 PMMU's eight-level privilege mechanism. The 68030 does not implement the 68851's breakpoint mechanism. Finally, most users do not require the full sophistication of the 68851. Consequently, the 68030's MMU has a simpler register set than the PMMU. As you can see from figure 7.68, the 68030's MMU registers comprise *two* root pointers (CRP and SRP), a translation control register (TC), a status register (MMUSR), and two *new* transparent translation registers (TT0 and TT1).

The 68030 does not support a DMA root pointer register because no access to the chip's logical address bus is possible. Moreover, as there is no root pointer table, the 68030 does not implement task aliasing. Note that the 68030's MMU status register is called MMUSR, but it is effectively the same as the PMMU's PSR. Although the 68030's actual logical-to-physical address translation mechanism is essentially the same as that of the PMMU, the 030's address translation cache holds only 22 entries, in comparison with the PMMU's 64. This reduction will *slightly*

FIGURE 7.68   The 68030 MMU registers

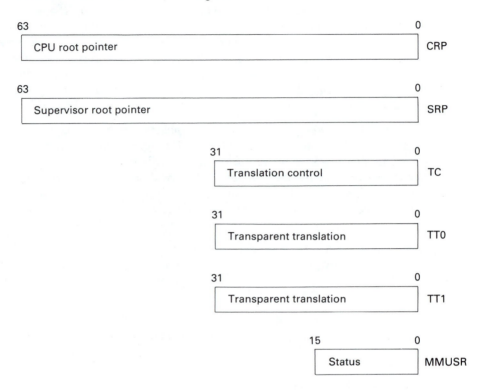

reduce the performance of the on-chip cache (but not as much as you might think). The reduction was made necessary in order to get the 68020 and the 68851 on the same piece of silicon. Moreover, you cannot lock descriptors in the ATC and prevent the 68030 from overwriting them by new descriptors. Finally, ATC entries cannot be defined as shared globally.

The 68030's MMU instruction set is rather smaller than that of the PMMU. The only instructions supported by the 68030 are PFLUSH, PLOAD, PMOVE, and PTEST. Note that conditional MMU instructions are not supported by the 68030 (i.e., PBcc).

The 68030's MMU does implement two new registers, called transparent translation register 0 and 1 (TT0 and TT1). Each of these two 32-bit registers specifies a block of at least 16M bytes of logical address space that does not take part in memory management. That is, the logical address is mapped into its identical physical address and access rights are not checked. Transparent translation bypasses the MMU and is useful, for example, when large blocks of data are to be moved during DMA or graphics applications. You could even use transparent translation to improve the efficiency of the MMU by bypassing a large block of program or data (doing this makes room for other entries in the ATC). For example, the Unix operating system can employ transparent translation because it is permanently loaded

FIGURE 7.69 Structure of the two transparent translation registers

31	24	23	16	15	14	13	12	11	10	9	8	7	6	5	4	3	2	1	0
Logical address base		Logical address mask		E	0	0	0	0	CI	R/W	RMW		0	FC base			0	FC mask	

into memory. Note that if a transparent translation cannot take place using TT0 or TT1, the address translation is handled by the MMU in the normal way by a table walk.

Figure 7.69 describes the structure of the two transparent translation registers (TT0 and TT1). The eight logical base address bits point to one of 2^8 possible logical blocks, each of 2^{24} bytes (i.e., 16M bytes), or 2^n times this size. If the transparent translation bit (i.e., the E-bit) is enabled, any logical address within this block is passed unchanged to the 68030's physical address pins.

The 8 logical address mask bits of the transparent translation register are used to mask out bits of the logical address base. Setting a bit in this field causes the corresponding bit in the logical address base to be ignored. Blocks of logical memory larger than 16M bytes can be transparently translated (up to the entire memory space of the 68030) by setting appropriate bits of the logical address mask. Note that this scheme is similar to the segmented memory management of the 68451 (the only difference is that the logical and physical blocks have the same address).

The meanings of the remaining fields of the two transparent registers TT0 and TT1 are as follows.

CI (Cache Inhibit)   If the cache inhibit bit is set, the address translated by TT0/TT1 is not cached by the 68030 and the cache inhibit output, CIOUT*, is asserted during the access.

R/W (Read/Write)   The R/W bit defines the type of accesses transparently translated. If R/W = 0, write accesses are transparent. If R/W = 1, read accesses are transparent.

RWM (Read/Write Mask)   The read/write mask masks the read/write bit. If RWM = 1, both read and write accesses are transparently translated. When RWM = 0, only read or write accesses are translated, depending on the state of the R/W bit. Note that RWM must be clear to enable read-modify-write cycles to be transparently translated.

FC Base (Function Code Base)   The 3-bit function code base defines the function code necessary for accesses to be transparently translated. That is, the function code and the contents of the FC base field must match before a transparent translation can take place. Accesses that do not match this function code are not transparently translated.

FC Mask (Function Code Mask)   The 3-bit function code mask field masks the function code base bits to permit more than one address space to be transparently translated. Setting a bit in the function code mask makes the corresponding function code base bit a don't care bit. You can use this field to force the TT register to translate addresses for more than one function code. In the extreme, setting the function code mask to 1,1,1 will make the TT register ignore the function code base.

Suppose we wish to program TT0 to perform transparent translation to any logical address in the 64M-byte range $2000 0000 to $23FF FFFF that reads from user data space (FC = 001) or supervisor data space (FC = 101). The fields to be packed in TT0 are

$$\text{Logical address base} = \$20 = 0001\ 0000$$

$$\text{Logical address mask} = \$03 = 0000\ 0011\ (\text{i.e., ignore } A_{24}, A_{25})$$

$$E = 1\ (\text{enable the transparent translation})$$

$$CI = 0\ (\text{permit caching})$$

$$R/W = 1\ (\text{transparent read accesses})$$

$$RWM = 0\ (\text{read-only access})$$

$$FC = 001\ (\text{or } 101)$$

$$FC\ \text{mask} = 100\ (\text{to mask out FC2})$$

The data to be loaded into TT0 is

000100000 0000011 1 0000 0 1 0 0 001 0 100   or $1003 8214

The data can be set up by the instructions

```
MOVE.L      #$10038314,D0
PMOVE       D0,TT0
```

If you wish to transparently translate user program space in the region $0000 0000– $0FFFF FFFF, you would set RMW = 1, FC base = 010, FC mask = 000, logical base address = $00, and logical address mask = $0F.

Example of Address Translation

Let's look at some examples of address translation (taken from the 68851 user's manual). Figure 7.70 describes a logical address with two address translation fields and a 1K-byte page-size that requires a simple two-level table. The root pointer points to the first-level (or A-level) table, which has 2^{12} entries. Each of these entries points to a second-level table containing 2^{10} entries. Finally, the B level tables point to the actual page descriptors. In this example, the logical address is $00A0 1A00 and you can see how the translation is performed.

Figure 7.71 demonstrates how function code lookup can be combined with address translation. The root pointer points to a first-level table containing eight entries—one for each of the function codes. When the processor accesses memory, the function code indexes into the function code table and selects a pointer to the next-level table. The table walk then continues exactly as for the previous example, using the hierarchical table structure. As you can see, this PMMU mode enables the

FIGURE 7.70 Example of address translation

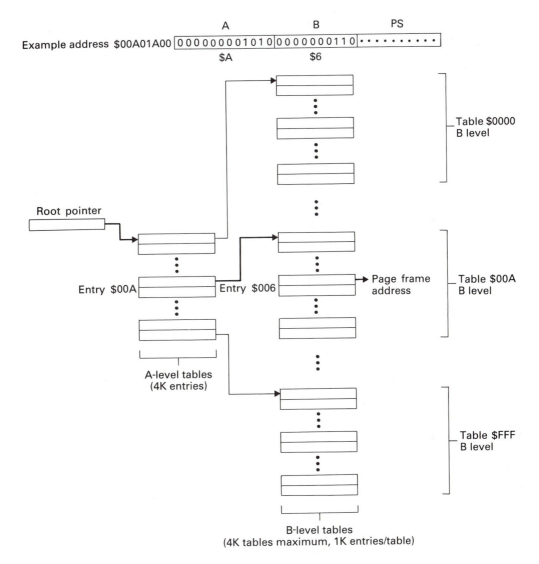

operating systems designer to separate entirely supervisor/user and program/data memory spaces.

The third example, in figure 7.72 is an extension of both figures 7.69 and 7.70. Two CPU root points are used (but not at the same time). One points to the first-level table for task A and the other to the first-level table for task B. In this case, some of the pages are shared. The operating system can change user tasks by simply reloading the CPU root pointer.

FIGURE 7.71 Example of address translation using function codes

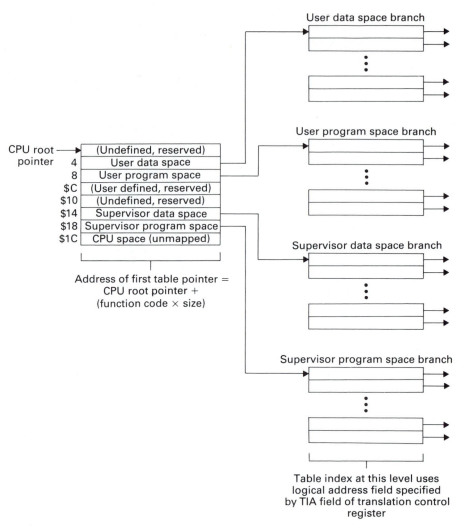

User data space branch

User program space branch

CPU root → (Undefined, reserved)
pointer 4 | User data space
         8 | User program space
       $C | (User defined, reserved)
      $10 | (Undefined, reserved)
      $14 | Supervisor data space
      $18 | Supervisor program space
      $1C | CPU space (unmapped)

Address of first table pointer =
CPU root pointer +
(function code × size)

Supervisor data space branch

Supervisor program space branch

Table index at this level uses
logical address field specified
by TIA field of translation control
register

7.5 CACHE MEMORIES

Cache memory provides system designers with a way of exploiting high-speed processors without incurring the cost of large high-speed main memory systems. The
word *cache* is pronounced "cash" or "cash-ay" and is derived from the French word
meaning *hidden*. Cache memory is hidden from the programmer and appears as part
of the system's memory space. There is nothing mysterious about cache memory. It
is simply a quantity of very high speed memory that can be accessed rapidly by the
processor. The element of magic comes from the ability of systems with cache
memory to employ a tiny amount of high-speed memory (e.g., 64K bytes of cache
memory in a system with 4M bytes of DRAM) and to have the processor make over

FIGURE 7.72 Example of address translation with shared pages

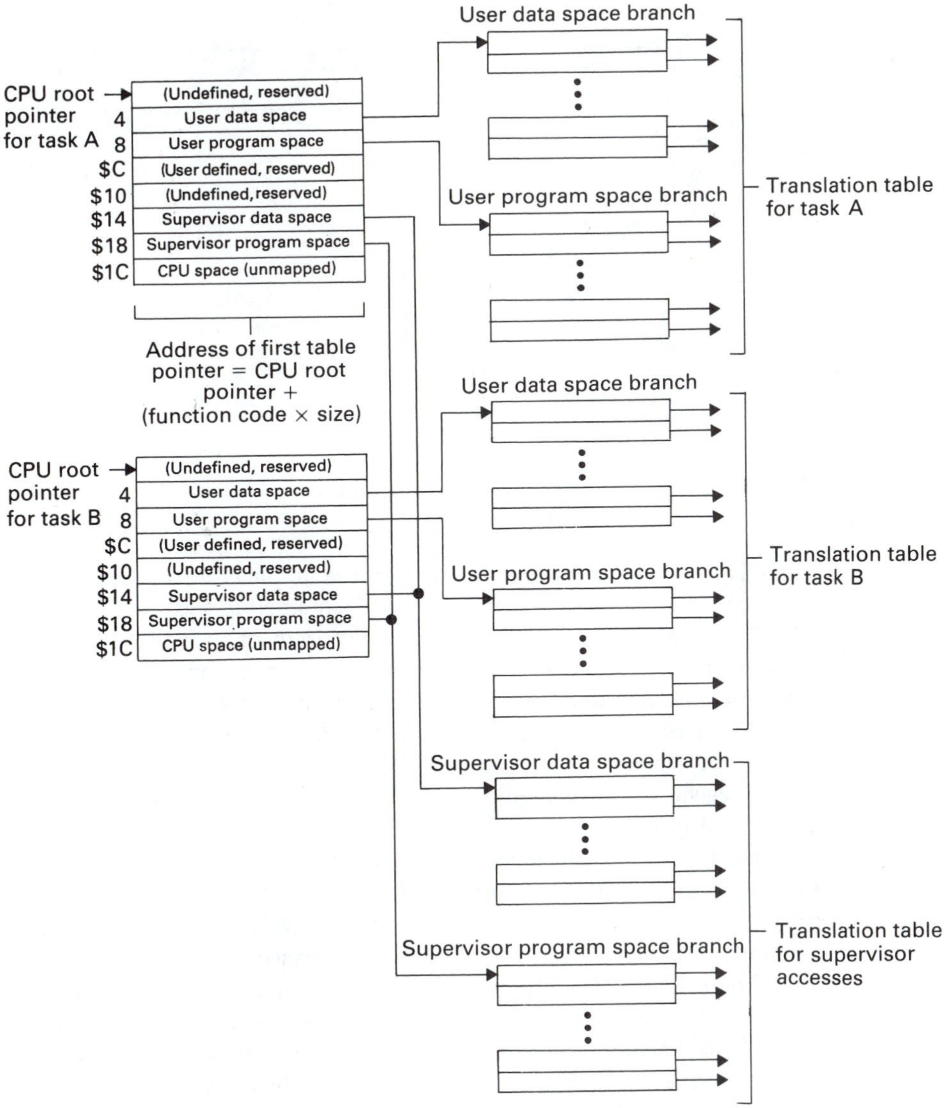

95 percent of its accesses to the cache rather than the slower DRAM. We are going to look at cache principles first and then at the way in which the 68020, the 68030, and the 68040 implement cache memory on-chip.

Cache memory locates frequently accessed information in high-speed SRAM (with access times in the region 15 ns to 45 ns) rather than in the much slower main memory. Unfortunately, the computer cannot know, a priori, what data is most likely to be accessed. Computer caches operate on a learning principle. By experience they learn what data is most frequently used and then transfer it to the cache.

The general structure of a cache memory is provided in figure 7.73. A block of cache memory sits on the processor's address and data buses in parallel with the

FIGURE 7.73   General structure of cache memory

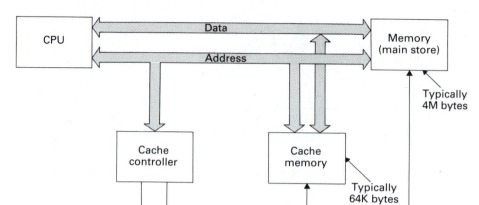

much larger main memory. Note that the implication of *parallel* in the previous sentence is that data in the cache is also maintained in the main memory.

The probability of accessing the next item of data in memory is not simply a random function. Because of the nature of programs and their attendant data structures, the data required by a processor is often highly clustered throughout memory. This aspect of memories is called the *locality of reference*; it makes the use of cache memory possible.

A cache memory requires a cache controller to determine whether or not the data currently being accessed by the CPU resides in the cache or whether it must be obtained from the main memory. When the current address is applied to the cache controller, the controller returns a signal called *hit*, which is asserted if the data is currently in the cache. Before we look at how cache memories are organized, we will demonstrate their effect on a system's performance.

The principal parameter of a cache system is its hit ratio H, which defines the ratio of hits to all memory accesses. The hit ratio is determined by statistical observations of the operation of a real system and cannot readily be calculated. Furthermore, the hit ratio is dependent on the actual nature of the programs being executed. It is perfectly possible to have some programs with very high hit ratios and others with very low hit ratios. Fortunately, the effect of locality of reference usually means that the hit ratio is very high—often in the region of 98 percent. Before calculating the effect of a cache memory on a processor's performance, we need to introduce some terms.

Access time of main store	t_m
Access time of cache memory	t_c
Hit ratio	H
Miss ratio	M
Speedup ratio	S

The speedup ratio is defined as the ratio of the memory system's access time without cache to its speed with cache. For N accesses to memory, the total access time of a memory without cache is given by Nt_m.

For N accesses to a memory with cache, the total access time is given by $N(Ht_c + Mt_m)$. We can express M in terms of H as $M = (1 - H)$ (i.e., if an access is not a hit, it must be a miss). Therefore, the total access time for a system with cache is given by $N(Ht_c + (1 - H)t_m)$.

The speedup ratio is therefore given by

$$S = \frac{Nt_m}{N(Ht_c + (1 - H)t_m)} = \frac{t_m}{(Ht_c + (1 - H)t_m)}$$

As we are not interested in the absolute speed of the main and cache memories, we can introduce a new parameter, k, that defines the ratio of the speed of cache memory to main memory. That is, $k = t_c/t_m$. Typical values for t_m and t_c might be 100 ns and 20 ns, respectively, which gives a value for k of .2.

Therefore,

$$S = \frac{t_m/t_m}{Ht_c/t_m + (1 - H)t_m/t_m} = \frac{1}{Hk + (1 - H)} = \frac{1}{1 - H(1 - k)}$$

Figure 7.74 provides a plot of S as a function of the hit ratio, H. As you might expect when $H = 0$ all accesses are made to the main memory and the speedup ratio is 1. When $H = 1$ all accesses are made to the cache and $S = 1/k$. The most important conclusion to be drawn from figure 7.73 is that the speedup ratio is a *sensitive*

FIGURE 7.74 Speedup ratio as a function of H (hit ratio)

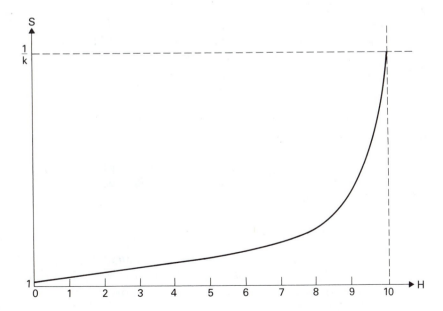

function of the hit ratio. Only when H approaches about 90 percent does the effect of the cache memory become really significant. This result is consistent with common sense. If H drops below about 90 percent, the accesses to main store take a disproportionate amount of time and the fast accesses to the cache have little effect on average system performance.

The actual speedup ratio achieved by practical microprocessors is nowhere near as optimistic as those just derived. The reason for this discrepancy is quite simple. A real microprocessor operates at a rate determined by its clock speed, the number of clock cycles per memory access, and the number of wait states introduced by the memory. These factors mean that there is little point in speeding up the cache memory beyond that needed to achieve zero wait states. Even if you use a very fast cache, you cannot reduce a memory access to less than that of a bus cycle without wait states. Consider the following example.

Microprocessor clock cycle time	20 ns
Minimum clock cycles per bus cycle	3
Memory access time	80 ns
Wait states introduced by memory	2 clock cycles
Cache memory access time	30 ns
Wait states introduced by cache	Zero

These figures tell us that an access to memory takes $(3 + 2) \times 20$ ns $= 100$ ns, whereas an access to the cache takes 3×20 ns $= 60$ ns. Note that the actual access times for the main memory and the cache do not appear in this calculation. In this case the speedup ratio is given by

$$\frac{100}{60h + 100(1 - H)} = \frac{100}{100 - 40H}$$

Assuming an average hit ratio of 95 percent, the speedup ratio is 1.61. This figure offers a modest improvement in performance but is considerably less than indicated by our original equations based only on the access time of the cache memory and the main store (i.e., 2.46).

Cache Organization

We are going to describe three ways of organizing a cache memory: direct mapped, associative mapped, and set associative mapped. Each of these has its own performance-to-cost trade-off.

Direct-Mapped Cache

The simplest way of organizing a cache memory is *direct mapping*, which employs a simple algorithm to map data line i from the main memory into data line i in the cache. For the purpose of this section we will regard the smallest unit of data held in a cache as a *line* (sometimes called a block), which is typically made up of 2 or 4 consecutive words. We employ the term *line* because it is used extensively in literature dealing with cache memories.

Figure 7.75 illustrates the structure of a highly simplified direct-mapped cache. The memory is composed of 32 words and is accessed by a 5-bit address bus. The cache memory itself holds 4 lines, and the main memory holds four *sets* each of 4 lines. For the purpose of this discussion we need consider only the *set* and *line* (as it does not matter how many words there are in a line). The address in this example has a 2-bit set field (A_3, A_4), a 2-bit line field (A_1, A_2) and a 1-bit word field (A_0). When the processor generates an address, the appropriate line in the cache is accessed. For example, if the processor generates the 5-bit address 01110, line 3 is accessed.

A glance at figure 7.75 reveals that there are four possible lines numbered 3 in

FIGURE 7.75 Structure of a direct-mapped cache system

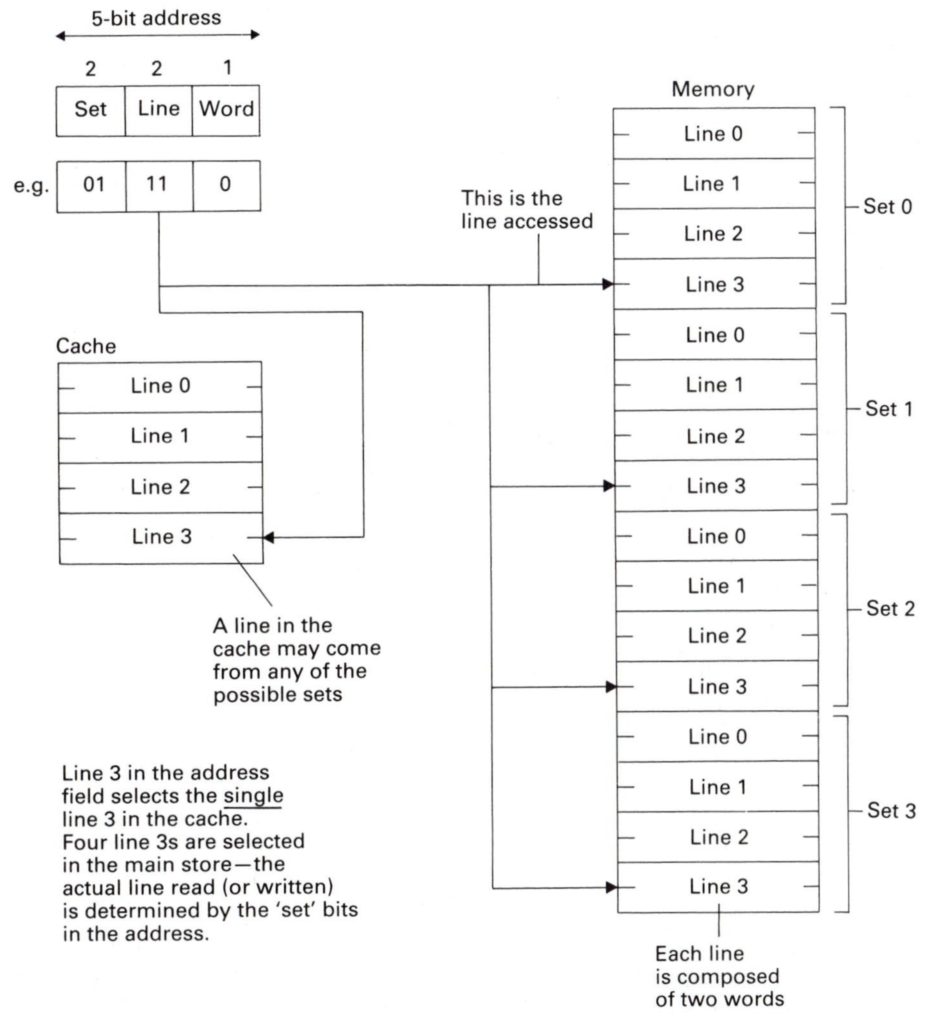

Line 3 in the address field selects the single line 3 in the cache. Four line 3s are selected in the main store—the actual line read (or written) is determined by the 'set' bits in the address.

A line in the cache may come from any of the possible sets

Each line is composed of two words

the main memory—a line 3 in set 0, a line 3 in set 1, a line 3 in set 2, and a line 3 in set 3. But, there is room only for one line 3 in the cache. In this example the processor accessed line 3 in set 1 with the address 01110. The obvious question to ask is, How does the system know whether the line 3 accessed in the cache is the line 3 from set 1 in the main memory?

Figure 7.76 demonstrates how the contention between lines is resolved by direct-mapped cache. Associated with each line in the cache memory is a *tag*, or *label*, that identifies to which set that particular line belongs. When the processor accesses line 3, the tag belonging to line 3 in the cache is sent to a comparator. At the same time the set field from the processor (i.e., higher-order address lines A_3, A_4) is also sent to the comparator. If they are the same, the line in the cache is the desired line and a hit occurs.

If they are not the same because the line in the cache is not from set 1, a miss occurs and the cache must be updated. The old line 3 in the cache is either simply discarded or rewritten back to main memory, depending on how the updating of main memory is organized.

FIGURE 7.76   Resolving contention between lines in a direct-mapped cache

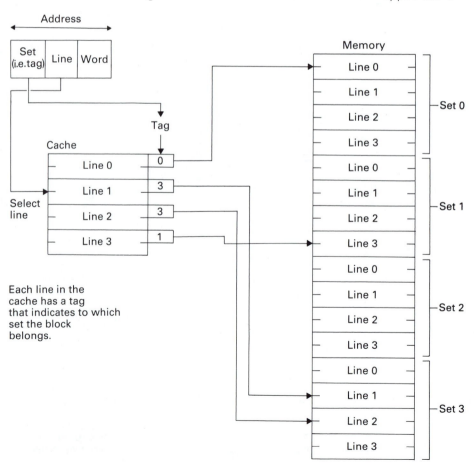

FIGURE 7.77 Simplified structure of direct-mapped cache memory system

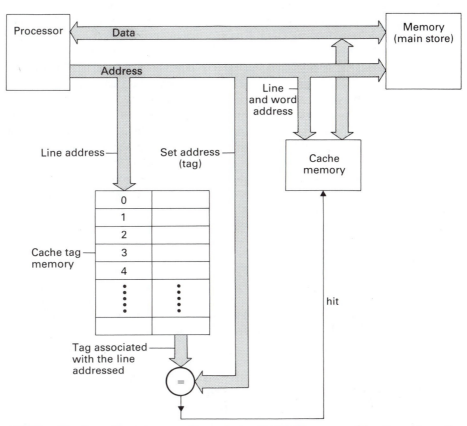

NOTE: The line address accesses the cache tag RAM. The accessed location returns the tag (i.e., set) corresponding to the accessed line. The tag is compared with the set address on the address bus. If they match, the line being accessed is in the cache.

Figure 7.77 provides a skeleton diagram of the structure of a direct-mapped cache memory system. The cache memory itself is nothing more than a block of very high speed random access read/write memory. The *cache tag RAM* is a fast combined memory and comparator circuit that receives both its address and data inputs from the processor's address bus. The cache tag RAM's address input is the line address from the processor; it is used to access a unique location (one for each of the possible lines) in the cache tag RAM. The data in the cache tag RAM at this location forms the tag associated with that location (i.e., it indicates to which set the line belongs). The cache tag RAM also has a data input, which is the tag field from the processor's address bus. If the tag field from the processor matches the contents of the tag (i.e., set) field being accessed, the cache tag RAM returns a hit signal. Otherwise, external logic must intervene to update the cache and the cache tag RAM.

The advantage of the direct-mapped cache is almost self-evident. Both the cache memory and the cache tag RAM are widely available devices that, apart from their speed, are no more complex than any other mainstream integrated circuit.

Moreover, the direct-mapped cache requires no complex line-replacement algorithm. If line x in set y is accessed and a miss takes place, then line x from set y in the main store is loaded into the frame for line x in the cache memory (a frame is a unit of memory employed to hold a page); that is, no decision has to be made concerning which line from the cache is to be rejected when a new line is to be loaded.

Another important advantage of direct-mapped cache is its inherent parallelism. Since the cache memory holding the data and the cache tag RAM are entirely independent, they can both be accessed simultaneously. Once the tag has been matched and a hit has occurred, the data from the cache will also be valid (assuming the two cache data and cache tag memories have approximately equal access times).

The disadvantage of direct-mapped cache is almost a corollary of its advantage. A cache with n lines has one restriction: At any instant it can hold only one line numbered x. What it cannot do is hold a line x from set p and a line x from set q. This restriction arises because there is one page frame in the cache for each of the possible lines. We now need to ask ourselves the question, Is this restriction important? To answer this, consider the following code:

```
REPEAT
      CALL Get_data
      CALL Compare
UNTIL match OR end_of_data
```

This innocuous fragment of code reads a string of data from a buffer and then matches it with another string until a match is found. Suppose that by bad luck the compiled version of this code is arranged so that part of the Get_data routine is in set x, line y and that part of the Compare routine is in set z, line y. Because the direct-mapped cache permits the loading of only one line y at a time, the frame corresponding to line y will have to be reloaded twice for each path through the loop. Consequently, a direct-mapped cache can have a very poor performance if the data is arranged in a certain way. However, statistical measurements on real programs indicate that the very poor worst-case behavior of direct-mapped caches has no significant impact on their average behavior.

To add insult to injury, you can imagine a situation in which a cache is almost empty (i.e., most of its page frames have not been loaded with active data) and yet two particular pages have to be swapped in and out frequently because two active lines in the main store just happen to have the same line numbers. In spite of these objections to direct-mapped cache, it is popular because of its low cost of implementation and high speed.

Associative Mapped Cache

An excellent way of organizing a cache memory that overcomes the limitations of direct-mapped cache is called an *associative mapped cache*. A simple associative mapped cache is described in figure 7.78. This cache organization places no restrictions on what data it can contain.

FIGURE 7.78 The associative mapped cache

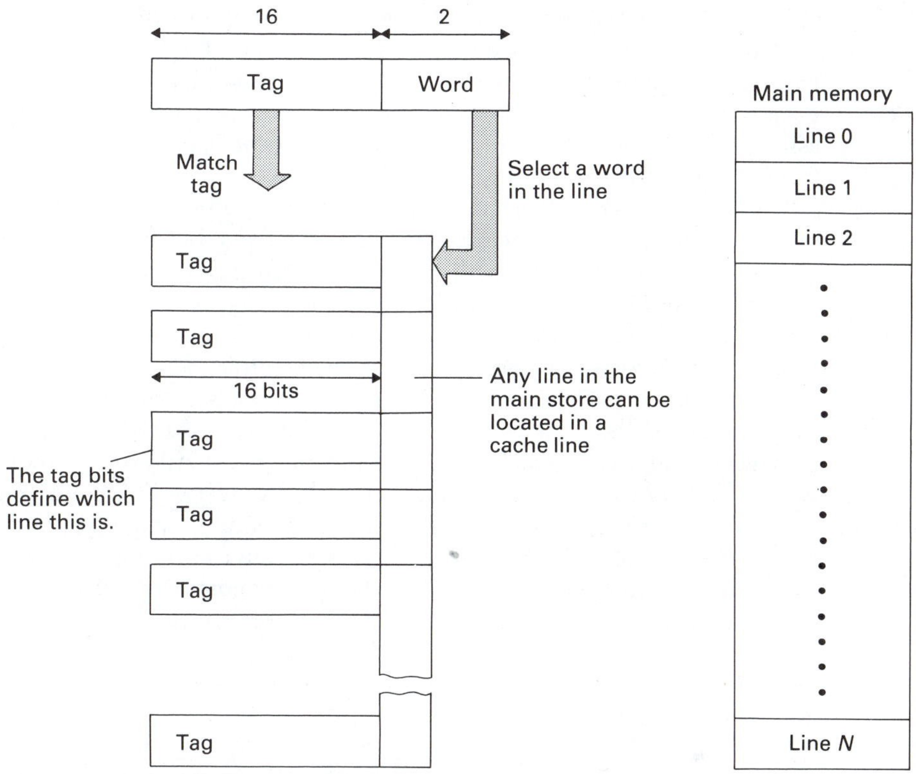

NOTE: The tag field is matched with all tags in the cache *simultaneously*. If there is a match, the corresponding line is in the cache and there is a hit. If there is no match, the corresponding line is in main store.

An address from the processor is divided into two fields: the tag and the word (figure 7.78). Like the direct-mapped cache, the smallest unit of data transferred into and out of the cache is the line. Unlike the direct mapped cache, the associative cache displays no relationship between the number of lines in the cache and the number of lines in the main memory. For example, consider a system with 1M bytes of main store and 64K bytes of associatively mapped cache. If a line comprises four 32-bit words (i.e., 16 bytes), the main memory is composed of $2^{20}/16 = 64$K lines and the cache is composed of $2^{16}/16 = 4,096$ lines. An associative cache permits any line in the main store to be loaded into one of its page frames. In this example, line *i* in the associative cache can be loaded with any one of the 64K possible lines in the main store. Therefore, line *i* requires a 16-bit tag to label it uniquely as being associated with line *i* from the main store. Note that, since the lines in the cache are not ordered, the tags are not ordered and cannot be stored in a simple look-up table, as for the direct-mapped cache.

When the processor generates an address, the word bits select a word location in both the main memory and the cache. The high-order address bits from the processor comprise the current line's tag and are sent to the cache. The cache memory in figure 7.78 can store any of the 64K memory lines in one of its frames, as it requires a 16-bit tag to identify each of its frames. If one of the 16-bit tags in the cache matches the tag from the processor, the corresponding line is in the cache. Otherwise a miss occurs and the cache must be updated. The cache tag memory must use associative memory because all stored tags are *simultaneously* matched with the tag from the processor.

Associative cache systems require a special type of memory called *associative memory*. An associative memory has an *n*-bit input but not necessarily 2^n unique internal locations. The *n*-bit address input is a tag that is matched with a tag field in each of its locations *simultaneously*. If the input tag matches a stored tag, the data associated with that location is output. Otherwise the associative memory produces a miss output. Associative memory is also called *content-addressable memory* (CAM).

Associative cache memories are efficient because they place no restriction on the location of the data they hold. Unfortunately, associative memories are very expensive, and large associative memories are not available. Moreover, once the cache is full, a new line can be brought in only by overwriting an existing line. As in the case of virtual memories, cache memories must use a page (i.e., line) replacement policy. Because of the high cost of fully associative memories, computer designers frequently employ an arrangement that is a compromise between direct-mapped caches and fully associative caches called set associative caches.

Set Associative Cache

A *set associative cache* memory is arranged as a direct-mapped cache, but each line in the cache is replicated. The simplest set associative cache is called a *two-way set associative cache* and duplicates each line in the cache. For example, there are two line 5s in the cache, so it is possible to hold one line 5 from set *x* and one line 5 from set *y*. When the processor accesses memory, the appropriate pair of lines in the cache are accessed. Since two lines respond to the access, a simple associative match between the set address from the processor and the two stored tags can be used to determine which (if either) of the lines in cache are to supply the data. Typical set associative caches employ four direct-mapped caches in parallel. See figure 7.79.

Apart from choosing the structure of a cache system and the line replacement policy (if it is an associative cache), the designer has to consider how write cycles are to be treated. Should write accesses be made only to the cache and then the main store updated when the line is written back? Should the main memory also be updated each time a word in the cache is modified? The latter policy is called a *write through* policy and is relatively efficient because the cache can be written to rapidly and the main memory can be updated over a longer span of time.

Another of the concepts often found in texts on cache systems is *cache coherency*. As we know, data in the cache also lives in the main memory. When the processor modifies data, it must modify both the copy in the cache and the copy in the main memory (although not necessarily at the same time). There are circum-

FIGURE 7.79 The set associative cache

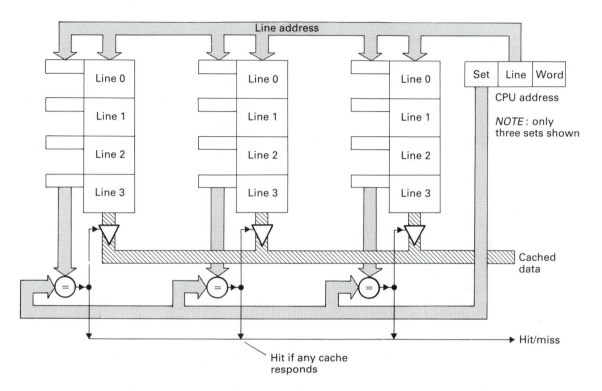

stances when the existence of two copies (which can differ) of the same item of data causes problems. For example, an I/O controller using DMA might attempt to move an old line of data from the main store to disk without knowing that the processor has just updated the copy of the data in the cache but has not yet updated the copy in the main memory. Cache coherency is also known as data consistency.

Cache Memory and the 68000 Family

From what we have already said about cache memory, it should be obvious that cache memory systems are both expensive and complex. Consequently, until recently they have not generally been associated with low-cost microprocessor systems. In any case, a 68000 running at 8 MHz can use low-cost DRAM with no wait states. Modern microprocessors such as the 68020 and 68030 operate at speeds sufficiently great to make cache memory worthwhile. On the other hand, the addition of a cache memory subsystem can greatly increase the cost of a microcomputer and can push its price into the minicomputer range.

Designers of the 68020 and 68030 have placed a rather modest quantity of cache on-chip and have therefore eliminated all the design and implementation overheads associated with cache memories. On the other hand, the on-chip cache gives these processors a boost in performance at no cost to the user. Modern technology has enabled the 68040 to implement an impressive on-chip cache system.

68020's Cache

The 68020 implements a 64-longword direct-mapped *instruction* cache. That is, instructions are cached but not data. By not caching data, the design of the cache is greatly simplified, since the problems of cache coherency and memory updating are eliminated. The 64 longwords (256 bytes) permit small loops to be run entirely from cache and thereby remove the need for instruction fetches from main store. As you can imagine, the 68020's cache has relatively little effect on the execution of pure in-line code with no loops. Note that the 68020's *internal* cache speeds up the processor more than an *external* cache, because the 68020 does not need to perform on external memory access if an instruction is cached.

The 68020's cache controller caches instructions automatically as they are prefetched. Instructions are stored in the 68020's direct-mapped cache, whose organization is described in figure 7.80. The logical address from the 68020 consists of a 32-bit address, A_{00} to A_{31}, and a 3-bit extension made up of the function code, FC0 to FC2. Address bits A_{02} to A_{07} select one of $2^6 = 64$ lines in the cache memory. Address bit A_{01} selects the lower or upper word of each line in the cache.

The 64-entry cache tag memory contains sixty-four 25-bit tags (i.e., 24 bits for address bits $A_{08}-A_{31}$ and FC2). FC2 is included to distinguish between a supervisor

FIGURE 7.80 Organization of the 68020's cache

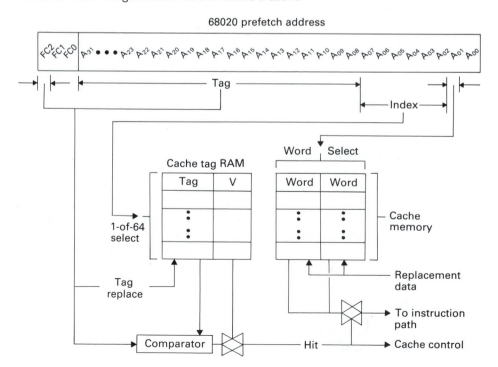

68020 prefetch address

space instruction and a user space instruction. When the 68020 prefetches an instruction, a line in the cache is accessed and the corresponding tag is read from the tag store. A hit is declared and the instruction is read from the cache if two conditions are met: (1) The tag matches both the high-order bits of the address bus and FC2 of the current function code, and (2) the V-bit is set (see later). Otherwise, the instruction is read from the external memory and the on-board cache is updated. Since the 68020 always prefetches instructions that are aligned at a longword boundary, both words of a line in the cache are always updated, regardless of which word caused the miss.

The tag memory contains just one control bit, a valid or V-bit. The function of the V-bit is to *validate* entries in the cache. During a processor reset (e.g., after power-up), all 64 V-bits are cleared to indicate that the cache is empty.

The 68020's instruction cache is almost, but not quite, invisible to the (systems) programmer. The 32-bit cache control register, CACR, in the 68020's supervisor space can be accessed by the privileged instructions MOVEC CACR,Rn and MOVEC Rn,CACR to control the operation of the cache. Although CACR is a 32-bit register, only 4 bits are defined (See figure 7.81).

During a reset the cache is cleared by resetting all V-bits to invalidate the entries in the cache, and the cache enable and cache freeze bits of the CACR are also cleared. The cache enable bit determines whether or not the cache is to be used. If CACR(E) = 0, the cache is disabled—it does not exist. You might wonder why it may be necessary to disable the cache. Suppose you are debugging the 68020 by observing the flow of information on the data and address buses. The internal cache suppresses the fetching of instructions that are already cached and therefore might make it difficult to debug the 68020.

Since the cache enable bit is cleared on reset, the supervisor program must explicitly set it to enable the cache. We can do this by

```
MOVE.L      #$01,D0    Set bit zero (i.e., the E bit)
MOVEC       D0,CACR    Load cache control register.
```

John Hodson (Motorola's former Northern Europe training manager) once described one of his 68020 courses. John was discussing the 68020's instruction cache and said, "Of course you can't benefit from the cache until you've first enabled it." An engineer in the audience turned an unattractive shade of grey, stood up and asked John if he could leave to make an important phone call. The engineer had attended the

FIGURE 7.81 Cache control register

31	• • •		8	7	6	5	4	3	2	1	0
0		0	0	0	0	0	0	C	CE	F	E

C = clear cache
CE = clear entry
F   = freeze cache
E   = enable cache

course because his company was using the 68020 and could not understand why it did not achieve the performance suggested by Motorola. They had simply forgotten to enable the cache.

By the way, if you disable the cache and then reenable it, the previously valid entries remain valid and can be used again.

The *cache freeze bit*, when set to 1, suspends the instruction cache's update mechanism. If a miss occurs, the line in the instruction cache is not updated. When F is cleared, the cache updates instructions normally. Do not confuse the E- and F-bits. The E-bit simply switches off the cache and prevents the 68020 from using it. The F-bit permits the cache to be accessed but not to be updated. Since the 68020's cache is rather small, it holds relatively few instructions and is refilled every time a section of in-line code is executed. Suppose you have a short task switcher that must be fast. You can unfreeze the cache on entry to the code and then freeze it again at the end of the code block. In this way, the cache is not updated and the task switcher's code will be cached next time it is called. Consider the following example:

```
SWITCH    MOVE.L    #%0001,D0         Enable and unfreeze cache
          MOVEC     D0,CACR           Setup cache control register
          :                 Code of task switcher
          MOVE.L    #%0011,D0         Enable and freeze cache
          MOVEC     D0,CACR           Setup cache control register
          RTS                         Return
```

The *clear cache bit*, CACR(3), is used to clear, or *flush*, all entries in the cache. Setting the C-bit of the CACR has the effect of clearing all V-bits. If the operating system performs a context switch or a new program is loaded into main store, it is necessary to clear (flush) the cache to remove old (i.e., *stale*) instructions.

The *clear entry bit*, CACR(2), can be used to clear a single entry (i.e., line) in the cache (as opposed to the C-bit, which clears all entries). We tell the 68020 which location is to be cleared by means of another supervisor space register, the CAR (cache address register). (See figure 7.82).

The index field of the CAAR determines which line of the cache is to be cleared when the CE-bit is set (the cache function address is not used by the 68020). The CE-bit is automatically reset to zero after it has been loaded with 1.

As you can see, it is easy to use the 68020's cache memory—you enable it and then forget about it. The use of cache memory has big implications, and problems caused by cache memories are some of the hardest to track down. The real danger lies in modifying data in main memory without changing the data in the cache. Suppose you load a new program from disk into main memory and begin executing it. The cache will contain old instructions even though corresponding information in main memory has now been overwritten by new instructions. You can avoid this

FIGURE 7.82 Cache address register

31	···	8 7	···	2 1	0
Cache function address			Index		

problem by *flushing* the cache each time you load new code. If you forget to flush the cache, you will spend weeks looking at the code byte by byte and wondering why the 68020 does not do what it was told. Incidentally, there is no explicit way in which you can examine the contents of the 68020's (or 68030's) cache.

The 68020 has one hardware cache control pin, CDIS* (cache disable), that can be asserted by external hardware to disable the cache. Asserting CDIS* disables the cache but does not flush it. CDIS* is used largely by external test equipment.

68030's Cache

The 68030 doubles the size of the 68020's cache by supporting a 256-byte *data cache* in addition to the 68020's 256-byte instruction cache. It is important to note that the 68030's two caches are *logical* caches, since they cache logical addresses from the 68030. This statement is necessary because the 68030 contains an on-chip memory management unit (as discussed in section 7.4), and therefore the address at its address pins is a *physical* address rather than a logical address. The point we must appreciate here is that if the mapping performed by the 68030's memory management unit is changed, both the 68030's caches must be flushed.

Since the 68030's instruction and data caches are entirely independent, they can operate autonomously. That is, the 68030 can perform an instruction fetch and a data fetch to its internal caches, perform a data fetch to external memory, and execute an instruction—all simultaneously. This degree of parallelism greatly enhances the performance of the 68030. Before looking at the details of the 68030's cache, we will briefly comment on the new pins (i.e., hardware interface) used to support the cache.

Since the 68000 family uses memory-mapped I/O, imagine the effect of caching data on a peripheral. The first time the CPU reads a peripheral, the data will be read correctly and then cached. The next time it reads the peripheral, it will read the old data from the cache even though the peripheral may have new data for the CPU. The 68030 has a cache disable input, CDIS*, that can be asserted to disable both caches. A separate cache inhibit input, CIIN*, can be asserted to disable the cache during certain memory accesses (e.g., I/O accesses to peripherals). CIIN* is ignored in write cycles.

An output signal, cache inhibit out, or CIOUT*, indicates to any external cache that the bus cycle should be ignored. Two signals are provided to permit an entire line of the cache to be filled in a burst of data transfers. Cache burst request, CBREQ*, is an output that requests a line of data, and cache burst acknowledge, CBACK*, is an input informing the cache that one more longword can be supplied.

You might expect the 68030's instruction cache to be organized in exactly the same way as the 68020's instruction cache. Figure 7.83 illustrates the organization of the 68030's cache. Note that it is organized as 16 lines of 4 longwords rather than 64 lines of 1 longword. The capacity of the 68030's instruction cache is the same as the 68020's, but its depth has been decreased and its width (i.e., line size) has been increased. If you look more closely at figure 7.83 (and compare it with figure 7.80), you will see another difference between the 68020's and 68030's instruction cache. The 68030's tag store has 4 V-bits per entry. Consequently, each longword in a line can be individually validated.

FIGURE 7.83   Organization of the 68030's instruction cache

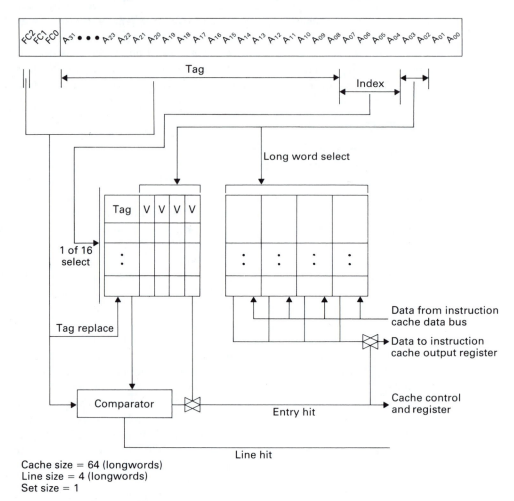

Cache size = 64 (longwords)
Line size = 4 (longwords)
Set size = 1

These modifications permit the 68030 to implement a burst mode cache refill. When a miss occurs, the appropriate longword in the current line is replaced. However, if the 68030's cache is programmed to operate in a burst mode (using CBREQ* and CBACK*), the entire line is refreshed and 4 longwords are transferred in a single burst. Of course, burst mode operation is possible only if the memory interface supports it.

The 68030's data cache is organized almost exactly like the instruction cache of figure 7.83. The only real difference between the instruction and data caches is that the instruction cache tag includes only FC2 (which indicates user/supervisor access), whereas the data cache tag stores the whole function code, FC0 to FC2. However, the operational details of the data cache are much more complex than those of the instruction cache, since a data cache can be written to by the 68030 as well as read from. The data cache stores any data references to address space (except to CPU space).

When a data access is made, the stored tag bits (A_{08} to A_{31} and FC0 to FC2) are indexed by A_{04} to A_{07}. Address bits A_{02} and A_{03} select the actual longword of the current line. If the tag field of the current address and function code match the stored tag bits, a hit occurs. Otherwise a miss occurs. As you know, the 68030 supports misaligned data operands, and a longword can be accessed at any boundary. In plain English, the 68030 can access a longword that has 1 byte cached and 3 bytes in main store only. When this happens, the 68030 treats the misaligned access as two accesses: one that results in a hit and one that results in a miss. These are then treated separately. I sometimes wonder whether the engineers who implemented the 68030's cache would like to strangle the engineers who implemented the 68020's dynamic bus sizing mechanism.

Read and write data accesses are treated differently by the 68030. Data reads are treated exactly like instruction reads. If a miss occurs, the operand is read from main store and the cache is updated. When data is written to memory, it is written to the cache and also written in parallel to the external main store. This approach means that the memory and cache remain always in step and that a line of the cache can be updated or flushed without having to write it back to memory. Such a cache is called a *write-through* cache.

The fine details of the 68030's data cache are rather complex, we will provide only an overview here by looking at the 68030's cache control register (figure 7.84). Although the 68030's CACR looks much more complex than that of the 68020's CACR, it is not. All the 68020's instruction cache control bits have been duplicated to refer to the instruction or the data cache explicitly. That is, you can enable or freeze either the instruction cache or the data cache independently.

The real additions to the 68030 are a write-allocate bit and data/instruction cache burst enable bits. The burst enable bits, DBE and IBE, are cleared after a reset and can be set to permit the 68030 to fill a line in its cache using the burst mode described before.

The write-allocate bit can be set to select the 68030's data cache write-allocate mode. The WA bit is cleared after a reset and is ignored if the data cache is frozen. When WA = 0, a write-around policy is implemented and write cycles resulting in a miss do not alter the data cache's contents. That is, the main store is updated but not the cache, because the cache is updated only during a write hit.

FIGURE 7.84 The 68030's cache control register, CACR

31 ···· 14	13	12	11	10	9	8	7	6	5	4	3	2	1	0
0 ··· 0	WA	DBE	CD	CED	FD	ED	0	0	0	IBE	CI	CEI	FI	EI

WA = write allocate	IBE = instruction burst enable
DBE = data burst enable	CI = clear instruction cache
CD = clear data cache	CEI = clear instruction cache entry
CED = clear entry in data cache	FI = freeze instruction cache
FD = freeze data cache	EI = enable instruction cache
ED = enable data cache	

When WA = 1, the 68030 operates in its write-allocation mode and the processor always updates the data cache on (cachable) write cycles but validates an updated entry only during hits or when the entry is a longword aligned at a longword boundary.

Once again, it is necessary to stress that the 68030's cache memory is invisible to the user programmer. Care has to be taken to inhibit the 68030's data cache by asserting CIIN* when a memory-mapped I/O port is accessed. If the supervisor mode implements task switching (using the on-chip MMU), the caches must be flushed before the new task runs, since the relationship between the cache and main memory will have altered.

68040's Cache

The 68040 represents a giant leap forward in cache technology over the 68020 and 68030. Like the 68030, the 68040 has independent and autonomous instruction and data caches. However, these caches are both four-way set associative caches with 64 sets of four 16-byte lines (i.e., 4K bytes each) and are located after the on-chip MMU to provide a physical cache. That is, the 68030 caches data before its logical address is passed to the MMU, whereas the 68040 caches data after the MMU has calculated the physical address. Figure 7.85 describes the structure of the 68040's cache.

The 68040's cache operates in conjunction with the on-chip MMU. Each page (i.e., unit of memory) is assigned a 2-bit code that determines how data on that page is to be cached. For example; pages containing I/O space can be marked as noncachable, or pages can be marked as *write-through* or *copy-back* pages.

Each entry in the instruction cache contains a V-bit, whereas each entry in the data cache contains both a V-bit and D-bit (dirty bit). Instead of relying on the write-back algorithm, the 68040 marks a line in the data cache as *dirty* if it has been written to. Lines with their D-bits set must be written back to memory. Write back occurs when the line is accessed and a miss results. The 68040 permits an explicit write back by means of the new CPUSH instruction, which pushes the selected dirty data cache lines to memory and then invalidates the lines in cache. All data transfers between the 68040's cache and external memory operate in a burst mode and transfer an entire line.

One of the great problems caused by cache memory in sophisticated microprocessor systems is called *cache coherency*. The same problem occurs when an author sends a chapter of a manuscript to a publisher, who edits it and makes changes. While this process is going on, the author looks through the chapter and decides to improve it. Before long, the author's copy and the publisher's copy are no longer the same. This situation causes endless confusion. In a similar way, a multiprocessor system (especially those with DMA) means that more than one processor can modify the contents of memory. For example, a disk drive might transfer a file to a block of RAM, or one processor might leave a message for another processor in a block of RAM. Noncached systems present no problems under these circumstances. However, a cached system must ensure that when the memory is updated, the cached copy of the data is also updated and not left "stale".

FIGURE 7.85 Internal organization of the 68040's caches

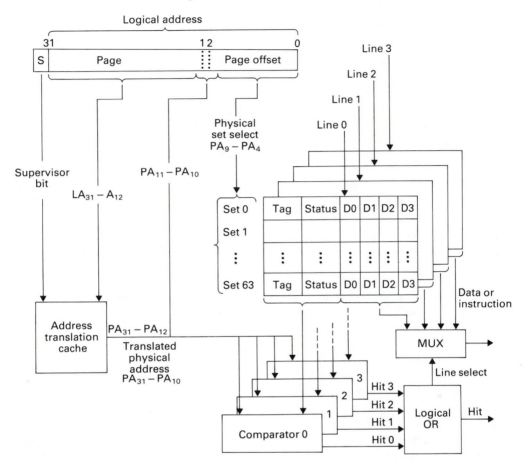

The 68040's cache solves the problem of cache coherency by means of a remarkable technique called *bus snooping*. Instead of only actively accessing the system bus like other members of the 68000 family, the 68040 also monitors the bus in a passive fashion. If an alternative bus master accesses the bus and writes data (when the 68040's snooping is enabled), the 68040 may then either invalidate the same line in its own cache or it may update its cache (depending on its programming).

Consider a read cycle by an alternate bus master. Suppose the alternate master performs a read cycle and accesses a location that is cached by the 68040 and that the location has its dirty bit set. In this case the alternate master will be in danger of reading stale data from memory. The 68040 can be programmed to prevent the memory from responding to the read access and to supply the correct data itself.

It should now be clear why the 68040 has a physical cache rather than a logical cache. If it had a logical cache, it would be able to deal only with addresses generated by the processor. By using a physical cache, the 68040 caches addresses that appear on the system address bus and are meaningful to the system memory. Without physical address caching, the 68040 would not be able to perform bus snooping and ensure cache coherency.

7.6 COPROCESSOR

The designer of any microprocessor would like to extend its instruction set almost infintely but is limited by the quantity of silicon available (not to mention the problems of complexity and testability). Consequently, a real microprocessor represents a compromise between what is desirable and what is acceptable to the majority of the chip's users. Having said this, there are many applications for which a given microprocessor lacks sufficient power. For example, even the powerful 68020 is not optimized for applications that require a large volume of scientific (i.e., floating-point) calculations. We are now going to look at one way in which the power of an existing microprocessor can be considerably enhanced by means of an external coprocessor.

Assume that we have a 68000 microprocessor and that it is necessary to increase its processing power. When we talk about increasing a microprocessor's processing power, we don't mean the addition of a few instructions, but the inclusion of a radically new facility. For example, we might require the processor to tackle floating-point arithmetic, or to handle high-speed graphics, or to perform memory management. Since it is not always practical to modify the 68000 die itself to include these new facilities, the obvious solution is to resort to parallel processing, in which an auxiliary processor takes on the burden of the new tasks.

In a general-purpose parallel processor, two or more processors operate concurrently. Such a system can be a very complex arrangement, since it requires communication paths between the various processors and special software to divide the task between them. A practical multiprocessing system based on a member of the 68000 family should be as simple as possible and require a minimum overhead in terms of both hardware and software.

An ideal coprocessor is designed to appear to the programmer as an extension of the CPU itself. For example, the 68882 floating-point coprocessor, FPC, can be employed in a 68020-based system to give the programmer an extended 68020 instruction set that is rich in floating-point operations. As far as the programmer is concerned, the architecture of the 68020 has just been expanded. A coprocessor like the FPC not only enhances the 68020's *instruction set* but also other components of its architecture. For example, the FPC "adds" eight 80-bit floating-point registers to the 68020's existing complement of registers. Moreover, the 68020's extended instruction set can cope with branches on FPC conditions and even handle exceptions initiated by the FPC.

We can choose several ways of arranging the coprocessor so that it can work alongside a microprocessor. One technique is to provide the coprocessor with an instruction interpreter and program counter. Each instruction fetched from memory is examined by both the microprocessor and the coprocessor. If it is a microprocessor instruction, the microprocessor executes it; otherwise the coprocessor executes it. As you might imagine, this solution is feasible but by no means easy, since it is difficult to keep the microprocessor and coprocessor in step. Another technique is to equip the microprocessor with a special bus to enable it to communicate with an external coprocessor. Whenever the microprocessor encounters an operation that requires the intervention of the coprocessor, the special bus provides a dedicated high-speed communication between the processor and its coprocessor. Once again, this solution is not simple.

The designers of 68000-family coprocessors decided to implement coprocessors that could work with *existing* and *future* generations of microprocessors with minimal hardware and software overhead. The actual approach adopted by 68000-family coprocessors is to tightly couple the coprocessor to the host microprocessor and to treat the coprocessor as a *memory-mapped peripheral* lying in *CPU address space*. In effect, the microprocessor fetches instructions from memory, and, if an instruction is a coprocessor instruction, the microprocessor passes it to the coprocessor by means of the microprocessor's asynchronous data transfer bus. By adopting this approach, the coprocessor does not have to fetch or interpret instructions itself.

The 68000-family coprocessors are of the *non-DMA* type because they never act as bus masters (greatly simplifying the design of the coprocessor interface). If a coprocessor requires data from memory, the host processor must fetch it. A corollary of this is that the coprocessor does not have to deal with, for example, bus errors, since all memory accesses are performed by 68000-series processor.

The 68000-series coprocessors are designed to work efficiently with the 68020, 68030, or 68040 because the microprocessor-coprocessor communication protocol is built into the *firmware* of the microprocessor itself. It is, in fact, perfectly possible to employ a coprocessor with a 68000, but the user must then emulate the 68020's coprocessor interface in software. In what follows, we will regard the 68020 as the host processor when describing coprocessors.

So, how can new coprocessor instructions be mapped onto the 68000's existing instruction set? In order to provide new instructions that can be interpreted by a coprocessor, a 68020 must allocate suitable op-code bit patterns. That is, a certain bit pattern must be interpreted by the processor as a coprocessor instruction. The 68020 uses its F-line op-codes to communicate with its coprocessors. You will remember from chapter 6 that the F-line op-codes are also *software exceptions*, or *traps*. We will soon see that the 68020 first treats an F-line bit pattern as a coprocessor instruction and then, if a coprocessor does not respond, treats it as a normal F-line exception.

The coprocessor is not located in conventional memory space but is memory-mapped in *CPU address space*, for which the function code on FC2,FC1,FC0 = 1,1,1. Since the coprocessor lies within CPU address space, it does not compete for program/data space and it is easy to detect accesses to the coprocessor (because FC2,FC1,FC0 = 1,1,1). Although it is theoretically possible to locate a coprocessor anywhere within the 68020's 2^{32}-byte address space, each coprocessor is restricted to a specific slice of CPU memory space. The 68000 family supports up to eight coprocessors in the region $20000 to $2E01F. Each coprocessor is allocated a 32-byte slice of memory space. The address of a coprocessor within CPU space is illustrated in figure 7.86.

FIGURE 7.86 Address of a coprocessor within CPU space

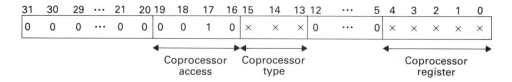

When the 68020 communicates with a coprocessor, address bits A_{31} to A_{20} are set to 0 (along with bits A_{12} to A_{05}) and the coprocessor identification code 0010 is placed on address bits A_{19} to A_{16}. This code indicates to the system hardware that a coprocessor access is taking place. Address bits A_{15} to A_{13} comprise the Cp-ID field and define one of eight possible coprocessor types from 000 to 111 (for example, the code $A_{15}, A_{14}, A_{13} = 0,0,1$ indicates a 68882 IEEE floating-point coprocessor). Currently, only two coprocessor types are defined (one is the FPC and the other is the MMU with a Cp-ID of 0,0,0). Finally, the least significant five address bits A_{04} to A_{00} can be used to access memory-mapped registers within the selected coprocessor. Figure 7.87 provides a memory map of the coprocessor-CPU memory space. The way in which a coprocessor is memory-mapped into a specific region of CPU address space is of interest to designers who have to interface coprocessors to microprocessors. It is of no interest to programmers, because the coprocessor appears as an extension of the microprocessor's instruction set. All that programmers need be aware of are the new instructions and registers provided by the coprocessor.

Coprocessor Interface

A 68020-series coprocessor employs an entirely conventional 68020 asynchronous bus interface, and absolutely no new signals whatsoever are required. Indeed, the coprocessor looks exactly like a typical 68020-series peripheral with a 4-bit register-select input (A_{01}–A_{04}), a 32-bit data bus, address and data strobe inputs, and DSACK0* and DSACK1* data transfer acknowledge strobe outputs. The 68020 CPU and the 68882 FPC may even employ entirely different clock frequencies without any problem.

The 68020-series coprocessors are very versatile and can be interfaced to 8-, 16-, or 32-bit data buses. The coprocessor uses its A_{00} and SIZE* input pins to configure it for 8-bit, 16-bit, or 32-bit data buses. Figure 7.88 demonstrates the interface between a 68020 and a 68882 FPC using the 32-bit data bus. Although the 68882 has a versatile interface and can be connected to a 32-bit, a 16-bit, or even an 8-bit data bus, it is reasonable to employ a 32-bit connection, as figure 7.88 illustrates, since the coprocessor is invariably used to improve the operation of a high-performance system.

Figure 7.89 demonstrates the FPC's connection to both a 68020 and a 68000 using a 16-bit data bus, and figure 7.90 demonstrates how you would connect the

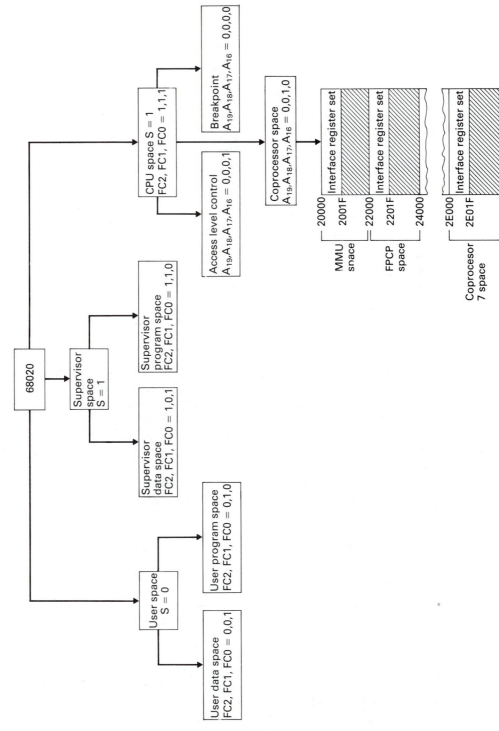

FIGURE 7.87 The coprocessor's memory space

FIGURE 7.88 Interfacing the 68882 coprocessor to a 32-bit data bus

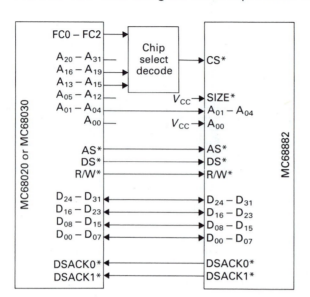

FPC to a 68020 or 68008 by means of an 8-bit bus. These diagrams are included to demonstrate both the versatility of the FPC's interface and the fact that it interfaces to the 68000 in almost exactly the same way that it interfaces to the 68020.

The 68882 FPC has a pin labeled SENSE* that is simply connected to the silicon die's ground. You can use this pin to detect the presence of a coprocessor in systems that may or may not have one. Suppose you sell two versions of a system: One is a low-cost model without a coprocessor that emulates coprocessor functions in software, and the other is a high-performance system with a coprocessor. You don't want to go to the trouble of designing two systems, so you fit a coprocessor socket with a 10 kΩ resistor between the SENSE* pin and V_{cc}. During the system's initialization routine you sample the state of the SENSE* line. If it is high, a coprocessor is absent and software emulation is necessary. If it is low, a coprocessor is present.

Coprocessor Instruction Set

All coprocessor instructions have an F-line format that *must* begin with the bit-pattern 1, 1, 1, 1. Coprocessor instructions must be at least one word long and multiword instructions are provided. A generic coprocessor instruction has the format illustrated in figure 7.91.

When the processor reads a coprocessor instruction, it uses the *Cp-ID field* to determine the particular coprocessor and the *instruction type field* to determine the class of the instruction. If the coprocessor ID field contains all zeros and the type field is nonzero, the processor treats the instruction as an F-line exception. The field labeled *type-dependent* is determined by the nature of the actual instruction, and other words may follow the first if the instruction requires it.

FIGURE 7.89 Interfacing the 68882 coprocessor to a 16-bit data bus

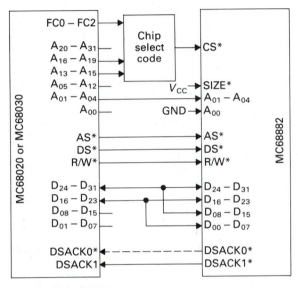

(a) Interface to 68020

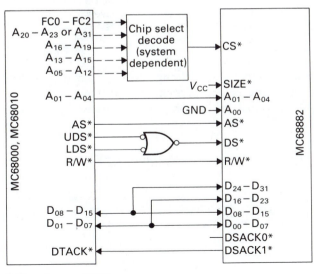

(b) Interface to 68000

FIGURE 7.90 Interfacing the 68882 coprocessor to an 8-bit data bus

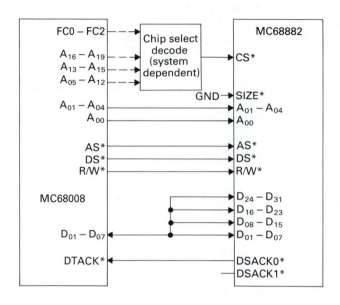

The 3-bit *type* field defines one of eight possible instruction classes.

TYPE BITS			MNEMONIC	MEANING
08	07	06		
0	0	0	cpGEN	General instruction
0	0	1	cpDBcc, cpScc, cpTRAPcc	DBcc, set, and TRAP on condition
0	1	0	cpBcc.W	Branch on condition cc
0	1	1	cpBcc.L	Branch on condition cc
1	0	0	cpSAVE	Save context
1	0	1	cpRESTORE	Restore context
1	1	0	Not defined	
1	1	1	Not defined	

Although these mnemonics look strange, they are really *generic* mnemonics and can be applied to any type of coprocessor, irrespective of its actual function. For example, when writing actual instructions, cp is replaced by the appropriate mne-

FIGURE 7.91 Generic coprocessor instruction format

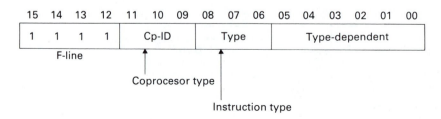

monic for the coprocessor, and GEN is replaced by the actual general instruction mnemonic. Thus, an MC68882 FPC would represent the instruction to calculate a tangent as FTAN (*F* indicates floating-point coprocessor and TAN is the general instruction to calculate a tangent). Similarly, an FPC cpTRAPcc instruction might be written as FTRAPEQ—trap on zero. Incidentally, the coprocessor condition codes represented by cc in the mnemonics are not necessarily the same as the 68020's conditions. A floating-point coprocessor can trap (or branch, etc.) on 32 conditions, including unordered condition, whereas a memory management unit can trap on 16 conditions, including write-protected.

Coprocessor instructions are used in exactly the same way as real 68020 instructions. For example, the FPC instruction FBEQ NEXT would cause a branch to the line labeled NEXT if the zero-bit of condition code register of the FPC (not the 68020) were set. Of course, you can write programs for coprocessors only if your assembler "knows" about coprocessors. Otherwise, you could always write your own macros.

A cpGEN instruction (i.e., a general coprocessor operation) may be monadic or dyadic and take one or two arguments, respectively. Typical 68882 cpGEN instructions are FCOSH (floating-point hyperbolic cosine), FACOS (floating-point arc cosine), and FADD (floating-point addition).

The cpSAVE and cpRESTORE instructions transfer an internal coprocessor *state frame* between a coprocessor and external memory. All this means is that you can save a coprocessor's status by means of a cpSAVE instruction and then restore it with a cpRESTORE. You would use these instructions in a multitasking or an interrupt-driven environment to save a coprocessor's status before it is used by another task. For example, FSAVE $-$(A7) would save a floating-point coprocessor's status on the system stack. Note that the FSAVE saves only the *invisible* status of the FPC and not its *visible* status, made up of FP0–FP7 and system control/status registers. You can save registers FP0 to FP7 with a FMOVEM FP0–FP7, $-$(A7) instruction.

As we have already stated, the coprocessor is allocated 32 bytes of CPU memory space, which is, effectively, arranged as 8 contiguous longwords. All coprocessors must have a specific arrangement of registers in order to communicate with the 68020 host. Figure 7.92 defines the register structure of a generic coprocessor. It is important to note that these registers are invisible to the programmer and are required to implement the 68020-coprocessor protocol.

Coprocessor Operation

When a 68020 detects an *F*-line instruction, it writes the instruction to the coprocessor's memory-mapped command register in CPU space. Remember, once again, that the programmer does not have to provide an address because the coprocessor instruction contains a Cp-ID field that identifies the type of coprocessor, and each type of coprocessor has its own unique region of CPU memory space. Having sent a command to the coprocessor, the processor reads the response from the coprocessor's response register. At this stage the coprocessor may use its response to request

FIGURE 7.92 The coprocessor interface register map

	31 16	15 00
00	Response	Control
04	Save	Restore
08	Operation word	Command
0C	Reserved	Condition
10		
14	Register select	Reserved
18	Instruction address	
1C	Operand address	

further actions such as "fetch an operand from the calculated effective address and store it in my operand register." Once the host processor has complied with the coprocessor's demands, it is free to continue normal computing. That is, processor and coprocessor may overlap their operations.

It is possible to use a coprocessor with a 68000 microprocessor—even though the 68000 does not support a coprocessor interface in its internal firmware. Since the coprocessor interface protocol is based solely on data transfers over the asynchronous bus, it is possible for a humble 68000 to emulate the 68020's coprocessor interface protocol entirely in software. When the 68000 reads a coprocessor F-line instruction, it takes the F-line exception and executes the appropriate coprocessor interface protocol. Motorola's application note AN947 indicates how you would go about writing such a coprocessor interface protocol for a 68000. While you might have to add a coprocessor to an existing 68000-based system to improve its performance, you would almost certainly use a 68020 if you were going to design a new system with a coprocessor.

Introduction to the MC68882 Floating-Point Coprocessor

In principle, the 68882 floating-point coprocessor is a very simple device, although in practice its full details are rather complex, as a glance at its extensive data manual will indicate—the 68882's manual is as large as that of the 68000 itself. Much of this complexity arises from the nature of IEEE floating-point arithmetic rather than from the nature of the FPC.

The 68882 extends the 68020's architecture to include the eight 80-bit floating-point data registers, FP0 to FP7, described in figure 7.93. It is important to understand that as far as the programmer is concerned, these registers magically appear

FIGURE 7.93 The 68882 FPC register model

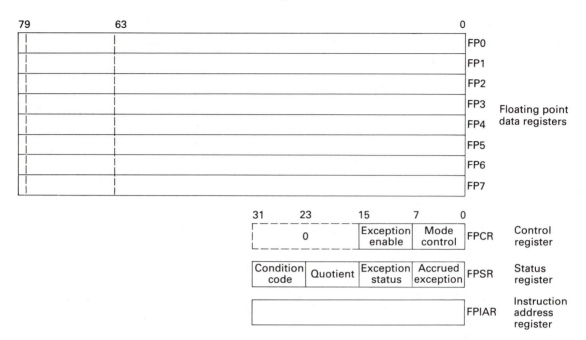

within the 68020's register space. In addition to the standard byte (.B), word (.W), and longword (.L) operands, the FPC supports four new operand sizes: single-precision real (.S), double-precision real (.D), extended-precision real (.X), and packed decimal string (.P). Figure 7.94 illustrates the structure of these operand formats. All on-chip calculations take place in extended precision format and all floating-point registers hold extended precision values. The single-real and double-real formats are used to input and output operands.

The three *real* floating-point formats support the corresponding IEEE floating-point numbers (single, double, and extended precision). The packed decimal real format holds a number in the form of a packed BCD 3-digit base-10 exponent and a 17-digit base-10 mantissa, as illustrated in figure 7.94. A complete packed decimal string value is 96 bits (packed in 3 longwards when stored in external memory) and is used to convert between decimal and binary floating-point values. For example, you can write the instruction

        FADD.P   # −4.123456E + 17,FP0

to add the literal decimal number -4.123456×10^{17} to the contents of floating-point register FP0. Note that the FPC *automatically* converts a packed decimal value to an extended precision value. You can convert from binary to decimal form by executing an instruction that has a packed decimal real as a destination operand. For example, the instruction

        FMOVE.P   FP6,Result{#6}

FIGURE 7.94 Data types supported by the 68882 FPC

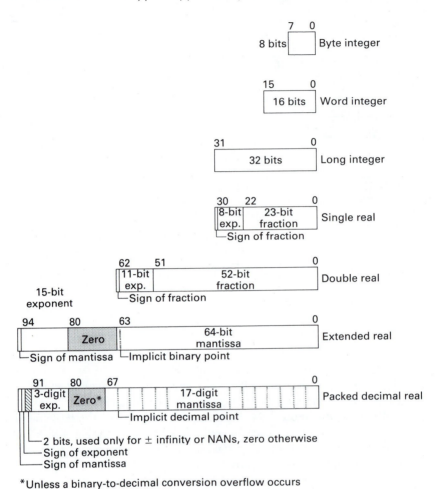

*Unless a binary-to-decimal conversion overflow occurs

has the effect of taking the value in FP6, converting it into packed decimal form, and moving it to main store at address "Result." Note that the {#6} field after the destination operand tells the FPC to store the number with six digits to the right of the decimal point.

The FPC's FMOVE ⟨source⟩,⟨destination⟩ instruction not only copies an operand from its source to its destination but also forces a data conversion. If the source is memory, the operand is converted to the internal extended precision format inside the FPC, and if the source is the FPC, the operand is converted into destination format before being stored in memory. The FPC also supports a move multiple register instruction, FMOVEM, that permits any set of the eight 80-bit floating-point registers to be moved to or from memory. Each of these floating-point registers is stored as 3 longwords, and no data conversion takes place when a FMOVEM is executed. A *new* move instruction is FMOVECR.X #data,FPn, which can move one of 22 floating-point constants to a floating-point register. These constants include 0, 1, e, $\log_2 e$, and π. Figure 7.95 illustrates some of the FPC's monadic and dyadic instructions.

In addition to the eight floating-point registers, the 68882 has a 32-bit control register, FPCR, and a 32-bit status register, FPSR. The FPCR is used by the pro-

FIGURE 7.95 Some of the 68882's monadic and dyadic instructions

FCOSH	Hyperbolic cosine
FETOX	e to the x power
FETOXM1	e to the x power -1
FGETEXP	Get exponent
FGETMAN	Get mantissa
FINT	Integer part
FINTRZ	Integer part (truncated)
FLOG10	Log base 10
FLOG2	Log base 2
FLOGN	Log base e
FLOGNP1	Log base e of $(x + 1)$
FNEG	Negate
FSIN	Sine
FSINCOS	Simultaneous sine and cosine
FSINH	Hyperbolic sine
FSQRT	Square root
FTAN	Tangent
FTANH	Hyperbolic tangent
FTENTOX	10 to the x power
FTST	Test
FTWOTOX	2 to the x power

DYADIC OPERATIONS

Dyadic operations have two input operands. The first input operand comes from a floating-point data register, memory, or an MPU data register. The second input operand comes from a floating-point data register. The destination is the same floating-point data register used for the second input. For example, the syntax for add is:

```
FADD.⟨fmt⟩    ⟨ea⟩,FPn or,
FADD.X        FPm,FPn
```

The MC68882 dyadic operations available are as follows:

FADD	Add
FCMP	Compare
FDIV	Divide
FMOD	Modulo remainder
FMUL	Multiply
FREM	IEEE remainder
FSCALE	Scale exponent
FSGLDIV	Single precision divide
FSGLMUL	Single precision multiply
FSUB	Subtract

grammer to determine which events cause floating-point exceptions and to specify how rounding is to be carried out (the *rounding* of inexact numbers in floating-point arithmetic is a very important consideration in numerical methods). The 32-bit FPSR provides a floating-point status (like a conventional CCR), the sign and least significant 7 bits of a quotient, the exception status, and the accrued exception status. Figure 7.96 illustrates the 68882's FPSR.

FIGURE 7.96   Structure of the 68882's status register

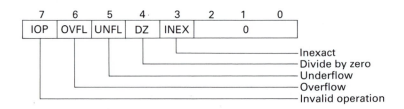

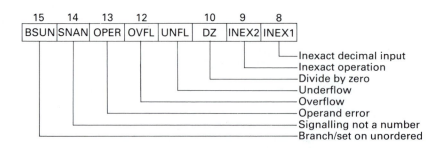

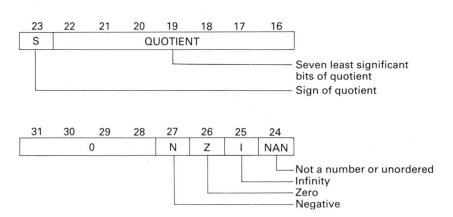

In order to see what coprocessor code looks like, consider the following fragment of code.

```
*               Calculate a vector times a constant plus a vector
*               FOR I = 1 to N
*                   X(i) := Y(i) × C + X(i)
*               ENDFOR
*
*               C = address of constant
*               XVec = address of vector X
*               YVec = address of vector Y

                MOVE.W      #N−1,D0         D0 contains the loop counter
                FMOVE.D     C,FP0           FP0 contains the constant
                LEA         XVec,A0         A0 points to XVec
                LEA         YVec,A1         A1 points to YVec
*
AGAIN           FMOVE.X     FP0,FP1
                FMUL.D      (A1)+,FP1       Calculate Y(i) × C
                FADD.D      (A0),FP1        Calculate Y(i) × C + X(i)
                FMOVE.D     FP1,(A0)+
                DBRA        D0,AGAIN        Repeat until all components formed
```

Exceptions and the FPC

The 68882 handles exceptions via the 68020 host processor; that is, 68020 and 68882 exceptions are treated in the same way—by the 68020. However, since the FPC can generate new exceptions (e.g., inexact result), the 68000's exception vector table is extended to include new exception vectors as described in table 7.7.

Now that we have introduced the cache, the memory management unit, and the floating-point coprocessor, we are going to provide a short introduction to the processor that includes all these facilities, the 68040.

Table 7.7   The 68020's exception vector table

Vector Number(s)	Vector Offset		Assignment
	Hex	Space	
0	000	SP	Reset initial interrupt stack pointer
1	004	SP	Reset initial program counter
2	008	SD	Bus error
3	00C	SD	Address error
4	010	SD	Illegal instruction
5	014	SD	Zero divide
6	018	SD	CHK, CHK2 instruction
7	01C	SD	cpTRAPcc, TRAPcc, TRAPV instructions
8	020	SD	Privilege violation

Table 7.7 — (continued)

Vector Number(s)	Vector Offset		Assignment
	Hex	Space	
9	024	SD	Trace
10	028	SD	Line 1010 emulator
11	02C	SD	Line 1111 emulator
12	030	SD	(Unassigned, reserved)
13	034	SD	Coprocessor protocol violation
14	038	SD	Format error
15	03C	SD	Uninitialized interrupt
16	040	SD	
⋮			Unassigned, reserved
23	05C	SD	
24	060	SD	Spurious interrupt
25	064	SD	Level 1 interrupt autovector
26	068	SD	Level 2 interrupt autovector
27	06C	SD	Level 3 interrupt autovector
28	070	SD	Level 4 interrupt autovector
29	074	SD	Level 5 interrupt autovector
30	078	SD	Level 6 interrupt autovector
31	07C	SD	Level 7 interrupt autovector
32	080	SD	
⋮			TRAP #0–15 instruction vectors
47	08C	SD	
48	0C0	SD	FPCP branch or set on unordered condition
49	0C4	SD	FPCP inexact result
50	0C8	SD	FPCP divide by zero
51	0CC	SD	FPCP underflow
52	0D0	SD	FPCP operand error
53	0D4	SD	FPCP overflow
54	0D8	SD	FPCP signaling NAN
55	0DC	SD	Unassigned, reserved
56	0E0	SD	PMMU configuration
57	0E4	SD	PMMU illegal operation
58	0E8	SD	PMMU access level violation
59	0EC	SD	
⋮			Unassigned, reserved
63	0FC	SD	
64	100	SD	
⋮			User-defined vectors (192)
255	3FC	SD	

SP   Supervisor Program Space   SD   Supervisor Data Space

7.7 INTRODUCTION TO THE 68040 MICROPROCESSOR

No text on the 68000 family would be complete without at least a mention of the 68040 microprocessor. The 68040 is an extension of the 68000, just as the Boeing 747 is an extension of the Sopwith Camel. Although it is true that the 68040 is indeed an extension of the 68000's user architecture, it is a radical departure from earlier members of the 68000 family—the 68000, 68020 and 68030. In this section we will describe some of the 68040's highlights, although you will have to consult its user's manual if you wish to explore it further. In short, the 68040 is a 68030 plus an internal floating-point processor with a 20-MIP integer performance and a 3.5-MFLOP floating-point performance.

If you maintain that the architecture of a processor represents the programmer's view of its structure, then the 68040 and the 68020 are virtually identical (except for the inclusion of an on-chip floating-point processor that implements a subset of the 68882's operations). As far as the user programmer is concerned, the 68040 looks like a 68020, with the addition of eight floating-point registers plus three registers used by the floating-point unit. Although the *user* models of the 68020 and the 68040 are virtually the same, the 68040 does implement changes to the *supervisor* model of the 68020. These changes mean that operating systems designed specifically for the 68020 would, ideally, require modification before running on the 68040. The difference between the 68020 and 68040's supervisor architectures is due to the inclusion of the 68040's instruction and data caches and its two memory management units. Incidentally, the 68040 does not implement a 68020-style coprocessor interface protocol.

The real power of the 68040 (four times that of the 68020) comes from its *internal organization* rather than its *architecture*. The 68040 has been designed to achieve a high throughput by a combination of parallelism and pipelining. Figure 7.97 illustrates the internal organization of the 68040. The two most striking things about figure 7.97 are the *pipelined* integer unit and the *separate* instruction and data memory units. These features constitute a *Harvard* architecture.

The integer unit, or IU, in figure 7.97 employs the type of pipelining more commonly associated with RISC (reduced instruction set computer) architectures and can execute many of the instructions in one clock cycle (the average time per instruction is 1.3 clock cycles). Instructions are fed to the IU from the *instruction memory unit* and are executed in stages as they flow through the pipeline.

In a conventional nonpipelined processor, an instruction is decoded, its operand effective address is calculated, the operand is fetched, the result is calculated, and the operand is stored before the next instruction is dealt with. Each of the functional parts of such a processor is idle for much of the time, which means that the silicon real estate is not being used efficiently. A pipelined processor overlaps the execution of instructions, so that while one instruction is being executed, the operand of the next instruction is being fetched, and the instruction after that is being decoded, and so on. Figure 7.98 illustrates the principles behind pipelining.

An illustration of the 68040's IU pipeline is given in figure 7.99 and is taken from an article by the 68040's design team (Edenfield et al., "The 68040 Processor,"

FIGURE 7.97 Internal organization of the 68040

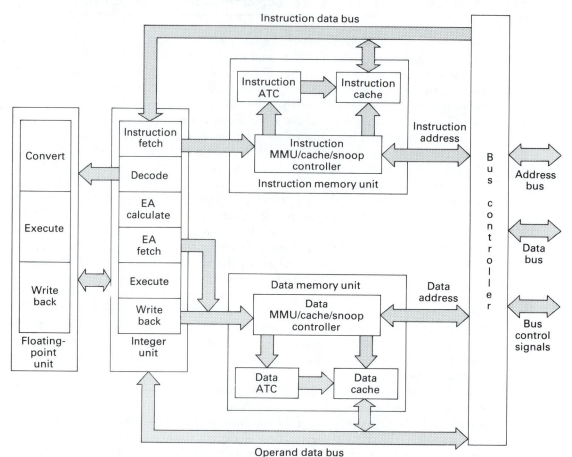

IEEE Micro (February 1990)). The figure demonstrates how each instruction in the sequence is executed in stages as it flows through the pipeline. Note that the entries labeled "dead" indicate that the corresponding unit is idle during that clock cycle.

Unfortunately, pipelining is not a perfect way of increasing a microprocessor's performance. As long as the pipeline is busy, with each stage executing part of an instruction, all is well. It is a fact of life that a remarkably large percentage of all instructions (in the region of 20 percent) are program-modification instructions (e.g., BRA, B_{cc}, BSR, RTS). Whenever such an instruction is encountered, the pipeline has to be *flushed*, since all the instructions prior to the branch have to be thrown away. Consequently, long pipelines are not practical.

The 68040 optimizes branch performance by calculating the target of the branch and fetching an instruction from that address after the instruction immediately following the branch—that is, the 68040 hedges its bets. Under these circumstances, the 68040 has only to decide whether to throw away the *next* instruction (branch taken) or the *target* instruction (branch not taken). Figure 7.100 is also taken from Edenfield et al. and illustrates how the 68040's pipeline handles branch instructions.

FIGURE 7.98 The principles of pipelining

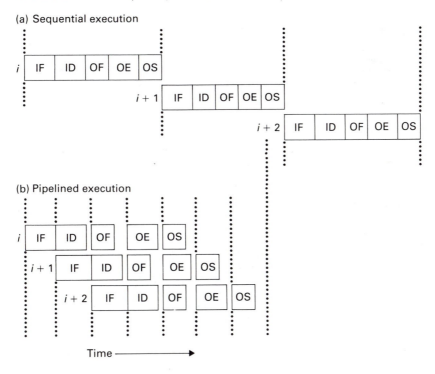

(a) Sequential execution

(b) Pipelined execution

Time ⟶

FIGURE 7.99 Example of data flow in the 68040's integer pipeline

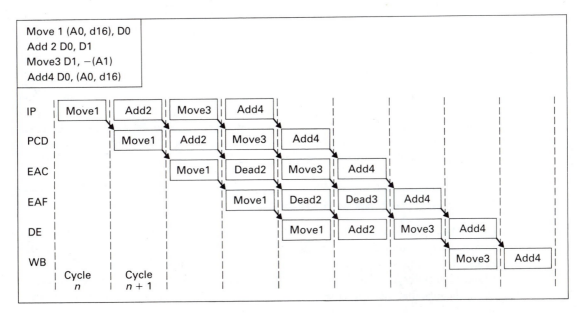

FIGURE 7.100 The effect of branch instructions on the 68040's pipeline

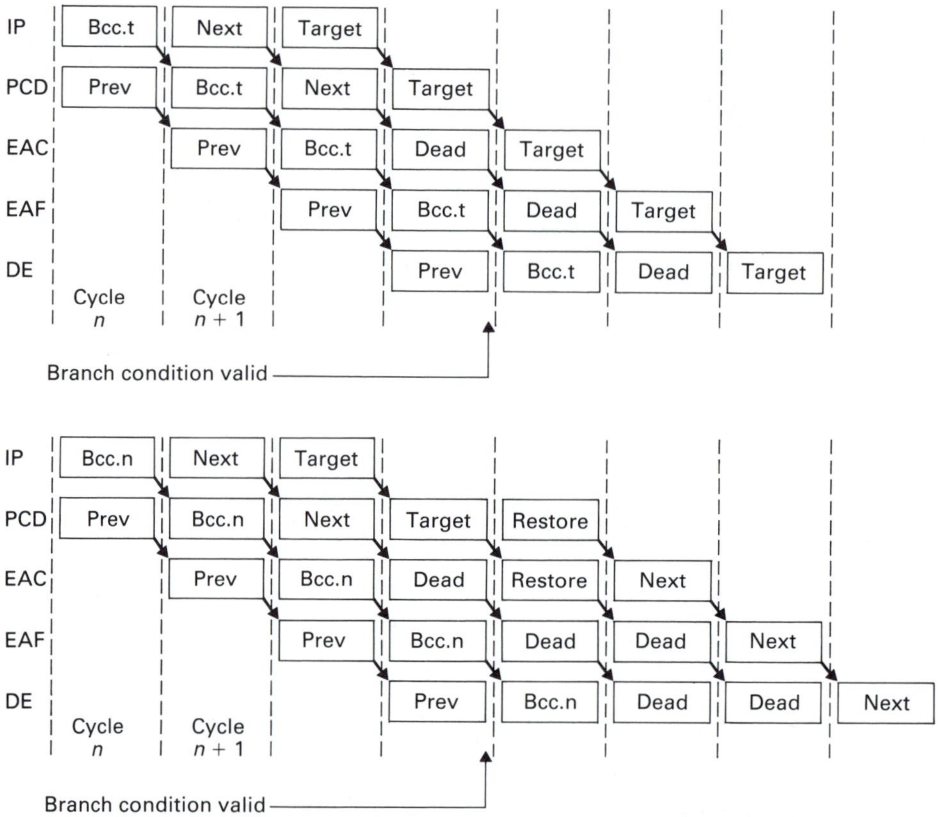

Bcc.n	Branch not taken	Next	Instruction after the branch
Bcc.t	Branch taken	Prev	Instruction before the branch
Dead	Inactive pipeline stage	Target	Taken instruction

The second radical approach to microprocessor design adopted by the 68040 is to implement two parallel and entirely independent memory units. An *instruction memory unit* is responsible for reading instructions from memory and a *data memory unit* is responsible for transferring data to and from memory. Each memory unit has its own controller, memory management unit, and cache. Consequently, the 68040 can fetch instructions and data simultaneously.

68040's Instruction and Data Caches

The 68040's two 4K-byte caches, one for data and one for instructions, increase its performance by reducing the traffic on the external bus. The chosen cache size, 4K bytes, represents a compromise between performance and the cost of silicon real-estate. Computer simulation of the performance of programs as a function of cache size indicates that performance (i.e., throughput) is an asymptotic function of size.

That is, once the cache size has reached about 2K to 4K bytes, increasing its size further does not radically improve system performance (on average).

We have already briefly described the 68040's cache earlier in this chapter, but it is worthwhile emphasizing that its caches are very sophisticated and implement bus snooping to ensure cache coherency. The 68040 can be programmed to monitor the external bus whenever another bus master (e.g., a DMA controller) is accessing memory. If the 68040 detects that memory locations that are currently cached have been modified, it can either update its own cache or invalidate the corresponding entries in its cache.

The 68040's two cache memories are located on the *physical* side of the address bus, which means that they cache *physical* data rather than *logical* data—that is, the cache lies between the on-chip MMU and physical memory rather than between the processor and the MMU. Does it really matter where the cache is located? If you place the cache between the processor and the MMU, the addresses used by the cache are all logical addresses, since they have not yet been translated by the MMU. The advantage of a logical data cache is that it is fast, since you can use an address straight from the CPU without having to wait for the MMU to perform a logical-to-physical translation.

Unfortunately, you cannot perform bus snooping if you cache logical addresses. A cache snooper must be able to monitor the physical addresses on the system bus if it is to detect any changes in physical memory. Therefore, the 68040 has adopted physical caching. We will soon see that the 68040 arranges its cache and MMU so that they can operate in parallel.

Cache memories can operate in two modes: write-through or copy-back. When data is written into a write-through cache, it is also written to external memory at the same time. Copying new data to external memory as well as the cache ensures cache coherency, since the data in the cache and memory do not diverge.

The copy-back policy does not update memory whenever data in the cache is modified. Instead, memory is updated only when a line in the cache is replaced and copied back to memory (usually following a write miss). Therefore, a copy-back strategy means that data in main store becomes stale, since it diverges from the copy of the data in the cache. A copy-back strategy is optimum in *single-processor* systems (where cache coherency is not a vital issue), as it offers the best use of the system bus. However, *multiprocessor* systems require a write-through strategy to prevent a new bus master from accessing stale data in memory. Imagine two processors in a multiprocessor system with very large caches and a copy-back policy. After a time they would each be accessing a main store full of stale data. Multiprocessors demand a write-through memory-updating strategy.

The 68040 supports both memory update strategies by permitting the supervisor mode programmer to specify each MMU page as write-through or copy-back.

68040's Memory Management Unit

The 68040's two on-chip memory management units operate like the 68851 PMMU and the 68030's on-chip MMU rather than the 68451 MMU; that is, they implement *paged* rather than *segmented* memory management. Indeed, the 68040's MMUs were designed to be as compatible with the 68030's MMU as possible.

The 68040's MMUs provide a minimum page size of 4K bytes (the 68851 supports a minimum page size as small as 256 bytes). Clearly, such a large page size increases the *granularity* of memory, but it does simplify the design of the two MMUs. In any case, as memory prices have dropped, operating systems designers have tended to use larger page sizes. Page sizes of 4K or 8K bytes are now the norm. 68040 systems programmers can specify a page size of either 4K or 8K bytes (the 68851 PMMU permits page sizes up to 32K bytes).

By making the page size as large as (or larger than) the cache memory, it is possible to perform a cache look-up concurrently with a logical to physical address translation, because the lower-order bits of the address bus can be applied to the cache's look-up table at the same time that the higher-order bits are applied to the MMU. If the page size were smaller than the cache, the cache would have to wait for the address translation to take place before it could begin its look-up.

Both MMUs employ a 64-entry cache to hold the most recent 64 logical-to-physical mappings. These entries are organized as a four-way set associative cache (do not confuse the 68040's data and instruction caches with its two MMU address translation caches).

The 68040's MMUs perform address translation and provide supervisor protection and write protection on a page-by-page basis. Each page descriptor has a 2-bit attribute field that is put out on the 68040's UPA1 and UPA0 pins (i.e., user-programmable attribute 1 and 0 pins); that is, the systems programmer can select the state of the UPA1 and UPA0 output pins for each logical page. These user-defined bits can be employed to control external logic.

Each MMU initiates a logical-to-physical address translation by searching its 64 cached descriptors for one that matches the current logical page. If a descriptor is not found in the cache, the MMU automatically performs external bus cycles to search the translation tables in external memory. After the descriptor has been located, it is cached by the MMU.

The address translation tables are arranged in the form of the three-level tree described in figure 7.101. The root pointer points at the first level of the tree in memory. The 68040 has two 32-bit root pointer registers: The URP (user root pointer) points at the tables to be used by user tasks, and the SRP (supervisor root pointer) points at the tables to be used by supervisor tasks.

The MMUs adopts two memory management options: a 7-7-6-12 and a 7-7-5-13 option. We will describe the 7-7-6-12 address table partitioning structure that provides a 4K page. The 7 most significant bits of the logical address access the first-level table pointed at by the root pointer. The selected entry then points to the second-level table, which is accessed by the next 7 most significant bits, A_{24} to A_{18}. The entry selected in the second level table points to the third-level table, which is accessed by the next 6 most significant bits, A_{17} to A_{12}. The entry selected in the third-level table is the appropriate descriptor and is used to perform the logical-to-physical address translation.

The 7-7-6-12 address table partitioning chosen by the 68040 is a compromise between the granularity of the memory, the levels of address translation, and the total size of the address translation tables. It would appear that the 68040 has been optimized to run Unix on the current generation of high-performance minicomputers.

FIGURE 7.101 The 68040's address translation mechanism

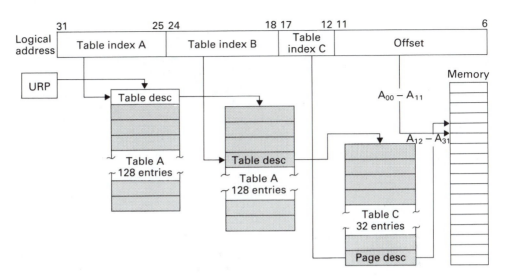

A 7-7-6-12 partitioning requires a total of $2^7 \times 4 + 2^7 \times 4 + 2^6 \times 4 = 512 + 512 + 256 = 1280$ bytes of storage for the translation data. The factor 4 in this equation is there because each entry in the translation tables is a longword of 4 bytes.

Each MMU has two *transparent translation registers* accessible from supervisor space (see figure 7.102) that define a one-to-one mapping for address space segments from 16M to 4G bytes. These registers permit a segment of logical address space to be mapped onto the corresponding segment of physical address space.

Programming Model of the 68040

Since the 68040 is the next link in the chain made up by the 68000, the 68020, and the 68030, it follows that it includes all the features of each of its predecessors. Its architecture is essentially that of the 68000 with the new addressing modes of the 68020, plus the basic facilities offered by the 68030's memory management unit, plus a subset of the 68882 floating-point coprocessor instructions. Figure 7.102 provides a programming model of the 68040, which is not as complex as you might expect, since the 68040's MMUs are not as complex as the 68851 PMMU.

The 68040 implements an internal pipelined and highly optimized floating-point unit, FPU, which (like the integer memory unit) is also pipelined. The FPU implements the IEEE 754 floating-point standard and is broadly compatible with the 68882 FPC (although it implements only the "most commonly used" subset of the 68882's instructions). The floating-point instructions implemented by the 68040 are

FIGURE 7.102 The 68040's programming model

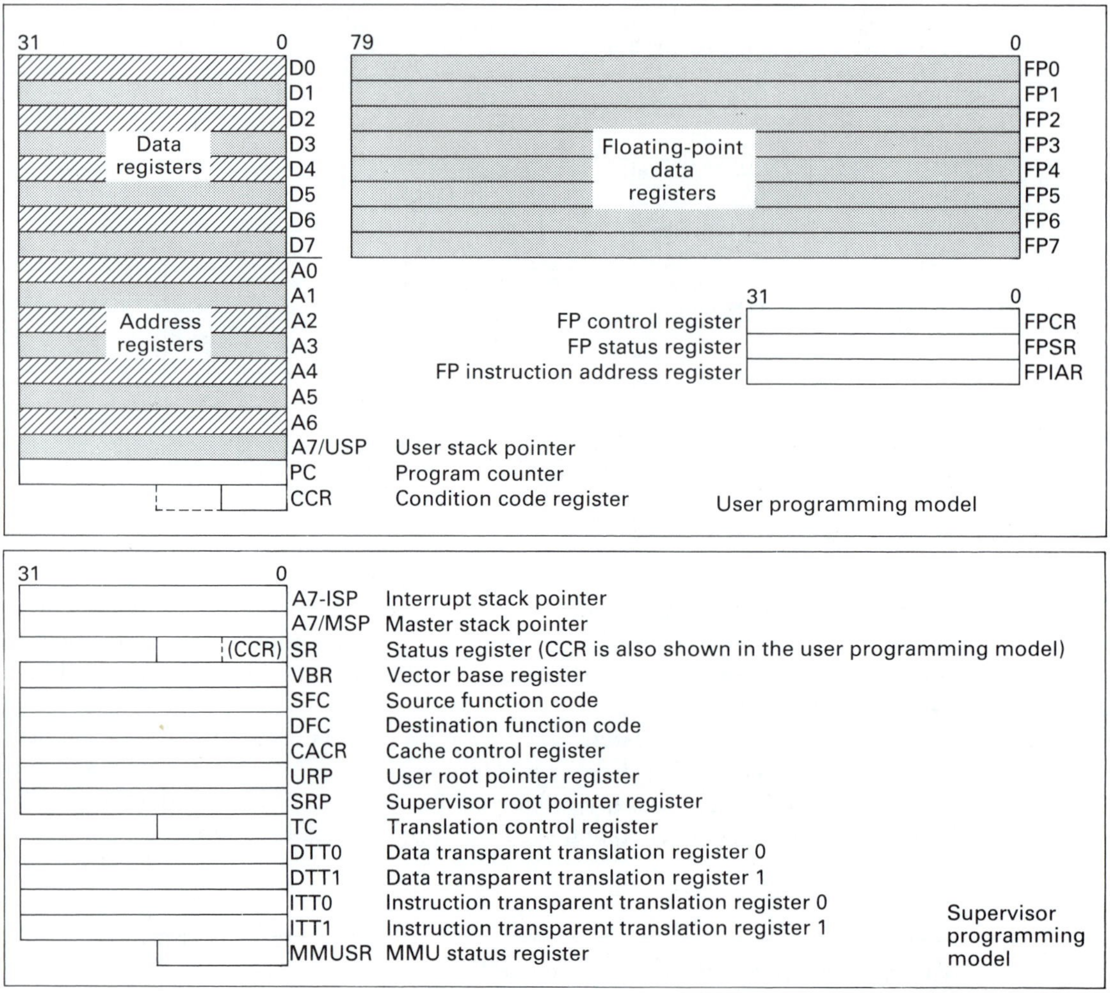

FABS	Absolute value	FNEG	Negate
FADD	Add	FRESTORE	Restore internal state
FBcc	Branch on condition	FSAVE	Save internal state
FCMP	Compare	FScc	Set according to condition
FDBcc	Decrement and branch	FSQRT	Square root
FDIV	Divide	FSUB	Subtract
FMOVE	Register move	FTRAPcc	Trap on condition
FMOVEM	Move multiple registers	FTST	Test
FMUL	Multiply		

Here's an interesting question. What happens if you are a 68882 user and decide to upgrade to the 68040? You cannot add a 68882 *directly* to the 68040, since the 68040 does not include a coprocessor interface protocol. The 68882 coprocessor instructions not implemented by the 68040 are trapped as unimplemented instructions, and it is up to the systems programmer to provide suitable code to execute the missing instructions in software. Interestingly enough, the 68040 is so fast that it can emulate 68882 instructions up to 130 percent faster than the 68882 itself. Motorola can supply a floating-point software emulation package to support all unimplemented 68882 instructions.

The 68040 with its internal floating-point unit is substantially faster than the 68030-68882 pair. Basic operations such as FADD, FSUB, and FMUL are about seven times faster on the 68040, and transcendental functions such as e^x, $\sin(x)$, and $\tan^{-1}(x)$ are about twice as fast (measured in terms of clock cycles).

The only real new user mode instruction implemented by the 68040 is the MOVE16, which permits the high-speed transfer of 16-byte data blocks between external devices such as memory or coprocessors. The MOVE16 instruction moves a cache line from one address to another. The source address may lie inside or outside the cache and the destination is in external memory. This instruction provides the fastest possible way of transferring a burst of 16 bytes. You can even use a MOVE16 instruction to transfer data between memory and a peripheral and thereby avoid the cost of a separate DMA controller.

68040's Interface

Although the 68040's architecture is essentially the same as that of earlier members of the 68000 family, its interface represents a radical departure. More specifically, the 68000's asynchronous data transfer bus has been abandoned in favor of a synchronous bus, and it does not include on-chip bus arbitration. However, separate non-multiplexed address and data buses have been retained.

The 68040 does not implement the 68020's dynamic bus sizing mechanism. Two outputs, SIZ0 and SIZ1, are employed in conjunction with address bits A_{00} and A_{01} to indicate the size of the current data bus transfer. These four signals can be used by the systems designer to generate individual byte-select signals for the memory system. Misaligned transfers are permitted and are treated by the 68040 as a series of aligned accesses of differing sizes.

Figure 7.103 describes the interface of the 68040, which is housed in a 179-pin pin-grid array. In order to increase the speed of 68040-based systems, the 68040 has a high address/control-pin drive capability to remove the need for separate address bus buffers. Address and control pins can be programmed to provide a conventional 5-mA drive or a much more substantial 55-mA drive (see chapter 10 for a discussion of bus drivers).

Address pins A_{00} to A_{31} and data pins D_{00} to D_{31} provide conventional 68000-series address and data buses. Transfer-type signals TT1 and TT2 indicate the type of the transfer taking place, and the three transfer modifier pins, TM0, TM1, and TM2, provide further information. These five pins perform a function similar to the FC0 to FC2 function code pins of a 68000.

FIGURE 7.103 The 68040's interface

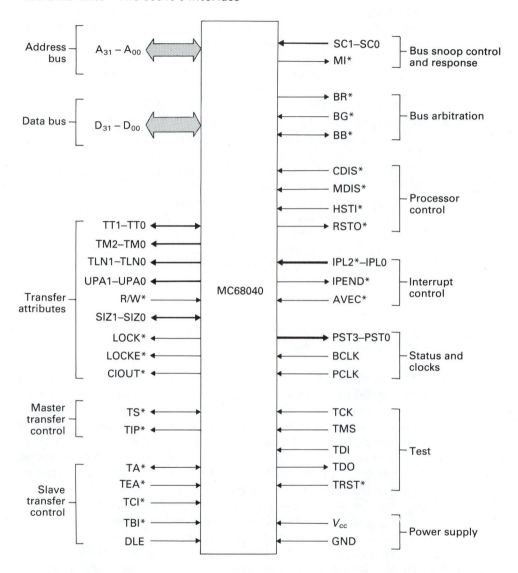

Eight new control signals (TS∗, TIP∗, TA∗, TEA∗, TCI∗, TBI∗, DLE) control data transfer on the synchronous address and data bus. The TS∗ (transfer start) bus master output is asserted by the 68040 to indicate that a valid address is on the address bus and the TA∗ (transfer acknowledge) input from a bus slave indicates the successful termination of the bus cycle. If the bus slave asserts TEA∗ (transfer error acknowledge) rather than TA∗, then a bus error is signaled. If *both* TA∗ and TEA∗ are asserted simultaneously, the 68040 reruns the bus cycle.

The transfer in progress (TIP∗) output is asserted by the 68040 for the duration of a bus transfer. The transfer cache inhibit (TCI∗) input indicates that the current bus

Table 7.8 Interpreting the 68040's PST0–PST3 status outputs

PST3–PST0	INTERNAL STATUS
0000	User start/continue current instruction
0001	User end current instruction
0010	User branch taken and end current instruction
0011	User branch not taken and end current instruction
0100	User table search
0101	Halted state—double bus fault
0110	Reserved
0111	Reserved
1000	Supervisor start/continue current instruction
1001	Supervisor end current instruction
1010	Supervisor branch taken and end current instruction
1011	Supervisor branch not taken and end current instruction
1100	Supervisor table search
1101	Stopped state—supervisor instruction
1110	RTS executed
1111	Exception stacking

cycle should not be cached. The transfer burst inhibit (TBI*) input indicates that the bus slave cannot receive a burst of data.

Four processor status outputs, PST0 to PST3, indicate to external hardware the internal status of the 68040. Table 7.8 provides an interpretation of these signals.

Since the 68040 provides a synchronous bus protocol, data is latched on the rising edge of the external bus clock input (BCLK). However, it is possible to inject a separate clock into the data latch enable, (DLE) pin and latch data on the rising edge of this clock. The 68040 has a separate processor clock input (PCLK) which runs at exactly twice the rate of BCLK. Figure 7.104 illustrates the 68040's synchronous access.

Instead of a bidirectional RESET* pin like the 68000, the 68040 has two reset pins. RSTI* is the reset input that is used to reset the 68040, whereas RSTO* is the reset output used by the 68040 to reset external peripherals when a RESET instruction is executed.

Two *unusual* pins supplied by the 68040 are the user programmable attributes UPA0 and UPA1. These are driven by 2 bits in the page descriptor corresponding to the page currently being accessed. You can use these pins in any way you want to provide special attributes (it's rather like providing your own "function codes"). For example, you might want to use UPA0 and UPA1 to control your own external cache memory. These pins can also be used to enable the 68040's bus snooping protocols. Two bus snoop control input pins, SC0* and SC1*, determine how the 68040 implements bus snooping.

The 68040's exception-handling interface offers no surprises for the 68000-systems designer. A seven-level encoded prioritized interrupt request is applied to

FIGURE 7.104 The 68040's synchronous memory access

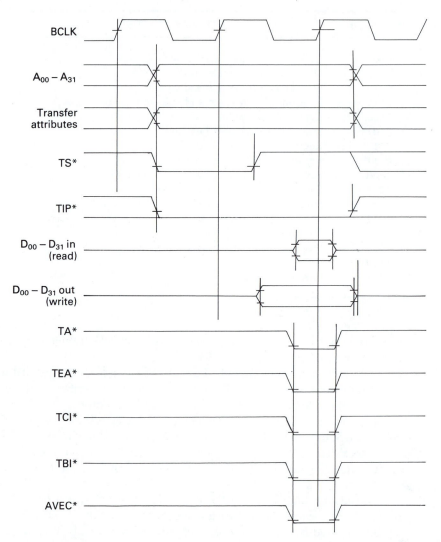

NOTE: Transfer Attribute Signals = UPA_N, SIZ_N, TT_N, TM_N, TLN_N, R/$\overline{W}$, LOCK$*$, LOCKE$*$, CIOUT$*$

IPL0$*$ to IPL2$*$ to request an interrupt. The AVEC$*$ (autovector pin) is asserted by a nonvectored peripheral to indicate an autovectored interrupt. The IPEND$*$ output indicates that an interrupt is pending.

The 68040 implements bus arbitration in a way very different from other members of the 68000 family. The 68000 uses its BR$*$, BG$*$, and BGACK$*$ pins to perform arbitration and permits an alternate master to take control of the bus at any time, unless a read-modify-write instruction is in progress. An external bus master asserts the 68000's BR$*$ *input* to force the 68000 off the bus. The 68000 responds by asserting its BG∗ *output* to indicate that it intends to give up the bus and then gives up the bus as long as its BGACK∗ *input* is asserted by the alternate bus master.

Designers of the 68040 found that some users wanted to be able to interrupt these *indivisible* read-modify-write cycles and have abandoned the 68000's bus arbitration mechanism. You might wonder why someone would like to interrupt an indivisible read-modify-write cycle. There is a good reason. Suppose that in a sophisticated system a processor begins a read-modify-write cycle to a block of on-card memory. Suppose also that another processor on a separate card also decides to access the same memory at approximately the same time. Under these circumstances, it is possible for deadlock to occur, with both processors locked out.

The 68040 has a bus request (BR∗) *output*, a bus grant (BG∗) *input*, and a bus busy (BB∗) *input/output* that is used by the current bus master to indicate that the bus is busy (BB∗ is essentially the same as BGACK∗). You should notice that the sense of these signals is reversed with respect to other member of the 68000 family. This is because the 68040 *does not* perform bus arbitration. It asserts its bus request output to request the bus from other bus masters and waits for BG∗ to be asserted in response. Once the 68040 has access to the bus, it maintains its BB∗ (bus busy) asserted until it gives up the bus. When you read about the VMEbus in chapter 10, you will see that the 68040 looks very much like a typical VMEbus master.

If the bus-granting input, BG∗, to the 68040 is removed, even an invisible bus-locked read-modify-write cycle will be interrupted. It is therefore left to the systems designer to construct his or her own external arbiter.

The 68040's LOCK∗ output is asserted to indicate to the external arbiter that the current bus cycle is an indivisible read-modify-write cycle. The arbiter may give the 68040 access to the bus for the duration of this cycle. However, the arbiter can, if it wishes, force the 68040 off the bus by negating its BG∗ input.

Unlike many other processors, the 68040 has a group of pins that are devoted entirely to its testing. These five pins (TCK, TMS, TDI, TDO, and TRST∗) are of interest only to those with automatic test equipment. These test pins support the *scan test methodology* used in board-level automatic testing.

In this section we have been able only to introduce the highlights of the 68040. Although, from the point of view of the user mode programmer, the 68040 is little more than a souped-up 68000, the 68040 marks a considerable departure from its predecessors. In the next chapter we return to the topic of input/output.

SUMMARY

In this chapter we have departed from the more bread-and-butter topics of microprocessor systems design and have looked at topics of interest to the designer of sophisticated microprocessor systems. We began with bus arbitration control, which is necessary in systems with two or more potential bus masters. Members of the 68000 family are well suited to applications in multimaster systems, because the 68000's three bus arbitration control pins make it possible for a potential bus master to request the bus and the 68000 to relinquish it in an orderly manner. We made use of the 68000's bus arbitration pins when we looked at the design of a DRAM memory refresh system.

One of the most significant features of today's general-purpose high-performance microcomputers is their very large memories. We have examined the characteristics of dynamic RAM devices and demonstrated that it is not as difficult to use a DRAM as its data sheet might sometimes suggest. Because DRAM must be refreshed periodically, the designer of DRAM memory arrays must decide whether to use the 68000's bus arbitration contol signals to request the bus and then perform a burst of refresh cycles.

Powerful microcomputers with large memories often use memory management techniques to map logical addresses onto physical addresses. We have briefly examined both the 68451 MMU and the more modern 68851 PMMU, which carry out the logical-to-physical address translation. These MMUs allow the operating system to associate blocks of memory with particular tasks.

As microprocessors have become faster and faster, the path between memory and the processor has grown into a bit of a bottleneck. In order to reduce the CPU-memory traffic, designers of the 68020, 68030, and 68040 have included on-chip caches. These caches keep a copy of recently used data on the chip and reduce the pressure on the CPU-memory highway. We have described the basic principles of the cache memory and have discussed some of the implications for the systems designer.

Systems designers using the 68020 and the 68030 often need to perform floating-point calculations. Performing floating-point arithmetic using the 68020's existing instruction set is time consuming. The floating-point coprocessor expands the 68020's architecture to include new floating-point registers and instructions. We have looked at how the 68020 uses its existing asynchronous bus interface to communicate with its coprocessor.

The final topic in this chapter provided an introduction to the 68040, which shares the user architecture of the 68020 but not its supervisor architecture. However, in spite of its compatability with the 68000, the 68040 is an immensely powerful member of the 68000 family. It includes a floating-point coprocessor on-chip plus two memory management units and two caches. By locating MMUs in both the instruction and data paths, the 68040 can access instructions and data in parallel.

Problems

1. What does *arbitration* mean and what is its function in a microprocessor system?

2. Describe the function carried out by each of the 68000's three bus arbitration pins.

3. The minimum number of bus arbitration lines required to implement a multiprocessor system is two (BR* and BG*). Why does the 68000 have an additional BGACK* line, and what are the advantages of a three-line bus control protocol over a two-line protocol?

4. In what way does the 68040's approach to bus arbitration differ from that of the 68000 to the 68030.

5. What is a TAS instruction and why is it needed in multiprocessor systems?

6. Suppose you are designing a 68000-based multiprocessor that uses TAS instructions. How does the use of a TAS instruction modify the design of your DTACK* generator?

7. In what way does the 68020's TAS instruction differ from that of the 68000's (from the point of view of the bus interface that implements the RMW cycle)?

8. Why does a DRAM require two "chip selects," RAS* and CAS*, whereas a static memory requires only a single CS* input?

9. It is much harder to design a DRAM system than a memory system using static RAM. What are the additional problems with which the designer of DRAM modules must contend?

10. What is a pseudomaximum? Why is the value of t_{RCD} (RAS* low to CAS* low) a pseudomaximum value? What factors determine the minimum and maximum values of t_{RCD}?

11. Why must DRAMs be refreshed and how may a refresh operation be carried out?

12. What are the maximum refresh periods for typical 64K-, 256K-, and 4M-bit DRAMs, respectively?

13. Why is there a limit of 10,000 ns on the time for which RAS* or CAS* can be active-low? What is the meaning of precharge time and what effect does it have on the design of DRAM systems?

14. The read cycle timing diagram of a 68000 microprocessor is given in figure 7.105 and the read cycle timing diagram of a 256K × 4-bit DRAM is shown in figure 7.106.
   a. Design a suitable interface between the 68000 and a DRAM array to provide 1M bytes of memory in the range $10 0000 to $1F FFFF. It is not necessary to consider arrangements for refreshing the memory. You may use SSI and MSI building blocks in your circuit.
   b. Using the timing diagrams provided, verify that neither the 68000's nor the DRAM's timing parameters are violated during a read cycle. You are not asked to consider a write cycle.

15. Why is RAS*-only refreshing now less popular than CAS* before RAS* refreshing? Why does the CAS*-only refresh make it easy to design a refreshing system?

16. Dynamic memory refresh can be carried out by either hardware or software techniques (in the latter a periodic interrupt or a periodically called procedure causes an access to each column address to carry out the access). What are the advantages and disadvantages of each type of refresh technique?

17. Some DRAMS support page, nibble, and static column modes (but not all three of these modes together). Explain what these modes do and how they may benefit the systems designer.

18. What (in the context of DRAMs) is a read/write cycle and how does it differ from a conventional memory access?

19. What is the difference between burst mode and distributed mode refreshing? What are the advantages and disadvantages of these modes from the point of view of the systems designer?

20. Given the data for a typical 256K × 4 DRAM in figure 7.106, analyze its read cycle timing when it is used in the ECB circuit described by figures 7.27 and 7.28. You may assume that the CPU is a 16-MHz device and a 32-MHz clock is available.

FIGURE 7.105 The 68000's read cycle timing diagram

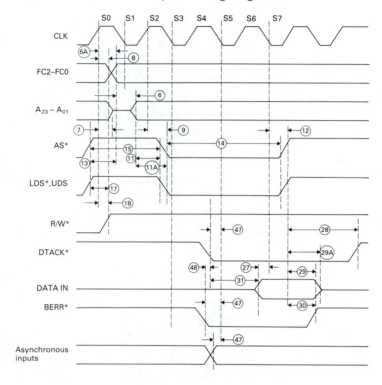

Num.	Characteristic	8 MHz		10 MHz		12.5 MHz		16.67 MHz '12F'		16 MHz		Unit
		Min	Max	Min	Max	Min	Max	Min	Max	Min	Max	
6	Clock Low to Address Valid	—	62	—	50	—	50	—	50	—	30	ns
6A	Clock High to FC Valid	—	62	—	50	—	45	—	45	0	30	ns
7	Clock High to Address, Data Bus High Impedance (Maximum)	—	80	—	70	—	60	—	50	—	50	ns
8	Clock High to Address, FC Invalid (Minimum)	0	—	0	—	0	—	0	—	0	—	ns
9	Clock High to AS*, DS* Asserted	3	60	3	50	3	40	3	40	3	30	ns
11	Address Valid to AS*, DS* Asserted (Read)/AS* Asserted (Write)	30	—	20	—	15	—	15	—	15	—	ns
11A	FC Valid to AS*, DS* Asserted (Read)/AS* Asserted (Write)	90	—	70	—	60	—	30	—	45	—	ns
12	Clock Low to AS*, DS* Negated	—	62	—	50	—	40	—	40	3	30	ns
13	AS*, DS* Negated to Address, FC Invalid	40	—	30	—	20	—	10	—	15	—	ns
14	AS* (and DS* Read) Width Asserted	270	—	195	—	160	—	120	—	120	—	ns
14A	DS* Width Asserted (Write)	140	—	95	—	80	—	60	—	60	—	ns
15	AS*, DS* Width Negated	150	—	105	—	65	—	60	—	60	—	ns
16	Clock High to Control Bus High Impedance	—	80	—	70	—	60	—	50	—	50	ns
17	AS*, DS* Negated to R/W* Invalid	40	—	30	—	20	—	10	—	15	—	ns
28	AS*, DS* Negated to DTACK* Negated (Asynchronous Hold)	0	240	0	190	0	150	0	110	0	110	ns
29	AS*, DS* Negated to Data-In Invalid (Hold Time on Read)	0	—	0	—	0	—	0	—	0	—	ns
29A	AS*, DS* Negated to Data-In High Impedance	—	187	—	150	—	120	—	90	—	90	ns
30	AS*, DS* Negated to BERR* Negated	0	—	0	—	0	—	0	—	0	—	ns
31	DTACK* Asserted to Data-In Valid (Setup Time)	—	90	—	65	—	50	—	40	—	50	ns
47	Asynchronous Input Setup Time	10	—	10	—	10	—	10	—	5	—	ns
48	BERR* Asserted to DTACK* Asserted	20	—	20	—	20	—	10	—	10	—	ns

FIGURE 7.106 A DRAM's read cycle timing diagram

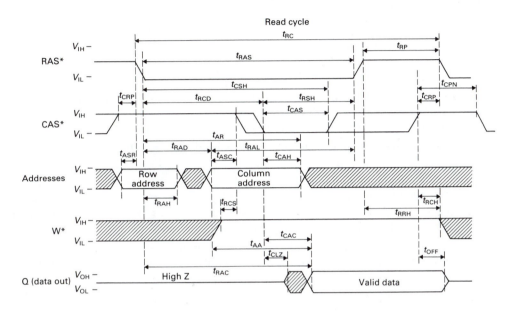

Parameter	Symbol		MCM511000A-70 MCM51L1000A-70		MCM511000A-80 MCM51L1000A-80		MCM511000A-10 MCM51L1000A-10		Unit
	Standard	Alternate	Min	Max	Min	Max	Min	Max	
Random Read or Write Cycle Time	t_{RELREL}	t_{RC}	130	—	150	—	180	—	ns
Access Time from RAS*	t_{RELQV}	t_{RAC}	—	70	—	80	—	100	ns
Access Time from CAS*	t_{CELQV}	t_{CAC}	—	20	—	20	—	25	ns
Access Time from Column Address	t_{AVQV}	t_{AA}	—	35	—	40	—	50	ns
CAS* to Output in Low-Z	t_{CELQX}	t_{CLZ}	0	—	0	—	0	—	ns
Output Buffer and Turn-Off Delay	t_{CEHQZ}	t_{OFF}	0	20	0	20	0	20	ns
RAS* Precharge Time	t_{REHREL}	t_{RP}	50	—	60	—	70	—	ns
RAS* Pulse Width	t_{RELREH}	t_{RAS}	70	10,000	80	10,000	100	10,000	ns
RAS* Hold Time	t_{CELREH}	t_{RSH}	20	—	20	—	25	—	ns
CAS* Hold Time	t_{RELCEH}	t_{CSH}	70	—	80	—	100	—	ns
CAS* Pulse Width	t_{CELCEH}	t_{CAS}	20	10,000	20	10,000	25	10,000	ns
RAS* to CAS* Delay Time	t_{RELCEL}	t_{RCD}	20	50	20	60	25	75	ns
RAS* to Column Address Delay Time	t_{RELAV}	t_{RAD}	15	35	15	40	20	50	ns
CAS* to RAS* Precharge Time	t_{CEHREL}	t_{CRP}	5	—	5	—	5	—	ns
Row Address Setup Time	t_{AVREL}	t_{ASR}	0	—	0	—	0	—	ns
Row Address Hold Time	t_{RELAX}	t_{RAH}	10	—	10	—	15	—	ns
Column Address Setup Time	t_{AVCEL}	t_{ASC}	0	—	0	—	0	—	ns
Column Address Hold Time	t_{CELAX}	t_{CAH}	15	—	15	—	20	—	ns
Column Address Hold Time Referenced to RAS*	t_{RELAX}	t_{AR}	55	—	60	—	75	—	ns
Column Address to RAS* Lead Time	t_{AVREH}	t_{RAL}	35	—	40	—	50	—	ns
Read Command Setup Time	t_{WHCEL}	t_{RCS}	0	—	0	—	0	—	ns
Read Command Hold Time Referenced to CAS*	t_{CEHWX}	t_{RCH}	0	—	0	—	0	—	ns
Read Command Hold Time Referenced to RAS*	t_{REHWX}	t_{RRH}	0	—	0	—	0	—	ns
CAS* Precharge Time	t_{CEHCEL}	t_{CPN}	10	—	10	—	15	—	ns

21. Modify the design of the DRAM system on the ECB to use 1M × 4-bit devices with an 8-MHz 68000. You must design the read/write access circuits and a suitable refresh generator.

22. Explain the meaning of the following DRAM parameters:

a. t_{RC} g. t_{RCD}
b. t_{RAC} h. t_{ASR}
c. t_{CAC} i. t_{RAH}
d. t_{RP} j. t_{ASC}
e. t_{RAS} k. t_{CAH}
f. t_{CAS} l. t_{CSH}

23. Microprocessors are widely used. DRAMs are widely used. There are virtually no microprocessors that support DRAM control and refresh on-chip. Equally, there are relatively few single-chip DRAM controllers. Explain why this is so.

24. Why does a designer have to take care when designing a DRAM's power supply?

25. Figures 7.107 and 7.108 give the write cycle details of a DRAM and a 68020, respectively. Design a DRAM controller that will provide the necessary signals (i.e., CAS∗, RAS∗, W∗, MPLX) to control a DRAM array during a write cycle. You may assume that any address multiplexers have a 10 ns delay from input change to output change and a 5 ns delay from row address to column address. You do not have to worry about refreshing.

26. What are the limitations of a conventional Hamming code (e.g. a 7, 3 Hamming code)? How does the addition of an overall parity bit improve the performance of a Hamming code?

27. What are the conditions necessary to implement each of the following?
   a. Single-error detection
   b. Single-error correction
   c. Double-error detection with single-error correction

28. Why do 16-bit microprocessors with byte read/write capabilities make life difficult for the designer of error-correcting memory systems? What can the designer do to overcome these difficulties?

29. If the MTBF for a given 1M-bit DRAM chip is 0.001 percent per thousand hours, what is the probability of failure in a 2M-byte memory system over a period of 48 hours?

30. What is the difference between FEC and ARQ (from the point of view or error correction)? Why can they both be applied to data transmission systems, but only an FEC can be used in a memory system?

31. What is the minimum number of check bits required to detect and correct a single error in a data block of 256 bits?

32. An 8-bit wide memory system is built with 4M × 1-bit chips. If the chips each have a probability of failure of 0.001 percent, what is the probability of the memory system failing? If the same chips are used to build a Hamming error detecting and correcting memory by adding four check bits, what is the probability of the memory system failing (i.e., what is the probability of two or more errors)?

33. Hamming codes become more efficient (efficiency is the ratio of check bits to total word length) as the number of bits in the source word increases. Consequently, a 16-bit 68000 with an ECC memory should be more efficient than an 8-bit 6800 with an ECC memory. A 68020 with its 32-bit data bus should be even more efficient. Why is this efficiency so hard to realize in practice?

FIGURE 7.107 Write cycle timing diagram of a 68020

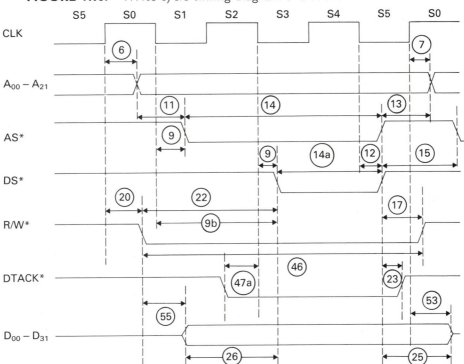

Num.	Characteristic	12.5 MHz		16.67 MHz		20 MHz		25 MHz		33.33 MHz		Unit
		Min	Max	Min	Max	Min	Max	Min	Max	Min	Max	
6	Clock High to Address, FC, Size, RMC* Valid	0	40	0	30	0	25	0	25	0	21	ns
7	Clock High to Address, Data, FC, Size, RMC*, High Impedance	0	80	0	60	0	50	0	40	0	30	ns
9	Clock Low to AS*, DS* Asserted	3	40	3	30	3	25	3	18	3	15	ns
9B	AS* Asserted to DS* Asserted (Write)	47	—	37	—	32	—	27	—	22	—	ns
11	Address, FC, Size, RMC* Valid to AS* (and DS* Asserted Read)	20	—	15	—	10	—	6	—	5	—	ns
12	Clock Low to AS*, DS* Negated	0	40	0	30	0	25	0	15	0	15	ns
13	AS*, DS* Negated to Address, FC, Size RMC* Invalid	20	—	15	—	10	—	10	—	5	—	ns
14	AS* width asserted	120	—	100	—	85	—	70	—	50	—	ns
14A	DS* Width Asserted Write	50	—	40	—	38	—	30	—	25	—	ns
15	AS*, DS* Width Negated	50	—	40	—	38	—	30	—	23	—	ns
17	AS*, DS* Negated to R/W* Invalid	20	—	15	—	10	—	10	—	5	—	ns
20	Clock High to R/W* Low	0	40	0	30	0	25	0	20	0	15	ns
22	R/W* Low to DS* Asserted (Write)	90	—	75	—	60	—	50	—	35	—	ns
23	Clock high to data out valid	—	40	—	30	—	25	—	25	—	18	ns
25	DS* Negated to Data Out Invalid	20	—	15	—	10	—	5	—	5	—	ns
26	Data Out Valid to DS* Asserted (Write)	20	—	15	—	10	—	5	—	5	—	ns
46	R/W* Width Valid (Write or Read)	180	—	150	—	125	—	100	—	75	—	ns
47A	Asynchronous Input Setup Time	10	—	5	—	5	—	5	—	5	—	ns
55	R/W* Valid to Data Bus Impedance Change	40	—	30	—	5	—	20	—	20	—	ns

FIGURE 7.108 Write cycle timing diagram of a DRAM

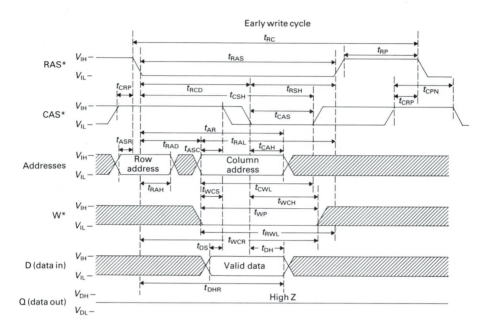

Parameter	Symbol		MCM511000A-70 MCM51L1000A-70		MCM511000A-80 MCM51L1000A-80		MCM511000A-10 MCM51L1000A-10		Unit
	Standard	Alternate	Min	Max	Min	Max	Min	Max	
Random Read or Write Cycle Time	t_{RELREL}	t_{RC}	130	—	150	—	180	—	ns
RAS* Precharge Time	t_{REHREL}	t_{RP}	50	—	60	—	70	—	ns
RAS* Pulse Width	t_{RELREH}	t_{RAS}	70	10,000	80	10,000	100	10,000	ns
RAS* Hold Time	t_{CELREH}	t_{RSH}	20	—	20	—	25	—	ns
CAS* Hold Time	t_{RELCEH}	t_{CSH}	70	—	80	—	100	—	ns
CAS* Pulse Width	t_{CELCEH}	t_{CAS}	20	10,000	20	10,000	25	10,000	ns
RAS* to CAS* Delay Time	t_{RELCEL}	t_{RCD}	20	50	20	60	25	75	ns
RAS* to Column Address Delay Time	t_{RELAV}	t_{RAD}	15	35	15	40	20	50	ns
CAS* to RAS* Precharge Time	t_{CEHREL}	t_{CRP}	5	—	5	—	5	—	ns
Row Address Setup Time	t_{AVREL}	t_{ASR}	0	—	0	—	0	—	ns
Row Address Hold Time	t_{RELAX}	t_{RAH}	10	—	10	—	15	—	ns
Column Address Setup Time	t_{AVCEL}	t_{ASC}	0	—	0	—	0	—	ns
Column Address Hold Time	t_{CELAX}	t_{CAH}	15	—	15	—	20	—	ns
Column Address Hold Time Referenced to RAS*	t_{RELAX}	t_{AR}	55	—	60	—	75	—	ns
Column Address to RAS* Lead Time	t_{AVREH}	t_{RAL}	35	—	40	—	50	—	ns
Write Command Hold Time Referenced to CAS*	t_{CELWH}	t_{WCH}	15	—	15	—	20	—	ns
Write Command Hold Time Referenced to RAS*	t_{RELWH}	t_{WCR}	55	—	60	—	75	—	ns
Write Command Pulse Width	t_{WLWH}	t_{WP}	15	—	15	—	20	—	ns
Write Command to RAS* Lead Time	t_{WLREH}	t_{RWL}	20	—	20	—	25	—	ns
Write Command to CAS* Lead Time	t_{WLCEH}	t_{CWL}	20	—	20	—	25	—	ns
Data In Setup Time	t_{DVCEL}	t_{DS}	0	—	0	—	0	—	ns
Data In Hold Time	t_{CELDX}	t_{DH}	15	—	15	—	20	—	ns
Data In Hold Time Referenced to RAS*	t_{RELDX}	t_{DHR}	55	—	60	—	75	—	ns
Write Command Setup Time	t_{WLCEL}	t_{WCS}	0	—	0	—	0	—	ns
CAS* Precharge Time	t_{CEHCEL}	t_{CPN}	10	—	10	—	15	—	ns

34. What is the alpha-particle problem and what is its effect on DRAM? Why does the alpha-particle problem not affect static RAM?

35. Design an error-logging system for an error detecting and correcting memory. Whenever the system detects an error, the error-detection circuit generates a low-priority interrupt, and the operating system then makes a note of the location of the word in error and of the bit in error. Running statistics of error locations are recorded so that systematic errors can be located and the faulty chip can be replaced. Design the hardware and software needed to carry out this function.

36. The following 7, 3 Hamming-coded words are read from memory. Each code word is constructed according to table 7.5 and the standard Hamming code. Which codes are in error and what should the correct data word be? The words are written in the order: I_4, I_3, I_2, C_3, I_1, C_2, C_1.

    a. 0 0 0 0 0 0 0   b. 1 1 1 1 1 1 1
    c. 1 0 1 1 1 1 1   d. 1 0 1 0 0 1 0
    e. 1 0 1 0 0 1 1

37. What are the main objectives of a memory management system?

38. What is the difference between logical and physical address space?

39. What is *bank switching* and under what circumstances is it used? Why do you never find it in 68000-based systems?

40. Describe how the 68451 MMU translates logical addresses into physical addresses.

41. The 68451 MMU employs a segmented memory-mapping scheme. What does this mean? List the advantages and disadvantages of the 68451's approach to memory management (in comparison to the PMMU).

42. A 68451 MMU has four descriptors in use. The contents of these descriptors (LBA, LAM, PBA) are given next. Using the information presented, draw an address map that illustrates the logical-to-physical address translation process.

    LBA_1 = $0000   LAM_1 = $FFF0   PBA_1 = $0000
    LBA_2 = $0010   LAM_2 = $FFFF   PBA_2 = $FFF0
    LBA_3 = $3000   LAM_3 = $F000   PBA_3 = $1000
    LBA_4 = $7000   LAM_4 = $FFF0   PBA_4 = $F0F0

43. Design a simple memory management system for a 68000 system without using a 68451 MMI. The logical memory space is to be divided up into 8,192 (i.e., 2^8) pages of 1K words by mapping the high-order logical address bits A_{11}–A_{23} into physical address bits PA_{11}–PA_{23}. Address bits A_{01}–A_{10} from the 68000 are passed unchanged to become PA_{01}–PA_{10}. All segments are of fixed size. The mapping is to be performed by two 8K by 8-bit static RAMs operated as a look-up table. These are connected in parallel so that a 13-bit logical address yields 16 bits from the memory. Thirteen bits form the desired physical address. The other 3 bits can be used to provide the M-bit, E-bit, and the WP bit. Design the circuitry required to support this arrangement, paying attention to the way in which the mapping RAM is loaded and to the reset function. (Following a reset the operating system must be able to set up the table). Calculate the overhead in terms of access time if the mapping RAM has an access time of 100 ns, the main memory has an access time of 200 ns, and the 68000 runs at 8 MHz.

44. Why cannot the 68000 be used (in conjunction with the 68451 MMU) to design true virtual memory systems?

45. Why can the 68010, the 68020, and so on be used to design virtual memory systems?

46. How does the 68451 MMU make use of the 68000's function code outputs during an address translation?

47. The 68851 PMMU uses paging. If the smallest page it can handle is 256 bytes, the total address space devoted to page descriptors will be several times that of the 68000's entire address space. How does the PMMU avoid the need for a vast number of page descriptors?

48. What is the difference between a PMMU *page* descriptor and a *table* descriptor? Why are both these descriptors available in long and short forms, and what effect does this have on the address translation process?

49. The PMMU can be programmed to treat the high-order 3 address bits from the 68020's address bus in a special way? What way and why?

50. How would you program the PMMU's TC register to implement a two-level paged system with a page size of 8K bytes and a 30-bit logical address? There is no unique answer to this question, and you must state your assumptions.

51. Why can some of the PMMU's cached page descriptors be locked and kept in the ATC permanently? What are the dangers of locking these descriptors, and how does the PMMU try to protect you?

52. In what ways are the 68030's transparent address translation register and the 68451 MMU alike and in what ways do they differ?

53. A cache memory may be operated in either a serial or a parallel mode. In the serial access mode, the cache is examined for data, and if a miss occurs the main store is accessed. In the parallel access mode both the cache and the main store are accessed simultaneously. If a hit occurs, the access to the main store is aborted.
   a. Assume that the system has a hit ratio h and that the ratio of cache memory access time to main store access time is k ($k < 1$). Derive expressions for the speedup ratio of both a parallel access cache and a serial access cache.
   b. If a serial mode cache is to be used and a 10 percent penalty in speedup ratio over the corresponding parallel access cache can be tolerated, what must the value of h be to achieve this? Assume that the main store access time is 150 ns and that the cache access time is 30 ns.

54. Why is it much harder to design a data cache than an instruction cache?

55. To what extent do the 68020's instruction cache and the 68030's instruction and data caches impinge on (a) the hardware systems designer, and (b) the operating systems programmer?

56. If my fairy godperson were to grant me the ability to design systems that lacked one class of faults (e.g., no timing errors, no logic errors, no faulty chip errors, no bus contention errors, etc.), I would without hesitation settle for freedom from errors due to cache systems. Why?

57. How does the 68020 implement a coprocessor interface—that is, how does it manage to use the 68020's existing hardware and software interface to incorporate the architecture of the coprocessor into its own architecture?

58. How does a coprocessor recognize a 68020 access?

59. In what ways is the 68040 a radical departure from the 68020 and the 68030?

THE MICROPROCESSOR INTERFACE

All microcomputers must be able to transfer information between themselves and an external system. The external system can vary from a CRT terminal to the valves and temperature or pressure sensors in an oil refinery. The three topics of greatest interest in this chapter are the microprocessor interface, the direct memory access interface controller (DMAC), which is able to control data transfers between a memory and a peripheral automatically, and the 68230 parallel interface and timer, which is a very versatile 8- or 16-bit parallel port. The term *parallel* indicates that the interface is able to input or output a byte or a word in one single operation. Chapter 9 examines interfaces that transfer information between computers and peripherals serially, one bit at a time.

8.1 INTRODUCTION TO MICROPROCESSOR INTERFACES

One of the greatest design limitations placed on the microprocessor, or on any other chip of similar complexity, is the number of connections between the chip and an external system. The 68000 requires 43 of its 64 pins just to communicate with memory. If the 68000 were to have a dedicated I/O interface, either a larger package would be required or some of the CPU's other features would have to be abandoned.

Fortunately, microprocessors do not require a dedicated I/O interface. The existing address, data, and control buses can handle I/O transactions as if they were normal memory accesses. This approach is called *memory-mapped input/output* and requires no overhead in the way of hardware or software (i.e., special I/O instructions).

Modest penalties have to be paid for the use of memory-mapped I/O. All data transfers must be of the same width as a normal memory access. More importantly, some of the address space must be dedicated to I/O space. This fact may be important where 8-bit microprocessors with their limited 64K-byte address spaces are concerned. Locating I/O space within memory space also runs the risk of errors due to spurious accesses to peripheral space. Imagine the effect of accidentally writing to the control register of a memory-mapped disk controller. Yet another disadvantage of memory-mapped I/O is the lack of the special-purpose I/O signals needed to control the operation of an external peripheral.

The 68000 has such a large memory space that the loss of a few bytes to I/O space is unimportant. The lack of special I/O control lines has been dealt with by locating I/O control functions within the I/O ports, rather than in the CPU itself. We shall soon see how a typical parallel I/O port, the 68230 PI/T, implements these control functions.

Figure 8.1 illustrates the essential components of a typical memory-mapped I/O system. The I/O port is the interface between the CPU and the actual peripheral

FIGURE 8.1 Essential components of a memory-mapped I/O port

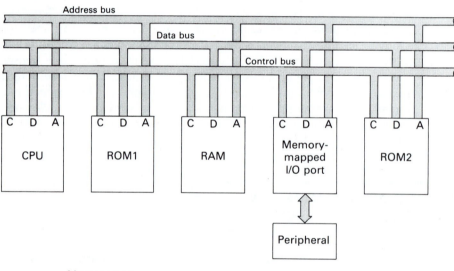

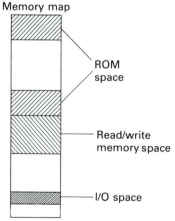

hardware. Really sophisticated ports, like disk controllers, are microcomputers in their own right. The host CPU communicates with such a port by transmitting commands along with I/O data.

Electrical Interface

Within the microcomputer, all well-behaved digital signals fall either below V_{OL} or above V_{OH}. When signals venture out of the CPU, they may be forced to abandon TTL levels and to conform with the signal levels in the peripheral equipment. Figure 8.2 illustrates the electrical interface between the interface port connected to the CPU and the external system proper.

One of the simplest input circuits is the switch of figure 8.3a. A switch connects a signal line to V_{cc} (though a pull-up resistor) or to ground. The switch may be a conventional device, a reed relay, a pressure switch, or a limit sensor. Mechanical switches suffer from bounce—the contacts do not make a clean connection, but bounce for a few milliseconds. To avoid spurious signals from a switch, a simple debounce circuit can be constructed from two cross-coupled NAND gates.

Sometimes the input is already in a binary form but is not TTL compatible. An example of this situation is given in chapter 9 where the two-level signals found on serial interfaces are described. These signals are typically -12 V or $+12$ V. In such circumstances, a level translator (figure 8.3b) is needed to convert the input signal to a TTL-level signal.

Often the signals in the equipment to be connected to the computer are in analog form and have an infinite number of values within a specified range; for example, a pressure transducer produces an output of from 9 to 12 V as the air pressure varies from 0 to 16 lb/in². In this case an analog-to-digital converter, ADC, is needed to transform the analog signal into an m-bit digital representation (figure 8.3c).

Consider now the output from a port. A TTL-level signal may control a relay as shown in figure 8.4a. This permits low-level signals in a digital system to switch high power loads in external systems. Although a TTL-level signal can operate some

FIGURE 8.2 Microprocessor interfaces

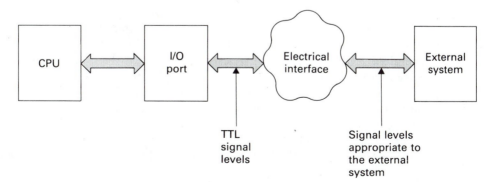

FIGURE 8.3 Examples of electrical interfaces—the input circuit

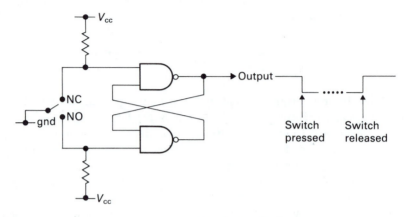

(a) Electrical interface to switch (debounced by RS flip-flop)

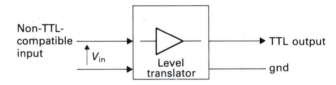

(b) Electrical interface between two-level non-TTL-compatible signal input and CPU input port

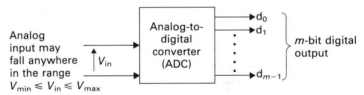

(c) Electrical interface between an infinitely variable input and a quantized 2^m-level output represented as an m-bit value (normally at TTL levels)

relays directly, it is more usual to buffer the TTL output from a port as shown. When the output is a logical zero, the transistor is in the off-state and the relay is not energized. When the output is in a logical one state, the transistor is turned on, the relay energized, and the switch closed. The switch may control an external system.

If the external circuit requires a two-level non-TTL signal, a level translator can be used as in figure 8.4b. Similarly, analog signals can be created by a digital-to-analog converter (figure 8.4c). The treatment of input/output circuits is beyond the scope of this text and is generally found in texts on "instrumentation and control." Here, we are more interested in the digital interface between the 68000 and the external system.

FIGURE 8.4 Examples of electrical interfaces—the output circuit

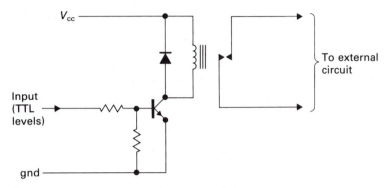

(a) The TTL input energizes a relay and closes a switch

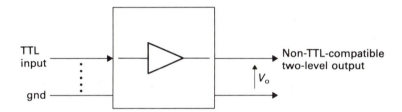

(b) A two-level TTL input is converted into a two-level non-TTL-compatible output

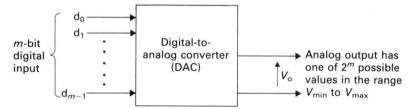

(c) An m-bit TTL-compatible input is converted to a 2^m-level analog signal

68000 Synchronous Interface

In chapter 4 we introduced the 68000's asynchronous bus, by which the CPU communicates with memory. It is perfectly possible to use this bus to communicate with memory-mapped peripherals. All we need is a peripheral that "looks like" a memory component, as far as the 68000 is concerned.

In the dark ages before the dawn of the 68000, we had the 6800 8-bit microprocessor and its 6800-series peripherals, such as the 6850 ACIA, the 6821 PIA, and the 68488 IEEE bus interface. All these peripherals interface easily to a 6800 bus through its synchronous interface. Unfortunately, they cannot be interfaced to the 68000's asynchronous bus directly. To permit designers to use the low-cost, tried, and tested 6800-series peripherals, the 68000 has been provided with three synchronous bus control signals—VPA∗, VMA∗, and E.

In order to understand why 6800-series peripherals cannot be used with the asynchronous bus, we must look at the timing diagram of one of these devices. Figure 8.5 gives the read and write cycle timing diagrams of the 6850 ACIA. Note that the timing parameters of its address inputs (i.e., RS = register select), its chip select, and its R/$\overline{\text{W}}$ input are all specified with respect to an enable clock input.

FIGURE 8.5 Timing diagram of a 6800-series I/O port

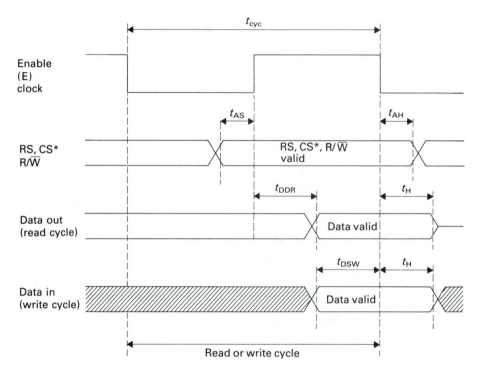

MNEMONIC	NAME	VALUE (ns)
t_{cyc}	Clock cycle time	1,000 typ.
t_{AS}	Address setup-time	160 min.
t_{AH}	Address hold-time	10 min.
t_{DDR}	Data delay-time (access time)	320 max.
t_H	Data hold-time (read)	10 max.
t_{DSW}	Data setup-time (write)	195 min.
t_H	Data hold-time (write)	10 min.

Microcomputers based on the 6800, 6809, or 6502 CPU all have some form of clock output, which is synchronized with their memory accesses and provides the enable input to the 6850 and similar peripherals. If these peripherals are interfaced to the 68000, an enable (or E) clock with the appropriate relationship between the 68000's address and data strobes needs to be provided. Although a suitable interface between the 68000 and a 6800-series peripheral is not difficult to design, it is messy. A handful of TTL devices would be required to satisfy the timing requirements of a 6800-series peripheral. Fortunately, the problem has been solved by putting the necessary synchronization circuitry on-chip.

The 68000 produces an E (enable) clock output, suitable for use with 6800-series peripherals. The E clock has a frequency of one tenth of the system clock and a low-to-high mark-space ratio of 6 : 4; that is, it is low for six CLK cycles and high for four CLK cycles. Equally importantly, the E clock is free running and bears no fixed relationship with any internal activity within the 68000.

Figure 8.6 gives the recommended interface between a 6800-series peripheral and a 68000. As these peripherals have bytewide data buses, they are interfaced

FIGURE 8.6 Interface between a 68000 CPU and a 6800-series port

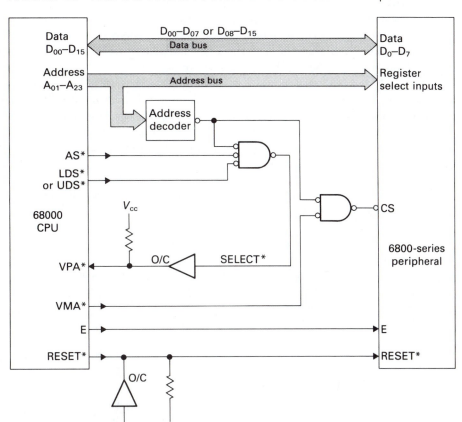

either to D_{00} to D_{07} or to D_{08} to D_{15} from the CPU. An address decoder detects an access to the peripheral's memory space and SELECT* goes active-low, following the assertion of AS* and LDS*/UDS*. A *conventional* peripheral would, of course, have its CS* input connected directly to SELECT*. However, in this case, SELECT* is connected to the 68000's valid peripheral address (VPA*) input via an open-collector buffer. Therefore, when the peripheral is selected, VPA* is asserted and the 68000 informed that the current bus cycle is to be a *synchronous* cycle. The cycle is not terminated by the assertion of DTACK*, and DTACK* must remain negated throughout the cycle.

On detecting the assertion of VPA*, the 68000 monitors its E clock and then asserts its valid memory address (VMA*) output at the appropriate point in the E cycle. VMA* is combined with the output of the address decoder to provide the necessary CS* input to the peripheral. Note that CS* should *not* be strobed with AS* or with UDS*/LDS*, as these signals are negated before the end of a synchronous access. Because the cycle is synchronous, it is terminated automatically on the falling edge of the E clock. Figure 8.7 gives the protocol flow diagram of a synchronous bus cycle. This protocol flow diagram is equally valid for read and write cycles.

Now let us examine the timing diagram of a synchronous bus cycle. Figure 8.8 gives two timing diagrams: One relates to the best case and one, to the worst case. In any synchronous cycle, the external decoder asserts VPA* within typically 100 ns of the assertion of AS*, which is recognized by the 68000 on the falling edge of S4 (just like DTACK*). No wait states are introduced (at this point) if VPA* meets its setup time, t_{ASI}, before the falling edge of S4.

In the best case (figure 8.8a), VPA* is recognized as being asserted three clock cycles before the rising edge of E. VMA* is synchronized with E and is asserted after the introduction of one wait cycle following S4. A clock state after the next falling edge of E, VMA*, is negated to end the bus cycle. Note that even in the best case, six wait cycles are introduced and the minimum cycle time is 20 states or 10 clock cycles.

In the worst case (figure 8.8b), VPA* is asserted less than 3 clock cycles before the rising edge of E, and an entire E cycle elapses before internal synchronization is achieved. We can see from figure 8.8b that a worst-case synchronous cycle requires 38 clock states.

The advantage of the 68000's synchronous cycle is the ease with which it permits interfacing with 6800-series peripherals. Its disadvantage is its excessively long bus cycle. A 68000 with an 8-MHz clock has worst-case synchronous bus cycles of 38×62.5 ns = 2375 ns = 2.375 μs. In many applications, this restriction is perfectly acceptable. However, some 6800-series peripherals (like the 68B54 advanced data-link controller) operate with an enable clock frequency of 2 MHz. These peripherals cannot be interfaced to the 68000's synchronous bus and still operate at their full rates. One solution is to abandon 6800-series peripherals and use the newer 68000-series peripherals that interface to the CPU's asynchronous bus. This solution is not always cost effective.

A second solution is to interface 6800-series peripherals to the 68000's *asynchronous* bus as outlined in Motorola's Application Note AN-808. The circuit diagram of this interface is given in figure 8.9. Two JK flip-flops, ICs 1a and 1b, are held in their clear state whenever LDS* from the CPU is negated.

FIGURE 8.7 Protocol flow diagram for the 68000 synchronous cycle

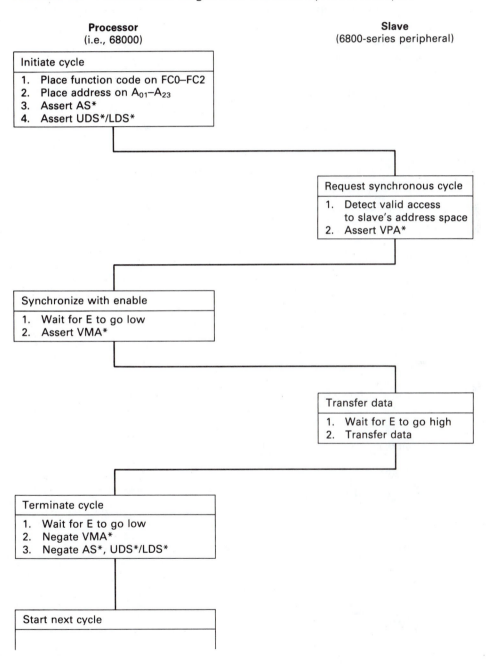

If the CPU executes a memory access to the peripheral's address space, the active-high CS from the address decoder enables AND gate IC7. When LDS* is asserted in the same cyle, the J_1 input to IC1a goes high along with its CLR* input.

Both flip-flops are clocked by E, which is generated by a user-supplied circuit

FIGURE 8.8 Timing diagram of a synchronous access cycle. (Reprinted by permission of Motorola Limited)

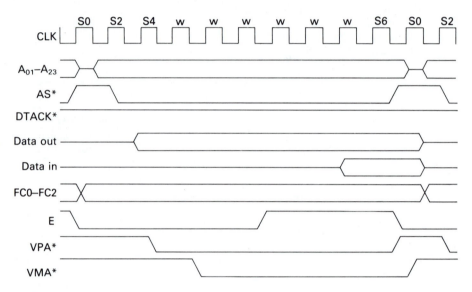

(a) Best-case synchronous access timing

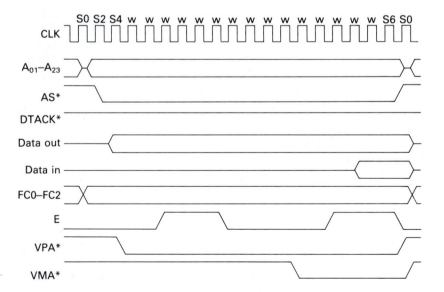

(b) Worst-case synchronous access timing

FIGURE 8.9 Circuit diagram of an interface between a 6800-series peripheral and the 68000's asynchronous bus. (Reprinted by permission of Motorola Limited)

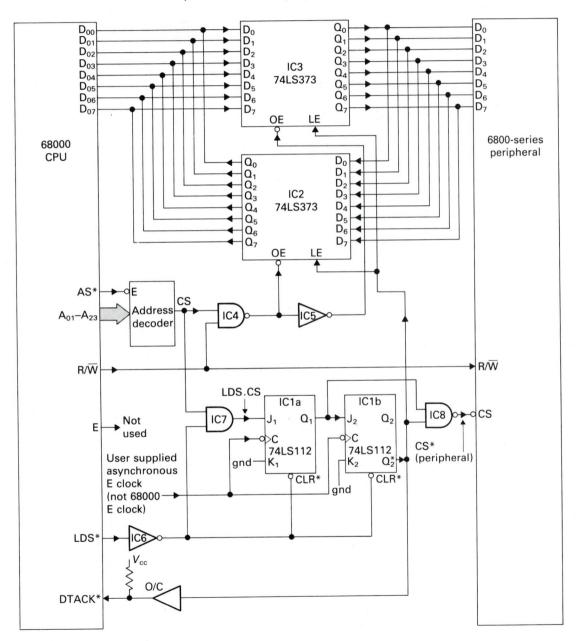

and is not obtained from the 68000. This circuit allows E clocks of a much greater frequency than the 68000's E clock operating at CLK/10. When E makes a negative transition, the Q_1 output of IC1a is asserted. Q_1 is NANDed with CS in IC8 to generate an active-low chip select for the peripheral. Figure 8.10 gives the timing diagram for figure 8.9.

The next falling edge of E clocks Q_1 into flip-flop 1b, forcing Q_2 high and Q_2* low. Q_2* is returned to the 68000 as DTACK* and therefore terminates the current bus cycle by negating AS* and LDS*, thereby, in turn, clearing flip-flops 1a and 1b to complete the access. This circuit reduces the bus cycle times by an average of 32 percent over the 68000's synchronous cycle and permits the E clock to operate at the maximum rate supported by the peripheral. Octal latches IC2 and IC3 synchronize the data bus with the peripheral access and provide the appropriate data setup and hold times.

The 68020 and 68030 do not have a synchronous bus interface, since it is assumed that they will be employed in systems using modern peripherals designed specifically for 16/32-bit microprocessors. Of course, you can always interface a 6800-series device to the 68020 using the circuit of figure 8.9. Although the 68020 does not implement the 68000's E, VMA*, and VPA* pins, it does have an AVEC* (autovector) input, which plays the same role as VPA* in autovectored interrupts.

FIGURE 8.10 Timing diagram for the interface of figure 8.9

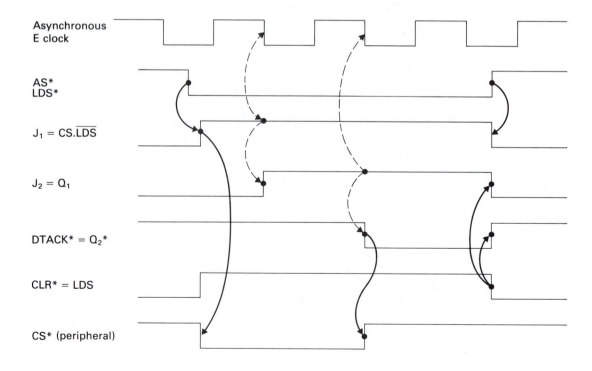

8.2 DIRECT MEMORY ACCESS

Both programmed I/O and interrupt-driven I/O require the CPU to take an active part in the input/output process. When a peripheral transfers data to a port, the CPU must read it and then store it in memory (if it is required later). Similarly, when a peripheral is ready for data, the CPU must read data from memory and transfer it to the output port. These actions both reduce the data transfer rate between the micro-computer and the peripheral, and occupy the CPU with house-keeping duties.

Direct memory access (DMA) bypasses the CPU-peripheral bottleneck and permits the transfer of data between a peripheral and the microcomputer's random access memory without the active intervention of the CPU. This process allows exceedingly high data transfer rates—often of the order of 10M bytes/s.

Figure 8.11 gives an idea of the hardware needed to support DMA. At the heart of the system is a direct memory access controller (DMAC), which can be obtained as a single LSI chip. In a sense, the DMAC is a coprocessor that shares similar privileges with the CPU—that is, it is able to take control of the system bus.

Most DMACs are connected to the processor's address, data, and control buses exactly like any other peripheral. The CPU can read from and write to their internal

FIGURE 8.11 Hardware needed to support DMA mode I/O

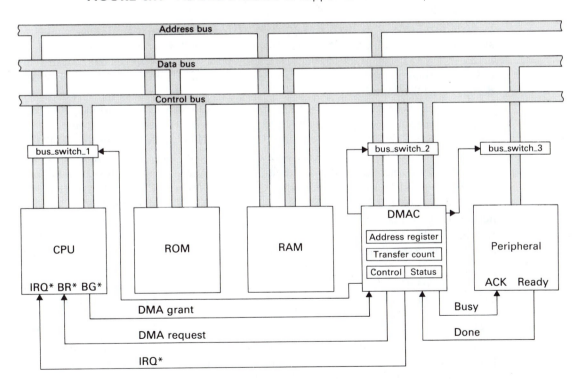

registers. A minimum DMAC register set includes an address register that points to the source/destination of data to be transferred from/to memory, a count register that contains the number of bytes to transfer, and status and control registers.

The CPU sets up a DMA operation by writing the appropriate parameters into the DMAC's registers. The DMAC then requests access to the system bus. When granted access by the CPU, the DMAC opens bus_switch_1 and closes bus_switch_2 and bus_switch_3 (figure 8.11). The DMAC puts out an address on the address bus and generates all the control signals necessary to move data between the peripheral and memory. Two signals, busy and done, synchronize data transfers between the DMAC and an external peripheral. When all the data has been transferred, the DMAC may interrupt the CPU, if it is programmed to do so.

The operating details of any DMAC vary from device to device and are closely related to the CPU with which it works. The DMAC must be able both to emulate CPU bus cycles and to request the bus from the CPU. Some DMACs interleave DMA operations with normal CPU memory accesses while others operate in a burst mode, carrying out a number of DMA cycles at a time (i.e., without intervening CPU cycles).

The SCB68430 DMAC

The 68450 is the *standard* DMAC intended for application in 68000-based systems. Because the 68450 is a rather complex and expensive device, the Signetics SCB68430 DMAC has been designed to provide a basic DMA facility in 68000 systems. The 68430 provides a *compatible* subset of the 68450's functions. Figure 8.12 gives a block diagram of the SCB68430 and its interface pins. Here, we provide only an outline of the operation of a DMAC and indicate the principles involved in designing a DMA interface.

The left-hand side of figure 8.12 shows the interface to the 68000 system bus. Note that many lines such as AS* and R/$\overline{W}$ are bidirectional. These lines act as inputs when the 68000 is accessing the DMAC and as outputs when the DMAC is accessing memory through the bus. Because the DMAC has a limited number of pins (48), its address and data buses are multiplexed. Figure 8.13 illustrates the additional hardware needed to interface the DMAC to a 68000 bus. OWN* is an active-low open-collector output from the DMAC that is asserted whenever the DMAC is a bus master. DBEN* (data bus enable) is also an active-low open-collector output that is asserted by the DMAC whenever it is being accessed by the CPU—that is, whenever CS* is asserted or when IACK* is asserted and the DMAC has an interrupt pending.

The 68430 communicates with a peripheral by means of five control lines. These lines allow the peripheral to request data transfers and the DMA controller to manage the data transfer between the peripheral and the memory. Some of the more sophisticated peripheral chips have pins that can directly be connected to a DMA controller. However, the systems designer often has to provide an interface between the *peripheral-side* pins of the DMAC and the interface pins of the peripheral. The five pins of the 68430 dedicated to peripheral control are as follows.

REQ* (Request)   This input to the DMAC from the peripheral requests service and causes the DMAC to request control of the bus from the current bus master (i.e., the 68000 CPU).

ACK* (Request Acknowledge)   ACK* is asserted by the DMAC to indicate that it has control of the bus and the cycle is now beginning. ACK* is asserted at the beginning of every bus cycle after AS* has been asserted and negated at the end of every bus cycle.

RDY* (Device Ready)   RDY* is asserted by the requesting device (i.e., peripheral) to indicate to the DMAC that valid data has either been stored or put on the bus. If negated, RDY* indicates that data has not been stored or presented, causing the DMAC to enter wait states.

FIGURE 8.12   Structure of the SCB68430 DMAC

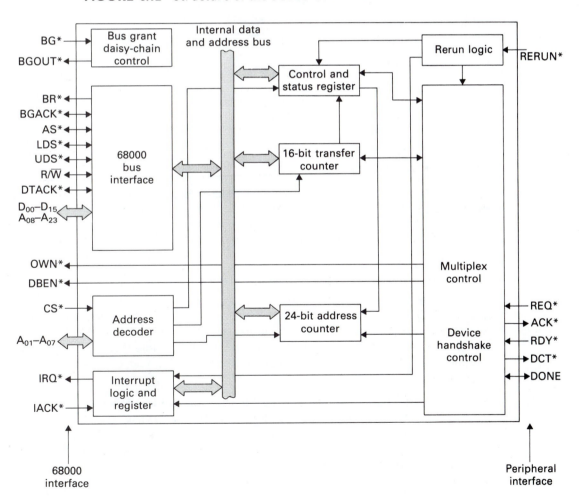

FIGURE 8.13 Interfacing the SCB68430 DMAC to a 68000 system

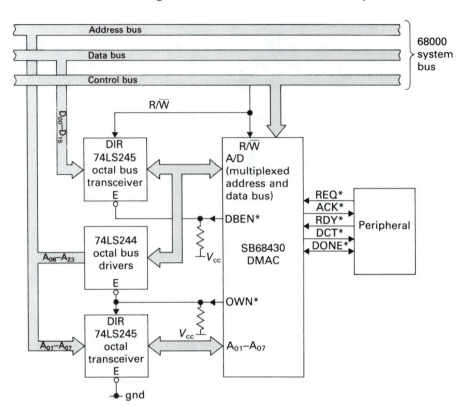

NOTE: Because the SCB6840 uses a multiplexed address and data bus, bus drivers and transceivers are used to interface the DMAC to the 68000's nonmultiplexed address and data buses. The DMAC generates two signals to control these buffers: DBEN* and OWN*. The data bus between the peripheral and the system bus is not shown here.

DTC* (Device Transfer Complete)   DTC* is asserted by the DMAC to indicate to the peripheral that the requested data transfer is complete. On a write to memory, DTC* indicates that the data from the peripheral has been successfully stored. On a read from memory, it indicates to the peripheral that the data from memory is present on the data bus and should be latched. Note that DTC* is asserted after *each* data bus transfer. DONE* is asserted at the end of a batch of data transfers.

DONE* (Done)   DONE* is a dual-function, active-low input or output pin. As an open-collector output, DONE* is asserted by the DMAC concurrently with the ACK* output to indicate that the transfer count is exhausted and that the DMAC's operation is complete. As an input, if DONE* is asserted by the peripheral before the transfer count reaches zero, it forces the DMAC to abort the operation and (if enabled) generate an interrupt request.

DMAC Operation

A DMA transfer takes place in a number of stages. The CPU first sets up the DMAC's registers (see the next section) to define the quantity of data to be moved, the type of DMA operation, and the direction of data transfer (to or from memory). During this phase, the DMAC behaves exactly like any other memory-mapped peripheral.

The DMAC is activated by a request for service from its associated peripheral. When the peripheral asserts REQ*, the DMAC requests control of the bus by asserting its BR* output, waiting for BG* from the bus master, and then asserting BGACK*.

Once the DMAC has control of the bus, it generates all the timing signals needed to transfer data between the peripheral and memory. DMA transfers take place in either the burst mode or in the cycle-stealing mode. In the burst mode several operands are transferred in consecutive bus cycles. In the cycle-stealing mode, the system bus may be relinquished between successive data transfers, allowing DMA and normal processing to be interleaved.

DMAC's Registers

The SCB68430 has seven register select inputs, A_{01} to A_{07}, permitting up to 128 internal registers to be uniquely specified. However, this device has only twelve registers, as it is a single-channel DMAC and supports only one peripheral at a time. The more complex 68450 DMAC supports up to four independent DMA channels. Table 8.1 gives the names and address offsets of the SCB68430's internal registers.

TABLE 8.1 The registers of the SCB68430

ADDRESS (A_{07}–A_{01})	OFFSET (hex)	MNEMONIC	NAME
0000000 (0)	00	CSR	Channel status register
0000000 (1)	01	CER	Channel error register
0000010 (0)	04	DCR	Device control register
0000010 (1)	05	OCR	Operation control register
0000011 (1)	07	CCR	Channel control register
0000101 (0)	0A	MTCH	Memory transfer counter high
0000101 (1)	0B	MTCL	Memory transfer counter low
0000110 (1)	0D	MACH	Memory address counter high
0000111 (0)	0E	MACM	Memory address counter middle
0000111 (1)	0F	MACL	Memory address counter low
0010010 (1)	25	IVR	Interrupt vector register

NOTE: The number in parentheses in the address column represents A_{00}. If $A_{00} = 0$, UDS* is asserted. If $A_{00} = 1$, LDS* is asserted. The SCB68430 data sheet treats registers as either the upper or lower half of a 16-bit word; for example, CSR and CER together form a word, with CSR having bits 8–15 and CER bits 0–7.

The registers do not have sequential addresses because this device is software compatible with the 68450 and its register set is, therefore, a subset of the 68450's.

A DMA operation is set up by loading the 24-bit memory address counter (MAC) with the location of the source/destination of the first operand. Once initialized, the MAC automatically increments after each data transfer. The increment is 1, 2, or 4, depending on whether the DMAC is programmed to transfer bytes, words, or longwords, respectively. The 16-bit memory transfer counter (MTC) is initialized by loading it with the number of transfers to be made during the current operation. The MTC is decremented after each transfer.

The interrupt vector register (IVR) is loaded with the vector to be placed on the data bus during an IACK cycle initiated by the CPU. Only the seven most significant bits of the IVR are gated onto the data bus during an IACK cycle. The least significant bit of the vector is set to zero by the DMAC if a normal termination occurred or to a one if the operation was terminated by an error condition. A reset to the DMAC presets the IVR to $0F, corresponding to the uninitialized vector exception.

The operating mode of the SCB68430 is determined by the device control register (DCR), the operation control register (OCR), and the channel control register (CCR). Bit 15 of the DCR determines whether the DMAC operates in a burst mode (DCR_15 = 0) or in a cycle steal mode (DCR_15 = 1). The burst mode allows a peripheral to request the transfer of multiple operands using consecutive bus cycles. In the cycle steal mode, the peripheral requests a single operand transfer at a time. Each request for service by the peripheral results in a request for bus arbitration by the DMAC.

The operation control register uses 3 bits, OCR_7, OCR_5, and OCR_4, to determine the direction and size of the data transfer. If OCR_7 is a logical zero, data is transferred from memory to the peripheral. If OCR_7 is a logical one, data is transferred from the peripheral to memory. OCR_5 and OCR_4 determine the size of each operand as illustrated in table 8.2.

Three bits of the channel control register, CCR_7, CCR_4, and CCR_3, are used by the DMAC. When CCR_7 makes a zero to one transition (under software control), the DMA operation is initiated. This should, of course, be done only when all the other registers have previously been initialized. Bit CCR_4 is a software abort and may be set to terminate the current data transfer and to place the DMAC in an idle state. Setting CCR_4 causes the channel_operation_complete and error bits in the CSR to be set, the channel_active_bit in the CSR to be reset, and the pending bit (CCR_7) to be reset. Bit CCR_3 is an interrupt enable bit. When clear, it disables DMAC interrupts. When set, it enables an interrupt request on the completion of a data transfer.

The channel status register (CSR), together with the channel error register (CER), provide the host CPU with an indication of the status of the DMAC. The status and error bits are defined as follows:

CSR_15 (Channel Operation Complete)   CSR_15 is set following the termination of an operation—whether that operation was successful or not. This bit must be cleared to start another operation.

TABLE 8.2 Bits OCR_4 and OCR_5 of the operation control register

OCR_5	OCR_4	OPERAND SIZE
0	0	Byte transfer. If the LSB of the MAC is 0, UDS* is asserted during the transfer. If the LSB of the MAC is 1, LDS* is asserted. The MAC is incremented by 1 after each transfer. The transfer count is decremented by 1 before each byte is transferred.
0	1	Word transfer. The transfer counter decrements by 1 before each word is transferred and the MAC increments by 2 after each transfer.
1	0	Longword transfer. The 32-bit operand is transferred as two 16-bit words. The transfer counter is decremented by 1 before the entire longword is transferred and the MAC is incremented by 2 after each transfer.
1	1	Double-word transfer. The operand size is 32 bits and is transferred as a *single* 32-bit word. The MAC is incremented by 4 after each operand transfer and the transfer counter is decremented by 1 before it. This mode is included for compatibility with the VME bus.

CSR_13 (Normal Device Termination) CSR_13 is set when the peripheral terminates the DMAC operation by asserting the DONE* line while the peripheral was being acknowledged. CSR_13 must be cleared to start another operation.

CSR_12 (Error) When set, CSR_12 indicates the termination of a DMA operation by an error. Its cause can be determined by reading the channel error register. CSR_12 must be cleared to start another channel operation. When cleared, the channel error register is also cleared.

CSR_11 (Channel Active) CSR_11 is set (by the DMAC) after the channel has been started and remains set until the channel operation terminates. It is automatically cleared by the DMAC.

CSR_8 (Ready Input State) CSR_8 reflects the state of the RDY* input at the time the CSR is read. CSR_8 = 0 if RDY* = 0 and CSR_8 = 1 if RDY* = 1.

CER_4 to CER_0 (Error Code) These five bits of the channel error register indicate the source of an error when CER_12 is set. Only three values are defined by the SCB68430:

00000 No error.

01001 Bus error. A bus error occurred during the last bus cycle generated by the DMAC.

10001 Software abort. The channel operation was terminated by a software abort.

Using the SCB68430 DMAC is relatively straightforward. All that is required is a peripheral conforming to the DMAC's data transfer control signals. The DMAC is programmed according to its control registers as described previously and, once

CCR_7 has been set, it executes the DMA operation as programmed and sets CSR_15 when the operation is complete. Figure 8.14 provides the protocol flowchart for a memory-to-peripheral operation. Further details of the SCB68430 are found in its data sheet.

FIGURE 8.14 Transfer protocol flowchart for a DMAC memory-to-peripheral data

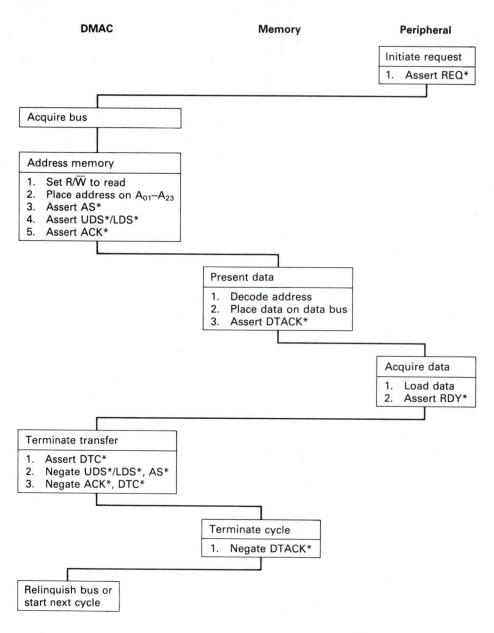

8.3 THE 68230 PARALLEL INTERFACE/TIMER

We are now going to look at the way in which a parallel I/O port can be constructed between a microcomputer and an external system. In particular, we will show how a sophisticated parallel interface can be implemented by the 68230. The 68230 parallel interface/timer (PI/T) is a general-purpose peripheral, with the primary function of an 8- or a 16-bit parallel interface between a computer and an external system, and a secondary function as a programmable timer. In this section, we examine the characteristics of the 68230 PI/T. The novice may be forgiven for wondering why a special parallel interface is necessary. After all, what is wrong with an octal D latch? In principle, the PI/T is little more than a set of latches which store I/O information. In practice, the PI/T has had many powerful attributes linked to its basic function, and all of these facilities are programmable.

Figure 8.15 provides an example of the simplest possible parallel output port based on the 74LS373 octal D latch. A small quantity of TTL is needed to trap a valid write access to the latch, clock it, and return DTACK* to the 68000. The poverty of this design lies in the absence of any two-way communication between

FIGURE 8.15 Basic output port using octal latches

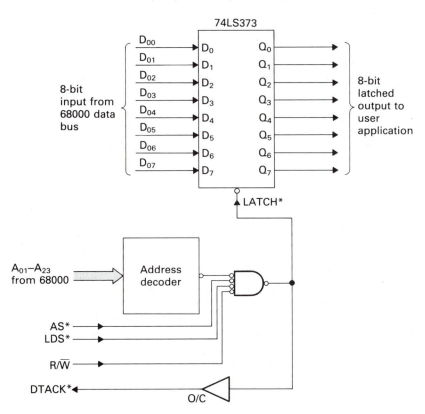

the CPU and the peripheral to which the port is connected. This interface leads to *open-loop* operation, in which the CPU transfers data to the latch by means of a write operation but lacks any feedback from the peripheral; for example, the peripheral cannot tell the CPU either that it has received the data or that it is ready for new data. These functions could be included in the arrangement of figure 8.15, but would require a large number of SSI and MSI chips. The 68230 PI/T provides all these functions in one 48-pin chip. Before we look at the 68230 itself, we need to define some important concepts related to input/output techniques.

I/O Fundamentals

The 68230 PI/T furnishes the systems designer with two things fundamental to all but the most primitive I/O ports. These facilities are handshaking and buffering. Handshaking permits data transfers to be interlocked with an external activity (e.g., a disk drive), so that data is moved at a rate in keeping with the peripheral's capacity. *Interlocked* means that the next action cannot go ahead until the current action has been completed. Buffering is a facility that permits an overlap in the transfer of data between the CPU and the PI/T, and between the PI/T and its associated peripheral; for example, the PI/T may be obtaining the next byte of data from a disk controller while the CPU is reading the last byte from the PI/T. Buffering requires some temporary storage.

Input Handshaking

Figure 8.16 illustrates the sequence of events taking place when the PI/T operates in its interlocked handshake input mode, and is similar to the asynchronous memory access discussed in chapter 4. For the time being, we are assuming that the PI/T has two data transfer control lines: an edge-sensitive input H1 and an output H2. These control lines are additional to the PI/T's parallel I/O bus. The PI/T also has internal status flags, readable by the CPU, that indicate the status of the handshake inputs; for example, H1S is set by an active transition on H1. The status flags can be read by the CPU at any time. The following description describes a single-buffered input operation. Note that the 68230 can operate in several different modes, which will be described shortly.

At state (a) in figure 8.16, control output H2 from the PI/T is in its asserted state, indicating to the peripheral that the PI/T is ready to receive data. The 68230 permits the sense of H1 and H2 to be programmed by the user. At state (b), the peripheral forces an active transition on the PI/T's H1 input, informing it that data is available on the PI/T's data input bus. Asserting H1 sets a status bit within the PI/T and generates an interrupt request if it is programmed to do so. At state (c), H2 is negated by the PI/T, informing the peripheral that the data has been accepted. Equally, the PI/T is no longer in a position to receive further data. At state (d), H1 is negated by the peripheral, informing the PI/T that the peripheral has acknowledged the data transfer. At state (e), the PI/T asserts H2 to indicate that it is once more ready to receive data from the peripheral. At this stage, the system is in the same condition as state (a), and a new cycle may commence.

FIGURE 8.16 Closed-loop data transfer and the input handshake

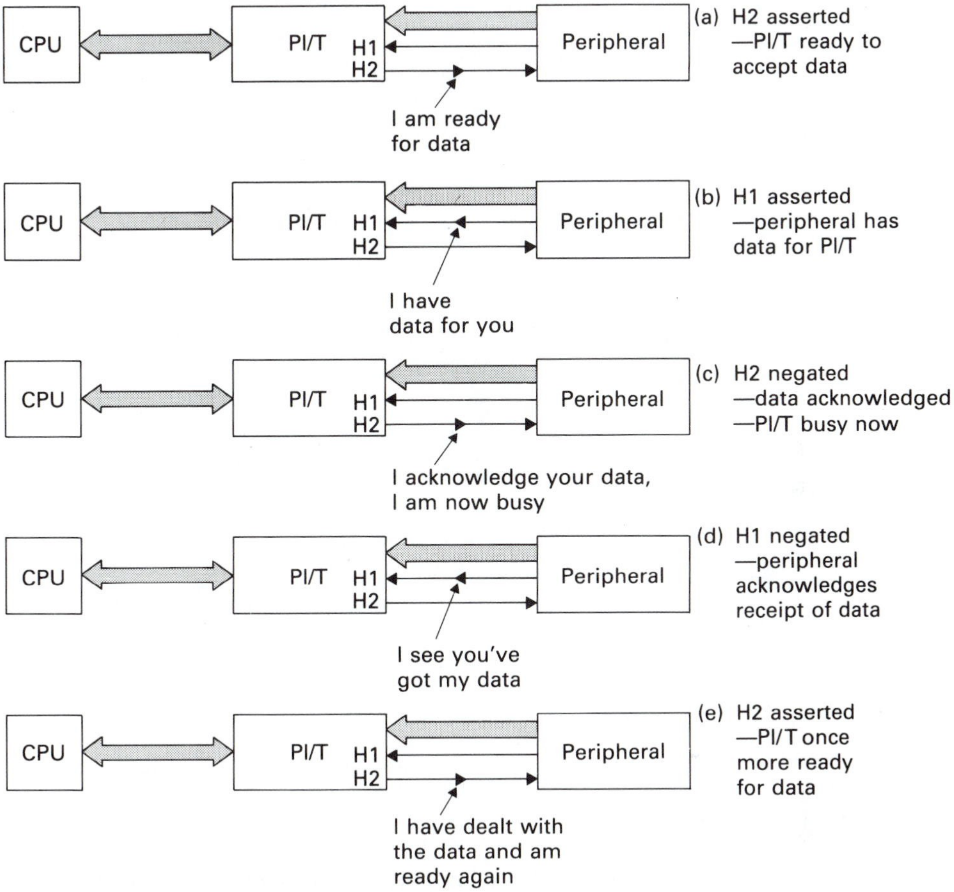

The timing diagram of two successive interlocked handshake input transfers is given in figure 8.17. Two cycles are shown because the PI/T is double-buffered. Double-buffering means that the PI/T can be receiving a new input while storing the previous input. In the first cycle, H2 is negated after H1 has been asserted and H2 is reasserted automatically after approximately four clock cycles because the input has been transferred from PI/T's *initial input latches* to its *final input latches* and the initial input latches are once more free to accept data. However, on the second input cycle, H2 remains inactive-high (i.e., it is not self-clearing) because both input buffers are full. Only when the CPU reads from the storage buffer does the H2 output reassert itself.

Double-buffering means that data can be transferred at almost the maximum rate at which the CPU can read the PI/T, without information being lost. Had the PI/T been supplied with many more buffers (making it a FIFO), instantaneous data rates of several times the host processor's transfer rate could have been supported."

FIGURE 8.17 Two consecutive input cycles using interlocked handshaking with double-buffered input

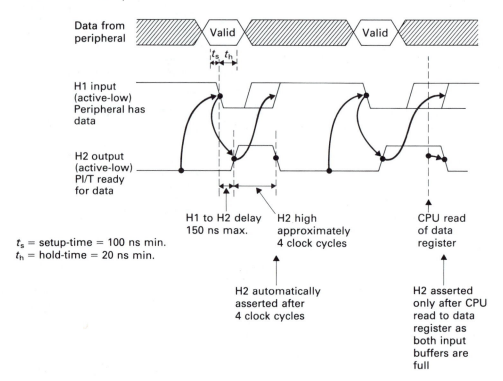

t_s = setup-time = 100 ns min.
t_h = hold-time = 20 ns min.

Output Handshaking

The PI/T implements double-buffered output transfers in very much the same fashion as the corresponding input transfers. Figure 8.18 shows the sequence of events taking place during an output transfer and figure 8.19 provides the corresponding timing diagram for two cycles of double-buffered output.

An output transfer starts at state (a) in figure 8.18, with the CPU loading data into the PI/T's output register, which causes H2 to be asserted after a delay of two clock cycles. This condition indicates to the peripheral that the data is available. At state (b), the peripheral asserts the PI/T's H1 input to indicate that it has read the data. The assertion of H1 causes the PI/T to negate H2 at state (c), indicating that the PI/T has acknowledged the peripheral's receipt of data. In turn, the peripheral negates H1 at state (d) to indicate that it is once more ready for data. Finally, the processor loads new data into the PI/T and H2 is asserted again to indicate a data-ready state at point (e).

The timing diagram of figure 8.19 also illustrates the effect of double-buffering on an output data transfer. Initially, both the PI/T's output buffers are empty. When data is first loaded into the PI/T by the CPU, the data is transferred to the chip's output terminals and H2 is asserted. At this point, one of the two output buffers is full. The buffer connected to the CPU is called the *initial O/P buffer* and that connected to the output pins is called the *final output buffer*.

When the next write to the PI/T's data register is made, the data is not immediately transferred to the output buffer, because the PI/T is in a busy state and cannot accept new data. Only when H1 is asserted by the peripheral does the PI/T transfer its latest data to the output register. Now the PI/T may once more accept data from the CPU. As in the case of input transfers, the CPU knows when the PI/T is ready for data by examining the state of the H1 flag bit, H1S.

Structure of the 68230 PI/T

The 68230 parallel interface and timer is available in a 48-pin package that interfaces to the 68000's asynchronous data bus via its 8 data pins and fully supports DTACK* and vectored interrupts. The key feature of the PI/T is its two independent 8-bit,

FIGURE 8.18 Closed-loop data transfer and the output handshake

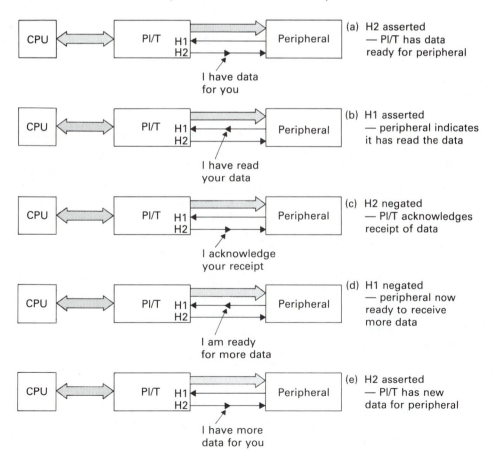

(a) H2 asserted
— PI/T has data
ready for peripheral

I have data
for you

(b) H1 asserted
— peripheral indicates
it has read the data

I have read
your data

(c) H2 negated
— PI/T acknowledges
receipt of data

I acknowledge
your receipt

(d) H1 negated
— peripheral now
ready to receive
more data

I am ready
for more data

(e) H2 asserted
— PI/T has new
data for peripheral

I have more
data for you

FIGURE 8.19 Two consecutive cycles of output using interlocked handshaking and double-buffering

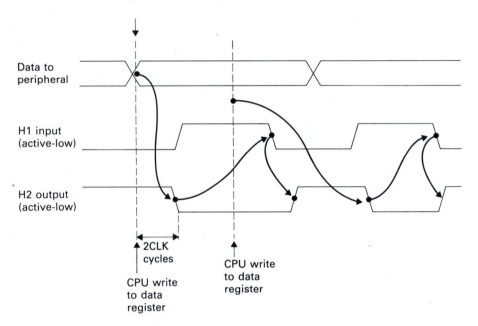

programmable I/O ports and its third dual-function C port. The C port can be programmed to operate as a simple I/O port without handshaking and double-buffering. However, the C port can also be programmed to act as a timer. In the latter mode, some of the port C pins perform other system functions. The PI/T also supports DMA operation in conjunction with a suitable DMA controller.

Figure 8.20 provides a block diagram of the internal structure of the PI/T and its pinout. The possible interface between a 68000 and a PI/T is presented in figure 8.21. The two 8-bit ports A and B and their handshake lines (H1, H2 for port A and H3, H4 for port B) are entirely application dependent and are unconnected in figure 8.21.

The interface between the 68230 PI/T and a 68000 CPU is simplicity itself. D_0 to D_7 from the PI/T is connected to D_{00} to D_{07} from the CPU (or alternatively to D_{08} to D_{15}, if the upper byte is to be used). Address lines A_{01} to A_{05} from the CPU are connected to the PI/T's five register select pins (RS_1 to RS_5) to enable the CPU to access any of the 32 addressable registers. The PI/T is accessed whenever its CS* input is active-low. CS* is derived from A_{06} to A_{23}, LDS*, and AS*. The active-low, open-drain DTACK* output from the PI/T is connected directly to the 68000's DTACK* input or to any suitable DTACK* point that is also driven by open-collector (or open-drain) outputs. The R/$\overline{W}$, RESET*, and CLK inputs to the PI/T are supplied by the 68000. Note that the CLK input does not have to be synchronized with the 68000's own clock.

Port C is a dual-function port and may be used as a simple 8-bit I/O port "without frills" or it may be used to support the chip's timer, interrupt, or DMA functions. The timer takes up three pins of port C ($PC2 = T_{IN}$, $PC3 = T_{OUT}$, $PC7 = TIACK*$). The PI/T supports a fairly simple 24-bit timer that has an optional (i.e., programmable) 5-bit input prescaler. The counter counts either clock pulses at the CLK input or pulses at the $PC2/T_{IN}$ pin. The timer output, $PC3/T_{OUT}$, generates the single or periodic pulses required by external equipment or it can be connected

FIGURE 8.20 Internal arrangement of the 68230 PI/T. (Reprinted by permission of Motorola Limited)

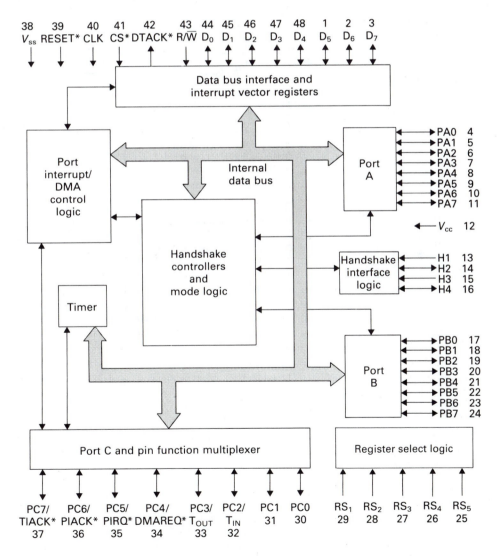

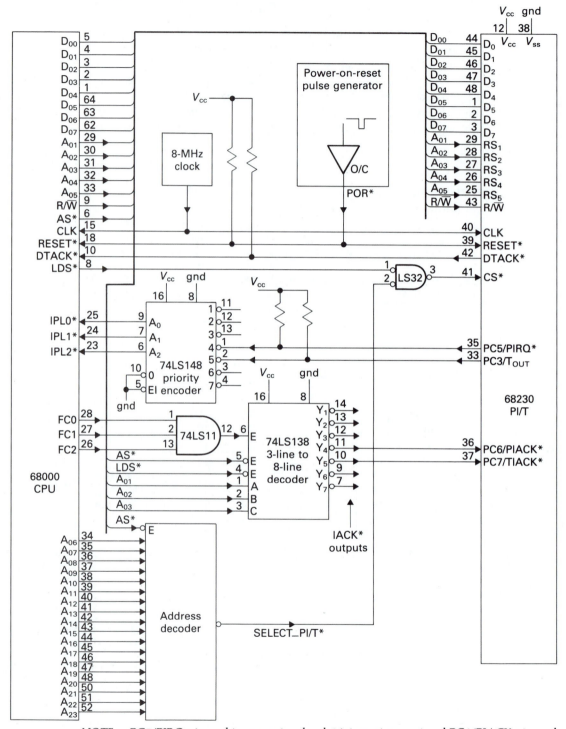

FIGURE 8.21 Interface between a 68000 CPU and a 68230 PI/T

NOTE: PC5/PIRQ* is used to generate a level 4 interrupt request and PC6/PIACK* is used to receive a level 4 IACK* from the 68000. PC3/T$_{OUT}$ and PC7/TIACK* provide a level 5 interrupt request and interrupt acknowledge from the timer part of the 68230.

to one of the CPU's interrupt request inputs to generate timed interrupts. The PC7/TIACK* input must be derived from the 68000's interrupt acknowledge circuitry and acknowledges an interrupt generated by an active transition on PC3/T_{OUT}. Of course, these three pins may be used as simple port C I/O pins if the timer function is not required.

Three other system functions are provided by port C. PC5/PIRQ* is a composite interrupt request output for parallel ports A and B. PC6/PIACK* is the corresponding interrupt acknowledge input to the PI/T. PC4/DMAREQ* is a DMA request output from the PI/T and may be used in conjunction with an external DMA controller to request a DMA operation between the peripheral connected to port A or port B and the system memory.

Table 8.3 defines the addresses and the names of the PI/T's 23 internal registers. Out of the 32 possible addresses on RS_1 to RS_5, nine values are not used and, when accessed, these null registers return the value $00. Although table 8.3 appears complex, the registers can be divided into functional groups: PI/T control and status, port A/B data and data direction registers, and counter registers. The functions of these registers are dealt with later.

TABLE 8.3 The 68230's internal register set

REGISTER SELECT BITS					REGISTER MNEMONIC	REGISTER DESCRIPTION	TYPE
RS_5	RS_4	RS_3	RS_2	RS_1			
0	0	0	0	0	PGCR	Port general control register	R/W
0	0	0	0	1	PSRR	Port service request register	R/W
0	0	0	1	0	PADDR	Port A data direction register	R/W
0	0	0	1	1	PBDDR	Port B data direction register	R/W
0	0	1	0	0	PCDDR	Port C data direction register	R/W
0	0	1	0	1	PIVR	Port interrupt vector register	R/W
0	0	1	1	0	PACR	Port A control register	R/W
0	0	1	1	1	PBCR	Port B control register	R/W
0	1	0	0	0	PADR	Port A data register	R/W
0	1	0	0	1	PBDR	Port B data register	R/W
0	1	0	1	0	PAAR	Port A alternate register	R only
0	1	0	1	1	PBAR	Port B alternate register	R only
0	1	1	0	0	PCDR	Port C data register	R/W
0	1	1	0	1	PSR	Port status register	R/W
1	0	0	0	0	TCR	Timer control register	R/W
1	0	0	0	1	TIVR	Timer interrupt vector register	R/W
1	0	0	1	1	CPRH	Counter preload register high	R/W
1	0	1	0	0	CPRM	Counter preload register middle	R/W
1	0	1	0	1	CPRL	Counter preload register low	R/W
1	0	1	1	1	CNTRH	Counter register high	R only
1	1	0	0	0	CNTRM	Counter register middle	R only
1	1	0	0	1	CNTRL	Counter register low	R only
1	1	0	1	0	TSR	Timer status register	R/W

Operating Modes of the PI/T

At first sight, the PI/T appears to be an exceedingly complex chip because it is a general-purpose device and supports so many different modes of operation. Each of the PI/T's operating modes is really very straightforward. The A and B ports of the PI/T can be configured to operate in seven ways. The ports are divided into four modes together with their submodes. The modes select the type of buffering (none/single/double) and whether the ports operate as independent 8-bit ports or as a combined 16-bit port. The operating modes of the PI/T are determined by bits 6 and 7 of its global control register (PGCR); that is, both port A and port B must operate in the same mode. However, the individual port control registers may be programmed to permit independent submodes of ports A and B.

Mode 0 Operation

Figure 8.22 illustrates the PI/T's mode 0 operation and its three submodes 00, 01, and 1X. These submodes are so called because they are chosen by setting bits 7 and 6 of the relevant port control register (PACR or PBCR). In figure 8.22, only port A is shown. Port B behaves in exactly the same way as port A, except that H3 acts like H1 and H4 acts like H2.

In modes 0 and 1, a data direction register (DDR) is associated with each port. PADDR controls port A and PBDDR independently controls port B. Each bit of the data direction register determines whether the corresponding bit of the port is an input or an output. A logical zero in bit i of a DDR defines the corresponding bit of a port as an input. A logical one defines the bit as an output. The contents of both DDRs are set to zero after a reset; for example, if PADDR is loaded with 00011111, PA5 to PA7 are inputs and PA0 to PA4 are outputs. This operating mode is called *undirectional* because it is changed only by resetting the PI/T or by reconfiguring the DDRs.

We can see from figure 8.22 that the submodes differ in terms of the buffering they permit on their inputs and outputs. The direction of data transfer that permits *double-buffering* is known as the primary data direction of the port. Data transfers in the primary direction are controlled by handshake pins H1, H2 for port A and H3, H4 for port B.

Mode 0, Submode 00   In submode 00, double-buffered input is provided in the primary direction. Output from the PI/T is single-buffered. Data is latched into the input register by the asserted edge of H1 and H2 behaves according to its programmed function defined in table 8.4. Up to now, we have discussed only the interlocked handshake mode offered by H1 and H2. It can be seen from table 8.4 that PACR_3 to PACR_5 may be used to define H2 as a simple output (i.e., an output at a logical 0 or logical 1 level), an interlocked handshake output, or a pulsed handshake output. In the latter case, the H2 output is asserted as in the interlocked mode of figures 8.17 and 8.19, but is negated *automatically* after approximately four clock cycles. In what follows, H1S and H2S are the status bits associated with H1 and H2, respectively, which are bits PSR_0 and PSR_1.

FIGURE 8.22 The PI/T in mode 0 (unidirectional 8-bit mode)

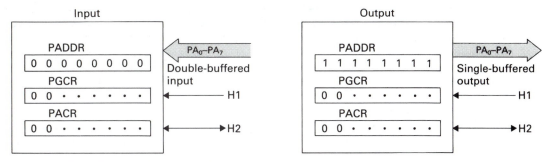

Submode 00   Pin-definable double-buffered input or single buffered output

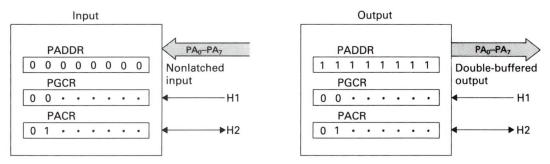

Submode 01   Pin-definable double-buffered output or single-buffered input

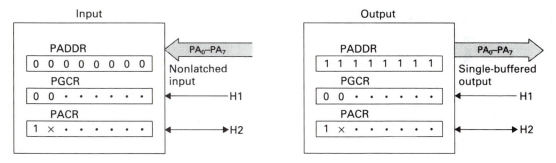

Submode 1X   Pin-definable single-buffered output or nonlatched input

NOTE: A dot in any register implies that the bit does not take part in the selection of either the mode or the submode. Only port A is shown here.

Mode 0, Submode 01   In submode 01, the primary data direction is from the PI/T and double-buffered output is provided. Input in this mode is nonlatched; that is, the input read by the CPU reflects the state of the input pin at the moment it is read. The programming of the port A control register in submode 01 is given in table 8.5 and is almost exacly the same as that of table 8.4, with the exception of the submode control field and the H1 status control bit, PACR_0. When PACR_0 = 0, the H1 status bit is set if *either* port A initial or final output latches can accept data and is clear otherwise. When PACR_0 = 1, the H1 status bit is set if *both* port A

TABLE 8.4 Port A control register (PACR) in mode 0, submode 00

Bit	PACR7 PACR6	PACR5 PACR4 PACR3	PACR2	PACR1 PACR0
Function	0 0	H2 control	H2 interrupt enable	H1 control

← Submode 00 →

PACR5	PACR4	PACR3	H2 CONTROL	
0	×	×	H2 edge-sensitive input	H2S set on asserted edge
1	0	0	H2 output—negated	H2S always clear
1	0	1	H2 output—asserted	H2S always clear
1	1	0	H2 output—interlocked handshake	H2S always clear
1	1	1	H2 output—pulsed handshake	H2S always clear

PACR2	H2 INTERRUPT ENABLE
0	H2 interrupt disabled
1	H2 interrupt enabled

PACR1	PACR0	H1 CONTROL
0	×	H1 interrupt and DMA request disabled
1	×	H1 interrupt and DMA request enabled
×	×	H1S status bit set if input data available

NOTE: H1S = H1 status bit of the port status register
            H2S = H2 status bit of the port status register

output latches are empty and is clear otherwise. In other words, we can program H1S to indicate the state *fully empty* or the state *half-empty*.

    Mode 0, Submode 1X In mode 0, submode 1X (see figure 8.22), simple bit I/O is available in both directions. Double-buffered I/O cannot be used in either direction. Data read from a pin programmed as an input is the instantaneous (i.e., nonlatched) signal at that pin. Data written to an output is single-buffered. H1 is an edge-sensitive input only and plays no part in any handshaking procedure related to the PI/T.

TABLE 8.5 Port A control register (PACR) in mode 0, submode 01

Bit	PACR7	PACR6	PACR5	PACR4	PACR3	PACR2	PACR1	PACR0
Function	0	1		H2 control		H2 interrupt enable	H1 control	

←——— Submode 01——→

PACR1	PACR0	
0	×	H2 interrupt and DMA request disabled
1	×	H2 interrupt and DMA request enabled
×	0	H1S indicates initial or final O/P latches empty
×	1	H1S indicates both O/P latches empty

H2 may be programmed as an edge-sensitive input that sets status bit H2S when asserted. As in the case of the other submodes described previously, H2 can be programmed as an output and set or cleared under program control. Table 8.6 defines the options available in this submode.

Mode I Operation

In mode 1, the two 8-bit ports are combined to act as a single 16-bit port. This port is still *unidirectional* in the sense that the primary direction of data transfer is associated with double-buffering and handshake control, and ports A and B data direction registers define whether the individual bits of the 16-bit port are to act as inputs or outputs. Figure 8.23 illustrates the possible configurations of the PI/T in mode 1 operation.

A combined port raises two problems—what do we do about the two pairs of handshake signals (H1, H2 and H3, H4) and what about the two port control registers (PACR and PBCR)? In this mode, port B supplies the handshake signals and the control register (PBCR). The port A control register is used in conjunction with H1 and H2 to provide the 16-bit port with additional facilities. Table 8.7 defines the effect of the PACR on the mode 1 operation of the PI/T.

In mode 1, the port A control register simply treats H1 as an edge-sensitive input and H2 as an edge-sensitive input or an output that may be set or cleared.

Mode 1, Submode X0 In mode 1, submode X0, double-buffered inputs or single-buffered outputs of up to 16 bits are possible. Note that the PI/T has only an 8-bit interface to the 68000 so that a 16-bit word must be transferred to the CPU as 2 bytes. Port A should be read before port B. For compatibility with the MOVEP instruction, port A should contain the most significant byte of data. The operation of the 16-bit port is determined by the port B control register, the structure of which is

TABLE 8.6 Port A control register (PACR) in mode 0, submode 1 ×

Bit	PACR7	PACR6	PACR5	PACR4	PACR3	PACR2	PACR1	PACR0
Function	1	×	H2 control			H2 interrupt enable	H1 control	

← Submode 1 × ⟶

PACR5	PACR4	PACR3	H2 FUNCTION	
0	×	×	Edge-sensitive input	H2S set on asserted edge
1	×	0	H2 output—negated	H2S always clear
1	×	1	H2 output—asserted	H2S always clear

PACR1	FUNCTION
0	H1 interrupt disabled
1	H1 interrupt enabled

PACR0	FUNCTION
×	H1 is an edge-sensitive input and H1S is set by an asserted edge of H1

given in table 8.8. The signal at each input is latched synchronously with the asserted edge of H3 and placed in either the initial input latch or the final input latch. As in mode 0 operation, H4 may be programmed to act as an input, a fixed output, or a pulsed/interlocked handshake signal.

TABLE 8.7 The format of port A control register during a mode 1 operation

Bit	PACR7	PACR6	PACR5	PACR4	PACR3	PACR2	PACR1	PACR0
Function	0	0	H2 control			H2 interrupt enable	H1 control	

Port A submode   PACR_0–PACR_5 behave exactly as in mode 0
                      submode 1 × —see table 8.6.

FIGURE 8.23 PI/T in mode I (unidirectional 16-bit mode)

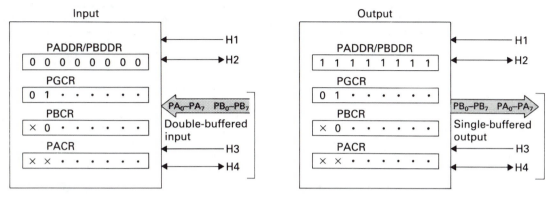

Submode X0   Pin-definable double-buffered input or single-buffered output

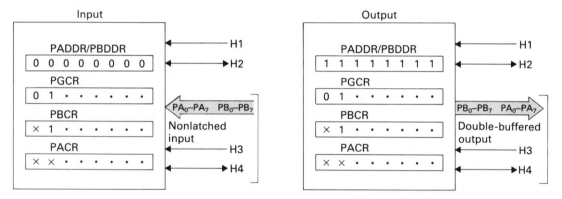

Submode X1   Pin-definable double-buffered output or nonlatched input

For pins programmed as outputs, the data path consists of a single latch driving the output buffer. Data written to this port's data register does not affect the operation of any handshake pin, status bit, or any other aspect of the PI/T.

Mode 1, Submode X1 In mode 1, submode X1, double-buffered outputs or nonlatched inputs of up to 16 bits are possible. Data is written to the PI/T as 2 bytes. The first byte (most significant) is written to the port A data register and the second byte is written to the port B data register (in that order). The PI/T then automatically transfers the 16-bit data to one of the output latches.

The port A control register and associated handshake signals (H1 and H2) behave exactly as in mode 1, submode X0, defined by tables 8.7 and 8.6. In a similar fashion, the port B control register behaves rather like the same register in mode 1, submode X0. The only differences are in bits PBCR_7 and PBCR_6, which are set to 0, 1 to select this mode, and in bit PBCR_0. When PBCR_0 is zero, the H3S status bit is set when either the initial or final output latch of ports A and B can accept new data. Otherwise it is clear. When PBCR_0 is 1, the H2S status bit is set when both the initial and final output latches of ports A and B are empty and is clear otherwise.

TABLE 8.8 Format of the port B control register during a mode I, submode X0 operation

Bit	PBCR7	PBCR6	PBCR5	PBCR4	PBCR3	PBCR2	PBCR1	PBCR0
Function	×	0	H4 control			H4 interrupt enable	H3 control	

Submode × 0

PBCR5	PBCR4	PBCR3	H4 FUNCTION	
0	×	×	Edge sensitive input	H4S set on asserted edge
1	0	0	Output—negated	H4S always cleared
1	0	1	Output—asserted	H4S always cleared
1	1	0	Output—interlocked handshake	H4S clear
1	1	1	Output—pulsed handshake	H4S clear

PBCR2	H4 INTERRUPT ENABLE
0	The H4 interrupt is disabled
1	The H4 interrupt is enabled

PBCR1	H3 SERVICE REQUEST ENABLE
0	The H3 interrupt and DMA request are disabled
1	The H3 interrupt and DMA request are enabled

PBCR0	HS STATUS CONTROL
×	The H3S status bit is set any time input data is present

Mode 2 Operation

Mode 2 offers bidirectional I/O and is illustrated in figure 8.24. Port A is used for bit I/O transfers with no associated handshake pins and provides unlatched input or single-buffered output. Individual pins can be programmed as inputs or outputs by the setting or clearing bits in the port A data direction register. Port B is the

FIGURE 8.24 PI/T configured for mode 2 (bidirectional 8-bit mode)

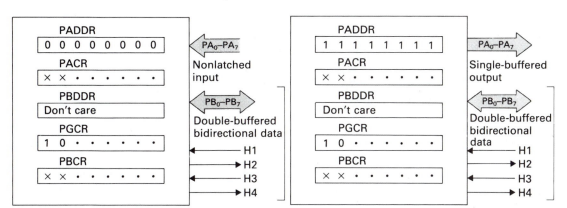

"workhorse" and acts as a bidirectional, 8-bit, double-buffered I/O port. The hand-shake pins are all associated with port B and operate in two pairs. H1, H2 control output transfers and H3, H4 control input transfers. The instantaneous direction of the data is determined by the H1 handshake pin; that is, the *external device* deter-mines the direction of data transfer. The port B data direction register has no effect in this mode because only byte I/O is permitted in mode 2. Similarly, ports A and B submode fields do not affect PI/T operation in mode 2.

The output buffers of port B are controlled by the level on the H1 input. When H1 is negated, the port B output buffers are enabled and its pins drive the output bus. Note that all eight buffers are enabled because individual pins cannot be pro-grammed as inputs or outputs in modes 2 and 3. Generally, H1 is negated by a peripheral in response to the assertion of H2, which indicates that new output data is present in the double-buffered latches. Following acceptance of the data, the periph-eral asserts H1, disabling the port B output buffers. H1 acts as an edge-sensitive input.

Double-Buffered Input Transfers Data at the port B input pins is latched on the asserted edge of H3 and deposited in one of the input latches. The corre-sponding H3 status bit, H3S, is set whenever input data, which has not been read by the host computer, is present in the input latches. As in all other modes, H4 is programmable. In modes 2 and 3, H4 can be programmed to perform two functions:

1. H4 may be an output pin in the interlocked handshake mode. H4 is assert-ed when the port is ready to accept new data and is negated asynchronously following the asserted edge of the H3 input. As soon as one of the input latches becomes ready to receive data, H4 is reasserted. Once both input latches are full, H4 remains negated until data is read by the host processor.

2. H4 may be an output pin in the pulsed input handshake mode. H4 is asserted when the input port is ready to receive new data exactly as in function 1 and is automatically cleared (negated) approximately four clock cycles later. Should a subsequent active transition on H3 take place while H4 is asserted,

H4 is negated asynchronously; that is, once the active edge of H4 has been detected by the peripheral, new data may be loaded into the double-buffered input latches.

Double-Buffered Output Transfers Data is written into one of the PI/T's output latches by the host processor. The peripheral connected to port B accepts data by asserting H1. The H1 status bit may be programmed to be set when either one or both output buffers are empty or only when both output buffers are empty. H2 may be programmed to act in one of two modes:

1. H2 may be an output pin in the interlocked output handshake mode and is asserted whenever the output latches are ready to accept new data from the host processor. H2 is negated asynchronously following the asserted edge of H1. As soon as one or both output latches become available, H2 is reasserted. When both output latches are full, H2 remains asserted intil at least one latch is emptied.

2. H2 may be an output pin in the pulsed output handshake protocol. H2 is asserted whenever the output latches are ready to accept new data, but is automatically negated approximately four clock cycles later. Should the asserted edge of H1 be detected while H2 is asserted, H2 is negated asynchronously.

The programming of port control registers A and B in mode 2 is illustrated in tables 8.9 and 8.10. Note that, in this mode, many bits are don't care functions.

Mode 3 Operation

Mode 3 operation is an extension of mode 2 and is illustrated in figure 8.25. In mode 3, both ports A and B are dedicated to 16-bit double-buffered input/output transfers. As in mode 2, H1 and H2 control output transfers and H3 and H4 control input transfers. The function of the port control registers (PACR, PBCR) in mode 3 is exactly the same as in mode 2 and therefore tables 8.9 and 8.10 also apply to mode 3.

TABLE 8.9 Format of the port A control register during a mode 2 operation

Bit	PACR7	PACR6	PACR5	PACR4	PACR3	PACR2	PACR1	PACR0
Function	×	×	×	×	H2 mode	H2 interrupt enable	H1 control	

Submode × × (don't care)

0 Interlocked handshake
1 Pulsed handshake

PACR_1, PACR_0 exactly as in table 8.5

TABLE 8.10 Format of the port B control register during a mode 2 operation

Bit	PBCR7	PBCR6	PBCR5	PBCR4	PBCR3	PBCR2	PBCR1	PBCR0
Function	×	×	×	×	H4 mode	H4 interrupt enable	H3 control	

Submode × × (don't care)

0 Interlocked handshake
1 Pulsed handshake

PBCR_1, PBCR_0 exactly as in table 8.4

Mode 3 provides a relatively convenient way of transferring data between the PI/T and 16-bit peripherals at high speed. The word "relatively" has been included because the PI/T interfaces to its host processor by an 8-bit data bus, allowing 16-bit peripheral-side data transfers but denying 16-bit CPU-side data transfers. However, by using the 68000's MOVEP (move peripheral instruction), 16-bit data transfers can be performed with only one instruction.

Registers of the PI/T

So far, we have looked at the PI/T from the point of view of its operational modes and discussed its internal registers only when the need arose. Here we provide further details on the PI/T's internal registers. Table 8.3 gives the names and mnemonics of the 68230's registers together with the register select lines (RS_1 to RS_5) needed to access them. In the following discussion, registers are also specified in

FIGURE 8.25 PI/T configured for mode 3 (bidirectional 16-bit mode)

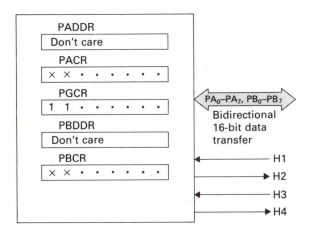

terms of their offset addresses. The offset address is given with respect to the device's base address, assuming that RS_1 to RS_5 are connected to A_{01} to A_{05}, respectively; for example, the address offset of PGCR is 0 and of TSR it is $34. Note that the offset is a *byte* value.

Port General Control Register (PGCR, offset = $00)

The PGCR is a global register that determines the operating mode of the PI/T. Bits 6 and 7 of the PGCR select the PI/T's operating mode (see table 8.11), bits 4 and 5 enable the handshake pairs H1, H2 and H3, H4, and bits 0 to 3 determine the sense of the four handshake lines; for example, setting PGCR_1 = 1 defines the active transition on H2 as a zero to one.

Port Service Request Register (PSRR, offset = $02)

The port service request register is also a global register and determines the circumstances under which the PI/T may request service. Table 8.12 gives the format of the PSRR, which is split into three logical fields: a service request field (SVCRQ) that determines whether the PI/T generates an interrupt or a DMA request when H1 or H3 are asserted, an operation select field that determines whether two of the dual-function pins belong to port C or perform special-purpose functions, and an interrupt priority control field. A DMA request is signified by an active-low pulse on the DMAREQ* (direct memory access request) pin for three clock cycles.

Port Data Direction Registers (PDDRA, offset = $04; PDDRB, offset = $06; PDDRC, offset = $08)

The port data direction registers determine the direction and buffering characteristics of each of the appropriate port pins. A logical one in a PDDR bit makes the corresponding port I/O pin act as an output, while a logical zero makes the pin an output. All DDRs are cleared to zero after a reset.

TABLE 8.11 Format of the port general control register (PGCR)

Bit	PGCR7	PGCR6	PGCR5	PGCR4	PGCR3	PGCR2	PGCR1	PGCR0
Function	Port mode control		H34 enable	H12 enable	H4 sense	H3 sense	H2 sense	H1 sense

00 Mode 0
01 Mode 1
10 Mode 2
11 Mode 3

0 = disable
1 = enable

Sense = 0 assertion level low
Sense = 1 assertion level high

NOTE: The H12 and H34 enable/disable fields enable or disable the operation of the H1, H2 or H3, H4 handshake lines. All bits of the PGCR are cleared after a reset operation. The handshake lines are enabled only when PGCR_5 and PGCR_4 are set under program control to avoid spurious operation of the handshake lines until the PI/T has been fully configured.

TABLE 8.12 Format of the port service request register (PSRR)

Bit	PSRR7	PSRR6	PSRR5	PSRR4	PSRR3	PSRR2	PSRR1	PSRR0
Function	×	SVCRQ		Operation select		Port interrupt priority		

0 ×   PC4/DMAREQ* = PC4 (DMA not used)

10   P4C/DMAREQ* = DMAREQ* and is associated with double-buffered transfers controlled by H1. H1 does not cause interrupts in this mode.

11   P4C/DMAREQ* = DMAREQ* and is associated with double-buffered transfers controlled by H3. H3 does not cause interrupts in this mode.

OPERATION SELECT		INTERRUPT PIN FUNCTION SELECT	
PSRR4	PSRR3		
0	0	PC5/PIRQ* = PC5	No interrupt support
		PC6/PIACK* = PC6	No interrupt support
0	1	PC5/PIRQ* = PIRQ*	Autovectored interrupts supported
		PC6/PIACK* = PC6	Autovectored interrupts supported
1	0	PC5/PIRQ* = PC5	
		PC6/PIACK* = PIACK*	
1	1	PC5/PIRQ* = PIRQ*	Vectored interrupts supported
		PC6/PIACK* = PIACK*	Vectored interrupts supported

PORT INTERRUPT PRIORITY			ORDER OF INTERRUPT PRIORITY			
PSRR2	PSRR1	PSRR0	HIGHEST ←			→ LOWEST
0	0	0	H1S	H2S	H3S	H4S
0	0	1	H2S	H1S	H3S	H4S
0	1	0	H1S	H2S	H4S	H3S
0	1	1	H2S	H1S	H4S	H3S
1	0	0	H3S	H4S	H1S	H2S
1	0	1	H3S	H4S	H2S	H1S
1	1	0	H4S	H3S	H1S	H2S
1	1	1	H4S	H3S	H2S	H1S

The port C PDDR behaves in the same way as the other two PDDRs and determines whether each dual-function pin chosen for port C operation is an input or an output pin.

Port Interrupt Vector Register (PIVR, offset = $0A)

The port interrupt vector register contains the upper-order 6 bits of the four port interrupt vectors. The contents of this register may be read in one of two ways: by an ordinary read cycle to the PIVR at offset address $0A or by a port interrupt acknowledge bus cycle.

During a normal read cycle, no "consequence" following the reading of this register exists. However, when the PIVR is read during an interrupt acknowledge cycle, the least two significant bits are determined by the source of the interrupt, as illustrated in table 8.13.

As an example, if the PIVR is loaded with 01101100 ($6C) during the PI/T's initialization phase, an interrupt initiated by H2 will yield an interrupt vector number of 01101101 ($6D). Similarly, an interrupt initiated by H4 will yield an interrupt vector number of 01101111 ($6F). This procedure is followed to avoid having four separate vector number registers. After the RESET* input to the PI/T has been asserted, the contents of the PIVR are initialized to $0F, which is, of course, the *uninitialized vector* number.

Port Control Registers (PCRA, offset = $0C; PCRB, offset = $0E)

The port control registers determine the submode operation of ports A and B and control the operation of the handshake lines. The programming of these two registers has already been dealt with when the operating modes of the PI/T were described.

Port Data Registers (PADR, offset = $10; PBDR, offset = $12; PCDR, offset = $18)

Port A and port B data registers are holding registers between the CPU-side bus of the PI/T and its port pins and internal buffer registers. These registers may be written to or read from at any time and are not affected by a reset on the PI/T.

TABLE 8.13 Relationship between PIVR0, PIVR1, and interrupt source

INTERRUPT SOURCE	PIVR1	PIVR0
H1	0	0
H2	0	1
H3	1	0
H4	1	1

TABLE 8.14 Accessing the port C data register

OPERATION	PORT C FUNCTION		ALTERNATE FUNCTION	
	PCDDR = 0	PCDDR = 1	PCDDR = 0	PCDDR = 1
Read PCDR	Read pin	Read output register	Read pin	Read output register
Write PCDR	Output register, buffer disabled	Output register, buffer enabled	Output register	Output register

The port C data register, PCDR (offset = $18), is a holding register for moving data to and from port C or its alternate-function pins. The exact nature of an information transfer depends on the type of cycle (read or write) and on the way in which port C is configured. Table 8.14 shows how the PCDR is affected by read/write accesses. Pins configured as port C functions offer single-buffered output or non-latched input.

Note that we are able to directly read the state of a dual-function pin even when it is used for non–port C functions. We are also able, of course, to generate non–port C functions "manually" by switching back to port C mode and writing to the PCDR. The port C data register is readable and writable at all times and is not affected by the state of the RESET* pin.

Port Alternate Registers (PAAR, offset = $14; PBAR, offset = $16)

Port A and port B alternate registers provide a way of reading the state of port A and port B pins, respectively. Both PAAR and PBAR are read-only and their contents reflect the actual instantaneous logic levels at the I/O pins. Writing to PAAR or PBAR results in a DTACK* handshake, but no data is latched by the PI/T and the bus cycle has no other effect on the PI/T. These registers are not affected by the operating modes of the PI/T.

Port Status Register (PSR, offset = $1A)

The port status register is the global register that reflects the activity of the handshake pins. The format of the PSR is given in table 8.15. Bits PSR_4 through PSR_7 reflect the instantaneous level at the respective handshake pin and are independent of handshake pin sense bits in the PGCR. Bits PSR_0 through PSR_3 are the handshake status bits and are set or cleared as specified by the appropriate operating mode. Each of these bits is active-high and is set when the appropriate handshake line is asserted.

TABLE 8.15   Format of the port status register

Bit	PSR7	PSR6	PSR5	PSR4	PSR3	PSR2	PSR1	PSR0
Function	H4 level	H3 level	H2 level	H1 level	H4S	H3S	H2S	H1S

Bits set or cleared by instantaneous level on handshake pin.	Bits set by assertion of handshake pin as programmed.

Timer Functions of the PI/T

The 68230 contains a single timer that interfaces to the host processor through the same CPU-side pins as the parallel interface and interfaces to external systems (or to the 68000 interrupt structure) through the alternate function pins of port C. Typical functions performed by the 68230 (or any other timer) are: the generation of square waves of programmable frequencies, the generation of single pulses of programmable duration, the production of single or periodic interrupts, and the measurement of frequency or elapsed time. A timer is, essentially, a simple device consisting of a counter that is clocked, typically, downward toward zero. By selecting the clock rate and the initial contents of the counter, a specific delay between starting the counter and the moment it reaches zero can be generated. If the counter is reloaded every time it reaches zero, we have a method of generating repetitive action.

Figure 8.26 shows the three peripheral-side interface lines of the timer together with its associated registers. The time contains a 24-bit synchronous down-counter (CNTR) that is loaded from three 8-bit counter preload registers (CPR). The synchronous counter is clocked either by the system clock (CLK) or by an external input applied to T_{IN}. The clock may, optionally, be prescaled by 32.

As the counter clocks downward, it eventually reaches zero and sets the zero-detect status bit (ZDS) of the timer status register (TSR). This event can be used to assert the T_{OUT} output from the timer. If T_{OUT} is connected to one of IRQ1* to IRQ7*, an interrupt may be generated.

The operating mode of the timer is determined by the timer control register, whose format is given in table 8.16. This table is rather complex but, in principle, it controls: (1) the choice between the port C option and the timer option on three dual-function pins; (2) whether the counter is loaded from the counter preload register or rolls over when zero detect is reached; (3) the source of the clock input; (4) whether the clock source is prescaled (i.e., divided by 32); and (5) whether the timer is enabled or disabled.

The 68230 has an independent timer interrupt vector register (TIVR) that supplies an 8-bit vector whenever the timer interrupt acknowledge pin, TIACK*, is asserted. The TIVR is automatically loaded with the unitialized interrupt vector, $0F, following a reset.

TABLE 8.16 Format of the timer control register

Bit	TCR7	TCR6	TCR5	TCR4	TCR3	TCR2	TCR1	TCR0
Function	T_{OUT}/TIACK* control			ZD control	×	Clock control		Timer enable

TCR7	TCR6	TCR5	T_{OUT}/TIACK* CONTROL
0	0	×	PC3/T_{OUT}, PC7/TIACK* are port C functions
0	1	×	PC3/T_{OUT} is a timer function. In the run state T_{OUT} provides a square wave that is toggled on each zero-detect. The T_{OUT} pin is high in the halt state. PC7/TIACK* is a port C function.
1	0	0	PC3/T_{OUT} is a timer function. In the run or halt state it is used as a timer interrupt request output. The timer interrupt is disabled—the pin is always three-stated. PC7/TIACK* is a timer function. Since interrupt request is negated, the PI/T produces no response to an asserted TIACK*.
1	0	1	PC3/T_{OUT} is a timer function and is used as a timer interrupt request output. The timer interrupt is enabled and T_{OUT} is low whenever the ZDS bit is set. PC7/TIACK* is a timer function and acknowledges interrupts generated by the timer. This combination supports vectored interrupts.
1	1	0	PC3/T_{OUT} is a timer function. In the run or halt state, it is used as a timer interrupt request output. The timer interrupt is disabled. PC7/TIACK* is a port C function.
1	1	1	TC3/T_{OUT} is a timer function and is used as a timer interrupt request output. The timer interrupt is enabled and T_{OUT} is low when the ZDS status bit is set. PC7/TIACK* is a port C function. Autovectored interrupts are supported.

TCR4	ZERO-DETECT CONTROL
0	The counter is loaded from the counter preload register on the first clock to the 24-bit counter after zero-detect; counting is then resumed.
1	The counter rolls over on zero-detect and then continues counting.

TABLE 8.16 (*continued*)

TCR2	TCR1	CLOCK CONTROL
0	0	PC2/T_{IN} is a port C function. Counter clock is from CLK prescaled by 32. The timer enable bit determines whether the timer is in the run or halt state.
0	1	PC2/T_{IN} is a timer input. The prescaler is decremented on the falling edge of the CLK input and the 24-bit counter is decremented when the prescaler rolls over from $00 to $1F. The timer is in the run state when the enable bit is one and the T_{IN} pin is high. Otherwise the timer is in the halt state.
1	0	PC2/T_{IN} is a timer input and is prescaled by 32. The prescaler is decremented following the rising transition of T_{IN} after being synchronized with the internal clock. The 24-bit counter is decremented when the prescaler rolls over from $00 to $1F. The timer enable bit determines whether the timer is in the run or halt state.
1	1	PC2/T_{IN} is a timer input and prescaling is not used. The 24-bit counter is decremented following the rising edge of the signal at the T_{IN} pin after being synchronized with the internal clock. The timer enable bit determines whether the timer is in the run or halt state.

TCR0	TIMER ENABLE BIT
0	Timer disabled
1	Timer enabled

Timer States

The timer is always in one of two states: running or halted. The control of the timer's state is effected by loading the appropriate value into the timer control register (table 8.16). The characteristics of the two states are as follows.

1. *Halt state*

    a. The contents of the counter are stable (do not change) and can be reliably and repeatedly read from the count registers.

    b. The prescaler is forced to $1F whether or not it is in use.

    c. The ZDS bit is forced to zero, regardless of the contents of the 24-bit counter.

2. *Run state*

    a. The counter is clocked by the source programmed in the timer control register.

b. The counter is not *reliably* readable.

c. The prescaler is allowed to decrement if it is so programmed.

d. The ZDS status bit is set when the counter makes a $00 0001 to $00 0000 transition.

Timer Applications

The timer section of the PI/T is not as complex as a first reading of table 8.16 would suggest. In order to illustrate the operation of the timer, some of its possible applications are now presented.

Real-Time Clock A real-time clock, RTC, generates an interrupt at periodic intervals and may be used by the operating system to switch between several tasks (see chapter 6) or to update a record of the time of day. In this configuration, the

FIGURE 8.26 Timer function of the 68230 PI/T

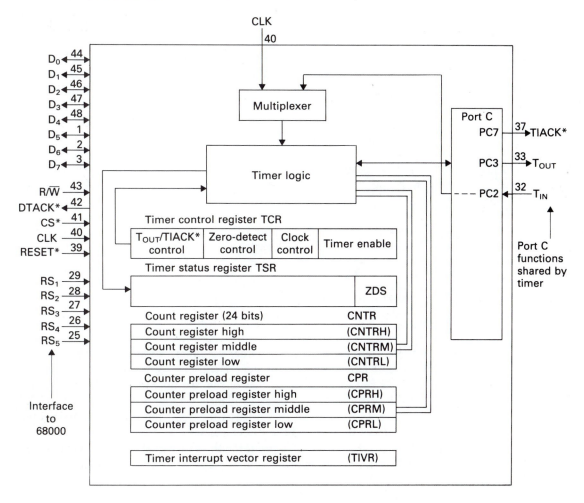

TABLE 8.17 Format of the TCR in the real-time mode

Bit	TCR7 TCR6 TCR5	TCR4	TCR3	TCR2 TCR1	TCR0
Value	1 × 1	0	0	0 0	1
Function	T_{OUT}/TIACK* control	ZD control		Clock control	Timber enable
	PC3/T_{OUT} = T_{OUT} timer interrupt enabled; T_{OUT} low when ZDS set	Counter reload on zero detect		Counter clock = CLK/32	

T_{OUT} pin is connected to one of the host processor's interrupt request inputs and the TIACK* input used as an interrupt request input to the timer. The T_{IN} pin may be used as a clock input or the system clock selected. The format of the TCR needed to select the real-time clock mode is given in table 8.17.

The host processor first loads the counter preload registers with a 24-bit value and then configures the TCR as described previously. The timer enable bit of the TCR may be set at any time counting is to begin—it need not be set during the timer initialization phase.

When the counter counts down from $00 0001 to $00 0000, the ZDS status bit is set and the T_{OUT} pin asserted to generate an interrupt request. At the next clock input to the 24-bit counter, the counter is loaded with the contents of the counter preload register. The host processor should clear the ZDS status bit to remove the source of the interrupt. The operation of the timer in this mode can be illustrated in PDL and by the timing diagram of figure 8.27:

```
REPEAT WHILE Timer_enable_bit = 1
        FOR I = Counter_preload_value DOWN_TO 0
                Clear ZDS_bit
                Negate T_OUT
        END_FOR
        Set ZDS_bit
        Assert T_OUT
END_REPEAT
```

Square-Wave Generator   In this mode, the timer produces a square wave at its T_{OUT} terminal and interrupts are not generated. The format of the TCR in the square-wave mode is almost identical to that of the real-time mode—the only major difference is that bit 7 of the TCR is clear. A glance at table 8.16 reveals that TCR7 controls T_{OUT}. When TCR7 is clear, the signal at the T_{OUT} pin is toggled every time the counter counts down to zero and the ZDS bit is set.

Figure 8.28 provides a timing diagram for the timer in the square-wave mode. Note that, as above, the timer counting source may be obtained from CLK or from T_{IN} and may be prescaled by 32.

FIGURE 8.27 Timing diagram of the PI/T as a real-time clock

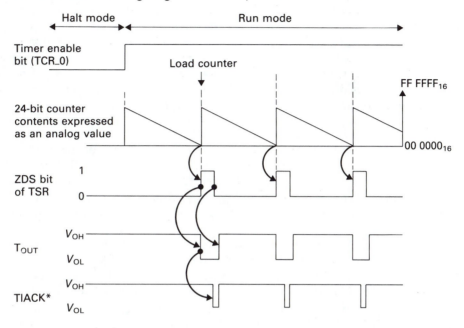

Interrupt After Time-OUT   In this mode, the timer generates an interrupt after a programmed period of time has elapsed. As in the case of the real-time clock, T_{OUT} is connected to the appropriate interrupt request line of the host processor and TIACK* may be used as an interrupt acknowledge input. The source of timing may be derived from the system clock or from T_{IN}.

FIGURE 8.28 Timing diagram of the PI/T as a square wave generator

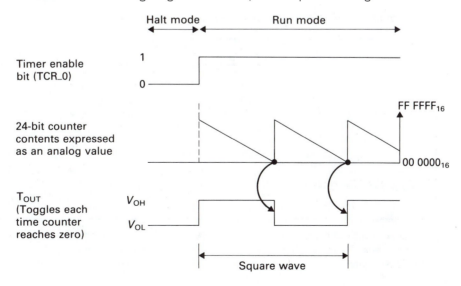

Table 8.18 gives the format of the TCR appropriate to this mode and the other operating modes discussed here. This configuration is similar to the real-time clock mode, except that the zero-detect bit of the TCR is set. Consequently, when the counter reaches zero it rolls over to its maximum value rather than being loaded from the counter load registers. Figure 8.29 illustrates this process. Once the interrupt has been serviced, the host processor can halt the timer and, if necessary, read the contents of the counter. At this point, the number in the counter gives an indication of the time elapsed between the interrupt request and its servicing.

Elapsed Time Measurement This configuration allows the host processor to determine the time that elapses between the triggering of the 68230 timer and its halting by clearing its enable bit. Table 8.18 (mode 4) gives the format of the TCR in this mode. The processor initializes the timer by loading its counter preload register with $FF FFFF (all ones) and setting its TCR and then enables it by setting TCR_0. The value $FF FFFF is selected because it provides the longest possible counting period.

Once TCR_0 has been set, the prescaler counts down toward zero and decrements the counter each time it rolls over from $00 to $1F. When the event, whose action signals the end of the timing period, takes place, the processor clears the enable bit (TCR_0) and halts the countdown. The processor determines the timing period by reading the contents of the counter registers. Note that we are able to program the TCR to permit an external clock to be connected to T_{IN}. When an external clock or pulse generator is connected to the T_{IN} input and the TCR initialized, as in table 8.18, mode 5, the timer can be used to count the number of pulses at the T_{IN} pin between the points at which TCR_0 is set and cleared.

By setting bits TCR_2, TCR_1 to 0, 1, respectively, the timer can be started and stopped by T_{IN} (table 8.18, mode 6). In this case, the timer requires that both TCR_0 *and* T_{IN} be high before counting may begin. Therefore, once TCR_0 has

TABLE 8.18 Format of the TCR in various operating modes of the PI/T

Bit	TCR7	TCR6	TCR5	TCR4	TCR3	TCR2	TCR1	TCR0	MODE
	1	×	1	0	0	00 or	1×	1	1
	0	1	×	0	0	00 or	1×	1	2
	1	×	1	1	0	00 or	1×	1	3
	0	0	×	1	0	0	0	1	4
	0	0	×	1	0	1	×	1	5
	1	×	1	1	0	0	1	1	6
Function	T_{OUT}/TIACK* control			ZD control		Clock control		Timer enable	

Mode 1 = real-time clock   Mode 4 = elapsed time measurement
Mode 2 = square-wave generator   Mode 5 = pulse counter
Mode 3 = interrupt after time-out   Mode 6 = period measurement

FIGURE 8.29 Timing diagram of the PI/T as a time-out interrupt generator

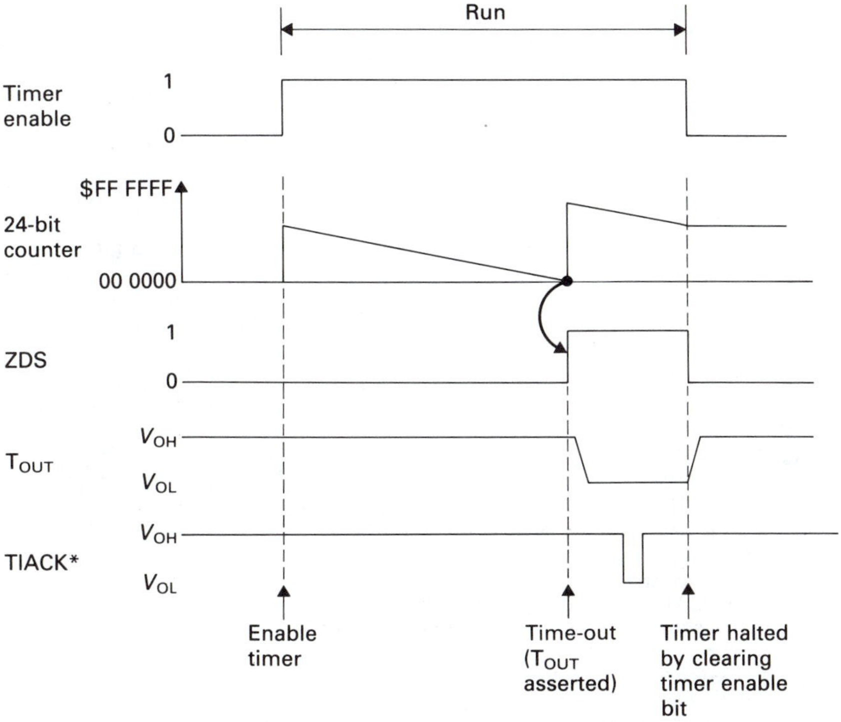

been set under software control, counting begins only when T_{IN} goes high and stops when T_{IN} returns low. If T_{IN} is controlled by external circuitry, the processor can determine the period between the positive and negative transition of T_{IN}.

Example of a PI/T Application

We are now going to describe how a PI/T can be used to interface a printer to a 68000 microprocessor with a minimum of hardware and software. This example is taken from Motorola's Application Note AN-854. Printers are usually connected to a host computer either by a serial RS232C data link or by a parallel data link conforming to the Centronics standard. In this application, a 68230 PI/T is used to implement a Centronics interface.

The Centronics interface is composed of an 8-bit data bus together with printer and data flow control signals. Figure 8.30 illustrates the Centronics interface and briefly describes its signal lines. For the purpose of this example we will assume that the printer we wish to interface uses a 7-bit data bus and prints ASCII-encoded characters from the host processor. Figure 8.31 provides a timing diagram for the transmission of a character to the printer.

FIGURE 8.30   Centronics interface

PIN No.	Signal	Direction
1	DSTB*	In
2	DATA 1	In
3	DATA 2	In
4	DATA 3	In
5	DATA 4	In
6	DATA 5	In
7	DATA 6	In
8	DATA 7	In
9	DATA 8	In
10	ACKNLG*	Out
11	BUSY	Out
12	PE	Out
13	SLCT	Out
14	AUTO LINE FEED*	In
15	NC	
16	gnd	
17	FG	
18	NC	

PIN No.	Signal	Direction
19	DSTB*-Return	
20	DATA 1-Return	
21	DATA 2-Return	
22	DATA 3-Return	
23	DATA 4-Return	
24	DATA 5-Return	
25	DATA 6-Return	
26	DATA 7-Return	
27	DATA 8-Return	
28	ACKNLG* Return	
29	BUSY-Return	
30	PE-Return	
31	INPRM*	In
32	FAULT*	Out
33	gnd	
34	NC	
35	+5 V	
36	SLCT-IN*	In

NOTES: 1. The-Return signal is always connected to gnd.
2. The level is raised to +5 V at 3.3 k Ω.

(a) DATA 1–8
- Used for character codes and image codes.
- Loaded to printer upon DSTB* signal.
- This signal should not be changed while DSTB* = low

(b) DSTB*
- Used for loading DATA 1–8.
- Becomes effective upon busy = low
- Do not send out another DSTB* signal before the output of ACKNLG*.

(c) INPRM*
- Used for initializing the printer.
- If this signal is received during operation, the printer immediately stops. Initialization is executed when the signal turns from low to high.

(d) ACKNLG*
- A response signal for DSTB*.
- Do not send out additional DSTB* signals before the output of this signal.
- This signal is sent out, irrespective of DSTB* signals, when the printer mode is changed from off-line to on-line, or INPRM* signal is entered.

(e) BUSY
- Indicates that the printer is BUSY when this signal is high. No code other than DC 1 will be accepted.
- Indicates that the printer is READY when this signal is low. Output of ACKNLG* occurs when changing to low.
- The signal changes to high when the printer mode is off-line.

(f) PE
- The signal changes to high when out of paper.
- The signal always changes to low when the paper select switch is set to CUT SHEET.

(g) SLCT
- Indicates the select mode when this signal is high.

(h) FAULT*
- The signal changes to low when:
  1) Detecting no-paper error.
  2) Printer is off-line.

(i) AUTO LINE FEED*
- When this signal is low, the printer will feed a line and return the carriage to the home position code.

(j) SLCT-IN*
- The printer switches to the select mode when this signal is low.

FIGURE 8.31 Timing of the data exchange across the Centronics interface

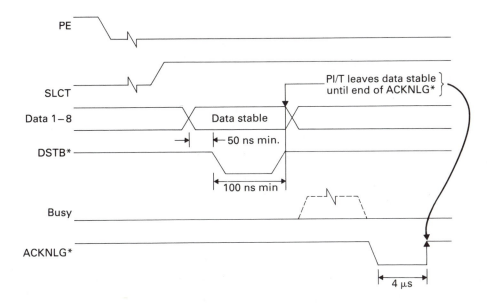

The DSTB∗ (data strobe) and ACKNLG∗ (acknowledge) lines perform a data transfer handshake between the printer and interface. DSTB∗ is an output from the interface informing the printer that data from the host computer is available on the data lines. The data must be set up for at least 50 ns before the falling edge of SSTB∗, and DSTB∗ must have a minimum duration of 100 ns. The printer acknowledges the receipt of data by pulsing its ACKNLG∗ output low for 4 μs.

Active-high output control lines BUSY, PE (printer error), and SLCT (select) reflect the status of the printer. When asserted, SLCT indicates that the printer is on-line, PE indicates that the printer is not operating correctly (e.g., because it has run out of paper), and BUSY indicates that the printer is busy and cannot accept further data. When the printer cannot accept further data, BUSY is asserted by the printer following a DSTB∗ pulse from the interface.

Figure 8.32 shows a possible interface implemented by a 68230 PI/T. Side A peripheral lines PA0–PA7 provide a data interface to the printer after suitable buffering by a 74LS244. The 74LS244 octal buffer is capable of driving a length of ribbon cable and isolates the PI/T from direct contract with the printer. Control lines H1 and H2 are connected to the handshake signals ACKNLG∗ and DSTB∗, respectively, after buffering. The three status lines from the printer, BUSY, PE, and SLCT, are connected to the PI/T's PA7, PC0, and PC1 pins, respectively. Note that since PA7 is not required as a data bus output, there is no reason why we cannot take advantage of the PI/T's ability to support both input and output lines and use PA7 as a control input (i.e., by connecting it to BUSY).

In this example, port A of the PI/T is to be configured as a double-buffered, unidirectional output port. PA7, which is used as an input, is unbuffered so that the state of the printer's BUSY output can be sampled at any time by the host processor.

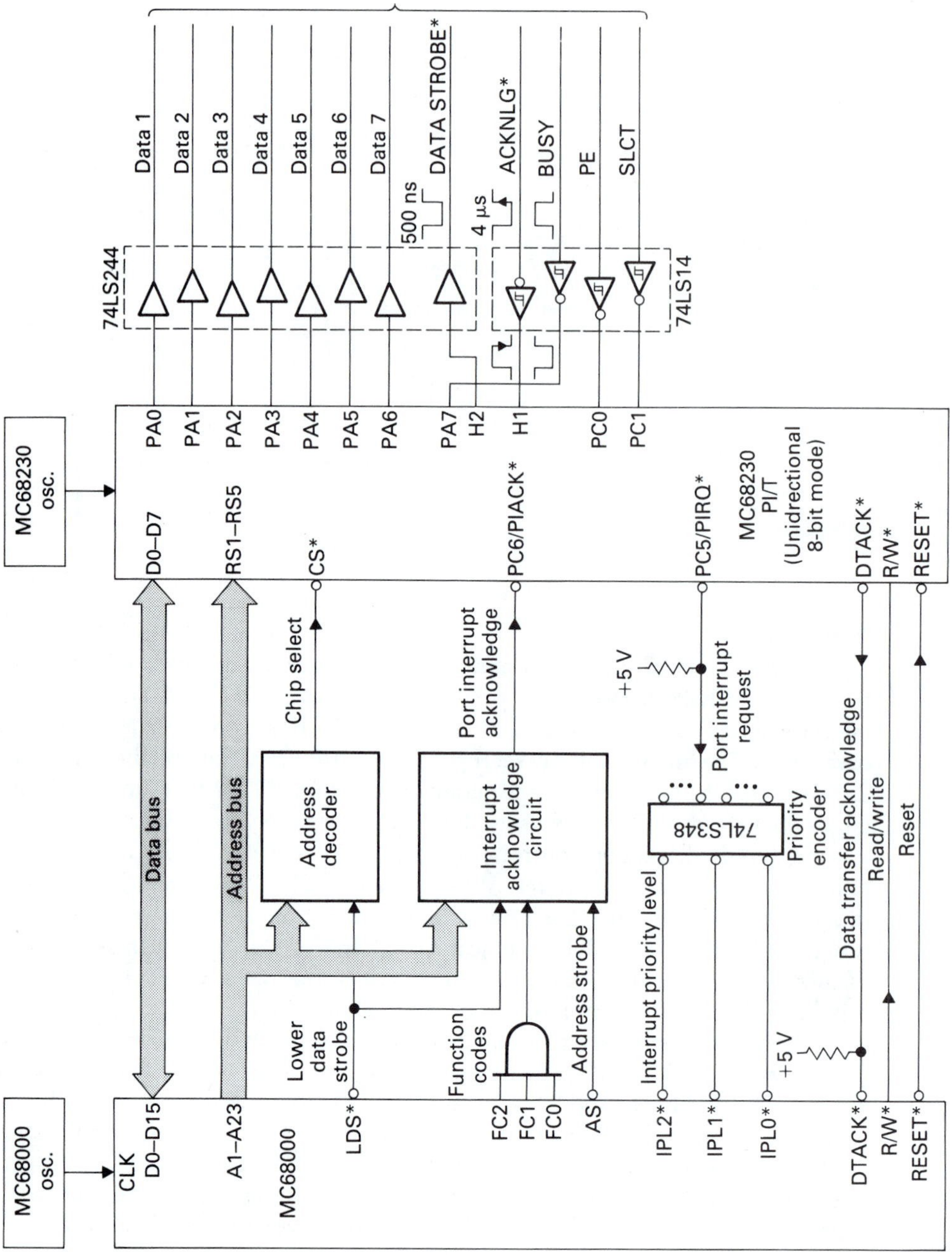

FIGURE 8.32 Centronics interface implemented by the 68230

Note that this application of the PI/T does not make use of port B and its associated control lines H3 and H4.

Since the Centronics interface requires a 100-ns minimum pulse on its DSTB* input, we can make good use of the PI/T's pulsed handshake mode, in which H2 is asserted for four clock cycles (i.e., 4×500 ns, or 2 μs, at a clock frequency of 8 MHz) to provide DSTB*. The printer responds to the DSTB* pulse by asserting ACKNLG* for 4 μs, which is detected at the PI/T's H1 input.

ACKNLG* from the printer is connected to the PI/T's H1 input after inversion by a buffer. A falling edge at the H1 pin indicates to the interface that the printer is ready to receive new data.

The PI/T's double-buffered output mode maximizes throughput because a new character can be sent to the printer as soon as the ACKNLG* strobe is received. Suppose that the PI/T's two buffers contain characters from the host processor. When ACKNLG* is detected by an active edge at H1, the next character is transferred to the PI/T's output pins, without the need to get a new character from the processor. At the same time, the PI/T can interrupt the processor to refill its empty buffers.

The software used to control the Centronics interface described by AN-854 consists of three parts: an initialization routine, LPOPEN, a buffer routine, LPWRITE, and a printer interrupt handler, LPINTR. The initialization routine is called once to configure the PI/T's internal registers. The LPWRITE routine is called by the application program and stores the data to be transmitted in a buffer. LPWRITE checks the status of the printer and enables the 68000's interrupt mechanism if the printer is ready for data.

The LPINTR routine performs the actual output operation. After each character has been transferred to the printer, the ACKNLG* strobe received by the PI/T on its H1 line is used to initiate the transfer of a new character to the PI/T's output pins (i.e., double-buffering) and the transfer of a new character into the PI/T from the buffer.

```
*
PIT       EQU      $0C0000      Base address of the PI/T
PGCR      EQU      PIT+1        Port general control register
PSRR      EQU      PIT+3        Port service request register
PADDR     EQU      PIT+5        Port A data direction register
PIVR      EQU      PIT+$B       Port interrupt vector register
PACR      EQU      PIT+$D       Port A control register
PADR      EQU      PIT+$11      Port A data register
PCDR      EQU      PIT+$19      Port C data register
PSR       EQU      PIT+$1B      Port status register
*
*
* LPOPEN —          This initialization routine is called to "open the printer channel" and
*                   is used to configure the PI/T before it can be used. Port A is set up to
*                   operate in an 8-bit, double buffered mode, and H2 is programmed to
*                   provide pulsed output handshake. FINFLAG is the "finished" flag and is
*                   set to all 1s to denote printing finished (all 0s otherwise).
```

```
*
LPOPEN      ST          FINFLAG         FINFLAG = $FF = finished, printer idle
            MOVE.B      #$7F,PADDR      Set up port A data direction for 7-bit output, 1-bit input
            MOVE.B      #$78,PACR       Port A = submode 01, pulsed H2
            MOVE.B      #$10,PGCR       Enable port A, mode 0
            MOVE.B      #$40,PIVR       Load the PIA's interrupt vector with $40 (handler at $100)
            MOVE.B      #$18,PSRR       Enable PIA's interrupt pins
            RTS                         Initialization complete—return
* LPWRITE —          This is called from a user program via a TRAP instruction (i.e., this is
*                    a trap handler). The byte count is in D0 and A0 points to the buffer that
*                    holds the data to be printed. If the printer is on-line, this routine just
*                    enables interrupts. On exit, D0 holds the status.
*
LPWRITE     CLR.B       FINFLAG         Reset finished flag to zero
            MOVE.L      D0,BYTECNT      Save user parameter (byte count)
            MOVE.L      A0,BUFFADDR     Save user parameter (buffer pointer)
            BTST        #0,PCDR         Test PrinterError status on pin PC0
            BEQ.S       PError          IF PE = 0 THEN  Printer error
            BTST        #1,PCDR         Test SLCT status on pin PC1 (SLCT = PrinterOnLine)
            BEQ.S       LPWG0           IF  SLCT = 0 THEN printer on-line, so continue
*
PError      ST          D0              Set D0 to 1s (to indicate error)
            RTS                         Return following error
*
LPWG0       BSET        #1,PACR         Enable H1S interrupt
LPW1        TST.B       FINFLAG         Wait until FINFLAG = $FF
            BEQ.S       LPW1            ... The PI/T interrupt routines clears FINFLAG ...
            CLR.B       D0              Clear returned status—no error
            RTS                         Return
* LPINTR —            Printer interrupt service routine. This gets characters from the buffer
*                     and sends them to the PI/T. When the printing is complete, the interrupts
*                     are disabled.
*
LPINTR      MOVE.L      A0,−(SP)        Save current A0 on stack
            MOVEA.L     BUFFADDR,A0     Get address of printer buffer
            TST.L       BYTECNT         Examine byte count
            BEQ.S       EMPTY           IF zero THEN all done
PRINT       MOVE.B      (A0)+,PADR      Send a character to the printer
            SUBQ.L      #1,BYTECNT      One less character to be printed
            BEQ.S       EMPTY           IF count zero THEN stop printing
            BTST        #0,PSR          See if room in PI/T for another character
            BNE         PRINT           IF room THEN print again
            BRA         NOTRDY                  ELSE printer PI/T not ready
EMPTY       BCLR        #1,PACR         Disable H1S interrupts
            ST          FINFLAG         Set finished status
NOTRDY      MOVE.L      A0,BUFFADDR     Save buffer address
            MOVE.L      (SP)+,A0        Restore original A0 from stack
            RTE                         Return from interrupt
*
BUFFADDR    DC.L        1               Space for buffer pointer
BYTECNT     DC.L        1               Space for byte counter
FINFLAG     DC.B        1               Space for finish flag
*
            END
```

SUMMARY

This chapter has introduced the interface between a microprocessor and an external device. We have looked at two aspects of interfaces in some detail—the DMA controller and the parallel interface/timer. Such peripherals are exceedingly complex because of their highly programmable modes of operation and their dependence on the system to which they are connected. However, the peripheral chip has done as much as any other component to make today's low-cost microcomputer a reality. Without serial and parallel interfaces, CRT controllers, and floppy disk controllers, the microcomputer would require large numbers of MSI chips to implement these interfaces.

Most of this chapter is devoted to the 68230 PI/T, one of the most sophisticated parallel ports available. The great advantage of the 68230 is its ability to operate as a 16-bit port or as two independent 8-bit ports. Each of these two modes has a wide range of handshaking procedures and can provide either single or double buffering of data. All these variations mean that the PI/T can easily interface with a very wide range of parallel ports with a minimum amount of user-supplied hardware or software.

Problems

1. What are the advantages and disadvantages of memory-mapped input/output?

2. Why does the 68000 have a synchronous interface in addition to its synchronous interface?

3. What is direct memory access, DMA, and how is it used?

4. Explain the meaning of the following terms:

   a. Nonbuffered input (or output)   b. Single-buffered input (or output)

   c. Double-buffered input (or output)   d. Open-loop data transfer

   e. Closed-loop data transfer   f. Handshaking

5. Design a simple, single-channel DMA controller for the 68000 using SSI and MSI logic. The DMAC has a two-wire interface to the peripheral, consisting of REQ* and ACK*. When REQ* is asserted by a peripheral, the DMAC requests the bus by using its bus arbitration control lines, BR*, BG*, and BGACK*. The DMAC executes a single cycle of DMA at a time. When the transfer is complete, the DMAC asserts ACK* to indicate that it is ready for the next transfer.

6. A printer has an 8-bit parallel Centronics interface that consists of an 8-bit parallel data bus and three control lines. DSTB* is an active-low pulsed data strobe that indicates to the printer that the data on the data bus is valid. BUSY* is an output from the printer that, when high. indicates to the computer that the printer cannot accept data. ACKNLG* is an active-low response from the printer to DSTB* and indicates that the printer has captured the data. Design a suitable interface between a computer and the printer using a 68230 PI/T and write a program to control the PI/T.

7. Why does the 68230 PI/T have both port A and B data registers *and* port A and B alternate registers?

8. What is the effect of loading the following data values into the stated registers of a 68230 PI/T?

a. $00 into PADDR (assume mode 0) b. $F0 into PADDR (assume mode 0)
c. $FA into PADDR (assume mode 0) d. $5A into PADDR (assume mode 3)
e. $00 into PGCR f. $F0 into PGCR
g. $18 into PSRR h. $6E into PACR (assume $1F in PGCR)

9. Write an initialization routine to set up a 68230 PI/T with port A as an 8-bit double-buffered input port and port B as an 8-bit double-buffered output port. The PI/T is to operate in an interrupt-driven mode and all control signals are active-low. Port A uses H2 in a fully interlocked handshake mode and port B uses H4 in a pulsed handshake mode.

10. Initializing a 68230 PI/T is no fun because of its complexity. We would be better advised to use a menu-driven initialization program that displays options on the screen and asks the user to select the desired option(s). Write a program in any suitable language that will provide a menu-driven initialization program for the 68230 PI/T. The user must be invited to supply the options in plain English; for example, "Do you require 8-bit on 16-bit data transfers?"

THE SERIAL INPUT/ OUTPUT INTERFACE

The vast majority of general-purpose microcomputers, except some entirely self-contained portable models, use a serial interface to communicate with remote peripherals such as CRT terminals. The serial interface, which moves information from point to point one bit at a time, is generally preferred to the parallel interface, which is able to move a group of bits simultaneously. This preference is not due to the high performance of a serial data link but to its low cost, simplicity, and ease of use. In this chapter we first describe how information is transmitted serially and then examine a typical parallel-to-serial and serial-to-parallel chip that forms the interface between a microprocessor and a serial data link. Because a serial data link can operate in one of two modes, asynchronous or synchronous, a separate section is devoted to each mode. We also take a brief look at some of the standards for the transmission of serial data. The chapter ends with the description of a suitable serial interface for a 68000-based system. Throughout this chapter, the word *character* refers to the basic unit of information transmitted over a data link. The term character has been chosen because many data links transmit information in the form of text, so that the unit of transmitted information corresponds to a printed character.

Figure 9.1 illustrates the basic serial data link between a computer and a CRT terminal. A CRT terminal requires a two-way data link because information from the keyboard is transmitted to the computer and information from the computer is transmitted to the screen. Note that the *transmitted* data from the computer becomes the *received* data at the CRT terminal. Although this statement is an elementary and self-evident observation, confusion between transmitted and received data is a common source of error in the linking of computers and terminals.

A more detailed arrangement of a serial data link in terms of its functional components is given in figure 9.2. The heart of the data link is the box labeled "serial interface," which translates data between the form in which it is stored within the computer and the form in which it is transmitted over the data link. The conversion of data between parallel and serial form is often performed by a single LSI device called an asynchronous communications interface adaptor (ACIA).

The line drivers in figure 9.2 have the function of translating the TTL level signals processed by the ACIA into a suitable form for sending over the transmission path. The transmission path itself is normally a twisted pair of conductors, which accounts for its very low cost. Some systems employ more esoteric transmission

FIGURE 9.1   Serial data link

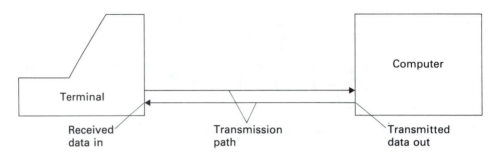

paths such as fiber optics or infrared (IR) links. The connection between the line drivers and transmission path is labeled *plug and socket* in figure 9.2 to emphasize that such mundane things as plugs become very important if interchangeability is required. International specifications cover this situation and other aspects of the data link.

The two items at the computer end of the data link enclosed in "clouds" in figure 9.2 represent the software components of the data link. The lower cloud contains the software that directly controls the serial interface itself by performing operations such as transmitting a single character or receiving a character and checking it for certain types of error. On top of this software sits the application-level software, which uses the primitive operations executed by the lower-level software to carry out actions such as listing a file on the screen.

FIGURE 9.2   Functional units of a serial data link

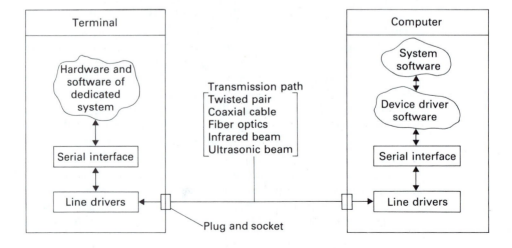

9.1 ASYNCHRONOUS SERIAL DATA TRANSMISSION

By far the most popular serial interface between a computer and its CRT terminal is the asynchronous serial interface. This interface is so called because the transmitted data and the received data are not synchronized over any extended period and therefore no special means of synchronizing the clocks at the transmitter and receiver is necessary. In fact, the asynchronous serial data link is a very old form of data transmission system and has its origin in the era of the teleprinter.

Serial data transmission systems have been around for a long time and are found in the telephone (human speech), Morse code, semaphore, and even the smoke signals once used by native Americans. The fundamental problem encountered by all serial data transmission systems is how to split the incoming data stream into individual units (i.e., bits) and how to group these units into characters. For example, in Morse code the dots and dashes of a character are separated by an intersymbol space, while the individual characters are separated by an intercharacter space, which is three times the duration of an intersymbol space.

First we examine how the data stream is divided into individual bits and the bits grouped into characters in an asynchronous serial data link. The key to the operation of this type of link is both simple and ingenious. Figure 9.3 gives the format of data transmitted over such a link.

An asynchronous serial data link is said to be character oriented, as information is transmitted in the form of groups of bits called characters. These characters are invariably units comprising 7 or 8 bits of "information" plus 2 to 4 control bits and

FIGURE 9.3 Format of asynchronous serial data

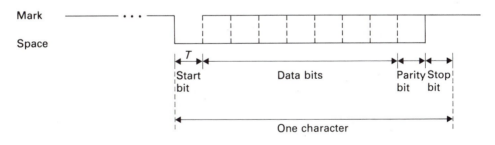

Example: Letter M = ASCII \$4D = 1001101_2 (even parity)

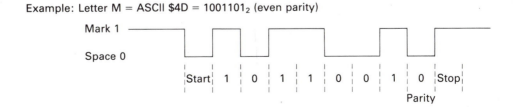

frequently correspond to ASCII-encoded characters. Initially, when no information is being transmitted, the line is in an idle state. Traditionally, the idle state is referred to as the *mark level*. By convention this corresponds to a logical 1 level.

When the transmitter wishes to send data, it first places the line in a *space* level (i.e., the complement of a mark) for one element period. This element is called the start bit and has a duration of T seconds. The transmitter then sends the character, 1 bit at a time, by placing each successive bit on the line for a duration of T seconds, until all bits have been transmitted. Then a single parity bit is calculated by the transmitter and sent after the data bits. Finally, the transmitter sends a stop bit at a mark level (i.e., the same level as the idle state) for one or two bit periods. Now the transmitter may send another character whenever it wishes. The only purpose of the stop bit is to provide a rest period for the receiver between consecutive characters. This bit is a relic of the days of electromechanical receivers and is not now strictly required for technical reasons, existing only for the purpose of compatibility with older equipment.

As the data wordlength may be 7 or 8 bits with odd, even, or no parity bits, plus either 1 or 2 stop bits, a total of 12 different possible formats can be used for serial data transmission—and this is before we consider that there are about seven commonly used values of T, the element duration. Connecting one serial link with another may therefore be difficult because so many options are available.

At the receiving end of an asynchronous serial data link, the receiver continually monitors the line looking for a start bit. Once the start bit has been detected, the receiver waits until the end of the start bit and then samples the next N bits at their centers, using a clock generated locally by the receiver. As each incoming bit is sampled, it is used to construct a new character. When the received character has been assembled, its parity is calculated and compared with the received parity bit following the character. If they are not equal, a parity error flag is set to indicate a transmission error.

FIGURE 9.4   Effect of unsynchronized transmitter and receiver clocks

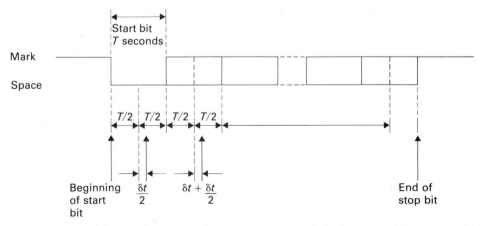

NOTE: Vertical lines with arrows indicate the points at which the received data is sampled.

The most critical aspect of the system is the receiver timing. The falling edge of the start bit triggers the receiver's local clock, which samples each incoming bit at its nominal center. Suppose the receiver clock waits $T/2$ seconds from the falling edge of a start bit and samples the incoming data every T seconds thereafter until the stop bit has been sampled. Figure 9.4 shows this situation. As the receiver's clock is not synchronized with the transmitter clock, the sampling is not exact.

Let us assume that the receiver clock is running slow, so that a sample is taken every $T + \delta t$ seconds. The first bit of the data is sampled at $(T + \delta t)/2 + (T + \delta t)$ seconds after the falling edge of the start bit. The stop bit is sampled at time $(T + \delta t)/2 + N(T + \delta t)$, where N is the number of bits in the character following the start bit. The total accumulated error in sampling the stop bit is therefore $(T + \delta t)/2 + N(T + \delta t) - (T/2 + NT)$, or $(2N + 1)\delta t/2$ seconds. For correct operation, the stop bit must be sampled within $T/2$ seconds of its center, so that:

$$\frac{T}{2} > \frac{(2N + 1)\delta t}{2}$$

or

$$\frac{\delta t}{T} < \frac{1}{2N + 1}$$

or

$$\frac{\delta t}{T} < \frac{100}{2N + 1} \qquad \text{as a percentage}$$

If $N = 9$ for a 7-bit character + parity bit + 1 stop bit, the maximum permissible error is $100/19 = 5$ percent. Fortunately, almost all clocks are now crystal controlled, and the error between transmitter and receiver clocks is likely to be a tiny fraction of 1 percent.

The most obvious disadvantage of asynchronous data transmission is the need for a start, parity, and stop bit for each transmitted character. If 7-bit characters are used, the overall efficiency is only $7/(7 + 3) \times 100 = 70$ percent. A less obvious disadvantage is due to the character-oriented nature of the data link. Whenever the data link connects a CRT terminal to a computer, few problems arise, as the terminal is itself character oriented. However, if the data link is being used to, say, dump binary data to a magnetic tape, problems arise. If the data are arranged as 8-bit bytes with all 256 possible values corresponding to valid data elements, it is difficult (but not impossible) to embed control characters (e.g., tape start or stop) within the data stream because the same character must be used both as pure data (i.e., part of the message) and for control purposes.

If 7-bit characters are used, pure binary data cannot be transmitted in the form of one character per byte. Two characters are needed to record each byte and this condition is clearly inefficient. We will see later how synchronous serial data links overcome this problem.

We have now described how information can be transmitted serially in the form of 7- or 8-bit characters. The next step is to show how these characters are encoded.

ASCII Code

Although computing generally suffers from a lack of standardization, the ASCII code is one of the few exceptions. Many microcomputers employ the ASCII code to represent information in character form internally, and for the exchange of information between themselves and CRT terminals. The ASCII code, or American Standard Code for Information Interchange, is one of several codes used to represent alphanumeric characters. As long ago as the 1920s, the Baudot or Murray code was designed for the teleprinter. This code, still used by the international telex service, represents characters by 5 bits. As this system provides only 2^5 unique values, one of the 32 possible values acts as a shift, affecting the meaning of the following characters. The effective number of characters available is thereby increased.

The ASCII code employs 7 bits to give a total of 128 unique values. These bits are sufficient to provide a full 96-character upper- and lowercase printing set, together with 32 characters to control the operation of the data link and the terminal itself. The ASCII code has now been adopted universally, and is almost identical to the International Standards Organization ISO-7 code.

Had the ASCII code been developed today, it would almost certainly be an 8-bit code. Unfortunately, the ASCII character set does not include "national" char-

TABLE 9.1 ASCII code

$b_6 b_5 b_4$ / $b_3 b_2 b_1 b_0$		0 / 000	1 / 001	2 / 010	3 / 011	4 / 100	5 / 101	6 / 110	7 / 111
0	0000	NUL	DLC	SP	0	@	P	'	p
1	0001	SOH	DC1	!	1	A	Q	a	q
2	0010	STX	DC2	"	2	B	R	b	r
3	0011	ETX	DC3	#	3	C	S	c	s
4	0100	EOT	DC4	$	4	D	T	d	t
5	0101	ENQ	NAK	%	5	E	U	e	u
6	0110	ACK	SYN	&	6	F	V	f	v
7	0111	BEL	ETB	'	7	G	W	g	w
8	1000	BS	CAN	(	8	H	X	h	x
9	1001	HT	EM	)	9	I	Y	i	y
A	1010	LT	SUB	*	:	J	Z	j	z
B	1011	VT	ESC	+	;	K	[	k	{
C	1100	FF	FS	,	<	L	\	l	|
D	1101	CR	GS	–	=	M	]	m	}
E	1110	SO	RS	.	>	N	∧	n	~
F	1111	SI	VS	/	?	O	_	o	DEL

acters such as the German umlaut or the French accents. Moreover, a graphical character set similar to that used by Teletex would have been very helpful. However, microcomputer manufacturers have tended to design their own graphics codes, leading to incompatibiliy.

Table 9.1 presents the ASCII code. The binary value of a character is obtained by reading the three most significant bits at the top of the column in which the character occurs and then taking the four least significant bits from its row. For example, the character m is in the column headed 110 and the row headed 1101; therefore, the binary code for m is 110 1101 (or \$6D in hexadecimal).

9.2 ASYNCHRONOUS COMMUNICATIONS INTERFACE ADAPTOR (ACIA)

One of the first general-purpose interface devices produced by the semiconductor manufacturers was the asynchronous communications interface adaptor, or ACIA. The ACIA relieves the system software of all the basic tasks involved in converting data between serial and parallel forms; that is, the ACIA contains almost all the logic necessary to provide an asynchronous data link between a computer and an external system.

One of the earliest and still popular ACIAs is the 6850 illustrated in figure 9.5. This particular ACIA will be described because it is much easier to understand than some of the newer ACIAs and is still widely used in microcomputers. Once the reader understands how the 6850 ACIA operates, he or she can read the data sheet of any other ACIA. Like any other digital device, the 6850 has a hardware model, a software model, and a functional model. We look at the hardware model first. Figure 9.6 gives the hardware model of the 6850 together with its timing diagram. From the designer's point of view, the 6850's hardware can be subdivided into three sections: the CPU side, the transmitter side, and the receiver side.

CPU Side

As far as the CPU is concerned, the 6850 behaves almost exactly like a static read/write memory; figure 9.6 shows the read and write cycles on the same diagram. However, one important difference exists between the 6850 and conventional RAM. The 6850's memory accesses are synchronized to an external E or enable clock. In 6800- or 6809-based systems, this situation presents no problem as the processor itself is also synchronized to the E (or $\phi2$ in 6800 terminology) clock that it provides. In the case of the 68000, the ACIA must be interfaced either by using VPA* and VMA* or asynchronously by means of additional logic.

The ACIA is a byte-oriented device and can be interfaced to D_{00} through D_{07} and strobed by LDS*, or to D_{08} through D_{15} and strobed by UDS*. The ACIA has a

FIGURE 9.5 The 6850 ACIA

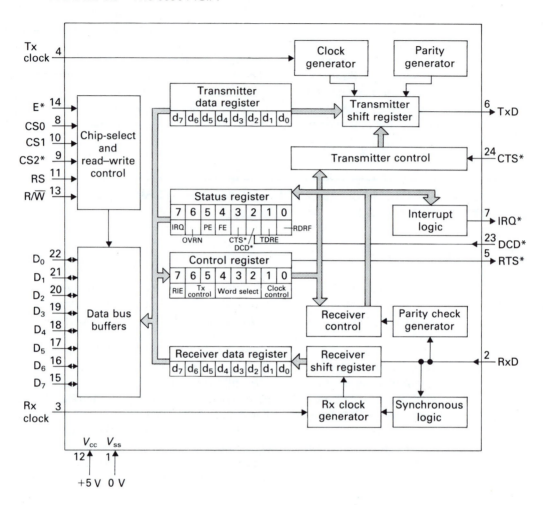

single register select line, RS, that determines the internal location (i.e., register) addressed by the processor. Typically, RS is connected to the processor's A_{01} address output, so that the lower location is selected by address X and the upper, by address X + 2.

Three chip-select inputs are provided, two of which are active-high and one, active-low. This spectacular display of overkill comes from the days when address decoders were relatively expensive and memories small. By using the chip selects alone, you can achieve partial address decoding without any additional components. In many modern systems just one of the chip-select inputs takes part in the address decoding process. The remaining two are permanently enabled. This situation is unfortunate, as the other two pins could have provided the ACIA with additional features—such as a RESET* input or an on-chip clock.

The 6850 has an interrupt request output, IRQ∗, that can be connected to any of the 68000's seven levels of interrupt request input. As the 6850 does not support vectored interrupts, autovectored interrupts must be used in the way described in chapter 6.

Unusually, the 6850 does not have a RESET∗ input because there were not

FIGURE 9.6 Hardware model of the ACIA and its timing diagram

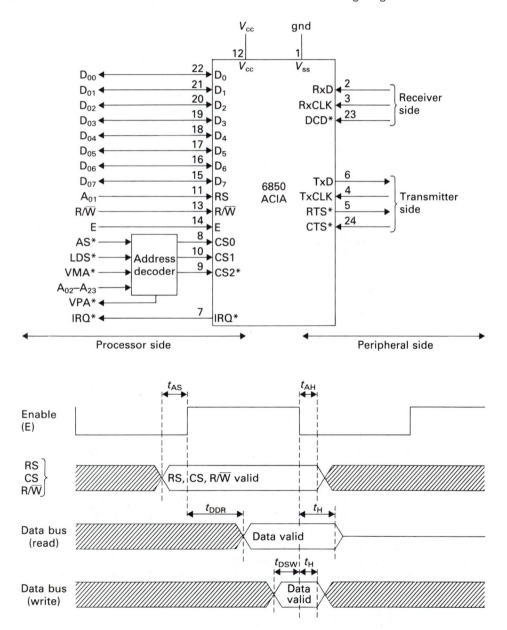

enough pins to provide the function and the manufacturer felt that RESET∗ was the most dispensable of functions. When power is first applied, some sections of the ACIA are reset automatically by an internal power-on-reset circuit. Afterwards, a secondary reset by software is performed, as we shall describe later.

The CPU side of the 6850 has a clock input labeled E (i.e., ENABLE). As with other 6800-series peripherals, the E input must be both free running and synchronized to read/write accesses between the ACIA and the processor. The simplest interface between the ACIA and a 68000 processor is given in figure 9.7. This circuit is entirely conventional and makes use of the CPU's synchronous bus control signals.

The lower byte of the 68000's data bus is connected to the ACIA's data input/output pins, D_0 to D_7, which locate all the ACIA's registers in the lower half of words at odd addresses. Remember that the 68000 address space is arranged so that lower-order bits (D_{00} to D_{07}) have odd addresses and higher-order bits (D_{08} to D_{15}) have even addresses. Whenever the 68000 addresses the ACIA, the address decoder detects the access and forces SELECT_ACIA∗ low. This signal drives the 68000's VPA∗ low via an OR gate, signaling that a synchronous bus cycle is to begin. The CPU then forces VMA∗ low and the ACIA is selected by SELECT_ACIA∗, VMA∗, and LDS∗ all being low simultaneously. During this access, R/$\overline{W}$ from the CPU determines the direction of data transfer and A_{01} the location of the internal register selected in the ACIA.

The lower portion of figure 9.7 is intended to show how the ACIA is operated in the autovectored interrupt mode. When the ACIA forces its IRQ∗ line low, a level 5 interrupt is signaled to the CPU. Assuming this level is enabled, IACK5∗ from the decoder goes low and is then ANDed with IRQ∗ from the ACIA and connected to VPA∗ via an OR gate. The purpose of ANDing IACK5∗ with IRQ∗ is to permit an interrupt acknowledge to the 68000 (via VPA∗) only when the ACIA is putting out an interrupt request while an interrupt acknowledge at the appropriate level is being indicated by the CPU.

Receiver and Transmitter Sides of the ACIA

One of the great advantages of peripherals such as the 6850 ACIA is that they isolate the CPU from the outside world both physically and logically. The *physical isolation* means that the engineer who is connecting a peripheral device to a microprocessor system does not have to worry about the electrical and timing requirements of the CPU itself. In other words, all the engineer needs to understand about the ACIA is the nature of its transmitter-side and receiver-side interfaces. Similarly, the peripheral performs a *logical isolation* by hiding the details of information transfer across it; for example, the operation of transmitting a character from an ACIA is carried out by the instruction MOVE.B D0,ACIA_DATA, where register D0 contains the character to be transmitted and ACIA_DATA is the address of the data register in the ACIA. All the actions necessary to actually serialize the data and append start, parity, and stop bits are carried out automatically (i.e., invisibly) by the ACIA.

FIGURE 9.7 Interface between a 6850 ACIA and a 68000 CPU

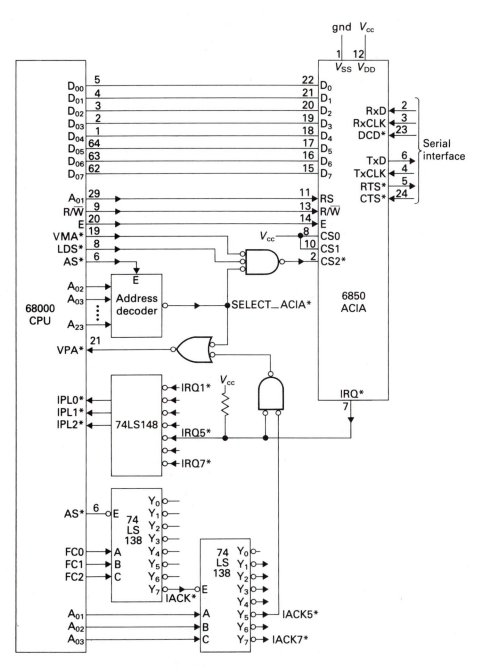

Here, only the essential details of the ACIA's transmitter and receiver sides are presented, because the way in which they function is described more fully when we come to the logical organization of the 6850. The peripheral-side interface of the 6850 is divided into two entirely separate groups—the receiver group, which forms the interface between the ACIA and a source of incoming data, and the transmitter group, which forms the interface between the ACIA and the destination for outgoing data. *Incoming* and *outgoing* are used with respect to the ACIA. The nature of these signals is strongly affected by one particular role of the ACIA—its role as an interface between a computer and the public switched telephone network via a modem.

Receiver Side

Incoming data to the ACIA is handled by three pins: RxD, RxCLK, DCD∗. Like all other inputs and outputs to the ACIA, these are TTL-level compatible signals. The RxD (receiver data input) pin receives serial data from the transmission path to which the ACIA is connected. The idle (mark) state at this pin is a TTL logical 1 level. A receiver clock is provided at the RxCLK (receiver clock) input pin by the systems designer. The RxCLK clock must be either the same, 16, or 64 times the rate at which bits are received at the data input terminal. Many modern ACIAs include on-chip receiver and transmitter clocks, relieving the system designer of the necessity of providing an additional external oscillator.

The third and last component of the receiver group is an active-low DCD∗ (data carrier detect) input. DCD∗ is intended for use in conjunction with a modem and, when low, indicates to the ACIA that the incoming data is valid. When inactive-high, DCD∗ indicates that the incoming data might be erroneous. This situation may arise if the level (i.e., signal strength) of the data received at the end of a telephone line drops below a predetermined value or the connection itself is broken.

Transmitter Side

The transmitter side of the ACIA comprises four pins: TxCLK, TxD, RTS∗, and CTS∗. The transmitter clock input (TxCLK) provides a timing signal from which the ACIA derives the timing of the transmitted signal elements. In most applications of the ACIA, the transmitter and receiver clocks are connected together and a common oscillator used for both transmitter and receiver sides of the ACIA. Serial data is transmitted from the TxD (transmit data) pin of the ACIA, with a logical one level representing the idle (mark) state.

An active-low request to send (RTS∗) output indicates that the ACIA is ready to transmit information. This output is set or cleared under software control and can be used to switch on any equipment needed to transmit the serial data over some data link. Some use it to switch on a cassette recorder when the ACIA is interfaced to a magnetic tape recording system.

An active-low clear to send (CTS∗) input indicates to the transmitter side of the ACIA that the external equipment used to transmit the serial data is ready. When negated, this input inhibits the transmission of data. CTS∗ is a modem signal that indicates that the transmitter carrier is present and that transmission may go ahead.

Operation of the 6850 ACIA

The software model of the 6850 has four user-accessible registers, as defined in table 9.2. These registers are a transmit data register (TDR), a receive data register (RDR), a system control register (CR), and a system status register (SR). As there are four registers and yet the ACIA has only a single register-select input, RS, a way must be found to distinguish between registers. The ACIA uses the R/W input to make this distinction. Two registers are read-only (i.e., RDR, SR) and two are write-only (TDR, CR). Although a perfectly logical, indeed an elegant, thing to do, I do not like it. I am perfectly happy to accept read-only registers, but I am suspicious of the write-only variety because the contents of a write-only register are impossible to verify. Suppose a program with a bug executed an unintended write to a write-only register. The change cannot be detected by reading back the contents of the register.

Table 9.2 also gives the address of each register, assuming that the base address of the ACIA is $00 E001 and that it is selected by LDS*. The purpose of this exercise is twofold: it shows that the address of the lower-order byte is odd and that the pairs of read-only and write-only registers are separated by two (i.e., $00 E001 and $00 E003).

Control Register

Because the ACIA is a versatile device and can be operated in any of several different modes, the control register permits the programmer to define its operational characteristics. This job can even be done dynamically if the need ever arises. Table 9.3 shows how the 8 bits of the control register are grouped into four logical fields.

Bits CR0 and CR1 determine the ratio between the transmitted or received bit rates and the transmitter and receiver clocks, respectively. The clocks operate at the same, 16, or 64 times the data rate. Most applications of the 6850 employ a receiver/transmitter clock at 16 times the data rate with CR1 = 0 and CR0 = 1. Setting CR1 = CR2 = 1 is a special case and serves as a software reset of the ACIA. A software reset clears all internal status bits, with the exception of CTS* and DCD*. A software reset to the 6850 is invariably carried out during the initialization phase of the host processor's reset procedures.

The *word-select* field, bits CR2, CR3, and CR4, determines the format of the

TABLE 9.2 Register-selection scheme of the 6850 ACIA

ADDRESS	RS	R/W	REGISTER TYPE	REGISTER FUNCTION	
00 E001	0	0	Write only	Control register	(CR)
00 E001	0	1	Read only	Status register	(SR)
00 E003	1	0	Write only	Transmit data register	(TDR)
00 E003	1	1	Read only	Receive data register	(RDR)

NOTE: Base address of ACIA = $00 E001.

TABLE 9.3 Structure of the ACIA's control register

BIT	CR7	CR6 CR5	CR4 CR3 CR2	CR1 CR0
Function	Receiver interrupt enable	Transmitter control	Word select	Counter division

CR1	CR0	DIVISION RATIO
0	0	1
0	1	16
1	0	64
1	1	Master reset

CR4	CR3	CR2	WORD SELECT			
			DATA WORD LENGTH	PARITY	STOP BITS	TOTAL BITS
0	0	0	7	Even	2	11
0	0	1	7	Odd	2	11
0	1	0	7	Even	1	10
0	1	1	7	Odd	1	10
1	0	0	8	None	2	11
1	0	1	8	None	1	10
1	1	0	8	Even	1	11
1	1	1	8	Odd	1	11

CR6	CR5	TRANSMITTER CONTROL	
		RTS*	TRANSMITTER INTERRUPT
0	0	Low (0)	Disabled
0	1	Low (0)	Enabled
1	0	High (1)	Disabled
1	1	Low (0)	Disabled and break

CR7	RECEIVER INTERRUPT ENABLE
0	Receiver may not interrupt
1	Receiver may interrupt

received or transmitted characters. The eight possible data formats are given in table 9.3. Note that these bits also enable the type of parity (if any) and the number of stop bits to be defined under software control, which is one of the nice features of a programmable peripheral. Possibly the most common data format for the transmission of information between a processor and a CRT terminal is: start bit + 7 data bits + even parity + 1 stop bit. The corresponding value of CR4, CR3, CR2 is 0, 1, 0.

The *transmitter control* field, CR5 and CR6, selects the state of the active-low request to send (RTS*) output and determines whether or not the transmitter section of the ACIA may generate an interrupt by asserting its IRQ* output. In most systems, RTS* is active-low whenever the ACIA is transmitting, because RTS* is used to activate equipment connected to the ACIA. The programming of the transmitter interrupt enable, and for that matter the receiver interrupt enable, is very much a function of the operating mode of the ACIA. If the ACIA is operated in a polled-data mode, interrupts are not necessary.

If the transmitter interrupt is enabled, an interrupt is generated by the transmitter whenever the transmit data register (TDR) is empty, signifying the need for new data from the CPU. When the ACIA's clear to send (CTS*) input is inactive-high, the TDR empty flag of the status register is held low, inhibiting any transmitter interrupt.

Setting both CR6 and CR5 to a logical 1 simultaneously creates a special case. When both these bits are high, a *break* is transmitted by the transmitter data output pin. A break is a condition in which the transmitter output is held at the active level (i.e., space or TTL logical zero) continuously. This state may be employed to force an interrupt at a distant receiver, because the asynchronous serial format precludes the existence of a space level for longer than about ten bit periods. The term *break* originates from the old current-loop data transmission system when a break was affected by disrupting (i.e., breaking) the flow of current round a loop.

The *receiver interrupt enable* field consists of 1 bit, CR7, which enables the generation of interrupts by the reviewer when it is set (CR7 = 1) and disables receiver interrupts when it is clear (CR7 = 0). The receiver asserts its IRQ* output, assuming CR7 = 1, when the receiver data register full (RDRF) bit of the status register is set, indicating the presence of a new data character ready for the CPU to read. Two other circumstances also force a receiver interrupt. An overrun (see later) sets the RDRF bit and generates an interrupt. Finally, a receiver interrupt can also be generated by a low-to-high transition at the active-low data carrier detect (DCD*) input, signifying a loss of the carrier from a modem. Note that CR7 is a composite interrupt enable bit and enables all the three forms of receiver interrupt described previously. To enable either an interrupt caused by the RDR being full or an interrupt caused by a positive transition at the DCD* pin alone is impossible.

Status Register

The 8 bits of the read-only status register are depicted in table 9.4 and serve to indicate the status of both the transmitter and receiver portions of the ACIA at any instant.

TABLE 9.4 Format of the status register

BIT	SR7	SR6	SR5	SR4	SR3	SR2	SR1	SR0
Function	IRQ	PE	OVRN	FE	CTS	DCD	TDRE	RDRF

SR0—Receiver Data Register Full (RDRF) When set, the RDRF bit indicates that the receiver data register (RDR) is full and a new word has been received. If the receiver interrupt is enabled by CR7 = 1, a logical one in SR0 also sets the interrupt status bit SR7 (i.e., IRQ). The RDRF bit is cleared either by reading the data in the receiver data register or by carrying out a software reset on the control register. Whenever the data carrier detect (DCD∗) input is inactive-high, the RDRF bit remains clamped at a logical zero, indicating the absence of any valid input.

SR1—Transmitter Data Register Empty (TDRE) The TDRE bit is the transmitter counterpart of the RDRF bit, SR0. A logical 1 in SR1 indicates that the contents of the transmit data register (TDR) have been sent to the transmitter and that the register is now ready to transmit new data. TDRE is cleared either by loading the transmit data register or by performing a software reset. If the transmitter interrupt is enabled, a logical one in bit SR1 (i.e., TDRE) also sets bit SR7 of the status word. Note again that SR7 is a composite interrupt bit because it is also set by an interrupt originating from the receiver side of the ACIA. If the clear to send (CTS∗) input is inactive-high, the TDRE bit is held low, indicating that the terminal equipment is not ready for data.

SR2—Data Carrier Detect (DCD) This status bit, associated with the receiver side of the ACIA, is normally employed when the ACIA is connected to the telephone network via a modem. Whenever the DCD∗ input to the ACIA is inactive-high, SR2 is set. A logical one on the DCD∗ line generally signifies that the incoming serial data is faulty, which also has the effect of clearing the SR0 (i.e., RDRF) bit, as possible erroneous input should not be interpreted as valid data.

When the DCD∗ input makes a low-to-high transition, not only is SR2 set but the composite interrupt request bit, SR7, is also set if the receiver interrupt is enabled. Note that SR2 remains set even if the DCD∗ input later returns active-low. This action traps any occurrence of DCD∗ high, even if it goes high only briefly. To clear SR2, the CPU must read the contents of the status register and then the contents of the data register.

SR3—Clear to Send (CTS) The CTS∗ bit directly reflects the status of the CTS∗ input on the ACIA's transmitter side. An active-low level on the CTS∗ input indicates that the transmitting device (modem, paper tape punch, teletype, cassette recorder, etc.) is ready to receive serial data from the ACIA. If the CTS∗ input and therefore the CTS∗ status bit are high, the transmit data register empty bit, SR1, is inhibited (clamped at a logical zero) and no data may be transmitted by the ACIA. Unlike the DCD∗ status bit, the logical value of the CTS∗ status bit is determined only by the CTS∗ input and is not affected by any software operation on the ACIA.

SR4—Framing Error (FE)   A framing error is detected by the absence of a stop bit and indicates a synchronization (i.e., timing) error, a faulty transmission, or a break condition. The framing error status bit, SR4, is set whenever the ACIA determines that a received character is incorrectly framed by a start bit and a stop bit. The framing error status bit is automatically cleared or set during the receiver data transfer time and is present throughout the time that the associated character is available. In other words, an FE bit is generated for each character received and a new character overwrites the old one's FE bit.

SR5—Receiver Overrun (OVRN)   The receiver overrun status bit is set when a character is received by the ACIA but is not read by the CPU before a subsequent character is received, overwriting the last character, which is now lost. Consequently, the receiver overrun bit indicates that one or more characters in the data stream have been lost. The OVRN status bit is set at the midpoint of the last bit of the second character received in succession without a read of the RDR having occurred. Synchronization of the incoming data is not affected by an overrun error— the error is due to the CPU not having read a character, rather than by any fault in the transmission and reception process. The overrun bit is cleared after reading data from the RDR or by a software reset.

SR6—Parity Error (PE)   The parity error status bit, SR6, is set whenever the received parity bit in the current character does not match the parity bit of the character generated locally in the ACIA from the received data bits. Odd or even parity may be selected by writing the appropriate code into bits CR2, CR3, and CR4 of the control register. If no parity is selected, then both the transmitter parity generator and receiver parity checker are disabled. Once a parity error has been detected and the parity error status bit set, it remains set as long as the erroneous data remains in the receiver register.

SR7—Interrupt Request (IRQ)   The interrupt request status bit, SR7, is a composite active-high (note!) interrupt request flag, and is set whenever the ACIA wishes to interrupt the CPU, for whatever reason. The IRQ bit is set active-high by any of the following events:

1. Receiver data register full (SR0 set) and receiver interrupt enabled.
2. Transmitter data register empty (SR1 set) and transmitter interrupt enabled.
3. Data carrier detect status bit (SR2) set and receiver interrupt enabled.

Whenever SR7 is active-high, the active-low open-drain IRQ* output from the ACIA is pulled low. The IRQ bit is cleared by a read from the RDR, or by a write to the TDR, or by a software master reset.

Using the 6850 ACIA

The most daunting thing about many microprocessor interface chips is their sheer complexity. Often this complexity is more imaginary than real, because such peripherals are usually operated in only one of the many different modes that are software selectable. This fact is particularly true of the 6850 ACIA. Figure 9.8 shows how the 6850 is operated in a minimal mode. Only its serial data input (RxD) and output (TxD) are connected to an external system. The request to send (RTS*) output is left unconnected and clear to send (CTS*) and data carrier detect (DCD*) are both strapped to ground at the ACIA.

In a minimal (and noninterrupt) mode, bits 2 to 7 of the status register can be ignored. Of course, the error-detecting facilities of the ACIA are therefore thrown away. The software necessary to drive the ACIA in this minimal mode consists of three subroutines: an initialization, an input, and an output routine:

```
ACIAC      EQU       $E0001              Address of control/status register
ACIAD      EQU       ACIAC+2             Address of data register
RDRF       EQU       0                   Receiver data register full
TDRE       EQU       1                   Transmitter data register empty

INITIALIZE MOVE.B    #%00000011,ACIAC    Reset the ACIA
           MOVE.B    #%00011001,ACIAC    Set up control word—disable
           RTS                           interrupts, RTS* low, 8 data
                                         bits, even parity, 1 stop bit,
                                         16 × clock

INPUT      BTST.B    #RDRF,ACIAC         Test receiver status
           BEQ       INPUT               Poll until receiver has data
           MOVE.B    ACIAD,D0            Put data in D0
           RTS

OUTPUT     BTST.B    #TDRE,ACIAC         Test transmitter status
           BEQ       OUTPUT              Poll until transmitter ready for data
           MOVE.B    D0,ACIAD            Transmit the data
           RTS
```

The INITIALIZE routine is called once before either input or output is carried out and has the effect of executing a software reset on the ACIA followed by setting up its control register. The control word %00011001 (see table 9.3) defines an 8-bit word with even parity and a clock rate (TxCLK, RxCLK) 16 times the data rate of the transmitted and received data.

The INPUT and OUTPUT routines are both entirely straightforward. Each tests the appropriate status bit and then reads data from or writes data to the ACIA's data register.

It is also possible to operate the ACIA in a minimal interrupt-driven mode. The IRQ* output is connected to one of the 68000's seven levels of interrupt request input and arrangements are made to supply the CPU with VPA* during an interrupt acknowledge cycle. Both transmitter and receiver interrupts are enabled by writing 1, 0, 1 into bits CR7, CR6, CR5 of the status register.

When a transmitter or receiver interrupt is initiated, it is still necessary to examine the RDRF and TDRE bits of the status register to determine that the ACIA did indeed request the interrupt and to separate transmitter and receiver requests for service. The effect of interrupt-driven I/O is to eliminate the time-wasting polling routines required by programmed I/O.

Figure 9.9 shows how the ACIA can be operated in a more sophisticated mode.

FIGURE 9.8 Minimal serial interface using the 6850 ACIA

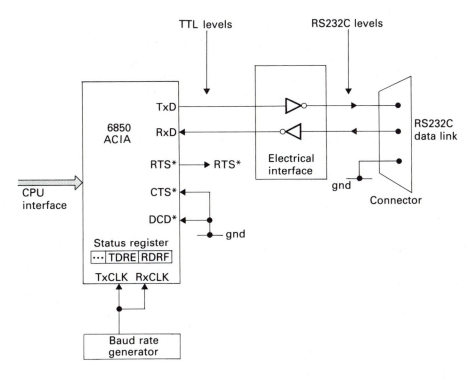

FIGURE 9.9 General-purpose serial interface using the 6850 ACIA

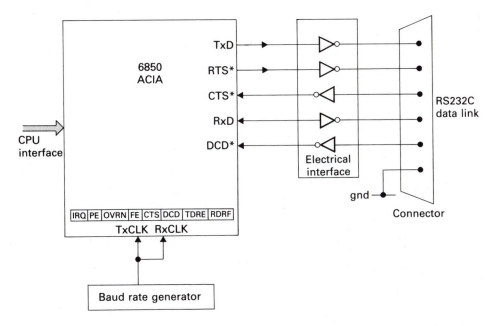

The reader may be tempted to ask, Why bother with a complex operating mode if the 6850 works quite happily in a basic mode? The answer is that the operating mode in figure 9.9 provides more facilities than the basic mode of figure 9.8.

In Figure 9.9 the transmitter side of the ACIA sends an RTS∗ signal and receives a CTS∗ signal from the remote terminal equipment. Now the remote equipment is able to say, "I am ready to receive your data," by asserting CTS∗. In the cut-down mode of figure 9.8, the ACIA simply sends data and hopes for the best.

Similarly, the receiver side of the ACIA uses the data carrier detect (DCD∗) input to signal the host computer that the receiver circuit us in a position to receive data. If DCD∗ is negated, the terminal equipment is unable to send data to the ACIA.

The software necessary to receive data when operating the 6850 in its more sophisticated mode is considerably more complex than that of the previous example. Provision for a full input routine is not possible here, as such a routine would include recovery procedures from the errors detected by the 6850 ACIA. These procedures are, of course, dependent on the nature of the system and the protocol used to move data between a transmitter and receiver. However, the following fragment of an input routine gives some idea of how the 6850's status register is used:

```
ACIAC         EQU           $<ACIA address>
ACIAD         EQU           ACIAC+2
RDRF          EQU           0                 Receiver_data_register_full
TDRE          EQU           1                 Transmitter_data_register_empty
DCD           EQU           2                 Data_carrier_detect
CTS           EQU           3                 Clear_to_send
FE            EQU           4                 Framing_error
OVRN          EQU           5                 Over_run
PE            EQU           6                 Parity_error
*
INPUT         MOVE.B        ACIAC,D0          Get status from ACIA
              BTST          #RDRF,D0          Test for received character
              BNE.S         ERROR_CHECK       If character received, then test SR
              BTST          #DCD,D0           Else test for loss of signal
              BEQ           INPUT             Repeat loop while CTS clear
              BRA.S         DCD_ERROR         Else deal with loss of signal
ERROR_CHECK   BTST          #FE,D0            Test for framing error
              BNE.S         FE_ERROR          If framing error, deal with it
              BTST          #OVRN,D0          Test for overrun
              BNE.S         OVRN_ERROR        If overrun, deal with it
              BTST          #PE,D0            Test for parity error
              BNE.S         PE_ERROR          If parity error, deal with it
              MOVE.B        ACIAD,D0          Load the input into D0
              BRA.S         EXIT              Return
DCD_ERROR                                     Deal with loss of signal
              BRA.S         EXIT
FE_ERROR                                      Deal with framing error
              BRA.S         EXIT
OVRN_ERROR                                    Deal with overrun error
              BRA.S         EXIT
PE_ERROR                                      Deal with parity error
EXIT                        RTS
```

9.3 THE 68681 DUART

The 6850 ACIA is a first-generation interface device designed in the 1970s to work with the 8-bit 6800 microprocessor and is now rather outdated (although it is still widely used). Today's designers would rather implement an asynchronous serial interface with a more modern component, such as the 68681 DUART (dual universal asynchronous receiver/transmitter). The 68681 (from now on we will just call it "DUART") performs the same basic functions as a *pair* of 6850s plus a baud-rate generator. Designers prefer to use the DUART for the following reasons:

1. The DUART provides two independent asynchronous serial channels and therefore replaces two 6850 ACIAs.

2. The DUART has a full 68000 asynchronous bus interface, which means that it supports asynchronous data transfers and can supply a vector number during an interrupt acknowledge cycle.

3. The DUART has an on-chip programmable baud-rate generator, which saves both the cost and board space of a separate baud-rate generator. Moreover, the DUART's baud-rate generator can be programmed simply by loading an appropriate value into a clock select register. This feature makes it very easy to connect a system with a DUART to a communications system with an unknown baud rate. Communications systems based on the 6850 have to change their baud rate by altering links on the board, making it tedious to change the baud rate frequently. Note that the DUART can receive and transmit at different baud rates (as can the 6850).

4. The DUART has a quadruple buffered input which means that up to four characters can be received in a burst before the host processor has to read the input stream. The host computer has to read each character from a 6850 as it is received (otherwise an overrun will occur and characters will be lost). Similarly, the DUART has a double-buffered output, permitting one character to be transmitted while another is being loaded by the CPU.

5. The DUART has 14 I/O pins (6 input, 8 output) that can be used as modem-control pins, clock input and outputs, or as general-purpose input/output pins.

6. The 6850 has just one operating mode. The DUART can support several modes (e.g., a self-test loopback mode).

Figure 9.10 illustrates the internal organization of a 68681 DUART.

Since the 68681 is so much better than the 6850, why, then, have we not used it to replace the 6850 in this chapter? The answer to this question is very simple. The 6850 is still widely used and is very much easier to understand than the more versatile DUART. However, since the DUART has become a standard in 68000-based systems, we cannot neglect it. The DUART is described in *Microprocessor Interfacing and the 68000* (see the bibliography), so the material is not repeated here.

FIGURE 9.10   The 68681 DUART

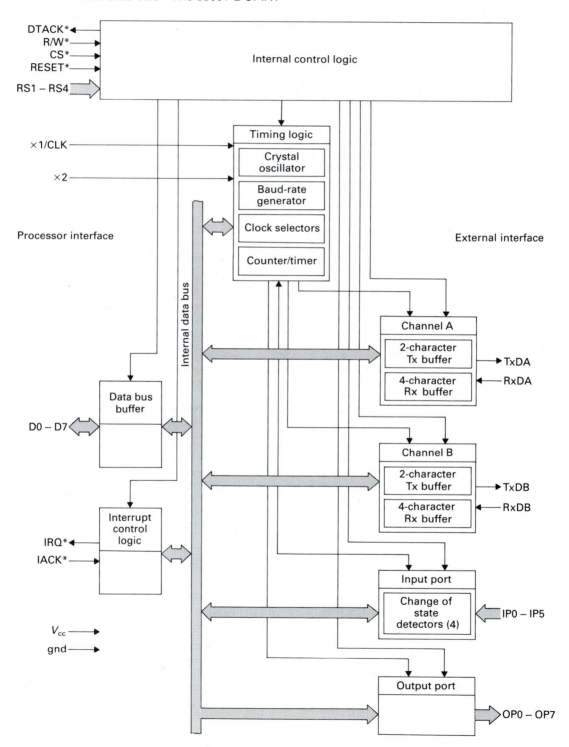

One way of approaching the DUART is to ignore its sophisticated functions and to treat it as an *advanced* ACIA. We will do this and describe how the DUART can be interfaced to a 68000 without going into fine detail. In short, we will treat the DUART as a black box.

The DUART's Registers

The DUART has 16 addressable registers, as illustrated in figure 9.11. Some registers are read-only, some are write-only, and some are read/write. For our current purposes, we concentrate on the DUART's five control registers that must be configured before it can be used as a transmitter or receiver (the 6850 has just a single control register). Note that some registers are *global* and affect the operation of both the DUART's serial channels, whereas others are *local* to channel A or to channel B (in what follows, we use channel A registers). These five control registers are: MR1A (master register 1), MR2A (master register 2), CSRA (clock select register), CRA (command register), and ACR (auxiliary control register). Note that MR1A and MR2A share the same address. After a reset, MR1A is selected at the base address of the DUART. When MR1A is loaded with data by the host processor, MR2A is automatically selected at the same address (you can access MR1A again only by resetting the DUART or by executing a special *select MR1A* command.

Figure 9.12 provides a simplified extract from the DUART's data sheet that describes the five control registers and the status register. Modes of no interest to us here, such as the DUART's parallel I/O capabilities, have not been included in figure 9.12. The following notes provide sufficient details about the DUART's registers to enable the reader to use it in its basic operating mode.

The *auxiliary control register*, ACR, selects the DUART's clock source (internal or external), selects its baud-rate set (there are two sets—setting ACR_7 to 0 selects set 1 and setting ACR_7 to 1 selects set 2), and controls certain parallel input pins. For our purposes, ACR can be loaded with $80 (to select baud-rate set 2) and ignored.

The *clock-select register(s)*, CSRA and CSRB, permit the programmer to select the DUART's baud rate (CSRA selects the channel A baud rate and CSRB the channel B baud rate). Figure 9.12 demonstrates that it is possible to select independent baud rates for transmission and reception. The values shown here can be loaded into the clock-select register to select the following popular baud rates (for both transmission and reception).

VALUE	BAUD RATE
44_{16}	300
55_{16}	600
66_{16}	1,200
88_{16}	2,400
99_{16}	4,800
BB_{16}	9,600
CC_{16}	19,200

FIGURE 9.11 The DUART's registers

RS4	RS3	RS2	RS1	Read (R/W* = 1)	Write (R/W* = 0)
0	0	0	0	Mode register A (MR1A, MR2A)	Mode register A (MR1A, MR2A)
0	0	0	1	Status register A (SRA)	Clock-select register A (CSRA)
0	0	1	0	Do not access*	Command register A (CRA)
0	0	1	1	Receiver buffer A (RBA)	Transmitter buffer A (TBA)
0	1	0	0	Input port change register (IPCR)	Auxiliary control register (ACR)
0	1	0	1	Interrupt status register (ISR)	Interrupt mask register (IMR)
0	1	1	0	Counter mode: current MSB of counter (CUR)	Counter/timer upper register (CTUR)
0	1	1	1	Counter mode: current LSB of counter (CLR)	Counter/timer lower register (CTLR)
1	0	0	0	Mode register B (MR1B, MR2B)	Mode register B (MR1B, MR2B)
1	0	0	1	Status register B (SRB)	Clock-select register B (CSRB)
1	0	1	0	Do not access*	Command register B (CRB)
1	0	1	1	Receiver buffer B (RBB)	Transmitter buffer B (TBB)
1	1	0	0	Interrupt-vector register (IVR)	Interrupt-vector register (IVR)
1	1	0	1	Input port (unlatched)	Output port configuration register (OPCR)
1	1	1	0	Start-counter command**	Output port Register (OPR) — Bit set command**
1	1	1	1	Stop-counter command**	Output port Register (OPR) — Bit reset command**

* This address location is used for factory testing of the DUART and should not be read. Reading this location will result in undesired effects and possible incorrect transmission or reception of characters. Register contents may also be changed.
** Address triggered commands.

Each baud-rate value loaded into a clock-select register consists of two 4-bit values (bits 0–3 select the transmitter baud rate and bits 4–7 select the receiver baud rate). For example, the instruction MOVE.B #$B8,CSRA selects a receive rate of 9,600 baud and a transmit rate of 2,400 baud.

The *channel mode control* registers define the operating mode of the DUART (MR1A, MR2A for channel A and MR1B, MR2B for channel B). Figure 9.12 provides a simplified account of these bits. To operate the DUART in its normal, 8-bit character mode with no parity, 1 stop bit, and no modem control functions activated, MR1A is loaded with 13_{16} and MR2A, with 07_{16}. Remember that these registers share the same address and that MR2A is selected automatically after MR1A has been loaded—that is,

```
MR1A    EQU    DUART_BASE
MR2A    EQU    MR1A             MR1A, MR2A have same address
               MOVE.B #$13,MR1A   Load MR1A—no parity, 8 bits
               MOVE.B #$07,MR2A   Now load MR2A—normal mode, 1 stop bit
```

The *command registers* (CRA and CRB) permit the programmer to enable and disable a channel's receiver or transmitter and to issue certain commands to the DUART. The command CRA(6 : 4) = 001 resets the master register pointer to MR1A (since MR2A is automatically selected after MR1A has been loaded). You can load CRA with $0A_{16}$ to disable both channels during its setting up phase and then load it with 05_{16} to enable its transmitter and receiver ports once its other registers have been set up.

The DUART's *status registers* (SRA and SRB) are very similar to their 6850 counterpart. The major additions are SRA_7, which detects that a break has been received, SRA_3 (TxEMT), which indicates that the transmitter buffer is empty (i.e., there are no characters in the DUART's buffer waiting to be transmitted), and SRA_1 (FFULL), which indicates that the receiver buffer is full (there are four received characters waiting to be read). You can, of course, forget about these new bits and operate the DUART exactly like the ACIA just by using the TxRDY and RxRDY bits of its status register.

The difference between the status bits FFULL and TxRDY is that the FFULL flag is applied to the whole receiver buffer, whereas RxRDY tells us that there is at least one free place in the receiver buffer. Similarly, TxEMT tells us that there is no character in the transmitter buffer and that the buffer is completely empty, whereas TxRDY tells us that the DUART is ready for another character.

The DUART has sophisticated *interrupt control* and handling facilities (figure 9.13). The *interrupt vector register*, IVR, provides a vector number when the DUART generates an interrupt and receives an IACK response from the 68000. If the IVR has not been loaded by the programmer since the last time the DUART was reset, the DUART supplies an uninitialized vector number during an IACK cycle.

The DUART has two interrupt control registers with identical formats: ISR is an *interrupt status register* whose bits are set when "interrupt-generating" activities take place. IMR is an *interrupt mask register* whose bits are set by the programmer to enable an interrupt or cleared to mask the interrupt. For example, ISR_0 is set if

FIGURE 9.12 The DUART's control registers

Clock-select register A (CSRA)

Receiver-clock select				Transmitter-clock select			
Bit 7	Bit 6	Bit 5	Bit 4	Bit 3	Bit 2	Bit 1	Bit 0
	Baud rate				Baud rate		
	Set 1 ACR bit 7 = 0	Set 2 ACR bit 7 = 1			Set 1 ACR bit 7 = 0	Set 2 ACR bit 7 = 1	
0 0 0 0	50	75		0 0 0 0	50	75	
0 0 0 1	110	110		0 0 0 1	110	110	
0 0 1 0	134.5	134.5		0 0 1 0	134.5	134.5	
0 0 1 1	200	150		0 0 1 1	200	150	
0 1 0 0	300	300		0 1 0 0	300	300	
0 1 0 1	600	600		0 1 0 1	600	600	
0 1 1 0	1200	1200		0 1 1 0	1200	1200	
0 1 1 1	1050	2000		0 1 1 1	1050	2000	
1 0 0 0	2400	2400		1 0 0 0	2400	2400	
1 0 0 1	4800	4800		1 0 0 1	4800	4800	
1 0 1 0	7200	1800		1 0 1 0	7200	1800	
1 0 1 1	9600	9600		1 0 1 1	9600	9600	
1 1 0 0	38.4k	19.2k		1 1 0 0	38.4k	19.2k	

Channel A mode register 1 (MR1A) and channel B mode register 1 (MR1B)

Rx RTS control	Rx IRQ* select	Error mode	Parity mode		Parity type	Bits-per-character	
Bit 7	Bit 6	Bit 5	Bit 4	Bit 3	Bit 2	Bit 1	Bit 0
					With parity 0 = even 1 = odd		
0 = Disabled 1 = Enabled	0 = RxRDY 1 = FFULL	0 = Char 1 = Block	0 0 = With parity 0 1 = Force parity 1 0 = No parity 1 1 = Multidrop mode*		Force parity 0 = Low 1 = High	0 0 = 5 0 1 = 6 1 0 = 7 1 1 = 8	
					Multidrop mode 0 = Data 1 = Address		

*The parity bit is used as the address/data bit in multidrop mode.

Channel A command register (CRA) and channel B command register (CRB)

Not used*	Miscellaneous commands			Transmitter commands		Receiver commands	
Bit 7	Bit 6	Bit 5	Bit 4	Bit 3	Bit 2	Bit 1	Bit 0
X	0 0 0 No command 0 0 1 Reset MR pointer to MR1 0 1 0 Reset receiver 0 1 1 Reset transmitter 1 0 0 Reset error status 1 0 1 Reset channel's break-change interrupt 1 1 0 Start break 1 1 1 Stop break			0 0 No action, stays in present mode 0 1 Transmitter enabled 1 0 Transmitter disabled 1 1 Don't use, indeterminate		0 0 No action, stays in present mode 0 1 Receiver enabled 1 0 Receiver disabled 1 1 Don't use, indeterminate	

*Bit seven is not used and may be set to either zero or one.

Channel A status register (SRA) and channel B status register (SRB)

Received break	Framing error	Parity error	Overrun error	TxEMT	TxRDY	FFULL	RxRDY
Bit 7*	Bit 6*	Bit 5*	Bit 4	Bit 3	Bit 2	Bit 1	Bit 0
0 = No 1 = Yes	0 = No 1 = Yes	0 = No 1 = Yes	0 = No 1 = Yes	0 = No 1 = Yes	0 = No 1 = Yes	0 = No 1 = Yes	0 = No 1 = Yes

* These status bits are appended to the corresponding data character in the receive FIFO and are valid only when the RxRDY bit is set. A read of the status register provides these bits (seven through five) from the top of the FIFO together with bits four through zero. These bits are cleared by a reset error status command. In character mode, they are discarded when the corresponding data character is read from the FIFO.

Auxiliary control register (ACR)

BRG set Select*	Counter/timer mode and source**			Delta*** IP3 IRQ*	Delta*** IP2 IRQ*	Delta*** IP1 IRQ*	Delta*** IP0 IRQ*
Bit 7	Bit 6	Bit 5	Bit 4	Bit 3	Bit 2	Bit 1	Bit 0
0 = Set 1 1 = Set 2	Mode 0 0 0 Counter 0 0 1 Counter 0 1 0 Counter 0 1 1 Counter 1 0 0 Timer 1 0 1 Timer 1 1 0 Timer 1 1 1 Timer		Clock source External (IP2)**** TxCA – 1X clock of channel A transmitter TxCB – 1X clock of channel B transmitter Crystal or external clock (X1/CLK) divided by 16 External (IP2)**** External (IP2) divided by 16**** Crystal or external clock (X1/CLK) Crystal or external clock (X1/CLK) divided by 16	0 = Disabled 1 = Enabled	0 = Disabled 1 = Enabled	0 = Disabled 1 = Enabled	0 = Disabled 1 = Enabled

* Should only be changed after both channels have been reset and are disabled.
** Should only be altered while the counter/timer is not in use (i.e., stopped if in counter mode, output and/or interrupt masked if in timer mode).
*** Delta is equivalent to change-of-state.
**** In these modes, because IP2 is used for the counter/timer clock input, it is not available for use as the channel B receiver-clock input.

Channel A mode register 2 (MR2A) and channel B mode register 2 (MR2B)

Channel mode		Tx RTS control	CTS enable transmitter	Stop bit length			
Bit 7	Bit 6	Bit 5	Bit 4	Bit 3	Bit 2	Bit 1 6–8 bits/ character	Bit 0 5 bits/ character
0 0 = Normal 0 1 = Automatic echo 1 0 = Local loopback 1 1 = Remote loopback		0 = disabled 1 = enabled	0 = disabled 1 = enabled	(0) 0 0 0 0 = (1) 0 0 0 1 = (2) 0 0 1 0 = (3) 0 0 1 1 = (4) 0 1 0 0 = (5) 0 1 0 1 = (6) 0 1 1 0 = (7) 0 1 1 1 = (8) 1 0 0 0 = (9) 1 0 0 1 = (A) 1 0 1 0 = (B) 1 0 1 1 = (C) 1 1 0 0 = (D) 1 1 0 1 = (E) 1 1 1 0 = (F) 1 1 1 1 =		0.563 0.625 0.688 0.750 0.813 0.875 0.938 1.000 1.563 1.625 1.688 1.750 1.813 1.875 1.938 2.000	1.063 1.125 1.188 1.250 1.313 1.375 1.438 1.500 1.563 1.625 1.688 1.750 1.813 1.875 1.938 2.000

NOTE:
If an external 1X clock is used for the transmitter, MR2 bit 3 = 0 selects one stop bit and MR2 bit 3 = 1 selects two stop bits to be transmitted.

FIGURE 9.13 The DUART's interrupt control registers

Interrupt vector register IVR

Interrupt vector bits D7 – D0

Interrupt status register ISR

7	6	5	4	3	2	1	0
Input port change	Delta break B	RxRDYB/ FFULLB	TxRDYB	Counter/ timer	Delta break A	RxRDYA/ FFULLA	TxRDYA

Interrupt mark register IMR

7	6	5	4	3	2	1	0
Input port change	Delta break B	RxRDYB/ FFULLB	TxRDYB	Counter/ timer	Delta break A	RxRDYA/ FFULLA	TxRDYA

NOTE:

RxRDY/FFULL   Interrupt if RxRDY bit in status register set
TxRDY   Interrupt if TxRDY bit in status register set

Bit 6 of the channel mode register (MR1A$_6$ or MR1B$_6$) determines whether interrupt status register bits ISR$_1$ and ISR$_5$ are set on RxRDY or FFULL. If MR1A$_6$ = 0, ISR$_1$ is set on RxRDY (i.e., at least one character received). If MR1A$_6$ = 1, ISR$_1$ is set on FFULL (i.e., receiver buffer full).

TxRDYA is asserted to indicate that the channel A transmitter is ready for a character. If IMR$_0$ is set to 1, the DUART will generate an interrupt when channel A is ready to transmit a character.

We said earlier that the DUART has multipurpose I/O pins, which can be used as simple I/O pins or to perform special functions. Input pins IP0–IP5 are configured by bits in the CSRA/B and ACR registers. These pins can be programmed to provide inputs for the DUART's timer/counter, its baud-rate generator, and its clear to send modem control. When MR2A$_4$ = 1, pin IP0 acts as a channel A active-low clear to send input. Similarly, IP1 can be configured as channel B's CTS* input by setting MR2B$_4$ to 1. If the DUART is programmed to use its CTS* pins (i.e., MRSA/B$_4$ = 1), data is not transmitted by the DUART whenever CTS* is high—that is, the remote receiver can negate CTS* to stop the DUART sending further data. Figure 9.14 demonstrates how CTS* is used in conjunction with RTS* (see the following discussion).

FIGURE 9.14 Flow control and the DUART

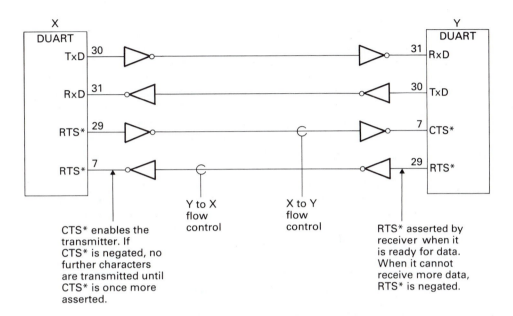

CTS* enables the
transmitter. If
CTS* is negated, no
further characters
are transmitted until
CTS* is once more
asserted.

RTS* asserted by
receiver when it
is ready for data.
When it cannot
receive more data,
RTS* is negated.

NOTE the following steps to configure the DUART to perform flow control:

```
*    Set MR2A4 to 1 and MR2A5 to 1 to configure OP0 as
*    RTS output
*
     MOVE.B      #$83,MR1A
     MOVE.B      #$27,MR2A
*
*
*    Note that RTS* must initially be asserted
*        manually—after that, RTS* is asserted
*        automatically whenever the receiver is ready to
*        receive more data. Note also that the contents
*        of the DUART's output port register are
*        inverted before they are fed to the output
*        pins. That is, to assert RTS* low, it is
*        necessary to load a one into the appropriate
*        bit of the OPR.
*
     MOVE.B      #$01,OPR     Set OPR0 to assert RTS*
```

The 8-bit output port is controlled by an output port configuration register
(OPCR) and certain bits of the ACR, MR1A, MR2A, MR1B, and MR2B registers.
Output bits can be programmed as simple outputs cleared and set under programmer
control, timer and clock outputs, and status outputs. Some of the output functions
that can be selected are the following:

PIN	FUNCTION	ACTION
OP0	RxRTSA*	Asserted if channel A Rx is able to receive a character
OP0	TxRTSA*	Negated if channel A Tx has nothing to transmit
OP1	RxRTSB*	Asserted if channel B Rx is able to receive a character
OP1	TxRTSB*	Negated if channel B Tx has nothing to transmit
OP4	RxRDYA	Asserted if channel A Rx has received a character
OP5	RxRDYB	Asserted if channel B Rx has received a character
OP6	TxRDYA	Asserted if channel A Tx ready for data
OP7	TxRDYB	Asserted if channel B Tx ready for data

Note the difference between the RxRTS* and TxRTS* functions. RxRTS* is used by a receiver to indicate to the remote transmitter that it (the receiver) is able to accept data. RxRTS* is connected to the transmitter's CTS* input to perform flow control (figure 9.14). The TxRTS* function is used to indicate to a modem that the DUART has further data to transmit.

Programming the 68681 DUART

Once the DUART has been configured, it can be used to transmit and receive characters exactly like the 6850. The following fragment of code provides basic initialization, receive, and transmit routines for the DUART.

```
*               DUART equates
MR1A      EQU    1                    Mode register 1
MR2A      EQU    1                    Mode register 2 (same address as MR1A)
SRA       EQU    3                    Status register
CSRA      EQU    3                    Clock select register
CRA       EQU    5                    Command register
RBA       EQU    7                    Receiver buffer register (i.e., serial in)
TBA       EQU    7                    Transmitter buffer register (i.e., data out)
IPCR      EQU    9                    Input port change register
ACR       EQU    9                    Auxiliary control register
ISR       EQU    11                   Interrupt status register
IMR       EQU    11                   Interrupt mask register
IVR       EQU    25                   Interrupt vector register
*
*               Initialize the DUART
*
INITIAL   LEA    DUART,A0             A0 points at DUART base address
*
*               Note the following three instructions are not necessary
*               after a hardware reset to the DUART. They are included to
*               show how the DUART is reset.
*
          MOVE.B  #$30,CRA(A0)        Reset port A transmitter
          MOVE.B  #$20,CRA(A0)        Reset port A receiver
          MOVE.B  #$10,CRA(A0)        Reset port A MR (mode register) pointer
*
*               Select baud rate, data format and operating modes by
*               setting up the ACR, MR1 and MR2 registers
```

```
*
                MOVE.B      #$00,ACR(A0)        Select baud-rate set 1
                MOVE.B      #$BB,CSRA(A0)       Set both Rx and Tx speeds to 9600 baud
                MOVE.B      #$93,MR1A(A0)       Set port A to 8 bits, no parity, 1 stop bit,
*                                              enable RxRTS output
                MOVE.B      #$37,MR2A(A0)       Select normal operating mode, enable
*                                              TxRTS, TxCTS, 1 stop bit
                MOVE.B      #$05,CRA(A0)        Enable port A transmitter and receiver
                RTS
*
*           Input a single character from port A (polled mode) into D2
*
PUT_CHAR        MOVEM.L     D0–D1/A0,–(SP)      Save working registers
                LEA         DUART,A0            A0 points to DUART base address
Input_poll      MOVE.B      SRA(A0),D1          Read the port A status register
                BTST        #RxRDY,D1           Test receiver ready status
                BEQ         Input_poll          UNTIL character received
                MOVE.B      RBA(A0),D2          Read the character received by port A
                MOVEM.L     (SP)+,D0–D1/A0      Restore working registers
                RTS
*
*           Transmit a single character in D0 from port A (polled mode)
*
PUT_CHAR        MOVEM.L     D0–D1/A0,–(SP)      Save working registers
                LEA         DUART,A0            A0 points to DUART base address
Output_poll     MOVE.B      SRA(A0),D1          Read port A status register
                BTST        #TxRDY,D1           Test transmitter ready status
                BEQ         Output_poll         UNTIL transmitter ready
                MOVE.B      D0,TBA(A0)          Transmit the character from port A
                MOVEM.L     (SP)+,D0–D1/A0      Restore working registers
                RTS
*
```

In spite of the DUART's complexity, you can see that it may be operated in a simple, noninterrupt-driven, character-by-character input/output mode, exactly like the ACIA, once its registers have been set up. On at least one occasion we have tested software written for a 6850-based system on a board with a DUART by making the following modifications to the 6850's I/O routines.

	6850 I/O			DUART I/O	
SETUP	LEA	ACIA,A0	SETUP	LEA	DUART,A0
	MOVE.B	#$03,(A0)		MOVE.B	#$13,(A0)
	MOVE.B	#$15,(A0)		MOVE.B	#$07,(A0)
	RTS			MOVE.B	#$BB,2(A0)
				MOVE.B	#$05,4(A0)
				RTS	
	LEA	ACIA,A0		LEA	DUART,A0
INPUT	BTST.B	#0,0(A0)	INPUT	BTST.B	#0,2(A0)
	BNE	INPUT		BNE	INPUT
	MOVE.B	2(A0),D0		MOVE.B	6(A0),D0
	RTS			RTS	
OUTPUT	BTST.B	#1,0(A0)	OUTPUT	BTST.B	#2,2(A0)
	BNE	OUTPUT		BNE	OUTPUT
	MOVE.B	D0,2(A0)		MOVE.B	D0,6(A0)

FIGURE 9.15 Interface between a DUART, a 68000 CPU, and two serial channels

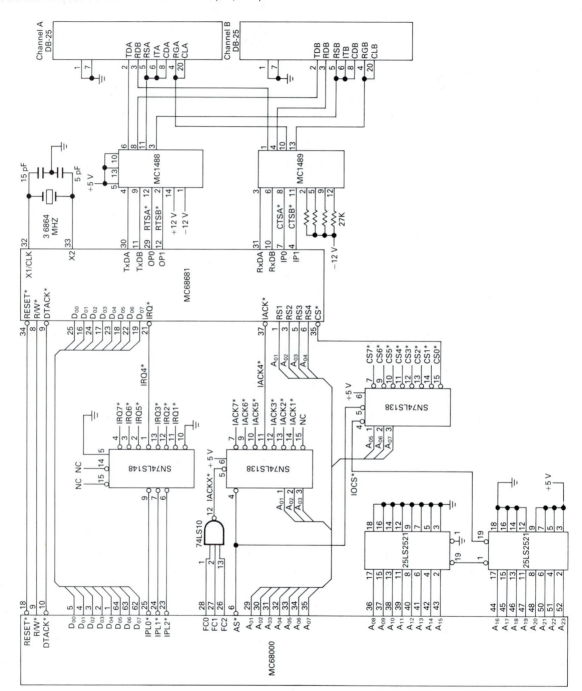

The interface between a DUART and both a 68000 and two serial data links is illustrated in figure 9.15. As you can see, the DUART's 68000 interface is entirely conventional and requires no further comment. In this case, the DUART uses its internal baud-rate generator to supply clocks to both serial channels. The baud-rate generator uses a 3.6864-MHz quartz crystal, which is widely available.

The serial data links each provide a request to send output and a clear to send input to provide flow control of both transmitted and received data.

9.4 SYNCHRONOUS SERIAL DATA TRANSMISSION

The type of asynchronous serial data link described in section 9.1 is widely employed to link relatively slow peripherals such as printers and VDTs with processors. Where information has to be transferred between the individual computers of a network, synchronous serial data transmission is a more popular choice. In a synchronous serial data transmission system, the information is transmitted continuously without gaps between adjacent groups of bits. We use the expression *groups of bits* because synchronous systems can transmit entire blocks of pure binary information at a time, rather than transmitting information as a sequence of ASCII-encoded characters. Before continuing, we need to point out that synchronous serial data links are often used in a much more sophisticated way than their asynchronous counterparts, which simply move data between a processor and its peripheral. This section covers only the basic details of a synchronous serial data link.

Two problems face the designer of a synchronous serial system. One is how to divide the incoming data stream into individual bits and the other is how to divide the data bits into meaningful units.

Bit Synchronization

As synchronous serial data transmission involves very long (effectively infinite) streams of data elements, the clocks at the transmitting and receiving ends of a data link must therefore be permanently synchronized. If a copy of the transmitter's clock were available at the receiver, no difficulty would be encountered in breaking up the data stream into individual bits. As this arrangement requires an extra transmission path between the transmitter and the receiver, it is not a popular solution to the problem of bit synchronization.

A better solution is found by encoding the data to be transmitted in such a way that a synchronizing signal is included with the data signal. Here, we do not delve deeply into the ways in which this situation may be achieved but show one popular arrangement.

The basic method of extracting a timing signal from synchronous serial data is illustrated in figure 9.16. The serial data stream is combined with a clock signal to give the encoded signal, which is actually transmitted over the data link. The encod-

FIGURE 9.16   Phase-encoded synchronous serial bit stream

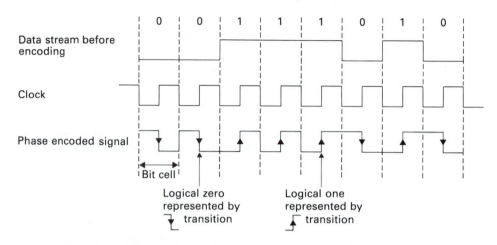

ing algorithm is simple. A logical one is represented by a positive transition in the center of a bit cell and a logical zero by a negative transition. This form of encoding is called phase encoding (PE) or Manchester encoding and is widely used. At the receiver, the incoming data can readily be split into a clock signal and a data signal. Integrated circuits that modulate or demodulate Manchester-encoded signals are readily available.

Word Synchronization

Having divided the incoming stream into individual data elements (i.e., bits), the next step is to group the bits together into meaningful units. We have called these *words*, although they may vary from 8 bits long to thousands of bits long. At first sight, dividing a continuous stream of bits into individual groups of bits might appear to be a most difficult task. Infactitisquiteaneasytasktoformbitsintowords. Here we have deleted interword spacing in the plain text, making it harder, but not impossible, to read. The reader examines the string of letters and looks for recognizable groups corresponding to valid words in English. A similar technique can be applied to continuous streams of binary data. Two basic modes of operation of synchronous serial data links exist: character oriented and bit oriented. In the former, the data stream is divided into separate characters and in the latter it is divided into much longer blocks of pure binary data.

Character-Oriented Data Transmission

In character-oriented data transmission systems, the information to be transmitted is encoded in the form of (usually) ASCII characters. One of the most popular character-oriented systems is called BISYNC, or binary synchronous data transmission. Take, for example, the four-character string "Alan;" it would be sent as the sequence of four 7-bit characters. The individual letters are ASCII encoded as:

A = $41

l = $6C

a = $61

n = $6E

Putting these together and reading the data stream from left to right with the first bit representing the least significant bit of the "A," we get:

1000001001101110000110111011

Some method is needed of identifying the beginning of a message. Once this has been done, the bits can be divided into characters by arranging them into groups of seven (or eight if a parity bit is used) for the duration of the message.

The ASCII code includes a number of characters specifically designed to control the flow of data over a synchronous serial data link. One such character is SYN (as in SYNchronization), whose code is $16, or 0010110. SYN is used to denote the beginning of a message. The receiver reads the incoming bits and looks for the string 0010110, representing a SYN and therefore the start of a message. Unfortunately, such a simple scheme is fatally flawed. The end of one character might be combined with the neginning of the following character to create a false SYN pattern. To avoid this situation, two SYN characters are transmitted sequentially. The receiver reads the first SYN and then looks for the second. If the receiver does not find another SYN, it assumes a false synchronization and continues looking for a valid SYN.

In addition to the synchronization character, the ASCII code provides other characters, such as STX (start of text), to help the user format data into meaningful units. However, character-oriented data transmission systems are not as popular as bit-oriented systems and are therefore not dealt with further here.

Bit-Oriented Data Transmission

Although the ASCII code is excellent for representing text, it is ill-fitted to the representation of pure binary data. Pure binary data can be anything from a core dump (a block of memory) or a program in binary form to floating-point numbers. When data is represented in character form, choosing one particular character (e.g., SYN) as a special marker is easy. When the data is in a pure binary form, choosing any particular data word as a reserved marker or flag is apparently impossible. Bit-oriented protocols (BOPs) have been devised to handle pure binary data.

Fortunately, a remarkably simple and very elegant technique can be used to solve this problem. The beginning of each new block of data, called a frame, is denoted by the special (i.e., unique) binary sequence 01111110. Whenever the receiver detects this pattern, it knows that it has found the start (or end) of a block of data. The special sequence 01111110 is called an *opening* or *closing flag*. Of course, we still have the problem of what to do if we wish to send the pattern 01111110 as part of the data stream to be transmitted. Clearly, it cannot be sent in the form it occurs naturally, as the receiver would regard it as an opening or closing flag.

The transmitter avoids the preceding problem by a process called *bit-stuffing*. Whenever the pattern 011111 is detected at the transmitter (i.e., five 1s in series), the transmitter says, "If the next two bits are a 1 followed by a 0, a spurious flag will be created." Therefore, the transmitter inserts (i.e., stuffs) a 0 after the fifth logical one in succession in order to avoid the generation of a flag pattern. In this way, a flag can never appear by accident in the transmitted data stream.

At the receiver, the incoming bit stream is examined and opening or closing flags are deleted from the data stream. If the sequence 0111110 is found, the 0 following the fifth logical 1 is deleted, as it *must* have been inserted at the transmitter. In this way, any bit pattern may be presented to the transmitter, as bit-stuffing prevents the accidental occurrence of the opening or closing flag. Figure 9.17 illustrates the process of bit-stuffing.

Modern bit-oriented synchronous serial data transmission systems have largely been standardized and use the HDLC data format. HDLC stands for *high-level data link control*. Information is transmitted in the form of packets or frames, with each packet separated by one or two flags as described previously. The format of a typical HDLC frame is given in figure 9.18. Following the opening flag is an address field of 8 bits, which defines the address of the secondary station (or slave) in situations

FIGURE 9.17   Process of bit-stuffing

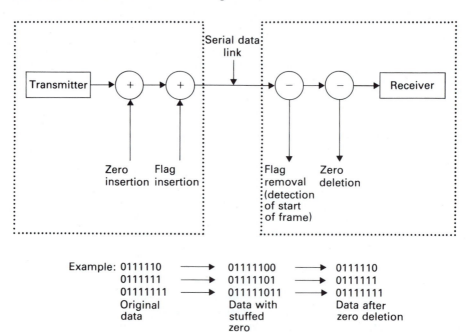

FIGURE 9.18 High-level data link control format

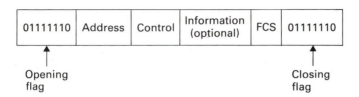

where a master station may be in communication with several slaves. An address field allows the master to send a message to one of its slaves without ambiguity. Any slave receiving a message whose address does not match that in the address field of a frame ignores that frame. Figure 9.19 shows the arrangement of a typical master-slave system. Remember that we have already stated that synchronous serial transmission systems are frequently used in more sophisticated ways than their asynchronous counterparts.

Following the address field is an 8-bit control field that controls the operation of the data link. The purpose of this field is to permit an orderly exchange of messages and to help detect and deal with lost messages. All that need be said here is that the control field provides the HDLC scheme with some very powerful facilities that are almost entirely absent in simple synchronous serial data links described earlier. The control field is followed by an optional data field (information field, or I field) containing the data to be transmitted. The I field is optional because frames may be transmitted for purely control purposes without an I field. Immediately after the I field (or control field if the I field is absent) comes the frame check sequence, FCS, which is a very powerful error-detecting code of 16 bits' length that is able to

FIGURE 9.19 Master-slave data transmission

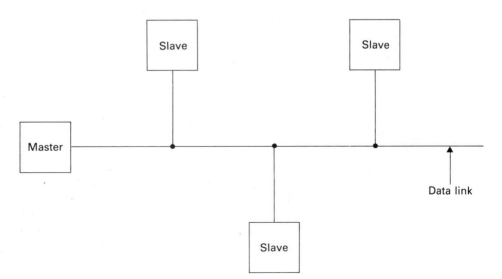

detect the vast majority of errors (single or multiple bit) in the preceding fields. We are not able to go into detail about the theory of the FCS here, but the following notes should help.

The p bits in the packet, or frame, between the opening flag and the FCS itself are regarded as forming the coefficients of a polynomial of degree p. This polynomial is divided by a standard polynomial using modulo two arithmetic to yield a quotient and a 16-bit remainder. The quotient is discarded and the remainder forms the 16-bit FCS. At the receiver, the bits between the opening flag and received FCS are divided by the same generator polynomial to yield a local FCS. If the local FCS is the same as the received FCS, we can assume that the frame is free from all transmission errors. If they differ, the current frame is rejected.

Following the FCS is a closing flag, 01111110. Some arrangements require a closing flag for the current frame to be followed by an opening flag for the next frame. Other systems use one flag both to close the current frame and to open the next frame.

Clearly, a synchronous system is more efficient than an asynchronous system because of the absence of start, parity, and stop bits for each transmitted character. However, the real advantage of a synchronous system combined with the HDLC frame structure is its ability to control a data transmission system.

9.5 SERIAL INTERFACE STANDARDS (RS-232 AND RS-422/RS-423)

Because of the low cost of a serial interface and transmission path, the serial data link is used to connect a very wide range of peripherals to computer equipment. Once only teletypes and modems were likely to be connected to a computer by serial data links. Today, almost any peripheral, from CRT terminals to graphics tablets to disk drives, may use serial data links. Therefore, the data link should be standardized so that a peripheral from one manufacturer can be plugged into the serial port of a computer from another manufacturer.

Such a serial interface standard has been created by the Electronic Industries Association (EIA) and is known as the RS-232C serial interface.

The RS-232C Serial Interface

The EIA RS-232C standard was largely intended to link data terminal equipment (DTE) with data communications equipment (DCE). The DCE corresponds to the computer or terminal and the DTE corresponds to the modem or similar line equipment. RS-232C specifies the electrical and mechanical aspects of the serial interface together with the functions of the signals forming the interface. In theory, any RS-232C-compatible DCE can be connected directly to any RS-232C-compatible DTE.

Alas, the life of the computer technician is filled with time-wasting requests by programmers to get their printer to work with their computer—both of which have "RS-232C" serial interfaces. We put "RS-232C" in quotation marks because many (the majority?) equipment suppliers implement their own subset of an RS-232C interface. The likelihood that the subset of the RS-232C interface in the printer is incompatible with the subset in the computer is highly probable.

Mechanical Interface

Fortunately, the vast majority of equipment suppliers adhere to the mechanical aspects of the RS-232C standard and use the D-type connectors illustrated in figure 9.20. This connector is available in 9-, 15-, 25-, 37-, and 50-pin versions, but only the 25-way D connector may be used with RS-232C standard serial data links. The pinout of this connector is given in table 9.5, although we must appreciate the fact that very few implementations of the RS-232C standard implement the full standard.

FIGURE 9.20 D-type connector

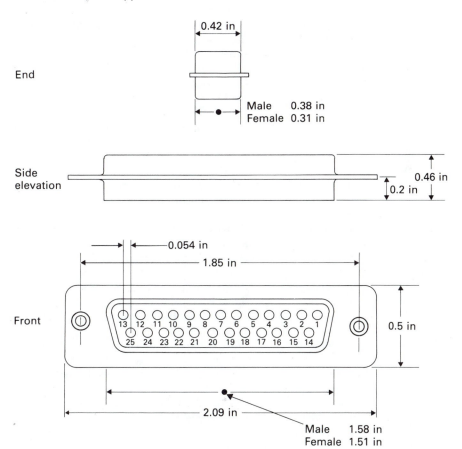

TABLE 9.5   Pinout of the RS-232C 25-way connector

PIN	NAME	FUNCTION
1	Protective ground	Electrical equipment frame and dc power ground
2	Transmitted data	Serial data generated by the DTE
3	Received data	Serial data generated by the DCE
4	Request to send	When asserted indicates that the DTE is ready to transmit primary data
5	Clear to send	When asserted indicates that the DCE is ready to transmit primary data
6	Data set ready	When asserted indicates that the DCE is not in a test, voice, or dial mode, that all initial handshake, answer tone, and timing delays have expired
7	Signal ground	Common ground reference for all circuits except protective ground
8	Received line signal detector	When asserted indicates that carrier signals are being received from the remote equipment
9	Reserved	
10	Reserved	
11	Unassigned	
12	Secondary received line signal detector	When asserted indicates that the secondary channel data carrier signals are being received from the remote equipment
13	Secondary clear to send	When asserted indicates that the DCE is ready to transmit secondary data
14	Secondary transmitted data	Low-speed secondary data channel generated by the DTE
15	Transmitted signal element timing	The signal on this line provides the DTE with signal element timing information
16	Secondary received data	Low-speed secondary channel data generated by the DCE
17	Receiver signal element timing	The signal on this line provides the DTE with signal element timing information
18	Unassigned	
19	Secondary request to send	When asserted indicates that the DTE is ready to transmit secondary channel data
20	Data terminal ready	When asserted indicates that the data terminal is ready
21	Signal quality detector	When asserted indicates that the received signal is probably error free; when negated indicates that the received signal is probably in error
22	Ring indicator	When asserted indicates that modem has detected a ringing tone on the telephone line
23	Data signal rate detector	Selects between two possible data rates
24	Transmit signal element timing	The signal on this line provides the DCE with signal element timing information
25	Unassigned	

Electrical Interface

The RS-232C standard is intended to provide serial communication facilities over relatively short distances and its electrical specifications reflect this. Table 9.6 gives the basic electrical parameters of the standard and figure 9.21 shows how the electrical interface may be implemented.

The circuit of figure 9.21 uses a single-ended bipolar unterminated circuit; that is, the circuit is single-ended (i.e., unbalanced) because the signal level to be transmitted is referred to ground and one of the signal-carrying conductors is grounded at both ends of the data link. The circuit is unterminated because no requirement exists in the RS-232C standard to match the characteristic impedance of the receiver to that of the transmission path.

One of the key parameters in table 9.6 is the receiver maximum input threshold of -3 to $+3$ V. A space is guaranteed to be recognized if the input is more positive than $+3$ V and a mark is guaranteed to be recognized if the input is more negative than -3 V. The threshold separating mark and space levels is truly massive. Unless a transmitter can produce a voltage swing at the end of a transmission path of greater than 6 V, the received signal falls outside the minimum requirements of RS-232C. However, most real receivers for RS-232C signals have *practical* input thresholds well below -3 to $+3$ V. Therefore, as most engineers have noticed, it is often possible to have much longer transmission paths than the standard stipulates.

Interfacing to RS-232C lines is now very easy, as the major semiconductor manufacturers have produced suitable line drivers and receivers. Figure 9.22 gives

TABLE 9.6 EIA RS-232C electrical interface characteristic

CHARACTERISTIC	VALUE
Operating mode	Single ended
Maximum cable length	15 m
Maximum data rate	20 kilobaud
Driver maximum output voltage (open-circuit)	-25 V $< V <$ $+25$ V
Driver minimum output voltage (loaded output)	-25 V $< V < -5$ V or $+5$ V $< V < +25$ V
Driver minimum output resistance (power off)	300 Ω
Driver maximum output current (short-circuit)	500 mA
Maximum driver output slew rate	30 V/μs
Receiver input resistance	3–7 kΩ
Receiver input voltage	-25 V $< V_i < +25$ V
Receiver output state when input open-circuit	Mark (high)
Receiver maximum input threshold	-3 to $+3$ V

FIGURE 9.21   RS-232C interface

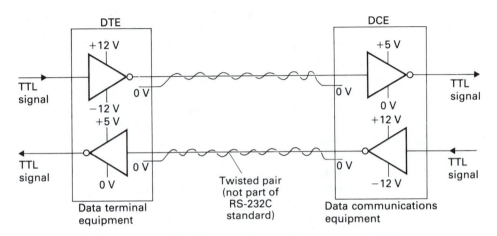

details of the 1488 quad RS-232C line driver and figure 9.23 gives details of the 1489 quad RS-232C line receiver. An example of the application of these chips is given in figure 9.24. Note that the 1489 receiver has an input control pin that can be used to define the amount of hysteresis at the input. We may leave this pin floating, in which case the input switching threshold is approximately 1 V.

RS-232C Interconnection Subset

So far we have seen three forms of RS-232C interconnection: the most basic arrangement of figure 9.8, the somewhat more complete circuit of figure 9.9, and the full RS-232C interface of table 9.5. A glance at table 9.5 makes it very clear that the RS-232C standard is aimed squarely at linking computer equipment with modems. Consequently, many of the facilities offered by the RS-232C standard are irrelevant to the engineer who wishes to connect a CRT terminal to a microcomputer.

Figure 9.25 shows a possible connection between two DTEs. As one DTE is the sink for the other's data, making the cross-connections shown in figure 9.25 is necessary. These cross-connections are frequently made at the junction of the cable and the connector. Such a cable linking a DTE to another DTE is called a *null modem* cable. When a DTE is connected to a modem (i.e., DCE) such a crossover is not necessary.

Many DTEs do not even use the subset of figure 9.25; for example, at the computer end of a computer-to-printer serial data link, no request to send (RTS) output may be provided. If the printer requires that its carrier detect input be driven by RTS, we need to strap the printer's carrier detect input to a logical one condition.

The 1987 Revision of RS-232C

The RS-232C was updated in 1987 because it no longer represented current practice. Engineers were using it in ways not anticipated by those who drew up the original standard. RS-232C has now been replaced by EIA232D. The new standard is, as we

FIGURE 9.22 The 1488 quad line driver

FUNCTIONAL DESCRIPTION

The 1488 is a quad line driver that conforms to EIA specification RS-232C. Each driver accepts one or two TTL/DTL inputs and produces a high-level logic signal on its output. The high and low logic levels on the output are defined by the positive and negative power supplies of plus and minus 9 volts, the output levels are guaranteed to meet the $\pm$ 6-volt specification with a 3 kΩ load. There is an internal 300 Ω resistor in series with the output to provide current limiting in both the high and low logic levels. The 1488 driver is intended for use with the 1489 or 1489A quad line receivers.

LOGIC SYMBOL

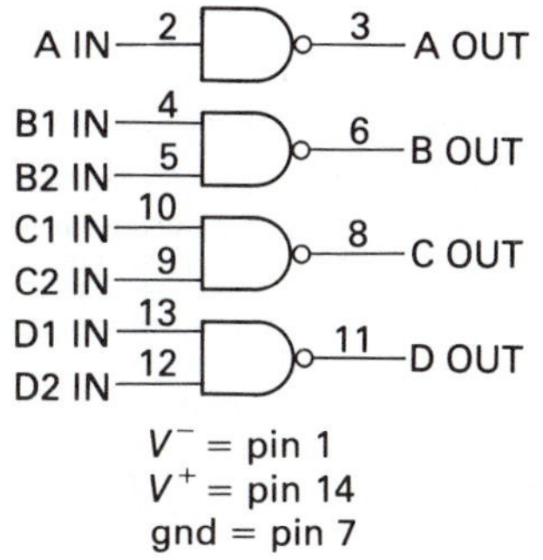

A IN — 2 — 3 — A OUT

B1 IN — 4
B2 IN — 5 — 6 — B OUT

C1 IN — 10
C2 IN — 9 — 8 — C OUT

D1 IN — 13
D2 IN — 12 — 11 — D OUT

$V^- = $ pin 1
$V^+ = $ pin 14
gnd = pin 7

CIRCUIT DIAGRAM
(one driver shown)

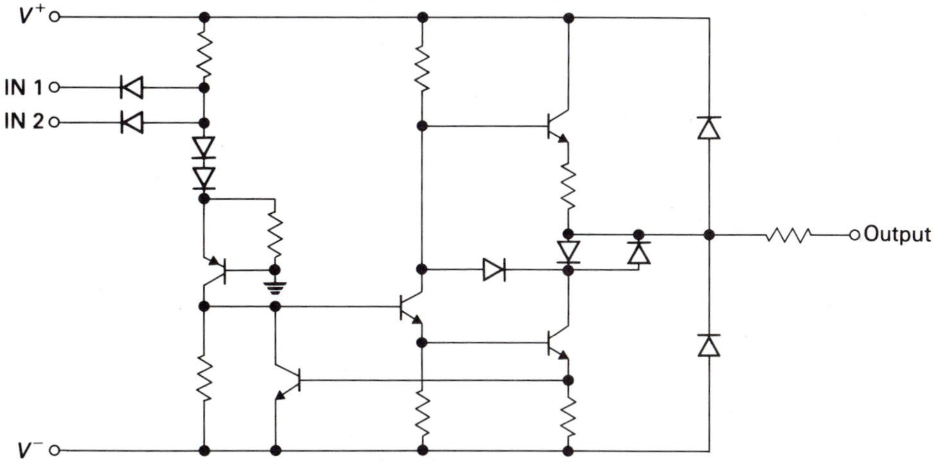

CONNECTION DIAGRAM
Top view

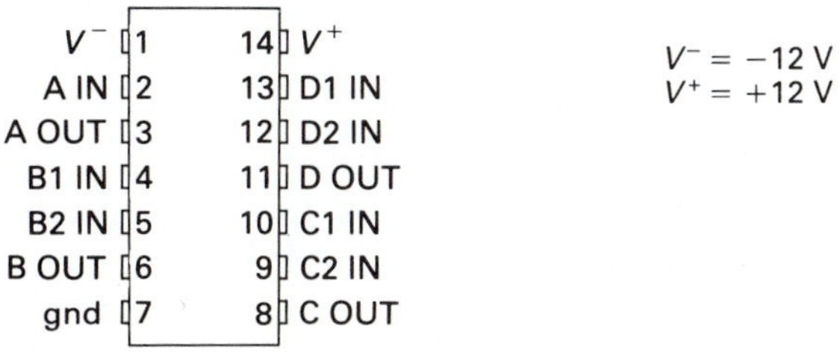

V^- 1	14 V^+
A IN 2	13 D1 IN
A OUT 3	12 D2 IN
B1 IN 4	11 D OUT
B2 IN 5	10 C1 IN
B OUT 6	9 C2 IN
gnd 7	8 C OUT

$V^- = -12$ V
$V^+ = +12$ V

FIGURE 9.23   The 1489 quad line receiver

FUNCTIONAL DESCRIPTION

The 1489 and 1489A are quad line receivers whose electrical characteristics conform to EIA specification RS-232C. Each receiver has a single data input that can accept signal swings of up to ± 30 V. The output of each receiver is TTL/DTL compatible, and includes a 2 kΩ resistor pull-up to V_{cc}. An internal feedback resistor causes the input to exhibit hysterisis so that a.c. noise immunity is maintained at a high level even near the switching threshold. For both devices, when a driver is in a low state on the output, the input may drop as low as 1.25 V without affecting the output. Both devices are guaranteed to switch to the high state when the input voltage is below 0.75 V. Once the output has switched to the high state, the input may rise to 1.0 V for the 1489 or 1.75 V for the 1489A without causing a change in the output. The 1489 is guaranteed to switch to a low output when its input reaches 1.6 V and the 1489A is guaranteed to switch to a low output when its input reaches 2.25 V. Because of the hysterisis in switching thresholds, the devices can receive signals with superimposed noise or with slow rise and fall times without generating oscillations on the output. The threshold levels may be offset by a constant voltage by applying a d.c. bias to the response control input. A capacitor added to the response control input will reduce the frequency response of the receiver for applications in the presence of high-frequency noise spikes. The companion line driver is the 1488.

LOGIC SYMBOL

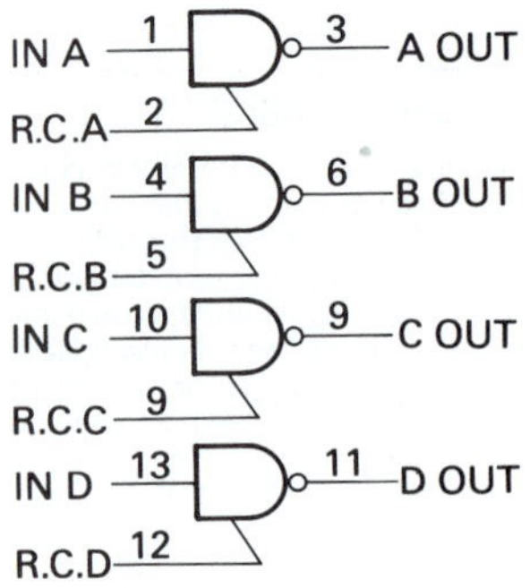

IN A —1— ▷o —3— A OUT
R.C.A —2—

IN B —4— ▷o —6— B OUT
R.C.B —5—

IN C —10— ▷o —9— C OUT
R.C.C —9—

IN D —13— ▷o —11— D OUT
R.C.D —12—

V_{cc} = pin 14
gnd = pin 7

CIRCUIT DIAGRAM
(one receiver)

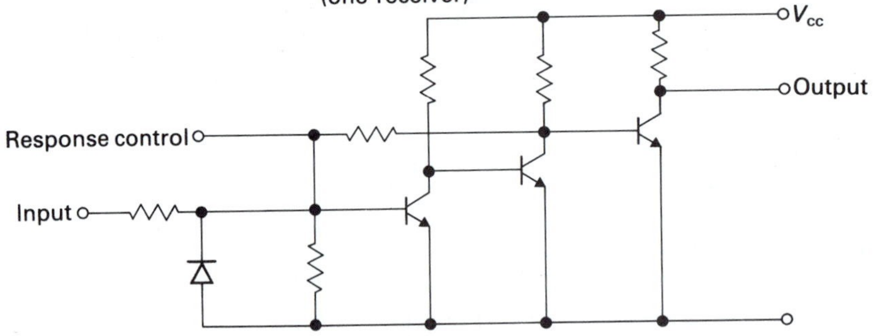

Response control
Input

V_{cc}
Output

CONNECTION DIAGRAM
Top view

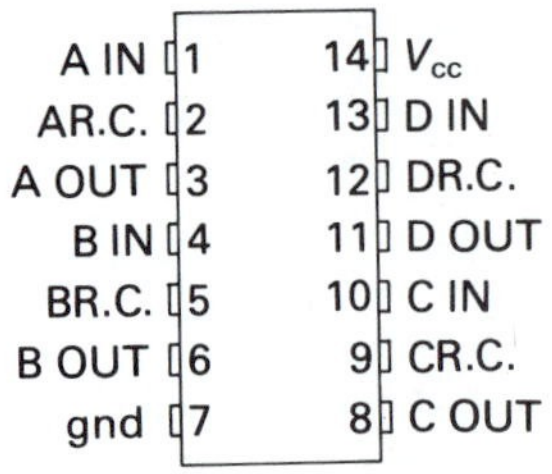

A IN 1	14 V_{cc}
AR.C. 2	13 D IN
A OUT 3	12 DR.C.
B IN 4	11 D OUT
BR.C. 5	10 C IN
B OUT 6	9 CR.C.
gnd 7	8 C OUT

might expect, compatible with the old standard, and only slight modifications have been made in order to match RS-232C more closely to its European equivalents CCITT V24 and V28 and to take account of actual current practice in linking DCEs to DTEs. The following are some of the changes to RS-232C:

1.   The RS-232C pin 1, a protective ground, has been replaced by a shield. Pin 1 may be used to connect the screen of an interface cable to the frame of the

FIGURE 9.24 Example of an RS-232C data link

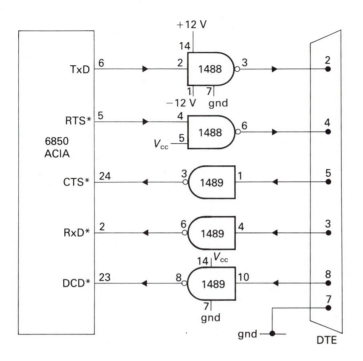

FIGURE 9.25 Connecting two DTEs together

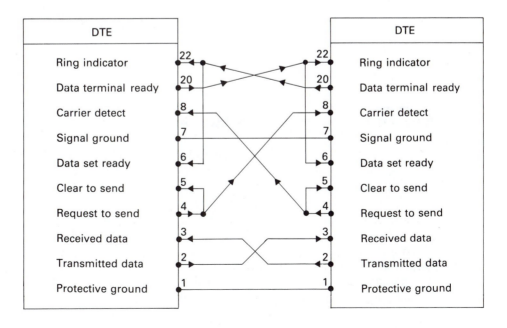

DTE; that is, pin 1 is connected to the screen at only one end of the data link (this avoids ground-loop problems).

2. EIA232D now specifies the mechanical characteristics of the 25-pin interfaces. The old RS-232C specification only *recommended* the use of 25-pin D connectors in an appendix.

3. Provision for local and remote loopback testing has been made by defining three new signals (on pins 21, 18, and 25 of the D connector). Pin 21 is RL (remote loopback) and is asserted by the DTE to tell the local DCE to instruct the remote DCE to go into its loopback mode, allowing the local DTE to test both DCEs and the channel linking them; that is, the remote DCE at the other end of the data link will return signals received from the local DCE via the communication channel. Since the data is echoed back, it is very easy to test the operation of the data link by comparing the transmitted data with that echoed back. Modern peripherals such as the 68681 DUART provide automatic echo modes to facilitate testing. In the old RS-232C standard, pin 21 was a signal-quality detector used by the modem to indicate when a signal was of such a poor quality that it was no longer reliable.

Pin 18 in the new EIA232D standard is called LL (local loopback) and acts like pin 21, except that it establishes a loopback path through the local DCE only. Local loopback permits the system to be tested from the local CPU to the local DCE and back.

Pin 25 is called TM (test mode) and is asserted by the DCE to inform the DTE that the DCE is in a test mode because it has received either RL or LL from the local DTE or a message from the remote DCE requesting a test mode.

4. The recommendation that the RS-232C cable length be restricted to no more than 15 m (50 ft) has been removed; EIA232D permits longer transmission paths, whose length is determined by the electrical loading on the cable. One of the reasons for including this modification is that many users of the RS232C standard have been tolerating longer transmission paths than the legal maximum of 15 m.

The RS-422 and RS-423 Serial Interfaces

The RS-232C interface is now thought to be rather limited because of its low bandwidth and its maximum transmission path of only 15 m. Two improved standards for serial data links have been approved by the EIA. These standards are the RS-422 and RS-423, which define the electrical characteristics of a data link. Unlike RS-232C, these standards refer only to the electrical aspects of a data link.

Table 9.7 gives the basic electrical parameters of the RS-423 and RS-422 standards and figure 9.26 shows how they are arranged. The RS-423 standard differs little from the RS-232C standard of table 9.6. Indeed, the only real difference is that the RS-423 standard specifies much smaller receiver thresholds, permitting both a longer cable length and a higher signaling rate.

TABLE 9.7 EIA RS-422 and RS-423 electrical interfaces

CHARACTERISTIC	RS-423 VALUE	RS-422 VALUE
Operating mode	Single ended	Differential
Maximum cable length	700 m (2,000 ft)	1,300 m (4,000 ft)
Maximum data rate	300 kilobaud	10 megabaud
Driver maximum output voltage (open-circuit)	$-6\,V < V_o < +6\,V$	6 V between outputs
Driver minimum output voltage (loaded output)	$-3.6\,V < V_o < +3.6\,V$	2 V between outputs
Driver minimum output resistance (power off)	100 mA between -6 and $+6$ V	100 μA between $+6$ and -0.25 V
Driver maximum output short-circuit current	150 mA	150 mA
Maximum driver output slew rate	Determined by cable length and modulation rate	No limit on slew rate necessary
Receiver input resistance	$>4\,k\Omega$	$>4\,k\Omega$
Receiver maximum input voltage	$-25\,V < V_i < +25\,V$	-12 to $+12$ V
Receiver maximum input threshold	-0.2 to 0.2 V	-0.2 to $+0.2$ V

FIGURE 9.26 RS-432 and RS-422 serial interfaces

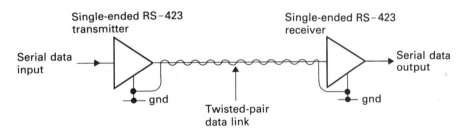

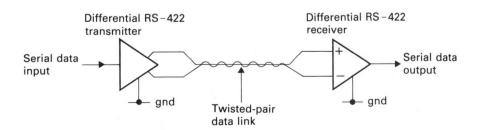

The RS-422 standard offers a significant improvement over both the RS-232C and RS-423 standards by adopting a balanced transmission mode. Balanced transmission requires two transmission paths per signal, because information is transmitted as a differential voltage between the two conductors. Noise voltages due to ground currents are introduced as common mode voltages that affect *both* transmission paths equally and have little effect on the differential signal between the lines. RS-422 systems can operate over distances of 15 m at 10 megabaud or over distances of 1300 m at 100 kilobaud.

9.6 SERIAL INTERFACE FOR THE 68000

Instead of providing a relatively sterile textbook example of a 68000 serial interface based on the ACIA, examining the serial interface in Motorola's MEX68KECB single board microcomputer is most instructive. After all, this board is designed by real engineers to provide the maximum functionality while minimizing the board area taken up by the interface components and keeping their cost low.

Figure 9.27 shows how the ECB is arranged with respect to a terminal and a host computer. In a minimal mode, only the terminal interface on port 1 is necessary. This interface permits the user to interact with the ECB and to develop and debug software. The ECB has a parallel printer interface and an audio cassette interface, allowing programs and data to be printed and stored on tape, respectively. Unfortunately, the serial interface is rather slow and does not have any file-handling capability.

By connecting the ECB to a host computer via the ECB's second serial interface, port 2, a moderately powerful 68000 development system can be created. During my own initial 68000 development work, I used a 6809-based system running under the Flex 09 operating system as a host computer. After a reset, the

FIGURE 9.27 Relationship between the ECB, its console terminal, and a host computer

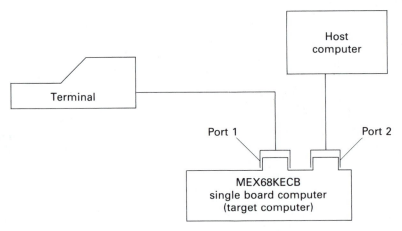

ECB communicates with the terminal through port 1. If the command TM ⟨exit character⟩ is entered, the ECB goes into its transparent mode. The expression ⟨exit character⟩ represents the character that must be entered from the terminal to leave the transparent mode. The default value is control A (ASCII $01).

Once the transparent mode has been entered, the terminal is effectively connected to port 2 and the 68000 on the ECB simply monitors the input from the terminal until it encounters the exit character. Therefore, the user can operate the host computer as if the ECB did not exist; for example, in the transparent mode I am able to edit a 68000 assembly language program on my host system and then assemble it into 68000 machine code using a cross-assembler. Once this process has been done, the exit character is entered and the terminal is once more "connected" to the ECB.

The next step is to transfer the machine code file on disk in the host system to the memory on the ECB. This step is performed by the load command, which moves object data in S-record format from an external device to the 68000's memory. S-record format is a way of representing machine code memory dumps. The syntax of the load command is

    LO[port number][;⟨options⟩][=text].

Square brackets enclose options. Port number 2 (the default port number) specifies the host computer. Other options are not of interest here. The "[=text]" field is available only with port 2 and causes the text following the " = " to be sent to port 2 before the loading is carried out, thus allowing the user to communicate with the host computer. Suppose, for example, that we have a program on disk in the host system in S-record format whose file name is PROG23.BIN. To transfer this program to the 68000's memory space, we enter

    LO2;=LIST PROG23.BIN

from the terminal, which sends the message "LIST PROG23.BIN" to the host computer. The message is interpreted by the operating system as meaning "list the file named PROG23.BIN." This file is then transmitted to port 2 and stored in the 68000's memory by its monitor software. When the loading is complete, the program can be executed and debugged using any of the ECB's facilities. Finally, the transparent mode may be entered and the source file, PROG23.TXT, reedited on the host system. This process is repeated until the software has been debugged.

Figure 9.28 gives the circuit diagram of the serial interface of port 1 and port 2 on the ECB. This diagram has been slightly simplified and redrawn from that appearing in the ECB user manual. Figure 9.28 includes both the serial interface between the ACIAs and the ports and the interface between the ACIAs and the 68000.

Interface between the ACIAs and the 68000 on the ECB

The CPU side of the ACIAs is fairly conventional as they are both connected directly to the 68000's data bus without additional buffering. Port 1 is connected to D_{08} to D_{15} and strobed by UDS*, whereas port 2 is connected to D_{00} to D_{07} and

FIGURE 9.28 Circuit diagram of the ECB's serial interface. (Reprinted by permission of Motorola Limited)

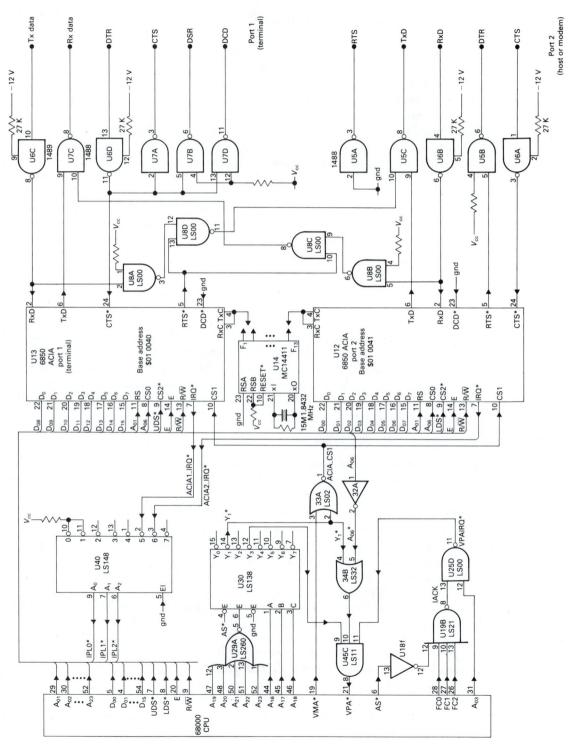

strobed by LDS*. Address line A_{01} is connected to the register select input, RS, of each ACIA and is used to distinguish between the control/status and data register.

To simplify address decoding, each of the three ACIA's chip-select inputs are pressed into service: CS2* is connected to UDS* or LDS* to select between ACIAs, CS1 is connected to the output of the primary address decoding network, ACIA_CS1, and CS0 is connected to A_{06} from the CPU. Primary address decoding is performed by IC29a, a five-input NOR gate, and IC30, a three-line-to-eight-line decoder. These two chips decode A_{16} to A_{23} to produce an active-low signal, Y1*, from IC30. Y1* is combined with VMA* from the 68000 in a NOR gate, IC33a, to give the active-high ACIA_CS1 signal. Note that the ACIAs are not fully address decoded and take up the half of the 64K-byte page of memory space from $01 0000 to $01 FFFF for which A6 = 1.

Because the ACIA is interfaced to the 68000's synchronous bus, VPA* must be asserted whenever an ACIA is accessed. ICs 32a, 34b, and 45c perform this function. Note that VPA* is also asserted when the VPAIRQ* input to IC45c is asserted.

The final aspect of the interface between the ACIA and the CPU to be dealt with is the interrupt-handling hardware. Both ACIAs have independent interrupt request outputs. IRQ* from port 1 is wired to the level 5 input of a 74LS148 priority encoder (IC40) and IRQ* from port 2 is wired to the level 6 interrupt input.

During an interrupt acknowledge cycle, the output of the four-input AND gate IC19b goes active-high to generate an IACK signal. When a level 4 through 7 interrupt is acknowledged, A_{03} is high during the IACK cycle. Therefore, by combining A_{03} with IACK in IC25d, an active-low VPAIRQ* signal generated. VPAIRQ* is fed back to VPA* via IC45c, as indicated earlier. This arrangement converts interrupt levels 4 to 7 into autovectored interrupts, greatly simplifying the hardware design at the cost of reducing the number of possible interrupt vectors.

Serial Interface Side of the ACIAs

IC14, an MC14411 baud-rate generator, provides the transmitter and receiver clocks of the two ACIAs with a source of element timing at 16 times the baud rate of the transmitted or received signal elements. Jumpers on the ECB must be positioned to select the appropriate clock output from the MC14411 for both ACIAs. We should note that if the terminal is to communicate with the host computer, both ACIAs must operate at the same baud rate.

Little comment need be made about the connection of the ACIA's RS-232C signals to their respective ports. However, one interesting feature has been added to the ECB. Whenever the RTS* output of the port 1 (i.e., terminal) ACIA is asserted active-low, both ports 1 and 2 operate independently. However, whenever RTS* from ACIA1 is negated, ICs 6c, 8a, 8d, and 5c route the incoming data from port 1 to the outgoing data on port 2. Incoming data on port 2 is routed to port 1 via ICs 6b, 8b, 8c, and 7c when RTS* is negated. Consequently, negating RTS* connects port 1 to port 2. This is, of course, exactly what happens when the ECB enters its transparent mode. The software required to operate this type of serial link is described in detail in the section dealing with a monitor in chapter 11.

SUMMARY

In this chapter we have looked at the serial interface used to link digital systems to video display terminals and to modems. The simplest method of transmitting serial data is based on a character-oriented asynchronous protocol. If microprocessors had to perform the task of controlling serial links themselves, a considerable part of their power would be lost. We have examined the 6850 ACIA, which performs all serial to parallel and parallel to serial conversion itself. Once a 6850 ACIA is interfaced to a microprocessor, all the microprocessor has to do is to read data from or write data to the appropriate port. Moreover, the ACIA also checks the received data for both transmission and framing errors, further reducing the burden placed on the host microprocessor. Part of the power of the 6850 lies in its ability to cater for a wide variety of serial formats that are selectable under program control. The 6850 also provides three modem control signals, which further simplifies the design of a serial data link between a microprocessor system and a terminal or modem.

We have looked at the electrical interface between the serial transmission path and the microprocessor system. As the TTL-level voltages found in digital equipment are not best suited to transmission paths longer than a few meters, we have described the RS-232C, the RS-422, and the RS423 standards for the transmission of serial data over transmission paths that extend to 1000 meters or more.

Problems

1. What is the difference between asynchronous and synchronous transmission systems? What are the advantages and disadvantages of each mode of transmission?

2. What are the functions of the DCD∗, CTS∗, and RTS∗ pins of the 6850 ACIA?

3. The control register of a 6850 ACIA is loaded with the value $B5. Define the operating characteristics of the ACIA resulting from this value.

4. The status register of the 6850 is read and is found to contain $43. How is this value interpreted?

5. Write an exception-handling routine for a 6850 ACIA to deal with interrupt-driven input. Each new character received is placed in a 4K-byte circular buffer. Your answer must include schemes to (a) deal with buffer overflow and (b) deal with transmission errors.

6. Is connection of the output of an RS-423 transmitter to the input of an RS-232C receiver possible without violating the parameters of either standard? Is connection of an RS-232C output to an RS-423 input possible?

7. An asynchronous transmission system employs unsynchronized transmitter and receiver clocks, both of which are controlled by quartz crystals. It is guaranteed that the worst-case frequency difference between the clocks will never exceed .01 percent. A designer wishes to transmit long bursts of data asynchronously over a serial data link. Each data burst employs a start bit, a single parity bit, and a stop bit. What is the maximum permitted burst length if the designer caters for a maximum frequency error between transmitter and receiver clocks of 80 percent of the stated worst case? (The designer cannot employ the stated worst case value. Why?)

8. Define the following errors associated with asynchronous serial transmission systems and state how each might occur in practice:

    a. Framing error   b. Receiver overrun error

    c. Parity error

9. Describe how bit-stuffing is employed by synchronous serial data links to ensure data transparency. Can you think of any other way in which data transparency can be achieved without resorting to bit-stuffing?

10. What are the advantages of the 68681 DUART over the 6850 ACIA?

11. Write a similar procedure to that of problem 5 for the 68681 DUART.

12. The DUART has a programmable baud-rate generator that is set by loading the appropriate value in clock-select register. This feature makes it possible to adapt to an "unknown" data rate. Write a subroutine that receives a string of carriage returns from a system (at an unknown speed) and adjusts the baud rate to match the incoming data. When the unknown baud rate has been determined, the DUART returns the string "ready."

MICROCOMPUTER BUSES

In this chapter we look at the design of the system bus, which acts as the computer's skeleton, holding all its other "organs" (the functional modules) together. We would not consider it unreasonable to say that the microprocessor bus has done as much to promote the growth of the microcomputer industry as the CPU or the memory chips themselves. We begin by introducing the bus and then describe its electrical characteristics and its interface to microprocessors and to memories or peripherals. At the end of this chapter, we include an introduction to the VMEbus, which is a standard bus and is closely associated with professional 68000-based microcomputers. We also look at the NuBus.

The bus is nothing more than a number of parallel conductors designed to transfer information between separate modules or cards in a microprocessor system. Although not an exciting or glamorous component, the bus serves two vital purposes. Firstly, it makes the production of complex systems with large quantities of memory and peripherals possible. If components were wired together on a point-to-point basis, without a bus, the sheer number of interconnections would be uneconomic. Secondly, once a standard for a bus has been promulgated, independent manufacturers can produce their own cards to plug into another's bus. The PC bus is a spectacular example of this process.

Figure 10.1 illustrates some of the concepts that must be dealt with when microcomputer buses are discussed. Although, strictly speaking, we have only one bus—the system bus—most engineers talk about the *data bus*, the *address bus*, and the *control bus*. These buses are really logically distinct subgroups of the system bus. We must also appreciate that the d.c. power fed to the various cards of a system is usually supplied by the system bus.

What is a bus? A bus is the electrical highway linking the modules of a computer system. This statement conceals far more than it reveals. Figure 10.2 shows why this is the case by illustrating the three factors influencing the design of a suitable bus, which are: its mechanical specification, its electrical specification, and its protocol. The mechanical specification governs the physical aspects of the bus (size, material, connectors), the electrical specification governs the requirements that must be met by the signals on the bus, and the protocol governs the sequence of signals that must be complied with to ensure an orderly exchange of data. Here we are most concerned with the electrical characteristics and the protocols of buses.

FIGURE 10.1 Microcomputer bus

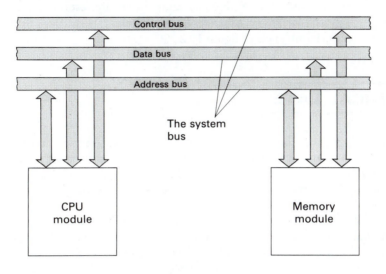

FIGURE 10.2 Components of a bus

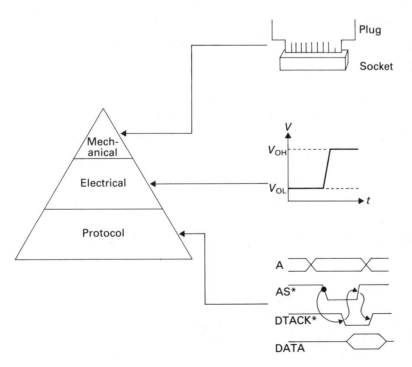

Possibly more than in any other area, economics plays a vital role in determining the type of bus used by any given system; that is, the economics of bus design and construction are its limiting factor. This statement is particularly true at the mechanical and electrical levels of figure 10.2.

10.1 MECHANICAL LAYER

The fact is sad but true that the mechanical nature of a bus largely determines its cost but has very little direct influence on its electrical performance. The mechanical aspects of a bus comprise those elements related to the physical structure of the bus, its mounting within the computer system, its physical dimensions and weight, its strength, and its reliability.

One of the simplest forms of bus is illustrated in figure 10.3 and consists of a motherboard, or backplane, into which a number of cards, or daughterboards, plug. Electrically, the bus is composed of parallel copper conductors running along the length of the motherboard. These conductors are generally arranged on a 0.1- or 0.15-in. pitch. At regular intervals along the motherboards are edge connectors into which the duaghterboards are plugged.

We can see from figure 10.3 that the *fingers* on the daughterboard make contact with spring-loaded connectors in the edge connector. Thus, the *i*th pin on one daughtercard is connected to the *i*th pin of all daughtercards. This arrangement is found in the Apple II, the IBM PC, and the once popular S100 bus, because it offers the cheapest form of bus mechanics.

FIGURE 10.3 Motherboard and daughterboard

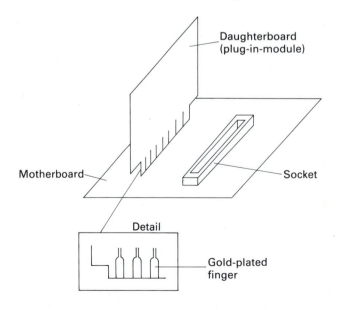

Unfortunately, the type of bus mechanics illustrated in figure 10.3 is relatively unreliable. Wear and tear due to repeated card insertion and removal or the gradual ingress of dirt or corrosive agents eventually lead to intermittent contact between the motherboard and daughterboards. Gold-plated connectors have a higher reliability than tin-plated connectors but are considerably more expensive. More than anywhere else, the reliability of the hardware is very much related to its cost.

In addition to the reliability of the electrical contacts, the edge connectors and daughterboards must be manufactured to quite tight physical tolerances; for example, if, say, 50 fingers are on a daughterboard and the pitch of the fingers varies by e from its nominal value, the accumulated error may be up to $50e$. Such a large error may make insertion of the daughterboard into an edge connector impossible.

The favored mechanical arrangement of the motherboard-daughterboard connection in many of today's professional and semiprofessional systems is the so-called two-piece connector. One part of the connector is attached to the motherboard and

FIGURE 10.4 Two-part connector

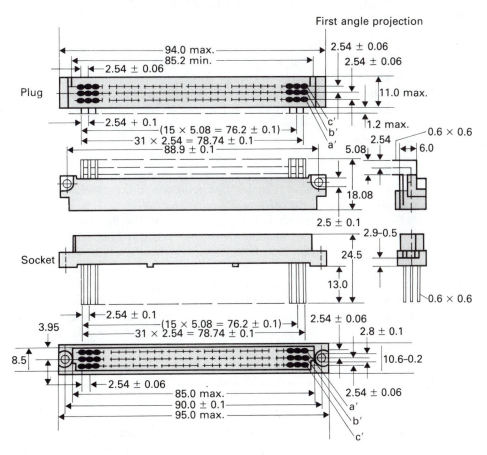

All dimensions in millimeters.

the other to the daughterboard. When a card is plugged into the motherboard, the physical connection takes place at the connector-connector level, rather than at the connector-card level in figure 10.3.

The arrangement of one of the most popular types of two-piece connector is given in figure 10.4. The connectors are designed to be mechanically compatible with the Eurocard System as defined by IEC 297 and DIN 41494. The connectors themselves are compatible with the DIN 41612 standard. By making both the connectors and the cards on which they are located an international standard, the designers know that if they buy a connector or a module conforming to the relevant standard, they will achieve a guaranteed level of compatibility; that is, the designer is freed from the tyranny of the single supplier. A DIN 41612 connector has 32 pins in one to three rows, providing a 32-, 64-, or 96-way bus. This arrangement is sufficient for many of the new 32-bit microprocessor systems.

10.2 ELECTRICAL CHARACTERISTICS OF BUSES

Each line of a bus distributes a digital signal from the card supplying it to all the other cards receiving it. Behind this seemingly trivial remark lie many complex design considerations. Even if we forget, for the time being, the problems of bus arbitration and signal timing protocols, three aspects are vital to the electrical characteristics of bus design: bus drivers, bus receivers, and bus transmission characteristics.

A bus driver is an active device that can change the logical level of the bus line it is driving. As each module capable of driving the bus has its own bus drivers, some mechanism must be provided to avoid bus contention. Bus contention is a situation in which two or more modules attempt to drive the bus simultaneously. Later we shall see that there are two solutions to the problem of bus contention: the tristate output and the open-collector output.

A bus receiver reads the logical level on the line to which it is connected. In principle, any TTL-compatible input may act as a bus receiver. In practice, special-purpose bus receivers have been designed whose characteristics have been optimized to provide high-speed, good noise immunity, and minimal bus loading.

Ideally, once a logic level is applied at one point on a bus, the same logic level should appear at all other points along the bus instantaneously. Unfortunately, real signals on real buses propagate along the bus at a finite speed and suffer reflections at the end of the bus or at any change in bus impedance. More is said about the flow of electrical energy down a conductor in the last part of this section.

Bus Drivers

Although digital systems operate with logical 0 and logical 1 levels, many different logic elements that both generate and detect these levels are evident. Figure 10.5 shows a bus line driven by an NMOS output stage (typical of a microprocessor

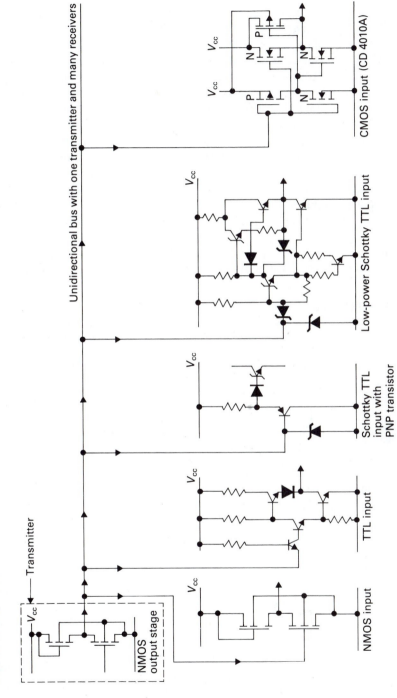

FIGURE 10.5 Bus driver

output). This bus is connected to receivers fabricated with NMOS, TTL, low-power Schottky TTL, and CMOS technology. The purpose of figure 10.5 is to demonstrate that many different types of receiver circuit exist. As may be imagined, these different input circuits also have a spread of electrical characteristics.

Figure 10.6 illustrates the connection between a driver and a single receiver. Figure 10.6a represents the system in a logical zero state and figure 10.6b the same system in a logical one state. To the right of the general diagrams are examples of NMOS drivers connected to low-power Schottky TTL receivers. When considering the interconnection of logic elements, two sets of characteristics must be satisfied—those relating to voltage levels and those relating to current levels.

One figure of merit often quoted for a logic element is its dc noise immunity. The dc noise immunity of a logic element is the amount of noise that can be tolerated at its input without exceeding V_{IL} (in a low state) or falling below V_{IH} (in a high state). The dc noise immunity of a logic element is defined as

High-level noise immunity $= V_{OH} - V_{IH}$

Low-level noise immunity $= V_{IL} - V_{OL}$

FIGURE 10.6   Connecting bus drivers to bus receivers

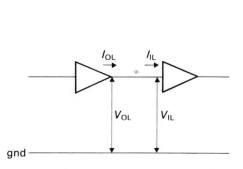
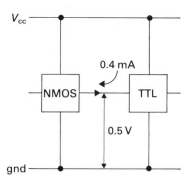

(a) Bus in a logical zero state

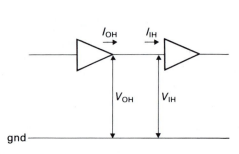
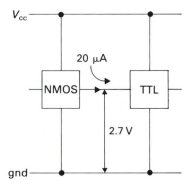

(b) Bus in a logical one state

TABLE 10.1 Characteristics of four types of logic element

CHARACTERISTIC	LOGIC FAMILY					UNITS
	LS TTL	S TTL	ALS TTL	NMOS	CMOS	
V_{OL}	0.5	0.5	0.4	0.4	0.01	V
V_{OH}	2.7	2.7	2.7	2.4	4.99	V
V_{IL}	0.8	0.8	0.8	0.8	1.5	V
V_{IH}	2.0	2.0	2.0	2.0	3.5	V
I_{OL}	8	20	4	1.6	0.4	mA
I_{OH}	-400	$-1,000$	-400	-200	-500	μA
I_{IL}	-0.4	-2.0	-0.4	2.5 μA	10 pA	mA
I_{IH}	20 μA	50 μA	20	2.5 μA	10 pA	mA
Propagation delay	9.5	3	4	25	35	ns
Input capacitance	3.5	—	—	10–160	5	pF

Obviously, the noise immunity of any device must be greater than zero, and the higher the figure the better. Table 10.1 gives the basic parameters of four of the most popular types of logic element found in today's microprocessor systems. The noise immunity for each possible combination of logic family to logic family is presented in table 10.2. For each combination, two values are given: the low-level noise immunity and the high-level noise immunity; for example, when LS TTL is connected to S TTL, the noise immunity (low level/high level) is 0.3/0.7 V. As the worst value has to be taken, the quoted noise immunity for this combination is 0.3 V.

Note that the high-level noise immunity for all logic families driving CMOS inputs (except CMOS outputs) is negative; that is, these families are not able to drive CMOS inputs because TTL V_{OH} values are too low. However, it is sometimes possible to drive CMOS inputs with TTL-compatible gates if the TTL outputs are pulled up to V_{cc} by means of a 2–6-kΩ resistor.

TABLE 10.2 Noise immunity of various gate combinations

OUTPUT LOGIC	INPUT LOGIC				
	LS TTL	S TTL	ALS TTL	NMOS	CMOS
LS TTL	0.3/0.7	0.3/0.7	0.3/0.7	0.3/0.7	1.0/ -0.8
S TTL	0.3/0.7	0.3/0.7	0.3/0.7	0.3/0.7	1.0/ -0.8
ALS TTL	0.4/0.7	0.4/0.7	0.4/0.7	0.4/0.7	1.1/ -0.8
NMOS	0.4/0.4	0.4/0.4	0.4/0.4	0.4/0.4	1.1/ -1.0
CMOS	0.79/2.99	0.79/2.99	0.79/2.99	0.79/2.99	1.45/1.45

NOTE: Each value is presented as "logical 0/logical 1" noise immunity. The effective value is the lower of this pair.

Current Levels and Digital Circuits

Not only do the voltage characteristics of bus drivers and bus receivers have to be matched, but their current characteristics must also be considered. As figure 10.6 demonstrates, a device driving a single LS TTL input must be able to source 20 μA in a logical one state and to sink 0.4 mA in a logical zero state. To a first approximation, a bus driver must be able to source (or sink) sufficient current to hold all the receivers on the bus in a logical one (or zero) state.

When the output of, say, an NMOS gate changes state, it attempts to drive the load to which it is connected from a logical one state to a logical zero state, or vice versa. Figure 10.7 shows an NMOS output driving an NMOS input. Suppose the upper transistor changes state from off to on, as the output switches from a logical zero to a logical one. Current flows both into the NMOS input and into the distributed capacitance in the circuit.

This capacitance is made up of the output capacitance of the driver, of the bus itself, and of all receivers connected to the bus. Sometimes, the total capacitive loading on the bus may be rather large. When the upper transistor runs on, the distributed capacitance charges through the resistance of the transistor as illustrated in figure 10.8. If the switching threshold of the input is V_T, the input does not change state until the capacitance charges from V_{OL} to V_T. Consequently, one of the limiting factors in designing microcomputer buses is the rate at which drivers can supply current to charge the bus capacitance and the number of highly capacitive inputs connected to the bus.

Because the capacitive loading of CMOS and NMOS inputs is so great, a common practice is to isolate them from the bus by means of bus drivers and bus receivers. Figure 10.9 illustrates a bus line driven by a bus driver on one card. Other

FIGURE 10.7   An NMOS output driving an NMOS input stage

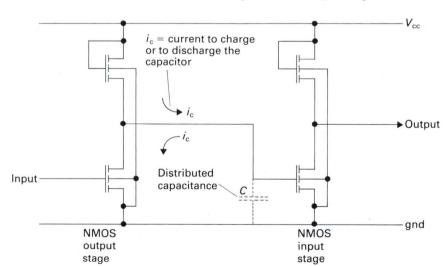

FIGURE 10.8 Charging the distributed-capacitance of an input circuit

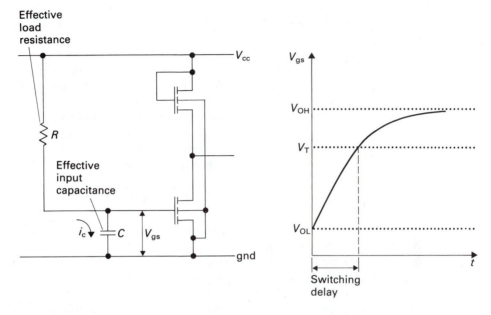

FIGURE 10.9 Bus drivers and receivers

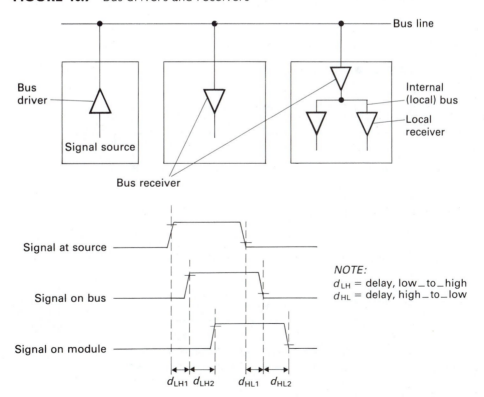

NOTE:
d_{LH} = delay, low_to_high
d_{HL} = delay, high_to_low

cards, which listen to this bus line, interface to it by means of receivers. As most buses are specified so that each card connected to them must not present more than one LS TTL load, we sometimes need to employ a local bus within a card and then buffer this with further buffers as illustrated in figure 10.9.

The great advantage of bus drivers and receivers is that the characteristics or behavior of the bus are made independent of the electrical properties of the modules connected to the bus. As we shall see, the propagation of signals on the bus is also determined by its transmission-line behavior rather than by the characteristics of the many NMOS or CMOS devices connected to it. Unfortunately, a price has to be paid for this bus isolation. This price is the signal delay incurred by the drivers and receivers. In figure 10.9, the timing diagram illustrates the delay caused by the bus driver and bus receiver in series. When a module has a local bus, a third delay is incurred.

Table 10.3 gives the properties of a typical bus driver and receiver. Note that the same device is frequently used in both roles; that is, the device has an input designed to lighly load a bus and an output designed to source or sink large bus currents. From this table we can see that the bus driver/receiver has low values for I_{IL} and I_{IH} (its loading effect) and large values of I_{OL} and I_{OH} (its driving capability). The signal transition delay of 18 ns maximum for a 74LS241 noninverting buffer is negligible in older microprocessor systems with clocks running at 1 or 2 MHz, but in today's high-speed systems with clocks of 8 MHz upward, this delay must be taken into account when designing a system.

TABLE 10.3   Characteristics of a bus driver/receiver

PARAMETER	UNITS	74LS241
V_{IH} min.	V	2.0
V_{IL} max.	V	0.8
V_{OH} min.	V	2.4
V_{OL} max.	V	0.5
I_{IH} max.	μA	2.0
I_{IL} max.	μA	-200
I_{OH} max.	mA	-15
I_{OL} max.	mA	24
I_{OS} max.	mA	-225
t_{PLH} max.	ns	18
t_{PHL} max.	ns	18

NOTES: I_{OS} = short-circuit output current
t_{PLH} = low-to-high signal propagation delay
t_{PHL} = high-to-low signal propagation delay

Passive Bus Drivers

Up to now, we have considered the situation depicted in figure 10.9, where a single bus driver is connected to the bus and communicates with many receivers. No provision has yet been added to permit several transmitters to share the bus. Simply connecting more than one TTL output stage to the bus is not a possible solution, as figure 10.10 demonstrates.

In figure 10.10 the outputs of two totem-pole stages are directly connected together via the bus. Suppose that output G1 is in a logical zero state. The lower transistor in its totem-pole, T2, conducts, pulling the bus down to ground level (i.e., V_{OL}). Suppose also that output G2 is in a logical one state with the upper transistor, T3, of its totem-pole conducting. Now G2 is trying to pull the bus upward to V_{OH}.

Clearly, this situation is contradictory. The bus cannot be at both V_{OL} and V_{OH} simultaneously. What actually happens is that a low-impedance path exists between V_{cc} and ground through T3 and T2 via the bus. The short-circuit current flowing along this path may burn out both gates. At best the state of the bus is undefined.

There are two basic approaches to allowing more than one output to control the bus. The first uses drivers with outputs that may only pull the bus down to V_{OL}, which avoids the situation where one driver pulls the bus down to V_{OL} while another attempts to pull it up to V_{OH}. The V_{OH} level on the bus is produced passively by means of a pull-up resistor between the bus and V_{cc}. The second solution involves the so-called tristate driver that can be electrically disconnected from the bus when it is not actively forcing the bus up to V_{OH} or down to V_{OL}. We will deal first with the passive pull-up solution.

FIGURE 10.10 Effect of connecting two TTL output stages together

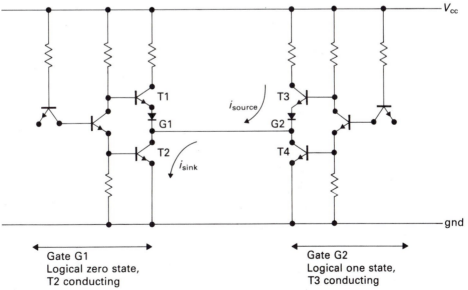

The totem-pole output circuit of gate G1 in figure 10.10 is always pulled up to V_{cc} when T1 is turned on or down to ground when T2 is turned on. An open-collector output dispenses with the upper transistor (T1). The term *open-collector* is used because the collector of the lower transistor has no path to V_{cc} within the gate itself. Therefore, this arrangement can only actively pull the output down to ground. The gate still has a two-state output, but instead of V_{OL}/V_{OH} states it has V_{OL}/floating states. In the floating state, the voltage level at the output is determined by the level of the signal on the bus.

Figure 10.11 shows how two open-collector outputs drive the same bus line. The key to understanding the open-collector bus driver is that, when a transmitter is not driving the bus, its output must be in a logical one state. Suppose that in figure 10.11 both transmitters are simultaneously in logical one states. Both output transistors will be turned off and the bus will be left to float. A pull-up resistor, R, defines the state of the bus whenever it is not actively pulled down to ground. This resistor pulls the bus up toward V_{cc}, so that an input connected to the bus sees a voltage not less than its required V_{IH}.

If transmitter 1 is currently controlling the bus, it will either be in a logical 1 state with R defining the level on the bus or it will be in a logical 0 state. In the latter case, T1 is turned on and the bus pulled down toward ground. The voltage at the collector of T1 (i.e., its output) is its saturation voltage (V_{cs}), and current can now flow out of the transistor in the receiver circuit (ie., T3) and into T1. If for any reason more than one bus driver is turned on, the current flowing through R and the receiver input is simply divided between the outputs in the logical zero state.

As any open-collector transmitter can pull down the bus to a logical zero state,

FIGURE 10.11 Open-collector bus driver

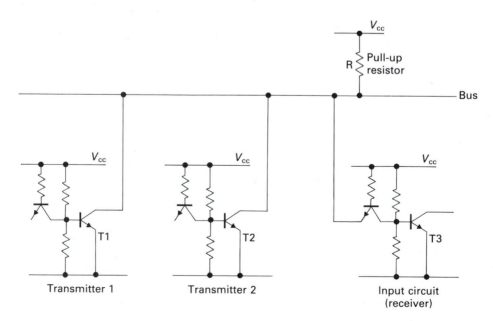

the arrangement is called *wired OR logic*. Of course, if a transmitter puts out a constant logical zero, the bus cannot be used by any other device. When this condition occurs, at least no potentially harmful situation exists.

Calculating the Pull-up Resistor Value

The value of the pull-up resistor required by a bus driver with open-collector outputs is obtained by considering the two limiting conditions—the maximum resistance that will guarantee a logical one on the bus and the minimum resistance that will keep dissipation inside the bus drivers within limits:

1. *Maximum value of R.* The maximum value of R is given by calculating the voltage drop across R when the bus is pulled up to V_{OH}, its minimum guaranteed high-level state. This state is shown by figure 10.12a. When the bus is at V_{OH}, the current flowing through R consists of two components: the input current flowing into any receiver connected to the bus and the leakage current flowing into each open-collector driving the bus.

   Suppose a system has m transmitters and n receivers. The total current flowing in R is $m \times I_{leakage} + n \times I_{IL}$. For LS TTL devices, $I_{leakage}$ is less than 250 μA and I_{IH} is 20 μA. Suppose we design the system to give a high-level dc noise margin of 0.4 V. The value of V_{IH} for TTL gates is 2.0 V, giving a required value of V_{OH} equal to 2.4 V.

   The voltage across R in a logical one state is $V_{cc} - 2.4 = 2.6$. Therefore, the maximum value of R is given by

$$R_{max} = \frac{2.6}{m \times 0.00025 + n \times 0.00002}$$

For ten receivers and ten transmitters, the maximum value of R is

$$R_{max} = \frac{2.6}{10 \times 0.00025 + 10 \times 0.00002} = \frac{2.6}{0.0025 + 0.0002}$$

$$= 1,000 \ \Omega$$

2. *Minimum value of R.* To calculate the minimum value of R, we consider the case in which a single bus driver is turned on to pull the bus down to no more than V_{OL}. This situation is illustrated in figure 10.12b. The current sunk by the active output dissipates energy in the output transistor and too large a current flow will physically destroy the transistor. To make matters worse, the active output must not only sink the current through R but also the current out of any receiver circuits connected to the bus (i.e., I_{IL}).

   The current flowing through R when the bus is in a logical zero state is given by $I_{OL} - n \times I_{IL}$. Therefore, the minimum value of R is given by

$$R_{min} = \frac{V_{cc} - V_{OL}}{I_{OL} - n \times I_{IL}}$$

FIGURE 10.12 Current flowing in open-collector circuits

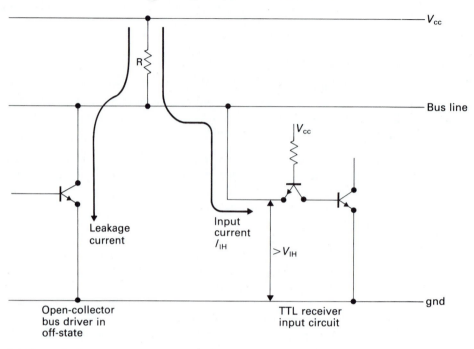

(a) Bus pulled up to V_{cc} passively

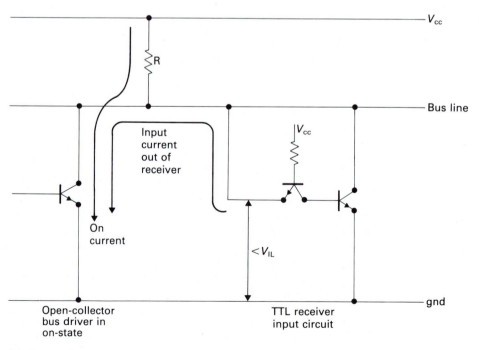

(b) Bus pulled down to ground actively

For one receiver with $I_{IL} = 0.4$ mA, $V_{OL} = 0.4$ V, and $I_{OL} = 16$ mA, we have

$$R_{min} = \frac{5 - 0.4}{0.016 - 0.0004} = \frac{4.6}{0.0156} = 300 \ \Omega$$

Had ten receivers been connected to the bus, the minimum value of R would be increased to 4.6/0.012 mA $= 380 \ \Omega$.

As the pull-up resistor may lie between 380 and 1,000 Ω, a compromise value of 470–680 Ω seems quite reasonable.

The NMOS technology equivalent of the open-collector output is the open-drain output circuit, whose output is the drain of an NMOS transistor without its usual active pull-up. In this case, the value of the pull-up resistor is normally recommended as typically 3 kΩ by the NMOS manufacturers.

Tristate Logic

Buses capable of being driven by more than one transmitter are almost invariably controlled by tristate (or three-state) buffers. We will now examine the characteristics of tristate bus drivers and show how they are used to implement microcomputer buses.

A tristate logic element is a device whose output circuit can assume one of three distinct states: a logical zero with the output actively pulled down to ground; a logical one with the output actively pulled up to V_{cc}; and a high-impedance state in which the output is floating and is electrically isolated from the buffer's circuitry. Figure 10.13a gives the logical representation of a tristate output circuit and figure 10.13b the circuit diagram of a typical gate. All tristate outputs have a control input labeled invariably: E (enable), CS (chip select), or OE (output enable). When this control input is asserted, the tristate output behaves *exactly* like the corresponding TTL output and provides a low-impedance path to ground or V_{cc}, depending only on the state of the output. When the control input, E, is negated, the tristate output goes into its high-impedance state, irrespective of any other activity within the device or of the state of its other inputs. Most tristate gates are arranged so that their enable inputs are active-low.

Because of the popularity of tristate bus drivers, semiconductor manufacturers have produced a range of bus drivers and transceivers to suit today's microprocessors. These devices are 4, 6, or 8 bits wide and are available with inverting or noninverting outputs. A transceiver is a transmitter-receiver and is able to drive the bus or to receive data from it—but not both activities at the same time. One of the most popular families of tristate buffer/transceiver is the 74LS240 series. The basic features of this series are given in table 10.4.

The buffers in table 10.4 differ largely in terms of their control arrangements and in whether they are inverting or noninverting; for example, the 74LS240 is organized as two independent quad buffers, each with its own active-low enable. The groups are able to operate entirely independently, or with the two enables strapped

FIGURE 10.13 *a*, Logical representation and *b*, the circuit diagram of a tristate output stage

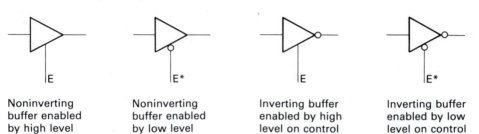

Noninverting buffer enabled by high level on control input

Noninverting buffer enabled by low level on control input

Inverting buffer enabled by high level on control input

Inverting buffer enabled by low level on control input

(a)

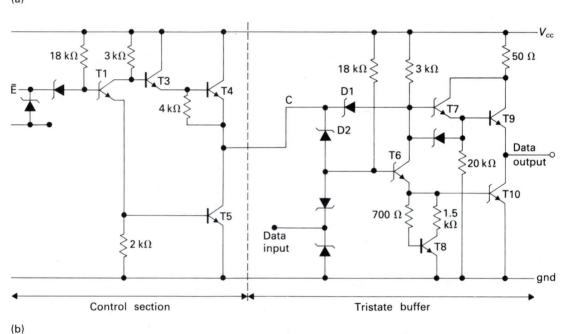

(b)

TABLE 10.4 Characteristics of the 74LS240 series tristate buffers

DEVICE	PINS	TYPE	POLARITY	FUNCTION	CONTROL
74LS240	20	2 × quad	Inverting	Driver	E∗ for each quad
74LS241	20	2 × quad	Noninverting	Driver	E∗, E
74LS242	14	Quad	Inverting	Transceiver	E∗ = read, E = write
74LS243	14	Quad	Noninverting	Transceiver	E* = read, E = write
74LS244	20	2 × quad	Noninverting	Driver	E* for each quad
74LS245	20	Octal	Noninverting	Transceiver	E*, DIR = direction

together and the whole device treated as a single octal buffer. Figure 10.14 shows how each of the buffers in table 10.4 is arranged internally.

The 74LS241 is organized as two separate quad buffers, with one group of four enabled by an active-high signal and the other by an active-low signal. By connecting the enables together and wiring the two pairs back to back (i.e., output of one

FIGURE 10.14 Logical arrangement of the buffers in table 10.4

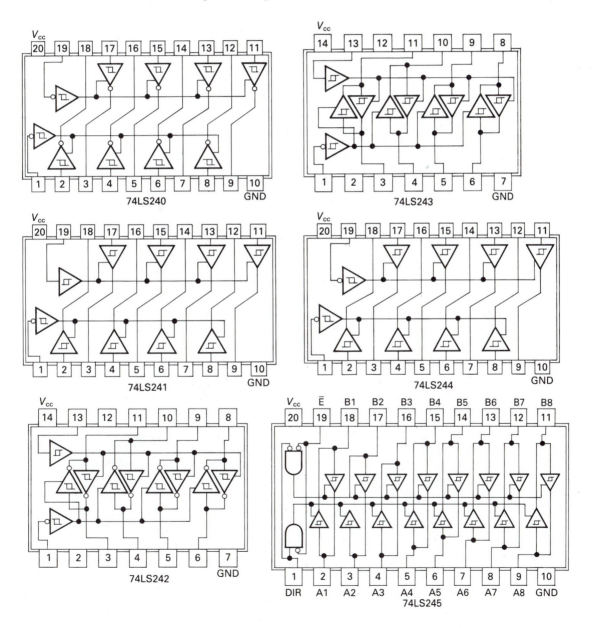

pair to input of the other), we can operate the device as a quad bidirectional transceiver. The 74LS241 is useful when a module has *separate* input and output data buses.

Tristate Bus Drivers in Microprocessor Systems

The next step in our consideration of bus drivers is to examine how tristate bus drivers can be applied to the design of microcomputer buses. Figure 10.15 shows how tristate bus drivers are used in a microprocessor system. Each bus driver or receiver is denoted by a four-letter code, as follows:

First letter	B	=	buffer
Second letter	A/D	=	address/data bus buffer
Third letter	I/O	=	data direction with respect to bus
	I	=	in from the bus (i.e., receiver)
	O	=	out to the bus (i.e., transmitter)
Fourth letter	C/M	=	CPU/memory (location of buffer).

For example, an address from the CPU is buffered onto the bus by BAOC. Each buffer is enabled by an active-low signal, labeled by the same name at the buffer itself. Thus, BAOC is enabled by BAOC∗.

The address buffers in figure 10.15, BAOC and BAIM, buffer an address from the CPU onto the bus and an address from the bus onto the memory module, respectively. A more detailed diagram of the address bus part of figure 10.15 is given in figure 10.16. Three 74LS244 noninverting bus drivers buffer the 23-bit address from the CPU, assumed to be a 68000, onto the system address bus.

FIGURE 10.15 Tristate bus driver in a microcomputer

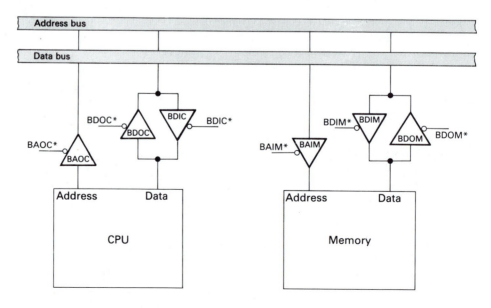

The 74LS244 is arranged as two groups of four buffers with active-low enable inputs. In figure 10.16 all six enable inputs of the address bus buffers on the computer card are connected together and enabled by BAOC*. If no device other than the CPU is ever to take control of the address bus, strapping BAOC* permanently to a logical zero becomes perfectly reasonable.

Many real systems have provision either for direct memory access or for multiprocessing. Both these modes of system operation allow a device other than the CPU to take control of the system bus and to access memory or peripherals. In this case, the 68000's address bus buffers must be turned off while another *bus master* is controlling the bus. One simple way of achieving this end is to connect BAOC* to AS* from the CPU. The 68000 asserts AS* only when it is actively accessing memory. Therefore, if the 68000 is not accessing memory, AS* is inactive-high and the address buffers are turned off, leaving the bus for another controller.

On the memory card, another three 74LS244s buffer the address from the

FIGURE 10.16 Controlling the tristate address buffers

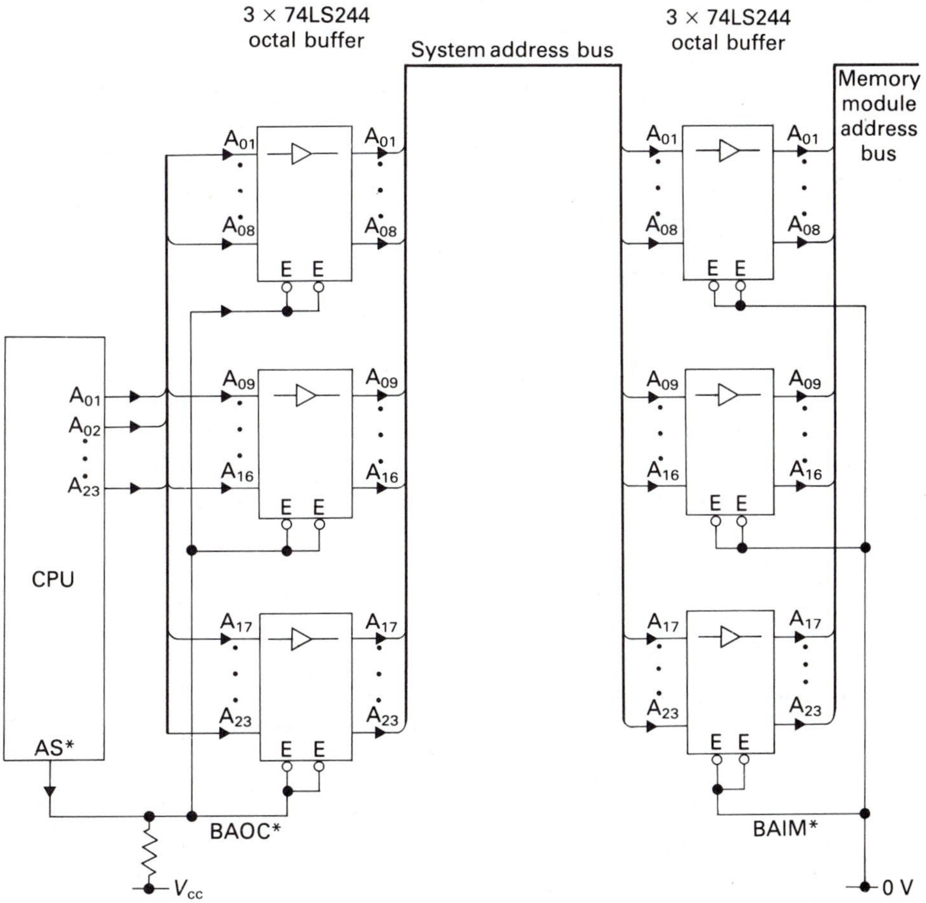

address bus and drive the address inputs of the memory components. As no device on this board will ever control either the system address bus or the address bus local to the card, enabling these buffers permanently becomes an entirely reasonable proposition.

The arrangement of the data buffers is rather more complex because the data bus is bidirectional. Figure 10.15 shows two pairs of buffers: BDOC and BDIC on the CPU card and BDOM and BDIM on the memory card. Unlike the address bus, the control of the data bus represents a reasonably difficult problem for the systems designer. Not least of the problems facing the designer is the restriction that no two data bus buffers must ever try to drive the data bus simultaneously.

A more detailed description of the data bus drivers and receivers is provided by figure 10.17. One of the most popular data bus driver-receivers is the 74LS245 octal transceiver, two of which are necessary to buffer the 68000's 16-bit data bus.

Each 74LS245 octal transceiver is controlled by two inputs: an active-low enable, E∗, and DIR. Whenever E∗ is inactive-high, both bus drivers and bus receivers are disabled and their outputs floated. Whenever E∗ is active-low, the transceiver either moves data from its A-side terminals to its B-side terminals, or vice versa. The actual direction of information flow is determined by the state of the transceiver's DIR (direction) input. A high level on DIR selects A-side to B-side transmission and a low level selects B-side to A-side transmission.

In figure 10.17, the buffers on the CPU card are connected with their B side to the 68000 data bus and their A side to the system data bus. DIR is connected directly to the 68000's $R/\overline{W}$ output. Whenever the 68000 sets $R/\overline{W} = 0$, the transceivers move data from the CPU to the system bus (i.e., side B to side A), and when $R/\overline{W} = 1$, they move data from the system data bus to the CPU data bus (i.e., side A to side B).

Similarly, the transceivers on the memory card are also controlled by the 68000's $R/\overline{W}$ output. In this case the transceivers are wired so that their B side is connected to the system bus and their A side to the local data bus on the memory card.

When designing a circuit to enable the data bus transceivers, all that need be remembered is that no two bus drivers may attempt to put data on the same bus at the same time. This restriction includes not only the system data bus but also the local data buses in both the CPU and memory cards. Local memory on the CPU card comprises memory components whose I/O data pins are connected to the CPU side of the system data bus buffers.

Let us consider first the control of the data bus transceivers on the CPU module. Seven states must be considered. These states are defined in table 10.5, where we can see that three possible control states exist for the transceivers: (1) enable, (2) disable, (3) don't care. When the 68000 is reading from or writing to external memory via the system bus, the transceivers must be enabled and their data direction controlled by $R/\overline{W}$ from the CPU.

When the CPU is reading from its local memory, the data bus transceivers on the CPU card *must* be disabled. If this action is not taken, the local memory will place its data on the local data bus while the bus transceivers are still controlling the local bus. Consequently, whenever a read to local memory is detected, the transceivers

FIGURE 10.17 Controlling the data bus buffers

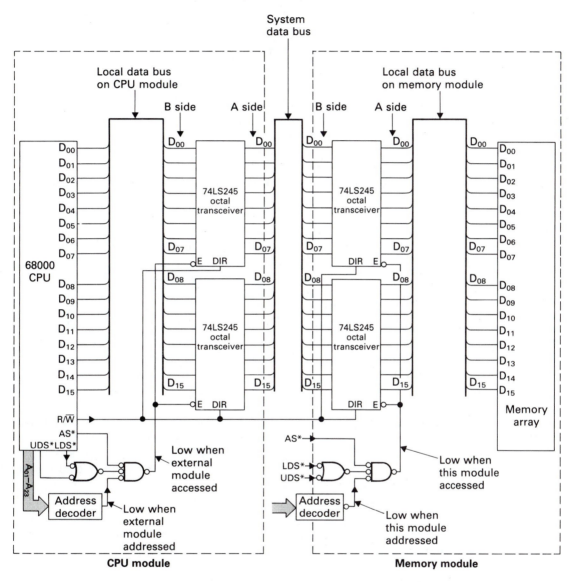

must be disabled. Similarly, if a card other than the CPU module becomes a bus master (e.g., for a DMA operation), the bus transceivers on the CPU card must not attempt to put data on the system bus during a write by the alternate bus master.

Some states in table 10.5 are labeled "don't care"—that is, actions of the data bus transceivers do not affect the operation of those don't care states. Therefore, during any period in which the CPU is idle, it does not matter whether the data bus transceivers are turned off, driving the system data bus, or driving the local data bus, as long as no other device is attempting to drive the same bus.

TABLE 10.5 Seven states of the CPU data bus transceivers

CASE	OPERATION	DATA BUS TRANSCEIVER
1	CPU idle (no bus activity)	Don't care
2	CPU read from memory card	Enable read
3	CPU write to memory card	Enable write
4	CPU read from local memory	Disable
5	CPU write to local memory	Don't care
6	DMA write to memory card	Disable
7	DMA read from memory card	Don't care

NOTE: A DMA operation implies that a device other than the CPU is controlling the system bus.

In the vast majority of well-designed systems, data bus transceivers have their outputs floated unless they are explicitly being used to drive data from one bus to another. Thus, all the *don't care* states in table 10.5 are replaced *disable* states.

As the 68000 asserts AS* and one or both data strobes during a memory access, these may be used to control the data bus transceivers on the CPU card. Unless AS* and UDS* or LDS* are asserted, the transceivers are turned off and their outputs floated.

When the 68000 performs a valid access to local memory on the CPU card, an address decoder must detect an address from the CPU falling in this range and employ it to turn off the data bus transceivers. Figure 10.17 shows how this action can be taken.

The control of the data bus transceivers on the memory card is almost identical to the control of the corresponding transceivers on the CPU card. The transceivers may be enabled only when the CPU card (or any other bus master) is generating the appropriate memory access signals (AS*, UDS*/LDS*) and when the memory being accessed falls within the memory card.

Bus Contention and Data Bus Transceivers

Having decided that the main criterion in designing data bus transceiver circuits is the avoidance of bus contention and having produced such a circuit, the reader may be forgiven for thinking that our problems are over. What we have avoided is static data bus contention; that is, for any given state, no two bus drivers attempt to drive the data bus at the same time.

Alas, another problem waits in the wings in the form of dynamic data bus contention. This contention is called dynamic because it is associated with changes of state on the bus and is due to overlap as the old bus driver switches off and the new one switches on; for example, when the processor is executing a write cycle to the memory card, buffer BDOC is placing data on the system bus and buffer BDIM is receiving this data. When the write cycle is followed by a read cycle, the transceivers must be turned around. Therefore, buffer BDOM must not turn on until BDOC has turned off.

In order to analyze the problems of dynamic data bus contention, we need a detailed diagram of the address and data bus buffering and control arrangements. Figure 10.18 provides us with this information. Note that the full analysis of such a system is rather complex because of the large number of buffers in the signal and control paths. The next step is to examine all possible modes of data bus contention in both the write-to-read and read-to-write changes of state. Here we provide two examples of possible bus contention.

FIGURE 10.18 Data bus contention and the bus transceiver

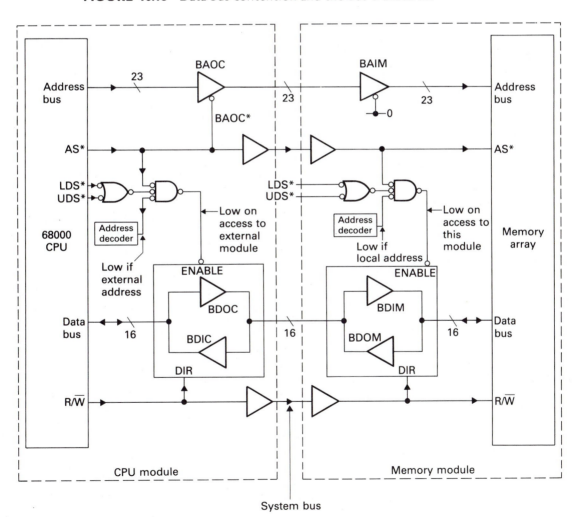

Write-to-Read Data Bus: Data Bus Contention

Suppose the 68000 is executing a write cycle to the memory card. Buffer BDOC on the CPU card is driving the data bus. At the end of the write cycle, BDOC is turned off by the negation of AS* and UDS*/LDS*. In the following read cycle, buffer BDOM on the memory card begins to drive the bus. We need to determine that BDOC-on does not overlap BDOM-on. Figure 10.19 gives the timing diagram for the switchover between a write and a read cycle.

At the end of the write cycle, AS* is negated and $E_{BDOC}*$ rises after a delay, $t_{decode1}$ seconds, due to the buffer control circuitry on the CPU card. After a further t_{OFF} seconds, the data bus buffer is turned off and the data bus floats. At the start of the read cycle, valid address and data strobes, AS* and UDS*/LDS*, cause $E_{BDOM}*$ on the CPU card to be asserted. This action takes place $t_{decode2}$ seconds after the

FIGURE 10.19 Write-to-read contention between BDOC and BDOM

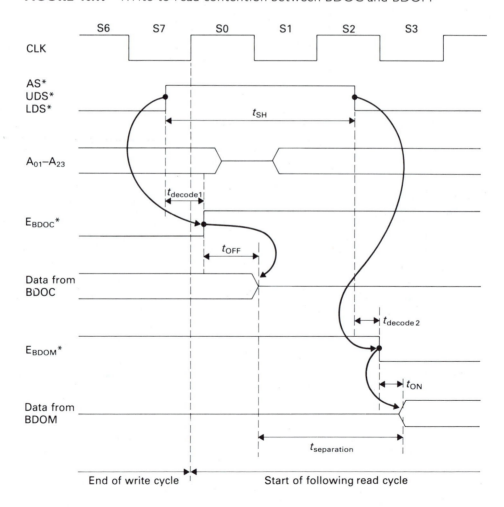

assertion of AS*. A further t_{ON} seconds later, the data bus transmitter on the memory card is turned on.

The time between BDOC turning off and BDOM turning on, $t_{separation}$, is

$$t_{separation} = t_{SH} - t_{decode1} - t_{OFF} + t_{decode2} + t_{ON}$$

Assume the following values:

$$t_{SH} \quad = 150 \text{ ns min. (8 MHz), 65 ns min. (12.5 MHz)}$$

$$t_{decode1} = \quad 30 \text{ ns max.}$$

$$t_{OFF} \quad = \quad 25 \text{ ns max.}$$

$$t_{decode2} = \quad 10 \text{ ns min.}$$

$$t_{ON} \quad = \quad 10 \text{ ns min.}$$

Therefore,

$$t_{separation} = 115 \text{ ns (8-MHz 68000)}$$

$$t_{separation} = \quad 30 \text{ ns (12.5-MHz 68000)}$$

As the separation is positive, one buffer is turned off before the other is turned on and no problem arises.

Write-to-Read CPU: Data Bus Contention

A second form of contention can take place between the data bus driver in the 68000 itself and the data bus receivers (BDIC) on the CPU card. Figure 10.20 shows this situation. The data bus drivers in the 68000 may be active for up to t_{CHADZ} seconds following the rising edge of S0 in the next cycle. As the data bus receivers (BDIC) are not turned on until $t_{decode2} + t_{ON}$ following the falling edge of AS*, the separation between the on-times of the two buffers is given by

$$t_{separation} = 2 \times t_c - t_{CHADZ} + t_{CHSL} + t_{decode2} + t_{ON}$$

$$= 2 \times 62.5 - 80 + 0 + 10 + 10$$

$$= 65 \text{ ns (8 MHz)}$$

$$t_{separation} = 2 \times 40 - 60 + 0 + 10 + 10$$

$$= 40 \text{ ns (12.5 MHz)}$$

These values are positive for both the 8- and 12.5-MHz versions of the 68000. Of course, other forms of dynamic bus contention also exist, but these are left as an exercise for the student.

Examples of Bus Contention in a 68030 Circuit

Unless you are unlucky, you might never run into bus contention problems when designing systems with 68000 microprocessors (especially at 8 MHz). If you are using a 68020 or 68030 at high clock rates, the picture is radically different.

FIGURE 10.20 Write-to-read contention between the CPU and BDIC

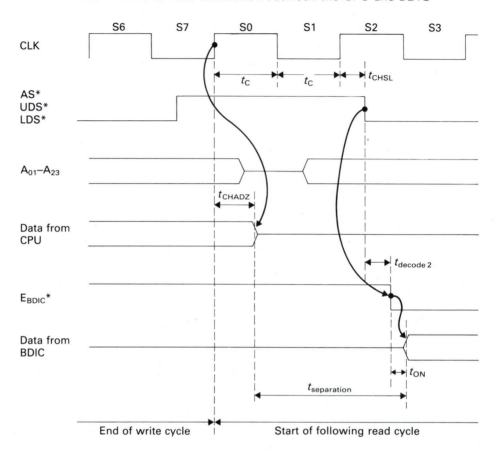

Example 1 Buffer-to-Buffer Contention in a Write-to-Read Transition
Consider again the example of buffer-to-buffer contention during the transition from
a write cycle to a read cycle. Figure 10.21 illustrates write-to-read bus contention. In
this example it is assumed that the 68030 is operating at 40 MHz and its memory
components are on the same board. Data bus buffers are necessary because there are
too many loads on the data bus for the 68030 to drive directly.

We will also assume that the data bus buffers are enabled by the address strobe
from the 68030. Note that we have shown the data bus buffer enable signals, $E_{BDOC}*$
and $E_{BDOM}*$, as two separate and different signals, even though they are both connected to the 68030's buffered AS*. In reality $E_{BDOC}*$ and $E_{BDOM}*$ are not the same
signals because they are also strobed by the 68030's R/$\overline{W}$ output (the additional
control circuitry has been omitted to simplify the description).

Assume first that the buffers are all traditional LS TTL devices (e.g., 74LS244
and 74LS245). The separation between the time at which buffer BDOC stops driving
the data bus and the time at which buffer BDOM begins to drive the data bus is
given by

$$t_{separation} = t_{SH} - t_{LH} - t_{OFF} + t_{HL} + t_{ON}$$

where t_{SH} is the minimum time for which AS* is negated, t_{LH} is the maximum low-to-high propagation time of the AS* buffer, t_{HL} is the minimum high-to-low propagation time of the AS* buffer, t_{OFF} is the maximum output disable time of the data bus buffer, and t_{ON} is its minimum output enable time.

If we use LS TTL data in this equation, we get

$$t_{separation} = t_{SH} - t_{LH} - t_{OFF} + t_{HL} + t_{ON}$$

$$= 18 - 18 - 25 + 9 + 12 = -4 \text{ ns}$$

FIGURE 10.21 Write-to-read contention in a 68030 system

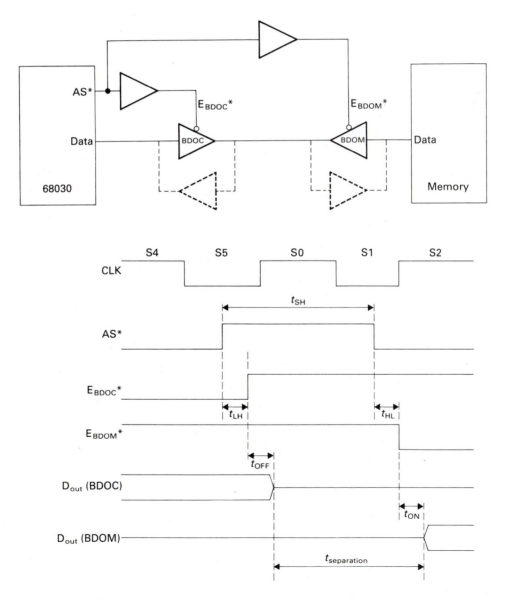

In this case we do get data bus contention for 4 ns and should, therefore, not use these components.

Suppose that we carry out the same calculation using FAST TTL logic (i.e., 74F244 and 74F245 buffers). In this case, $t_{\text{separation}}$ is given by

$$t_{\text{separation}} = t_{\text{SH}} - t_{\text{LH}} - t_{\text{OFF}} + t_{\text{HL}} + t_{\text{ON}}$$

$$= 18 - 5.2 - 6.5 + 2.5 + 3.5 = 12.3 \text{ ns}$$

We have now eliminated bus contention by a small margin.

Now suppose we use mixed logic families on the same board. The memory's data bus buffer and its control are implemented by FAST logic, whereas the 68030's buffer and its control are implemented by LS TTL logic. We now have the situation described by

$$t_{\text{separation}} = t_{\text{SH}} - t_{\text{LH}} - t_{\text{OFF}} + t_{\text{HL}} + t_{\text{ON}}$$

$$= 18 - 18 - 25 + 2.5 + 3.5 = -19 \text{ ns}$$

In this case we have severe bus contention, because the LS TTL devices are slow to release the data bus at the end of a write cycle, whereas the 74F devices grab the bus early in the following read cycle. You might be tempted to think that this is an entirely unreasonable example, since nobody would be stupid enough to make such a mistake. Well I did, and it took me some time to discover why my system would not work. Even if the designer specifies 74F-series devices throughout, the acquisition department might use a few 74LS-series devices (because they have some in stock and assume these chips do the same job).

Example 2 Memory-to-CPU Contention in a Read-to-Write Transition We will now consider bus contention in a circuit using a 68030 at 40 MHz with fast MCM6264 8K × 8 static RAM. Figure 10.22 illustrates the connection between a 68030 and MCM6264 memory and provides the timing diagram of a read-to-write transition. Assume that the RAM is connected directly to the 68030's data bus and that the RAM is enabled by the 68030's address strobe.

At the end of the read cycle, AS* from the 68030 is negated in bus state S5, and the memory's CS* input is negated t_{CS} seconds later. The MCM6264 memory stops driving the data bus t_{GHOZ} seconds after the rising edge of CS*, and the 68030 puts its own data on the data bus t_{CHDO} seconds after the start of state S2 in the following write cycle.

The period from the falling edge of state S4 to the point at which the 68030 drives its data bus in the following write cycle is given by

$$t_{\text{CLSH}} + t_{\text{CS}} + t_{\text{GHOZ}} + t_{\text{separation}}$$

The same period of time is equal to three clock states plus t_{CHDO}. If we equate these two values, $t_{\text{separation}}$ is given by

$$t_{\text{separation}} = 3t_{\text{cyc}} + t_{\text{CHDO}} - t_{\text{CLSH}} - t_{\text{CS}} - t_{\text{GHOZ}}$$

$$= 3 \times 12.5 + 0 - 10 - t_{\text{CS}} - 15 = 12.5 - t_{\text{CS}}$$

FIGURE 10.22 Read-to-write contention in a 68030 circuit

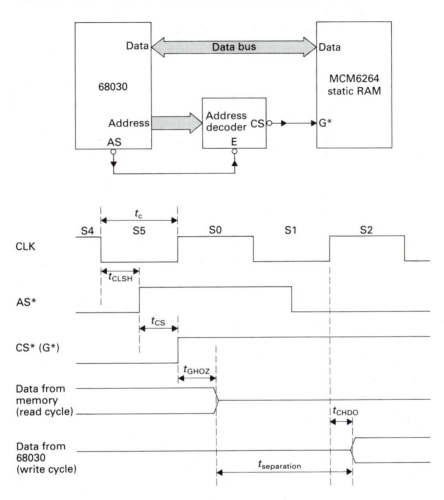

This equation tells us that the address decoder that generates CS* from AS* must have an AS* high to CS* high delay of no more than 12.5 ns if data bus contention is to be avoided.

Bus Contention and Power Dissipation

Bus contention is more than a just a source of errors in digital systems. Under certain circumstances it can lead to damage, as we shall now demonstrate. One of the effects of an electric current is to heat the material through which it flows. Consequently, a bus conflict increases the power dissipation within the gates through which the short-circuit current flows. Since the temperature of the chip depends on the power dissipated within it, bus contention can increase chip temperature. Should the temperature rise to 150°C, the reliability of the chip is dramatically reduced, because the ions used to dope the silicon become mobile and start to migrate through the silicon—that is, if you overheat the silicon, you can destroy the chip.

The temperature of a chip, T_j, is given by

$$T_j = T_a + P_{TOT} \times R_{thja},$$

where T_a is the ambient temperature of the chip, P_{TOT} is the total dissipation in the chip, and R_{thja} is the thermal resistance of the chip (typically 80 to 100°C/W). What this equation tells us is that the chip's temperature rises as the power dissipated within it increases.

The total power dissipated by a chip is given by

$$P_{TOT} = P_Q + (P_S \times t_S + P_o \times 2t_p + P_C \times t_C) \times f \times n$$

where

P_Q = quiescent power dissipation of the chip (i.e., standby power)
P_S = average power dissipation caused by current spikes
t_S = average duration of current spikes
P_o = power dissipation during charging of bus capacitance
t_p = propagation time of gate
P_C = power dissipation during a bus conflict
t_C = duration of bus conflict
f = operating frequency
n = number of outputs involved in bus conflict

As you can see, the dissipation of the chip is determined by the frequency and is a function of the frequency of bus conflicts, their duration, and the current taken during bus contention. Severe bus contention (caused by poor systems design) could possibly destroy a chip, although this is unlikely if the logic is low-power logic (e.g., LS TTL).

Buses as Transmission Lines

Now that we have struggled with bus drivers and crept through the minefield of bus contention, we have one last hurdle to overcome—the transmission line properties of the bus. In other words, having determined that only one output is actively driving the bus and with enough current to clamp the bus at V_{OH} or V_{OL}, things can still go wrong. In short, a pulse propagated along the bus can be sufficiently distorted to produce misleading effects. Why this happens and what can be done about it is the subject of this section.

The term *transmission line* may be new to computer scientists with little background in electrical engineering. As a matter of fact, the origin of what is now called electronics evolved from a study of the effect of transmission lines on digital data! In the 1850s engineers had already observed that signals received at the ends of long submarine telegraph cables were noticeably distorted. A cleanly switched signal at the transmitting end was received as a slowly changing signal at the far end of the cable.

FIGURE 10.23 Idealized view of a transmission line

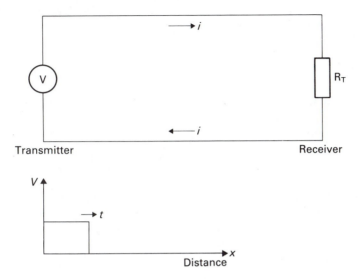

The sponsors of the project to lay the first transatlantic cable linking North America with Europe asked Professor Thomson (later Lord Kelvin) to investigate the problem. In May 1855 he presented a paper to the Royal Society that was to become the cornerstone of modern transmission line theory.

Figure 10.23 illustrates an idealized form of a transmission line. This transmission line is made up of a bus (conductor) and a ground return path. At one end a voltage source (the driver) can place a step voltage on the line and at the far end the line is terminated by a resistor, R_T.

Suppose the voltage between the ends of the line is initially everywhere zero and a step voltage of V is applied at the transmitter. This step corresponds to a zero-to-one transition of a bus driver in a digitial system. Because of the fundamental electrical properties of matter, it is not possible for an electrical disturbance to travel down a circuit instantaneously. The resistance, inductance, and capacitance of the transmission line affect the way the pulse flows down it.

The physics of transmission lines forms an entire branch of electronic engineering. Here we can only mention some of the implications of transmission-line theory. One way of dealing with the transmission line is to model it in terms of conventional components (e.g., R, L, C). Figure 10.24 illustrates a simple RC (i.e., resistor-capacitor) model of a transmission line. This model is also called a low-pass circuit.

FIGURE 10.24 Representing a line as a simple RC circuit

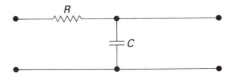

An RC circuit can be described by two equations, which express its *cutoff frequency*, f_c, and its rise time, t_r. The cutoff frequency is the frequency at which the RC circuit attenuates a sine wave by 3 dB.

$$f_c = \frac{1}{2\pi RC} \quad \text{and} \quad t_r = 2.2RC$$

The rise time can also be expressed as $t_r = 2.2/2\pi f_c = 1/2.86 f_c = 1/3T$, where T is the propagation delay of the circuit.

The cutoff frequency of typical PCB tracks is in the region of 1 GHz, implying a rise time of less than 1 ns. Consequently, the PCB tracks do not significantly distort signals. However, longer lines (i.e., tracks) cannot be described in terms of a simple RC model, and a more complex model must be adopted.

Figure 10.25a illustrates the classical model of a section of a transmission line. The line is considered to be made up of an infinite number of infinitely short sections, whose characteristics are defined in figure 10.25a. Since the series resistance of a PCB track is very small and its conductance (i.e., leakage between adjacent tracks) is very large, we can simplify the model of the transmission line to that of figure 10.25b (we do not include their derivation here, since partial differential equations tend to frighten even the most masochistic readers).

The propagation delay of a pulse on a transmission line is given by $\sqrt{LC}$ per unit length, where L is the inductance and C is the capacitance of the bus (both per unit length). In free space (i.e., a vacuum), a signal propagates at the speed of light, 3×10^8 m/s (i.e., 30 cm/ns). Practical buses have values of $\sqrt{LC}$ greater than that of free space and signals propagate at about 15–20 cm/ns. This level of propagation delay may not seem much, but buses longer than tens of centimeters can incur a signal delay of 10 ns or more. Such a value is of the order of a gate delay and cannot be ignored in timing calculations for high-speed processors.

FIGURE 10.25 The model of a transmission line

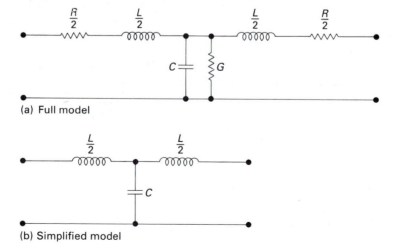

(a) Full model

(b) Simplified model

TABLE 10.6 Characteristics of typical transmission media

MEDIUM	CHARACTERISTIC			
	L (nH/cm)	c (pF/cm)	Z (Ohm)	T (ns/m)
Single wire (far from ground)	20	0.06	600	4
Free space (i.e., a vacuum)	μ_0	ℓ_0	370	3.3
Twisted-pair cable	5–10	0.5–1	80–120	5
Flat cable	5–10	0.5–1	80–120	5
Track on a PCB	5–10	0.5–1.5	70–100	5
Coaxial cable	2.5	1.0	50	5
Back plane bus line	5–10	1.0–3.0	20–40	10–20

Various types of transmission lines modeled by figure 10.25b are described by parameters in table 10.6.

We can say (as a rule of thumb) that transmission-line theory has to be applied to any system in which the rise time of the signal is shorter than twice its propagation time. Some engineers use a ratio of 4. For example, consider the case in which the propagation delay per meter is 5 ns, the rise time is 2 ns, and the signal path length is 4 cm. The ratio of rise time to propagation delay is given by

$$\frac{2 \text{ ns}}{5 \text{ ns/m} \times 4/100 \text{ m}} = 10$$

As you can see, the rise time dominates the ratio and we do not have to worry about transmission-line effects.

Now consider a 50-cm bus with a 2-ns rise time and a propagation delay of 20 ns/m. The ratio of the rise time to propagation delay is now

$$\frac{2 \text{ ns}}{20 \text{ ns/m} \times 50/100 \text{ m}} = 0.25$$

This value is less than 0.5, and therefore the bus should be treated as a transmission line.

It would be bad enough if the transmission line were only to introduce a pure delay into our model for the propagation of signals along a bus. We have not yet thought about the current flowing in the transmission line. When the transmitter in figure 10.23 creates a voltage step of V, what current flows in the bus? Ohm's law provides us with the formula $I = V/R$, where R is the resistance of the bus plus its load R_T. If we have learned one thing from this book, it is that nothing in microcomputer systems design is simple. At the time the pulse is generated, the current does not "know" the value of R_T. How could it? A disturbance travels down the line at $1/\sqrt{LC}$ and cannot "see ahead" of itself.

Therefore, what current actually flows down the line? Because of the electrical nature of matter, the transmission line has a characteristic impedance, Z_0, whose value is determined by the geometry of the bus and the properties of the dielectric separating the signal path of the bus and its ground return. From table 10.6 it can be seen that the characteristic impedance of typical backplanes is approximately 100 Ω. The characteristic impedance of a transmission line is given by $\sqrt{L/C}$, where L and C are the distributed inductance and capacitance per unit length.

We now have a step voltage of V moving down the line at $1/\sqrt{LC}$ meters per second with a current of V/Z_o amperes. All good things must come to an end, and the pulse eventually reaches the end of the line. If the bus is terminated by a load equal to its characteristic impedance, the pulse is dissipated in the load and the voltage at all points along the bus remains at V.

If the bus is not terminated by Z_o, the pulse is reflected from the termination back toward its source. To understand why this happens, consider the effect of the pulse arriving at the termination R_T. Immediately before the pulse reaches R_T, the relation between the pulse and current in the bus is given by $V = I/Z_o$. On reaching the termination, Ohm's law must be obeyed, and $I_T = V_T/R_T$, where $I_T =$ the current in the termination and $V_T =$ the voltage across the terminator.

The current flowing in the load can be defined as the sum of two components:

$$I_T = I_i + I_r$$

where $I_i =$ incident current and $I_r =$ reflected current. Therefore,

$$I_T = \frac{V_T}{R_T} = \frac{V_i + V_r}{R_T}$$

However,

$$I_i = \frac{V_i}{Z_o} \quad \text{and} \quad I_r = -\frac{V_r}{Z_o}$$

The minus sign indicates that the reflected voltage, V_r, is moving away from the terminator and back toward the generator. These equations can be rearranged to give

$$\frac{V_i}{Z_o} - \frac{V_r}{Z_o} = \frac{V_i + V_r}{R_T} = \frac{V_i}{R_T} + \frac{V_r}{R_T}$$

that is,

$$V_i\left(\frac{1}{Z_o} - \frac{1}{R_T}\right) = V_r\left(\frac{1}{Z_o} + \frac{1}{R_T}\right)$$

or

$$V_r = V_i \frac{R_T - Z_o}{R_T + Z_o} = V_i\Gamma \quad (\Gamma \text{ is used to represent the reflection coefficient})$$

Thus, the reflected voltage can be calculated in terms of the load impedance R_T and the characteristic impedance of the line Z_o. Consider the three limiting cases: $R_T = Z_o$ (line terminated by characteristic impedance), $R_T = 0$ (short circuit), and $R_T =$ infinity (open circuit):

1. Matched line $V_r = V_i \dfrac{R_T - R_T}{R_T + R_T} = 0$ (no reflected wave).

2. Short-circuit $V_r = V_i \dfrac{0 - Z_o}{0 + Z_o} = -V_i$ (inverted pulse reflected).

3. Open-circuit $V_r = V_i \dfrac{\infty - Z_o}{\infty + Z_o} = V_i$ (pulse reflected).

Note that when the termination is zero (short-circuit), the reflected wave has an equal amplitude but opposite polarity to the incident wave. These waves cancel to produce a zero voltage, which is reassuringly in line with common sense. When the incident wave reaches the short circuit, the voltage across the lines must fall to zero, which travels back down the line to the generator.

Real transmission lines fall between the extremes. Suppose that $Z_o = 100\ \Omega$ and the line is terminated by $150\ \Omega$. The reflection coefficient is therefore

$$\frac{R_T - Z_o}{R_T + Z_o} = \frac{150 - 100}{150 + 100} = \frac{50}{250} = \frac{1}{5}$$

In this case, the reflected voltage is $V_i/5$. When the reflected voltage has traveled back down the line, it reaches the generator and is reflected, exactly as it was at the terminator. In this case, the new reflected voltage is given by

$$V_r = V_i \frac{R_G - Z_o}{R_G + Z_o}$$

where $R_G =$ resistance of generator. A pulse is therefore reflected to and fro between the generator and transmission line termination until, in the limit, the voltage on the bus reaches its steady-state value of $VR_T/(R_G + R_T)$.

Figure 10.26 illustrates the effect of applying a pulse to a transmission line with mismatches at both the generator and terminator. In any real system the waveform will be much more complex as the impedance of the transmission line is not uniform, the transmission line contains nonlinear elements (i.e., semiconductors), and reflections occur at discontinuities in the transmission line such as connectors and circuits connected to the line (so-called stubs).

Example of a Bus with Reflections at Both Ends

Before leaving the subject of bus reflections, we will look at an example in which a pulse is transmitted down a bus from a generator and reflected at the far end. When the reflected pulse returns to the generator, it too is reflected (since the generator does not have the same impedance at the transmission line). Figure 10.27 illustrates this arrangement.

FIGURE 10.26   Effect of a mismatch on a pulse on a transmission line

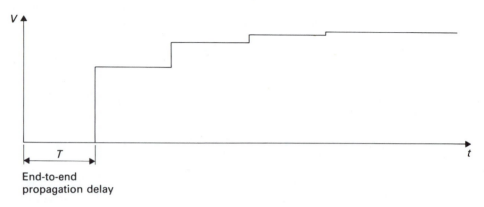

End-to-end
propagation delay

The transmission line has an impedance of 75 Ω, the generator has an impedance of 30 Ω, and the far-end load has an impedance of 100 Ω. The initial step applied by the generator is 3.7 V. The reflection coefficients at the generator and load ends of the transmission line are given by

$$\frac{30 - 75}{30 + 75} = -0.429 \quad \text{(generator)} \qquad \text{and} \qquad \frac{100 - 75}{100 + 75} = +0.143 \quad \text{(far end)}$$

The initial step of 3.7 V at the generator produces a pulse of 3.7 V $\times$ 75/(30 + 75) = 2.64 V, which travels down the line. Note that the initial pulse is scaled by 75/105 because the pulse is applied to the end of a potentiometer formed by the generator impedance (30 Ω) and the transmission line impedance (75 Ω).

When the pulse reaches the load end, part of it is reflected back toward the generator. The reflected part is given by 2.64 V $\times$ 0.143 = 0.378 V. At the moment the forward-going pulse hits the far end, the far-end voltage rises to 2.64 + 0.378 = 3.02 V (i.e., the incident signal plus the reflected signal). The pulse reflected by the load end of the transmission line travels back to the generator end, where it too is reflected, and so on.

The system can, therefore, be viewed as a pulse traveling from one end of the line to the other and back, suffering a positive reflection of +0.143 at one end and a negative reflection of −0.429 at the other end. In theory, the reflections take an infinite time to decay, but in practice the line reaches its final value (within a percent or so) after a few reflections. We can write the first few reflections as follows:

1. Initial step: 3.7 $\times$ 75/105 = 2.64 V

2. $V_{r1} = 2.64 \times 0.143 = 0.378$ V

3. $V_{r2} = 0.378 \times -0.429 = -0.16$ V

4. $V_{r3} = -0.16 \times 0.143 = -0.02$ V

5. $V_{r4} = -0.02 \times -0.429 = 0.0099$ V

6. $V_{r5} = 0.0099 \times 0.143 = 0.00142$ V

Figure 10.27 also demonstrates the *lattice diagram* that can be used to keep track of

FIGURE 10.27 Example of multiple reflections

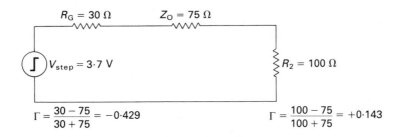

$$\Gamma = \frac{30 - 75}{30 + 75} = -0.429 \qquad\qquad \Gamma = \frac{100 - 75}{100 + 75} = +0.143$$

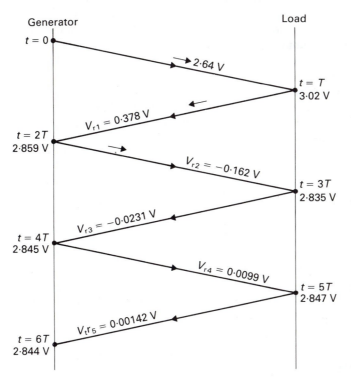

the successive reflections. Each reflection is added to the current level of the signal on the transmission line.

These theories are very nice, but where are they getting us? The answer is that unless a bus is terminated by a reasonable approximation to its characteristic impedance, fast pulses will suffer from reflections that may play havoc with the system's operation. In the early days of the microcomputer (mid-1970s), little attention was paid to bus design. Fortunately, clock speeds were rather low and relatively long times were available for signals to settle and reflections to die down. Today clock rates of 8 to 50 MHz or more make bus design critical.

Bus designers now attempt to define the characteristic impedance of a bus by controlling its geometry. They also provide ground return paths as close as possible to each signal line.

Terminating the Bus

To reduce reflections, the ends of a transmission line should be terminated by connecting a resistance, equal to Z_o, across the line. The value of Z_o is approximately 100 Ω for a typical PCB. Unfortunately, if a 100-Ω resistor is wired between a bus line and ground, the upper logic level will be pulled down and the noise immunity reduced. Equally, connecting a 100-Ω resistor between the bus and V_{cc} will pull up the lower logic level and reduce the low-level noise immunity. Remember that the terminator can be connected between the bus and ground *or* V_{cc}, as a low impedance exists between ground and V_{cc} as far as transients are concerned.

The classic solution to bus termination is illustrated in figure 10.28. Two resistors are connected to the bus, one to ground and one to V_{cc}. A typical resistor pair is 330 Ω to V_{cc} and 470 Ω to ground. The bus termination circuit can be reduced to its Thevinin equivalent of a single resistance of 194 Ω connected to a voltage source of 2.94 V. Thus, the line is terminated without being pulled down to ground or up to V_{cc}.

A refinement of this circuit, illustrated in figure 10.29, is rather misleadingly called an *active termination* circuit. A voltage regulator produces the desired termination voltage and a single resistor of 120 Ω is connected between this voltage source and each of the bus lines to be terminated.

Designing a Microcomputer Bus

It should be apparent by now that the design of a bus is no trivial matter. If a relatively low clock rate is anticipated, conventional wisdom may help the designer. The three pillars of conventional wisdom are these:

1. Allow each signal time for all reflections to have died away to vanishingly small proportions; for example, the contents of address and data buses should not be sampled until, say, 100 ns after they are nominally valid.

FIGURE 10.28 Terminating the bus

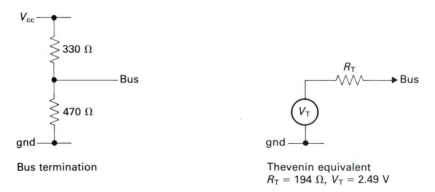

Bus termination

Thevenin equivalent
R_T = 194 Ω, V_T = 2.49 V

FIGURE 10.29 Active bus termination

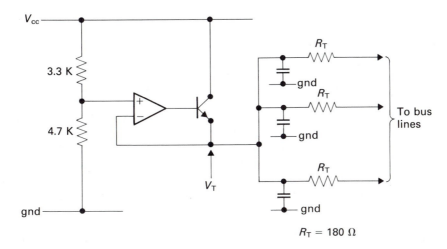

$R_T = 180 \ \Omega$

2. Apply matched terminations to each end of the bus. A perfect match is impossible to obtain, so a termination of approximately 100 Ω should severely attenuate reflections.

3. Load the bus as little as possible and avoid long stubs. A stub is an extension to the bus rather like a T junction.

A better approach to bus design is to buy a bus off the shelf. This course may not be a particularly adventurous path, I'll admit, but sometimes caution should be exercised in systems design. Few areas of microprocessor systems design are as complex as the design of reliable buses. Many manufacturers subcontract the analysis of their buses to noted academics, but this is not always helpful. Academics differ more on their attitude to bus design than on almost any other area. Today ready-made buses can be obtained that are constructed to international standards from reputable manufacturers. Such buses are produced by state-of-the-art technology to minimize reflections and cross talk. You should not design a bus for a high-speed microcomputer without facilities to test and modify the design in the light of experience. Necessary equipment for such testing includes oscilloscopes with rise times an order of magnitude better than the pulses you might wish to observe.

10.3 VMEbus

Now that we have described the way in which a bus operates, we are going to look at a bus designed specifically for 68000 systems. The VMEbus is intended to support the 68000-series microcomputers and is a backplane bus. The VMEbus is asynchronous only in the sense that the 68000 is asynchronous; that is, memory accesses of a variable duration are possible by means of the 68000's DTACK* input. The VMEbus is, to some extent, a synchronous bus, as the latching of addresses and data is controlled by a system clock.

Motorola developed its own bus, the VERSAbus, for use on its EXORmacs 68000 development system. The VERSAbus is a particularly large (physically) bus, and a version suited to the popular Eurocard was developed by Motorola in 1981. The Eurocard is available as a single card (160 × 100 mm) or as a double card (160 × 233.4 mm). The bus developed for these cards was called the *Versa Module Europe* bus and is now known as the VMEbus.

The VMEbus was rapidly adopted as a standard by industry and is now supported by Motorola, Signetics, Mostek, Philips, and Thomson-EFCIS. Although the Eurocard originated in Europe, it is also popular in the United States and is displacing other, more conventional, formats. Today, many independent manufacturers produce a wide range of Eurocard modules for the VMEbus.

In September 1983 the IEEE Standards Board gave their approval to the VMEbus standards group and assigned project number P1014 as the IEEE reference number to this bus during its development phase. The VMEbus was approved by the IEEE in 1984 and is therefore known officially as the IEEE 1014 bus. Similarly, the International Electrochemical Commission, IEC, started formal standardization of the VMEbus in 1982 and called it the IEC 821 bus. The American National Standards Institute approved the VMEbus standard in September 1987 as ANSI/IEEE STD 1014-1987.

A definitive treatment of the VMEbus is impossible to provide here. The VME system architecture manual that defines the bus is several hundred pages long. Only an overview of the bus and its characteristics can be given. The purpose of the VMEbus is to allow the systems designer to put together a microprocessor system by buying off-the-shelf hardware and software components. This approach can be very economical, particularly as it frees the designer to spend more time on those parts of the system that must be custom-made. The formal objectives of the VMEbus are

1. To provide communication facilities between two devices (i.e., cards) on the VMEbus without disturbing the internal activities of other devices interfaced to it. In plain English, we can plug in a new card without "harming" the existing system.

2. To specify the electrical and mechanical system characteristics required to design devices that will reliably and unambiguously communicate with other devices interfaced to the VMEbus.

3. To specify protocols that precisely define the interaction between the VMEbus and devices interfaced to it.

4. To provide terminology and definitions that precisely describe system protocols.

5. To allow a broad range of design latitude so that the designer can optimize cost and/or performance without affecting system compatibility.

6. To provide a system where performance is primarily device limited, rather than system interface limited.

The VMEbus has been designed with flexibility in mind and can operate with data widths of 8, 16, or 32 bits, and with 24- or 32-bit address buses. The VMEbus complements all the powerful features of the 68000 (except for its synchronous bus) and has important facilities of its own.

VMEbus Electrical Characteristics

The VMEbus is, essentially, a TTL-logic-compatible bus that employs the logic levels you would expect to find in any conventional digital system. The VMEbus specification defines the characteristics of signals explicitly and states what is and what is not permissible. For example, the maximum length of a VMEbus is specified as no longer than 19.68 in. (500 mm) with no more than 21 slots.

Not only does the VMEbus specification define the obvious (e.g., $V_{cc} = 5$ V), it covers subtleties (e.g., the upper limit on V_{cc} is $5 + 0.25$ V, the lower limit is $5 - 0.125$ V, and the maximum permitted ripple on V_{cc} is 50 mV peak-to-peak). Reading a standard like that of the VMEbus is an interesting experience, not the least because it illustrates just how many factors you have to take into account when formally specifying a standard.

The worst-case VMEbus signal levels are defined as

$$V_{OH}(\text{min.}) = 2.4 \text{ V}$$

$$V_{IH}(\text{min.}) = 2.0 \text{ V}$$

$$V_{IL}(\text{max.}) = 0.8 \text{ V}$$

$$V_{OL}(\text{max.}) = 0.6 \text{ V}$$

These values provide a minimum high-state dc noise immunity of 0.4 V and a minimum low-state dc noise immunity of 0.2 V.

Rule 6.10 in the VMEbus specification document states that all boards shall provide input clamping on each signal line they monitor, to prevent negative excursions below -1.5 V; that is, all receivers connected to a VMEbus line should employ diode clamping. All this means is that the input circuit should have a reverse-biased diode connected between it and ground. If the input falls below ground level (usually due to a negative reflection or ringing on the bus), the clamping diode conducts and limits the negative-going input voltage excursion to about 1 V. Diode clamping avoids negative voltages on signal lines, which might otherwise damage inputs. Having laid down rule 6.10 (which might frighten the VMEbus user into buying a load of diodes for input clamping), observation 6.4 in the VMEbus specification tells us that standard 74LS-series and 74F-series logic elements have internal clamping diodes on their inputs, and therefore we don't have to worry about rule 6.10 after all.

Some VMEbus lines (i.e., strobes) might have a heavy loading, and the specification states that AS*, DS0*, and DS1* should have the following drive capability:

I_{OL} = 64 mA min.

I_{OH} = 3 mA min.

I_{OS} = 50 mA min.

I_{OS} = 225 mA max.

I_{OZL} = 450 μA max.

I_{OZH} = 100 μA max.

C_T = 20 pF max. (output capacitance including PCB tacks)

Note that I_{OS} is the short-circuit current of a gate and is the maximum output current when its output is clamped at ground. I_{OZL} and I_{OZH} represent the maximum current leakage of the drivers when their outputs are floating (i.e., tristated).

The address, data, address modifier, IACK*, LWORD*, and WRITE* control lines have similar characteristics to the address/data strobes, except that I_{OL} needs to be no more than 48 mA and I_{OZL}/I_{OZH} is 700/150 μA.

Some of the VMEbus lines (the bus arbitration/IACK lines to be described later) do not drive more than one input, and these have relatively modest drive specifications:

I_{OL} = 8 mA max.

I_{OH} = 400 μA max.

V_{OH} = 2.7 V (min.) at I_{OH} = 400 μA

The VMEbus termination network described in figure 10.28 is mandatory (a termination network must be applied to each end of a signal line). In addition to reducing reflections, the termination network provides a high state for open-collector outputs and restores a signal line to a high-level state when all three-state drives connected to it are disabled.

The VMEbus specification does not specify the actual impedance of the signal lines, but it does recommend that the backplane be designed to produce an impedance as close to 100 Ω as possible.

VMEbus Mechanics

VMEbus cards are designed to slot into a 19-in. (482.6 mm) rack that may be either 3U (132.5 mm) or 6U (265.9 mm) high. Figure 10.30 shows a card frame that supports both the single Eurocard (3U height) and double Eurocard (6U height) formats.

FIGURE 10.30 Card frame. (Reprinted by permission of Motorola Limited)

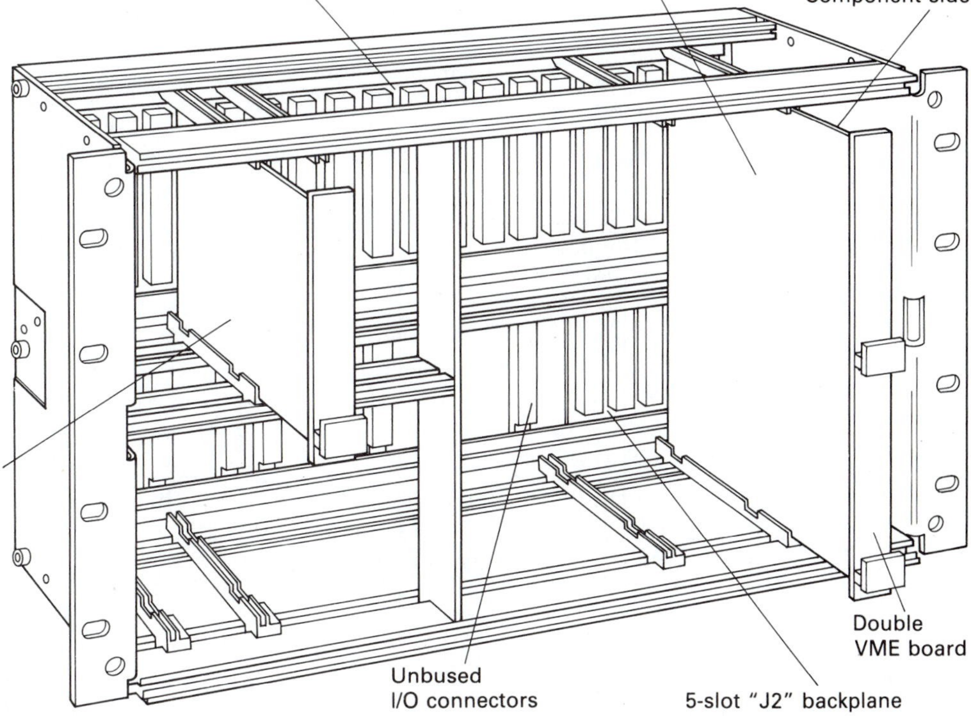

The cards themselves are either single-height boards 100 mm (3.937 in.) by 160 mm (6.299 in.) deep or double-height boards 233.35 mm (9.187 in.) high and 160 mm deep. Figure 10.31 illustrates a double-height board. A clever feature of the VMEbus is its two connectors, called P1 and P2. If the VMEbus had all the facilities it needed on one connector, it would be unwieldy. Moreover, the cut-down version (i.e., 3U size) of a Eurocard would be impossible to construct.

The approach adopted by the VMEbus is to define a primary connector P1 and a secondary connector P2. All the functions necessary to implement a basic VMEbus are provided by P1, which permits the construction of a system based entirely on standard Eurocards. Connector P2 provides expansion facilities, permitting the bus to be extended from 24 to 32 address bits and from 16 to 32 data bits. The connector on the card is referred to as P1 (or P2) and the connector on the backplane is referred to as J1 (or J2).

Both connectors are two-piece devices with three rows of 32 pins (96 in all) and conform to DIN 41612 standard. A VME backplane may be implemented as a single backplane (for P1) or a double backplane for P1 and P2. Invariably, separate backplanes are used for P1 and P2 rather than a double backplane.

FIGURE 10.31 Dimensions of a double-height VME card. (Reprinted by permission of Motorola Limited)

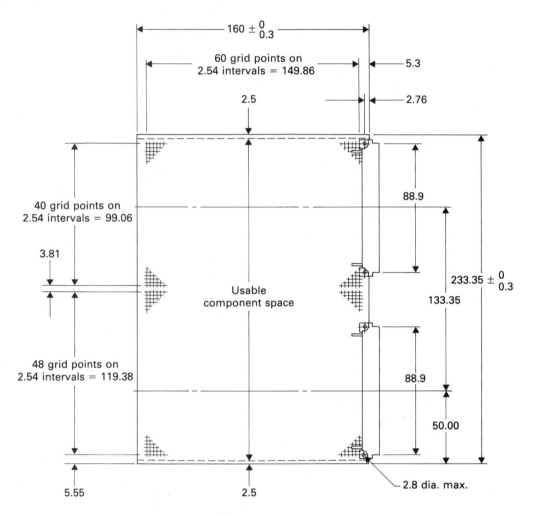

NOTES: Board thickness 1.6 ± 0.2 reference IEC 249–2.
All dimensions are shown in millimeters.

Functions Provided by the VMEbus

Although the VMEbus is a single entity and is not physically subdivisible, its specification logically divides the bus into four distinct subbuses, as illustrated in figure 10.32. The positions along the VMEbus into which cards are plugged are called *slots*. In VMEbus terminology, a module is a collection of electronic components with a single functional purpose. More than one such module may exist on the same card.

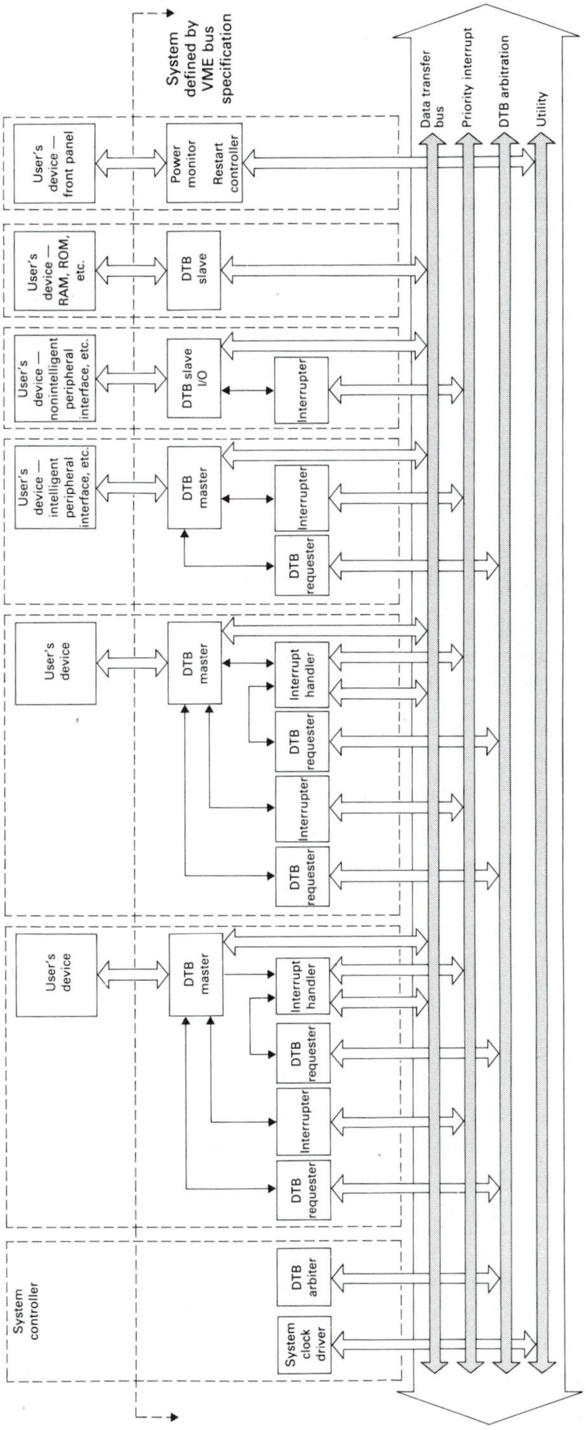

FIGURE 10.32 VMEbus. (Reprinted by permission of Motorola Limited)

From figure 10.32, we can see that the VMEbus system definition specifies a number of modules that form the interface between the VMEbus backplane and the various user modules making up the microcomputer. The functional modules forming part of the VMEbus specification are

1. *DTB requester.* "DTB" stands for data transfer bus and includes the address and signal paths necessary to execute a data transfer (8, 16, or 32 bits) between a DTB master and a DTB slave. A DTB requester is a module on the same board as a master or interrupt handler and is capable of requesting control of the data transfer bus whenever its master or interrupt handler needs it.

2. *Interrupter.* An interrupter is a functional module capable of requesting service from a master subsystem by generating an interrupt request. The interrupter must also provide status information when the interrupt handler requests it.

3. *Interrupt handler.* An interrupt handler is a functional module capable of detecting interrupt requests and initiating appropriate responses.

4. *DTB arbiter.* A data transfer bus arbiter is a functional module that receives requests for the DTB from other modules, prioritizes them, and grants the bus to the appropriate requester.

5. *DTB slave.* A DTB slave, or simply slave, is a functional module capable of responding to a data transfer operation initiated by a master; for example, a memory module is a typical DTB slave.

6. *DTB master.* A DTB master, or simply master, is a functional module capable of initiating bus transfers. A 68000 is a prime example of a DTB master. Note that all 68000s are not necessarily DTB masters. A 68000 on a card may operate entirely locally and may not be able to access the VMEbus itself.

Now that we have defined the functions of some of the modules forming part of the VMEbus specification, we can look at the four groups of signals making up the VMEbus:

1. *Data transfer bus.* The data transfer bus is the data and address pathways and their associated control signals, and is employed for the purpose of transferring data from a DTB master to a DTB slave. Of all the buses, the DTB most closely matches the corresponding pin functions (i.e., the asynchronous bus) of the 68000.

2. *DTB arbitration bus.* At any instant a VMEbus can be configured with only one master that is capable of transferring data between itself and one or more slaves. The DTB arbitration bus and its associated DTB arbiter module provide a means of transferring control of the DTB between two or more masters in an orderly manner.

3. *Priority interrupt bus.* The priority interrupt bus and its associated modules extend the interrupt-handling capabilities of the 68000 microprocessor (see chap. 6). The priority interrupt capability of the VMEbus provides a means by which devices can request interruption of normal bus activity and can be serviced by an interrupt handler. These interrupt requests can be prioritized into a maximum of seven levels.

4. *Utilities bus.* The utilities bus is a "miscellaneous functions bus" by another name and includes the system clock, a system reset line, a system fail line, and an ac fail line.

The pin assignments of the P1 connector of the VMEbus are given in table 10.7. These pins provide all the functionality of the four subbuses. Table 10.8 gives

TABLE 10.7 P1 pin assignments on the J1 VMEbus

PIN NUMBER	SIGNAL MNEMONIC		
	ROW A	ROW B	ROW C
1	D_{00}	BBSY*	D_{08}
2	D_{01}	BCLR*	D_{09}
3	D_{02}	ACFAIL*	D_{10}
4	D_{03}	BG0IN*	D_{11}
5	D_{04}	BG0OUT*	D_{12}
6	D_{05}	BG1IN*	D_{13}
7	D_{06}	BG1OUT*	D_{14}
8	D_{07}	BG2IN*	D_{15}
9	GND	BG2OUT*	GND
10	SYSCLK	BG3IN*	SYSFAIL*
11	GND	BG3OUT*	BERR*
12	DS1*	BR0*	SYSRESET*
13	DS0*	BR1*	LWORD*
14	WRITE*	BR2*	AM5
15	GND	BR3*	A_{23}
16	DTACK*	AM0	A_{22}
17	GND	AM1	A_{21}
18	AS*	AM2	A_{20}
19	GND	AM3	A_{19}
20	IACK*	GND	A_{18}
21	IACKIN*	SERCLK	A_{17}
22	IACKOUT*	SERDAT*	A_{16}
23	AM4	GND	A_{15}
24	A_{07}	IRQ7*	A_{14}
25	A_{06}	IRQ6*	A_{13}
26	A_{05}	IRQ5*	A_{12}
27	A_{04}	IRQ4*	A_{11}
28	A_{03}	IRQ3*	A_{10}
29	A_{02}	IRQ2*	A_{09}
30	A_{01}	IRQ1*	A_{08}
31	−12 V	+5 standby	+12 V
32	+5 V	+5 V	+5 V

TABLE 10.8 P2 pin assignments on the J2 VMEbus

PIN NUMBER	SIGNAL MNEMONIC		
	ROW A	ROW B	ROW C
1	User I/O	+5 V	User I/O
2	User I/O	GND	User I/O
3	User I/O	Reserved	User I/O
4	User I/O	A_{24}	User I/O
5	User I/O	A_{25}	User I/O
6	User I/O	A_{26}	User I/O
7	User I/O	A_{27}	User I/O
8	User I/O	A_{28}	User I/O
9	User I/O	A_{29}	User I/O
10	User I/O	A_{30}	User I/O
11	User I/O	A_{31}	User I/O
12	User I/O	GND	User I/O
13	User I/O	+5 V	User I/O
14	User I/O	D_{16}	User I/O
15	User I/O	D_{17}	User I/O
16	User I/O	D_{18}	User I/O
17	User I/O	D_{19}	User I/O
18	User I/O	D_{20}	User I/O
19	User I/O	D_{21}	User I/O
20	User I/O	D_{22}	User I/O
21	User I/O	D_{23}	User I/O
22	User I/O	GND	User I/O
23	User I/O	D_{24}	User I/O
24	User I/O	D_{25}	User I/O
25	User I/O	D_{26}	User I/O
26	User I/O	D_{27}	User I/O
27	User I/O	D_{28}	User I/O
28	User I/O	D_{29}	User I/O
29	User I/O	D_{30}	User I/O
30	User I/O	D_{31}	User I/O
31	User I/O	GND	User I/O
32	User I/O	+5 V	User I/O

the pin assignments of the P2 connector. We can see that the J2 bus is divided between user-defined I/O pins and an extension of the J1 address and data buses to 32 bits. The J2 bus is not considered further here.

Data Transfer Bus

As stated previously, the DTB is little more than an extension of the 68000's asynchronous bus. Moreover, the specification of the DTB is given in terms of the timing diagrams and protocol flow diagrams introduced in chapter 4. The signals of the DTB are defined in table 10.9.

TABLE 10.9 Signals of the data transfer bus (DTB)

VMEBUS MNEMONIC	68000 MNEMONIC	NAME
$A_{01}–A_{31}$	$A_{01}–A_{23}$	Address bus
$D_{00}–D_{31}$	$D_{00}–D_{15}$	Data bus
AS*	AS*	Address strobe
LWORD*	None	Longword
DS1*	UDS*	Data strobe 1
DS0*	LDS*	Data strobe 0
WRITE*	R/$\overline{W}$	Write
DTACK*	DTACK*	Data acknowledge
BERR*	BERR*	Bus error
AM0–AM5	None	Address modifier bus

Eight- or 16-bit data transfers are controlled exactly as in the 68000 itself, with DS1* replacing UDS* and DS0* replacing LDS*. A new function is provided by LWORD*, which, when asserted, permits a 32-bit longword data transfer on D_{00} to D_{31}. Note that longword transfers require that both P1 and P2 connectors be present. Longword data transfers can, of course, be implemented by the 68020 CPU, but not by a 68000 or a 68010. In systems that do not support 32-bit data transfers, the DTB master must put an inactive-high level on LWORD*.

A special feature of the DTB is the 6-bit address modifier, bits AM0 to AM5. The purpose of the address modifier bits is to allow the master to pass up to six bits of additional information to a slave during a data transfer. The modifier bits may provide the information present on FC0 to FC2 from the 68000. To a great extent, the way in which this information is encoded and actually employed is left up to the user. Some possible applications of the address modifier bits are now given:

1. *System partitioning.* Slaves may be programmed with an address modifier value that must match the value on AM0 to AM5 if they are to take place in a valid data exchange with a master. In this way, a slave may be assigned to a given master even though other masters generate addresses falling within the slave's address range. Accesses by other masters will be ignored. The effect of this arrangement is to partition the slaves among the masters.

2. *Memory map manipulation.* Slaves may be designed to respond to more than one range of addresses, the actual range depending on the current address modifier received from the master. Thus the master places the system resources in selected map locations by providing different address modifier codes.

3. *Privileged access.* Because slaves may be designed to respond to some address modifier values and not to others, different levels of privilege may be established. A master executing a data bus transfer (DBT) puts out a level of privilege on AM0 to AM5 and a slave responds only if it is able to operate at that level of privilege.

TABLE 10.10 Address modifier codes

HEX CODE	ADDRESS MODIFIER						FUNCTION
	5	4	3	2	1	0	
3F	H	H	H	H	H	H	Standard supervisory block transfer
3E	H	H	H	H	H	L	Standard supervisory program access
3D	H	H	H	H	L	H	Standard supervisory data access
3C	H	H	H	H	L	L	Reserved
3B	H	H	H	L	H	H	Standard nonprivileged block transfer
3A	H	H	H	L	H	L	Standard nonprivileged program access
39	H	H	H	L	L	H	Standard nonprivileged data access
38	H	H	H	L	L	L	Reserved
37	H	H	L	H	H	H	Reserved
36	H	H	L	H	H	L	Reserved
35	H	H	L	H	L	H	Reserved
34	H	H	L	H	L	L	Reserved
33	H	H	L	L	H	H	Reserved
32	H	H	L	L	H	L	Reserved
31	H	H	L	L	L	H	Reserved
30	H	H	L	L	L	L	Reserved
2F	H	L	H	H	H	H	Reserved
2E	H	L	H	H	H	L	Reserved
2D	H	L	H	H	L	H	Short supervisory access
2C	H	L	H	H	L	L	Reserved
2B	H	L	H	L	H	H	Reserved
2A	H	L	H	L	H	L	Reserved
29	H	L	H	L	L	H	Short nonprivileged access
28	H	L	H	L	L	L	Reserved
27	H	L	L	H	H	H	Reserved
26	H	L	L	H	H	L	Reserved
25	H	L	L	H	L	H	Reserved
24	H	L	L	H	L	L	Reserved
23	H	L	L	L	H	H	Reserved
22	H	L	L	L	H	L	Reserved
21	H	L	L	L	L	H	Reserved
20	H	L	L	L	L	L	Reserved
1F	L	H	H	H	H	H	User-defined
1E	L	H	H	H	H	L	User-defined
1D	L	H	H	H	L	H	User-defined
1C	L	H	H	H	L	L	User-defined
1B	L	H	H	L	H	H	User-defined
1A	L	H	H	L	H	L	User-defined
19	L	H	H	L	L	H	User-defined
18	L	H	H	L	L	L	User-defined

TABLE 10.10 (continued)

HEX CODE	ADDRESS MODIFIER						FUNCTION
	5	4	3	2	1	0	
17	L	H	L	H	H	H	User-defined
16	L	H	L	H	H	L	User-defined
15	L	H	L	H	L	H	User-defined
14	L	H	L	H	L	L	User-defined
13	L	H	L	L	H	H	User-defined
12	L	H	L	L	H	L	User-defined
11	L	H	L	L	L	H	User-defined
10	L	H	L	L	L	L	User-defined
0F	L	L	H	H	H	H	Extended supervisory block transfer
0E	L	L	H	H	H	L	Extended supervisory program access
0D	L	L	H	H	L	H	Extended supervisory data access
0C	L	L	H	H	L	L	Reserved
0B	L	L	H	L	H	H	Extended nonpriviledge block transfer
0A	L	L	H	L	H	L	Extended nonprivileged program access
09	L	L	H	L	L	H	Extended nonprivileged data access
08	L	L	H	L	L	L	Reserved
07	L	L	L	H	H	H	Reserved
06	L	L	L	H	H	L	Reserved
05	L	L	L	H	L	H	Reserved
04	L	L	L	H	L	L	Reserved
03	L	L	L	L	H	H	Reserved
02	L	L	L	L	H	L	Reserved
01	L	L	L	L	L	H	Reserved
00	L	L	L	L	L	L	Reserved

L = low-signal level
H = high-signal level

4. *Address range determination.* As a full implementation of the VMEbus offers 32 address bits, each slave requires a rather large amount of address decoding circuitry (see chapter 5) to determine whether the current address in the 4G-byte range falls within its own range. By using address modifier bits to specify a short address on A_{01} to A_{15}, a standard address on A_{01} to A_{23}, or an extended address on A_{01} to A_{31}, slaves with rather simpler address decoding circuits may be built.

Table 10.10 lists the address modifier codes specified in ANSI/IEEE STD 1014-1987. Address modifier codes can be provided by hard-wired logic or generated by DIP switches (set by the user). However, a much more flexible way of generating address modifier codes is to use a PROM (which can easily be changed if the system is modified or updated). Figure 10.33 illustrates this application of a PROM, in which

FIGURE 10.33   Using a PROM to generate address modifier codes

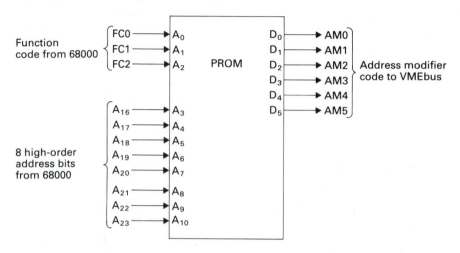

the function code and address lines A_{23} to A_{16} from a 68000 are employed to generate a 6-bit address modifier code.

Data Transfer on the VMEbus   When a module wishes to transfer data to or from a slave, it must first acquire control of the bus (if it is not already a bus master) via its bus requester module, to be described later. Once a module is a bus master, it executes a data transfer in very much the same way as a 68000. Of course, adaptation of a 68000 CPU to the VMEbus is very easy: Only the appropriate buffering between the chip and the bus is needed. Other CPUs can be interfaced to the VMEbus but not necessarily as conveniently as the 68000.

Figure 10.34 defines the VMEbus protocol for a DTB byte read cycle. This diagram is essentially the same as the protocol diagram for a 68000 read cycle in chapter 4, apart from its greater detail and the inclusion of LWORD* and IACK*. The setting of IACK* high indicates that the master is not executing an interrupt acknowledge cycle.

The VMEbus specification also provides a number of timing diagrams to augment the protocol diagrams. Once more, these diagrams mirror the 68000's read and write cycle timing diagrams.

Although the 68000's bus cycles and the VMEbus's data transfer cycles are effectively identical, there is one significant difference. The VMEbus standard states that a master shall assert its address strobe, AS*, 35 ns after the address on the address bus has stabilized. An 8-MHz 68000 specifies t_{AVSL}, address valid to AS*, asserted as 30 ns, and all faster members of the 68000 family specify much shorter values for t_{AVSL}. Consequently, you can interface all the 68000's asynchronous bus signals directly to the VMEbus (after suitable buffering), but the 68000's AS* output must be delayed. Figure 10.35 demonstrates how AS* can be delayed with a simple D flip-flop.

The address strobe, AS*, from the 68000 goes low t_{CHSL} seconds after the rising edge clock at the start of the S2 state. Consider an 8-MHz version of the

68000. The quoted value for t_{CHSL} is 3–60 ns. This figure means that AS∗ may go low at any time in state S2. Indeed, additional circuit delays external to the 68000 may even cause AS∗ to go low at the beginning of bus state S3.

By using a positive-edge-triggered D flip-flop to latch the 68000's address strobe, the VMEbus's AS∗ will not be asserted until the beginning of state S4.

FIGURE 10.34 Protocol flowchart for a DTB single-byte read cycle

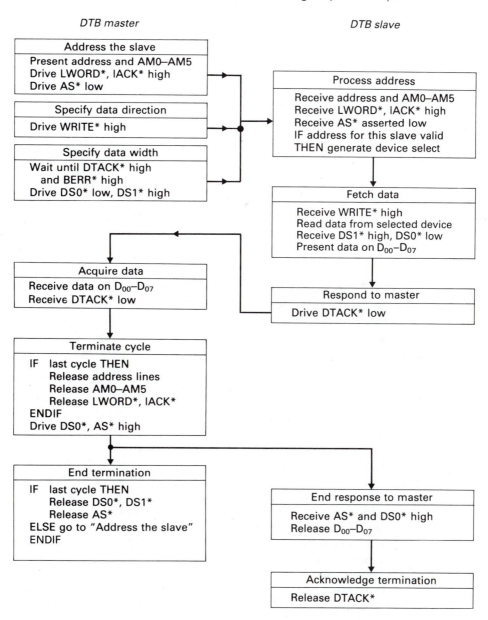

FIGURE 10.35   Generating AS∗ for the VMEbus from the 68000's AS∗

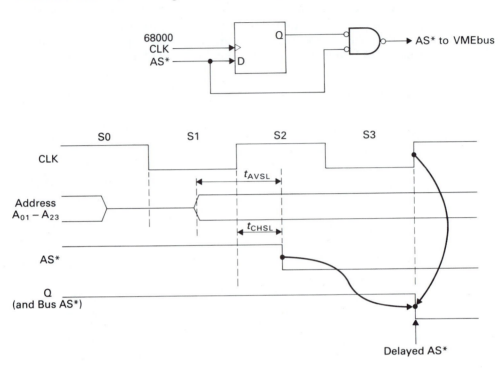

Therefore, VMEbus AS∗ will not be asserted until at least 60 ns after the address from the 68000 has become stable. Even a 16-MHz 68000 will guarantee an address setup time of better than 30 ns.

In addition to normal 68000-style data transfers, the DTB part of the VMEbus supports both block transfers and address-only cycles. A *block read* (or write) cycle is used to transfer 1 to 256 bytes of data to or from a slave. The difference between a conventional sequence of data transfers and a block transfer is that the address (and address modifier code) is presented at the beginning of the block transfer and is held constant throughout the data transfer. The slave itself locally increments the address after each byte is transferred.

Note that block transfers always take place in ascending order. The master puts out the lowest address in the block and the slave counts upward as successive bytes are transferred. A block transfer is initiated exactly like a VMEbus read/write cycle. Once the first transfer has taken place, the master maintains AS∗ asserted and toggles the data strobe(s) each time a transfer takes place.

Since AS∗ is asserted low throughout the block transfer, no other device should attempt to access the VMEbus while a master is driving AS∗ low. In other words, AS∗ is more than a simple address strobe; it is an indication that the VMEbus is being actively used.

The *address-only cycle* is a DTB cycle in which an address is broadcast but no data is transferred. That is, an address is placed on the address bus and AS∗ is asserted, but neither of the data strobes is asserted. Slaves do not acknowledge an address-only cycle, and therefore the cycle is terminated when the master negates AS∗. Handshaking between the master and slave does not take place and DTACK∗ and BERR∗ are not asserted at the end of an address-only cycle. Incidentally, the address-only cycle is the only cycle on the DTB that does not transfer data.

Address-only cycles can be used to trigger user-defined activity within slaves. They can be transmitted to slaves that do not (or cannot) respond without forcing a bus error exception. The VMEbus specification makes the observation that address-only cycles can be used to allow a slave to decode an address concurrently with the CPU board.

One final comment on the DTB should be made concerning address pipelining. Since separate address and data strobes are used to control a bus cycle, it is perfectly possible for a master to broadcast the next address while data is still being transferred in the current cycle.

Since neither the 68000 nor the VMEbus employ a multiplexed address and data bus, the address bus becomes redundant after the slave has latched (or otherwise "used") the address from the master. However, an address pipelining mechanism cannot be imposed on the VMEbus retrospectively by simply saying that, for example, the master's address should be disregarded 60 ns after the assertion of AS∗. Address pipelining has to fit in with the 68000's existing mode of operation.

Once the master has detected that the slave taking part in the current cycle has acknowledged the cycle by asserting DTACK∗ or BERR∗, it may change the address on the address bus. This restriction makes sense, because an acknowledgment from the slave implies that the current address has been accepted and acted upon. After driving AS∗ high for its minimum negated time, the master may drive AS∗ low again. The master must, of course, negate AS∗ before asserting it again; otherwise the slave would not be able to recognize the new address (since an address is captured on the falling edge of AS∗).

You might wonder what the essential difference is between address pipelining and a normal DTB cycle. In a pipelined cycle, the master may issue a new address and address strobe as soon as it detects DTACK∗ (or BERR∗), even though the master has not yet read data from the data bus or negated its data strobe. In a normal cycle, a new address is not issued until AS∗, the data strobe(s), and DTACK∗ have all been negated.

A corollary of address pipelining is that a slave must be designed to take account of the fact that address pipelining might take place (even if the slave is not designed to respond to address pipelining). This restriction (rule 2.18 in the VMEbus specification) may seem a little strange, but it is both logical and necessary. If a slave were to be designed on the assumption that the address on the address bus were valid throughout the cycle, a pipelined address might cause its incorrect operation. Rule 2.18 of the VMEbus specification states that "... slaves MUST NOT be designed on the assumption that they will never encounter pipelined cycles."

It is not difficult to comply with rule 2.18. For example, a slave may be designed to latch the current address on the falling edge of AS∗ from the master. The address is now captured for the duration of the current cycle. If the address latch is prevented from latching a new address until after the data strobe(s) and DTACK∗

have been negated, any pipelined address appearing during the current bus cycle will be ignored.

The VMEbus specification suggests that problems caused by address pipelining can be eliminated simply by ensuring that the slave initiates data transfers only on the falling edge of the data strobe(s), rather than a simultaneous low level on both the address and data strobes. Doing this may, of course, slow the system down during write cycles, since the data strobe is not asserted until two clock states after the address strobe.

DTB Arbitration

Bus arbitration is a mechanism that enables control of the DTB to be passed to one master in a group of masters, all of which are requesting use of the DTB. Systems with only one processor and no other *processorlike* modules such as DMA devices do not require the VMEbus's DTB arbitration facilities. Here we discuss only the arbitration facilities offered by the VMEbus (i.e., we do not discuss the way in which the various modules implement arbitration).

Before we look at the arbitration subbus of the VMEbus, we need to form a mental picture of what is really happening on the VMEbus. I used to think of the VMEbus as just another backplane bus (i.e., like the PC bus). I was wrong. The VMEbus belongs in an entirely different category, and it might be more reasonable to call it a *tightly coupled, high-bandwidth, low-latency local area network* than a backplane bus.

A traditional backplane bus is used by a microprocessor to communicate with memory and peripherals (usually on another card). The increase in memory capacity over the past decade coupled with the development of sophisticated peripherals has made it possible to put together very powerful systems with large memories on a *single board*. Consequently, a microprocessor might need to communicate with other cards in the system relatively infrequently.

The VMEbus is well suited to systems in which several modules (each of which contains a processor and memory) operate in parallel. These modules communicate with each other via the VMEbus relatively infrequently. In such an environment each module (i.e., potential bus master) requires some mechanism that permits it to request access to the VMEbus. Equally, a mechanism must be implemented to determine which module is to get control of the bus and to perform an orderly transfer of VMEbus ownership from the current bus master to the would-be bus master.

In what follows, *requester* refers to the mechanism employed by a master to gain control of the VMEbus's DTB. Similarly, *arbiter* refers to the module in slot 1 that decides which requester is to get control of the bus. A VME system may have only one arbiter but several requesters. The design of the arbiter is left up to the designer of the VME system and is not laid down by the VMEbus specification.

We have just said that requesters can compete for mastership of the DTB, which implies that there is more than one request line (just as there is more than one interrupt request line). The VMEbus supports four levels of bus request, called BR0* (lowest priority) to BR3* (highest priority). Four levels of priority are a compromise between the desire to have as many priority levels as possible and the need to restrict the number of VMEbus tracks to 96.

TABLE 10.11 VME arbitration bus

PIN/ROW	MNEMONIC	NAME	GROUP	FUNCTION
12b	BR0*	Bus request 0	Requester	Used by requester to
13b	BR1*	Bus request 1	Requester	gain access to the
14b	BR2*	Bus request 2	Requester	DTB
15b	BR3*	Bus request 3	Requester	
5b	BG0OUT*	Bus grant out 0	Requester	Used by requester to
7b	BG1OUT*	Bus grant out 1	Requester	pass on BGIN* from the
9b	BG2OUT*	Bus grant out 2	Requester	arbiter
11b	BG3OUT*	Bus grant out 3	Requester	
1b	BBSY*	Bus busy	Requester	Indicates bus busy
2b	BCLR*	Bus clear	Arbiter	Informs master that the DTB is needed
4b	BG0IN*	Bus grant in 0	Arbiter	Used by arbiter to
6b	BG1IN*	Bus grant in 1	Arbiter	indicate level of
8b	BG2IN*	Bus grant in 2	Arbiter	DTB access
10b	BG3IN*	Bus grant in 3	Arbiter	

The VMEbus arbitration subbus employs a total of 14 lines, which are arranged into two groups. One group is made up of lines driven by requester modules in DTB masters and the other group is made up of lines driven by the arbiter, which, as we have already stated, must be physically located in slot 1. Table 10.11 gives the names and mnemonics of the arbitration bus lines.

The VMEbus supports four levels of arbitration (the 68000 itself supports only one level). In general, the systems designer or "integrator" must decide (for each module with a 68000) which of the four VMEbus levels of arbitration is to be connected to the 68000's arbitration pins. Two types of module take part in an arbitration process: the DTB requesters, forming part of a bus master, and the DTB arbiter, which belongs to the system controller and acts on the bus globally.

Before we look at how the arbitration lines are used, we will describe three of the ways in which a VMEbus arbiter might operate. Whenever a situation arises in which a number of entities are competing for limited resources (be they people or bus masters), an algorithm must be devised to deal with the distribution of the resources. In human terms, we can adopt "fair shares for all" policies, "first-come, first-served" policies, or "survival of the fittest" policies. Similar strategies have been applied to the VMEbus and are known as arbiter options. The following three options are available:

1. *Option RRS (round robin select).* The RRS option assigns priority to the DTB masters on a rotating basis. Each of the four levels of bus request has a turn at being the highest level. The four levels of bus request, BR0*–BR3*, are treated cyclically with BR3* following BR0*; that is, the sequence of successive highest levels of priority is BR0*–BR3*–BR2*–BR1*–BR0*–BR3*–BR2* . . . .

At any instant, one of the four levels is made the highest level so that a requester at that level may gain control of the bus. If a requester at the current highest level does not wish to use the bus, the next level downwards is made the new highest level, and so on. For example, if the current highest level is BR2*, in the next cycle the highest level will be BR1*.

Suppose a requester is granted control of the bus. After the bus has been released, the next level downwards is made the new highest level and the cycle continues. Consequently, all levels of bus request become the highest priority in turn and no level is ever left out. Round robin select is a fair method of arbitration as all the masters are granted equal access to the bus.

2. *Option PRI (prioritized)*. The PRI option assigns a level of priority to each of the bus request lines from BR3* (highest) to BR0* (lowest). Whenever a master requests access to the DTB bus, the arbiter deals with the request by comparing the new level of priority with the old (i.e., current) level. A higher-priority request always defeats a lower-priority request. Option PRI is similar to the 68000's own interrupt request facilities. It is not a fair strategy, as a low-level request may, theoretically, never be serviced if higher priority devices are "greedy."

3. *Single level (SGL)*. The SGL option provides a minimal bus arbitration facility using bus request line BR3* only. The priority of individual modules is determined by "daisy-chaining," so that the module next to the arbiter module in slot 1 of the VMEbus rack has the highest priority. As the position of a module moves further away from the arbiter, its priority becomes lower.

Scheduling algorithms other than these three types of arbitration may be used on the VMEbus. The arbitration algorithm may be selected by the user.

Arbitration Lines   The arrangement of the VMEbus's arbitration lines is illustrated in figure 10.36, and the relationship of the lines and the arbiter and requester module is given in figure 10.37. Note that in figure 10.36, the bus_grant_in and the bus_grant_out lines are broken and run only from slot to slot rather than from end to end. A bus_grant_in from a left-hand module (i.e., higher-priority requester module) is passed out on its right as a bus_grant_signal. By convention, the level of a request is written as x (where $x = 0$, 1, 2, or 3). Therefore, the BGxOUT* of one module is connected to the BGxIN* of its right-hand neighbor.

The arrangement of figure 10.36 is called daisy-chaining because of its *head-to-tail* nature; it adds a special feature to a bus line. A normal, continuous bus line transmits a signal in both directions to all devices connected to it. The daisy-chained line is unidirectional, transmitting a signal from one specific end to the other. Moreover, each module connected to (i.e., receiving from and transmitting to) a daisy-chained line may either pass a signal on down the line or inject a signal of its own onto the line.

As you can see in figure 10.36, the arbiter in slot 1 sends a bus grant input to the card in slot 2. The card in slot 2 takes this bus grant input and passes it on as a bus grant output to the card in shot 3, and so on. In this way, a card receives a bus grant input from its left hand neighbor and passes it on as a bus grant output to its

FIGURE 10.36 Arrangement of VMEbus arbitration lines on the JI bus

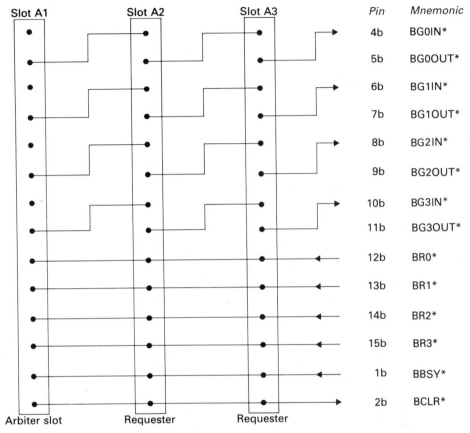

			Pin	Mnemonic
Slot A1	Slot A2	Slot A3	4b	BG0IN*
			5b	BG0OUT*
			6b	BG1IN*
			7b	BG1OUT*
			8b	BG2IN*
			9b	BG2OUT*
			10b	BG3IN*
			11b	BG3OUT*
			12b	BR0*
			13b	BR1*
			14b	BR2*
			15b	BR3*
			1b	BBSY*
			2b	BCLR*

Arbiter slot   Requester   Requester

NOTE:
The bus_grant_in lines of slot 1 are driven
by the arbiter which is plugged into slot
1; that is, the BGxIN* pins of slot 1 are *not*
driven from the VMEbus

right-hand neighbor. A card might choose to end the daisy-chain signal-passing sequence and not transmit a bus grant signal to its right-hand neighbor, as we shall soon see. If a slot is empty (i.e., no card is plugged into it), bus jumpers must be provided to route the appropriate bus_grant_in signals to the corresponding bus_grant_out terminals.

A DTB requester module makes a bid for control of the system's data transfer bus by asserting one of the bus request lines, BR0* to BR3*. Note that only one line is asserted and that the actual line is chosen by assigning a given priority to the requester. This priority may be assigned by on-board user-selectable jumpers or dynamically by software.

The arbiter in slot 1, on receiving a request for the bus, may (depending on the arbitration option in force) assert one of its bus_grant_out lines (BG0OUT* to

FIGURE 10.37 Relationship between VMEbus, arbiter, and requester

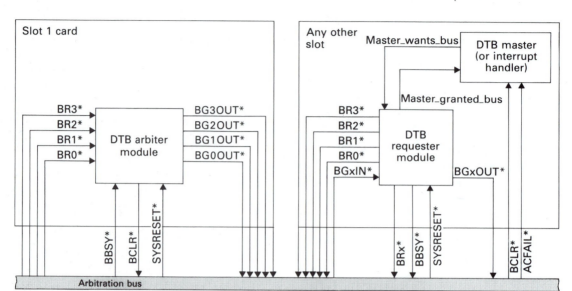

BG3OUT*). The bus_grant_out signal then propagates down the daisy-chain. Each BGxOUT* arrives at the BGxIN* of the next module. If that module does not require access to the bus, it passes along the request on its BGxOUT* line. If, however, the module does wish to request the bus, it does not assert its BGxOUT* signal. Daisy-chaining provides automatic prioritization, because bus requesters further down the line do not receive a bus grant. Figure 10.38 provides a protocol flowchart of the VMEbus arbitration procedure.

Once a bus requester has been granted control of the data transfer bus by an active-low level on its BGxIN* input, it asserts bus busy (BBSY*) active-low. BBSY* performs roughly the same function in a VME system as BGACK* in a 68000 system. BBSY* is an input to the arbiter and is not daisy-chained but runs the length of the VMEbus. By asserting BBSY*, a requester signifies its possession of the bus and control may not be taken back until the requester releases BBSY*. Note that this situation contrasts with the prioritization of interrupt requests. A low-priority interrupt may be serviced if no other interrupt is pending. A higher-level interrupt will always interrupt one with a lower priority. The arbitration bus functions differently. An active DTB master cannot be forced off the bus.

BBSY* must be asserted by a requester for at least 90 ns and remain asserted for at least 30 ns after the requester has released the bus. Furthermore, BBSY* must be asserted until the requester's bus grant is negated in order to assure that the arbiter has seen the BBSY* transition.

The bus clear line, BCLR*, from the arbiter informs the current DTB master that another master with a higher priority now wishes to access the bus. As we said before, the current master does not have to relinquish the bus within a prescribed time limit. Typically, it will release the bus at the first convenient instant. It releases the bus by negating BBSY*. There is no 68000 equivalent of the BCLR* line.

Bus clear is driven only by *option PRI* arbiters. Because the bus request lines have no fixed priority in a round-robin arbitration scheme, an RRS arbiter does not drive bus clear, BCLR*. In this case BCLR* will be driven inactive-high by the bus termination network.

Figure 10.37 shows how the arbiter module communicates with requester modules. The task of the arbiter is to prioritize incoming bus requests and to grant access to the appropriate requester. When operating in the fixed priority mode (PRI),

FIGURE 10.38 Protocol flowchart for VMEbus arbitration

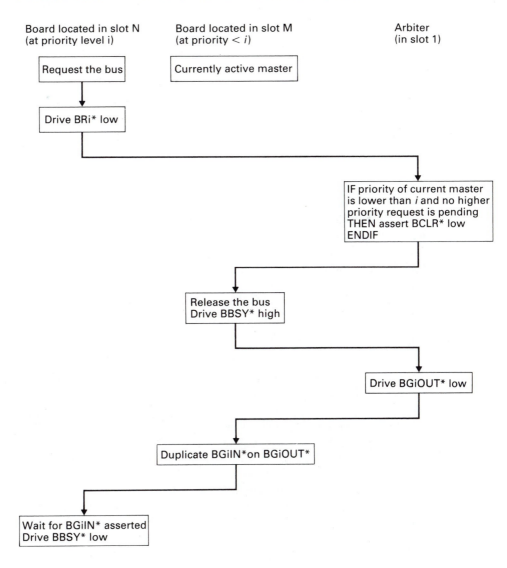

the arbiter also informs any master currently in control of the DTB that a higher level of request is pending by asserting BCLR*.

Bus Requester Operation   A bus requester module receives an indication from its DTB master (via the master_wants_bus signal in figure 10.37) that the latter wishes to access the DTB. The requester then asserts the appropriate bus request.

After the arbitration has taken place, the requester reads the incoming bus grant signals, BGxIN*, and passes them unchanged on BGxOUT*. If the on-board master wants the DTB and the requester's priority is equal to that on the BGxIN* inputs, the bus grant is latched internally and a master_granted_bus signal is passed to the master. BBSY* is asserted by the requester as long as the master indicates its intention to use the bus.

The requester may implement one of two options for releasing the DTB. One is called option RWD (release when done), and the other is called option ROR (release on request). The simpler of the options is RWD, which requires the requester to release the bus as soon as the on-board master stops indicating bus busy. In other words, the master remains in control of the bus until its task has been completed. This situation can, of course, lead to undue *bus hogging*. The ROR option is more suitable in systems in which it is unreasonable to grant unlimited bus access to a master. The ROR requester monitors the four bus request lines. If it sees that another requester has requested service, it releases its BBSY* output and defers to the other request. The ROR option also reduces the number of arbitrations requested by a master, since the bus is frequently cleared voluntarily.

The Arbitration Process   We will now look briefly at an example of the arbitration sequence. Figure 10.39 demonstrates the sequence of events taking place during arbitration between two requesters at different levels of priority. Further examples are found in the VME system manual.

The sequence of events begins when both requester A and requester B assert their request outputs simultaneously. Requester A asserts BR1* and requester B asserts BR2*. Assuming that the arbiter detects BR1* and BR2* low simultaneously, the arbiter will assert only BG2IN* on slot 1, because BR2* has a higher priority than BR1*. When this signal has propagated down the daisy-chain to requester B, requester B will respond to BG2IN* low by asserting BBSY*. Requester B then releases BR2* and informs its own master that the DTB is now available.

After detecting that BBSY* has been asserted, the arbiter negates BG2IN*. At this point, both BR2* and BG2IN* are inactive-high, because BBSY* and the bus grants are interlocked, as shown in figure 10.39. The arbiter is not permitted to negate a bus grant until it detects BBSY* low. When master B completes its data transfer or transfers, requester B releases BBSY*. The negation of BBSY* is conditional on BG2IN* remaining high and at least 30 ns having elapsed since the release of BR2*. The 30-ns delay ensures that the arbiter will not interpret the old active-low value of BR2* as another request. Requester B will wait until the 30-ns interval has elapsed and will then release BBSY*.

The arbiter interprets the release of BBSY* as a signal to arbitrate bus requests once more. Since BR1* is still active-low and is the only bus request line asserted,

FIGURE 10.39 Arbitrating between two requests on different levels

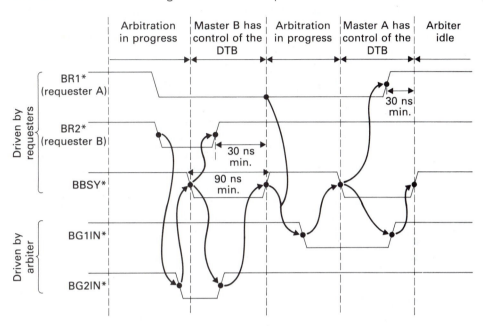

the arbiter grants access of the DTB to requester A by asserting BG1IN*. Requester A responds by asserting BBSY*. When master A has completed its data transfer, requester A releases BBSY*, provided BG1IN* has been received and 30 ns have elapsed since the release of BR1*. Since no bus request lines remain asserted when requester A releases BBSY*, the arbiter remains idle until a new request is made.

The preceding description is equally valid for both PRI and RRS option arbiters. The arbitration bus has been dealt with only superficially, and the reader is directed to the VMEbus manual for a definitive treatment. The 68000 systems designer must provide logic to interface between the 68000 bus master and the arbitration bus.

The final aspect of arbitration we are going to introduce before moving on to the VMEbus's priority interrupt bus is the design of the bus arbitration control circuits and the arbiter itself. This topic could be expanded to fill an entire text because it is more complex than you might first think. Here, we will just provide an overview. The reason for this complexity is due to a problem called *metastability*, which plagues the design of asynchronous systems.

Essentially, metastability is a state into which a clocked circuit such as a latch can fall if its data input changes at the same time it is clocked. That is, metastability can result when a latch's data setup or hold times with respect to its clock are violated. The metastable state results in uncertainty in the output of the flip-flip until the metastable state is *resolved*. This resolution might take a few tens of nanoseconds. Consequently, we cannot guarantee that the output of a latch will be valid until some time after it has been clocked if we cannot guarantee that the latch's data setup and hold times have been met. In everyday terms, metastability is similar to the effect you would observe if you tossed a coin into the air and it landed on its edge—the coin might wobble for a short time before falling heads or tails up.

Synchronous systems can be designed to be free from the dangers of metastability, because the designer controls when flip-flops are clocked with respect to data at their inputs. Unfortunately, the designer has no such control over asynchronous systems. In the case of the VMEbus, the arbiter might be confronted with several asynchronous and near-simultaneous requests for service on BR0* to BR3*. If the arbiter is clocked at the instant its inputs are changing, it will enter a metastable state and its output will not be immediately valid. Metastability is not a trivial problem, since it could lead to an unspecified state in which two masters are granted access to the bus simultaneously. (While a flip-flop is in a metastable state, its Q and Q* outputs might even be the same.)

The VMEbus specification for ANSI/IEEE STD 1014-1987 provides several sample circuits that can be used to perform arbitration. Here we will describe just two of them.

Figure 10.40 illustrates an asynchronous arbitration circuit that can be used on

FIGURE 10.40 Asynchronous arbitration and the BGxIN* = BGxOUT* daisy-chain

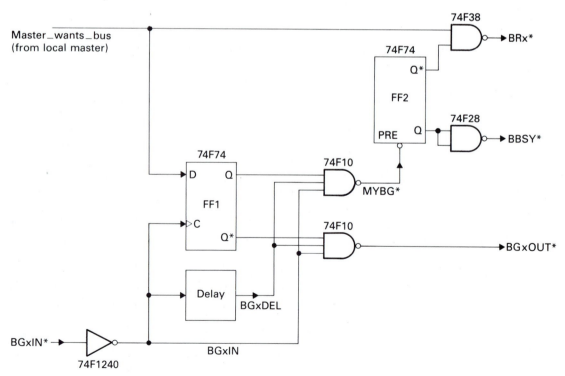

a card in slot *i* to control the BGxIN∗–BGxOUT∗ arbitration daisy-chain. As we shall soon see, this circuit arbitrates between an incoming bus grant on the BGxIN∗–BGxOUT∗ daisy-chain and a local request for the bus. Address decoding logic associated with the on-board master detects that the master has generated an address that is off-card. That is, the address corresponds to a slave on another card and this slave must be accessed via the DTB. The local (would-be) master generates a master_wants_bus signal—which, when high, indicates that the local master wishes to access the VMEbus—by driving the VMEbus request signal BRx∗ active-low.

Since the master_wants_bus signal has no particular timing relationship with BGxIN∗ from the daisy-chain, the circuit must perform an asynchronous arbitration between BGxIN∗ and master_wants_bus. The arbiter has to determine whether to drive BBSY∗ low and assume control of the bus on behalf of its local master or whether to pass the low level on its BGxIN∗ input down the daisy-chain by driving BGxOUT∗ low.

The falling edge of BGxIN∗ from the VMEbus bus grant daisy-chain is used to clock FF1, a 74F74 positive-edge-triggered D flip-flop, and therefore capture the current value of master_wants_bus. BGxIN is also applied to a delay line, whose output, BGxDEL (bus grant delay) is used to enable two 74F10 three-input NAND gates. BGxDEL is delayed to allow time for the output of FF1 to settle should it go into a metastable state. A typical delay might be 100 ns. By the time BGxDEL goes high, the output of FF1 will be valid and BGxOUT∗ will be asserted low if the local master does not want the bus.

If master_wants_bus is high when it is sampled, MYBG∗ (my_bus_grant) will go low to set a second D flip-flop, FF2. The outputs of FF2 assert BBSY∗ to claim the VMEbus and negate BRx∗ (since the local master now has control). Note that the circuit does not provide logic to release BBSY∗ once the on-board master has relinquished the bus. That logic must be user-supplied.

We will now look at a second example of a metastability-free arbitration circuit. Figure 10.41 illustrates an asynchronous bus arbiter that can be used by the arbiter in slot 1 to arbitrate between bus request inputs on BR0∗ to BR3∗. Asynchronous inputs BR0∗ to BR3∗, BBSY∗, and SYSRESET∗ from the VMEbus are buffered by two 74F1244 buffers. The bus request inputs are ORed together by a 74F20 to produce a signal, BR, that is asserted if any BRi∗ is asserted. BR is ANDed with BBSY∗ and SYSRESET∗ by a 74F11 to produce ARBGO, which is high if a bus request is asserted while both BBSY∗ and SYSRESET∗ are negated.

ARBGO is used to clock the bus request inputs into the 74F175 quad latch. The outputs of the latches are fed to a block called *grant logic* that determines which bus request is to win the arbitration. The exact nature of this logic depends on the arbitration procedure adopted by the particular VMEbus. In figure 10.41 we illustrate the logic necessary to implement arbitration option PRI.

The outputs of the bus grant logic circuit are transmitted through four 74F00 NAND gates and passed to the respective BGxIN∗ pins for feeding down the daisy chain. These NAND gates are all enabled by ARBGO after it has been delayed in a delay line (or a similar delay circuit). As in figure 10.40, the delay is provided to give the 74F175 latch time to settle should it be triggered as its inputs are changing.

FIGURE 10.41   An asynchronous arbiter

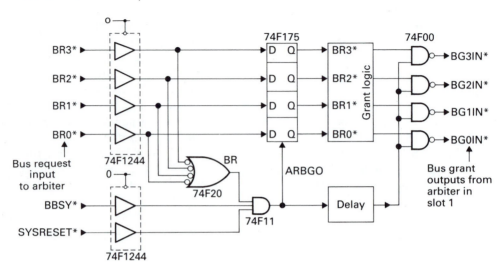

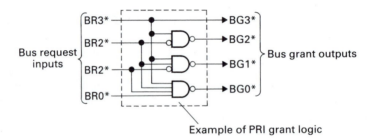

Example of PRI grant logic

Priority Interrupt Bus

In general, the VMEbus's interrupt-handling structure is closely associated with the 68000's interrupt-handling scheme described in chapter 6. That is, the VMEbus supports a seven-level prioritized, vectored interrupt system. The priority interrupt bus enables modules connected to the VMEbus to request service from a DTB master. Note that we say *a* DTB master rather than *the* DTB master, because more than one potential bus master may handle interrupts.

Two type of module are associated with the interrupt bus: the interrupt requester (*interrupter*), which requests service, and the *interrupt handler*, which receives the request and later processes it. Note that not all interrupt handling in a VMEbus system involves the VMEbus. A microprocessor can handle interrupts generated on the same card locally, without the use of the VMEbus. The VMEbus interrupt subbus allows an interrupter on one card to request service from an interrupt handler on another card. Figure 10.42 illustrates the components of a VMEbus system that contribute to the VMEbus's interrupt handling capability.

FIGURE 10.42 Relationship between the VMEbus, interrupters, and interrupt handlers. (Reprinted by permission of Motorola Limited)

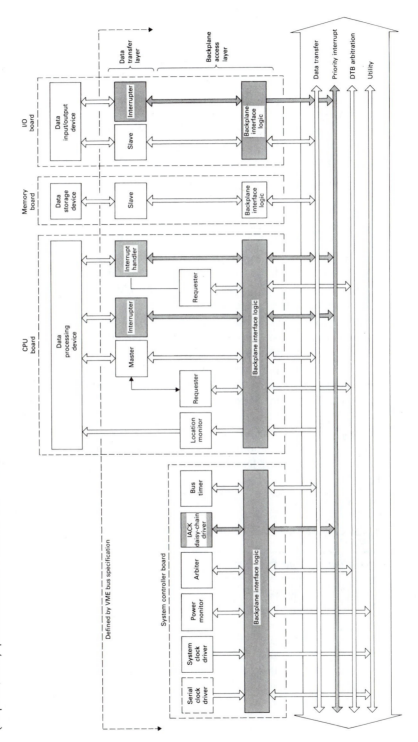

Interrupt Bus Lines

The ten lines of the priority interrupt bus are illustrated in table 10.12. Seven of the lines, IRQ1* to IRQ7*, are assigned to interrupt requests from the interrupters. The IRQ* output of an interrupt requester can be connected to any one of the VMEbus's seven interrupt request lines. In chapter 6, we explained that the 68000 responds to an interrupt by initiating an IACK cycle, during which it reads the interrupting device's vector number. The vector number is used by the 68000 to locate the appropriate exception-processing routine. The 68000 employs the function code FC2, FC1, FC0 = 1, 1, 1 to indicate an IACK cycle.

Since the VMEbus does not have function code lines, an explicit IACK* line from interrupt handlers indicates that an IACK cycle is being executed. Note that the IACK* line runs the full length of the VMEbus and is connected to the IACK daisy-chain driver in slot 1. Any interrupt handler can assert IACK* in response to an interrupt request, as illustrated in figure 10.43.

When IACK* is asserted by an interrupt handler, the negative transition is detected by the IACK daisy-chain driver (figure 10.43), which asserts its own IACKOUT* pin in response. Because the IACKOUT* and IACKIN* pins form a daisy-chain running the length of the VMEbus, the active transition of IACK* initiates a low-going transition that propagates down the interrupt acknowledge daisy-chain. Consider the sequence of events taking place when an interrupter signals attention. The interrupter asserts IRQi* to indicate its desire for attention.

The interrupt handler detects IRQi* asserted and asserts its IACK* output in turn. Interrupters are not connected to the IACK* line. (How would two requesters know which was to respond if both were connected to the same IACK* line?) The IACK* line is used by the daisy-chain driver in slot 1 to pass an IACK message from interrupt handler to interrupt handler along the IACKOUT*–IACKIN* daisy-chain. When an interrupt handler that generated an interrupt at the same level as the IACK detects that its IACKIN* has been asserted, it executes the appropriate interrupt-handling routine.

TABLE 10.12 VMEbus J1 priority interrupt bus

PIN/ROW	MNEMONIC	NAME
24b	IRQ7*	Interrupt request 7
25b	IRQ6*	Interrupt request 6
26b	IRQ5*	Interrupt request 5
27b	IRQ4*	Interrupt request 4
28b	IRQ3*	Interrupt request 3
29b	IRQ2*	Interrupt request 2
30b	IRQ1*	Interrupt request 1
20a	IACK*	Interrupt acknowledge
21a	IACKIN*	Interrupt acknowledge input
22a	IACKOUT*	Interrupt acknowledge output

FIGURE 10.43 Structure of the IACKIN*–IACKOUT* daisy-chain

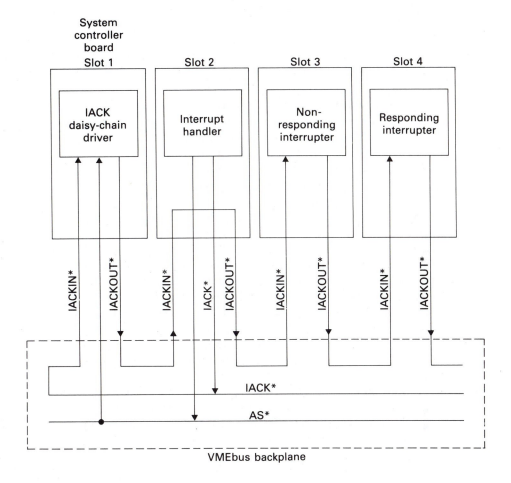

VMEbus backplane

Each of the seven interrupt request lines, IRQ1* to IRQ7*, may be shared by two or more interrupter modules. Therefore, when IRQx* is asserted, the interrupt handler cannot positively identify the source of the interrupt. However, the interrupt acknowledge daisy-chain solves the problem of interrupt contention by making certain that only one interrupter receives an acknowledgment.

When an interrupt is acknowledged, IACKIN* is asserted at slot 1. Each module driving an interrupt request line low must wait for the low level to arrive at its own slot (i.e., be propagated down the chain) before accepting the acknowledge. The module accepting the acknowledge does not pass the active-low level on down the daisy-chain, guaranteeing that only one interrupt requester will be acknowledged. We will now describe the features of the two types of module taking part in the interrupt process: the interrupter and the interrupt handler.

Interrupter

The interrupter is the source of interrupt requests and is used to accomplish three tasks.

1. It requests service from the interrupt handler. The interrupt handler monitors the interrupt request line from the interrupter.

2. It supplies a status/ID byte to the interrupt handler when its interrupt request is acknowledged. We use the term *status/ID* byte rather than the 68000 term *vector number*, because the interrupt handler does not necessarily process interrupts in exactly the same way as the 68000; that is, the 68000 multiplies the vector number by 4 and uses the result to find the address of the interrupt handler in the 68000's exception vector table. An interrupt handler may use the status/ID byte in any way the designer wishes.

3. It passes the signal at its IACKIN* pin on to its IACKOUT* pin to propagate down the interrupt acknowledge daisy-chain signal if the interrupter is not requesting that level of interrupt.

Interrupts can be identified by the level of the interrupt request line they assert to request service. The VMEbus option notation is I(n), where n is the interrupt request line number. For example, I(4) means that the interrupter asserts IRQ4*. Note that more than one interrupter module may be located on any given card. Each of these interrupters may have a different option number.

An interrupter monitors the address bus of the DTB and the IACKIN*-IACKOUT* daisy-chain to determine when its interrupt is being acknowledged. When the acknowledgment is received, it places a status/ID byte on the lower 8 lines (D_{00} to D_{07}) of the data bus and signals the byte's validity to the interrupt handler via the DTACK* line. As we can see, the response of a interrupter to an IACK cycle is essentially the same as that of an interrupting peripheral in a 68000 system.

Interrupt Handler

The interrupt handler is responsible for dealing with the interrupts originating from interrupters and performs the following four functions.

1. It prioritizes the incoming interrupt requests within its assigned range (maximum range IRQ1* to IRQ7*) from the interrupt bus—that is, the interrupter responds to the highest level of interrupt that is assigned to it. Suppose that an interrupter is designed to handle interrupts IRQ1* to IRQ5*. If IRQ2*, IRQ4*, and IRQ6* are asserted simultaneously, the interrupt handler will respond to IRQ4* (since IRQ4* is the highest level of interrupt within its range).

2. It uses its associated requester to access the DTB and, when granted use of the DTB, it acknowledges the interrupt. In other words, the interrupt handler cannot just put a status byte on the DTB in response to an IACK cycle. Like any other part of a VMEbus system, the interrupt handler (or rather its own DTB requester) must arbitrate for the VMEbus before it can use the DTB.

3. It reads the status (i.e., ID byte) from the interrupter being acknowledged.

4. Based on the information received in the status/ID byte, it initiates the appropriate interrupt-servicing routine.

Once more we must stress that the VMEbus specification has nothing to say about the way in which the interrupt handler actually services the interrupts. This action is device-dependent, and the users are left to write their own appropriate interrupt-handling routines.

Interrupt handlers may be identified by an option code IH(a-b), where a is the lowest interrupt request level serviced and b is the highest. For example, an IH(3-5) option interrupt handler may service interrupts only on IRQ3*, IRQ4*, and IRQ5*. Notice that this option demands that the interrupt handler service a contiguous sequence of interrupt levels.

When the interrupt handler receives more than one interrupt request on IRQ1* to IRQ7*, it uses the DTB requester to service the highest-priority interrupter.

Interrupt Request and Interrupt-Handling Sequence

Now that we have introduced all the components required to implement an interrupt-handling system, we will put them together and describe an interrupt sequence. Figure 10.44 provides a simplified protocol flowchart for a typical interrupt sequence. The interrupter requests service and the interrupt handler responds by requesting the DTB. When granted control of the DTB, the interrupt handler begins an IACK cycle.

The interrupt acknowledge output from the interrupt handler is transmitted to slot 1 on IACK* and then to the other slots on the IACKOUT*-IACKIN* daisy-chain. When the interrupter receives IACKIN* asserted, it puts a status/ID byte (i.e., a vector number in 68000 terminology) on D_{00} to D_{07} and completes the IACK cycle by asserting DTACK*. The interrupt handler terminates the IACK cycle and then deals with the interrupt request in whatever (user-defined) fashion is necessary.

Slaves and Interrupts

Although both a DTB slave and an interrupter may access the system data transfer bus, there are 3 important distinctions between a DTB slave and an interrupter. Unless all three of these conditions are met, the interrupter does not respond to the acknowledge sequence. If IACKIN* is asserted but the other conditions are not met, the interrupter passes on the low level on IACKIN* to the next module in the daisy-chain via IACKOUT*. The three distinctions between interrupter and slave are

1. The interrupter ignores the contents of the address modifier bus, AM0 to AM5. Equally, the DTB slave ignores the IACK* signal and never responds when IACK* is asserted.

2. The slave decodes the contents of the address bus (A_0 to A_{15}, A_{01} to A_{23}, or A_{01} to A_{31}) together with AM0 to AM5 and responds accordingly. The interrupter decodes only the lowest-order address lines, A_{01} to A_{03}, when it detects an interrupt acknowledge. However, the interrupter responds only if it has an interrupt request pending, if the interrupt acknowledge on A_{01} to A_{03}

FIGURE 10.44 Protocol flowchart of interrupt-processing sequence

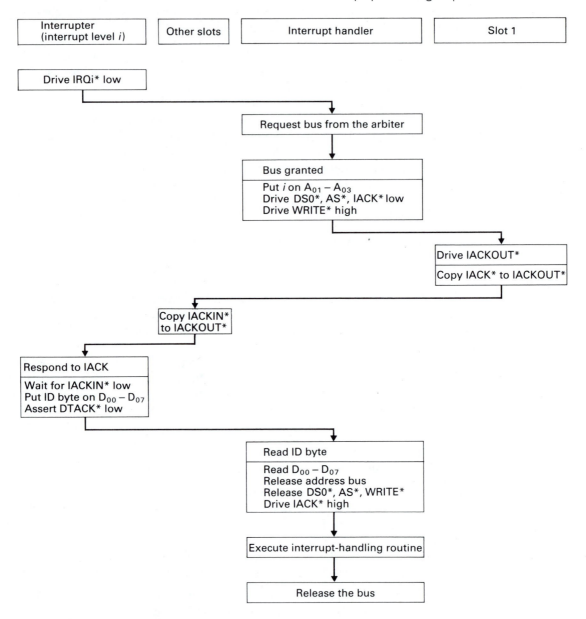

matches the level of the pending request, and if the interrupter is receiving a low level on its IACKIN∗.

3. The interrupter is required to drive only the lowest 8 data bits in response to an acknowledgment. Therefore, it is not required to monitor DS1∗. It is also not required to monitor the state of the WRITE∗ line, as the interrupter is never written to.

Centralized Distributed Interrupt Handlers

Engineers who begin their careers by designing single-board microcomputers are well aware of systems in which a single processor handles interrupts from two or more peripherals on the same board. The VMEbus goes one step further and supports *multiple interrupt handlers*. In single-handler VMEbus systems (Figure 10.45a), the seven interrupt request lines are all monitored by a single interrupt-handler module. Interrupts are prioritized by level, 1 to 7, and by position along the interrupt daisy-chain.

In a distributed interrupt-handling system (Figure 10.45b), up to seven independent interrupt handlers may be allocated to interrupt processing. No real problems are introduced by distributed interrupt-handling. Each interrupt handler is assigned one or more interrupt request lines. An interrupt handler in a distributed system operates exactly as in its single-handler counterpart.

Should two or more interrupt handlers respond to interrupts at the same time, a bus contention problem exists. Which handler is going to process the interrupt? This problem is resolved by an arbiter module. Each interrupt handler asserts a bus request

FIGURE 10.45 Single and distributed interrupt handlers

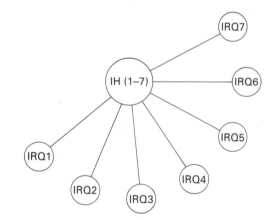

(a) Single interrupt handler

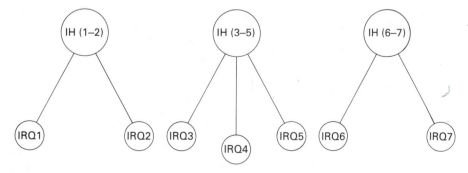

(b) Distributed interrupt handlers

line (one of BR0* to BR3*) in order to gain control of the DTB, and the arbiter determines which handler gets access to the DTB, according to the priority option in force.

Details of Interrupt Processing

We will now look at the interrupt-handling sequence in more detail. The full protocol flowchart for a single interrupt handler is given in figure 10.46. This single interrupt handler, located in slot 2, deals with interrupts on IRQ1* to IRQ7* (i.e., there is no other interrupt handler on the VMEbus). The interrupting device is located in slot 5 and asserts IRQ4* to request attention. We will assume that the current bus master, located in slot 3, has control of the VMEbus (at level BR2*) prior to the interrupt.

When the interrupt handler in slot 2 detects that IRQ4* has been asserted, it asserts a local "device_wants_bus" signal to its bus requester to request mastership of the DTB. The requester responds by asserting BR3* and the arbiter in slot 1 asserts BCLR*, in turn, to inform the current bus master that it should give up the bus.

When the bus master in slot 3 detects that another device wants the bus, it informs its local bus requester (by negating the local signal device_wants_bus) that is giving up the DTB. The requester releases BBSY* to permit the new bus master to take control. Of course, the requester does not have to respond to BBSY* asserted (according to the VMEbus specification), but it is likely that any system implementer would use BBSY* to force the current bus master off the bus to give priority to the interrupt.

The arbiter in slot 1 detects that BBSY* has been negated and asserts BG3OUT*, which ripples down the bus grant daisy-chain to the interrupt handler (we are assuming that the arbiter is implementing a PRI scheme). The interrupt handler places a 3-bit code on A_{01} to A_{03} (i.e., 100) to indicate that it is acknowledging a level 4 interrupt and drives IACK* and AS* active-low.

The low level on IACK* is detected by the IACK daisy-chain controller in slot 1, and a falling edge is propagated down the IACKIN*–IACKOUT* daisy-chain. When the interrupter in slot 5 detects IACKIN* asserted, it checks the value on A_{01} to A_{03}. If the interrupt being acknowledged is at level 4, the interrupter places a status byte on D_{00} to D_{07} and asserts DTACK*. Finally, the interrupt handler detects DTACK* low, reads the status byte, and negates the data strobe. At this stage, the interrupt handler is in control of the bus, and it carries out the appropriate user-defined actions in response to the interrupt request.

Interrupt Interface

Systems designers can choose one of several ways of implementing VMEbus interface circuits. You can exploit one of the modern powerful LSI VMEbus interface circuits that perform all VMEbus functions. These began to appear in the mid-1980s. If, perhaps, the use of sophisticated LSI circuits is a bit of an overkill, you can employ an MSI device that performs an individual VMEbus function (e.g., bus controller, interrupter, interrupt handler). Three Philips VMEbus interface circuits are described in *Microprocessor Interfacing* (see the bibliography). Finally, you can design your own

FIGURE 10.46 Protocol flowchart for interrupt processing

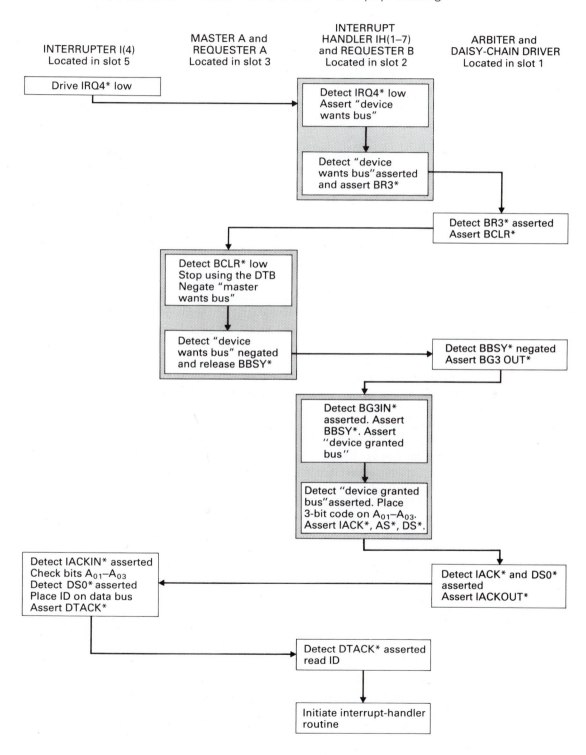

specific VMEbus interface. We will conclude our discussion of interrupts and the VMEbus by describing two basic interrupter circuits.

The first interrupter circuit (figure 10.47) is programmable and can issue an interrupt request at any level. Eight-bit latch L1 is loaded with a 3-bit code on D_{00} to D_{02} by the local processor. This code is decoded into one of seven levels by the 74LS138 interrupt decoder, and the outputs of the decoder are placed on the VMEbus via seven open-collector buffers.

Bit D_{03} in latch L1 is used by the local master as a strobe to set RS flip-flop FF1 and enable the interrupt level encoder (a 74LS138). For example, if we wish to generate a level 5 interrupt, L1 must be loaded with $XXXX1101_2$.

When the interrupt handler (in another slot) recognizes the interrupt, it begins an IACK cycle. The interrupt requester uses a 4-bit comparator (e.g., 74LS85) to detect the IACK cycle (i.e., the IRQ level on $A_{01}-A_{03}$ corresponds to the level being requested and AS* is asserted). If both these conditions are met, the active-high signal LIACK (local IACK) is asserted. LIACK is qualified with IACKIN* from the VMEbus and is used to enable the 74LS244 VME data bus buffer. This buffer places the 8-bit status byte in latch 2, L2 onto the DTB. The same enable signal resets flip-flop FF1 and removes the interrupt request from the VMEbus.

LIACK is used also to control the IACKIN*–IACKOUT* daisy-chain. If LIACK is negated, the VMEbus signal on IACKIN* is copied directly to IACKOUT*. If LIACK is active-high, the IACKIN* signal is not copied onto IACKOUT* but is used to trigger the circuit's IACK response, as we have just described.

The second interrupter circuit (figure 10.48) is taken from the VMEbus specification and is designed to control the IACKIN*–IACKOUT* daisy-chain. In this example, only one level of hard-wired interrupt request, IRQx*, is catered for (MY_IRQ from a local interrupter). Furthermore, this circuit does not provide a status value during the IACK cycle.

The function of the circuit in figure 10.48 is to arbitrate between MY_IRQ and IACKIN* from the VMEbus. AS* from the VMEbus is buffered and inverted and then applied to a delay line to produce two delayed versions of itself, ASDL1 and ASDL2 (see the timing diagram of figure 10.49). ASDL1 triggers the D flip-flop and latches MY_LEVEL (i.e., the level of the interrupt being acknowledged). The Q* output of the 74F74 D flip-flop is used to determine whether IACKIN* is copied to IACKOUT* or not. Note that Q* is enabled by ASDL2, which gives time for the output of the flip-flop to settle should it be clocked at the moment MY_LEVEL is changing.

The Q output of the flip-flop is used to generate a local MY_IACK* signal to indicate to the on-board master that it should go ahead with an IACK cycle. Note that this circuit can be used in figure 10.47 to control the IACKIN*–IACKOUT* daisy-chain and remove any problems caused by metastability.

Releasing the Interrupt Request

As you know, an interrupter requests service by asserting one of IRQ1*–IRQ7*, and, eventually, an interrupt handler responds to this request. However, we still have to answer one question: When does the interrupter release (i.e., negate) the interrupt

FIGURE 10.47 Circuit diagram of an interrupter

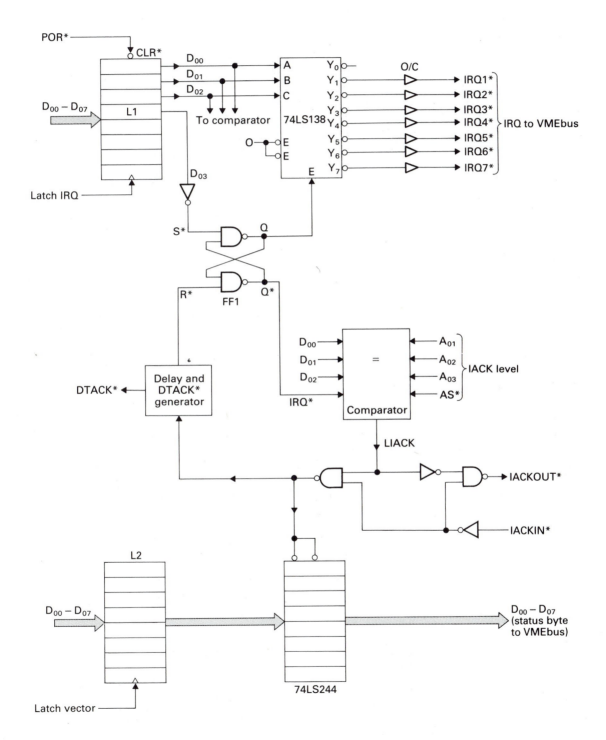

FIGURE 10.48 Circuit diagram of the IACKIN∗–IACKOUT∗ daisy-chain controller

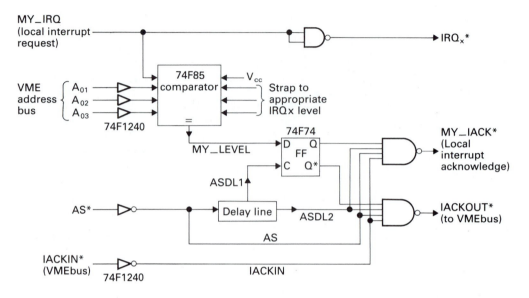

FIGURE 10.49 Timing diagram of the daisy-chain controller in figure 10.48

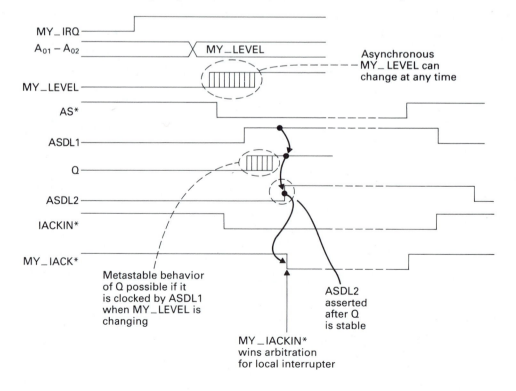

line that it has asserted? If IRQx* is released too soon, the interrupt might not be processed. On the other hand, if IRQ* is released too late, the same interrupt may be processed more than once. The problem of deciding when to release IRQx* is compounded by the fact that the various peripherals capable of generating interrupt requests do not behave in a standard way.

The VMEbus specification describes two types of interrupt release policy. A RORA (*release on register access*) interrupter maintains its IRQx* output asserted until after a register within the interrupter has been read. The RORA interrupter detects that data strobe DS0* has been asserted during a register access, and it releases interrupt request output within 2 μs of the rising edge of DS0* at the end of the bus cycle.

A second IRQx* release strategy is called ROAK, *release on acknowledge*. The ROAK interrupter releases its interrupt request output after it detects the falling edge of data strobe DS0* during the IACK cycle. A ROAK interrupter must release IRQx* within 500 ns of the rising edge of DS0* at the end of the IACK cycle. Note that the interrupter in figure 10.47 uses the ROAK algorithm, since the IACK cycle is used to release IRQx*.

Now that we have covered the VMEbus's data bus, arbitration bus, and interrupt bus, we shall briefly mention some of its other signals.

VMEbus Utilities

The VMEbus supports the following four miscellaneous signal lines:

PIN/ROW	MNEMONIC	NAME
10a	SYSCLK	System clock at 16 MHz
3b	ACFAIL*	AC failure (loss of power)
12c	SYSRESET*	System reset
10c	SYSFAIL*	System fail—indicate failure mode

SYSCLK   The system clock is a master timing signal that is free running at 16 MHz and that serves as the source of timing for all VME modules. It is not necessarily of the same frequency or phase as the processor's own clock—that is, the SYSCLK has no fixed phase relationship with any other VMEbus signal. Moreover, the provision of SYSCLK does not preclude the use of a locally generated clock in any VMEbus module.

SYSRESET*   System reset is an open-collector line driven by either a power monitor module or a manual reset switch. This signal is normally identical to the 68000's RESET* input/output (see chapter 6), and its effect is to place the processor and/or all other modules in a known state on initial power-up or following a manual reset. SYSRESET* must be asserted for a minimum period of 200 ms.

SYSFAIL* System fail is a general-purpose signal whose function is to indicate that the system has failed in some sense. What constitutes a failure and what action is to be taken when SYSFAIL* is asserted is not defined by the VMEbus standard.

We recommend that all cards within the VMEbus system drive SYSFAIL* active-low on power-up and maintain SYSFAIL* low until they have all passed their self-tests. In systems with nonintelligent cards (i.e., without an on-board processor), we recommend that they hold SYSFAIL* low until a master on another card has completed a test on them. SYSFAIL* can be asserted at any instant during normal (i.e., nonpower-up) operation of the system when the failure of one of the modules is detected.

SYSFAIL* and SYSRESET* are related. After SYSRESET* is negated (i.e., released), all cards enter their self-test mode and hold SYSFAIL* low until each test has been successfully completed. Once SYSFAIL* has been negated, the system enters its normal operating mode.

ACFAIL* The ACFAIL* line is driven by open-collector circuits and, when asserted, indicates that the ac power supply to the VMEbus cardframe has either failed or is no longer within its specified operating range. Once ACFAIL* has been asserted, the master may use the period of time between the negative transition of ACFAIL* and the point at which the system's 5-V supply falls below its minimum specification to force an orderly power-down sequence. For example, there may be sufficient time to carry out a core dump and store all working memory on disk so that an orderly restart may be made later.

Interfacing the 68020 to the VMEbus

The VMEbus grew out of the need to engineer sophisticated 68000-based microcomputer systems and to exploit fully the 68000's capabilities. However, the VMEbus was designed before the introduction of the 68020 and 68030. Consequently, some of the special features of the 68020 are not directly exploited by the VMEbus. In particular, the VMEbus does not make use of the 68020's dynamic bus sizing mechanism. Fortunately, the 68020's interrupt-handling and bus arbitration systems are the same as those of the 68000 and, therefore, present no problems when interfacing the 68020 to the VMEbus. It is the 68020's data bus we have to worry about.

The first problem caused by the 68020 is due to its AS* timing with respect to address valid. Since the VMEbus requires that AS* go low 35 ns after the address is stable, it is necessary to delay AS* from the 68020, as we have already demonstrated in figure 10.35.

A more serious problem concerns mapping the 68020's data bus sizing control signals onto the VMEbus control signals. Using the 68020's DS*, A_{00}, SIZ1, and SIZ0 control outputs, we have to manufacture DS0* and DS1* signals for the VMEbus. Figure 10.50 (from Motorola's engineering bulletin EB114R1) demonstrates how this is done. As you can see, the data strobes are delayed in the same way AS*

FIGURE 10.50 Generating DS0* and DS1* in a 68020 system

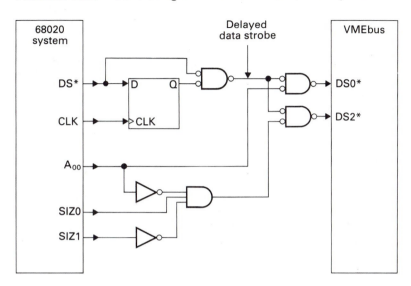

is delayed, and a little logic is necessary to synthesize DS0* and DS1* from the 68020's control signals (chapter 4 describes the 68020's bus interface in more detail).

The VMEbus's LWORD* signal is asserted whenever the 68020 executes a 32-bit aligned longword operation on D_{00}–D_{31}. The 68020 indicates a 32-bit aligned data transfer by $A_{01}, A_{00}, SIZ1, SIZ0 = 0, 0, 0, 0$. Note that the VMEbus permits only aligned longword operations. In other words, you can interface a 68020 to the VMEbus only by operating the 68020 in a 68000 *data transfer mode*.

In addition to creating DS0* and DS1* for the VMEbus, we have to create DSACK0* and DSACK1* data acknowledge strobes for the 68020 from the VMEbus's single DTACK* signal (see figure 10.51). DTACK1* is a copy of DTACK* and DSACK0* is the logical AND (in negative logic terms) of LWORD* and DTACK*. That is, DSACK0* is asserted if DTACK* is asserted and the VMEbus is executing a 32-bit data transfer (i.e., LWORD* is asserted low).

FIGURE 10.51 Generating DSACK1* and DSACK0* for a 68020 in a VMEbus system

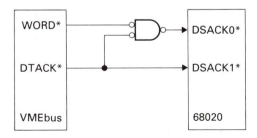

Another problem posed by the 68020's dynamic bus sizing mechanism is its data alignment. You will remember from chapter 4 that the 68020 places its least significant byte on D_{24} to D_{31} when it accesses an 8-bit port. Consequently, 68020-based systems have to locate 8-bit ports on D_{24} to D_{31}. Since we cannot carry this restriction over to the VMEbus when we interface it to a 68020, we must make the 68020's data bus look like a 68000's data bus. Figure 10.52 (from EB114R1) shows how a 68020 can be connected to the VMEbus by using bus transceivers to multiplex the data.

The 74ALS640 bus transceivers perform the necessary multiplexing and are controlled by A_{01}, A_{00}, SIZ1, and SIZ0. Extra logic is required to detect bus 68020

FIGURE 10.52 Interfacing the 68020 to the VMEbus

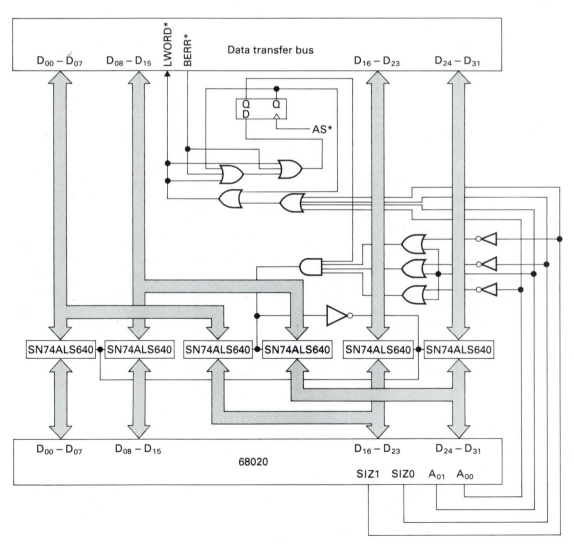

cycles that cannot be mapped onto a VMEbus cycle and to assert BERR*. When an 8-bit or a 16-bit port receives LWORD* asserted (i.e., misaligned data transfer), BERR* is asserted to terminate the bus cycle. Suppose that the interface logic also asserts HALT* to force a rerun of the bus cycle. The second time the system attempts the transfer, it blocks LWORD* and the interface logic attempts a 16-bit data transfer. If the second attempt leads to the assertion of BERR*, then an exception will take place; otherwise a third bus cycle will be executed by the 68020 to complete the longword transfer. Therefore, an aligned longword transfer to a 16-bit port will occur in three bus cycles instead of the normal two cycles.

10.4 NuBus

The VMEbus is not the only general-purpose high-performance backplane bus used by the designers of 68000 systems. We are now going to describe briefly an alternative to VMEbus called NuBus. NuBus is a processor-independent bus that was originally conceived at MIT in 1970 and was later supported by Western Digitial and Texas Instruments (1983). NuBus is now a registered trademark of TI. Today, over 100 companies use NuBus, the most important of which is Apple. Apple implements a subset of NuBus in its Macintosh II. NuBus now has a IEEE specification: ANSI/ IEEE STD 1186-1988.

The obvious questions to ask at this stage are, What is the basic difference between the VMEbus and NuBus? and Why should anyone adopt it when there are so many other buses from which to choose? The short answer to the first question is that the NuBus is a *synchronous bus* with *multiplexed* address and data lines, whereas the VMEbus is an *asynchronous bus* with separate address and data buses. The equally short answer to the second question is that NuBus is highly cost-effective.

One consequence of multiplexing the address and data lines is that the NuBus is cheaper than the VMEbus and has only 51 signal lines (it employs the same connectors as the VMEbus and uses all nonsignal lines for either power or ground). The two principal reasons that Apple and other companies have adopted the NuBus are that (1) it is as processor-independent as you could realistically expect, and (2) it is a powerful but inexpensive bus (its protagonists argue that it has the performance of buses in the VMEbus class, whereas its cost falls within the PC AT bus class).

NuBus employs 366.7-mm × 280-mm (14.44-in. × 11.02-in.) triple-Eurocards, providing more board area than the corresponding VMEbus card. As we have said, both NuBus and VMEbus employ the same type of two-part DIN 41612-C96 connectors. NuBus also supports a 4-in. × 12.875-in. card, which gives NuBus users two very different form factors from which to choose (i.e., the *professional Eurocard* format or the *desktop personal computer* format).

Like any other new bus (or microprocessor, etc.), the NuBus has its own terminology, the most annoying aspect of which is the use of *word* to indicate a 32-bit value. Since the VMEbus and 68000 world reserves *word* to refer to a 16-bit value, there is a lot of room for confusion. *Master* and *slave* are used in their VMEbus sense, but *tenure* is "NuBus-speak" for *bus mastership*. A NuBus read or write bus cycle is called a *transaction*.

Philosophy of NuBus

Probably the most visible of NuBus's attributes is its processor independence. If we were to show the VMEbus specification to engineers who knew all about microprocessors and nothing about buses, they would all immediately identify the VMEbus as a 68000 product. Of course, you can interface an 80386 to the VMEbus, just as you can order a hamburger at the Ritz. The point is that the VMEbus fits the 68000 like a glove. It would be an impossible task to associate NuBus with a particular microprocessor, since NuBus has no special signals or timing requirements. Of course, that does not mean that it is equally easy to interface any microprocessor to the NuBus.

Unlike VMEbus, which has a special location within the bus (i.e., slot 1), NuBus is entirely decentralized, and all slots are of equal importance. However, each slot has its own unique ID (identification) code that is *hard-wired* into the backplane. Thus, the ID value of each slot is *fixed* and is provided by the backplane connector itself (and not by the card). For example, if you plug a card into slot 11, active-low backplane lines ID3*–ID0* are hard-wired at the connector to provide the value 0100 (i.e., the inverse of $1011_2 = 11_{10}$). The identification lines ID3*–ID0* do not run along the backplane and are simply connected to ground or to V_{cc} at each connector to provide the appropriate slot number.

It is worth noting that NuBus's lines are not daisy-chained, and therefore you do not have to worry about using jumpers to bypass daisy-chain links in empty slots.

The NuBus implements a philosophy called *geographic addressing*; that is, the address space is partitioned, and each card is allocated a unique slice of the address space. A 32-bit NuBus address can be expressed in hexadecimal form as $YXXX\ XXXX_{16}$. The 256M-byte address space for which Y = 1111 (i.e., $F000\ 0000_{16}$ to $FFFF\ FFFF_{16}$) is reserved and is called *slot space*. This slot space is divided into 16 blocks of 16M bytes, and each of these blocks is allocated to a slot, as described in figure 10.53. Each slot therefore has a 16M-byte block of address space associated with it. It follows that the NuBus cannot support more than 16 slots.

Data Formats and NuBus

One of the fundamental differences between the 68000 and the 8086 series of microprocessors is the way in which these two families arrange data. The 68000 family takes what is called a *big-endian* approach to data. That is, the 68000 stores data with the most significant unit at the lowest address. For example, if a 68000 system were to execute the instruction MOVE.L #$12345678,1000, memory location 1000 would be loaded with $12, 1001 with $34, 1002 with $56, and 1003 with $78.

FIGURE 10.53 NuBus slot space

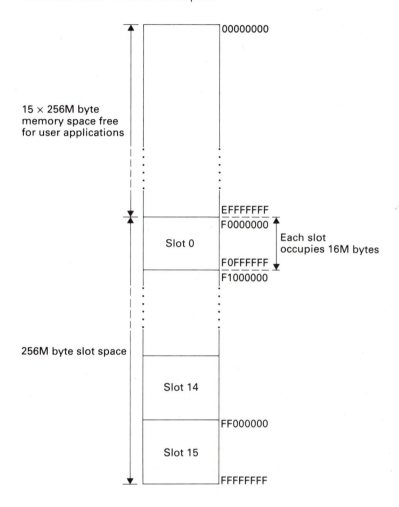

Members of the Inter family take a *little-endian* approach and store the least significant unit of data at the lowest address. In other words, Intel and Motorola chips store data in memory in opposite ways. This difference makes life terribly difficult if you attempt to interface, say, an 80386 to a 68020 system (because one processor stores $12.34\ 5678$ in memory and the other reads it back as $7856\ 3412$).

The terms big endian and little endian are now used to describe the relationship between byte ordering and byte storage. Jonathan Swift coined these terms in Gulliver's Travels to describe two groups of Lilliputians: Little endians break eggs at their narrow ends, and big endians break eggs at their wide ends. A law was passed to force everyone to break eggs at the big end. A rebel uprising by little endians led to a civil war in which 11,000 Lilliputians perished. Swift was, in fact, satirizing the religious wars of his day. The terms *little endian* and *big endian* were used by Danny

Cohen ("On holy wars and a plea for peace," *IEEE Computer*, October 1981) to call for the way in which we store data to be standardized.

The NuBus takes a little-endian approach to data storage and organizes data in the form illustrated by figure 10.54. For example, the NuBus stores bits 0–7 of a longword in the same location that the 68000 stores bits 24–31. NuBus literature refers to 32-, 16-, and 8-bit entities as words, halfwords, and bytes, respectively. Some authors describe 32 bits as a *NuBus word* to distinguish it from the 68000's 16-bit word. A 16-bit NuBus halfword is stored in memory with the most significant byte at the lower address.

FIGURE 10.54 NuBus and little-endian data

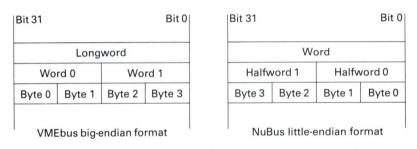

(a) NuBus/VMEbus data-size terminology and the arrangement of 8-bit, 16-bit and 32-bit units of data in memory.

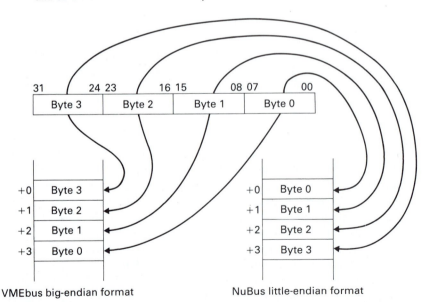

(b) Storing a 32-bit word in memory. The NuBus stores the least significant byte at the lowest address; the VMEbus stores the most significant byte at the highest address.

Before looking at some of the NuBus's details, we will list its signals, together with their grouping (table 10.13). As you can see, the NuBus uses negative logic for all its signals, including the address/data and even the clock. If you compare NuBus's signals with those of VMEbus, you cannot fail to be impressed by the minimalist approach taken by NuBus. Of the 96 signals paths on a DIN connector, 45 are dedicated to the power and ground rails. The mnemonics and signal groupings in table 10.13 are fairly self-expananatory, with a few exceptions. The *transaction control signals* regulate the flow of data in read and write cycles and perform a similar function to the VMEbus's asynchronous bus control signals.

The slot identification signals are used to identify each of the slots on a NuBus (from 0 to the maximum of 15), and there is no VMEbus equivalent to these signals. Two parity signals are provided: SP* indicates the parity on the address/data bus and SPV* (parity valid) indicates whether the parity bit is valid or not. Note that the Macintosh II version of NuBus does not implement parity checking, and SP*/SPV* are passively pulled up to their inactive levels. PFW* is a power failure warning signal that is asserted to indicate that the power has failed. The RESET* line should be asserted for no less than 1 ms to comply with STD 1196.

Data Transfer on the NuBus

NuBus does not support a range of different types of data transfer (e.g., normal read/write cycle, I/O processing, interrupt processing) normally associated with other high-performance buses. Instead, NuBus implements a basic bus cycle called a *NuBus transaction*. Figure 10.55a describes a read cycle on the NuBus's 32-bit data transfer bus (remember that VMEbus is a 16-bit bus unless both J1 and J2 backplanes are implemented). The NuBus's 10-MHz clock controls the operation of the bus. This clock has a 72 : 25 duty ratio and is used directly to control data transfer. At point A, the bus master drives the address and data bus, AD00*–AD31*, with an address and asserts the control signal START* to indicate a new bus cycle. At the same time, the master places a code on TM1*, TM0* to indicate the type of the bus cycle. Instead of using a conventional R/$\overline{W}$ signal like the VMEbus, the NuBus employs TM1* and TM0* to pass a message between the master and the slave.

The slave samples the data on AD00*–AD31* and TM1*, TM0* at point B, and the master stops driving these buses at point C 25 ns later (the master also negates START*). The slave then puts its data on the address/data bus at D and supplies a status code on TM1*, TM0*. At the same time, the slave asserts the acknowledgment line, ACK*, to complete the cycle. The master samples the bus at E to read both the data and status from the slave. The clock, CLK*, is asserted for 25 ns, which provides a 25-ns data setup time and which helps avoid bus skew problems.

As you can see from figure 10.55a, the NuBus cycle is synchronous, but wait states can be introduced by delaying the slave's ACK* response. The number of data paths is reduced by multiplexing the address and data, and the number of control paths is reduced by employing the same lines to pass messages from the master to the slave and from the slave to the master.

TABLE 10.13   NuBus signals

PIN	CONNECTOR ROW		
	A	B	C
1	− 12 V	− 12 V	RESET*
2	Gnd	Gnd	Gnd
3	SPV*	Gnd	+ 5 V
4	SP*	+ 5 V	+ 5 V
5	TM1*	+ 5 V	TM0*
6	AD1*	+ 5 V	AD0*
7	AD3*	+ 5 V	AD2*
8	AD5*	− 5.2 V	AD4*
9	AD7*	− 5.2 V	AD6*
10	AD9*	− 5.2 V	AD8*
11	AD11*	− 5.2 V	AD10*
12	AD13*	Gnd	AD12*
13	AD15*	Gnd	AD14*
14	AD17*	Gnd	AD16*
15	AD19*	Gnd	AD18*
16	AD21*	Gnd	AD20*
17	AD23*	Gnd	AD22*
18	AD25*	Gnd	AD24*
19	AD27*	Gnd	AD26*
20	AD29*	Gnd	AD28*
21	AD31*	Gnd	AD30*
22	Gnd	Gnd	Gnd
23	Gnd	Gnd	PFW*
24	ARB1*	− 5.2 V	ARB0*
25	ARB3*	− 5.2 V	ARB2*
26	ID1*	− 5.2 V	ID0*
27	ID3*	− 5.2 V	ID2*
28	ACK*	+ 5 V	START*
29	+ 5 V	+ 5 V	+ 5 V
30	RQST*	Gnd	+ 5 V
31	NMRQ*	Gnd	Gnd
32	+ 12 V	+ 12 V	CLK*

NOTE: These signals fall into six groups:

Address/data/parity (AD0*–AD31*, SPV*, SP*)	34 lines
Transaction control signals (TM0*, TM1*, START*, ACK*)	4 lines
Arbitration bus (ARB0*–ARB3*, RQST*)	5 lines
Slot identification bus (ID0*–ID3*)	4 lines
Utility bus (NMRQ*, RESET*, CLK*, PWF*)	4 lines
Power and ground (+ 5 V, − 5.2 V, + 12 V, − 12 V, Gnd)	45 lines

Signal lines are driven by three-state bus drivers, except CLK*, which is driven by a TTL totem-pole output, and ARB0*–ARB3*, RQST*, RESET*, and PFW*, which are all driven by open-collector outputs.

FIGURE 10.55 Data transfer on the NuBus

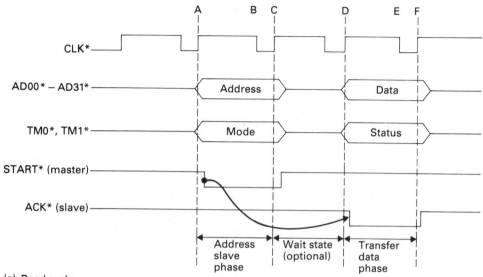

(a) Read cycle

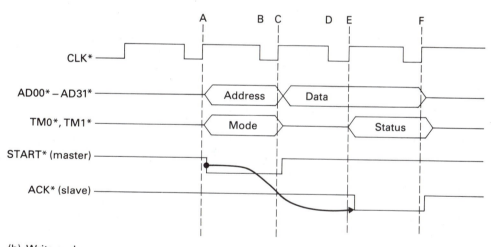

(b) Write cycle

Figure 10.55b describes the corresponding NuBus write cycle. In a write cycle the master supplies data on AD00*–AD31* at point C (i.e., at the start of the clock cycle following the setting up of the address). At the end of the write cycle (point E), the slave supplies an ACK* signal and a status code.

Since the NuBus is synchronous, uses a minimum of two clock cycles to perform a data transfer, and is 32 bits (i.e., 4 bytes) wide, its maximum data rate is 20M bytes/s. The VMEbus has a slightly faster maximum data rate of 25.0M bytes/s.

Table 10.14 interprets the meaning of the TM0* and TM1* lines. When used to indicate the *mode* of the cycle, TM1*, TM0* signify whether the cycle is to be a read or write cycle and whether it is to involve a byte or a 32-bit word (i.e., TM1* and TM0* perform the function of the VMEbus's R/$\overline{\text{W}}$ and data strobes).

When used to convey the status at the end of a bus cycle, TM1*, TM0* tell us rather more than DTACK* and BERR* in a VMEbus system. The four status messages include normal termination and *bus error* plus a timeout message and a *try again later* message.

In addition, NuBus's two status lines provide four messages, as table 10.14 indicates. These are not implemented by the Macintosh, but they can be used in multiprocessing environments. For example, the *attention-resource-lock* cycle can be issued to force a slave to lock out local CPU accesses while a *locked* NuBus transfer is in operation. The locked state is released when the master sends an *attention-null* cycle. The VMEbus implements locked activity by maintaining AS* asserted.

Table 10.15 relates NuBus's transaction control signals and its two least significant address bits to the type and size of the data transfer. The same table also provides the corresponding signals on the VMEbus for similar data transfer cycles. In this example, the VMEbus is assumed to have a single connector, so that a VMEbus longword is transferred as two separate words. Note in table 10.15 that the terminology (word, etc.) is the VMEbus/68000 terminology and the entries 0 and 1 in the tables refer to the electrical level on the bus ($0 = V_{OL}$, $1 = V_{OH}$). Since the NuBus takes a little-endian approach to the storage of data, it is necessary to use multiplexers to map the 68000's big-endian data onto the NuBus.

In addition to single data transfers, the NuBus supports a block transfer mode. A block of 2 to 16 words (i.e., 32-bit values) can be transferred at a time. The master

TABLE 10.14 Interpreting TM1*, TM0*

MESSAGE TYPE	TM1*	TM0*	CYCLE	COMMENT
Mode	0	0	Write	Write a byte.
Mode	0	1	Write	Write a word (32 bits).
Mode	1	0	Read	Read a byte.
Mode	1	1	Read	Read a word (32 bits).
Status	0	0	Transfer complete	Normal bus cycle termination.
Status	0	1	Error	Bus cycle error (e.g., parity).
Status	1	0	Timeout	The slave failed to respond in 256 cycles.
Status	1	1	Retry	The slave cannot complete the bus cycle. The slave might be able to complete the cycle later.
Transmit	0	0	Attention-null	These messages are used in multiprocessing systems.
Transmit	0	1	Reserved	
Transmit	1	0	Attention-resource-lock	
Transmit	1	1	Reserved	

TABLE 10.15 TM1*, TM0*, and data bus transfer sizes

CYCLE	TRANSFER	NuBus					VMEbus			
		TM1*	TM0*	AD1*	AD0*	DATA BUS	A_{01}	UDS*	LDS*	DATA BUS
Write	Byte 0	0	0	1	1	D_{31}–D_{24}	0	0	1	D_{15}–D_{08}
Write	Byte 1	0	0	1	0	D_{23}–D_{16}	0	1	0	D_{07}–D_{00}
Write	Byte 2	0	0	0	1	D_{15}–D_{08}	1	0	1	D_{15}–D_{08}
Write	Byte 3	0	0	0	0	D_{07}–D_{00}	1	1	0	D_{07}–D_{00}
Write	Word 0/1	0	1	1	0	D_{31}–D_{16}	0	0	0	D_{15}–D_{00}
Write	Word 2/3	0	1	0	0	D_{15}–D_{00}	1	0	0	D_{15}–D_{00}
Write	Longword	0	1	1	1	D_{31}–D_{00}	0	0	0	D_{15}–D_{00} 1
							1	0	0	D_{15}–D_{00} 2
Read	Byte 0	1	0	1	1	D_{31}–D_{24}	0	0	1	D_{15}–D_{08}
Read	Byte 1	1	0	1	0	D_{23}–D_{16}	0	1	0	D_{07}–D_{00}
Read	Byte 2	1	0	0	1	D_{15}–D_{08}	1	0	1	D_{15}–D_{08}
Read	Byte 3	1	0	0	0	D_{07}–D_{00}	1	1	0	D_{07}–D_{00}
Read	Word 0/1	1	1	1	0	D_{31}–D_{16}	0	0	0	D_{15}–D_{00}
Read	Word 2/3	1	1	0	0	D_{15}–D_{00}	1	0	0	D_{15}–D_{00}
Read	Longword	1	1	1	1	D_{31}–D_{00}	0	0	0	D_{15}–D_{00} 1
							1	0	0	D_{15}–D_{00} 2

indicates to the slave the length of the block by putting the length (i.e., 2, 4, 8, or 16) on AD5* to AD2* before the transfer takes place. AD6* to AD31* provide the address of the block in the normal way. As each word in the block is transferred, the slave responds by asserting TM0* as an intermediate acknowledge and then asserts ACK* only after the entire block has been transmitted. Block transfers can take place at up to 100 ns/word, corresponding to 40M bytes/s.

Bus Arbitration

Like the VMEbus, NuBus supports multiprocessing and multiple bus masters. NuBus has four arbitration lines, ARB0* to ARB3*. A would-be master begins arbitration by asserting its bus request line, RQST*, which it continues to assert until it gets control of the bus. Unlike the VMEbus, no special arbiter exists to decide which master will get the bus. All potential masters join in the arbitration process, which is called *distributed arbitration*. As we stated earlier, each slot has a unique number (from 0_{16} to F_{16}) and this number plays a key role in the arbitration process.

Any potential master that wants to use the bus places its arbitration level on the 4-bit arbitration bus. The arbitration bus is driven by open-collector circuits and can therefore be driven by more than one card simultaneously without creating bus contention. Since NuBus uses negative logic, the arbitration number is *inverted*, so that the highest level of priority is 0000 and the least is 1111. If a competing master sees a higher level on the bus than its own level, it ceases to compete for the bus. That is, each master simultaneously drives the arbitration bus and observes the bus.

Consider the case in which three masters numbered 0100 (four), 0101 (five), and 0010 (two) simultaneously put the codes 1011, 1010, and 1101, respectively, onto the arbitration bus. Because the arbitration lines are open-collector, any output at a 0 level will pull the bus down to 0. In this example, the bus will be forced into the state 1000. The master at level 2 that puts the code 1101 on the arbitration bus will detect that ARB2* is being pulled down and will therefore leave the arbitrating process; the arbitration bus will then have the value 1010. The master with the code 1011 will detect that ARB1* is being pulled down and will leave the arbitration process. The value on the arbitration bus will then be 1010, and the master with that value will have gained control. In VMEbus terms, this is PRI arbitration (the only strategy permitted by NuBus).

Figure 10.56 illustrates the type of logic that a NuBus master might use to gain control of the bus. Lines ID0*–ID3* define the slot location of the master (i.e., its priority) and lines ARB0*–ARB3* are NuBus's arbitration lines. The signal labeled ARB* permits the master to arbitrate for the bus, and the output GRANT is asserted if the master wins the arbitration. NuBus arbitration uses simple nonsequential arbitration logic (and arbitration can take place in parallel with normal bus activity), so the NuBus is well suited to multiprocessing applications.

Since NuBus implements a prioritized arbitration system, there is a danger that a high-priority slot will monopolize the bus and stop a low-priority slot from ever using the bus. Such bus hogging is eliminated by a *deferral mechanism*. Once a slot has gained bus mastership and then relinquished it, that slot will not attempt to reestablish bus mastership until all pending bus requests have been dealt with. However, the NuBus does have special mechanisms that permit a bus master to maintain its bus mastership. A special bus cycle called an *attention cycle* can be executed to request continuing bus ownership. An attention cycle is indicated by asserting both START* and ACK* simultaneously.

FIGURE 10.56 Example of NuBus arbitration logic

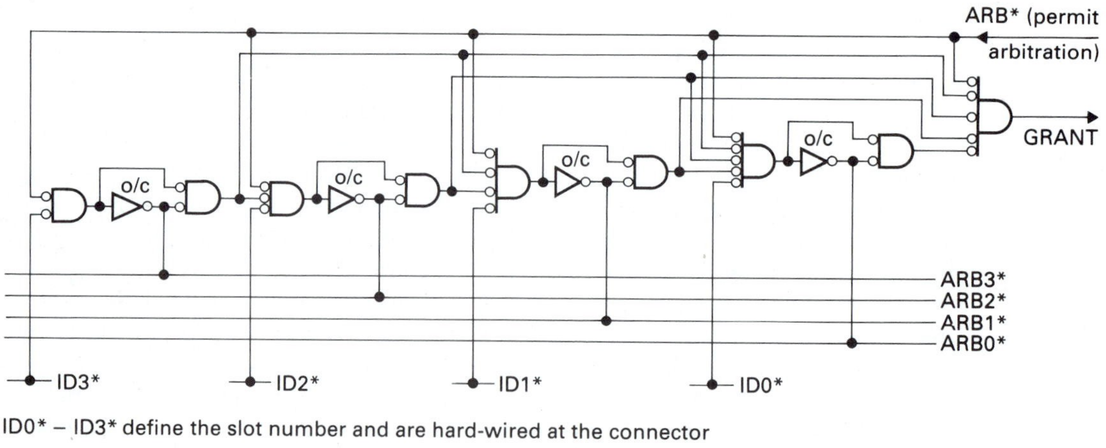

ID0* – ID3* define the slot number and are hard-wired at the connector

NuBus and Interrupts

NuBus does not implement an interrupt structure as the VMEbus does. Instead, it turns an interrupt into a *write cycle*. Consequently, a complex interrupt structure (both hardware and protocol) is not required. A device requiring attention can interrupt a processor simply by writing a message into a region of memory space monitored by that processor. As you can see, interrupt processing is replaced by a message-passing mechanism. However, a very primitive device (i.e., one not capable of becoming a bus master) might wish to indicate that it needs servicing and yet be too primitive to take part in a message exchange sequence. Such devices are catered for by the provision of a *nonmaster request* (NMRQ∗) line that operates like a conventional IRQ.

Miscellaneous Comments

The NuBus provides the same power rails as the VMEbus ($+5$ V, $+12$ V, -12 V), together with an additional rail at -5.2 V. Note that the Apple version of the NuBus does not support a -5.2-V rail.

NuBus cards use a configuration ROM to provide information concerning the function of the card and pointers to the various resources on the card. This ROM is located so that its highest address is the highest address of the card's slot space. The detailed purpose of the ROM (i.e., its contents and how they are used) is not defined by the NuBus specification.

SUMMARY

In this chapter, we have looked at the bus that distributes information between the various parts of a computer. We have examined the electrical characteristics of buses and the bus drivers and receivers needed to interface modules to the bus. We have looked at some of the practical problems, such as transmission-line effects, faced by bus designers and bus users alike. We have provided an overview of the VMEbus, which adds value to the 68000's own bus. In particular, the VMEbus provides all the facilities necessary to implement a multiprocessor system. By means of its SYSFAIL∗, SYSRESET∗, and ACFAIL∗ lines, it is able to provide a limited measure of automatic self-test and recovery from certain forms of failure.

The only feature of the 68000 CPU lacking in the VMEbus is the 68000's synchronous bus. Fortunately, this is not absolutely necessary, as a pseudo-synchronous bus can be derived from the existing VMEbus signals. Although the data transfer bus part of the VMEbus is almost identical to that of the 68000 itself, the system designer must provide his or her own interface to the arbitration and interrupt buses.

At the end of this chapter we looked at another bus that can be used in 68000-based systems but that differs from the VMEbus in many important ways.

This bus is the NuBus, which is found in the Macintosh range of computers. A particularly interesting feature of NuBus is its ability to support distributed arbitration.

Problems

1. Explain the meaning of the following terms as they are applied to bus technology.
   a. Bus driver
   b. Bus receiver
   c. Static bus contention
   d. Dynamic bus contention
   e. Bus protocol
   f. Passive bus driver
   g. Active bus driver
   h. Bus arbiter
   i. Reflection
   j. Bus termination

2. A manufacturer decides to produce a 68000-based personal computer. All on-board memory, interface, and peripherals are located on a single card. The designer wishes to keep the cost to an absolute minimum and decides not to include an external bus. However, the designer does wish to offer a range of add-on peripherals that do not require to take advantage of the 68000's full operating speed. Therefore, a compromise is chosen that exploits the 68000's asynchronous interface but is very cheap to implement.

   The proposed bus uses an 8-bit parallel interface (plus ground return paths) to take advantage of low-cost connectors. This bus uses a 4-bit control word and 4-bit data word to move data between the computer and a peripherals module.
   a. Design a 4-bit control bus, C0, C1, C2, C3, to control the flow of data on the 4-bit data bus, I0 to I3. The control signals should permit the bus to mimic the 68000's asynchronous bus.
   b. For the 8-bit bus you have designed, construct a protocol flowchart and a timing diagram for both a read (word) cycle and a write (word) cycle.

3. For the diagram of figure 10.18, calculate:
   a. The write-to-read bus to memory contention time
   b. The write-to-read CPU to data bus driver contention time
   c. The read-to-write bus to memory contention time

4. Like most microprocessors, the 68000 uses special-purpose control lines to augment the data transfer bus. These control lines include FC0 to FC2, BR*, etc., IPL0* to IPL2*, etc. An alternative approach is to employ a *message bus* that carries encoded system control messages; for example, a device requiring attention may inject a suitable message onto this bus.

   Devise a method of implementing such an arrangement in a 68000-based environment. Note that a suitable protocol is needed to determine which device may access the message bus. This situation can be achieved by using a special-purpose control line or by time-division multiplexing. The IEEE-488 bus has some of the characteristics of this type of control bus.

5. Derive a general expression for the maximum rate (bits per second) at which a bus can operate stating any assumptions you make. HINT: Consider the time taken for a message (e.g., address) to be transmitted from a master to a slave and the time taken for the transmitter to receive a reply.

6. Design an interface between a 68000 module and the VMEbus. Assume that the module is to go in slot 2 of the bus.

7. We stated previously that the SYSFAIL* line of the VMEbus is driven low by modules during their "self-test" mode immediately following the initial application of power. Describe how a 68000 module can be forced into a self-test mode and indicate the hardware and software necessary to perform this task.

8. What does *fairness* mean in the context of bus arbitration?

9. The VMEbus does not employ multiplexing in order to use a single address/data bus. What are the advantages and disadvantages of the VMEbus's nonmultiplexed address and data buses?

10. What is the difference between a TAS timing cycle and the timing cycle of other 68000 instructions? How is TAS used?

11. Why are address and data bus buffers required to implement a bused system?

12. To what extent is the following statement correct: Selecting a good computer bus is considerably more important than selecting a good microprocessor.

13. What are the three components, or elements, necessary to define a computer bus?

14. Why and in what sense are the DIN 4164 standard connectors specified by the VMEbus standard better than the connectors used by the IBM PC bus?

15. What is the significance of the term *noise immunity* when applied to a logic element?

16. Since the input current taken by the input stage of an NMOS or a CMOS gate is negligible, it follows that a single output can drive hundreds or thousands of NMOS/CMOS inputs. Why is this statement not true?

17. Calculate the limiting (i.e., minimum and maximum) values of the pull-up resistor for an open-collector bus, given the following data:

Number of bus drivers	10
Number of bus receivers	10
V_{OH}	2.7 V
V_{OL}	0.4 V
I_{IL}	0.5 mA
I_{IH}	50 μA
I_{OL}	8 mA
$I_{OH} = I_{leakage}$	100 μA

18. What is tristate logic and why is it so popular with the designers of computer buses?

19. Why is it necessary to terminate a bus with its *characteristic impedance*?

20. What is the reflection coefficient for a bus with a characteristic impedance of 100 Ω and a termination of 200 Ω?

21. What is the difference between a positive and a negative reflection coefficient?

22. What is meant by *active termination*?

23. What are the differences between the VMEbus's data transfer bus and the 68000's own data transfer bus?

24. The VMEbus has six address modifier lines, AM0 to AM5. What is their significance and how are they used?

25. A VMEbus system has several modules that can act as a bus master. Explain how one of these modules (which is not a current bus master) goes about requesting the VMEbus and receiving it.

26.   What does *daisy-chaining* mean in the context of the VMEbus?

27.   Why does the designer of backplane buses have to be aware of transmission-line theory?

28.   For the following bus, calculate the voltage at the near and far ends of the transmission line (i.e., bus) from the time at which a step function is applied (i.e., $t = 0$) to the near end to four units of delay later (one unit of delay, t, is the end-to-end propagation time of the bus).

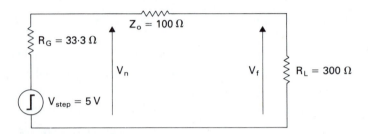

29.   A backplane has a 150 Ω far-end termination network. The bus has an inductance of 2.5 nH/cm, a capacitance of 1 pF/cm, and a length of 50 cm. What is the reflection coefficient of the far end of the bus and what is the end-to-end propagation time for the bus?

30.   What are the principal differences between the VMEbus and the NuBus?

31.   Compare and contrast the ways in which the VMEbus and NuBus carry out arbitration.

DESIGNING A MICROCOMPUTER SYSTEM

We are now going to apply some of the lessons learned in earlier chapters and design a modest 68000-based microcomputer. We also look at the design of a 68030 system. This chapter is divided into three parts:

1. An examination of some of the ways in which microcomputers are designed in order to make their testing relatively easy. A brief discussion then follows of the equipment and techniques commonly used in debugging digital systems.

2. The design of a 68000-based microcomputer with an expansion bus.

3. The design of a monitor and loader that permits information to be transferred to the 68000 system from a host computer. This monitor is written in assembly language and includes the routines necessary to drive a serial interface, to convert between ASCII strings and numeric quantities, and to interpret commands input from the console.

11.1 DESIGNING FOR RELIABILITY AND TESTABILITY

Once upon a time, an engineer designed a system to work, and to work as well as possible within its economic limitations. Design engineers were very important people and lived in castles (or at least mansions). Sometimes, due to faulty components or to the general perversity of nature, the apparatus that the designer had created stopped working. When this happened, the repairperson or troubleshooter was called in to pinpoint the source of the problem and to repair it. Unlike the designer, the troubleshooter lived in a cottage in the grounds of the castle, if he or she was lucky, or in a hut if he or she was not.

We now live in the age of realism and have banished such fairy stories. Design is no longer the only major factor in the production of complex digital equipment. Today, the cost of debugging and testing a microcomputer can be more than designing or building it in the first place. In other words, we can see little point in designing a system if it is almost impossible to test when it fails.

The designer and the test engineer now have to form an equal partnership and work together. The designer must produce a system with testability as one of its main criteria. To make this situation possible, designers must understand the limitations of the components they use, know their failure modes, and include facilities to help the test engineer to pinpoint the source of failure.

Before we can consider how to design testable equipment and examine fault-finding procedures, we need to look at some of the reasons why equipment fails. These reasons are as follows:

1. *The blunder.* A blunder is an act of folly that could have been avoided; for example, a designer may specify an OR gate instead of a NOR gate or may connect two or more gates with totem-pole outputs to the DTACK* line. Such blunders should not occur and are entirely due to human error. Blunders can be eliminated by double-checking circuits and systems before they reach production.

Sometimes blunders can be discovered at the prototype, or "breadboard," stage or when the system is emulated in software. The latter approach is popular with the designer of digital circuits, and software packages are available for the emulation of digital systems.

2. *The subtlety.* A subtlety is a sort of "gentle blunder" and is a human design error that does not stick out like the blunder. A subtlety may be missed when the circuit is double-checked or when it is emulated in software. A typical subtlety is a timing error, which appears when a system is expanded by, say, the addition of a memory module. Without the module, the system works and passes its initial tests. With the module, the system fails because the additional signal delays through bus transmitters and receivers cannot be tolerated.

3. *The "stuck-at" fault.* A logic element may fail internally so that its output becomes independent of its input and remains either stuck-at-one or stuck-at-zero. A stuck-at fault is relatively easy to locate, provided that the inputs of the logic elements can be modified to permit outputs to toggle between states. We will return to this topic when we consider the design of testable systems. Note that a stuck-at fault may also be due to a short circuit between adjacent tracks of a printed circuit.

4. *The faulty IC.* Sometimes an IC is faulty in some other way than the stuck-at fault described previously. This fault can always be detected or cleared by substituting the suspected device with another of the same type that is known to function. Unfortunately, we are no longer able to test all ICs before they are used as the number of internal states in a complex device is so great. Moreover, a faulty IC sometimes produces spectacularly obscure faults. For example, a microprocessor may execute an instruction correctly unless it is preceded by a particular instruction that causes it to fail; such a fault might be discovered only after the device has been in production for several months.

5. *The faulty PCB.* Many faults are caused by defects in the mechanical components of a digital system. In particular, the tracks of a printed circuit board may be shorted together or left open circuit because of a break in a trace. The worst type of fault is a blob of solder under an integrated circuit socket that short-circuits two pins but cannot be detected by optical inspection. These defects often produce symptoms identical to the "stuck-at" fault already mentioned.

6. *The intermittent fault.* The intermittent fault is one of the nastiest of faults because it is difficult to find, apparently inexplicable, and seemingly malicious. The origin of this fault may lie in any of the mechanisms already mentioned, although intermittent faults are most closely associated with mechanical problems. The intermittent fault is there one minute and gone the next, making it very difficult to trace. I have heard of engineers dealing with particularly frustrating intermittent faults by placing the board on the floor, jumping on it, and then reporting the defect as "too extensive to warrant further investigation."

Testability

Influence of Testability at the Design Stage

Testing a system is like detective work—the source of the fault has to be deduced from the evidence it leaves behind. The test engineer is, however, better off than the detective. A system can be designed in such a way that a fault either automatically "points to itself" or is made to "show up" (with a little prompting). We will soon see how this objective is achieved.

Two things are necessary to achieve testability: the ability to monitor activity within a system and to influence this activity. Figure 11.1 illustrates both these points with a simple circuit, a logic module driven by an on-card clock.

In figure 11.1a, the clock is connected directly to the system on the card. No convenient way exists to observe the action of the clock or to determine whether it is working according to its specification. At best, a probe can be attached to some pin carrying the clock signal.

In figure 11.1b, the clock is buffered and brought off the card via one of the pins on the PCB. We can now use external equipment to monitor the state of the clock without disturbing (i.e., loading) the system on the PCB. Note that two prices have been paid: an economic price (i.e., the cost of the additional buffer and pin on the connector) and a reduction in overall reliability (an extra component is needed on the board).

In figure 11.1c, not only has the buffered clock been brought off the board but two inputs have been provided to modify the operation of the system. The clock is gated through an AND gate and a control signal, local_clock_inhibit*, can be forced low to inhibit the on-card clock, effectively disconnecting it from the logic system. An external clock may be injected into the clock_in terminal once the local clock has been inhibited, thus permitting the system to be synchronized with an external clock that may be slowed or stopped for test purposes.

FIGURE II.I Example of a testable circuit

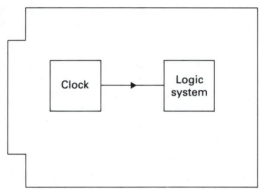

(a) Clock generator connected directly to the logic circuit

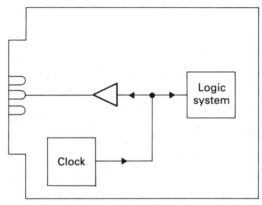

(b) Buffered clock brought off-card

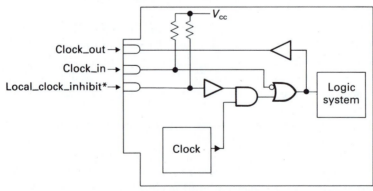

(c) On-card clock with facility for external clock input

FIGURE 11.2 Increasing testability by making control signals accessible

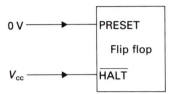

(a) Control signal permanently connected to a logical zero or a one

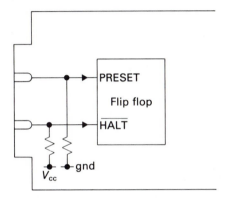

(b) Control signals brought out to pins on a connector for testing

Figure 11.2 illustrates another path to enhanced testability. In figure 11.2a, the two control signals, PRESET and HALT*, are permanently strapped to V_{cc} or ground, as appropriate, because their particular control functions are not required by the system. By using pull-up or pull-down resistors, these control lines can be brought out to test pins and employed to test the system by operating it in special modes (e.g., halted). This procedure can be extended to permit the injection of other stimuli and goes a long way to making stuck-at faults easier to detect.

In general, a digital system should be designed so that as many internal signals as possible (address, data, control) are available for examination off-card. Sometimes an additional connector can be added to the card and used solely for test purposes. Similarly, test inputs for the generation of control stimuli should also be provided. Apart from making testing digital equipment easier, such provisions facilitate automatic testing. In an automatic testing system, a digital computer applies preprogrammed signals via the test inputs and computer-controlled test equipment monitors the response at the test outputs.

Testability and Feedback Paths

The greatest obstacle to the effective testability of a computer system is its closed-loop nature. Consider first the open-loop circuit of figure 11.3a. An input is successively operated on by a number of processes to yield an output. At no stage does

FIGURE II.3 Open-loop and closed-loop systems

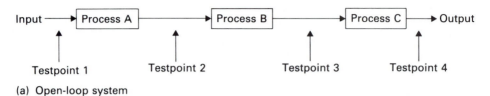

(a) Open-loop system

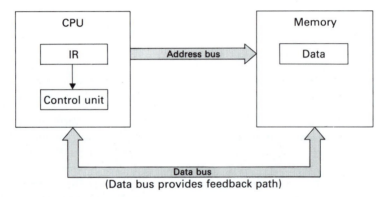

(b) Closed-loop system

(c) Computer as a closed-loop system

any feedback path exist between the output and input—hence the term *open-loop*. To debug such a system, a known signal is injected into the input and traced through the various processes. If the input to a process is known and the nature of the process is known, the output of a process can be calculated. Whenever the measured output of a process differs from its expected value, we can confidently say that the process is faulty. Televisions, radios, and hifi equipment are largely examples of open-loop systems (although even these use closed-loop techniques in some circuits).

Consider now the closed-loop system of figure 11.3b. The input is combined with some function of the output and applied to process A. The output of process A is fed to processes B and C, and the output of process C fed back to process A. Such a system is called a *closed-loop* system because information circulates round the loop, from output to input. Localizing the source of a fault is now very difficult. The effects of a fault propagate through the system, are fed back to the input, and flow round the circuit. Thus, the effect of a fault at one point may be detected at an entirely different point—even "upstream" of the actual fault.

Figure 11.3c shows the computer as a closed-loop system. The computer calculates an address internally and uses it to read the next instruction from memory. The op-code flows along the data bus to the computer, where it is interpreted. The result of this interpretation eventually leads to the generation of a new address and the sequence repeats ad infinitum.

Suppose the contents of a memory cell have been corrupted. The data read from the cell may cause the address of the next instruction to differ from that intended, and, say, a spurious jump made, leading to unpredictable behavior. Because all aspects of the behavior of the system appear faulty, we find great difficulty in pinpointing the source of the failure; for example, all the faults below could have produced the same observed effect:

1. The address from the computer is wrong because of a fault in the CPU. Equally, the CPU may be functionally perfect but its support circuitry (reset, clock, interrupt, etc.) may be faulty.

2. The address at the memory is wrong due to a fault on the address bus. Included here are the effects of errors in the address decoding circuitry.

3. The data on the data bus is wrong due to a fault in the memory (the actual fault in our example).

4. The data on the data bus is wrong due to a fault on the data bus. Data bus errors also arise from faults in the data bus buffers and their associated control.

There are two ways of dealing with this problem. One method is to monitor the operation of the system at several points over a period of time and then analyze its behavior to determine the cause of the fault. Complex and expensive test equipment (the logic analyzer) is involved, a subject dealt with later in this chapter. The other method is to break the feedback path. Without feedback, the digital system can be tested by the same techniques available to the television repairperson.

Good Circuit Design

In a book on microprocessor systems we are not able to become involved with the detailed design of digital circuits. However, some of the points that the engineer should bear in mind are worth listing when designing a circuit:

1. *Maximize access to the inputs and outputs.* Whenever possible, the inputs and outputs of circuits and submodules within the system under test should be accessible; that is, test points and test input paths should be liberally provided in the system. This is particularly true of systems with complex integrated circuits with inaccessible internal test points. The next best thing is to provide facilities to monitor the input and output pins of complex ICs.

2. *Adopt a modular approach to systems design.* As we have already seen, when we decompose a system into a number of subsystems, we can easily test the whole system by testing the subsystems one by one. Note that this approach

generally requires that feedback paths be broken for the duration of the test. Sometimes modularity conflicts with economics. We may find expediency in using several components, spread throughout the card, to carry out a given logical operation; for example, the reset function (power-on-reset, manual reset, etc.) may be implemented by using a gate here, a gate there. In other words, unused gates at various points in the system are combined to provide a reset circuit for very little additional cost. Although this procedure may be good from an economic point of view, it runs counter to the principle expressed previously. Using dedicated circuits to perform a single task makes it easier to test the circuit in isolation. Sometimes the function can be carried out by test hardware external to the system under test.

3. *Avoid asynchronous logic.* Asynchronous circuits are arranged so that the output of one element triggers the next element, and so on. An asynchronous circuit does not have a global clock to determine the instant when each element changes state. Because the behavior of an asynchronous circuit may change if the signal delay incurred by an element is greater or less than that expected and because the asynchronous circuit is prone to race conditions, this type of circuit is not popular with some designers. Therefore, designers should, wherever possible, choose fully clocked synchronous circuits.

4. *Avoid monostables.* The monostable is a classic digital circuit that generates a pulse of fixed duration whenever it is triggered. A resistor-capacitor network (figure 11.4a) determines the duration of the pulse. The monostable is not popular with either the test engineer or the designer. The test engineer is unhappy because observation of the output of a monostable is often difficult—especially if the pulse is very short. The designer does not like the monostable because it is inflexible (the timing delay is determined by analog components) and is also prone to trigger from noise on the power lines or other spurious inputs. An alternative to the conventional monostable is the purely digital circuit of figure 11.4b. When the RS flip-flop is set, the AND gate is enabled and the counter counts up. When the output of the counter reaches a preset value, the output of the comparator resets the flip-flop and the counter is cleared. The advantage of this circuit is its absence of analog components and its flexibility—the pulse width is modified by changing the clock rate or by reprogramming the counter. These facilities are very useful in testing the system.

5. *Place "important" devices in sockets.* Some designers believe that important (or even *all*) digital elements should be plugged into sockets rather than solder-ed directly to the board. Testing ICs becomes very easy as they can be unplugged and tested off-card. This approach is very controversial, because integrated circuit sockets are relatively unreliable and intermittent faults due to poor pin-socket contacts are common. However, without sockets, the testing of integrated circuits by substitution is a rather difficult task. Unsoldering ICs often damages PC boards.

6. *Do not use marginal design.* Never design a system that operates outside its guaranteed parameters. We all know that a gate with a fanout of 10 will drive 12 loads because worst-case parameters are just that. In "normal" operation

FIGURE 11.4 Monostable

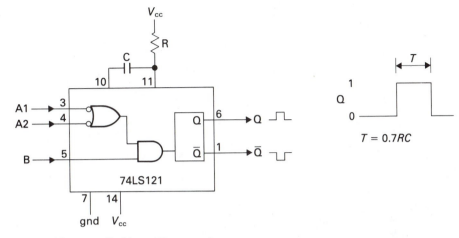

(a) Monostable controlled by a CR network

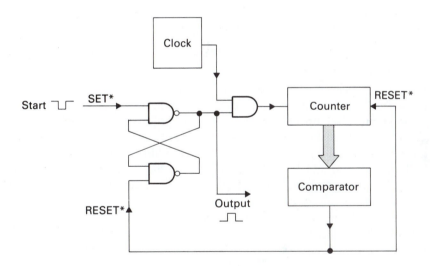

(b) Monostable controlled entirely by digital components

these parameters can be exceeded, but no designer should ever rely on this fact. Cases have been reported where manufacturers have had a working model on the test bed and then gone into full-scale production, with disastrous results.

Static Testing

Static testing is the name given to tests on a digital system that break the feedback loop between input and output. Basically, in a static test the result of one operation is not permitted to lead to the next operation; that is, only one cycle at a time is permitted.

FIGURE II.5 Static emulator

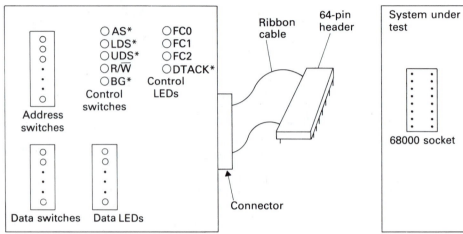

Emulator control panel

A static tester replaces the CPU by a *CPU emulator* and is illustrated in figure 11.5. The CPU emulator mimics certain aspects of a CPU. A test circuit has a number of switches enabling address, data, and control signals to be set up manually. From this module, a 64-pin header at the end of a length of ribbon cable is plugged into the 68000 socket of the system under test.

The circuit details of a possible static emulator are given in figure 11.6. Address information is set up on 13 switches and applied to the inputs of 23 tristate bus drivers. When enabled, these bus drivers place a user-selectable address onto the address bus of the system under test. As this address is static, we are able to examine address lines and address decoder outputs on the module under test with an oscilloscope or even a voltmeter. Of course, static emulation is not very helpful in debugging dynamic memory circuits!

In a similar way, control lines AS*, UDS*, LDS*, BG*, FC0 to FC2 can all be set up from switches on the emulator. Note that control lines should all be debounced by RS flip-flops, as shown in figure 11.6. The emulator must be able to both provide a source of data to emulate a write cycle and to read data from memory to emulate a read cycle. In figure 11.6, 16 bus drivers place data from switches on the data bus whenever $R/\overline{W} = 0$ and another 16 bus receivers drive LEDs to indicate the state of the data bus.

To use the static emulator in the read mode, the appropriate address is set up, $R/\overline{W}$ set to a logical 1, and AS* plus UDS* or LDS* set to a logical 0. The contents of the data bus are then displayed on the data LEDs. As the DTACK* pulse from the system under test may be very short, a latch is necessary to catch its leading edge.

If the system has a ROM with known data, we are able to test the address and data buses by applying an address and examining the resulting data. Should this data not be the one expected, the address and data paths can be traced until the fault has been located.

FIGURE 11.6 Circuit diagram of a simple static emulator

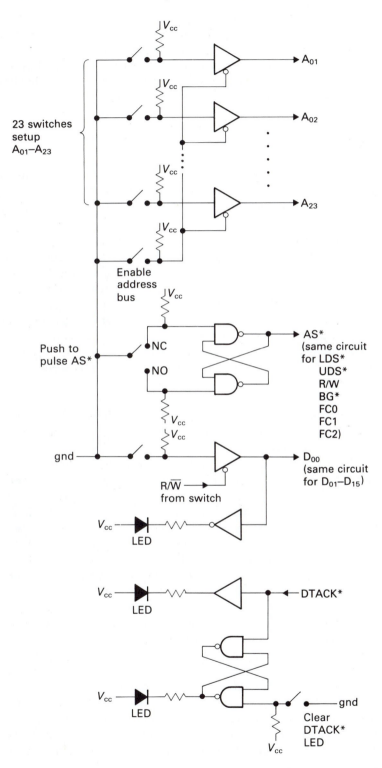

It would be wrong to suggest that this type of static emulation is anything other than a relatively crude error detector/locator; for example, the static emulator might not detect two address lines shorted together if the test address includes both the faulty address lines at the same level. However, such a circuit should prove quite effective for use in university and polytechnic laboratories.

Logic Analyzer

The logic analyzer continues from where the static emulator left off. The analyzer examines and displays the operation of a digitial system dynamically and in realtime. In principle, a logic analyzer is a digital-domain oscilloscope. An oscilloscope displays one, two (or up to about eight) analog signals on a CRT as a function of time. Most oscilloscopes are able to display only a periodic waveform.

Figure 11.7 illustrates the basic principles of a logic analyzer, which operates in one of two modes: an acquisition mode and a display mode. In the acquisition mode (figure 11.7a), a number of channels from the system under test are sampled and the samples stored in consecutive locations in the analyzer's memory. The samples are digital quantities and have the logical values 0 or 1. This fact is important, because signals at a test point at an indeterminate level are always recorded as a logical 0 or a logical 1 by the analyzer. The analyzer is triggered by a start signal and stops on receipt of a stop signal. During this time, the counter counts successively upward as each sample is taken and stored.

In the display mode (figure 11.7b), the counter free-runs and periodically steps through each memory location containing data collected during the acquisition phase. This data is fed to the inputs of a multichannel oscilloscope and displayed as a series of traces, one for each channel.

From the preceding comments, we can clearly see that the logic analyzer provides a snapshot of the state of the system under test over a period of time. If sufficient channels exist, we are able to sit down and analyze the activity on the buses and therefore to determine whether or not the system is functioning correctly. From the observed data, the cause of a fault can frequently be localized. Note that the logic analyzer can debug both hardware and software, and that most logic analyzers are able to deal with asynchronous events such as interrupts.

Many logic analyzers offer a number of display options. Figure 11.8 shows four popular display formats. Figure 11.8a and b illustrates the waveform and binary modes, respectively. The waveform is reconstructed from the digital data stored in the analyzer memory. Unlike the oscilloscope, these waveforms are purely digital and are *idealized* versions of the waveforms from the actual system under test. Because the original signals are sampled periodically, timing relationships cannot *accurately* be measured from the logic analyzer display. The binary mode displays the data as a table (octal and hexadecimal modes are also common).

Figure 11.8c illustrates the disassembly mode, in which the information from the system under test comes from the processor's address and data buses and is displayed in mnemonic form. In order to do this, the logic analyzer must include a "personality module" to disassemble the code of the particular microprocessor under

FIGURE 11.7 Logic analyzer

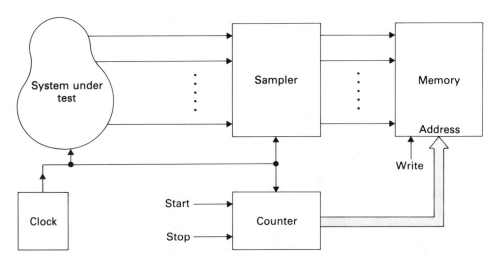

(a) Signal acquisition mode

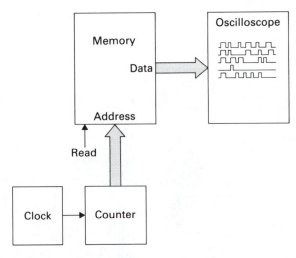

(b) Signal display mode

test. The display mode is very effective in software debugging—particularly for real-time and interrupt-driven systems.

Figure 11.8d illustrates the rather curious looking point-plotting display. An n-channel logic analyzer plots each n-bit sample as a single point on the screen, by dividing the screen into 2^n points. Therefore, as the digital system changes from one state to another, a sequence of points is displayed on the screen. Interpreting such a display is an art form! Some people would say that reading the display is comparable to reading the future in tea leaves.

FIGURE 11.8 Logic analyzer display formats

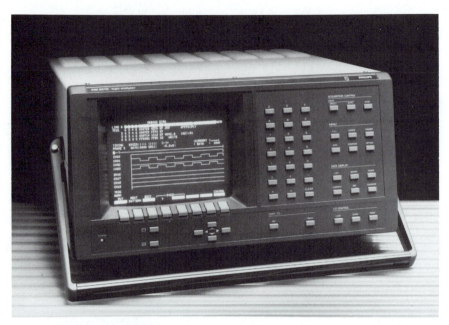

(a) The waveform display mode

(b) The binary display mode

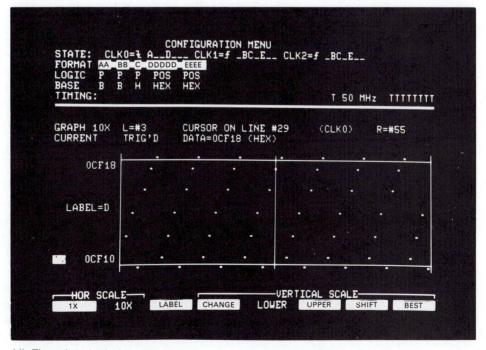

```
                      STATE TRIGGER MENU
   W1  0  X X X X 0000XX XXXX


      FINAL DELAY = 100    STATES OF CLK 0

   M68000 DISA CURRENT                                          TRIG
   LINE # I H M S  ADDRES  CODE  FC        MNEMONIC    . EXCEPTION ASYN
   TRIG    4 3 4 5 00006C 0000 SD                    LEVEL 3 INTERRUPT
       1   4 3 4 5 00006E 0500 SD
       2   4 3 4 6 000500 103C SP  MOVE.B   #$00FF,D0
       3   4 3 4 6 000502 00FF SP

       4   4 3 4 6 000504 4281 SP  CLR.L    D1
       5   4 3 4 6 000506 4DF9 SP  LEA.L    $00008010,A6
       6   4 3 4 6 000508 0000 SP
       7   4 3 4 6 00050A 8010 SP
       8   4 3 4 6 00050C 49F9 SP  LEA.L    $00008200,A4

       9   4 3 4 6 00050E 0000 SP
      10   4 3 4 6 000510 8200 SP
      11   4 3 4 6 000512 3CC0 SP  MOVE.W   D0,(A6)+
      13   4 3 0 5 008010 FFFF SD  WRITE

   DISPLAY   SYNC    SELECT  ┌────────DISPLAY POSITION────────┐
    FORMAT   MANUAL   STATE    CURSOR   FIND   NEXT   PAGE+   PAGE-
```

(c) The disassembly display mode

```
                   CONFIGURATION MENU
   STATE:   CLK0=↧ A__D___ CLK1=ƒ _BC_E__ CLK2=ƒ _BC_E__
   FORMAT  AA  BB  C  DDDDD  EEEE
   LOGIC   P   P   P  POS  POS
   BASE    B   B   H  HEX  HEX
   TIMING:                               T 50 MHz  TTTTTTTT

   GRAPH 10X   L=#3       CURSOR ON LINE #29    (CLK0)   R=#55
   CURRENT     TRIG'D     DATA=0CF18 (HEX)

   0CF18


   LABEL=D


   0CF10

   ┌─── HOR SCALE ───┐              ┌────── VERTICAL SCALE ──────┐
     1X     10X    LABEL   CHANGE   LOWER   UPPER   SHIFT   BEST
```

(d) The point-plotting display mode

Many logic analyzers can operate in a so-called state mode. In the state mode, the logic analyzer takes its clock from the system under investigation and is therefore able to operate in synchronism with the system. For example, a state mode logic analyzer investigating the 68000 would use the 68000's own clock to capture the contents of the address, data, and control buses as instructions are executed. By capturing the processors' address, data, and status signals, it is possible to display the current state of the processor and its past history on a screen. You can frequently display information in mnemonic form by disassembling the instructions. By doing this you can investigate the code that the processor actually executes (the listing gives you the code you think that the processor is executing). Moreover, the state mode is very useful in investigating the relationship between the program and external events such as interrupts. You can see how the processor responds to an actual interrupt.

Logic Analyzer Characteristics

Before we look at how logic analyzers help us to debug microprocessor systems, we need to think about their characteristics and limitations. Possibly the key part of a logic analyzer is its signal-acquisition circuit, which is shown in block diagram form in figure 11.9. The input signals, typically 16 to 64 or more channels, are applied to analog signal comparators that generate a logical one or zero output depending on whether the signal is above or below some threshold. This threshold may be switch selectable to suit TTL or ECL logic levels, or continuously variable to permit the acquisition of an arbitrary binary signal in the range (typically) -3 to $+12$ V.

Because the input goes through a comparator, it is always interpreted as a true or false level. Therefore, a logic analyzer cannot readily be used to detect faults due to incorrect signal levels. That procedure is the province of the oscilloscope.

The outputs of the comparators are then captured by a latch. Figure 11.10 provides a timing diagram of a synchronously clocked logic analyzer operating from, say, the system clock. To keep things simple, only two channels of input are shown. Three points are noted from this diagram:

1. At point A, the signals on channels 1 and 2 change state. The change is not displayed until the signals have been sampled by the falling edge of the clock. Therefore, the logic analyzer does not record input changes instantaneously.

2. At point B, channel 1 makes a negative transition before channel 2. As both channels are not sampled until the next falling edge of the clock, the displayed data shows them making a negative transition simultaneously; that is, the relative delay between traces is not preserved.

3. At point C, a short pulse, or glitch, occurs on channel 2. Because this pulse falls between two successive sampling clocks, it is not recorded and does not appear on the display. Therefore, short-term events that play havoc with the system under test may go unnoticed when subjected to investigation by a logic analyzer.

FIGURE 11.9 Signal acquisition circuit of a logic analyzer

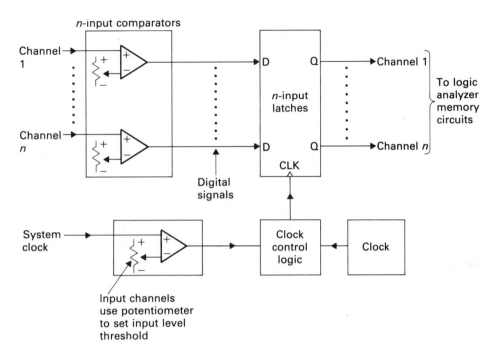

FIGURE 11.10 Effect of sampling a signal synchronously

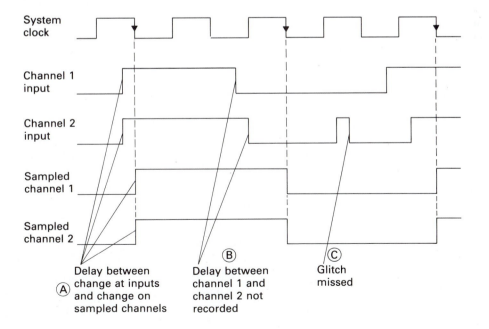

The above points present a more gloomy picture than is actually the case. Remember that a real microprocessor-based system is itself clocked synchronously. Moreover, the setup and hold times of the logic analyzer are likely to be smaller than those of many components in a microprocessor system. However, if a glitch does occur in a microprocessor system, *some* logic analyzers will most probably miss it.

Logic analyzers with asynchronously clocked data-acquisition latches are also available. Here the latch is triggered by the analyzer's own clock. The phase relationship between channels becomes easier to observe, as the number of sampling clock pulses per system clock is increased. The sampling clock frequency should be at least four times the system clock frequency and a factor of 10 to 20 is not uncommon. The use of such a high clock ratio also makes the capture of glitches much easier.

Figure 11.11 illustrates the advantages of asynchronous clocking.

Some logic analyzers have a special glitch-detection feature. As a glitch is missed entirely if it falls between two sampling points, a latch can be used to detect a glitch by applying the channel input to its clock. A glitch triggers the latch and is displayed when the analyzer enters its playback mode. Generally speaking, some logic analyzers are dedicated almost exclusively to detecting hardware faults while other analyzers are aimed more at debugging software errors. The engineer should be aware of this difference when buying a logic analyzer.

FIGURE 11.11 Effect of sampling a signal asynchronously

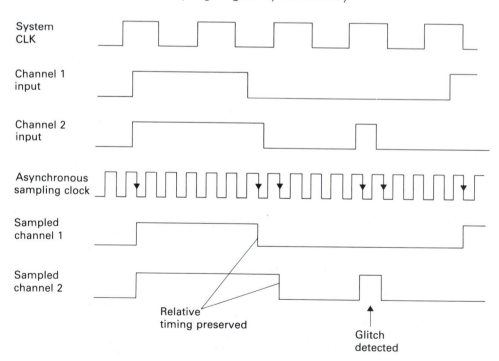

Triggering

Modern microprocessor systems, running at clock rates of 8 MHz and with 24-bit address buses, 16-bit data buses, and 10-bit (or more) control buses, generate about 400,000,000 bits of information per second. Clearly, no reasonably priced logic analyzer can record such a data flow for more than a few microseconds. A method of starting the recording process and terminating it at suitable times must be found.

Logic analyzers are invariably triggered by an *event* that is defined by the operator. The event corresponds to some pattern of data on the input channels and/or a particular data pattern at the *qualifier inputs*. A logic analyzer has inputs (qualifier inputs) that are employed by the analyzer to trigger the recording of data, but are not themselves displayed. We should note that a signal employed for the purposes of triggering the display may be defined as 1, 0, or X (i.e., don't care); for example, an address in the range $4500 to $450F may be used as a trigger by selecting the trigger to be 0100 0101 0000 XXXX. Early logic analyzers had their trigger conditions entered from front panel switches. Modern devices use a keyboard or a keypad to enter the trigger conditions.

The analyzer may also be triggered *before* the specified event. This apparent exercise in time travel is achieved by letting the analyzer record input data freely; that is, at any instant the analyzer's buffer contains the last N samples recorded, where N is the length of the buffer. The input is stored in a circular buffer and, once the buffer is full, new data overwrites the oldest data in the buffer. The recording is halted by the trigger, at which point the analyzer memory contains a record of the data flow from the system under test leading up to the trigger event.

Signature Analyzer

Signature analysis offers a quick, effective, low-cost troubleshooting technique and is aimed at the manufacturer of digital systems who produces large numbers of similar systems rather than at the engineer who builds a single prototype. A logic analyzer produces a snapshot of the operation of a system, which can later be examined to deduce the cause of a fault. Clearly, complex equipment and highly trained personnel are required. Signature analysis simply gives the equipment under test a task to perform and then offers a "go/no-go" analysis of the results.

At any point, or node, in a digital circuit, the signal level switches between logic states as information passes through the node. If the equipment executes a given task, the sequence of pulses observed at the node is the same each time the task is carried out. The node may be an address, data, or control line or an intermediate point such as a chip select.

The signal analysis condenses the complex sequences of pulses at a node into a single *fingerprint*, or *signature*, which is similar to the cyclic redundancy code found in data transmission systems.

Figure 11.12 illustrates the principle and the great simplicity of the signature analyzer. A 16-bit shift register is clocked by the system under test. The data moved

FIGURE 11.12 Principle of the signature analyzer

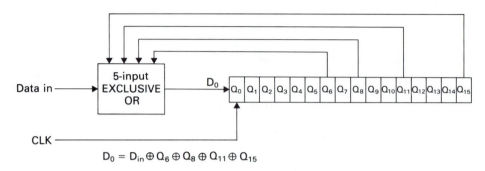

$$D_0 = D_{in} \oplus Q_6 \oplus Q_8 \oplus Q_{11} \oplus Q_{15}$$

into the least significant bit of the shift register is the EXCLUSIVE OR of the data input from the system under test and four outputs from the shift register; that is,

$$D_0 = D_{in} \oplus Q_6 \oplus Q_8 \oplus Q_{11} \oplus Q_{15}$$

Sixteen stages are used because a 16-bit shift register detects a multibit error in the data stream with a probability of 99.998 percent and a single-bit error with a probability of 100 percent. Fewer stages do not give such a large probability of detecting an error and more stages increase the cost of the signature analyzer.

Figure 11.13 provides an idea of the circuitry of a signature analyzer (albeit a highly simplified circuit). Four inputs to the analyzer are required: a clock, a data input, a start, and a stop. To use the analyzer, the four leads are attached to the

FIGURE 11.13 Highly simplified circuit of a signature analyzer

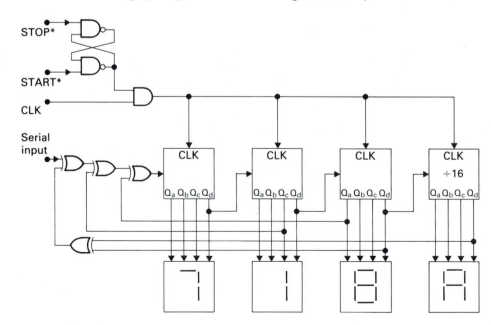

system under test. Suppose that the start signal makes an active transition. Data is clocked into the shift register until the stop signal is asserted. The displayed signature then represents the outcome of the test.

The signature is totally arbitrary and is not amenable to analysis. The operator compares the measured signature with that recorded in the troubleshooting manual for the system under test. If the signatures are the same, the equipment has passed the test. If they differ, it has failed.

In order to apply signature analysis usefully, the system under test should be placed in some cyclical free-running mode; for example, the operating system ROM may be replaced by a test ROM that executes a software loop. The start and stop signals are obtained from suitable points within the system under test; for example, A_{16} may be toggled after every hundredth cycle of the loop to generate a start signal. The data input may come from any point in the system that is not at a static signal level. For each node at which the probe is placed, the expected signature is recorded in the system manual—just as voltages are provided at test points in analog circuits.

As the system is operating in a cyclic mode, the signature is constantly updated—each new signature being the same as the previous signature. If the signature is unstable, an intermittent fault or a problem caused by asynchronous events (e.g., interrupts, DMA, DRAM refresh cycles) interfering with the test may result. Sometimes an unstable signature is caused by the choice of unsuitable clock, data, start, or stop signals. Clearly, the data input should make its transitions when the clock is stable.

Although the preceding method of forcing the system under test into a repeatable, known, cyclic mode involves the substitution of a test ROM, other techniques are available. It is possible to break the feedback path within a digital system by disconnecting the CPU's data inputs from the data bus by means of some test fixture. Then the data lines can be pulled up or down to force the CPU to execute an infinite series of NOPs or some other operation code. As with the test ROM, the signature is measured at various points in the circuit and compared with the published values.

Signature analysis is obviously suited to production line testing, where fault finding is turned into a flowchart procedure that test staff can learn in a very short time.

Microprocessor Development System

The microprocessor development system, MDS, is designed to facilitate both the design and production of a microprocessor system and to debug it. Inside the MDS lies a general-purpose digital computer, frequently (but now always) based on the microprocessor in the system to be developed. The general-purpose computer is disk based and offers comprehensive software development facilities. Combined with this computer is a built-in logic analyzer to test the hardware of the system under development. Unlike all other test equipment, the MDS can be used to debug a system from its paper implementation phase to its production-line testing.

Software Development

The software development system runs under an operating system, which may be a proprietary operating system such as UNIX or may originate from the manufacturer of the MDS. Software running under the operating system normally includes an editor, assembler, linker, emulator, and possibly one or more compilers.

As a microcomputer consists of two fundamental components (its hardware and its software) and one component is useless without the other, a microprocessor system is inherently difficult to develop. The MDS solves this dilemma by providing a framework within which the software can be constructed and debugged, entirely independently of the target system on which it will eventually run.

Once the hardware environment of the target system has been specified, the software development can begin. The MDS offers an editor and an assembler so that the necessary object code can be produced. Alternatively, a compiler may be provided that generates the object code for the appropriate CPU in the system being developed. If this system were all an MDS offered in the way of software development, it would hardly be worth the large price tag attached to it.

The MDS is also able to run the software under its emulation mode. Usually three levels of emulation are offered—levels 0, 1, and 2. In level 0, the target hardware is not available and the software runs entirely on the MDS system. All the usual debug-package software tools are supplied and we are able to examine and modify memory locations, to insert breakpoints, and to trace through the program; for example, keyboard input can be simulated as a file that returns a character wherever it is interrogated.

Under level 1 emulation, the target hardware is present and an emulation probe from the MDS is inserted in the socket of the CPU in the target hardware. The actual CPU is, of course, in the MDS. Partitioning the processor's memory space is possible between the target and MDS system.

FIGURE 11.14   Partitioning the memory space between the CPU and the MDS

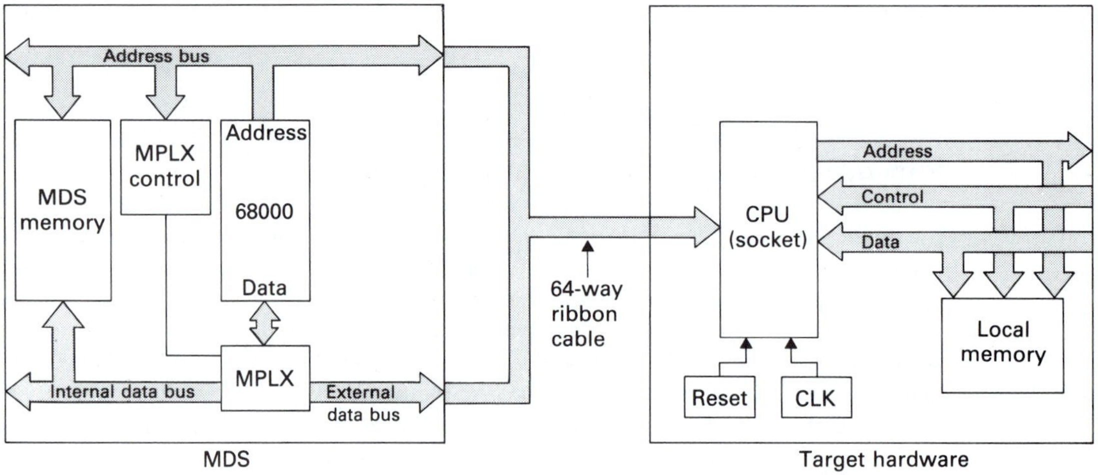

Figure 11.14 shows how the memory is partitioned between the MDS and target hardware. The 68000 itself is part of the MDS hardware. Its address bus is connected to both the MDS and the external target hardware. Consequently, the same location is accessed in both systems. The data bus and associated control signals are multiplexed between the MDS memory and the target hardware. Whenever the CPU generates an address, it is applied to a mapping table to determine whether that address belongs in the MDS or the target hardware. The output of the mapping table controls the multiplexer and routes signals between the CPU and the MDS or between the CPU and the target hardware.

During the software emulator initialization phase, the address table is set up by the programmer when he or she allocates address space to the MDS or to the target hardware. Suppose that the software has been successfully debugged and that the target hardware contains a serial I/O port. We are now able to assign all memory to the MDS and map only the serial interface address space to the target hardware. In this state, only the I/O port together with its address and data paths on the target system are being tested. The I/O port can, of course, be used normally even though its associated ROM and RAM are all in the MDS.

If the preceding test works, more features of the target hardware can be mapped onto the CPU address space—including the RAM. At this stage, all peripherals are operating in the target hardware rather than being emulated in the MDS and all data is located in the target's own memory. All the MDS's debug features can still be used.

By now, only the CPU and the ROM portion of the target system are in the MDS. The last stage in the development process is to transfer the program developed on the MDS to EPROM or even to mask-programmed ROM. When this stage is completed and the EPROM plugged into the target hardware, the MDS runs in emulation mode 2.

In emulation mode 2, the MDS monitors the operation of the target hardware. Indeed, we can safely say that the MDS is now operating as a logic analyzer. MDS offering this facility have additional channel inputs (usually as an option to the basic model) that can be connected to various points of the target system to permit the usual logic analyzer triggering modes.

The MDS was one of the most expensive and complex pieces of test equipment available (neglecting the computer-controlled automatic test station found on a production line) and was intended for use by system design and development engineers who follow the design of equipment from its original concept to prototype. Today a basic MDS can be bought for as little as $5000.

11.2 DESIGN EXAMPLE USING THE 68000

In this section we examine the design of a modest microcomputer based on the 68000 CPU. Before we consider the design of this computer, called TS2, we need to provide it with a specification.

Specification of the TS2

1. The TS2 uses a 68000 CPU. (I'd get funny looks if I used an 8086 in a book about the 68000.)

2. The TS2 is built on an extended, double Eurocard (233.4 × 220 mm), which provides ample room for the CPU, bus control, local memory, and interface circuitry.

3. The CPU card is capable of operating on its own. System testing is thus facilitated because other modules are not required to operate the CPU card in a stand-alone mode.

4. An external bus is used to connect other modules to the CPU card. An interface between the 68000 on the CPU module and the backplane is essential.

5. The VMEbus itself is not used, as the full functionality of the VMEbus is not necessary. Therefore, a relatively low cost backplane can be implemented.

6. The memory on the CPU card is static RAM and EPROM to avoid the difficulty in debugging the CPU system together with its DRAM. This author does not have a microprocessor development system that would permit the debugging of DRAM *independently* of the CPU and its software.

7. Full seven-level interrupt facilities are provided.

8. No on-card facility limits the capability of the system.

9. Full address decoding is provided. The address space is to be compatible with the Motorola MEX68KECB (ECB for short) development system in order to facilitate the transport of software between TS2 and the ECB.

10. The vector number table at $00 0000 to $00 03FF is implemented in RAM. Therefore, the reset vectors are overlaid as described in chapter 6.

11. The RAM is implemented by 8K × 8 CMOS devices to minimize the component count.

12. The ROM is implemented by 2764 type 8K × 8 EPROMs.

13. The terminal (console) interface is through a serial port. A secondary serial port is also provided. Configuration is to be the same as in the ECB development system.

The block diagram of the arrangement of a single-board computer satisfying the design criteria is given in figure 11.15, which is a very general diagram and serves as a "checklist" for the various parts of the system to be elaborated. Only one design decision has been taken at this stage. For the sake of simplicity, the module's local address and data buses have not been buffered. This fact implies that care should be taken not to load the local address and data buses too heavily.

FIGURE 11.15 Block diagram of a single-board computer

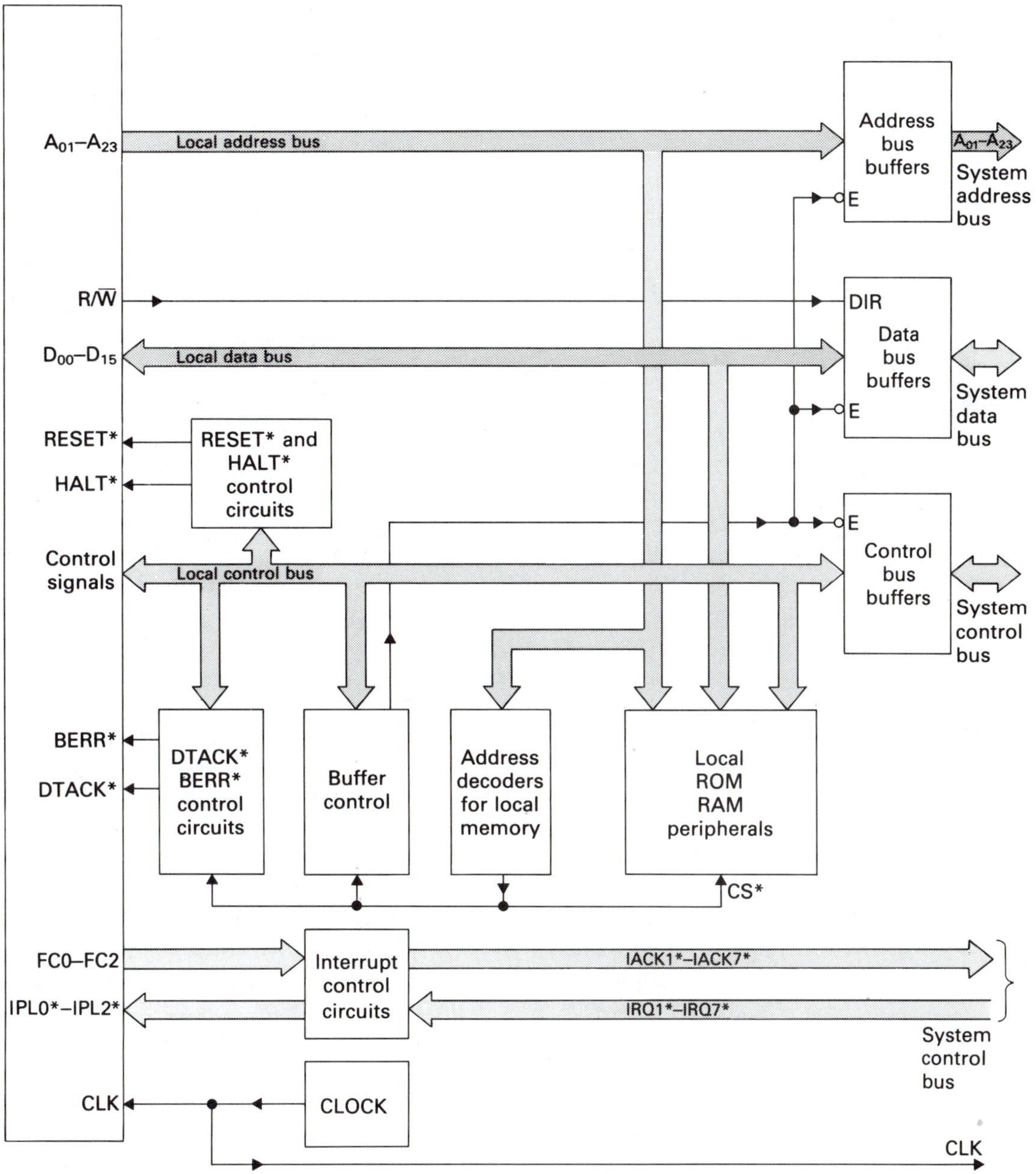

Basic CPU Control Circuitry

Every CPU requires a certain amount of basic control circuitry to enable it to operate—this circuitry includes its clock, reset, halt, and similar functions. Such circuitry can be designed largely independently of the rest of the system and is needed to perform even the simplest tests on the CPU. Therefore, we will design these circuits first.

Figure 11.16 gives the diagram of the control circuitry surrounding the 68000, excluding the interrupt request inputs. The control of HALT* and RESET* is conventional (see chapter 4). At power-on, a 555 timer configured as a monostable, L7, generates a single active-high pulse.

The position of each integrated circuit, or to be more precise each DIP package, on the double Eurocard is indicated by a letter and a number. The letter denotes the row in which the IC is found and the number denotes its position in that row. We have used this method to allow spaces on the board to be populated with ICs later, without altering the numbering (i.e., sequence) of existing ICs. Had we called them 1, 2, ..., the addition of a new IC in a previously empty location would have given it an out-of-sequence number.

Open-collector inverting buffers, L6a and L6b, apply the reset pulse to the 68000's RESET* and HALT* inputs, respectively. A manual reset generator is formed from two cross-coupled NAND gates, J7a and J7b, and applied to the reset lines by a further two open-collector buffers, L6c and L6d. An inverting buffer, G2b, gates the reset pulse from L7 onto the system bus as the active-low POR* (power_on_reset). This signal can be used by other modules to clear circuits on power-up.

An LED is connected to the HALT* pin via a buffer. This LED is fitted to the front panel and confirms the reset operation. It also shows if the CPU has asserted HALT* because of a double bus fault.

The 68000 clock input is provided by an 8-MHz crystal-controlled clock in a DIP package. L3, a 74LS93 divide-by-16 counter, provides submultiples of the basic clock frequency for other functions. When performing initial tests, we usually run the CPU from a 4-MHz clock.

The high impedance control inputs shown in figure 11.16 are pulled up to the V_{cc} by resistors. Although pull-up resistors are necessary on BR*, DTACK*, etc., the reader may be surprised to find them on AS*, UDS*, LDS*, and R/$\overline{W}$. They are required here because these pins are driven by tristate outputs in the 68000. When the 68000 relinquishes the bus, all tristate lines are floated. To leave the state of these bus lines undefined is unwise, as a spurious bus cycle might possibly be generated in certain circumstances. A better course is to be safe rather than sorry. During the testing phase, some of the pull-up resistors were temporary and used only for test purposes, because they were connected to lines that will later be pulled up or down by totem-pole outputs. They appear in figure 11.16 so that the circuit can be tested independently of the rest of the system.

Testing the CPU control circuitry is very easy. The power_on_reset circuit is tested by attaching an oscilloscope probe to test point 1 (TP1), switching on the V_{cc}

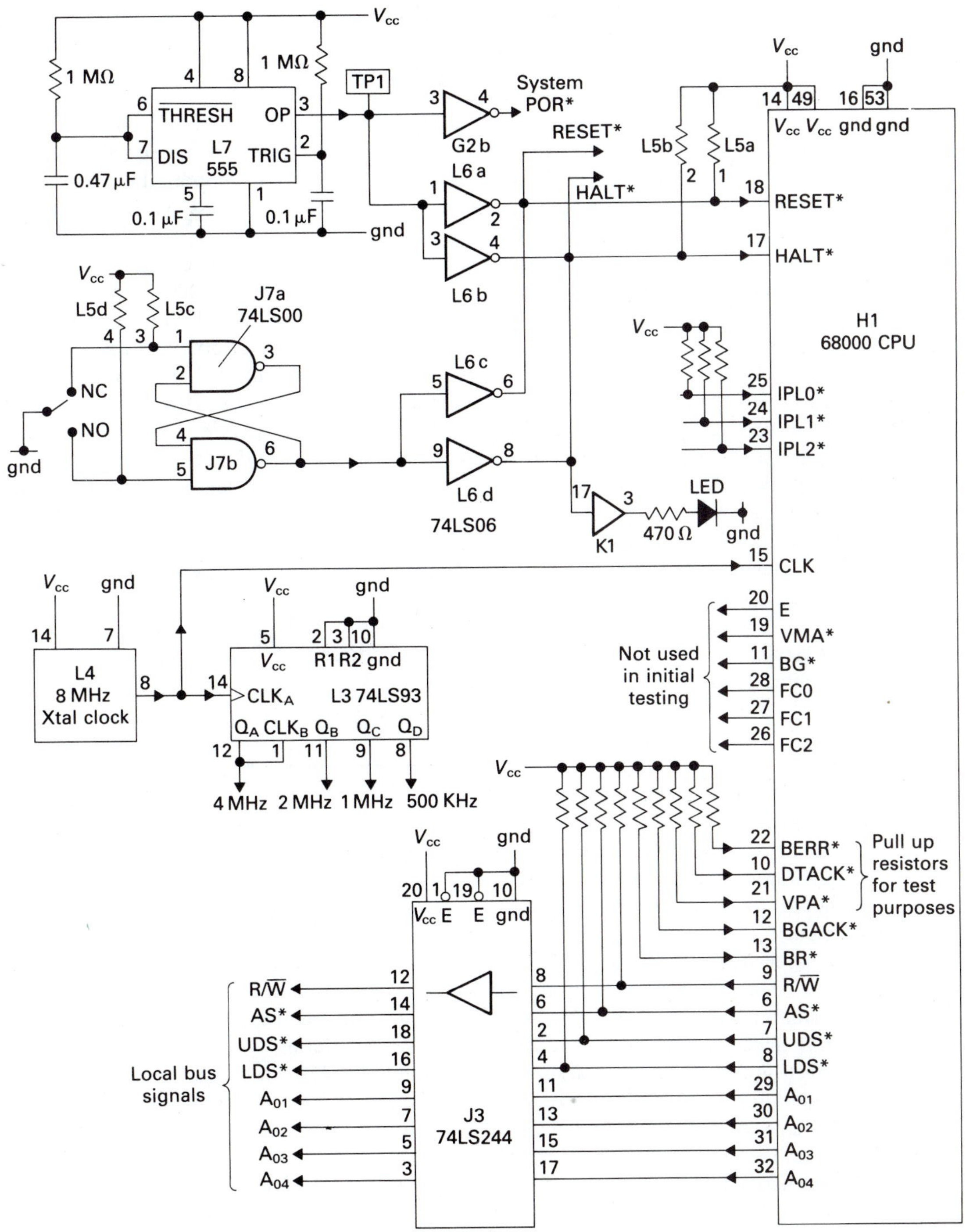

FIGURE 11.16 Circuit diagram of the 68000 reset, halt, and clock control circuits

power supply, and observing the positive-going pulse. A negative-going pulse should be observed at the CPU's RESET* and HALT* pins. The manual reset pulse generator should force HALT* and RESET* low whenever the reset button is pushed.

An 8-MHz square wave should be observed at the CLK input to the CPU. All inputs pulled up to a logical 1 should be at a logical 1 state.

The next step is to install the 68000 and to force the CPU to free-run. As no memory components have yet been fitted, the CPU must be fooled into thinking that it is executing valid bus cycles. To do this DTACK* is temporarily connected to the AS* output. Whenever the 68000 starts a memory access by asserting AS*, DTACK* is automatically asserted to complete the cycle.

The 68000 is a tricky beast to test in a free-running mode, because it generates an exception if a nonvalid op-code is detected. Should the 68000 then generate a second exception, the resulting bus fault will cause it to halt. Therefore, the 68000 must always see a valid op-code on its data bus. One way of doing this is to pull up (or down) the data bus lines with resistors to V_{cc} (or ground). Traditionally, CPUs are tested by placing a NOP (no operation) op-code on the data bus. The 68000 NOP code is $4E71 (i.e., %0100 1110 0111 0001). If this code is jammed onto the data bus, the 68000 will also use it for the stack pointer and reset vectors during the reset exception processing. Sadly, this code will lead to an address error. When the CPU reads the stack pointer from addresses $00 0000 and $00 0002 at the start of its reset exception processing, it obtains $4E71 4E71. Unfortunately, this value is *odd* and generates an address exception. We need to use a dummy op-code that is even to allow the CPU to free-run.

When a suitable op-code has been jammed onto D_{00} to D_{15}, the 68000 should free-run and a square wave be observed on address pins A_{01} to A_{23}. The frequency at pin A_i should be one half that at pin A_{i-1}.

Interrupt Circuit

The interrupt control circuitry surrounding the 68000 is entirely conventional and does not depart from that described in chapter 6. In figure 11.17 a 74LS148 eight-line-to-three-line priority encoder, J4, converts the seven levels of interrupt request input into a 3-bit code on IPL0* to IPL2*. Note that each interrupt request input must have a pull-up resistor.

The function code from the 68000 is decoded by J5, a 74LS138, and the resulting IACK* output used to enable a second decoder, J6. J6 is also strobed by AS* and converts the information on A_{01} to A_{03} during an IACK cycle into one of seven levels of interrupt acknowledge output (IACK1* to IACK7*). Other function code information supplied by J5 that may be useful in debugging the system is the "user/supervisor" memory access codes and the "program/data" bus cycle codes.

The interrupt control circuitry can be tested by periodically pulsing IRQ7* (the nonmaskable interrupt) and observing the response on IACK7*. A suitable source of

FIGURE 11.17 The 68000 interrupt control circuitry

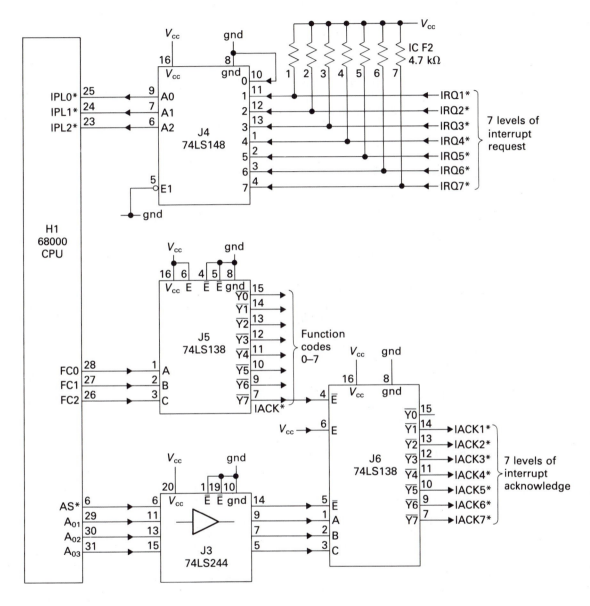

pulses for IRQ7* can be found on the address lines during the free-running mode—for example, A_{05}. Note that other levels of interrupt cannot be tested yet as the 68000 sets its mask bits to level 7 during its reset. In general, detailed testing of interrupt control circuits is not possible until exception-handling routines have been written.

Address Decoder Circuitry on the CPU Module

The specification of the TS2 CPU module calls for up to 32K bytes of static RAM and up to 32K bytes of EPROM at the bottom of the processor's 16M-byte memory space. Immediately above the RAM memory space sits the peripheral address space, permitting up to eight memory-mapped components, each occupying 64 bytes. Table 11.1 gives the memory map of the TS2 CPU module. The address decoding table corresponding to the memory map of table 11.1 is given in table 11.2.

The diagram of a possible implementation of table 11.2 is given in figure 11.18. A five-input NOR gate, K3a, generates an active-high output, G1, whenever A_{19} to A_{23} are all low. Together with $A_{18}*$ and $A_{17}*$, this gate enables a three-line-to-eight-line decoder, K5, that divides the lower 128K bytes of memory space from $00\ 0000 to $01\ FFFF into eight blocks of 16K. The first four blocks decode the address space for the read/write memory and ROM. We deal with the selection of the reset vector memory space in ROM later.

The active-low peripherals_group_select* output of K5 (i.e., the address range $01\ 0000 to $01\ 3FFF) enables a second three-line-to-eight-line decoder, K6. K6 is a 25LS2548 that has two active-low and two active-high enable inputs. It also has an active-low open-collector output, ACK*, that is asserted whenever the device is enabled *and* is strobed by a negative-going pulse on its RD* or WR* inputs.

K6 is also enabled by G3 from K4a, which is high when A_{09} to A_{13} are all low, and by AS* from the CPU. Thus, whenever a valid address in the range $01\ 0000 to $01\ 01FF appears on the address bus, one of K6's active-low outputs is asserted. When either UDS* or LDS* go low in the same cycle, the ACK* of the decoder is asserted, indicating a synchronous access to a peripheral by asserting the processor's VPA* input. Note that this arrangement is intended to be used in conjunction with 6800-series peripherals.

TABLE 11.1   Memory map of the TS2 CPU module

	SIZE (bytes)	DEVICE	ADDRESS SPACE
1	8	EPROM1	00 0000–00 0007
2	16K	RAM1	00 0008–00 3FFF
3	16K	RAM2	00 4000–00 7FFF
4	16K	EPROM1	00 8000–00 BFFF
5	16K	EPROM2	00 C000–00 FFFF
6	64	Peripheral 1	01 0000–01 003F
7	64	Peripheral 2	01 0040–01 007F
8	64	Peripheral 3	01 0080–01 00BF
9	64	Peripheral 4	01 00C0–01 00FF
10	64	Peripheral 5	01 0100–01 013F
11	64	Peripheral 6	01 0140–01 017F
12	64	Peripheral 7	01 0180–01 01BF
13	64	Peripheral 8	01 01C0–01 01FF

TABLE 11.2 Address decoding table for the TS2 CPU module memory map

	DEVICE	A_{23}	A_{22}	...	A_{16}	A_{15}	A_{14}	A_{13}	A_{12}	A_{11}	A_{10}	A_{09}	A_{08}	A_{07}	A_{06}	A_{05}	A_{04}	A_{03}	A_{02}	A_{01}
1	EPROM1	0	0	...	0	0	0	0	0	0	0	0	0	0	0	0	0	0	X	X
2	RAM1	0	0	...	0	0	0	X	X	X	X	X	X	X	X	X	X	X	X	X
3	RAM2	0	0	...	0	0	1	X	X	X	X	X	X	X	X	X	X	X	X	X
4	EPROM1	0	0	...	0	1	0	X	X	X	X	X	X	X	X	X	X	X	X	X
5	EPROM2	0	0	...	0	1	1	X	X	X	X	X	X	X	X	X	X	X	X	X
6	PERI1	0	0	...	1	0	0	0	0	0	0	0	0	0	0	X	X	X	X	X
7	PERI2	0	0	...	1	0	0	0	0	0	0	0	0	0	1	X	X	X	X	X
8	PERI3	0	0	...	1	0	0	0	0	0	0	0	0	1	0	X	X	X	X	X
9	PERI4	0	0	...	1	0	0	0	0	0	0	0	0	1	1	X	X	X	X	X
10	PERI5	0	0	...	1	0	0	0	0	0	0	0	1	0	0	X	X	X	X	X
11	PERI6	0	0	...	1	0	0	0	0	0	0	0	1	0	1	X	X	X	X	X
12	PERI7	0	0	...	1	0	0	0	0	0	0	0	1	1	0	X	X	X	X	X
13	PERI8	0	0	...	1	0	0	0	0	0	0	0	1	1	1	X	X	X	X	X

FIGURE 11.18 Arrangement of TS2's address decoding circuits on the CPU module

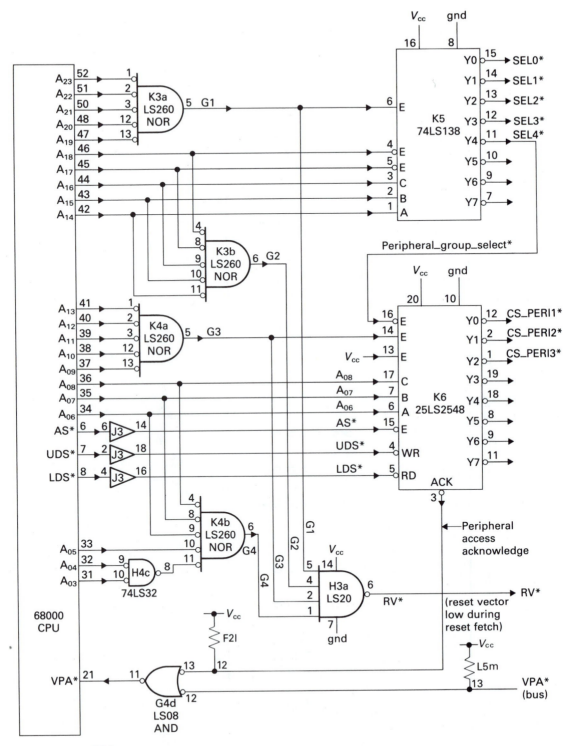

An access to the reset vectors in the range $00\ 0000$ to $00\ 0007$ is detected by gates K3a, K3b, K4a, K4b, H4c, and H3a. When the output of each NOR gate is high, signifying a zero on A_{03} to A_{23}, the output of the NAND gate H3a, RV*, goes active-low; that is, RV* is low whenever a reset vector is being accessed and is used to overlay the exception table in read/write memory with the reset vectors in ROM.

The address decoder of figure 11.18 can be tested to a limited extent by free-running the CPU and detecting decoding pulses at the outputs of the address decoder. A better technique is to insert a test ROM and to execute an infinite loop which periodically accesses the reset vector space. This makes it easy to observe the operation of the circuit on an oscilloscope.

The selection of the individual RAM and EPROM components from the address decoder outputs is carried out by the circuit of figure 11.19. Two-input NOR gates combine one of the four device-select signals (SEL0* to SEL3*) from the address decoder with the appropriate data strobe (UDS* or LDS*) to produce the actual active-low chip-select inputs to the eight memory components on the CPU module.

The circuit of figure 11.19 is also responsible for overlaying the reset vector space onto the ROM memory space. The technique used in figure 11.19 is exactly as described in chapter 6. When the RV* signal goes active-low while a reset vector is being fetched, the read/write memory at $00\ 0000$ to $00\ 3FFF$ is disabled and the EPROM at $00\ 8000$ to $00\ BFFF$ substituted.

Memory on the CPU Module

The use of $8K \times 8$ memory components permits the design of a memory with a very low component count and virtually no design effort. Figure 11.20 gives the circuit diagram of half the memory components on the CPU module—the others are arranged in exactly the same fashion but are enabled by different chip-select signals from the address decoder.

No further comment is required other than to point out that the EPROMs have their active-low output enables (OE*) driven by $R/\overline{W}$ from the processor via an inverter. This action is necessary to avoid a bus conflict if a write access is made to EPROM memory space.

DTACK* and BERR* Control

Each memory access cycle begins with the assertion of AS* by the 68000 and ends with the assertion of DTACK* (or VPA*) by the addressed device or with the assertion of BERR* by a watchdog timer. Figure 11.21 gives the diagram of the DTACK* and BERR* control circuitry on the CPU module.

Whenever a block of 16K bytes of memory is selected on the CPU module, one of the four select signals, SEL0* to SEL3*, goes active-low. The output, MSEL, of the NAND gate H3b is then forced active-high. MSEL becomes the ENABLE/LOAD* control input of a 74LS161 4-bit counter, H2. When MSEL = 0 (i.e., on-

FIGURE 11.19 Selecting RAM and ROM on the CPU module

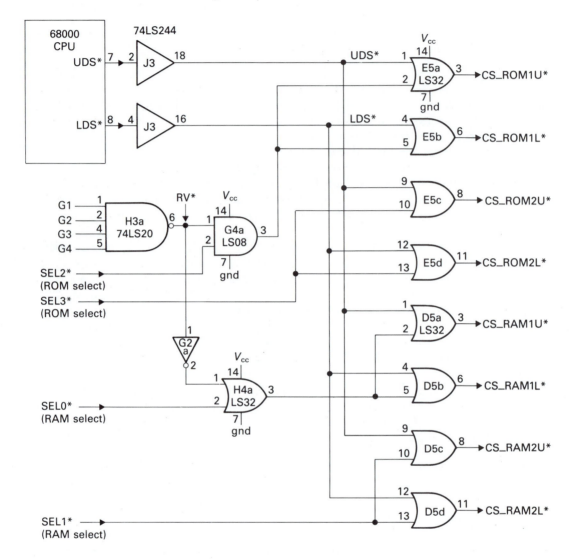

board memory not accessed), the counter is held in its load state and the data inputs on D_a to D_d are preloaded into the counter—in this example 1100. The Q_d output from the counter is gated, uninverted, through G4b, H4b, and G4c to form the processor's DTACK∗ input.

When MSEL goes high the counter is enabled. The counter is clocked from the 68000's clock and counts upward from 1100. After four clock pulses, the counter folds over from 1111 to 0000 and Q_d (and therefore DTACK∗) goes low to provide the handshake required by the 68000 CPU. At the end of the cycle, AS∗ is negated and MSEL goes low to preload the counter with 1100 and negate DTACK∗. Figure 11.22a gives the timing diagram of this circuit.

FIGURE 11.20 RAM and ROM on the CPU module

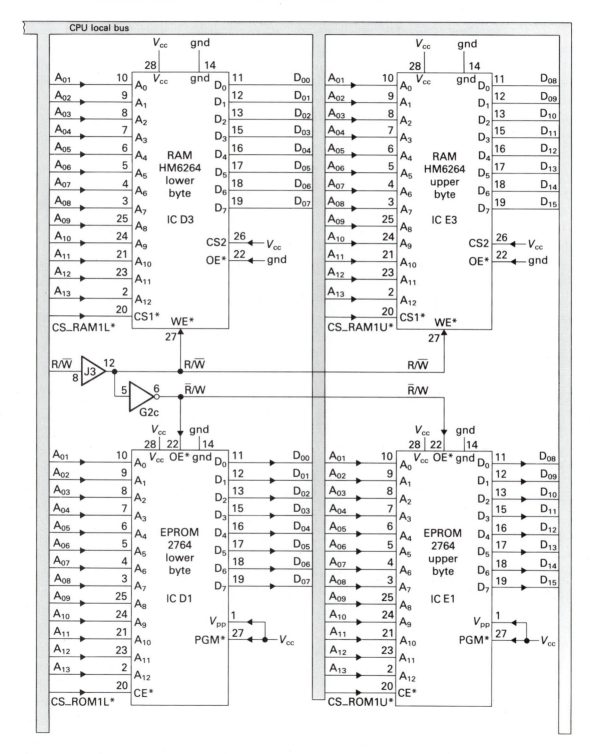

FIGURE II.21   DTACK∗ and BERR∗ control circuitry

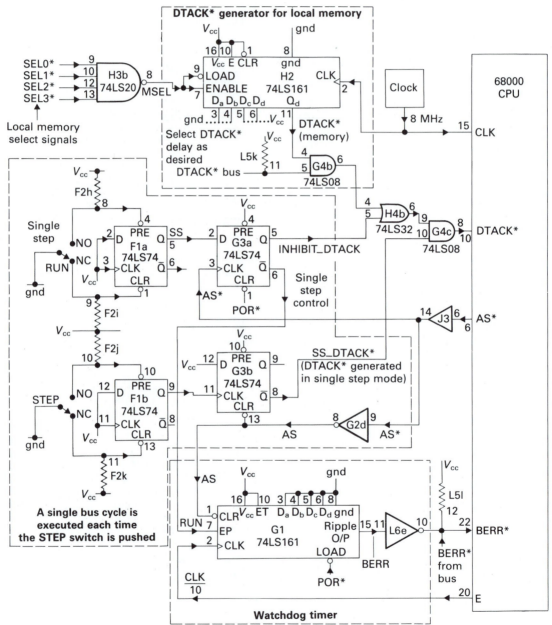

At the same time that H2 begins counting, a second timer, G1 (another 74LS161), also begins to count upward. The count clock is taken from the 68000's E output which runs at CLK/10. This counter is cleared to zero whenever AS* is negated. The ripple output from the counter goes high after the fifteenth count from zero and is inverted by the open-collector gate L6e to provide the CPU with a BERR* input. Therefore, unless AS* is negated within 15 E-clock cycles of the start of a bus cycle, BERR* is forced low to terminate the cycle. Note that the counter is disabled (EP = 0) in the single-step mode (discussed later) to avoid a spurious bus error exception. Figure 11.22b provides a timing diagram of the watchdog timer.

A useful feature of the DTACK* circuit is the addition of a single-step mode, allowing the execution of a single bus cycle (note bus cycle *not* instruction) each time a button is pushed. This facility can be used to debug the system by freezing the state of the processor.

One of the inputs to the OR gate H4b is INHIBIT_DTACK. If this is active-high, the output of the OR gate is permanently true and the generation of DTACK* by the DTACK* delay circuit (or from the system bus) is inhibited. Therefore, a bus cycle remains frozen with AS* asserted, forcing the CPU to generate an infinite stream of wait states.

Two positive-edge triggered D flip-flops, F1a and G3a, control INHIBIT_DTACK. F1a acts as a debounced switch and produces an SS/RUN* signal from its Q output, depending only on the state of the single-step/run switch. Unfortunately, it would be unwise to use the output of F1a to inhibit DTACK*, because changing from run to single-step mode in mid bus cycle might lead to unpredictable results. Instead, the output of F1a is synchronized with AS* from the processor by a second flip-flop, G3a. Figure 11.22c shows how the INHIBIT_DTACK signal from G3a is forced high only when AS* is negated at the end of a bus cycle. The 68000 always enters its single-step mode at the start of a new cycle before AS* is asserted.

In the single-step mode, DTACK* pulses are generated manually by depressing the "step" switch. The output of this switch is debounced by flip-flop F1b. A second flip-flop, G3b, generates a single, active-low pulse, SS_DTACK*, each time the step button is pushed. SS_DTACK* is gated in G4c to produce the DTACK* input needed to terminate the current bus cycle. Figure 11.22d gives the timing diagram of the SS_DTACK* generator.

There are two simple ways of testing the DTACK* control circuits. One is in the free-run mode and is done by connecting, say, SEL0* to AS*, so that a delayed DTACK* is produced for each bus cycle. The single-step circuit can also be tested in this mode. Another procedure is to construct a special test rig for the circuit, which simulates the behavior of the 68000 by providing AS*, CLK, and SEL0* signals.

Buffering and Bus Control on the CPU Module

The interface between the CPU module and other modules is via its backplane bus. This bus can be divided into three components: the address bus, the control bus, and the data bus. Figure 11.23 gives the circuit of the address and control signal paths and figure 11.24 gives the circuit of the data bus buffers and their control.

FIGURE 11.22 Timing diagrams for DTACK* and BERR* control

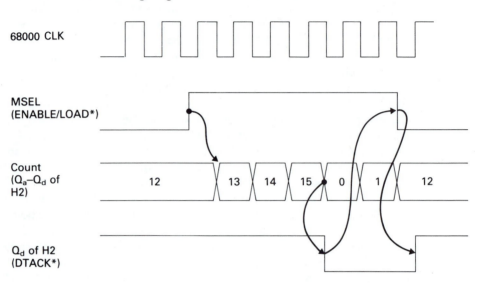

68000 CLK

MSEL
(ENABLE/LOAD*)

Count
(Q_a–Q_d of
H2)

Q_d of H2
(DTACK*)

(a) DTACK* delay generator

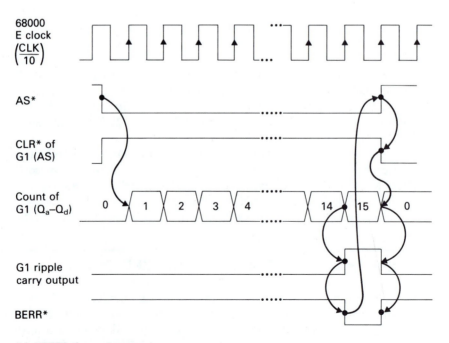

68000
E clock
$\left(\dfrac{CLK}{10}\right)$

AS*

CLR* of
G1 (AS)

Count of
G1 (Q_a–Q_d)

G1 ripple
carry output

BERR*

(b) BERR* timeout generator

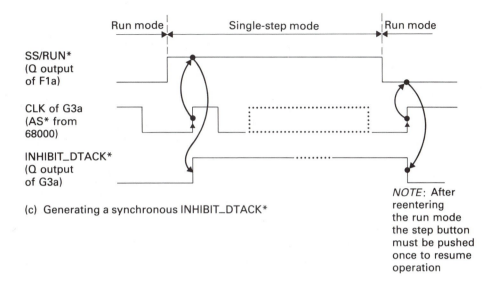

(c) Generating a synchronous INHIBIT_DTACK*

NOTE: After reentering the run mode the step button must be pushed once to resume operation

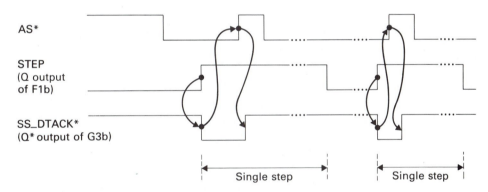

(d) Producing a single-step DTACK* signal

Three 74LS244 octal tristate bus drivers buffer the address from the 68000 onto the system bus (figure 11.23). The address buffers are all enabled by the complement of the BGACK* input to the 68000. BGACK* is pulled up to V_{cc} by a resistor and the address bus buffers are normally enabled (even when the 68000 is addressing local memory). Whenever a module on another card wishes to use the system bus, it asserts BR*, waits for BG* to be asserted, and then asserts BGACK*. This situation causes the address bus buffers on the CPU module to float, leaving the bus free for the new bus master.

The asynchronous bus control signals from the 68000 (AS*, UDS*, LDS*, R/$\overline{W}$) are buffered in exactly the same way as the address by one half of an octal bus transceiver. Three control signals, CLK, VMA*, and E, from the CPU are buffered by permanently enabled bus drivers, as we do not anticipate that another CPU will implement synchronous bus cycles from the system bus.

FIGURE 11.23 Address and control buffers on the CPU module

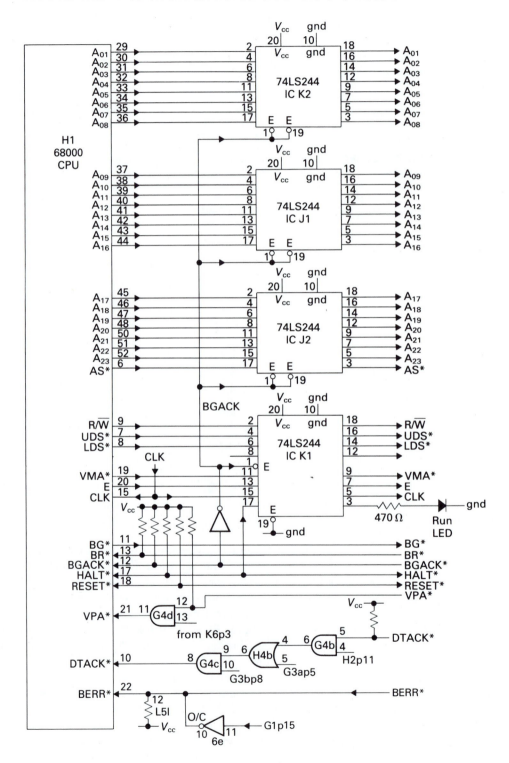

FIGURE 11.24 Data buffers and their control on the CPU module

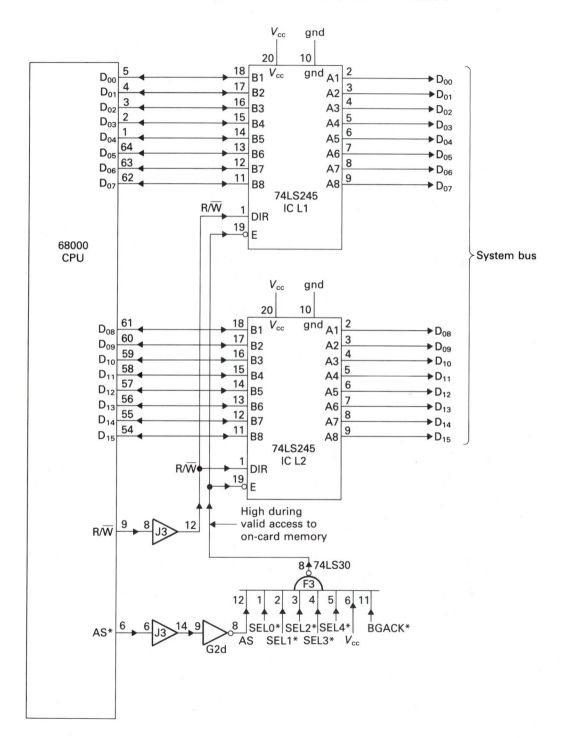

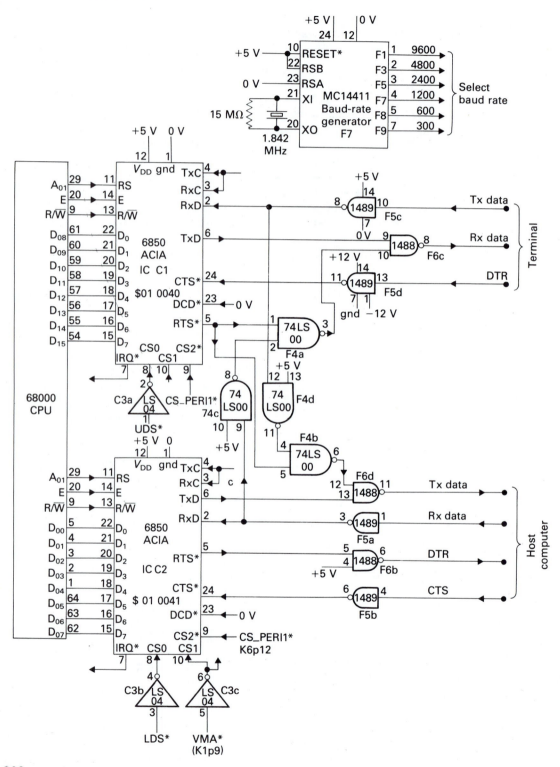

FIGURE 11.25 ACIAs on the CPU module

The remaining bus control signals have been dealt with elsewhere and are included in figure 11.23 for the sake of completeness. All inputs to the 68000 have pull-up resistors.

The data bus buffers of figure 11.24 are implemented by two 74LS245 octal bus transceivers. Both transceivers have their data direction controlled by R/$\overline{\text{W}}$ from the CPU and are enabled only when the 68000 executes a valid bus cycle to nonlocal memory. If local memory is accessed, one of SEL0* to SEL4* goes active-low and the logical one at the output of the eight-input NAND gate, F3, disables the data bus transceivers.

If the 68000 gives up the bus in response to the assertion of BR*, execution of the bus cycles is stopped and AS* is not asserted until the 68000 is once more in control. By making the BGACK* input to the 68000 a necessary condition to enable the data bus transceivers, the transceivers are automatically turned off whenever the on-board CPU has relinquished the bus.

I/O Ports on the CPU Module

The only I/O ports implemented on the TS2 CPU module are the two 6850 ACIAs illustrated in figure 11.25. This circuit is almost identical to that found in the ECB modulue described in chapter 9. One port is dedicated to the terminal (IC C1 at address $01 0040) and the other (IC C2 at address $01 0041) is dedicated to the host computer interface. Whenever RTS* from C1 is made electrically high, the terminal interface is connected directly to the host port.

11.3　DESIGN EXAMPLE USING THE 68030

To conclude this chapter on the design of 68000 systems, we will look at a simple, but powerful, 68030-based single-board microcomputer. This design is taken from Motorola's Application Note ANE426 written by David McCartney and Tommy Kelly. The microcomputer uses a 68030 with an 8-bit EPROM port, a 32-bit static RAM port, a 68681 DUART and a 68230 PI/T configured as a Centronics printer port. This circuit has been included to demonstrate that the design of a 68030 (or 68020) system is similar to its 68000 counterpart.

The read/write memory is implemented by eight MCM6164 8K by 8-bit high-speed static RAMs. These are interfaced to a 32-bit data port to make best use of the 68030's high-speed data path. The read-only memory is supplied by two 27512 512K-bit (i.e., 64K by 8-bit) EPROMs, although only one EPROM is strictly necessary. By exploiting the dynamic bus sizing capabilities of the 68030, the system firmware can be loaded into a single EPROM and the design of the system can be greatly simplified.

The 68030 systems designer is probably more concerned about the CPU-memory interface than any other aspect of the microcomputer. It is this interface that determines the overall performance of the system, because it dictates how fast the 68030 can run. The design we are going to discuss here has been optimized to allow for zero-wait-state operation using the 68030's special synchronous access mode.

The desire to implement no-wait-state operation strongly influences the design of the address decoder, since the address decoder must supply the handshake signal (called STERM*) needed to terminate the memory access very early in the bus cycle. Figure 11.26 illustrates the essential details of a 68030 synchronous read cycle (signals not of immediate interest are excluded). If the 68030 is to operate in its four- (yes, four) clock-state synchronous mode without the introduction of wait states, the synchronous termination handshake, STERM*, must be returned to the processor before the start of clock state S2.

FIGURE II.26   68030 synchronous read cycle

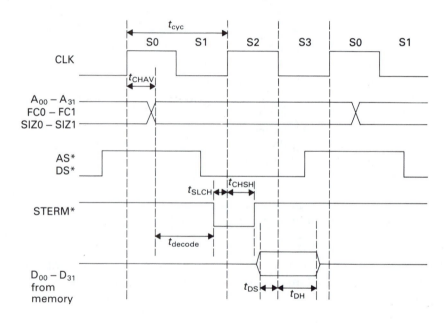

		Parameters (ns)	
		20 MHz	40 MHz
t_{cyc}	Clock cycle time	50	25
t_{CHAV}	Clock high to address valid	0–25	0–14
t_{SLCH}	STERM* low to clock high (setup time)	4	2
t_{CHSH}	Clock high to STERM* high (hold time)	12	6
t_{DS}	Data setup time	4	1
t_{DH}	Data hold time	12	6

t_{decode} = time available to generate STERM* = $t_{cyc} - t_{CHAV} - t_{SLCH}$

The 68030's interface to the slower EPROMs and I/O devices also includes buffer delays and, therefore, does not operate with zero wait states; that is, no attempt is made to optimize the speed of non-RAM accesses. This approach to microcomputer design is entirely reasonable. If code within EPROM is to be accessed frequently, it is better to copy code from EPROM to static RAM and run it in the RAM. Copying firmware into RAM is called *shadowing*.

Address Decoder

The address decoder must not only select the RAM, EPROM, and I/O devices, it must perform its task fast enough to permit the 68030 to execute synchronous memory accesses. In general, the speed of address decoders in 68000-based systems (especially at 8 MHz) is not critical. The same thing cannot be said for 68020-, 68030- or 68040-based systems. If we examine the timing parameters of the 68030 synchronous read cycle shown in figure 11.26, the following constraints are evident:

1. Address lines A_{00}–A_{31}, function code lines FC0–FC2, and SIZ0, SIZ1 become valid at time t_{CHAV} after the rising edge of bus state zero.

2. The synchronous handshake input to the 68030, STERM$*$, requires a setup time t_{SLCH} before the rising edge of S2 to ensure zero-wait-state operation.

Therefore, the time between the address valid (i.e., maximum value of the parameter t_{CHAV} from the rising edge of S0) and the falling edge of STERM$*$ (i.e., maximum value of the parameter t_{SLCH} before the rising edge of S2) is the time available to perform address decoding and to generate the STERM$*$ handshake signal. Generating STERM$*$ is not a mysterious process, since STERM$*$ is produced just like DTACK$*$ in a 68000 system. The problem faced by the 68030 systems designer is simply that STERM$*$ has to be asserted very early in a bus cycle. An 68030 operating at 20 MHz has a cycle time of 50 ns, yielding an address decode and handshake generation time of

$$50 - t_{CHAV} - t_{SLCH} = 50 - 25 - 4 = 21 \text{ ns}$$

In this application, address decoding is carried out by three 74F521s (i.e., octal comparators) and two 74F138s (i.e., IC14, IC15, IC16, IC12, and IC13), as illustrated by figure 11.27. This figure is taken from Application Note ANE426 and has been redrawn to clarify the operation of the circuit. Note that the original ANE426 circuit from Motorola, figure 11.30, employs conventional positive logic symbols for gates, whereas figure 11.27 uses negative logic symbols.

Address lines A_{31}–A_{24} are decoded to produce the signal BOARD_SEL$*$ that allocates the top 16M Bytes of the 68030's address space (i.e., 00000000– $00FF FFFF$) to this application board. Therefore, this board can be interfaced to other boards to add extra memory and peripherals in the address space above the top 16M bytes.

IC15 and IC16 produce two active-low select signals based on the state of address lines A_{23}–A_{19}. These are I/O.ROM_SEL$*$ for the top 512K Bytes of the

FIGURE 11.27   Circuit of the address decoder

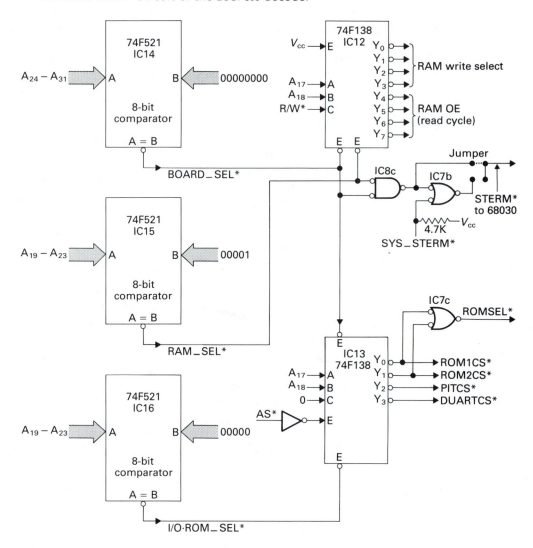

memory map ($0000 0000–$0007 FFFF) and RAM_SEL* for the next 512K Bytes ($0008 0000–$000F FFFF). These two select signals are combined with BOARD_SEL* and further decoded by IC12 and IC13. The BOARD_SEL*, RAM_SEL*, and I/O.ROM_SEL* signals are generated in parallel. Note that *two-level* address decoding is performed by IC14/IC15/IC16 in series with IC13/IC12. Three-level (or greater) address decoding would introduce an unacceptable delay.

ICs 12 and 13 provide individual block-select signals for 4 × 128K RAM banks, two separate 128K-byte ROM selects, and two separate 128K-byte I/O selects used for the 68681 and 68230 peripherals. The memory map for the board is shown in figure 11.28.

The 68030's synchronous termination signal, STERM∗, is generated by ANDing (ORing in positive logic terms) BOARD_SEL∗ and RAM_SEL∗ in IC8c. Jumper J6 allows the output of this gate to be fed directly to the 68030 processor or via an OR gate to allow a system STERM∗ signal to be injected from some other off-board source via IC7b.

The decode and handshake generation logic meets the timing constraints just discussed, since, for zero-wait states, the signal path consists of two 74F521s in

FIGURE 11.28 Memory map for the 68030-based SBC

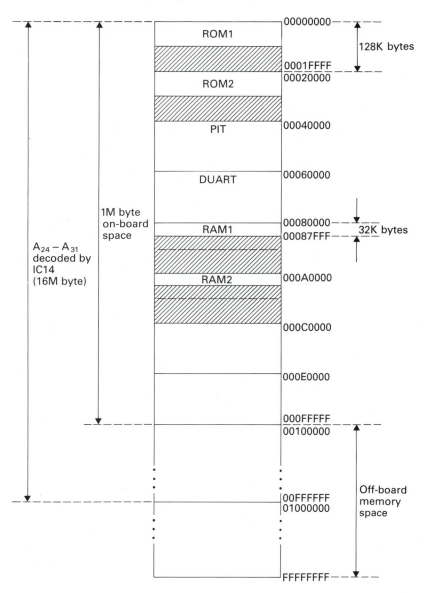

parallel (i.e., ICs 14 and 15) followed by a 74F32 (i.e., IC 8c). These devices introduce a maximum delay of 11 ns + 6.3 ns = 17.3 ns, which is within the 21-ns constraint.

Firmware EPROM

The firmware is located in one or two 27512 EPROMs. As we stated earlier, only one device is strictly necessary to hold reset vectors and a bootstrap loader. The two EPROMs, IC25 and IC26, are both connected to bits D_{31}–D_{24} of the processor's data bus and therefore form an 8-bit-wide data port. This 8-bit port permits the processor to be bootstrapped from a single EPROM, since the dynamic bus sizing capability of the 68030 will automatically handle all sizes of access to the program ROM.

Figure 11.29 describes the EPROM selection logic. The two EPROM chip-select signals, ROM1CS* and ROM2CS*, from the address decoder section are used to select the appropriate device. Since the interface port is 8 bits wide, the required handshake to the 68030 processor is DSACK0* = 0 and DSACK1* = 1. Most code will be run from RAM on the board; therefore, the ROM access time is not critical. Remember that the bootstrap program in the EPROM will copy itself to RAM and execute the monitor from there. Consequently, no attempt has been made to minimize the number of wait states for each ROM access.

The two EPROM chip-selects, ROM1CS* and ROM2CS*, are ORed by IC7c and then this combined signal is ANDed with a delay signal to produce the required DSACK0* handshake input to the 68030 (figure 11.29). The delay signal comes from a wait-state generator implemented by IC4 and provides a delay of two cycles of a 2.5-MHz clock. This clock is one eighth of the processor clock. A two-cycle delay guarantees enough access time to the EPROMs, even if the clock frequency of the processor is increased beyond the design speed of 20 MHz.

The wait-state generator implemented by IC4 is also used to implement a bus error time-out for faulty memory accesses (indicated by dotted lines in figure 11.29). If AS* is not negated to reset IC4 during a bus cycle, output Q3 from the counter will eventually rise to V_{cc} and BERR* will be asserted. If the processor does not receive either a STERM* or the DSACKx* handshake within four clock cycles of the slowest clock available on the board, then IC4 will generate a BERR* signal to the 68030 and force exception processing.

Read/Write Memory

As we have said, the address decode logic produces four separate RAM chip-select signals, each covering a 128K-Byte address range. Due to the capacitive loading of the RAM chips, only two banks have been incorporated on the board.

The RAM is organized as a 32-bit-wide port and uses the new synchronous interface on the 68030, which yields a two-clock-cycle read and write bus cycle. The decode time and access time of these RAMs are part of the critical timing path. To improve the access time, the MCM6164 RAM chips have separate chip enable and output enable inputs. The chip enable inputs are tied to a logic zero to enable them

FIGURE 11.29 EPROM selection logic

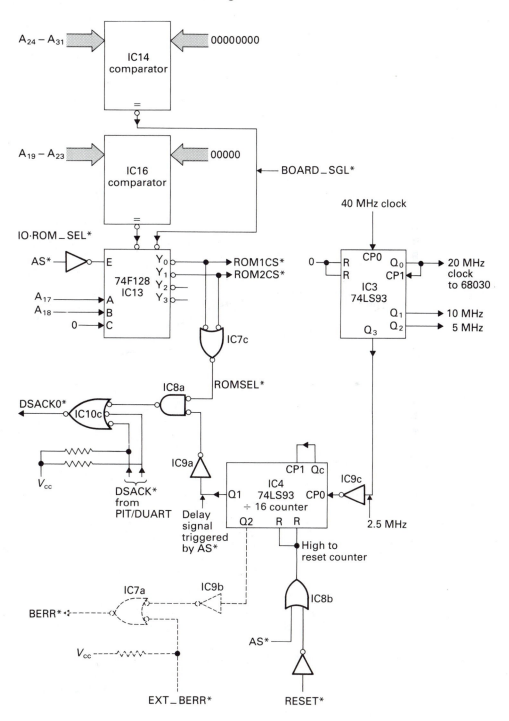

permanently, and the output enables are connected to the chip-select signals generated by the address decode logic.

On a read operation, all four chips of one bank are always enabled (i.e., ICs 17–20 or ICs 21–24, see figure 11.30). Enabling all four chips drives the full 32 bits of the data bus, irrespective of the size of the operation being carried out by the 68030. The processor will then latch the appropriate sections of the data bus, depending on the size of the access. A processor write operation is more complex, since only the relevant sections of the data bus contain valid data.

The 16L8 PAL IC38 uses the A_{00}, A_{01}, SIZ0, and SIZ1 lines from the 68030 to generate four separate byte-select signals, UUD*, UMD*, LMD*, and LLD*. These signals are gated with the RAM1 write-select, the RAM2 write-select, and the address strobe in IC39, IC40, IC41, and IC42 to produce the individual RAM chip write enable signals. The equations used to code the PAL of IC38 are those given in the "Applications" section of the 68030 User's Manual and are as follows:

$$UUD* = \overline{A_{00}} \cdot \overline{A_{01}} \qquad \text{Upper byte, directly addressed, any size}$$

$$
\begin{aligned}
UMD* = {} & A_{00} \cdot \overline{A_{01}} && \text{Upper middle, directly addressed, any size} \\
& + \overline{A_{01}} \cdot \overline{\text{SIZ0}} && \text{Word aligned, size} = 1 \text{ or } 3 \text{ bytes} \\
& + \overline{A_{01}} \cdot \text{SIZ1} && \text{Word aligned, size} = \text{word or longword}
\end{aligned}
$$

$$
\begin{aligned}
LMD* = {} & \overline{A_{00}} \cdot A_{01} && \text{Lower middle, directly addressed, any size} \\
& + \overline{A_{01}} \cdot \overline{\text{SIZ0}} \cdot \overline{\text{SIZ1}} && \text{Word aligned, size} = \text{longword} \\
& + \overline{A_{01}} \cdot \text{SIZ0} \cdot \text{SIZ1} && \text{Word aligned, size} = 3 \text{ bytes} \\
& + \overline{A_{01}} \cdot A_{00} \cdot \overline{\text{SIZ0}} && \text{Word aligned, size} = \text{word or longword}
\end{aligned}
$$

$$
\begin{aligned}
LLD* = {} & A_{00} \cdot A_{01} && \text{Lower byte, directly addressed, any size} \\
& + A_{00} \cdot \text{SIZ0} \cdot \text{SIZ1} && \text{Odd aligned, size} = 3 \text{ bytes} \\
& + \overline{\text{SIZ0}} \cdot \overline{\text{SIZ1}} && \text{Any address, size} = \text{longword} \\
& + A_{01} \cdot \text{SIZ1} && \text{Word aligned, size} = \text{word or 3 bytes}
\end{aligned}
$$

As no data bus buffering is used between the RAM chips and the processor, no extra delay is introduced into the system. The address decode section generates additional RAM selects for banks 3 and 4. These signals can be employed to select additional RAM. However, if you add extra RAM to the board, you will require buffered address and data buses.

Serial and Parallel Communications

Figure 11.30 gives the full circuit diagram of the 68030-based SBC and shows the arrangement of the I/O ports. The 68681 DUART, IC28, provides two independent serial communications ports, which are normally connected to local terminals or used to provide links to host computers. The driver chips, IC36 and IC37, provide the TTL-to-RS232 voltage-level conversions between the DUART and the serial ports. These ports are brought out to standard DB25 way connectors J4 and J5.

A Centronics-type parallel printer port has been implemented using a 68230 PI/T IC27. The output of this device is buffered by a 74LS244 octal driver IC29 and brought out to a standard 50-way connector J3.

FIGURE 11.30 Circuit diagram of a 68030-based SBC

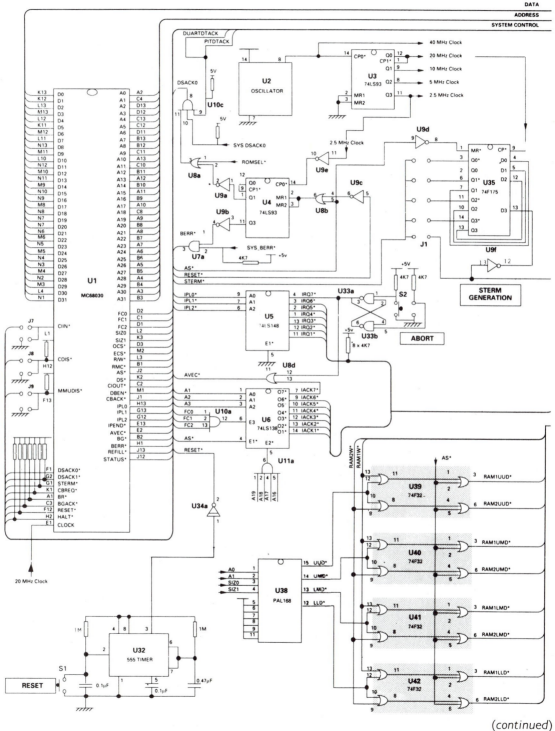

(continued)

FIGURE 11.30 (continued)

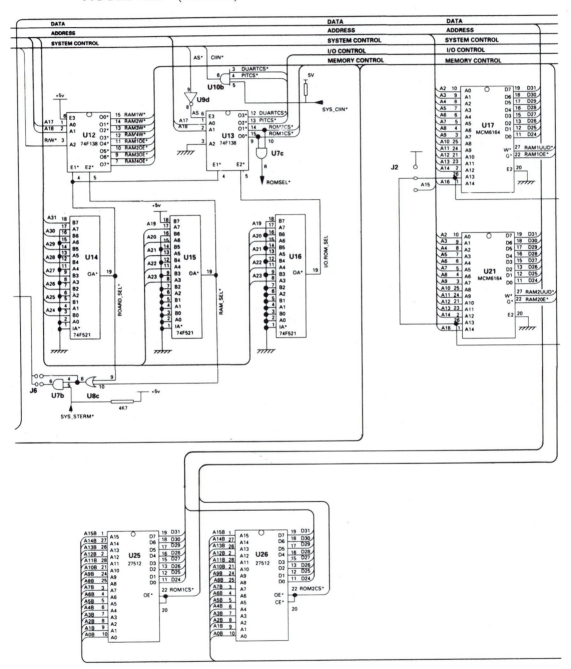

FIGURE 11.30 (continued)

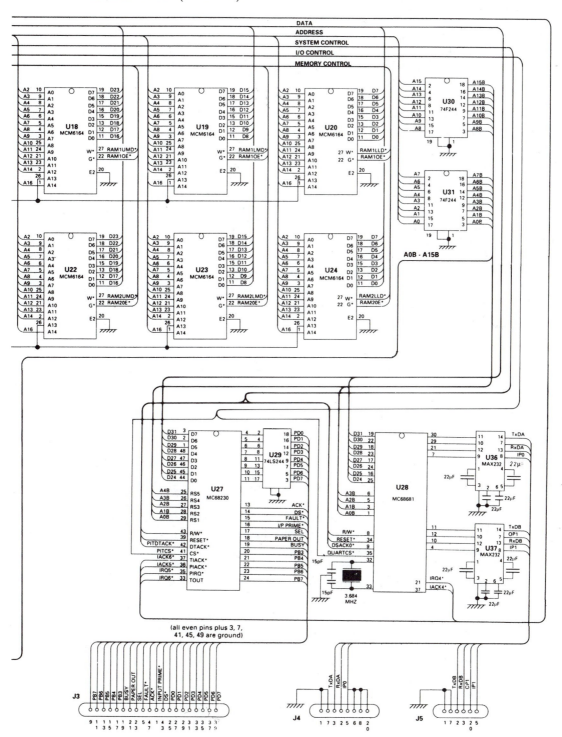

System Control and Interface

The time-critical paths to the RAM are unbuffered to allow full zero-wait-state operation. Two 74F244 devices, IC30 and IC31, buffer address lines A_{00}–A_{15}. These buffered address lines are then connected to the EPROMs and I/O devices in the system and could also be used for additional RAM banks. The basic system contains no data bus buffers due to the need for high-performance zero-wait-state operation. Four 74F245 bidirectional buffers could be added to the data bus if extra RAM is required.

The remaining functional circuits in figure 11.30 are all virtually identical to their 68000 counterparts. The power-on-reset circuit provided by IC32 and IC34a is exactly like a 68000 reset generator, except that HALT* need not be asserted concurrently with RESET*. A manual reset is provided by retriggering the 555 timer.

The interrupt control circuits use IC5 to encode IRQ1*–IRQ7*, IC10a, IC6, and IC11a to provide the corresponding IACK*s. The function of IC11a is to detect a 68030 IACK access to CPU space (it is not necessary to decode A_{16} to A_{19} in a 68000 system as the 68000 supports only one type of CPU space—IACK space). Note that IC33a/b provides a manual level 7 interrupt request (i.e., abort). When IRQ7* is asserted and IACK7* is asserted by the 68030 in response, AND gate IC8d (negative logic) generates an AVEC* input to the 68030 to trigger an autovectored interrupt.

As this example demonstrates, the design of a basic 68030 system is essentially the same as that of a 68000 system. The only critical difference is in the CPU-RAM path, which must be optimized if the 68030 is to run at anything like its full speed.

11.4 MONITORS

When microprocessors first began to appear in the mid-1970s, their initial markets were relatively small because few engineers, designers, or even academics had any practical experience with programmable digital systems. Semiconductor manufacturers were quick to realize that the key to microprocessor sales was education. Accordingly, they produced a number of single-board microcomputers that could be connected to a VDT or teletype. These microcomputers also provided the user with an introduction to programming and to the control of external systems, because all boards had some form of parallel I/O port.

A whole generation of engineers learned to program—often in machine code, as some engineers did not even have access to an assembler. Early SBCs had tiny memories. My first microcomputer had less than 1,024 bytes of read/write memory. In order to get such a primitive system to do anything useful, the single-board computers were supplied with a program in read-only memory, called a monitor. The monitor was the microcomputer's "operating system." The words are in quotes because the operating system of the SBC was often so primitive that few computer scientists would recognize it as such. A monitor provided the user with at least three basic functions:

1. The ability to input data or instructions (normally in machine code form) into the computer's memory.

2. The ability to execute a program starting at a given address.

3. The ability to read the contents of a memory location and to display them on the terminal.

The preceding facilities did at least enable many engineers to come to terms with the microprocessor. As time passed, the monitor grew more sophisticated and many functions were added. However, the monitor did not grow and grow without limit. Once the cost of read/write memory had declined to a small fraction of its original price and the floppy disk system became affordable, the monitor was not often needed and was replaced by the *bootstrap loader*. Following a reset, the bootstrap loader in ROM reads the operating system off a disk, places it in RAM, and runs it. Now, instead of the primitive monitor, the engineer has access to a full operating system with assemblers, compilers, editors, debuggers, and system utilities.

Today, the monitor has not disappeared entirely. It is not found in expensive, industrial microprocessor development systems. Nor is it found in low-cost, high-volume personal computers, where the scale of production permits a more comprehensive operating system and its associated secondary storage. The monitor is, however, found in some educational systems, where the cost of a disk-based operating system is prohibitive and the volume of production does not warrant a tailor-made operating system in ROM. One such monitor is called TUTOR and occupies 16K bytes of ROM in Motorola's ECB. We will examine some of TUTOR's facilities later.

There have been several noticeable trends in the facilities offered by modern monitors. The greatest change reflects the environment in which they are now used. Originally, the single board development system was the only computer in the laboratory or classroom. Now it is often surrounded by many relatively sophisticated systems. Therefore, monitors are frequently designed to allow software to be developed on a larger machine (the host computer) and then transferred (or downline-loaded) into the single board computer (the target machine).

Such monitors place more emphasis on the debugging of programs than on their creation. If the initial editing and assembly is done on the host computer, only the machine code (i.e., object code) needs to be transferred to the SBC. Then the programmer can set up initial conditions and run the code in the environment for which it was intended.

The TS2 microcomputer described earlier in this chapter needs a monitor to permit it to be tested and to allow programs to be entered and executed. Development of a monitor similar to TUTOR would be unrealistic here as that would represent a major design effort and consume a book of its own. A more tractable approach is to design a monitor capable of performing the three basic functions listed earlier in this section. Such a monitor is able to transfer programs and data between itself and an external host. The monitor to be presented also gives the reader an introduction to 68000 assembly language programming and a collection of useful subroutines dealing with input/output transactions. Before designing this monitor, we look at the structure of a monitor and at some of the facilities it offers.

Structure of a Monitor

A monitor is a program, resident in ROM, whose minimal function is to enter a program into the microprocessor's RAM and then transfer control to that program. In order to do this, the monitor must provide procedures to control the micro-computer's input/output port, which is connected to a display device (the console or terminal). Most monitors communicate with a VDT by means of an asynchronous serial interface.

When we consider the design of a practical monitor, three points must be appreciated:

1. The monitor must be in read-only memory and located in the CPU's memory space where it is activated following a reset.

2. The monitor must control at least the terminal I/O interface. Therefore, the type of I/O port and its physical location within the processor's memory space is fixed.

3. A monitor should provide some facility for dealing with interrupts and exceptions. The structure of a monitor can be described in terms of PDL.

```
Module_Monitor
                    Initialize_system_constants
                    Set_up_console_I/O_port
                    Set_up_interrupt_vector_table
                    Display_heading_on_VDT
                    REPEAT
                        Clear_input_buffer
                        Get Command
                        CASE Command OF
                                Memory: Display/modify_memory_contents
                                Breakpoint: Set/clear_breakpoint
                                GO: GO_and_execute instructions
                                LOAD: Load formatted data into memory
                                DUMP: Output formatted data from memory
                                TM: Enter transparent mode
                        END CASE
                    UNTIL system reset
        END module
```

Although many possible commands may be included in a monitor, a few of the most important commands and their effects can be identified. Some of these commands are now detailed. As the TUTOR monitor on the MEX68KECB is so widely available and is also based on debugging facilities offered by Motorola's EXORMACS MDS, the following commands are related to TUTOR where necessary.

Memory Display/Modify   The display/modify command allows the contents of a selected memory location to be examined and, if necessary, modified. Although found on all monitors, the memory display/modify function exhibits a wide spread of facilities from system to system. A primitive monitor permits the examination of a byte at a specified address and its replacement by a new value, if desired. More sophisticated monitors allow byte, word, or longword operations and may even permit data to be displayed or entered in mnemonic form; that is, the contents of the specified location are disassembled.

TUTOR has two separate memory display/modify functions: MD (memory display), which displays the contents of one or more memory locations, and MM (memory modify), which displays the contents of a memory location and permits new data to be entered. Both commands allow data display and data entry to be in mnemonic form.

Memory Block Move   The block move command allows a block of memory to be moved (i.e., copied) from one part of the memory space to another. It is frequently used in conjunction with EPROM programmers, where the data to be written into the EPROM is copied from its source destination to the EPROM buffer, or vice versa.

Memory Block Fill   The block fill command presets the contents of a region of memory space to a given value, which is frequently done to initialize data storage areas before running a program. Clearly, if the data is initialized to, say $00, before executing a program, any change from $00 (after the execution of the program) can be detected by means of the memory display functions.

Memory Search   Sometimes we need the location of a particular data item in a region of memory space. The memory search command allows the user to seek the first occurrence of a given byte (word, longword) or a given string within a region of memory.

Load a Block of Data into Memory   The load command transfers a block of data from a terminal, secondary storage device, or host processor into the read/write memory space of the computer. The data to be loaded must be formatted in the exact way expected by the monitor; that is, the data is in the form of records with a header, byte count, data field, and error-detecting code.

TUTOR's load function has the form LO[⟨port number⟩][;options][=text]. The port number defines the source of the data to be loaded; the options (specified by the contents of the square brackets) permit checksums to be ignored or the data to be echoed back to the console terminal. The " =text" field causes the message "text" to be transmitted to the specified port before the data is loaded; for example, the command LO2; X = LIST MYPROG.BIN causes the string LIST MYPROG.BIN to be transmitted to the host computer via port 2 and the resulting data from the host to be displayed on the console terminal as well as stored in memory.

The data loaded by the LO function of TUTOR must conform to the S-record format that is widely used to record binary data. S-record data represents binary data in packets with five fields, as shown in figure 11.31. All binary data is represented in ASCII-encoded hexadecimal form. This representation is, of course, very inefficient,

FIGURE 11.31 S-record format

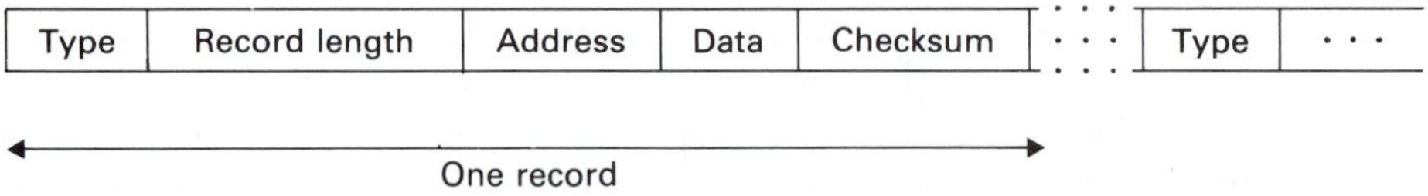

Type	Record length	Address	Data	Checksum	⋮⋮⋮	Type	⋯

◄――――――――――――――――――――――――――――――►

One record

Field	Field width	Field contents
Type	2 characters	S-record type – S0, S1, . . . , S9
Record length	2 characters	Total number of character pairs in the record, excluding type and record length fields
Address	4, 6, 8 characters	The 2-, 3-, or 4-byte address at which the data field is to be loaded into memory
Data	0–2n	From 0 to n bytes of data to be loaded in memory, starting at the specified address
Checksum	2 characters	The least-significant byte of the one's complement sum of the values represented by the pairs of characters making up the record length, address, and data fields

Although there are ten types of S record, the records of most interest are:

S0 An optional record defining succeeding records. The address field is normally all zeros.

S1 A record with a data field and a 2-byte address at which the data is to be loaded.

S2 As S1 but with a 3-byte address field.

S3 As S1 but with a 4-byte address field.

S7 A termination record for a block of S records. The address field contains the 4-byte address to which control is to be passed. There is no data field.

S8 Same as S7. The S8 record provides a terminator for S records, but the address field is 3 bytes.

S9 Same as S7. The S9 record provides a terminator for S records, but with a 2-byte address field.

and originates from the time when many data-storage devices (e.g., papertape) did not support pure binary, 8-bit characters.

Dump a Block of Data This command is the complementary function of the load command. A specified block of data is formed into S records and transmitted to a storage device (e.g., a cassette in low-cost systems) or to a host processor. Some monitors call this the PUNCH command—a term belonging to the days of papertape.

TUTOR's dump command has the form

DU[⟨port number⟩]⟨address 1⟩⟨address 2⟩[⟨text⟩].

If present, the optional text field is used to create an S0 header; for example, the command DU4 86C0 87FF CLEMENTS1 transmits to port 4 (the cassette port) an S0 message containing the data field CLEMENTS1, and a sequence of S1 messages with the data in memory locations $86C0 through $87FF. TUTOR follows this with an S9 or an S8 termination record.

Set/Remove Breakpoints This command allows users to place (or to remove) breakpoints in their programs. A breakpoint is a memory location that, when accessed by the processor, forces some specific action to take place. Normally, a breakpoint is a software exception operation code that is inserted in a program in place of a normal instruction. TUTOR uses the illegal op-code $4AFB to force an exception whenever it is encountered. Once the resulting exception has been raised, the contents of the processor's registers can be displayed on the console device. Execution is continued by replacing the illegal op-code with the saved instruction originally at that address. Some systems permit the use of one breakpoint, whereas others allow multiple simultaneous breakpoints. All monitors supporting a breakpoint also provide a command to clear existing breakpoints.

Execute a Program Once a program has been entered and, if necessary, modified, it can be executed, which is done by loading the program counter with the address of the first instruction to be executed. Sometimes, this address is called the program's transfer address (TA). Simple monitors provide a single EXECUTE or GO function. TUTOR has three variations on this command: GD, GO, GT. Before continuing, note that the term PC in what follows does not mean the 68000's program counter. PC is a "synthetic register" and contains the next address in the *user program* that is to be executed when this program is run.

GD The GD (GO direct) command has the form GD ⟨address⟩ and causes a program to be executed, starting at the specified address. If an address is not provided, execution begins at the point specified by the current contents of the PC.

GO The GO command has the form GO ⟨address⟩ and is similar to GD, thereby causing a program to be executed from the specified address. However, the GO command starts by tracing one instruction, setting any breakpoints, and then continuing.

GT The GT (GO until breakpoint) command has the form GT ⟨breakpoint⟩, thereby causing the processor to continue executing from the address currently in the PC until it encounters the temporary breakpoint address specified by the GT command.

Set Trace Mode When a program runs in the trace mode, an exception is raised after each instruction has been executed and the contents of the CPU's internal registers dumped on the console display. This action permits a program to be monitored line by line. TUTOR supports a trace command with the format TR [⟨count⟩]. The optional parameter, count, determines the number of instructions that are to be executed before the registers are dumped.

Transparent Mode The so-called transparent mode permits the console device to communicate directly with the host computer, thereby bypassing the target machine entirely. TUTOR supports this command with the syntax TM [⟨exit character⟩], where ⟨exit character⟩ specifies the character to be used to force a return to the TUTOR command level.

This command is used when the target computer is connected to both a console (on one port) and a host computer (on another port). By entering the transparent mode, the programmer is able to edit and assemble a program on the host machine. Then the exit character is entered (the default is control A) and a return to TUTOR control made. The DU (dump) function of TUTOR can then be employed to transfer the object program from the host to the target computer. Without the transparent mode, we would need to physically switch the console device from the target to the host processor and back.

Monitor Input/Output

Most monitors communicate with the outside world through a serial data link. How this communication is actually achieved varies widely from monitor to monitor. Such diversity is necessary if the functions provided by the monitor are to be made as versatile as possible. Here we consider four variations on the theme of input/output: I/O procedure, parameter-driven procedures, input/output and the device control block, and channel I/O.

I/O Procedure

Some of the earliest and most primitive monitors simply provided a subroutine (procedure) that either transmitted a character to the console (output) or received a character from the console (input). What more could one ask of an input/output routine? The answer is, "A lot." Predefined I/O routines are reasonable only when the nature of the I/O, its associated data path, and far-end (i.e., remote) terminal are all known in advance and never change.

Such an approach is very inflexible; for example, if the predefined I/O routine works with 7-bit ASCII characters, it can never be used to read 8-bit characters. Moreover, simple I/O routines do not have sophisticated built-in error-recovery procedures. What happens if the input data is faulty? What happens if the I/O is to be directed to a different port? In some of the early systems, the secondary storage device needed to be connected to the SBC by unplugging the console terminal and then plugging in the storage unit.

Parameter-Driven Procedures

The parameter-driven procedure still performs I/O transactions via predefined subroutines, but it permits the operational characteristics to be modified by changing parameters stored in read/write memory. During the running of the monitor's initialization routine, these parameters are set up to reflect the expected characteristics of the console terminal. A user can later alter them to modify the I/O characteristics of the SBC and to redirect I/O to an auxiliary port if necessary.

Consider an example of parameter-driven input expressed in PDL. The console input device is an ACIA whose address is stored in a variable called Console_ACIA.

Another pointer, Secondary_ACIA, holds the address of an alternative ACIA through which I/O can be directed. A flag bit, User_ACIA, is clear when I/O is performed by the console ACIA, and is set when it is performed by the secondary ACIA. By setting or clearing User_ACIA, we are able to switch I/O between ACIAs under software control.

Four other single-bit flags control the operation of the procedure. If Input_direction is set, the normal ACIA driver routine is not used and a jump is made to the subroutine, whose address is in the variable User_routine. This procecure permits the redirection of input to any device driver.

A parity strip flag is tested and, if clear, the input is to be regarded as 7-bit character-encoded data and the eighth (parity) bit stripped. Another flag (Case_conversion) determines whether lowercase characters should be converted to their uppercase equivalents. If Case_conversion is clear, upper- to lowercase conversion is carried out; for example, a lowercase "f" (110 0110) is converted into its uppercase equivalent "F" (100 0110) if Case_conversion is clear. Finally, the Echo_mode flag determines whether the input character is to be echoed on the console output device. If Echo_mode is clear, any input is echoed on the display device. All pointers and parameters are set up during the initialization phase of the monitor. Note that the default state of all single-bit flags is zero.

```
Module Input
            DEFINE POINTER: Console_ACIA {Points to console ACIA}
            DEFINE POINTER: Secondary_ACIA {Points to alternate ACIA}
            DEFINE POINTER: User_routine {Points to address of
                                          alternate I/O routine}
            DEFINE BYTE: Input_flag {Composite byte of 5 control bits of
                                     user supplied input}
            DEFINE BIT: Input_direction {If clear get data from console
                                         else from user routine}
            DEFINE BIT: User_ACIA {If set use Secondary_ACIA}
            DEFINE BIT: Parity_strip {If clear strip parity from input}
            DEFINE BIT: Case_conversion {If clear convert lower-case to
                                         upper-case}
            DEFINE BIT: Echo_mode {If clear echo input character on console}

            IF Input_direction = 0    THEN InConsole ELSE InUser END_IF
            IF Parity_strip = 0       THEN Strip_parity_bit END_IF
            IF Case_conversion = 0    THEN Convert_LC_to_UC END_IF
            IF Echo_mode = 0          THEN Output END_IF
END Input
InConsole
                MOVE.B      Input_flag,D1         Get input flag byte
                BTST        #User_ACIA,D1         Input from console?
                BEQ.S       InC1                  If clear then console
                MOVEA.L     Secondary_ACIA,A0     Load address of secondary
                BRA.S       InC2                  Skip load console ACIA
        InC1    MOVEA.L     Console_ACIA,A0       Load console ACIA address
        InC2    BTST        #0,(A0)               Test ACIA status
                BNE         InC2                  Loop until ACIA ready
                MOVE.B      2(A0),D0              Read input
                RTS

END InConsole
```

InUser

```
          MOVEA.L    User_routine,A0        Call user-supplied input
          JMP        (A0)
```

End InUser

Parameter-driven I/O is a great step forward over I/O provided by the rigid, embedded I/O procedures described previously. Unfortunately, this is a rather ad hoc approach to I/O and does not lend itself to generality; that is, the operating modes still have to be built into the monitor's software. If a radically different mode of operation were necessary, parameter-driven I/O would probably not prove sufficiently flexible. A better technique involves the more general concept of the device control block (DCB).

Input/Output and the Device Control Block

As its name suggests, a device control block (DCB) is a collection of parameters that completely defines the characteristics of an input/output transaction; that is, all the I/O procedure needs to know is the address of the appropriate DCB. All device-dependent information is stored in the DCB. Figure 11.32 illustrates a possible DCB for a console input device. The DCB is a data structure that, in the example of figure 11.32, has five fields.

The *header* supplies the name of the DCB. The header may be a logical device number or an ASCII string. The *device driver address* is a pointer to the subroutine that actually performs the input or output transaction. The *device address* provides the device driver with the location of the physical I/O device. The device *error code* is a status word returned by the device driver and reflects, for example, device_not_ready or parity errors. Finally, the DCB includes a *parameter block* that contains other information associated with the actual type of I/O being performed.

I/O by means of the DCB allows a greater degree of device independency;

FIGURE 11.32 Device control block (DCB)

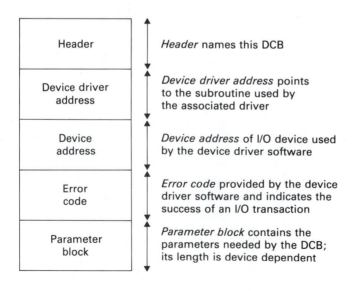

Header	*Header* names this DCB
Device driver address	*Device driver address* points to the subroutine used by the associated driver
Device address	*Device address* of I/O device used by the device driver software
Error code	*Error code* provided by the device driver software and indicates the success of an I/O transaction
Parameter block	*Parameter block* contains the parameters needed by the DCB; its length is device dependent

that is, the programmer can write programs that need to know nothing about the nature of the actual I/O devices. During the monitor's initialization phase, the console DCBs are set up in RAM using a table of default parameters held in ROM or on disk. The user can later redirect I/O by writing to the appropriate DCB, or a pointer can be set up to a new DCB.

The data structure forming the DCB is employed by a generalized I/O procedure. The programmer requests an input or an output transaction and passes the address of the DCB to the procedure; for example, in 68000 terminology, I/O may be performed by the following four steps:

1. Load D0 with the code of the operation to be performed (e.g., input, output, get_device_status, etc.).

2. Load D1 with any parameter needed to perform the desired operation.

3. Load A0 with the address of the DCB.

4. Call the generalized I/O handler (e.g., BSR IO_REQUEST).

Consider the following example of the use of a device control block in inputting data. To keep matters simple, the IO_REQUEST automatically inputs a byte. A real system would require the passing of a parameter to determine the nature of the operation to be performed. The example is in four parts: a call to the input handler (IO_REQUEST), the device control block appropriate to the input, the IO_REQUEST routine, and the device handler (IN_CON) used to obtain a character from the console ACIA.

```
CON_ACIA      EQU         $010040         Physical address of console ACIA
              :
*             Perform I/O here
              LEA         CON_DCB,A0      A0 points at the DCB to be used
              BSR         IO_REQUEST      Perform the input
              :                           Continue
              :                           :
*             Device control block for the console ACIA
CON_DCB       DC.B        'CON_ACIA'      8-byte header
              DC.L        CON_DRIVER      Address of console driver
              DC.L        CON_ACIA        Address of console ACIA
              DC.B        ERROR1          Error status 1 (logical error)
              DC.B        ERROR2          Error status 2 (physical error)
              DC.W        PARAM           Parameters needed by driver
              :
*             Entry point for standard I/O request
IO_REQUEST    JMP         8(A0)           Call input handler in DCB
              :
*             Actual device driver routine for the console ACIA
CON_DRIVER    MOVE.L      A1,-(SP)        Save A1 on stack
              LEA         12(A0),A1       Get address of ACIA from DCB
              CLR.B       16(A0)          Clear ERROR1 status
LOOP          MOVE.B      (A1),D0         Get ACIA status
              BTST        #0,D0           Test RDRF bit of status
              BEQ         LOOP            If RDRF clear then repeat
              MOVE.B      D0,17(A0)       Store device status in ERROR2 of DCB
              ANDI.B      #$70,D0         Mask to error bits of device status
```

```
                    BEQ.S      READ_DATA      If remaining bits clear, get data
                    MOVE.B     #1,16(A0)      Else set logical error flag in DCB
    READ_DATA       MOVE.B     2(A1),D0       Get input data from ACIA
                    MOVEA.L    (SP)+,A1       Restore A1
                    RTS
```

In this example, two error status bytes are associated with the DCB. ERROR1 is a logical error message and may be assigned codes to indicate: no_error, device_not_ready, etc. ERROR2 is a physical error message and is the status returned by the actual I/O device. The meaning of the bits in ERROR2 varies from DCB to DCB, whereas the bits of ERROR1 indicate one of a number of preassigned device-independent messages. For the sake of simplicity, ERROR2 is clear if there is no error and is set to $01 otherwise.

Note that no processing is performed on the input (e.g., parity stripping, lower-to uppercase conversion). This processing could be done by using the PARAM field of the DCB to determine the type of processing to be applied to the input.

It should now be clear that the application-level programmer does not have to know about the details of the actual I/O routines and their associated hardware. Furthermore, simply by altering the DCB address, the I/O can easily be redirected to some other channel.

Channel I/O

For the purpose of this discussion, channel I/O is considered as an application-level form of I/O using device control blocks. Channel I/O is built on the DCB mechanism and offers the programmer an even greater degree of freedom than that provided by the DCB alone.

Each I/O device is given a logical name, such as CON, PRNTR, MODEM, DISK, etc. This name is used by the programmer to form the header of the appropriate DCB. Channel I/O does not require the programmer to know the address of the DCB. When I/O is executed, the DCBs are searched until the DCB whose name matches that supplied by the programmer is found. To do this, each DCB contains a pointer to the next DCB in the chain. Figure 11.33 illustrates such a linked list.

As an example, in one possible arrangement the programmer simply creates an ASCII string in memory or provides a pointer to it and then calls a trap. The trap-handling routine searches each DCB for a header that matches the one provided. When the appropriate DCB has been located, the information in it is used to execute the appropriate I/O transaction. In order to avoid searching for a DCB each time a particular I/O transaction is executed, an alternative procedure is to "open" a channel. In this case, the DCB chain is searched once for the location of the appropriate DCB and the address of this DCB is "attached" to the current channel. The monitor written for the TS2 uses this form of I/O.

Monitor for the TS2

Now that we have designed the hardware of a 68000 system, the next step is to provide it with a monitor. The monitor presented here is a very simple monitor and is intended to achieve only two objectives: it provides an extended example of a

FIGURE 11.33 Linked list of DCBs used by I/O channels

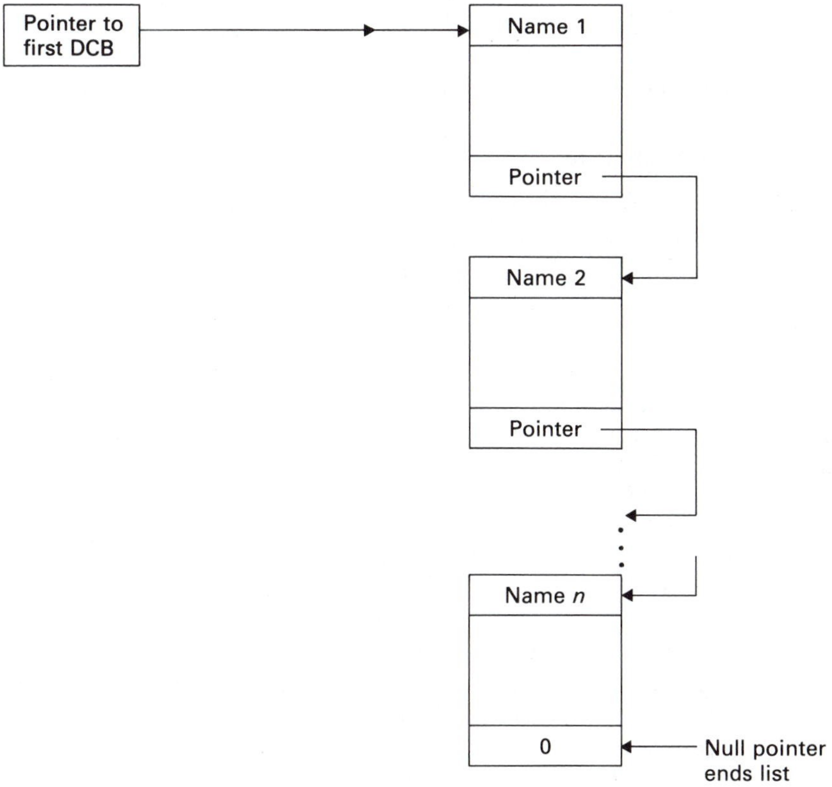

68000 assembly language program and it allows the hardware described earlier to be tested and downline-loaded from a host processor.

In designing such a monitor, the author is faced with conflicting goals—the monitor should be as simple as possible, yet it should illustrate a number of interesting or important features. Consequently, the monitor to be described is somewhat lopsided and, although very primitive, includes facilities normally associated with more sophisticated monitors. The monitor described is called TS2MON.

Specification of TS2MON

1. TS2MON is an EPROM-based monitor for a 68000 system and supports three functions: memory modify/examine, load a program from a host processor using S-formatted data, execute a program from some specified address.

2. TS2MON is a flexible monitor whose subroutines are capable of being used by other programs easily and efficiently. TS2MON is constructed so that additional commands may be added to its repertoire with little difficulty.

3. The command input is assembled into a buffer and then interpreted. A command line interpreter of the type described in chapter 3 is required.

4. Input/output is handled by means of device control blocks. Following a reset, two DCBs are set up by TS2MON: a DCB for the console and a DCB for the host processor. Both I/O devices are ACIAs.

5. TS2MON implements a very basic form of breakpoint mechanism. A program may be executed to a breakpoint and then run from the breakpoint.

Design of TS2MON

We are now going to discuss some of the features of TS2MON before presenting its listing. A detailed design is not given, as the listing is well endowed with comments. The basic structure of TS2MON is presented in PDL form.

```
Module: TS2MON
        Setup all pointers
        Setup ACIAs
        Setup exception table
        Setup DCB table
        Display heading
        REPEAT
            Get_command
            Execute_command
        END_REPEAT
End TS2MON
```

Following a reset, TS2MON sets up its operating environment, which involves creating device control blocks for the console ACIA and the auxiliary ACIA and loading the exception vector table with the addresses of all appropriate exception-handling routines.

Once the DCBs have been set up, the programmer is free to modify them in order to redirect I/O. Similarly, the exception vector table can be modified to provide alternative exception routines. Note that any exception not explicitly required by TS2MON is treated as an uninitialized interrupt.

In what follows, the names of subroutines are given in uppercase characters (usually in parentheses). The main part of the program is an infinite loop which assembles a line of text into a buffer (GETLINE), removes leading and multiple embedded spaces from the input (TIDY), and then matches the first string in the buffer with commands in the command table (EXECUTE). Before the built-in command table is searched, a user command table pointed at by a longword in UTAB is examined. This feature enables user-supplied commands to be added to TS2MON's instruction set.

The commands provided by TS2MON are self-explanatory (see the listing in figure 11.34). Only two features are worthy of special mention: the DCB structure and exception-handling facilities.

During the initialization process, the monitor sets up the appropriate DCBs in RAM (SET_DCB). Input/output is performed by loading register A0 with a pointer to the name of the desired DCB and then calling IO_OPEN. This searches the linked list of DCBs for the one whose name matches that pointed at by A0. On returning from IO_OPEN, A0 contains the address of the DCB itself.

Actual I/O is performed by calling IO_REQ, which reads the address of the device handler routine from the DCB pointed at by A0 and executes that routine. As all I/O carried out by TS2MON is in character form, two routines have been included to control the console device (still using DCBs). GETCHAR reads a character, strips the parity bit, converts lowercase to uppercase, and echoes the input to the console. Similarly PUTCHAR displays a character on the console.

Exceptions handled by TS2MON are: illegal instructions, bus error, address errors, and breakpoints. The breakpoint exception uses the TRAP #14 vector.

Group 1 exceptions (bus and address errors) are handled by displaying the appropriate error message and then calling GROUP1, which reads the program counter from the stack and the instruction being executed at the time of the exception. As the PC on the stack is not the actual value of the PC at the time of the exception (due to the 68000's prefetch facility), a search is made in the area pointed at by the saved PC until the op-code corresponding to the saved instruction is located. The address of this instruction is taken as the "correct" value of the PC. The GROUP1 stack frame is then cleaned up to make it look like a group 2 exception and the group 2 exception-handling routine is called.

GROUP2 handles all group 2 exceptions and group 1 exceptions after "preprocessing" by GROUP1. The action carried out by GROUP2 is to make a copy of all the 68000's registers and the program counter/status register in a data structure called TSK_T. Two commands operate on this data structure. EX_DIS displays the contents of these registers and REG permits any register to be updated within the table; for example, the command REG PC FF0A has the effect of altering the program counter stored in the data structure to $FF0A.

Up to eight breakpoints are set up by the command BRGT ⟨address⟩. This command only stores the user-supplied address in the breakpoint table (BP_TAB). Similarly, NOBR ⟨address⟩ deletes the appropriate breakpoint from the table. A NOBR command without an address clears all breakpoints.

A program may be executed by the command GB ⟨address⟩ or by GB. If an address is supplied, a jump to that address is made; otherwise the program counter is loaded from the value stored in TSK_T. In the latter case, all address and data registers (except SSP) are loaded from the TSK_T. In both cases, the breakpoints are set prior to execution. A TRAP #14 is placed at the address pointed at by each breakpoint and the instruction that was at the address is saved in the breakpoint table.

When a breakpoint is encountered, the volatile environment is displayed and all breakpoints are cleared. Execution can be continued from the breakpoint by entering the command GB.

FIGURE 11.34 Listing of TS2MON

Source file: MONITOR.X68
Assembled on: 91-02-08 at: 21:23:59
            by: X68K PC-1.7 Copyright (c) Teesside Polytechnic 1989
Defaults: ORG $0/FORMAT/OPT A,BRL,CEX,CL,FRL,MC,MD,NOMEX,NOPCO

```
  1                        *       TSBUG2 - 68000 monitor - version of 23 July 1986
  2                        *                               Symbol equates
  3          00000008      BS:      EQU    $08              ;Back_space
  4          0000000D      CR:      EQU    $0D              ;Carriage_return
  5          0000000A      LF:      EQU    $0A              ;Line_feed
  6          00000020      SPACE:   EQU    $20              ;Space
  7          57000000      WAIT:    EQU    'W'              ;Wait character (to suspend output)
  8          0000001B      ESC:     EQU    $1B              ;ASCII escape character (used by TM)
  9          00000001      CTRL_A:  EQU    $01              ;Control_A forces return to monitor
 10                        *                               Device addresses
 11          00000800      STACK:   EQU    $00000800        ;Stack_pointer
 12          00010040      ACIA_1:  EQU    $00010040        ;Console ACIA control
 13          00010041      ACIA_2:  EQU    ACIA_1+1         ;Auxilary ACIA control
 14          00000008      X_BASE:  EQU    $08              ;Start of exception vector table
 15          00004E4E      TRAP_14: EQU    $4E4E            ;Code for TRAP #14
 16          00000040      MAXCHR:  EQU    64               ;Length of input line buffer
 17                        *
 18          00000C00      DATA:    EQU    $00000C00        ;Data origin
 19 00000000 00000040      LNBUFF:  DS.B   MAXCHR           ;Input line buffer
 20          0000003F      BUFFEND: EQU    LNBUFF+MAXCHR-1   ;End of line buffer
 21 00000040 00000004      BUFFPT:  DS.L   1                ;Pointer to line buffer
 22 00000044 00000004      PARAMTR: DS.L   1                ;Last parameter from line buffer
 23 00000048 00000001      ECHO:    DS.B   1                ;When clear this enable input echo
 24 00000049 00000001      U_CASE:  DS.B   1                ;Flag for upper case conversion
 25 0000004A 00000004      UTAB:    DS.L   1                ;Pointer to user command table
 26 0000004E 00000004      CN_IVEC: DS.L   1                ;Pointer to console input DCB
 27 00000052 00000004      CN_OVEC: DS.L   1                ;Pointer to console output DCB
 28 00000056 0000004A      TSK_T:   DS.W   37               ;Frame for D0-D7, A0-A6, USP, SSP, SW, PC
 29 000000A0 00000030      BP_TAB:  DS.W   24               ;Breakpoint table
 30 000000D0 00000200      FIRST:   DS.B   512              ;DCB area
 31 000002D0 00000100      BUFFER:  DS.B   256              ;256 bytes for I/O buffer
 32                        *
```

```
33                     ***********************************************************************
34                     *
35                     *   This is the main program which assembles a command in the line
36                     *   buffer, removes leading/embedded spaces and interprets it by matching
37                     *   it with a command in the user table or the built-in table COMTAB
38                     *   All variables are specified with respect to A6
39                     *
40   00008000                      ORG      $00008000              ;Monitor origin
41   00008000 00000800             DC.L     STACK                  ;Reset stack pointer
42   00008004 00008008             DC.L     RESET                  ;Reset vector
43            00008008  RESET:      EQU      *                      ;Cold entry point for monitor
44   00008008 4DF80C00             LEA      DATA,A6                ;A6 points to data area
45   0000800C 42AE004A             CLR.L    UTAB(A6)               ;Reset pointer to user extension table
46   00008010 422E0048             CLR.B    ECHO(A6)               ;Set automatic character echo
47   00008014 422E0049             CLR.B    U_CASE(A6)             ;Clear case conversion flag (UC<-LC)
48   00008018 6136                 BSR.S    SETACIA                ;Setup ACIAs
49   0000801A 610005D8             BSR      X_SET                  ;Setup exception table
50   0000801E 6100044E             BSR      SET_DCB                ;Setup DCB table in RAM
51   00008022 49FA09D0             LEA      BANNER(PC),A4          ;Point to banner
52   00008026 6164                 BSR.S    HEADING                ;and print heading
53   00008028 207C0000C000         MOVE.L   #$0000C000,A0          ;A0 points to extension ROM
54   0000802E 2010                 MOVE.L   (A0),D0                ;Read first longword in extension ROM
55   00008030 0C80524F4D32         CMP.L    #'ROM2',D0             ;If extension begins with 'ROM2' then
56   00008036 6604                 BNE.S    NO_EXT                 ;call the subroutine at EXT_ROM+8
57   00008038 4EA80008             JSR      8(A0)                  ;else continue
58   0000803C 4E71     NO_EXT:     NOP                             ;Two NOPs to allow for a future
59   0000803E 4E71                 NOP                             ;call to an initialization routine
60   00008040 4287     WARM:       CLR.L    D7                     ;Warm entry point - clear error flag
61   00008042 6128                 BSR.S    NEWLINE                ;Print a newline
62   00008044 614C                 BSR.S    GETLINE                ;Get a command line
63   00008046 61000080             BSR      TIDY                   ;Tidy up input buffer contents
64   0000804A 610000BE             BSR      EXECUTE                ;Interpret command
65   0000804E 60F0                 BRA      WARM                   ;Repeat indefinitely
66                     *
67                     ***********************************************************************
68                     *
69                     *   Some initialization and basic routines
70                     *
71            00008050  SETACIA:    EQU      *                      ;Setup ACIA parameters
72   00008050 41F900010040         LEA      ACIA_1,A0              ;A0 points to console ACIA
73   00008056 10BC0003             MOVE.B   #$03,(A0)              ;Reset ACIA1
74   0000805A 117C00030001         MOVE.B   #$03,1(A0)             ;Reset ACIA2
75   00008060 10BC0015             MOVE.B   #$15,(A0)              ;Set up ACIA1 constants (no IRQ,
76   00008064 117C00150001         MOVE.B   #$15,1(A0)             ;RTS* low, 8 bit, no parity, 1 stop)
77   0000806A 4E75                 RTS                             ;     Return
78                     *
79            0000806C  NEWLINE:    EQU      *                      ;Move cursor to start of newline
80   0000806C 48E70008             MOVEM.L  A4,-(A7)               ;Save A4
81   00008070 49FA099C             LEA      CRLF(PC),A4            ;Point to CR/LF string
82   00008074 6106                 BSR.S    PSTRING                ;Print it
83   00008076 4CDF1000             MOVEM.L  (A7)+,A4               ;Restore A4
84   0000807A 4E75                 RTS                             ;     Return
```

```
85                      *
86           0000807C   PSTRING:  EQU      *              ;Display the string pointed at by A4
87  0000807C 2F00                 MOVE.L   D0,-(A7)       ;Save D0
88  0000807E 101C       PS1:      MOVE.B   (A4)+,D0       ;Get character to be printed
89  00008080 6706                 BEQ.S    PS2            ;If null then return
90  00008082 6100050C             BSR      PUTCHAR        ;Else print it
91  00008086 60F6                 BRA      PS1            ;Continue
92  00008088 201F       PS2:      MOVE.L   (A7)+,D0       ;Restore D0 and exit
93  0000808A 4E75                 RTS
94                      *
95  0000808C 61DE       HEADING:  BSR      NEWLINE        ;Same as PSTRING but with newline
96  0000808E 61EC                 BSR      PSTRING
97  00008090 60DA                 BRA      NEWLINE
98                      *
99                      ********************************************************************
100                     *
101                     *   GETLINE  inputs a string of characters into a line buffer
102                     *           A3 points to next free entry in line buffer
103                     *           A2 points to end of buffer
104                     *           A1 points to start of buffer
105                     *           D0 holds character to be stored
106                     *
107 00008092 43EE0000   GETLINE:  LEA      LNBUFF(A6),A1  ;A1 points to start of line buffer
108 00008096 47D1                 LEA      (A1),A3        ;A3 points to start (initially)
109 00008098 45E90040             LEA      MAXCHR(A1),A2  ;A2 points to end of buffer
110 0000809C 610004C0   GETLN2:   BSR      GETCHAR        ;Get a character
111 000080A0 0C000001             CMP.B    #CTRL_A,D0     ;If control_A then reject this line
112 000080A4 671E                 BEQ.S    GETLN5         ;and get another line
113 000080A6 0C000008             CMP.B    #BS,D0         ;If back_space then move back pointer
114 000080AA 660A                 BNE.S    GETLN3         ;Else skip past wind-back routine
115 000080AC B7C9                 CMP.L    A1,A3          ;First check for empty buffer
116 000080AE 67EC                 BEQ      GETLN2         ;If buffer empty then continue
117 000080B0 47EBFFFF             LEA      -1(A3),A3      ;Else decrement buffer pointer
118 000080B4 60E6                 BRA      GETLN2         ;and continue with next character
119 000080B6 16C0       GETLN3:   MOVE.B   D0,(A3)+       ;Store character and update pointer
120 000080B8 0C00000D             CMP.B    #CR,D0         ;Test for command terminator
121 000080BC 6602                 BNE.S    GETLN4         ;If not CR then skip past exit
122 000080BE 60AC                 BRA      NEWLINE        ;Else new line before next operation
123 000080C0 B7CA       GETLN4:   CMP.L    A2,A3          ;Test for buffer overflow
124 000080C2 66D8                 BNE      GETLN2         ;If buffer not full then continue
125 000080C4 61A6       GETLN5:   BSR      NEWLINE        ;Else move to next line and
126 000080C6 60CA                 BRA      GETLINE        ;repeat this routine
127                     *
```

```
128                          ******************************************************************************
129                          *
130                          *   TIDY cleans up the line buffer by removing leading spaces and multiple
131                          *       spaces between parameters. At the end of TIDY, BUFFPT points to
132                          *       the first parameter following the command.
133                          *       A0 = pointer to line buffer. A1 = pointer to cleaned up buffer
134                          *
135   000080C8 41EE0000   TIDY:    LEA      LNBUFF(A6),A0      ;A0 points to line buffer
136   000080CC 43D0                LEA      (A0),A1            ;A1 points to start of line buffer
137   000080CE 1018       TIDY1:   MOVE.B   (A0)+,D0           ;Read character from line buffer
138   000080D0 0C000020            CMP.B    #SPACE,D0          ;Repeat until the first non-space
139   000080D4 67F8                BEQ      TIDY1              ;character is found
140   000080D6 41E8FFFF            LEA      -1(A0),A0          ;Move pointer back to first char
141   000080DA 1018       TIDY2:   MOVE.B   (A0)+,D0           ;Move the string left to remove
142   000080DC 12C0                MOVE.B   D0,(A1)+           ;any leading spaces
143   000080DE 0C000020            CMP.B    #SPACE,D0          ;Test for embedded space
144   000080E2 660A                BNE.S    TIDY4              ;If not space then test for EOL
145   000080E4 0C180020   TIDY3:   CMP.B    #SPACE,(A0)+       ;If space skip multiple embedded
146   000080E8 57FA                BEQ      TIDY3              ;spaces
147   000080EA 41E8FFFF            LEA      -1(A0),A0          ;Move back pointer
148   000080EE 0C00000D   TIDY4:   CMP.B    #CR,D0             ;Test for end_of_line (EOL)
149   000080F2 66E6                BNE      TIDY2              ;If not EOL then read next char
150   000080F4 41EE0000            LEA      LNBUFF(A6),A0      ;Restore buffer pointer
151   000080F8 0C10000D   TIDY5:   CMP.B    #CR,(A0)           ;Test for EOL
152   000080FC 6706                BEQ.S    TIDY6              ;If EOL then exit
153   000080FE 0C180020            CMP.B    #SPACE,(A0)+       ;Test for delimiter
154   00008102 66F4                BNE      TIDY5              ;Repeat until delimiter or EOL
155   00008104 2D480040   TIDY6:   MOVE.L   A0,BUFFPT(A6)      ;Update buffer pointer
156   00008108 4E75                RTS
157                          *
158                          ******************************************************************************
159                          *
160                          *   EXECUTE matches the first command in the line buffer with the
161                          *   commands in a command table. An external table pointed at by
162                          *   UTAB is searched first and then the in-built table, COMTAB.
163                          *
164   0000810A 4AAE004A   EXECUTE: TST.L    UTAB(A6)           ;Test pointer to user table
165   0000810E 670C                BEQ.S    EXEC1              ;If clear then try built-in table
166   00008110 266E004A            MOVE.L   UTAB(A6),A3        ;Else pick up pointer to user table
167   00008114 6120                BSR.S    SEARCH             ;Look for command in user table
168   00008116 6404                BCC.S    EXEC1              ;If not found then try internal table
169   00008118 2653                MOVE.L   (A3),A3            ;Else get absolute address of command
170   0000811A 4ED3                JMP      (A3)               ;from user table and execute it
171                          *
172   0000811C 47FA0A46   EXEC1:   LEA      COMTAB(PC),A3      ;Try built-in command table
173   00008120 6114                BSR.S    SEARCH             ;Look for command in built-in table
174   00008122 6508                BCS.S    EXEC2              ;If found then execute command
175   00008124 49FA09CF            LEA      ERMES2(PC),A4      ;Else print ''invalid command''
176   00008128 6000FF52            BRA.L    PSTRING            ;and return
177   0000812C 2653       EXEC2:   MOVE.L   (A3),A3            ;Get the relative command address
178   0000812E 49FA0A34            LEA      COMTAB(PC),A4      ;pointed at by A3 and add it to
179   00008132 D7CC                ADD.L    A4,A3              ;the PC to generate the actual
180   00008134 4ED3                JMP      (A3)               ;command address. Then execute it.
181                          *
```

```
182           00008136      SEARCH:  EQU      *              ;Match the command in the line buffer
183  00008136 4280                   CLR.L    D0             ;with command table pointed at by A3
184  00008138 1013                   MOVE.B   (A3),D0        ;Get the first character in the
185  0000813A 6734                   BEQ.S    SRCH7          ;current entry. If zero then exit
186  0000813C 49F30006               LEA      6(A3,D0.W),A4  ;Else calculate address of next entry
187  00008140 122B0001               MOVE.B   1(A3),D1       ;Get number of characters to match
188  00008144 4BEE0000               LEA      LNBUFF(A6),A5  ;A5 points to command in line buffer
189  00008148 142B0002               MOVE.B   2(A3),D2       ;Get first character in this entry
190  0000814C B41D                   CMP.B    (A5)+,D2       ;from the table and match with buffer
191  0000814E 6704                   BEQ.S    SRCH3          ;If match then try rest of string
192  00008150 264C        SRCH2:     MOVE.L   A4,A3          ;Else get address of next entry
193  00008152 60E2                   BRA      SEARCH         ;and try the next entry in the table
194  00008154 5301        SRCH3:     SUB.B    #1,D1          ;One less character to match
195  00008156 670E                   BEQ.S    SRCH6          ;If match counter zero then all done
196  00008158 47EB0003               LEA      3(A3),A3       ;Else point to next character in table
197  0000815C 141B        SRCH4:     MOVE.B   (A3)+,D2       ;Now match a pair of characters
198  0000815E B41D                   CMP.B    (A5)+,D2
199  00008160 66EE                   BNE      SRCH2          ;If no match then try next entry
200  00008162 5301                   SUB.B    #1,D1          ;Else decrement match counter and
201  00008164 66F6                   BNE      SRCH4          ;repeat until no chars left to match
202  00008166 47ECFFFC    SRCH6:     LEA      -4(A4),A3      ;Calculate address of command entry
203  0000816A 003C0001               OR.B     #1,CCR         ;point. Mark carry flag as success
204  0000816E 4E75                   RTS                     ;and return
205  00008170 023C00FE    SRCH7:     AND.B    #$FE,CCR       ;Fail - clear carry to indicate
206  00008174 4E75                   RTS                     ;command not found and return
207                       *
208                       **********************************************************************
209                       *
210                       *  Basic input routines
211                       *  HEX    = Get one   hexadecimal character  into D0
212                       *  BYTE   = Get two   hexadecimal characters into D0
213                       *  WORD   = Get four  hexadecimal characters into D0
214                       *  LONGWD = Get eight hexadecimal characters into D0
215                       *  PARAM  = Get a longword from the line buffer into D0
216                       *  Bit 0 of D7 is set to indicate a hexadecimal input error
217                       *
218  00008176 610003E6    HEX:       BSR      GETCHAR        ;Get a character from input device
219  0000817A 04000030               SUB.B    #$30,D0        ;Convert to binary
220  0000817E 6B0E                   BMI.S    NOT_HEX        ;If less than $30 then exit with error
221  00008180 0C000009               CMP.B    #$09,D0        ;Else test for number (0 to 9)
222  00008184 6F0C                   BLE.S    HEX_OK         ;If number then exit - success
223  00008186 5F00                   SUB.B    #$07,D0        ;Else convert letter to hex
224  00008188 0C00000F               CMP.B    #$0F,D0        ;If character in range ''A'' to ''F''
225  0000818C 6F04                   BLE.S    HEX_OK         ;then exit successfully
226  0000818E 00070001    NOT_HEX:   OR.B     #1,D7          ;Else set error flag
227  00008192 4E75        HEX_OK:    RTS                     ;and return
228                       *
229  00008194 2F01        BYTE:      MOVE.L   D1,-(A7)       ;Save D1
230  00008196 61DE                   BSR      HEX            ;Get first hex character
231  00008198 E900                   ASL.B    #4,D0          ;Move it to MS nybble position
232  0000819A 1200                   MOVE.B   D0,D1          ;Save MS nybble in D1
233  0000819C 61D8                   BSR      HEX            ;Get second hex character
234  0000819E D001                   ADD.B    D1,D0          ;Merge MS and LS nybbles
235  000081A0 221F                   MOVE.L   (A7)+,D1       ;Restore D1
236  000081A2 4E75                   RTS
237                       *
```

```
238  000081A4  61EE      WORD:     BSR       BYTE              ;Get upper order byte
239  000081A6  E140                ASL.W     #8,D0             ;Move it to MS position
240  000081A8  60EA                BRA       BYTE              ;Get LS byte and return
241                      *
242  000081AA  61F8      LONGWD:   BSR       WORD              ;Get upper order word
243  000081AC  4840                SWAP      D0                ;Move it to MS position
244  000081AE  60F4                BRA       WORD              ;Get lower order word and return
245                      *
246                      *  PARAM reads a parameter from the line buffer and puts it in both
247                      *  PARAMTR(A6) and D0. Bit 1 of D7 is set on error.
248                      *
249  000081B0  2F01      PARAM:    MOVE.L    D1,-(A7)          ;Save D1
250  000081B2  4281                CLR.L     D1                ;Clear input accumulator
251  000081B4  206E0040             MOVE.L   BUFFPT(A6),A0     ;A0 points to parameter in buffer
252  000081B8  1018      PARAM1:   MOVE.B    (A0)+,D0          ;Read character from line buffer
253  000081BA  0C000020            CMP.B     #SPACE,D0         ;Test for delimiter
254  000081BE  6720                BEQ.S     PARAM4            ;The permitted delimiter is a
255  000081C0  0C00000D            CMP.B     #CR,D0            ;space or a carriage return
256  000081C4  671A                BEQ.S     PARAM4            ;Exit on either space or C/R
257  000081C6  E981                ASL.L     #4,D1             ;Shift accumulated result 4 bits left
258  000081C8  04000030            SUB.B     #$30,D0           ;Convert new character to hex
259  000081CC  6B1E                BMI.S     PARAM5            ;If less than $30 then not-hex
260  000081CE  0C000009            CMP.B     #$09,D0           ;If less than 10
261  000081D2  6F08                BLE.S     PARAM3            ;then continue
262  000081D4  5F00                SUB.B     #$07,D0           ;Else assume $A - $F
263  000081D6  0C00000F            CMP.B     #$0F,D0           ;If more than $F
264  000081DA  6E10                BGT.S     PARAM5            ;then exit to error on not-hex
265  000081DC  D200      PARAM3:   ADD.B     D0,D1             ;Add latest nybble to total in D1
266  000081DE  60D8                BRA       PARAM1            ;Repeat until delimiter found
267  000081E0  2D480040  PARAM4:   MOVE.L    A0,BUFFPT(A6)     ;Save pointer in memory
268  000081E4  2D410044            MOVE.L    D1,PARAMTR(A6)    ;Save parameter in memory
269  000081E8  2001                MOVE.L    D1,D0             ;Put parameter in D0 for return
270  000081EA  6004                BRA.S     PARAM6            ;Return without error
271  000081EC  00070002  PARAM5:   OR.B      #2,D7             ;Set error flag before return
272  000081F0  221F      PARAM6:   MOVE.L    (A7)+,D1          ;Restore working register
273  000081F2  4E75                RTS                         ;Return with error
274                      *
275                      *************************************************************************
276                      *
277                      *  Output routines
278                      *  OUT1X    = print one   hexadecimal character
279                      *  OUT2X    = print two   hexadecimal characters
280                      *  OUT4X    = print four  hexadecimal characters
281                      *  OUT8X    = print eight hexadecimal characters
282                      *  In each case, the data to be printed is in D0
283                      *
284  000081F4  1F00      OUT1X:    MOVE.B    D0,-(A7)          ;Save D0
285  000081F6  0200000F            AND.B     #$0F,D0           ;Mask off MS nybble
286  000081FA  06000030            ADD.B     #$30,D0           ;Convert to ASCII
287  000081FE  0C000039            CMP.B     #$39,D0           ;ASCII = HEX + $30
288  00008202  6302                BLS.S     OUT1X1            ;If ASCII <= $39 then print and exit
289  00008204  5E00                ADD.B     #$07,D0           ;Else ASCII := HEX + 7
290  00008206  61000388  OUT1X1:   BSR       PUTCHAR           ;Print the character
291  0000820A  101F                MOVE.B    (A7)+,D0          ;Restore D0
292  0000820C  4E75                RTS
293                      *
```

```
294   0000820E E818    OUT2X:  ROR.B   #4,D0           ;Get MS nybble in LS position
295   00008210 61E2            BSR     OUT1X           ;Print MS nybble
296   00008212 E918            ROL.B   #4,D0           ;Restore LS nybble
297   00008214 60DE            BRA     OUT1X           ;Print LS nybble and return
298                    *
299   00008216 E058    OUT4X:  ROR.W   #8,D0           ;Get MS byte in LS position
300   00008218 61F4            BSR     OUT2X           ;Print MS byte
301   0000821A E158            ROL.W   #8,D0           ;Restore LS byte
302   0000821C 60F0            BRA     OUT2X           ;Print LS byte and return
303                    *
304   0000821E 4840    OUT8X:  SWAP    D0              ;Get MS word in LS position
305   00008220 61F4            BSR     OUT4X           ;Print MS word
306   00008222 4840            SWAP    D0              ;Restore LS word
307   00008224 60F0            BRA     OUT4X           ;Print LS word and return
308                    *
309                    ***********************************************************************
310                    *
311                    * JUMP causes execution to begin at the address in the line buffer
312                    *
313   00008226 6188    JUMP:   BSR     PARAM           ;Get address from buffer
314   00008228 4A07            TST.B   D7              ;Test for input error
315   0000822A 6608            BNE.S   JUMP1           ;If error flag not zero then exit
316   0000822C 4A80            TST.L   D0              ;Else test for missing address
317   0000822E 6704            BEQ.S   JUMP1           ;field. If no address then exit
318   00008230 2040            MOVE.L  D0,A0           ;Put jump address in A0 and call the
319   00008232 4ED0            JMP     (A0)            ;subroutine. User to supply RTS!!
320   00008234 49FA08A1 JUMP1: LEA     ERMES1(PC),A4   ;Here for error - display error
321   00008238 6000FE42        BRA     PSTRING         ;message and return
322                    *
323                    ***********************************************************************
324                    *
325                    * Display the contents of a memory location and modify it
326                    *
327   0000823C 6100FF72 MEMORY: BSR    PARAM           ;Get start address from line buffer
328   00008240 4A07            TST.B   D7              ;Test for input error
329   00008242 6634            BNE.S   MEM3            ;If error then exit
330   00008244 2640            MOVE.L  D0,A3           ;A3 points to location to be opened
331   00008246 6100FE24 MEM1:  BSR     NEWLINE
332   0000824A 612E            BSR.S   ADR_DAT         ;Print current address and contents
333   0000824C 6140            BSR.S   PSPACE          ;update pointer, A3, and O/P space
334   0000824E 6100030E        BSR     GETCHAR         ;Input char to decide next action
335   00008252 0C00000D        CMP.B   #CR,D0          ;If carriage return then exit
336   00008256 6720            BEQ.S   MEM3            ;Exit
337   00008258 0C00002D        CMP.B   #'-',D0         ;If ''-'' then move back
338   0000825C 6606            BNE.S   MEM2            ;Else skip wind-back procedure
339   0000825E 47EBFFFC        LEA     -4(A3),A3       ;Move pointer back 2+2
340   00008262 60E2            BRA     MEM1            ;Repeat until carriage return
341   00008264 0C000020 MEM2:  CMP.B   #SPACE,D0       ;Test for space (= new entry)
342   00008268 66DC            BNE.S   MEM1            ;If not space then repeat
343   0000826A 6100FF38        BSR     WORD            ;Else get new word to store
344   0000826E 4A07            TST.B   D7              ;Test for input error
345   00008270 6606            BNE.S   MEM3            ;If error then exit
346   00008272 3740FFFE        MOVE.W  D0,-2(A3)       ;Store new word
347   00008276 60CE            BRA     MEM1            ;Repeat until carriage return
348   00008278 4E75    MEM3:   RTS
349                    *
```

```
350  0000827A 2F00         ADR_DAT: MOVE.L   D0,-(A7)           ;Print the contents of A3 and the
351  0000827C 200B                  MOVE.L   A3,D0              ;word pointed at by A3.
352  0000827E 619E                  BSR      OUT8X              ;and print current address
353  00008280 610C                  BSR.S    PSPACE             ;Insert delimiter
354  00008282 3013                  MOVE.W   (A3),D0            ;Get data at this address in D0
355  00008284 6190                  BSR      OUT4X              ;and print it
356  00008286 47EB0002              LEA      2(A3),A3           ;Point to next address to display
357  0000828A 201F                  MOVE.L   (A7)+,D0           ;Restore D0
358  0000828C 4E75                  RTS
359                        *
360  0000828E 1F00         PSPACE:  MOVE.B   D0,-(A7)           ;Print a single space
361  00008290 103C0020              MOVE.B   #SPACE,D0
362  00008294 610002FA              BSR      PUTCHAR
363  00008298 101F                  MOVE.B   (A7)+,D0
364  0000829A 4E75                  RTS
365                        *
366                        *************************************************************************
367                        *
368                        *  LOAD  Loads data formatted in hexadecimal ''S'' format from Port 2
369                        *        NOTE - I/O is automatically redirected to the aux port for
370                        *        loader functions. S1 or S2 records accepted
371                        *
372  0000829C 2F2E0052     LOAD:    MOVE.L   CN_OVEC(A6),-(A7)  ;Save current output device name
373  000082A0 2F2E004E              MOVE.L   CN_IVEC(A6),-(A7)  ;Save current input device name
374  000082A4 2D7C00008C22          MOVE.L   #DCB4,CN_OVEC(A6)  ;Set up aux ACIA as output
              0052
375  000082AC 2D7C00008C10          MOVE.L   #DCB3,CN_IVEC(A6)  ;Set up aux ACIA as input
              004E
376  000082B4 522E0048              ADD.B    #1,ECHO(A6)        ;Turn off character echo
377  000082B8 6100FDB2              BSR      NEWLINE            ;Send newline to host
378  000082BC 6100015A              BSR      DELAY              ;Wait for host to ''settle''
379  000082C0 61000156              BSR      DELAY
380  000082C4 286E0040              MOVE.L   BUFFPT(A6),A4      ;Any string in the line buffer is
381  000082C8 101C         LOAD1:   MOVE.B   (A4)+,D0           ;transmitted to the host computer
382  000082CA 610002C4              BSR      PUTCHAR            ;before the loading begins
383  000082CE 0C00000D              CMP.B    #CR,D0             ;Read from the buffer until EOL
384  000082D2 66F4                  BNE      LOAD1
385  000082D4 6100FD96              BSR      NEWLINE            ;Send newline before loading
386  000082D8 61000284     LOAD2:   BSR      GETCHAR            ;Records from the host must begin
387  000082DC 0C000053              CMP.B    #'S',D0            ;with S1/S2 (data) or S9/S8 (term)
388  000082E0 66F6                  BNE.S    LOAD2              ;Repeat GETCHAR until char = ''S''
389  000082E2 6100027A              BSR      GETCHAR            ;Get character after ''S''
390  000082E6 0C000039              CMP.B    #'9',D0            ;Test for the two terminators S9/S8
391  000082EA 6706                  BEQ.S    LOAD3              ;If S9 record then exit else test
392  000082EC 0C000038              CMP.B    #'8',D0            ;for S8 terminator. Fall through to
393  000082F0 662A                  BNE.S    LOAD6              ;exit on S8 else continue search
394           000082F2     LOAD3:   EQU      *                  ;Exit point from LOAD
395  000082F2 2D5F004E              MOVE.L   (A7)+,CN_IVEC(A6)  ;Clean up by restoring input device
396  000082F6 2D5F0052              MOVE.L   (A7)+,CN_OVEC(A6)  ;and output device name
397  000082FA 422E0048              CLR.B    ECHO(A6)           ;Restore input character echo
398  000082FE 08070000              BTST     #0,D7              ;Test for input errors
399  00008302 6708                  BEQ.S    LOAD4              ;If no I/P error then look at checksum
400  00008304 49FA07D1              LEA      ERMES1(PC),A4      ;Else point to error message
401  00008308 6100FD72              BSR      PSTRING            ;Print it
402  0000830C 08070003     LOAD4:   BTST     #3,D7              ;Test for checksum error
403  00008310 6708                  BEQ.S    LOAD5              ;If clear then exit
404  00008312 49FA07F3              LEA      ERMES3(PC),A4      ;Else point to error message
405  00008316 6100FD64              BSR      PSTRING            ;Print it and return
406  0000831A 4E75         LOAD5:   RTS
407                        *
```

```
408  0000831C 0C000031   LOAD6:   CMP.B    #'1',D0             ;Test for S1 record
409  00008320 671E                BEQ.S    LOAD6A              ;If S1 record then read it
410  00008322 0C000032            CMP.B    #'2',D0             ;Else test for S2 record
411  00008326 66B0                BNE.S    LOAD2               ;Repeat until valid header found
412  00008328 4203                CLR.B    D3                  ;Read the S2 byte count and address,
413  0000832A 613C                BSR.S    LOAD8               ;clear the checksum
414  0000832C 5900                SUB.B    #4,D0               ;Calculate size of data field
415  0000832E 1400                MOVE.B   D0,D2               ;D2 contains data bytes to read
416  00008330 4280                CLR.L    D0                  ;Clear address accumulator
417  00008332 6134                BSR.S    LOAD8               ;Read most sig byte of address
418  00008334 E180                ASL.L    #8,D0               ;Move it one byte left
419  00008336 6130                BSR.S    LOAD8               ;Read the middle byte of address
420  00008338 E180                ASL.L    #8,D0               ;Move it one byte left
421  0000833A 612C                BSR.S    LOAD8               ;Read least sig byte of address
422  0000833C 2440                MOVE.L   D0,A2               ;A2 points to destination of record
423  0000833E 6012                BRA.S    LOAD7               ;Skip past S1 header loader
424  00008340 4203      LOAD6A:   CLR.B    D3                  ;S1 record found - clear checksum
425  00008342 6124                BSR.S    LOAD8               ;Get byte and update checksum
426  00008344 5700                SUB.B    #3,D0               ;Subtract 3 from record length
427  00008346 1400                MOVE.B   D0,D2               ;Save byte count in D2
428  00008348 4280                CLR.L    D0                  ;Clear address accumulator
429  0000834A 611C                BSR.S    LOAD8               ;Get MS byte of load address
430  0000834C E180                ASL.L    #8,D0               ;Move it to MS position
431  0000834E 6118                BSR.S    LOAD8               ;Get LS byte in D2
432  00008350 2440                MOVE.L   D0,A2               ;A2 points to destination of data
433  00008352 6114      LOAD7:    BSR.S    LOAD8               ;Get byte of data for loading
434  00008354 14C0                MOVE.B   D0,(A2)+            ;Store it
435  00008356 5302                SUB.B    #1,D2               ;Decrement byte counter
436  00008358 66F8                BNE      LOAD7               ;Repeat until count = 0
437  0000835A 610C                BSR.S    LOAD8               ;Read checksum
438  0000835C 5203                ADD.B    #1,D3               ;Add 1 to total checksum
439  0000835E 6700FF78            BEQ      LOAD2               ;If zero then start next record
440  00008362 00070008            OR.B     #%00001000,D7       ;Else set checksum error bit,
441  00008366 608A                BRA      LOAD3               ;restore I/O devices and return
442                    *
443  00008368 6100FE2A  LOAD8:    BSR      BYTE                ;Get a byte
444  0000836C D600                ADD.B    D0,D3               ;Update checksum
445  0000836E 4E75                RTS                          ;and return
446                    *
```

```
447                          *********************************************************************
448                          *
449                          *   DUMP    Transmit S1 formatted records to host computer
450                          *          A3 = Starting address of data block
451                          *          A2 = End address of data block
452                          *          D1 = Checksum, D2 = current record length
453                          *
454    00008370 61000096     DUMP:    BSR      RANGE               ;Get start and end address
455    00008374 4A07                  TST.B    D7                  ;Test for input error
456    00008376 6708                  BEQ.S    DUMP1               ;If no error then continue
457    00008378 49FA075D              LEA      ERMES1(PC),A4       ;Else point to error message,
458    0000837C 6000FCFE              BRA      PSTRING             ;print it and return
459    00008380 B08B         DUMP1:   CMP.L    A3,D0               ;Compare start and end addresses
460    00008382 6A08                  BPL.S    DUMP2               ;If positive then start < end
461    00008384 49FA07D1              LEA      ERMES7(PC),A4       ;Else print error message
462    00008388 6000FCF2              BRA      PSTRING             ;and return
463    0000838C 2F2E0052     DUMP2:   MOVE.L   CN_OVEC(A6),-(A7)   ;Save name of current output device
464    00008390 2D7C00008C22          MOVE.L   #DCB4,CN_OVEC(A6)   ;Set up Port 2 as output device
                0052
465    00008398 6100FCD2              BSR      NEWLINE             ;Send newline to host and wait
466    0000839C 617A                  BSR.S    DELAY
467    0000839E 286E0040              MOVE.L   BUFFPT(A6),A4       ;Before dumping, send any string
468    000083A2 101C         DUMP3:   MOVE.B   (A4)+,D0            ;in the input buffer to the host
469    000083A4 610001EA              BSR      PUTCHAR             ;Repeat
470    000083A8 0C00000D              CMP.B    #CR,D0              ;Transmit char from buffer to host
471    000083AC 66F4                  BNE      DUMP3               ;Until char = C/R
472    000083AE 6100FCBC              BSR      NEWLINE
473    000083B2 6164                  BSR.S    DELAY               ;Allow time for host to settle
474    000083B4 528A                  ADDQ.L   #1,A2               ;A2 contains length of record + 1
475    000083B6 240A         DUMP4:   MOVE.L   A2,D2               ;D2 points to end address
476    000083B8 948B                  SUB.L    A3,D2               ;D2 contains bytes left to print
477    000083BA 0C8200000011          CMP.L    #17,D2              ;If this is not a full record of 16
478    000083C0 6502                  BCS.S    DUMP5               ;then load D2 with record size
479    000083C2 7410                  MOVEQ    #16,D2              ;Else preset byte count to 16
480    000083C4 49FA064C     DUMP5:   LEA      HEADER(PC),A4       ;Point to record header
481    000083C8 6100FCB2              BSR      PSTRING             ;Print header
482    000083CC 4201                  CLR.B    D1                  ;Clear checksum
483    000083CE 1002                  MOVE.B   D2,D0               ;Move record length to output register
484    000083D0 5600                  ADD.B    #3,D0               ;Length includes address + count
485    000083D2 612E                  BSR.S    DUMP7               ;Print number of bytes in record
486    000083D4 200B                  MOVE.L   A3,D0               ;Get start address to be printed
487    000083D6 E158                  ROL.W    #8,D0               ;Get MS byte in LS position
488    000083D8 6128                  BSR.S    DUMP7               ;Print MS byte of address
489    000083DA E058                  ROR.W    #8,D0               ;Restore LS byte
490    000083DC 6124                  BSR.S    DUMP7               ;Print LS byte of address
491    000083DE 101B         DUMP6:   MOVE.B   (A3)+,D0            ;Get data byte to be printed
492    000083E0 6120                  BSR.S    DUMP7               ;Print it
493    000083E2 5302                  SUB.B    #1,D2               ;Decrement byte count
494    000083E4 66F8                  BNE      DUMP6               ;Repeat until all this record printed
495    000083E6 4601                  NOT.B    D1                  ;Complement checksum
496    000083E8 1001                  MOVE.B   D1,D0               ;Move to output register
497    000083EA 6116                  BSR.S    DUMP7               ;Print checksum
498    000083EC 6100FC7E              BSR      NEWLINE
499    000083F0 B7CA                  CMP.L    A2,A3               ;Have all records been printed?
500    000083F2 66C2                  BNE      DUMP4               ;Repeat until all done
501    000083F4 49FA0622              LEA      TAIL(PC),A4         ;Point to message tail (S9 record)
502    000083F8 6100FC82              BSR      PSTRING             ;Print it
503    000083FC 2D5F0052              MOVE.L   (A7)+,CN_OVEC(A6)   ;Restore name of output device
504    00008400 4E75                  RTS                         ;and return
505                          *
```

```
506  00008402 D200          DUMP7:  ADD.B   D0,D1               ;Update checksum, transmit byte
507  00008404 6000FE08              BRA     OUT2X               ;to host and return
508                         *
509           00008408      RANGE:  EQU     *                   ;Get the range of addresses to be
510  00008408 4207                  CLR.B   D7                  ;transmitted from the buffer
511  0000840A 6100FDA4              BSR     PARAM               ;Get starting address
512  0000840E 2640                  MOVE.L  D0,A3               ;Set up start address in A3
513  00008410 6100FD9E              BSR     PARAM               ;Get end address
514  00008414 2440                  MOVE.L  D0,A2               ;Set up end address in A2
515  00008416 4E75                  RTS
516                         *
517           00008418      DELAY:  EQU     *                   ;Provide a time delay for the host
518  00008418 48E78008              MOVEM.L D0/A4,-(A7)         ;to settle. Save working registers
519  0000841C 203C00004000          MOVE.L  #$4000,D0           ;Set up delay constant
520  00008422 5380          DELAY1: SUB.L   #1,D0               ;Count down        (8 clk cycles)
521  00008424 66FC                  BNE     DELAY1              ;Repeat until zero  (10 clk cycles)
522  00008426 4CDF1001              MOVEM.L (A7)+,D0/A4         ;Restore working registers
523  0000842A 4E75                  RTS
524                         *
525                         ************************************************************************
526                         *
527                         *  TM  Enter transparant mode (All communication to go from terminal to
528                         *  the host processor until escape sequence entered). End sequence
529                         *  = ESC, E. A newline is sent to the host to ''clear it down''.
530                         *
531  0000842C 13FC00550001 TM:     MOVE.B  #$55,ACIA_1         ;Force RTS* high to re-route data
              0040
532  00008434 522E0048              ADD.B   #1,ECHO(A6)         ;Turn off character echo
533  00008438 61000124      TM1:    BSR     GETCHAR             ;Get character
534  0000843C 0C00001B              CMP.B   #ESC,D0             ;Test for end of TM mode
535  00008440 66F6                  BNE     TM1                 ;Repeat until first escape character
536  00008442 6100011A              BSR     GETCHAR             ;Get second character
537  00008446 0C000045              CMP.B   #'E',D0             ;If second char = E then exit TM
538  0000844A 66EC                  BNE     TM1                 ;Else continue
539  0000844C 2F2E0052              MOVE.L  CN_OVEC(A6),-(A7)   ;Save output port device name
540  00008450 2D7C00008C22          MOVE.L  #DCB4,CN_OVEC(A6)   ;Get name of host port (aux port)
              0052
541  00008458 6100FC12              BSR     NEWLINE             ;Send newline to host to clear it
542  0000845C 2D5F0052              MOVE.L  (A7)+,CN_OVEC(A6)   ;Restore output device port name
543  00008460 422E0048              CLR.B   ECHO(A6)            ;Restore echo mode
544  00008464 13FC00150001          MOVE.B  #$15,ACIA_1         ;Restore normal ACIA mode (RTS* low)
              0040
545  0000846C 4E75                  RTS
546                         *
```

```
547                         *************************************************************************
548                         *
549                         *    This routine sets up the system DCBs in RAM using the information
550                         *    stored in ROM at address DCB_LST. This is called at initialization.
551                         *    CN_IVEC contains the name ''DCB1'' and IO_VEC the name ''DCB2''
552                         *
553   0000846E 48E7F0F0     SET_DCB:  MOVEM.L  A0-A3/D0-D3,-(A7)    ;Save all working registers
554   00008472 41EE00D0               LEA      FIRST(A6),A0         ;Pointer to first DCB destination in RAM
555   00008476 43FA0774               LEA      DCB_LST(PC),A1       ;A1 points to DCB info block in ROM
556   0000847A 303C0005               MOVE.W   #5,D0                ;6 DCBs to set up
557   0000847E 323C000F     ST_DCB1:  MOVE.W   #15,D1               ;16 bytes to move per DCB header
558   00008482 10D9         ST_DCB2:  MOVE.B   (A1)+,(A0)+          ;Move the 16 bytes of a DCB header
559   00008484 51C9FFFC               DBRA     D1,ST_DCB2           ;from ROM to RAM
560   00008488 3619                   MOVE.W   (A1)+,D3             ;Get size of parameter block (bytes)
561   0000848A 3083                   MOVE.W   D3,(A0)              ;Store size in DCB in RAM
562   0000848C 41F03002               LEA      2(A0,D3.W),A0        ;A0 points to tail of DCB in RAM
563   00008490 47E80004               LEA      4(A0),A3             ;A3 contains address of next DCB in RAM
564   00008494 208B                   MOVE.L   A3,(A0)              ;Store pointer to next DCB in this DCB
565   00008496 41D3                   LEA      (A3),A0              ;A0 now points at next DCB in RAM
566   00008498 51C8FFE4               DBRA     D0,ST_DCB1           ;Repeat until all DCBs set up
567   0000849C 47EBFFFC               LEA      -4(A3),A3            ;Adjust A3 to point to last DCB pointer
568   000084A0 4293                   CLR.L    (A3)                 ;and force last pointer to zero
569   000084A2 2D7C00008BEC           MOVE.L   #DCB1,CN_IVEC(A6)    ;Set up vector to console input DCB
               004E
570   000084AA 2D7C00008BFE           MOVE.L   #DCB2,CN_OVEC(A6)    ;Set up vector to console output DCB
               0052
571   000084B2 4CDF0F0F               MOVEM.L  (A7)+,A0-A3/D0-D3    ;Restore registers
572   000084B6 4E75                   RTS
573                         *
574                         *************************************************************************
575                         *
576                         *    IO_REQ handles all input/output transactions. A0 points to DCB on
577                         *    entry. IO_REQ calls the device driver whose address is in the DCB.
578                         *
579   000084B8 48E700C0     IO_REQ:   MOVEM.L  A0-A1,-(A7)          ;Save working registers
580   000084BC 43E80008               LEA      8(A0),A1             ;A1 points to device handler field in DCB
581   000084C0 2251                   MOVE.L   (A1),A1              ;A1 contains device handler address
582   000084C2 4E91                   JSR      (A1)                 ;Call device handler
583   000084C4 4CDF0300               MOVEM.L  (A7)+,A0-A1          ;Restore working registers
584   000084C8 4E75                   RTS
585                         *
```

```
586                    **************************************************************************
587                    *
588                    *   CON_IN handles input from the console device
589                    *   This is the device driver used by DCB1. Exit with input in D0
590                    *
591   000084CA 48E74040   CON_IN:   MOVEM.L   D1/A1,-(A7)          ;Save working registers
592   000084CE 43E8000C             LEA       12(A0),A1            ;Get pointer to ACIA from DCB
593   000084D2 2251                 MOVE.L    (A1),A1              ;Get address of ACIA in A1
594   000084D4 42280013             CLR.B     19(A0)               ;Clear logical error in DCB
595   000084D8 1211       CON_I1:   MOVE.B    (A1),D1              ;Read ACIA status
596   000084DA 08010000             BTST      #0,D1                ;Test RDRF
597   000084DE 67F8                 BEQ       CON_I1               ;Repeat until RDRF true
598   000084E0 11410012             MOVE.B    D1,18(A0)            ;Store physical status in DCB
599   000084E4 020100F4             AND.B     #%011110100,D1       ;Mask to input error bits
600   000084E8 6706                 BEQ.S     CON_I2               ;If no error then skip update
601   000084EA 117C00010013         MOVE.B    #1,19(A0)            ;Else update logical error
602   000084F0 10290002   CON_I2:   MOVE.B    2(A1),D0             ;Read input from ACIA
603   000084F4 4CDF0202             MOVEM.L   (A7)+,A1/D1          ;Restore working registers
604   000084F8 4E75                 RTS
605                    *
606                    **************************************************************************
607                    *
608                    *   This is the device driver used by DCB2. Output in D0
609                    *   The output can be halted or suspended
610                    *
611   000084FA 48E76040   CON_OUT:  MOVEM.L   A1/D1-D2,-(A7)       ;Save working registers
612   000084FE 43E8000C             LEA       12(A0),A1            ;Get pointer to ACIA from DCB
613   00008502 2251                 MOVE.L    (A1),A1              ;Get address of ACIA in A1
614   00008504 42280013             CLR.B     19(A0)               ;Clear logical error in DCB
615   00008508 1211       CON_OT1:  MOVE.B    (A1),D1              ;Read ACIA status
616   0000850A 08010000             BTST      #0,D1                ;Test RDRF bit (any input?)
617   0000850E 6716                 BEQ.S     CON_OT3              ;If no input then test output status
618   00008510 14290002             MOVE.B    2(A1),D2             ;Else read the input
619   00008514 0202005F             AND.B     #%01011111,D2        ;Strip parity and bit 5
620   00008518 0C020057             CMP.B     #WAIT,D2             ;and test for a wait condition
621   0000851C 6608                 BNE.S     CON_OT3              ;If not wait then ignore and test O/P
622   0000851E 1411       CON_OT2:  MOVE.B    (A1),D2              ;Else read ACIA status register
623   00008520 08020000             BTST      #0,D2                ;and poll ACIA until next char received
624   00008524 67F8                 BEQ       CON_OT2
625   00008526 08010001   CON_OT3:  BTST      #1,D1                ;Repeat
626   0000852A 67DC                 BEQ       CON_OT1              ;until ACIA Tx ready
627   0000852C 11410012             MOVE.B    D1,18(A0)            ;Store status in DCB physical error
628   00008530 13400002             MOVE.B    D0,2(A1)             ;Transmit output
629   00008534 4CDF0206             MOVEM.L   (A7)+,A1/D1-D2       ;Restore working registers
630   00008538 4E75                 RTS
631                    *
632                    **************************************************************************
633                    *
634                    *   AUX_IN and AUX_OUT are simplified versions of CON_IN and
635                    *   CON_OUT for use with the port to the host processor
636                    *
637   0000853A 43E8000C   AUX_IN:   LEA       12(A0),A1            ;Get pointer to aux ACIA from DCB
638   0000853E 2251                 MOVE.L    (A1),A1              ;Get address of aux ACIA
639   00008540 08110000   AUX_IN1:  BTST      #0,(A1)              ;Test for data ready
640   00008544 67FA                 BEQ       AUX_IN1              ;Repeat until ready
641   00008546 10290002             MOVE.B    2(A1),D0             ;Read input
642   0000854A 4E75                 RTS
643                    *
```

```
644   0000854C 43E8000C   AUX_OUT:  LEA     12(A0),A1          ;Get pointer to aux ACIA from DCB
645   00008550 2251                 MOVE.L  (A1),A1            ;Get address of aux ACIA
646   00008552 08110001   AUX_OT1:  BTST    #1,(A1)            ;Test for ready to transmit
647   00008556 67FA                 BEQ     AUX_OT1            ;Repeat until transmitter ready
648   00008558 13400002             MOVE.B  D0,2(A1)           ;Transmit data
649   0000855C 4E75                 RTS
650                       *
651                       ***********************************************************************
652                       *
653                       *  GETCHAR gets a character from the console device
654                       *  This is the main input routine and uses the device whose name
655                       *  is stored in CN_IVEC. Changing this name redirects input.
656                       *
657   0000855E 2F08       GETCHAR:  MOVE.L  A0,-(A7)           ;Save working register
658   00008560 206E004E             MOVE.L  CN_IVEC(A6),A0     ;A0 points to name of console DCB
659   00008564 6154                 BSR.S   IO_OPEN            ;Open console (get DCB address in A0)
660   00008566 08070003             BTST    #3,D7              ;D7(3) set if open error
661   0000856A 6620                 BNE.S   GETCH3             ;If error then exit now
662   0000856C 6100FF4A             BSR     IO_REQ             ;Else execute I/O transaction
663   00008570 0200007F             AND.B   #$7F,D0            ;Strip msb of input
664   00008574 4A2E0049             TST.B   U_CASE(A6)         ;Test for upper -> lower case conversion
665   00008578 660A                 BNE.S   GETCH2             ;If flag not zero do not convert case
666   0000857A 08000006             BTST    #6,D0              ;Test input for lower case
667   0000857E 6704                 BEQ.S   GETCH2             ;If upper case then skip conversion
668   00008580 020000DF             AND.B   #%11011111,D0      ;Else clear bit 5 for upper case conv
669   00008584 4A2E0048   GETCH2:   TST.B   ECHO(A6)           ;Do we need to echo the input?
670   00008588 6602                 BNE.S   GETCH3             ;If ECHO not zero then no echo
671   0000858A 6104                 BSR.S   PUTCHAR            ;Else echo the input
672   0000858C 205F       GETCH3:   MOVE.L  (A7)+,A0           ;Restore working register
673   0000858E 4E75                 RTS                        ;and return
674                       *
675                       ***********************************************************************
676                       *
677                       *  PUTCHAR sends a character to the console device
678                       *  The name of the output device is in CN_OVEC.
679                       *
680   00008590 2F08       PUTCHAR:  MOVE.L  A0,-(A7)           ;Save working register
681   00008592 206E0052             MOVE.L  CN_OVEC(A6),A0     ;A0 points to name of console output
682   00008596 6122                 BSR.S   IO_OPEN            ;Open console (Get address of DCB)
683   00008598 6100FF1E             BSR     IO_REQ             ;Perform output with DCB pointed at by A0
684   0000859C 205F                 MOVE.L  (A7)+,A0           ;Restore working register
685   0000859E 4E75                 RTS
686                       *
687                       ***********************************************************************
688                       *
689                       *  BUFF_IN and BUFF_OUT are two rudimentary input and output routines
690                       *  which input data from and output data to a buffer in RAM. These are
691                       *  used by DCB5 and DCB6, respectively.
692                       *
693   000085A0 43E8000C   BUFF_IN:  LEA     12(A0),A1          ;A1 points to I/P buffer
694   000085A4 2451                 MOVE.L  (A1),A2            ;A2 gets I/P pointer from buffer
695   000085A6 1022                 MOVE.B  -(A2),D0           ;Read char from buffer and adjust A2
696   000085A8 228A                 MOVE.L  A2,(A1)            ;Restore pointer in buffer
697   000085AA 4E75                 RTS
698                       *
```

```
699   000085AC 43E8000C   BUFF_OT:  LEA      12(A0),A1         ;A1 points to O/P buffer
700   000085B0 24690004             MOVE.L   4(A1),A2          ;A2 gets O/P pointer from buffer
701   000085B4 14C0                 MOVE.B   D0,(A2)+          ;Store char in buffer and adjust A2
702   000085B6 228A                 MOVE.L   A2,(A1)           ;Restore pointer in buffer
703   000085B8 4E75                 RTS
704                      *
705                      ***************************************************************************
706                      *
707                      *  Open - opens a DCB for input or output. IO_OPEN converts the
708                      *  name pointed at by A0 into the address of the DCB pointed at
709                      *  by A0. Bit 3 of D7 is set to zero if DCB not found
710                      *
711   000085BA 48E7F870   IO_OPEN:  MOVEM.L  A1-A3/D0-D4,-(A7) ;Save working registers
712   000085BE 43EE00D0             LEA      FIRST(A6),A1      ;A1 points to first DCB in chain in RAM
713   000085C2 45D1       OPEN1:    LEA      (A1),A2           ;A2 = temp copy of pointer to DCB
714   000085C4 47D0                 LEA      (A0),A3           ;A3 = temp copy of pointer to DCB name
715   000085C6 303C0007             MOVE.W   #7,D0             ;Up to 8 chars of DCB name to match
716   000085CA 181A       OPEN2:    MOVE.B   (A2)+,D4          ;Compare DCB name with string
717   000085CC B81B                 CMP.B    (A3)+,D4
718   000085CE 6608                 BNE.S    OPEN3             ;If no match try next DCB
719   000085D0 51C8FFF8             DBRA     D0,OPEN2          ;Else repeat until all chars matched
720   000085D4 41D1                 LEA      (A1),A0           ;Success - move this DCB address to A0
721   000085D6 6016                 BRA.S    OPEN4             ;and return
722            000085D8   OPEN3:    EQU      *                 ;Fail - calculate address of next DCB
723   000085D8 32290010             MOVE.W   16(A1),D1         ;Get parameter block size of DCB
724   000085DC 43F11012             LEA      18(A1,D1.W),A1    ;A1 points to pointer to next DCB
725   000085E0 2251                 MOVE.L   (A1),A1           ;A1 now points to next DCB
726   000085E2 B3FC00000000         CMP.L    #0,A1             ;Test for end of DCB chain
727   000085E8 66D8                 BNE      OPEN1             ;If not end of chain then try next DCB
728   000085EA 00070008             OR.B     #8,D7             ;Else set error flag and return
729   000085EE 4CDF0E1F   OPEN4:    MOVEM.L  (A7)+,A1-A3/D0-D4 ;Restore working registers
730   000085F2 4E75                 RTS
731                      *
```

```
732                       ****************************************************************************
733                       *
734                       *  Exception vector table initialization routine
735                       *  All vectors not setup are loaded with uninitialized routine vector
736                       *
737   000085F4 41F80008   X_SET:    LEA      X_BASE,A0              ;Point to base of exception table
738   000085F8 303C00FD             MOVE.W   #253,D0               ;Number of vectors - 3
739   000085FC 20FC000089E4 X_SET1: MOVE.L   #X_UN,(A0)+           ;Store uninitialized exception vector
740   00008602 51C8FFF8             DBRA     D0,X_SET1             ;Repeat until all entries preset
741   00008606 91C8                 SUB.L    A0,A0                 ;Clear A0 (points to vector table)
742   00008608 217C000087B4         MOVE.L   #BUS_ER,8(A0)         ;Setup bus error vector
               0008
743   00008610 217C000087C2         MOVE.L   #ADD_ER,12(A0)        ;Setup address error vector
               000C
744   00008618 217C0000879E         MOVE.L   #IL_ER,16(A0)         ;Setup illegal instruction error vect
               0010
745   00008620 217C00008898         MOVE.L   #TRACE,36(A0)         ;Setup trace exception vector
               0024
746   00008628 217C00008652         MOVE.L   #TRAP_0,128(A0)       ;Setup TRAP #0 exception vector
               0080
747   00008630 217C000087D0         MOVE.L   #BRKPT,184(A0)        ;Setup TRAP #14 vector = breakpoint
               00B8
748   00008638 217C00008040         MOVE.L   #WARM,188(A0)         ;Setup TRAP #15 exception vector
               00BC
749   00008640 303C0007             MOVE.W   #7,D0                 ;Now clear the breakpoint table
750   00008644 41EE00A0             LEA      BP_TAB(A6),A0         ;Point to table
751   00008648 4298        X_SET2:  CLR.L    (A0)+                 ;Clear an address entry
752   0000864A 4258                 CLR.W    (A0)+                 ;Clear the corresponding data
753   0000864C 51C8FFFA             DBRA     D0,X_SET2             ;Repeat until all 8 cleared
754   00008650 4E75                 RTS
755                       *
```

```
756                          *********************************************************************
757                          *
758            00008652      TRAP_0:   EQU      *          ;User links to  TS2BUG via TRAP #0
759   00008652 0C010000                CMP.B    #0,D1      ;D1 = 0 = Get character
760   00008656 6606                    BNE.S    TRAP1
761   00008658 6100FF04                BSR      GETCHAR
762   0000865C 4E73                    RTE
763   0000865E 0C010001     TRAP1:     CMP.B    #1,D1      ;D1 = 1 = Print character
764   00008662 6606                    BNE.S    TRAP2
765   00008664 6100FF2A                BSR      PUTCHAR
766   00008668 4E73                    RTE
767   0000866A 0C010002     TRAP2:     CMP.B    #2,D1      ;D1 = 2 = Newline
768   0000866E 6606                    BNE.S    TRAP3
769   00008670 6100F9FA                BSR      NEWLINE
770   00008674 4E73                    RTE
771   00008676 0C010003     TRAP3:     CMP.B    #3,D1      ;D1 = 3 = Get parameter from buffer
772   0000867A 6606                    BNE.S    TRAP4
773   0000867C 6100FB32                BSR      PARAM
774   00008680 4E73                    RTE
775   00008682 0C010004     TRAP4:     CMP.B    #4,D1      ;D1 = 4 = Print string pointed at by A4
776   00008686 6606                    BNE.S    TRAP5
777   00008688 6100F9F2                BSR      PSTRING
778   0000868C 4E73                    RTE
779   0000868E 0C010005     TRAP5:     CMP.B    #5,D1      ;D1 = 5 = Get a hex character
780   00008692 6606                    BNE.S    TRAP6
781   00008694 6100FAE0                BSR      HEX
782   00008698 4E73                    RTE
783   0000869A 0C010006     TRAP6:     CMP.B    #6,D1      ;D1 = 6 = Get a hex byte
784   0000869E 6606                    BNE.S    TRAP7
785   000086A0 6100FAF2                BSR      BYTE
786   000086A4 4E73                    RTE
787   000086A6 0C010007     TRAP7:     CMP.B    #7,D1      ;D1 = 7 = Get a word
788   000086AA 6606                    BNE.S    TRAP8
789   000086AC 6100FAF6                BSR      WORD
790   000086B0 4E73                    RTE
791   000086B2 0C010008     TRAP8:     CMP.B    #8,D1      ;D1 = 8 = Get a longword
792   000086B6 6606                    BNE.S    TRAP9
793   000086B8 6100FAF0                BSR      LONGWD
794   000086BC 4E73                    RTE
795   000086BE 0C010009     TRAP9:     CMP.B    #9,D1      ;D1 = 9 = Output hex byte
796   000086C2 6606                    BNE.S    TRAP10
797   000086C4 6100FB48                BSR      OUT2X
798   000086C8 4E73                    RTE
799   000086CA 0C01000A     TRAP10:    CMP.B    #10,D1     ;D1 = 10 = Output hex word
800   000086CE 6606                    BNE.S    TRAP11
801   000086D0 6100FB44                BSR      OUT4X
802   000086D4 4E73                    RTE
803   000086D6 0C01000B     TRAP11:    CMP.B    #11,D1     ;D1 = 11 = Output hex longword
804   000086DA 6606                    BNE.S    TRAP12
805   000086DC 6100FB40                BSR      OUT8X
806   000086E0 4E73                    RTE
807   000086E2 0C01000C     TRAP12:    CMP.B    #12,D1     ;D1 = 12 = Print a space
808   000086E6 6606                    BNE.S    TRAP13
809   000086E8 6100FBA4                BSR      PSPACE
810   000086EC 4E73                    RTE
811   000086EE 0C01000D     TRAP13:    CMP.B    #13,D1     ;D1 = 13 = Get a line of text into
812   000086F2 6606                    BNE.S    TRAP14     ;the line buffer
813   000086F4 6100F99C                BSR      GETLINE
814   000086F8 4E73                    RTE
```

```
815  000086FA 0C01000E   TRAP14:  CMP.B    #14,D1          ;D1 = 14 = Tidy up the line in the
816  000086FE 6606                BNE.S    TRAP15          ;line buffer by removing leading
817  00008700 6100F9C6            BSR      TIDY            ;leading and multiple embeded spaces
818  00008704 4E73                RTE
819  00008706 0C01000F   TRAP15:  CMP.B    #15,D1          ;D1 = 15 = Execute the command in
820  0000870A 6606                BNE.S    TRAP16          ;the line buffer
821  0000870C 6100F9FC            BSR      EXECUTE
822  00008710 4E73                RTE
823  00008712 0C010010   TRAP16:  CMP.B    #16,D1          ;D1 = 16 = Call RESTORE to transfer
824  00008716 6606                BNE.S    TRAP17          ;the registers in TSK_T to the 68000
825  00008718 6100015A            BSR      RESTORE         ;and therefore execute a program
826  0000871C 4E73                RTE
827  0000871E 4E73       TRAP17:  RTE
828                      *
829                      ***********************************************************************
830                      *
831                      * Display exception frame (D0 - D7, A0 - A6, USP, SSP, SR, PC)
832                      * EX_DIS prints registers saved after a breakpoint or exception
833                      * The registers are saved in TSK_T
834                      *
835  00008720 4BEE0056   EX_DIS:  LEA      TSK_T(A6),A5    ;A5 points to display frame
836  00008724 49FA0313            LEA      MES3(PC),A4     ;Point to heading
837  00008728 6100F962            BSR      HEADING         ;and print it
838  0000872C 3C3C0007            MOVE.W   #7,D6           ;8 pairs of registers to display
839  00008730 4205                CLR.B    D5              ;D5 is the line counter
840  00008732 1005       EX_D1:   MOVE.B   D5,D0           ;Put current register number in D0
841  00008734 6100FABE            BSR      OUT1X           ;and print it
842  00008738 6100FB54            BSR      PSPACE          ;and a space
843  0000873C 5205                ADD.B    #1,D5           ;Update counter for next pair
844  0000873E 2015                MOVE.L   (A5),D0         ;Get data register to be displayed
845  00008740 6100FADC            BSR      OUT8X           ;from the frame and print it
846  00008744 49FA0311            LEA      MES4(PC),A4     ;Print string of spaces
847  00008748 6100F932            BSR.L    PSTRING         ;between data and address registers
848  0000874C 202D0020            MOVE.L   32(A5),D0       ;Get address register to be displayed
849  00008750 6100FACC            BSR      OUT8X           ;which is 32 bytes on from data reg
850  00008754 6100F916            BSR      NEWLINE
851  00008758 4BED0004            LEA      4(A5),A5        ;Point to next pair (ie Di, Ai)
852  0000875C 51CEFFD4            DBRA     D6,EX_D1        ;Repeat until all displayed
853  00008760 4BED0020            LEA      32(A5),A5       ;Adjust pointer by 8 longwords
854  00008764 6100F906            BSR      NEWLINE         ;to point to SSP
855  00008768 49FA02C6            LEA      MES2A(PC),A4    ;Point to ''SS =''
856  0000876C 6100F90E            BSR      PSTRING         ;Print it
857  00008770 201D                MOVE.L   (A5)+,D0        ;Get SSP from frame
858  00008772 6100FAAA            BSR      OUT8X           ;and display it
859  00008776 6100F8F4            BSR      NEWLINE
860  0000877A 49FA02A2            LEA      MES1(PC),A4     ;Point to 'SR ='
861  0000877E 6100F8FC            BSR      PSTRING         ;Print it
862  00008782 301D                MOVE.W   (A5)+,D0        ;Get status register
863  00008784 6100FA90            BSR      OUT4X           ;Display status
864  00008788 6100F8E2            BSR      NEWLINE
865  0000878C 49FA0299            LEA      MES2(PC),A4     ;Point to 'PC ='
866  00008790 6100F8EA            BSR      PSTRING         ;Print it
867  00008794 201D                MOVE.L   (A5)+,D0        ;Get PC
868  00008796 6100FA86            BSR      OUT8X           ;Display PC
869  0000879A 6000F8D0            BRA      NEWLINE         ;Newline and return
870                      *
```

```
871                   ************************************************************************
872             *
873             *   Exception handling routines
874             *
875          0000879E  IL_ER:    EQU       *                      ;Illegal instruction exception
876  0000879E 2F0C               MOVE.L    A4,-(A7)               ;Save A4
877  000087A0 49FA02DF           LEA       MES10(PC),A4           ;Point to heading
878  000087A4 6100F8E6           BSR       HEADING  .             ;Print it
879  000087A8 285F               MOVE.L    (A7)+,A4               ;Restore A4
880  000087AA 6176               BSR.S     GROUP2                 ;Save registers in display frame
881  000087AC 6100FF72           BSR       EX_DIS                 ;Display registers saved in frame
882  000087B0 6000F88E           BRA       WARM                   ;Abort from illegal instruction
883             *
884          000087B4  BUS_ER:   EQU       *                      ;Bus error (group 1) exception
885  000087B4 2F0C               MOVE.L    A4,-(A7)               ;Save A4
886  000087B6 49FA02A9           LEA       MES8(PC),A4            ;Point to heading
887  000087BA 6100F8D0           BSR       HEADING                ;Print it
888  000087BE 285F               MOVE.L    (A7)+,A4               ;Restore A4
889  000087C0 602C               BRA.S     GROUP1                 ;Deal with group 1 exception
890             *
891          000087C2  ADD_ER:   EQU       *                      ;Address error (group 1) exception
892  000087C2 2F0C               MOVE.L    A4,-(A7)               ;Save A4
893  000087C4 49FA02A9           LEA       MES9(PC),A4            ;Point to heading
894  000087C8 6100F8C2           BSR       HEADING                ;Print it
895  000087CC 285F               MOVE.L    (A7)+,A4               ;Restore A4
896  000087CE 601E               BRA.S     GROUP1                 ;Deal with group 1 exception
897             *
898          000087D0  BRKPT:    EQU       *                      ;Deal with breakpoint
899  000087D0 48E7FFFE           MOVEM.L   D0-D7/A0-A6,-(A7)      ;Save all registers
900  000087D4 61000180           BSR       BR_CLR                 ;Clear breakpoints in code
901  000087D8 4CDF7FFF           MOVEM.L   (A7)+,D0-D7/A0-A6      ;Restore registers
902  000087DC 6144               BSR.S     GROUP2                 ;Treat as group 2 exception
903  000087DE 49FA02B7           LEA       MES11(PC),A4           ;Point to heading
904  000087E2 6100F8A8           BSR       HEADING                ;Print it
905  000087E6 6100FF38           BSR       EX_DIS                 ;Display saved registers
906  000087EA 6000F854           BRA       WARM                   ;Return to monitor
907             *
908             *    GROUP1 is called by address and bus error exceptions
909             *    These are ''turned into group 2'' exceptions (eg TRAP)
910             *    by modifying the stack frame saved by a group 1 exception
911             *
```

```
912  000087EE 48E78080   GROUP1:    MOVEM.L  D0/A0,-(A7)          ;Save working registers
913  000087F2 206F0012              MOVE.L   18(A7),A0            ;Get PC from group 1 stack frame
914  000087F6 302F000E              MOVE.W   14(A7),D0            ;Get instruction from stack frame
915  000087FA B060                  CMP.W    -(A0),D0             ;Now backtrack to find the ''correct PC''
916  000087FC 670E                  BEQ.S    GROUP1A              ;by matching the op-code on the stack
917  000087FE B060                  CMP.W    -(A0),D0             ;with the code in the region of the
918  00008800 670A                  BEQ.S    GROUP1A              ;PC on the stack
919  00008802 B060                  CMP.W    -(A0),D0
920  00008804 6706                  BEQ.S    GROUP1A
921  00008806 B060                  CMP.W    -(A0),D0
922  00008808 6702                  BEQ.S    GROUP1A
923  0000880A 5588                  SUBQ.L   #2,A0
924  0000880C 2F480012   GROUP1A:   MOVE.L   A0,18(A7)            ;Restore modified PC to stack frame
925  00008810 4CDF0101              MOVEM.L  (A7)+,D0/A0          ;Restore working registers
926  00008814 4FEF0008              LEA      8(A7),A7             ;Adjust stack pointer to group 1 type
927  00008818 6108                  BSR.S    GROUP2               ;Now treat as group 1 exception
928  0000881A 6100FF04              BSR      EX_DIS               ;Display contents of exception frame
929  0000881E 6000F820              BRA      WARM                 ;Exit to monitor - no RTE from group 2
930                     *
931           00008822   GROUP2:    EQU      *                    ;Deal with group 2 exceptions
932  00008822 48E7FFFF              MOVEM.L  A0-A7/D0-D7,-(A7)    ;Save all registers on the stack
933  00008826 303C000E              MOVE.W   #14,D0               ;Transfer D0 - D7, A0 - A6 from
934  0000882A 41EE0056              LEA      TSK_T(A6),A0         ;the stack to the display frame
935  0000882E 20DF       GROUP2A:   MOVE.L   (A7)+,(A0)+          ;Move a register from stack to frame
936  00008830 51C8FFFC              DBRA     D0,GROUP2A           ;and repeat until D0-D7/A0-A6 moved
937  00008834 4E6A                  MOVE.L   USP,A2               ;Get the user stack pointer and put it
938  00008836 20CA                  MOVE.L   A2,(A0)+             ;in the A7 position in the frame
939  00008838 201F                  MOVE.L   (A7)+,D0             ;Now transfer the SSP to the frame,
940  0000883A 04800000000A          SUB.L    #10,D0               ;remembering to account for the
941  00008840 20C0                  MOVE.L   D0,(A0)+             ;data pushed on the stack to this point
942  00008842 225F                  MOVE.L   (A7)+,A1             ;Copy TOS (return address) to A1
943  00008844 30DF                  MOVE.W   (A7)+,(A0)+          ;Move SR to display frame
944  00008846 201F                  MOVE.L   (A7)+,D0             ;Get PC in D0
945  00008848 5580                  SUBQ.L   #2,D0                ;Move back to current instruction
946  0000884A 20C0                  MOVE.L   D0,(A0)+             ;Put adjusted PC in display frame
947  0000884C 4ED1                  JMP      (A1)                 ;Return from subroutine
948                     *
949                     *********************************************************************
950                     *
951                     *  GO executes a program either from a supplied address or
952                     *  by using the data in the display frame
953  0000884E 6100F960   GO:        BSR      PARAM                ;Get entry address (if any)
954  00008852 4A07                  TST.B    D7                   ;Test for error in input
955  00008854 6708                  BEQ.S    GO1                  ;If D7 zero then OK
956  00008856 49FA027F              LEA      ERMES1(PC),A4        ;Else point to error message,
957  0000885A 6000F820              BRA      PSTRING              ;print it and return
958  0000885E 4A80       GO1:       TST.L    D0                   ;If no address entered then get
959  00008860 670A                  BEQ.S    GO2                  ;address from display frame
960  00008862 2D40009C              MOVE.L   D0,TSK_T+70(A6)      ;Else save address in display frame
961  00008866 3D7C2700009A          MOVE.W   #$2700,TSK_T+68(A6)  ;Store dummy status in frame
962  0000886C 6006       GO2:       BRA.S    RESTORE              ;Restore volatile environment and go
963                     *
964  0000886E 6100007A   GB:        BSR      BR_SET               ;Same as go but presets breakpoints
965  00008872 60DA                  BRA.S    GO                   ;Execute program
966                     *
```

```
967                        *         RESTORE moves the volatile environment from the display
968                        *         frame and transfers it to the 68000's registers. This
969                        *         re-runs a program suspended after an exception
970                        *
971  00008874 47EE0056  RESTORE: LEA    TSK_T(A6),A3        ;A3 points to display frame
972  00008878 47EB004A           LEA    74(A3),A3           ;A3 now points to end of frame + 4
973  0000887C 4FEF0004           LEA    4(A7),A7            ;Remove return address from stack
974  00008880 303C0024           MOVE.W  #36,D0             ;Counter for 37 words to be moved
975  00008884 3F23       REST1:  MOVE.W  -(A3),-(A7)        ;Move word from display frame to stack
976  00008886 51C8FFFC           DBRA    D0,REST1           ;Repeat until entire frame moved
977  0000888A 4CDF00FF           MOVEM.L  (A7)+,D0-D7       ;Restore old data registers from stack
978  0000888E 4CDF7F00           MOVEM.L  (A7)+,A0-A6       ;Restore old address registers
979  00008892 4FEF0008           LEA    8(A7),A7            ;Except SSP/USP - so adjust stack
980  00008896 4E73               RTE                        ;Return from exception to run program
981                        *
982           00008898  TRACE:   EQU     *                  ;TRACE exception (rudimentary version)
983  00008898 287A020B           MOVE.L  MES12(PC),A4       ;Point to heading
984  0000889C 6100F7EE           BSR     HEADING            ;Print it
985  000088A0 6100FF4C           BSR     GROUP1             ;Save volatile environment
986  000088A4 6100FE7A           BSR     EX_DIS             ;Display it
987  000088A8 6000F796           BRA     WARM               ;Return to monitor
988                        *
989                        *****************************************************************************
990                        *  Breakpoint routines: BR_GET gets the address of a breakpoint and
991                        *  puts it in the breakpoint table. It does not plant it in the code.
992                        *  BR_SET plants all breakpoints in the code. NOBR removes one or all
993                        *  breakpoints from the table. KILL removes breakpoints from the code.
994                        *
995  000088AC 6100F902  BR_GET:  BSR     PARAM              ;Get breakpoint address in table
996  000088B0 4A07               TST.B   D7                 ;Test for input error
997  000088B2 6708               BEQ.S   BR_GET1            ;If no error then continue
998  000088B4 49FA0221           LEA     ERMES1(PC),A4      ;Else display error
999  000088B8 6000F7C2           BRA     PSTRING            ;and return
1000 000088BC 47EE00A0  BR_GET1: LEA     BP_TAB(A6),A3      ;A6 points to breakpoint table
1001 000088C0 2A40               MOVE.L  D0,A5              ;Save new BP address in A5
1002 000088C2 2C00               MOVE.L  D0,D6              ;and in D6 because D0 gets corrupted
1003 000088C4 3A3C0007           MOVE.W  #7,D5              ;Eight entries to test
1004 000088C8 201B       BR_GET2: MOVE.L  (A3)+,D0          ;Read entry from breakpoint table
1005 000088CA 660C               BNE.S   BR_GET3            ;If not zero display existing BP
1006 000088CC 4A86               TST.L   D6                 ;Only store a non-zero breakpoint
1007 000088CE 6710               BEQ.S   BR_GET4
1008 000088D0 274DFFFC           MOVE.L  A5,-4(A3)          ;Store new breakpoint in table
1009 000088D4 3695               MOVE.W  (A5),(A3)          ;Save code at BP address in table
1010 000088D6 4286               CLR.L   D6                 ;Clear D6 to avoid repetition
1011 000088D8 6100F944  BR_GET3: BSR     OUT8X              ;Display this breakpoint
1012 000088DC 6100F78E           BSR     NEWLINE
1013 000088E0 47EB0002  BR_GET4: LEA     2(A3),A3           ;Step past stored op-code
1014 000088E4 51CDFFE2           DBRA    D5,BR_GET2         ;Repeat until all entries tested
1015 000088E8 4E75               RTS                        ;Return
1016                        *
```

```
1017              000088EA  BR_SET:   EQU    *                   ;Plant any breakpoints in user code
1018  000088EA 41EE00A0             LEA    BP_TAB(A6),A0       ;A0 points to BP table
1019  000088EE 45EE009C             LEA    TSK_T+70(A6),A2     ;A2 points to PC in display frame
1020  000088F2 2452                 MOVE.L (A2),A2             ;Now A2 contains value of PC
1021  000088F4 303C0007             MOVE.W #7,D0               ;Up to eight entries to plant
1022  000088F8 2218      BR_SET1:   MOVE.L (A0)+,D1            ;Read breakpoint address from table
1023  000088FA 670A                 BEQ.S  BR_SET2            ;If zero then skip planting
1024  000088FC B28A                 CMP.L  A2,D1              ;Don't want to plant BP at current PC
1025  000088FE 6706                 BEQ.S  BR_SET2            ;location, so skip planting if same
1026  00008900 2241                 MOVE.L D1,A1              ;Transfer BP address to address reg
1027  00008902 32BC4E4E             MOVE.W #TRAP_14,(A1)      ;Plant op-code for TRAP #14 in code
1028  00008906 41E80002  BR_SET2:   LEA    2(A0),A0           ;Skip past op-code field in table
1029  0000890A 51C8FFEC             DBRA   D0,BR_SET1         ;Repeat until all entries tested
1030  0000890E 4E75                 RTS
1031                     *
1032              00008910  NOBR:    EQU    *                   ;Clear one or all breakpoints
1033  00008910 6100F89E             BSR    PARAM              ;Get BP address (if any)
1034  00008914 4A07                 TST.B  D7                 ;Test for input error
1035  00008916 6708                 BEQ.S  NOBR1              ;If no error then skip abort
1036  00008918 49FA01BD             LEA    ERMES1(PC),A4      ;Point to error message
1037  0000891C 6000F75E             BRA    PSTRING            ;Display it and return
1038  00008920 4A80      NOBR1:     TST.L  D0                 ;Test for null address (clear all)
1039  00008922 6720                 BEQ.S  NOBR4              ;If no address then clear all entries
1040  00008924 2240                 MOVE.L D0,A1              ;Else just clear breakpoint in A1
1041  00008926 41EE00A0             LEA    BP_TAB(A6),A0      ;A0 points to BP table
1042  0000892A 303C0007             MOVE.W #7,D0              ;Up to eight entries to test
1043  0000892E 2218      NOBR2:     MOVE.L (A0)+,D1           ;Get entry and
1044  00008930 41E80002             LEA    2(A0),A0           ;skip past op-code field
1045  00008934 B289                 CMP.L  A1,D1              ;Is this the one?
1046  00008936 6706                 BEQ.S  NOBR3              ;If so go and clear entry
1047  00008938 51C8FFF4             DBRA   D0,NOBR2           ;Repeat until all tested
1048  0000893C 4E75                 RTS
1049  0000893E 42A8FFFA  NOBR3:     CLR.L  -6(A0)             ;Clear address in BP table
1050  00008942 4E75                 RTS
1051  00008944 41EE00A0  NOBR4:     LEA    BP_TAB(A6),A0      ;Clear all 8 entries in BP table
1052  00008948 303C0007             MOVE.W #7,D0              ;Eight entries to clear
1053  0000894C 4298      NOBR5:     CLR.L  (A0)+              ;Clear breakpoint address
1054  0000894E 4258                 CLR.W  (A0)+              ;Clear op-code field
1055  00008950 51C8FFFA             DBRA   D0,NOBR5           ;Repeat until all done
1056  00008954 4E75                 RTS
1057                     *
1058              00008956  BR_CLR:  EQU    *                   ;Remove breakpoints from code
1059  00008956 41EE00A0             LEA    BP_TAB(A6),A0      ;A0 points to breakpoint table
1060  0000895A 303C0007             MOVE.W #7,D0              ;Up to eight entries to clear
1061  0000895E 2218      BR_CLR1:   MOVE.L (A0)+,D1           ;Get address of BP in D1
1062  00008960 2241                 MOVE.L D1,A1              ;and put copy in A1
1063  00008962 4A81                 TST.L  D1                 ;Test this breakpoint
1064  00008964 6702                 BEQ.S  BR_CLR2            ;If zero then skip BP clearing
1065  00008966 3290                 MOVE.W (A0),(A1)          ;Else restore op-code
1066  00008968 41E80002  BR_CLR2:   LEA    2(A0),A0           ;Skip past op-code field
1067  0000896C 51C8FFF0             DBRA   D0,BR_CLR1         ;Repeat until all tested
1068  00008970 4E75                 RTS
1069                     *
```

```
1070                             *  REG_MOD modifies a register in the display frame. The command
1071                             *  format is REG <reg> <value>. E.g. REG D3 1200
1072                             *
1073   00008972 4281    REG_MOD:  CLR.L    D1                     ;D1 to hold name of register
1074   00008974 41EE0040         LEA      BUFFPT(A6),A0          ;A0 contains address of buffer pointer
1075   00008978 2050             MOVE.L   (A0),A0                ;A0 now points to next char in buffer
1076   0000897A 1218             MOVE.B   (A0)+,D1               ;Put first char of name in D1
1077   0000897C E159             ROL.W    #8,D1                  ;Move char one place left
1078   0000897E 1218             MOVE.B   (A0)+,D1               ;Get second char in D1
1079   00008980 41E80001         LEA      1(A0),A0               ;Move pointer past space in buffer
1080   00008984 2D480040         MOVE.L   A0,BUFFPT(A6)          ;Update buffer pointer
1081   00008988 4282             CLR.L    D2                     ;D2 is the character pair counter
1082   0000898A 41FA0122         LEA      REGNAME(PC),A0         ;A0 points to string of character pairs
1083   0000898E 43D0             LEA      (A0),A1                ;A1 also points to string
1084   00008990 B258    REG_MD1:  CMP.W    (A0)+,D1               ;Compare a char pair with input
1085   00008992 6712             BEQ.S    REG_MD2                ;If match then exit loop
1086   00008994 5282             ADD.L    #1,D2                  ;Else increment match counter
1087   00008996 0C8200000013     CMP.L    #19,D2                 ;Test for end of loop
1088   0000899C 66F2             BNE      REG_MD1                ;Continue until all pairs matched
1089   0000899E 49FA0137         LEA      ERMES1(PC),A4          ;If here then error
1090   000089A2 6000F6D8         BRA      PSTRING                ;Display error and return
1091   000089A6 43EE0056 REG_MD2: LEA     TSK_T(A6),A1           ;A1 points to display frame
1092   000089AA E582             ASL.L    #2,D2                  ;Multiply offset by 4 (4 bytes/entry)
1093   000089AC 0C8200000048     CMP.L    #72,D2                 ;Test for address of PC
1094   000089B2 6602             BNE.S    REG_MD3                ;If not PC then all is OK
1095   000089B4 5582             SUB.L    #2,D2                  ;else dec PC pointer as Sr is a word
1096   000089B6 45F12000 REG_MD3: LEA     (A1,D2),A2             ;Calculate address of entry in disptable
1097   000089BA 2012             MOVE.L   (A2),D0                ;Get old contents
1098   000089BC 6100F860         BSR      OUT8X                  ;Display them
1099   000089C0 6100F6AA         BSR      NEWLINE
1100   000089C4 6100F7EA         BSR      PARAM                  ;Get new data
1101   000089C8 4A07             TST.B    D7                     ;Test for input error
1102   000089CA 6708             BEQ.S    REG_MD4                ;If no error then go and store data
1103   000089CC 49FA0109         LEA      ERMES1(PC),A4          ;Else point to error message
1104   000089D0 6000F6AA         BRA      PSTRING                ;print it and return
1105   000089D4 0C8200000044 REG_MD4: CMP.L #68,D2               ;If this address is the SR then
1106   000089DA 6704             BEQ.S    REG_MD5                ;we have only a word to store
1107   000089DC 2480             MOVE.L   D0,(A2)                ;Else store new data in display frame
1108   000089DE 4E75             RTS
1109   000089E0 3480    REG_MD5:  MOVE.W   D0,(A2)                ;Store SR (one word)
1110   000089E2 4E75             RTS
1111                             *
1112                             ********************************************************************
1113                             *
1114           000089E4 X_UN:     EQU      *                      ;Uninitialized exception vector routine
1115   000089E4 49FA0157         LEA      ERMES6(PC),A4          ;Point to error message
1116   000089E8 6100F692         BSR      PSTRING                ;Display it
1117   000089EC 6100FD32         BSR      EX_DIS                 ;Display registers
1118   000089F0 6000F64E         BRA      WARM                   ;Abort
1119                             *
```

```
1120                     *************************************************************************
1121                     *
1122                     *   All strings and other fixed parameters here
1123                     *
1124  000089F4 545342554720 BANNER:  DC.B      'TSBUG 2 Version 23.07.86',0,0
               322056657273
               696F6E203233
               2E30372E3836
               0000
1125  00008A0E 0D0A3F00     CRLF:    DC.B      CR,LF,'?',0
1126  00008A12 0D0A53310000 HEADER:  DC.B      CR,LF,'S','1',0,0
1127  00008A18 533920200000 TAIL:    DC.B      'S9 ',0,0
1128  00008A1E 20535220203D MES1:    DC.B      ' SR = ',0
               202000
1129  00008A27 20504320203D MES2:    DC.B      ' PC = ',0
               202000
1130  00008A30 20535320203D MES2A:   DC.B      ' SS = ',0
               202000
1131  00008A39 202044617461 MES3:    DC.B      '  Data reg        Address reg',0,0
               207265672020
               202020202041
               646472657373
               207265670000
1132  00008A57 202020202020 MES4:    DC.B      '          ',0,0
               20200000
1133  00008A61 427573206572 MES8:    DC.B      'Bus error   ',0,0
               726F72202020
               0000
1134  00008A6F 416464726573 MES9:    DC.B      'Address error   ',0,0
               73206572726F
               722020200000
1135  00008A81 496C6C656761 MES10:   DC.B      'Illegal instruction ',0,0
               6C20696E7374
               72756374696F
               6E200000
1136  00008A97 427265616B70 MES11:   DC.B      'Breakpoint ',0,0
               6F696E742020
               0000
1137  00008AA5 547261636520 MES12:   DC.B      'Trace   ',0
               202000
1138  00008AAE 443044314432 REGNAME: DC.B      'D0D1D2D3D4D5D6D7'
               443344344435
               44364437
1139  00008ABE 413041314132          DC.B      'A0A1A2A3A4A5A6A7'
               413341344135
               41364137
1140  00008ACE 53535352              DC.B      'SSSR'
1141  00008AD2 5043202000            DC.B      'PC  ',0
1142  00008AD7 4E6F6E2D7661 ERMES1:  DC.B      'Non-valid hexadecimal input   ',0
               6C6964206865
               786164656369
               6D616C20696E
               707574202000
1143  00008AF5 496E76616C69 ERMES2:  DC.B      'Invalid command  ',0
               6420636F6D6D
               616E64202000
```

```
1144   00008B07 4C6F6164696E ERMES3:   DC.B      'Loading error',0
                67206572726F
                7200
1145   00008B15 5461626C6520 ERMES4:   DC.B      'Table full  ',0,0
                66756C6C2020
                0000
1146   00008B23 427265616B70 ERMES5:   DC.B      'Breakpoint not active  ',0,0
                6F696E74206E
                6F7420616374
                697665202020
                0000
1147   00008B3D 556E696E6974 ERMES6:   DC.B      'Uninitialized exception ',0,0
                69616C697A65
                642065786365
                7074696F6E20
                0000
1148   00008B57 2052616E6765 ERMES7:   DC.B      ' Range error',0
                206572726F72
                00
1149                          *
1150                          *   COMTAB is the built-in command table. All entries are made up of
1151                          *      a string length + number of characters to match + the string
1152                          *      plus the address of the command relative to COMTAB
1153                          *
1154   00008B64 0404         COMTAB:   DC.B      4,4              ;JUMP <address> causes execution to
1155   00008B66 4A554D50               DC.B      'JUMP'           ;begin at <address>
1156   00008B6A FFFFF6C2               DC.L      JUMP-COMTAB      ;n
1157   00008B6E 0803                   DC.B      8,3              ;MEMORY <address> examines contents of
1158   00008B70 4D454D4F5259           DC.B      'MEMORY          ;<address> and allows them to be changed
                2020
1159   00008B78 FFFFF6D8               DC.L      MEMORY-COMTAB
1160   00008B7C 0402                   DC.B      4,2              ;LOAD <string> loads S1/S2 records
1161   00008B7E 4C4F4144               DC.B      'LOAD'           ;from the host. <string> is sent to host
1162   00008B82 FFFFF738               DC.L      LOAD-COMTAB
1163   00008B86 0402                   DC.B      4,2              ;DUMP <string> sends S1 records to the
1164   00008B88 44554D50               DC.B      'DUMP'           ;host and is preceeded by <string>.
1165   00008B8C FFFFF80C               DC.L      DUMP-COMTAB
1166   00008B90 0403                   DC.B      4,3              ;TRAN enters the transparent mode
1167   00008B92 5452414E               DC.B      'TRAN'           ;and is exited by ESC,E.
1168   00008B96 FFFFF8C8               DC.L      TM-COMTAB
1169   00008B9A 0402                   DC.B      4,2              ;NOBR <address> removes the breakpoint
1170   00008B9C 4E4F4252               DC.B      'NOBR'           ;at <address> from the BP table. If
1171   00008BA0 FFFFFDAC               DC.L      NOBR-COMTAB      ;no address is given all BPs are removed.
1172   00008BA4 0402                   DC.B      4,2              ;DISP displays the contents of the
1173   00008BA6 44495350               DC.B      'DISP'           ;pseudo registers in TSK_T.
1174   00008BAA FFFFFBBC               DC.L      EX_DIS-COMTAB
1175   00008BAE 0402                   DC.B      4,2              ;GO <address> starts program execution
1176   00008BB0 474F2020               DC.B      'GO              ;at <address> and loads regs from TSK_T
1177   00008BB4 FFFFFCEA               DC.L      GO-COMTAB
1178   00008BB8 0402                   DC.B      4,2              ;BRGT puts a breakpoint in the BP
1179   00008BBA 42524754               DC.B      'BRGT'           ;table - but not in the code
1180   00008BBE FFFFFD48               DC.L      BR_GET-COMTAB
```

```
1181  00008BC2 0402                     DC.B    4,2                  ;PLAN puts the breakpoints in the code
1182  00008BC4 504C414E                 DC.B    'PLAN'
1183  00008BC8 FFFFFD86                 DC.L    BR_SET-COMTAB
1184  00008BCC 0404                     DC.B    4,4                  ;KILL removes breakpoints from the code
1185  00008BCE 4B494C4C                 DC.B    'KILL'
1186  00008BD2 FFFFFDF2                 DC.L    BR_CLR-COMTAB
1187  00008BD6 0402                     DC.B    4,2                  ;GB <address> sets breakpoints and
1188  00008BD8 47422020                 DC.B    'GB '                ;then calls GO.
1189  00008BDC FFFFFD0A                 DC.L    GB-COMTAB
1190  00008BE0 0403                     DC.B    4,3                  ;REG <reg> <value> loads <value>
1191  00008BE2 52454720                 DC.B    'REG '               ;into <reg> in TASK_T. Used to preset
1192  00008BE6 FFFFFE0E                 DC.L    REG_MOD-COMTAB       ;registers before a GO or GB
1193  00008BEA 0000                     DC.B    0,0
1194                          *
1195                          *************************************************************************
1196                          *
1197                          *  This is a list of the information needed to setup the DCBs
1198                          *
1199           00008BEC  DCB_LST:  EQU   *
1200  00008BEC 434F4E5F494E DCB1:      DC.B    'CON_IN '            ;Device name (8 bytes)
               2020
1201  00008BF4 000084CA0001            DC.L    CON_IN,ACIA_1        ;Address of driver routine, device
               0040
1202  00008BFC 0002                    DC.W    2                    ;Number of words in parameter field
1203  00008BFE 434F4E5F4F55 DCB2:      DC.B    'CON_OUT '
               5420
1204  00008C06 000084FA0001            DC.L    CON_OUT,ACIA_1
               0040
1205  00008C0E 0002                    DC.W    2
1206  00008C10 4155585F494E DCB3:      DC.B    'AUX_IN '
               2020
1207  00008C18 0000853A0001            DC.L    AUX_IN,ACIA_2
               0041
1208  00008C20 0002                    DC.W    2
1209  00008C22 4155585F4F55 DCB4:      DC.B    'AUX_OUT '
               5420
1210  00008C2A 0000854C0001            DC.L    AUX_OUT,ACIA_2
               0041
1211  00008C32 0002                    DC.W    2
1212  00008C34 425546465F49 DCB5:      DC.B    'BUFF_IN '
               4E20
1213  00008C3C 000085A00000            DC.L    BUFF_IN,BUFFER
               02D0
1214  00008C44 0002                    DC.W    2
1215  00008C46 425546465F4F DCB6:      DC.B    'BUFF_OUT'
               5554
1216  00008C4E 000085AC0000            DC.L    BUFF_OT,BUFFER
               02D0
1217  00008C56 0002                    DC.W    2
1218                          *
```

```
1219        ************************************************************************
1220        *
1221        *   DCB structure
1222        *
1223        *                    --------------------
1224        *         0 ->    | DCB  name         |
1225        *                    |--------------------|
1226        *         8 ->    | Device driver     |
1227        *                    |--------------------|
1228        *        12 ->    | Device address    |
1229        *                    |--------------------|
1230        *        16 ->    |Size of param block |
1231        *                    |--------------------| --
1232        *        18 ->    |      Status        |    |
1233        *                    | logical  | physical |   | S
1234        *                    |--------------------|   |
1235        *                    .                    .  .
1236        *                    |--------------------| --
1237        *      18+S ->    | Pointer to next DCB |
1238        *
1239
1240
1241
```

Lines: 1241, Errors: 0, Warnings: 0.

SUMMARY

Throughout this book, we have looked at various aspects of the 68000, from its programming model to its exception-handling mechanism. In this chapter, we have considered some of the practical problems of systems design and test, and have applied the lessons learned elsewhere to design a basic single board, 68000-based microcomputer. This microcomputer, the TS2, has a monitor that permits it to receive input from a terminal, modify the input, debug it, and then run it.

The monitor presented in this chapter is not intended as an optimum monitor for the 68000 microprocessor but as an extended tutorial in 68000 assembly language programming. In particular, it is designed to demonstrate one of the more interesting ways of executing input or output transactions. The monitor provides a strong measure of device-independent I/O by means of device control blocks, DCBs.

Problems

1. What facilities or attributes make a microcomputer easy to test?

2. What facilities or attributes make a microcomputer difficult to test?

3. Why are closed-loop systems harder to test and debug than open-loop systems?

4. What additional logic and circuitry would be required to make a single-board 68000 microcomputer testable by turning it into an open-loop system? Assume that the SBC has a 96-pin connector that can be connected to external test equipment.

5. A microcomputer card can be designed to be "self-testing." For example, the CPU runs a program that tests its instruction set by executing a section of code in ROM, using only internal read/write storage, and then compares the result with a prestored value. Then the CPU tests the read/write memory and finally any peripherals.
   a. Design a program to test the 68000's instruction set.
   b. Write a program to test read/write memory.
   c. How do you think that peripherals may be self-tested?
   d. What additional hardware is required for these tests?

6. Describe how a logic analyzer is used to test and debug a 68000-based single-board computer. What are the limitations of a logic analyzer?

7. Signature analysis is used in production-line testing because it provides only a go/no-go result. Show how a signature analyzer can be built into a single-board computer to perform a test following a reset. If the test fails, it is repeated twice (by reasserting RESET*). If the test continues to fail, a front panel LED is lit.

8. Suppose that the single-board computer, TS2, is to be mass produced and that it is necessary to reduce the parts count. If the single-step feature is omitted and fusible logic (PAL, etc.) used wherever possible, how far can the chip count of TS2 be reduced?

9. A designer wishes to produce an entirely general-purpose 68000 board and one of the design specifications requires that the memory (both ROM and read/write) and peripherals should have software programmable addresses; that is, all addressible devices must be capable of being relocated under software control. To do this, the address decoders must be reprogrammable. Of course, ROM must be assigned to the reset vector area following a reset in order to set up the system. Design the logic required to implement this system.

10. A watchdog circuit generates an interrupt every T seconds. However, the system executes a program that resets the watchdog timer at least once every T seconds. Because the timer is reset before it times-out and generates an interupt, the processor is never interrupted by the watchdog timer. If, however, the processor hangs up for any reason, the timer times-out and the processor is reset. In this way, it is forced out of its hang-up state. Design a watchdog timer for a 68000 system.

11. The bus interface of TS2 is designed to make TS2 a master in a 68000 system. Consequently, the TS2 module can access the bus or it can be forced off the bus whenever its BGACK* input is asserted. TS2 cannot act as a slave and be accessed by another master. Redesign TS2 so that it can be a slave and its ROM/RAM and peripherals accessed from the system bus.

12. Some of the functions not implemented by TS2MON are:
   (1) A move command that copies a block of memory from Address_1 through Address_2 to start at Address_3.
   (2) A fill command that fills a block of memory from Address_1 through Address_2 with a constant. The command should be able to use byte, word, or longword constants.
   (3) A test command that tests read/write memory in the region from Address_1 through Address_2.
Design a subroutine to implement these functions.

13. Design a simple monitor that runs itself continually as a background task and will also execute a task as a background job; that is, executing GO⟨address⟩ invokes a multitasking kernel that switches between the monitor itself and the task invoked by the GO command. Assume that a constant stream of interrupts is available from a timer.

14. Why is design for test a desirable feature of a computer system?

15. Describe some of the ways in which a 68000 single-board computer can be designed to make it easier to test.

16. What is signature analysis and under what circumstances is it used?

17. Under what circumstances is it a good idea to free-run the 68000?

18. How can the 68000 be made to free-run?

19. What type of errors are likely to occur during the breadboard construction of a basic single-board computer?

20. Describe tests that can be carried out to detect the errors described in problem 8.

21. Suppose you wanted to design a highly reliable computer system that uses three 68000s operating in parallel. If the output of one disagrees with that of the others, it is disregarded. This is called *triple modular redundancy* and is widely used as the basis of reliable systems. In addition to the 68000 CPU themselves, control, memory, and I/O must be duplicated.

Unfortunately, the preceding system contains a logical flaw and cannot operate as described. What is the flaw? (HINT: It involves time). How could a reliable system be designed to overcome this flaw?

SUMMARY OF THE 68000 INSTRUCTION SET

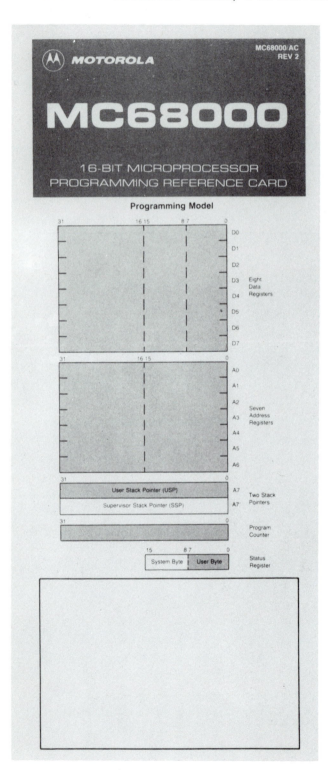

Effective Addressing Mode Categories

Type	Mode	Register	Generation	Assembler Syntax
Data Register Direct	000	reg. no.	EA = Dn	Dn
Address Register Direct	001	reg. no.	EA = An	An
Register Indirect	010	reg. no.	EA = (An)	(An)
Postincrement Register Indirect	011	reg. no.	EA = (An), An ← An + N	(An) +
Predecrement Register Indirect	100	reg. no.	An ← An − N, EA = (An)	− (An)
Register Indirect With Offset	101	reg. no.	EA = (An) + d_{16}	d_{16}(An)
Indexed Register Indirect With Offset	110	reg. no.	EA = (An) + (Xn) + d_8	d_8(An, Xn)
Absolute Short	111	000	EA = (Next Word)	xxx
Absolute Long	111	001	EA = (Next Two Words)	xxxxxxx
PC Relative With Offset	111	010	EA = (PC) + d_{16}	d_{16}(PC)
PC Relative With Index and Offset	111	011	EA = (PC) + (Xn) + d_8	d_8(PC + Xn)
Immediate	111	100	Data = Next Word(s)	#xxx
Quick Immediate	—	—	Inherent Data	#xxx (1–8)
Implied Register	—	—	EA = SR, USP, SP, PC	—

NOTES:
EA = Effective Address
An = Address Register
Dn = Data Register
Xn = Address or Data Register used as Index Register
SR = Status Register
PC = Program Counter

d_8 = Eight bit Offset (displacement)
d_{16} = Sixteen bit Offset (displacement)
N = 1 for Byte, 2 for Words and 4 for Long Words
( ) = Contents of
← = Replaces

Condition Code Computations

Operations	X	N	Z	V	C	Special Definition
ABCD	*	U	?	U	?	C = Decimal Carry; Z = Z · $\overline{Rm}$ · ... · $\overline{R0}$
ADD, ADDI, ADDQ	*	*	*	?	?	V = Sm·Dm·$\overline{Rm}$ + $\overline{Sm}$·$\overline{Dm}$·Rm; C = Sm·Dm + $\overline{Rm}$·Dm + Sm·$\overline{Rm}$
ADDX	*	*	?	?	?	V = Sm·Dm·$\overline{Rm}$ + $\overline{Sm}$·$\overline{Dm}$·Rm; C = Sm·Dm + $\overline{Rm}$·Dm + Sm·$\overline{Rm}$; Z = Z · $\overline{Rm}$ · ... · $\overline{R0}$
AND, ANDI, EOR, EORI, MOVEQ, MOVE, OR, ORI, CLR, EXT, NOT, TAS, TST	—	*	*	0	0	
CHK	—	*	U	U	U	
SUB, SUBI, SUBQ	*	*	*	?	?	V = $\overline{Sm}$·Dm·$\overline{Rm}$ + Sm·$\overline{Dm}$·Rm; C = Sm·$\overline{Dm}$ + Rm·$\overline{Dm}$ + Sm·Rm
SUBX	*	*	?	?	?	V = $\overline{Sm}$·Dm·$\overline{Rm}$ + Sm·$\overline{Dm}$·Rm; C = Sm·$\overline{Dm}$ + Rm·$\overline{Dm}$ + Sm·Rm; Z = Z · $\overline{Rm}$ · ... · $\overline{R0}$
CMP, CMPI, CMPM	—	*	*	?	?	V = $\overline{Sm}$·Dm·$\overline{Rm}$ + Sm·$\overline{Dm}$·Rm; C = Sm·$\overline{Dm}$ + Rm·$\overline{Dm}$ + Sm·Rm
DIVS, DIVU	—	*	*	?	0	V = Division Overflow
MULS, MULU	—	*	*	0	0	
SBCD, NBCD	*	U	?	U	?	C = Decimal Borrow; Z = Z · $\overline{Rm}$ · ... · $\overline{R0}$

Operations	X	N	Z	V	C	Special Definition
NEG	*	*	*	?	?	V = Dm · Rm, C = Dm + Rm
NEGX	*	*	?	?	?	V = Dm · Rm, C = Dm + Rm; Z = Z · $\overline{Rm}$ · ... · $\overline{R0}$
BTST, BCHG, BSET, BCLR	—	?	—	—		Z = $\overline{Dn}$
ASL	*	*	*	?	?	V = Dm·($\overline{D_{m-1}}$ + ... + $\overline{D_{m-r}}$) + $\overline{Dm}$·(D_{m-1} + ... + D_{m-r}); C = D_{m-r+1}
ASL (r=0)	—	*	*	0	0	
LSL, ROXL	*	*	*	0	?	C = D_{m-r+1}
LSR (r=0)	—	*	*	0	0	
ROXL (r=0)	—	*	*	0	?	C = X
ROL	—	*	*	0	?	C = D_{m-r+1}
ROL (r=0)	—	*	*	0	0	
ASR, LSR, ROXR	*	*	*	0	?	C = D_{r-1}
ASR, LSR (r=0)	—	*	*	0	0	
ROR	—	*	*	0	?	C = D_{r-1}
ROXR (r=0)	—	*	*	0	?	C = X
ROR (r=0)	—	*	*	0	0	

NOTES:
Sm = Source operand — most significant bit
Dm = Destination operand — most significant bit
Rm = Result operand — most significant bit
n = bit number
r = shift count
? = See Special Definition

Addressing Modes

Mnemonic	Size	Address Mode	Dn	An	(An)	(An)+	-(An)	d16(An)	d8(An,Xn)	Abs.W	Abs.L	d16(PC)	d8(PC,Xn)	s = Imm/d = SR,CC	Opcode Bit Pattern 1111 11 5432 1098 7654 3210	Boolean	Condition Codes X N Z V C
ABCD	B	s = −(An) d = −(An)	2 6				18								1100 RRR1 0000 0rrr 1100 RRR1 0000 1rrr	d10+s10+X → d	* U * U *
ADD	BW	s = Dn	4	ADDA	2	2	2	4	4	4	4				1101 DDD1 SSEE EEEE	d+Dn → d	* * * * *
ADDA	L																
ADDI	BW	s = Imm	8		6	6	6	6	6	6	6			8	0000 0110 SSEE EEEE	d+# → d	* * * * *
ADDQ	BW	s = Imm3	4	ADDA	4	4	4	4	4	4	4				0101 QQQ0 SSEE EEEE	d+# → d	* * * * *
ADDX	BW	s = Dn s = −(An)	2				18								1101 RRR1 SS00 0rrr 1101 RRR1 SS00 1rrr	d+s+X → d	* * * * *
AND	BW	s = Dn d = Dn	4		2	2	2	4	4	4	4				1100 DDD1 SSEE EEEE	d∧Dn → d	− * * 0 0
ANDI	BW	s = Imm s = Imm	8		6	6	6	6	6	6	6			8 16	0000 0010 SSEE EEEE	d∧# → d	− * * 0 0
ANDI CCR	B													20	0000 0010 0011 1100	s∧CCR → CCR	* * * * *
ANDI SR	W													20	0000 0010 0111 1100	s∧SR → SR	* * * * *
ASL, ASR		count = Dn count = #1–8 count = #1–8 count = 1	2+2n 2+2n 2+2n								6*				1110 CCC1 SSI0 0DDD 1110 QQQ1 SSI0 0DDD 1110 0010 10ee eeee		* * * * *
Bcc	B W	d8 = d16 =	<8 <12									branch taken branch not taken		10 8 10 12	0110 CCCC PPPP PPPP	If cc true, then PC+disp → PC	− − − − −
BCHG	B L	bit# = Dn bit# = Imm bit# = Dn bit# = Imm	<10 <14		4 6	4 6	4 6	4	4	4	4			2 4	0000 rrr1 01EE EEEE 0000 1000 01EE EEEE	!(bit#) of d → Z, !(bit#) of d → d	− − * − −
BCLR	B L	bit# = Dn bit# = Imm bit# = Dn bit# = Imm	<10 <14		4 6	4 6	4 6	4	4	4	4			2 4	0000 rrr1 10EE EEEE 0000 1000 10EE EEEE	!(bit#) of d → Z, 0 → (bit#) of d	− − * − −
BRA	B W	d8 = d16 =	<6											10 12	0110 0000 PPPP PPPP	PC+disp → PC	− − − − −
BSET	B L	bit# = Dn bit# = Imm bit# = Dn bit# = Imm	<8 <12		4 6	4 6	4 6	4	4	4	4			2 4	0000 rrr1 11EE EEEE 0000 1000 11EE EEEE	!(bit#) of d → Z, 1 → (bit#) of d	− − * − −
BSR	W	d8 = d16 =	<44											18	0110 0001 PPPP PPPP	PC → −(SP), PC+disp → PC	− − − − −
BTST	B L	bit# = Dn bit# = Imm	<6 <10		4 6	4 6	4 6	4	4	4	4			2 4	0000 rrr1 00EE EEEE 0000 1000 00EE EEEE	!(bit#) of d → Z	− − * − −
CHK	W	s = Dn s = Imm	10	trap no trap											0100 DDD1 10ee eeee	If Dn<0 or Dn>(bound), then trap	− * U U U
CLR	BW	d = Dn	4		4	4	4	4	4	4	4				0100 0010 SSEE EEEE	0 → d	− 0 1 0 0
CMP	BW	d = Dn	4		4	4	4	4	4	4	4				1011 DDD0 SSee eeee	Dn−s	− * * * *
CMPA	BW	d = An		6											1011 AAA0 11ee eeee	An−s	− * * * *
CMPI	BW	s = Imm	8		6	6	6	6	6	6	6				0000 1100 SSEE EEEE	d−#	− * * * *
CMPM	BW	s = (An)+													1011 RRR1 SS00 1rrr	d−s	− * * * *

Mnemonic	Size	Address Mode		Dn	#	An	#	(An)	#	(An)+	#	-(An)	#	d16(An)	#	d8(An,Xn)	#	Abs.W	#	Abs.L	#	d16(PC)	N	d8(PC,Xn)	#	s = immed / d = SR/CC	#	N	Opcode Bit Pattern	Boolean	Condition Codes X N Z V C

DBcc — W — d16=Imm — counter= ... — If cc true, then NOP else Dn - 1 → Dn; If Dn ≠ -1, then PC + disc → PC — - - - - -

DIVS — W — s = Dn; d = Dn — Dn32÷s16 → Dnr :q — - * * * 0

DIVU — W — s = Dn; d = Dn — Dn32÷s16 → Dnr :q — - * * * 0

EOR — B/W/L — d=Dn; s=Dn; s=Imm — d ⊕ Dn → d — - * * 0 0

EORI — B/L — d=Dn; s=Imm — d ⊕ # → d — - * * 0 0

EORI CCR — B — s=Imm — s ⊕ CCR → CCR — * * * * *

EORI SR — W — s=Imm — s ⊕ SR → SR — * * * * *

EXG — L — s=An — s → d — - - - - -

EXT — W/L — d=Dn — bit 7 → bits 15 : 8 / bit 15 → bits 31 : 16 — - * * 0 0

ILLEGAL — — — (illegal vector) → PC — - - - - -

JMP — — d — d → PC — - - - - -

JSR — — d — PC → - (SSP); SP → (SSP); (SP)/d → PC — - - - - -

LEA — L — d=An — PC → - (SSP)/d → PC — - - - - -

LINK — W — s=An — An → - (SP); SP → An; SP + disp → SP — - - - - -

LSL, LSR — B/W/L — count = Dn; count = #1-8; count = #1-8; count = 1 — s → d — * * * 0 *

Memory — — — — s → d — * * * 0 *

MOVE — B/W/L — d=Dn; d=An; d=(An); d=-(An); d=(An)+; d=d16(An); d=d8(An,Xn); d=Abs.W; d=Abs.L — s → d — - * * 0 0

MOVE — L — ... — s → d — - * * 0 0

MOVE CCR — W — d=CCR — s → CCR — * * * * *
MOVE SR — W — d=SR — s → SR — * * * * *
MOVE SR — W — s=SR — SR → d — - - - - -
MOVE USP — L — d=USP; s=USP — USP → An; An → USP — - - - - -

MOVEA — W/L — d=An; s=An — Xn → d — - - - - -

MOVEM — W/L — s=Xn; d=Xn — Xn → d — - - - - -

MOVEP — W/L — s=Dn; s=d16(An); s=Dn; s=d16(An) — Dn→d by bytes; Dn→d by bytes; Dn→d by bytes; Dn→d by bytes — - - - - -

Mnemonic	Size	Address Mode	Dn	An	(An)	(An)+	-(An)	d16(An)	d8(An,Xn)	Abs.W	Abs.L	d16(PC)	d8(PC,Xn)	s = Imm, d = SR,CC	Opcode Bit Pattern 1111 5432 1098 7654 3210	Boolean	Condition Codes X N Z V C
MOVEQ	L	s = Imm8, d = Dn													0111 DDD0 QQQQ QQQQ	# → Dn	- * * 0 0
MULS	W	d = Dn													1100 DDD1 11ee eeee	Dn × s → Dn	- * * 0 0
MULU	W	d = Dn													1100 DDD0 11ee eeee	Dn × s → Dn	- * * 0 0
NBCD	B	d = Dn													0100 1000 00ee eeee	0 - d10 - X → d	* U * U *
NEG	B,W														0100 0100 SSee eeee	0 - d → d	* * * * *
NEGX	B,W														0100 0000 SSee eeee	0 - d - X → d	* * * * *
NOP															0100 1110 0111 0001	none	- - - - -
NOT	B,W														0100 0110 SSee eeee	⌐d → d	- * * 0 0
OR	B,W	s = Dn													1000 DDD0 SSee eeee	d·or·s → Dn	- * * 0 0
		d = Dn													1000 DDD1 SSee eeee	Dn·or·d → d	
ORI	B,W	s = Imm													0000 0000 SSee eeee	d·or·# → d	- * * 0 0
ORI CCR	B	s = Imm													0000 0000 0011 1100	s·or·CCR → CCR	* * * * *
ORI SR	W	s = Imm													0000 0000 0111 1100	s·or·SR → SR	* * * * *
PEA	L														0100 1000 01ee eeee	d → -(SP)	- - - - -
RESET															0100 1110 0111 0000	assert RESET pin	- - - - -
ROL, ROR	B,W	count = Dn													1110 RRR1 SS11 1DDD		- * * 0 *
		count = #1-8													1110 QQQ1 SS01 1DDD		
	L	count = Dn													1110 RRR1 10ee eeee		
		count = #1-8													1110 QQQ1 10ee eeee		
	W	count = 1															
ROXL, ROXR	B,W	count = Dn													1110 RRR1 SS01 0DDD		* * * 0 *
		count = #1-8													1110 QQQ1 SS01 0DDD		
	L	count = Dn													1110 RRR1 1011 1DDD		
		count = #1-8													1110 QQQ1 1011 1DDD		
	W	count = 1													1110 010r 11ee eeee		
RTE															0100 1110 0111 0011	(SP)+ → SR, SP+ → PC	* * * * *
RTR															0100 1110 0111 0111	(SP)+ → CC, SP+ → PC	* * * * *
RTS															0100 1110 0111 0101	(SP)+ → PC	- - - - -
SBCD	B	s = Dn													1000 RRR1 0000 0rrr	d10 - s10 - X → d	* U * U *
	B	s = -(An)													1000 RRR1 0000 1rrr		
Scc	B	cc = True													0101 CCCC 11ee eeee	If cc true then 1's else 0's → d	- - - - -
	B	cc = False															
STOP	W														0100 1110 0111 0010	# → SR, wait for interrupt	* * * * *
SUB	B,W	s = Dn													1001 DDD0 SSee eeee	d - s → Dn	* * * * *
		d = Dn													1001 DDD1 SSee eeee	Dn - d → d	
SUBA	W	d = An													1001 AAA0 11ee eeee	An - s → An	- - - - -
	L	d = An													1001 AAA1 11ee eeee		
SUBI	B,W	s = Imm													0000 0100 SSee eeee	d - # → d	* * * * *
SUBQ	B,W	s = Imm3													0101 QQQ1 SSee eeee	d - # → d	* * * * *
SUBX	B,W	s = Dn													1001 RRR1 SS00 0rrr	d - s - X → d	* * * * *
		s = -(An)													1001 RRR1 SS00 1rrr		
SWAP	W														0100 1000 0100 0DDD	Dn(31:16) ←→ Dn(15:0)	- * * 0 0
TAS	B														0100 1010 11ee eeee	test d - cc, 1 → bit 7 of d	- * * 0 0
TRAP		trap taken													0100 1110 0100 VVVV	PC → -(SSP), SR → -(SSP), (1 of 16 trap vectors) → PC	- - - - -
TRAPV		trap taken													0100 1110 0111 0110	If V = 1, then PC → -(SSP), SR → -(SSP), (TRAPV vector) → PC, else NOP	- - - - -
		trap not taken															
TST	B,W														0100 1010 SSee eeee	test d → cc	- * * 0 0
UNLK															0100 1110 0101 1AAA	An → SP, (SP)+ → An	- - - - -

General Notes:

* Word Only
\# Number of Bytes in Instruction
s Execution Time in Clock Periods
d Destination (d10 = base 10 operand)
√ Value is Maximum Number

Complement (invert)
d8 8-Bit Displacement
d16 16-Bit Displacement
Imm Immediate Data
Imm3 Immediate Data, 3 Bits
Imm8 Immediate Data, 8 Bits

A Address Register Number
C Test Condition
D Data Register Number
E Destination Effective Address
e Source Effective Address

f Direction
  0 = Right 1 = Left
M Destination EA Mode
P Displacement
Q Quick Immediate Data

R Destination Register
S Source Register
S Size
  00 - Byte 01 - Word
  10 - Word 11 - Word
V Vector Number
XX Move size:
  01 - Byte 00 - Byte
  10 - Long 11 - Word

Condition Code Notation:
* Set according to result of operation
- Not affected by operation
0 Cleared
1 Set
U Undefined after operation
? Other - See Special Definition

Opcode Bit Pattern Codes:

Status Register

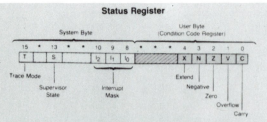

System Byte

User Byte
(Condition Code Register)

15	*	13	*	*	10	9	8	*	*	*	4	3	2	1	0
T		S			I_2	I_1	I_0				X	N	Z	V	C

- Trace Mode
- Supervisor State
- Interrupt Mask
- Extend
- Negative
- Zero
- Overflow
- Carry

*Denotes reserved bits. read only as "0"

Interrupt Encoding

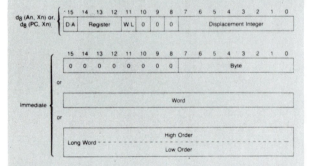

Priority	IPL2 1 0 Control Lines	Requested Interrupt Level	Status Reg. Int. Mask I2/I1/I0	Recognized Interrupt Level
Highest	LLL	7	111	7
•	LLH	6	110	7
•	LHL	5	101	6,7
•	LHH	4	100	5-7
•	HLL	3	011	4-7
•	HLH	2	010	3-7
•	HHL	1	001	2-7
Lowest	HHH	None	000	1-7

Instruction Operation Word General Format

15	14	13	12	11	10	9	8	7	6	5	4	3	2	1	0
Operation Word (First Word Specifies Operation and Modes)																
Immediate Operand (If Any. One or Two Words)																
Source Effective Address Extension (If Any. One or Two Words)																
Destination Effective Address Extension (If Any. One or Two Words)																

d_8 (An, Xn) or,
d_8 (PC, Xn)

15	14	13	12	11	10	9	8	7	6	5	4	3	2	1	0
D/A	Register	W L	0	0	0	Displacement Integer										

Immediate

15	14	13	12	11	10	9	8	7	6	5	4	3	2	1	0
0	0	0	0	0	0	0	0	Byte							

or

Word

or

High Order
Long Word
Low Order

Single-Effective-Address Instruction Operation Word

15	14	13	12	11	10	9	8	7	6	5	4	3	2	1	0
X	X	X	X	X	X	X	X	X	X	Effective Address Mode			Register		

Double-Effective-Address Instruction Operation Word

15	14	13	12	11	10	9	8	7	6	5	4	3	2	1	0
X	X	X	X	Destination Register		Mode			Source Mode			Register		

Reference Classification

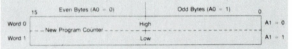

Function Code Output			Reference Class
FC2	FC1	FC0	
0	0	0	(Unassigned)
0	0	1	User Data
0	1	0	User Program
0	1	1	(Unassigned)

Function Code Output			Reference Class
FC2	FC1	FC0	
1	0	0	(Unassigned)
1	0	1	Supervisor Data
1	1	0	Supervisor Program
1	1	1	Interrupt Acknowledge

Exception Vector Format

15	Even Bytes (A0 = 0)		Odd Bytes (A0 = 1)	0	
Word 0	New Program Counter — High				A1 = 0
Word 1	Low				A1 = 1

Peripheral Vector Number Format

D15		D8	D7							D0
Ignored			v7	v6	v5	v4	v3	v2	v1	v0

Where
v7 is the MSB of the Vector Number
v0 is the LSB of the Vector Number

Address Translated from 8-Bit Vector Number

A31		A10	A9	A8	A7	A6	A5	A4	A3	A2	A1	A0
All Zeroes			v7	v6	v5	v4	v3	v2	v1	v0	0	0

Supervisor Stack Order for Bus or Address Error Exception

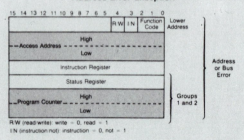

15	14	13	12	11	10	9	8	7	6	5	4	3	2	1	0	
											R/W	I/N	Function Code			Lower Address
Access Address — High															
Low															
Instruction Register															
Status Register															
Program Counter — High															
Low															

Address or Bus Error

Groups 1 and 2

R/W (read/write): write = 0, read = 1
I/N (instruction/not): instruction = 0, not = 1

Exception Grouping and Priority

Group	Exception	Processing
0	Reset Address Error Bus Error	Exception processing begins within two clock cycles.
1	Trace Interrupt Illegal Privilege	Exception processing begins before the next instruction.
2	TRAP. TRAPV. CHK Zero Divide	Exception processing is started by normal instruction execution.

Exception Vector Assignment

Vector Number(s)	Address			Assignment
	Dec	Hex	Space[6]	
0	0	000	SP	Reset: Initial SSP[2]
1	4	004	SP	Reset: Initial PC[2]
2	8	008	SD	Bus Error
3	12	00C	SD	Address Error
4	16	010	SD	Illegal Instruction
5	20	014	SD	Zero Divide
6	24	018	SD	CHK Instruction
7	28	01C	SD	TRAPV Instruction
8	32	020	SD	Privilege Violation
9	36	024	SD	Trace
10	40	028	SD	Line 1010 Emulator
11	44	02C	SD	Line 1111 Emulator
12[1]	48	030	SD	(Unassigned, Reserved)
13[1]	52	034	SD	(Unassigned, Reserved)
14	56	038	SD	Format Error[5]
15	60	03C	SD	Uninitialized Interrupt Vector
16-23[1]	64	040	SD	(Unassigned, Reserved)
	95	05F		—
24	96	060	SD	Spurious Interrupt[3]
25	100	064	SD	Level 1 Interrupt Autovector
26	104	068	SD	Level 2 Interrupt Autovector
27	108	06C	SD	Level 3 Interrupt Autovector
28	112	070	SD	Level 4 Interrupt Autovector
29	116	074	SD	Level 5 Interrupt Autovector
30	120	078	SD	Level 6 Interrupt Autovector
31	124	07C	SD	Level 7 Interrupt Autovector
32-47	128	080	SD	TRAP Instruction Vectors[4]
	191	0BF		
48-63[1]	192	0C0	SD	(Unassigned, Reserved)
	255	0FF		—
64-255	256	100	SD	User Interrupt Vectors
	1023	3FF		—

NOTES:
1. Vector numbers 12, 13, 16 through 23, and 48 through 63 are reserved for future enhancements by Motorola. No user peripheral devices should be assigned these numbers.
2. Reset vector (0) requires four words, unlike the other vectors which only require two words, and is located in the supervisor program space.
3. The spurious interrupt vector is taken when there is a bus error indication during interrupt processing.
4. Trap #n uses vector number 32 + n.
5. MC68010/MC68012 only.
   This vector is unassigned, reserved on the MC68000, and MC68008.
6. SP denotes supervisor program space, and SD denotes supervisor data space.

Exception Processing Execution Times

Exception	Periods
Address Error	50(4/7)
Bus Error	50(4/7)
CHK Instruction	44(5/4) +
Divide by Zero	42(5/4)
Illegal Instruction	34(4/3)
Interrupt	44(5/3)*
Privilege Violation	34(4/3)
RESET**	40(6/0)
Trace	34(4/3)
TRAP Instruction	38(4/4)
TRAPV Instruction	34(4/3)

+ Add effective address calculation time
*The interrupt acknowledge cycle is assumed to take four clock periods.
**Indicates the time from when $\overline{RESET}$ and $\overline{HALT}$ are first sampled as negated to when instruction execution starts.

Conditional Tests

Mnemonic	Condition	Encoding	Test
T	true	0000	1
F	false	0001	0
HI	high	0010	$\overline{C}\cdot\overline{Z}$
LS	low or same	0011	C + Z
CC(HS)	carry clear	0100	$\overline{C}$
CS(LO)	carry set	0101	C
NE	not equal	0110	$\overline{Z}$
EQ	equal	0111	Z
VC	overflow clear	1000	$\overline{V}$
VS	overflow set	1001	V
PL	plus	1010	$\overline{N}$
MI	minus	1011	N
GE	greater or equal	1100	$N\cdot V + \overline{N}\cdot\overline{V}$
LT	less than	1101	$N\cdot\overline{V} + \overline{N}\cdot V$
GT	greater than	1110	$N\cdot V\cdot\overline{Z} + \overline{N}\cdot\overline{V}\cdot\overline{Z}$
LE	less or equal	1111	$Z + N\cdot\overline{V} + \overline{N}\cdot V$

Pin Assignments

64-Pin Dual-in-Line Package

Left	Pin		Pin	Right
D4	1		64	D5
D3	2		63	D6
D2	3		62	D7
D1	4		61	D8
D0	5		60	D9
$\overline{AS}$	6		59	D10
$\overline{UDS}$	7		58	D11
$\overline{LDA}$	8		57	D12
R/$\overline{W}$	9		56	D13
$\overline{DTACK}$	10		55	D14
$\overline{BG}$	11		54	D15
$\overline{BGACK}$	12		53	GND
$\overline{BR}$	13		52	A23
V_{CC}	14		51	A22
CLK	15		50	A21
GND	16		49	V_{CC}
$\overline{HALT}$	17		48	A20
$\overline{RESET}$	18		47	A19
$\overline{VMA}$	19		46	A18
E	20		45	A17
$\overline{VPA}$	21		44	A16
$\overline{BERR}$	22		43	A15
$\overline{IPL2}$	23		42	A14
$\overline{IPL1}$	24		41	A13
$\overline{IPL0}$	25		40	A12
FC2	26		39	A11
FC1	27		38	A10
FC0	28		37	A9
A1	29		36	A8
A2	30		35	A7
A3	31		34	A6
A4	32		33	A5

Operation Code Map

Bits 15 through 12	Operation
0000	Bit Manipulation/MOVEP/Immediate
0001	Move Byte
0010	Move Long
0011	Move Word
0100	Miscellaneous
0101	ADDQ/SUBQ/Scc/DBcc
0110	Bcc/BSR
0111	MOVEQ

Bits 15 through 12	Operation
1000	OR/DIV/SBCD
1001	SUB/SUBX
1010	(Unassigned)
1011	CMP/EOR
1100	AND/MUL/ABCD/EXG
1101	ADD/ADDX
1110	Shift/Rotate
1111	(Unassigned)

Pin Assignments
68-Pin Grid Array

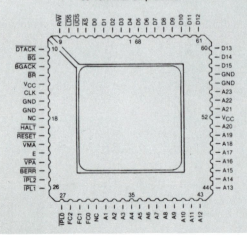

	1	2	3	4	5	6	7	8	9	10
K	NC	FC2	FC0	A1	A3	A4	A6	A7	A9	NC
J	$\overline{BERR}$	$\overline{IPL0}$	FC1	NC	A2	A5	A8	A10	A11	A14
H	E	$\overline{IPL2}$	$\overline{IPL1}$					A13	A12	A16
G	$\overline{VMA}$	$\overline{VPA}$						A15		A17
F	$\overline{HALT}$	$\overline{RESET}$			BOTTOM				A18	A19
E	CLK	GND			VIEW				V$_{CC}$	A20
D	$\overline{BR}$	V$_{CC}$							GND	A21
C	$\overline{BGACK}$	$\overline{BG}$	R/$\overline{W}$					D13	A23	A22
B	$\overline{DTACK}$	$\overline{LDS}$	$\overline{UDS}$	D0	D3	D6	D9	D11	D14	D15
A	NC	$\overline{AS}$	D1	D2	D4	D5	D7	D8	D10	D12

68-Terminal Chip Carrier

(68-terminal chip carrier diagram)

Top edge pins: R/$\overline{W}$, $\overline{LDS}$, $\overline{UDS}$, $\overline{AS}$, D0, D1, D2, D3, D4, D5, D6, D7, D8, D9, D10, D11, D12

Left side (top to bottom): $\overline{DTACK}$ 10, $\overline{BG}$, $\overline{BGACK}$, $\overline{BR}$, V$_{CC}$, CLK, GND, GND, NC 18, $\overline{HALT}$, $\overline{RESET}$, $\overline{VMA}$, E, $\overline{VPA}$, $\overline{BERR}$, $\overline{IPL2}$, $\overline{IPL1}$ 26

Bottom (left to right): $\overline{IPL0}$, FC2, FC1, FC0, NC, A1, A2, A3, A4, A5, A6, A7, A8, A9, A10, A11, A12

Right side (top to bottom): D13 60, D14, D15, GND, GND, A23, A22, A21, V$_{CC}$ 52, A20, A19, A18, A17, A16, A15, A14, A13

Corner numbers: 9, 1, 68, 61, 44, 43, 35, 27

MOTOROLA MC68000

Powers of 16, Powers of 2

16^m m =	2^n n =	Value
0	0	1
	1	2
	2	4
	3	8
1	4	16
	5	32
	6	64
	7	128
2	8	256
	9	512
	10	1,024
	11	2,048
3	12	4,096
	13	8,192
	14	16,384
	15	32,768

16^m m =	2^n n =	Value
4	16	65,536
	17	131,072
	18	262,144
	19	524,288
5	20	1,048,576
	21	2,097,152
	22	4,194,304
	23	8,388,608
6	24	16,777,216
	25	33,554,432
	26	67,108,864
	27	134,217,728
7	28	268,435,456
	29	536,870,912
	30	1,073,741,824
	31	2,147,483,648
8	32	4,294,967,296

ASCII Character Set (7-Bit Code)

LS Dig. \ MS Dig.	0	1	2	3	4	5	6	7
0	NUL	DLE	SP	0	@	P	`	p
1	SOH	DC1	!	1	A	Q	a	q
2	STX	DC2	"	2	B	R	b	r
3	ETX	DC3	#	3	C	S	c	s
4	EOT	DC4	$	4	D	T	d	t
5	ENQ	NAK	%	5	E	U	e	u
6	ACK	SYN	&	6	F	V	f	v
7	BEL	ETB	'	7	G	W	g	w
8	BS	CAN	(	8	H	X	h	x
9	HT	EM	)	9	I	Y	i	y
A	LF	SUB	*	:	J	Z	j	z
B	VT	ESC	+	;	K	[	k	{
C	FF	FS	,	<	L	\	l		
D	CR	GS	–	=	M	]	m	}
E	SO	RS	.	>	N	^	n	~
F	SI	US	/	?	O	_	o	DEL

Hexadecimal and Decimal Conversion

How to use:

Conversion to Decimal: Find the decimal weights for corresponding hexadecimal characters beginning with the least significant character. The sum of the decimal weights is the decimal value of the hexadecimal number.

Conversion to Hexadecimal: Find the highest decimal value in the table which is lower than or equal to the decimal number to be converted. The corresponding hexadecimal character is the most significant. Subtract the decimal value found from the decimal number to be converted. With the difference repeat the process to find subsequent hexadecimal characters.

	Byte			Byte			Byte	
23		16	15		8	7		0
23 Char 20	19 Char 16	15 Char 12	11 Char 8	7 Char 4	3 Char 0
Hex	Dec	Hex	Dec	Hex	Dec	Hex	Dec	Hex	Dec	Hex	Dec
0	0	0	0	0	0	0	0	0	0	0	0
1	1,048,576	1	65,536	1	4,096	1	256	1	16	1	1
2	2,097,158	2	131,072	2	8,192	2	512	2	32	2	2
3	3,145,728	3	196,608	3	12,288	3	768	3	48	3	3
4	4,194,304	4	262,144	4	16,384	4	1024	4	64	4	4
5	5,242,880	5	327,680	5	20,480	5	1280	5	80	5	5
6	6,291,456	6	393,216	6	24,576	6	1536	6	96	6	6
7	7,340,032	7	458,752	7	28,672	7	1792	7	112	7	7
8	8,388,608	8	524,288	8	32,768	8	2048	8	128	8	8
9	9,437,184	9	589,824	9	36,864	9	2304	9	144	9	9
A	10,485,760	A	655,360	A	40,960	A	2560	A	160	A	10
B	11,534,336	B	720,896	B	45,056	B	2816	B	176	B	11
C	12,582,912	C	786,432	C	49,152	C	3072	C	192	C	12
D	13,631,488	D	851,968	D	53,248	D	3328	D	208	D	13
E	14,680,064	E	917,504	E	57,344	E	3584	E	224	E	14
F	15,728,640	F	983,040	F	61,440	F	3840	F	240	F	15

BIBLIOGRAPHY

Bacon, J. *The Motorola MC68000: An Introduction to Processor, Memory and Interfacing.* Englewood Cliffs, N.J.: Prentice-Hall, 1986.

Beaston, J., and Tetrick, R.S. "Designers Confront Metastability in Boards and Buses." *Computer Design* (March 1, 1986): 67–71.

Borrill, P.L. "MicroStandards Special Feature: A Comparison of 32-bit Buses." *IEEE Micro*, Vol. 5 (December 1985): 71–79.

Borrill, P.L. "Objective Comparison of 32-bit Buses." *Microprocessors and Microsystems*, Vol. 10, No, 2 (March 1986): 94–100.

Bramer, B., and Bramer, S. *MC68000 Assembly Language Programming*, 2d ed. Edward Arnold, 1991.

Brown, G., and Harper, K. *MC68008 Minimum Configuration System.* Application Note AN897, Motorola, Inc., 1984.

Carter, E.M., and Bonds, A.B. "A 68000-Based System for Only $200." *Byte* (January 1984): 403–416.

Clements, A. "A Microprocessor for Teaching Computer Technology." *Computer Bulletin*, Vol. 2, Part 1 (March 1986): 14–16.

Clements, A. *Microcomputer Design and Construction.* Englewood Cliffs, N.J.: Prentice-Hall, 1982.

Clements, A. *Microprocessor Interfacing and the 68000: Peripherals and Systems.* New York: Wiley, 1989.

Clements, A. *Microprocessor Support Chips Sourcebook.* New York: McGraw-Hill, 1992.

Clements, A. *68000 Sourcebook.* New York: McGraw-Hill, 1990.

Coffron, J.W. *Using and Troubleshooting the MC68000.* Reston, Va.: Reston Publishing, 1982.

Coombs, T. "The VMEbus Specification—A Critique." *Electronic Product Design* (August 1987): 39–41; (September 1987): 75–76; (November 1987): 74.

Cornejo, C., and Lee, R. "Comparing IBM's Micro Channel and Apple's NuBus." *Byte* (1986 Extra Edition): 83–92.

Davies, R. *Prioritized Individually Vectored Interrupts for Multiple Peripheral Systems with the MC68000.* Application Note AN819, Motorola Inc., 1981.

Del Corso, D.; Kirrman, H.; and Nicoud, J.D. *Microcomputer Buses and Links.* New York: Academic Press, 1986.

Dr. Dobb's Journal. *Dr. Dobb's Toolbook of 68000 Programming.* Englewood Cliffs, N.J.: Prentice-Hall, 1986.

Eccles, W.J. *Microprocessor Systems: A 16-Bit Approach.* Reading, Mass.: Addison-Wesley, 1985.

Edenfield, R.W., et al. "The 68040 Processor: Part 1, Design and Implementation." *IEEE Micro*, Vol. 10, No. 1 (February 1990): 66–78.

Edenfield, R.W., et al. "The 68040 Processor: Part 2, Memory Design and Chip Verification." *IEEE Micro*, Vol. 10, No. 3, (June 1990): 22–35.

Fischer, W. "IEEE P1014—A Standard for the High-Performance VME Bus." *IEEE Micro* (February 1985): 31–41.

Ford, W., and Topp, W. *MC68000 Assembly Language and Systems Programming*. Lexington, Mass.: Heath, 1987.

Foster, C.C. *Real-Time Programming—Neglected Topics*. Reading, Mass.: Addison-Wesley, 1981.

Gillet, W.D. *An Introduction to Engineered Software*. Orlando, Fla.: Holt, Rinehart and Winston, 1982.

Gorsline, G.W. *Assembly and Assemblers: The Motorola MC68000 Family*. Englewood Cliffs, N.J.: Prentice-Hall, 1988.

Groves, S. "Balancing RAM Access Time and Clock Rate Maximizes Microprocessor Throughput." *Computer Design* (July 1980): 118–126.

Harman, L.T. *The Motorola MC68020 and MC68030 Microprocessors. Assembly Language, Interfacing and Design*. Englewood Cliffs, N.J.: Prentice-Hall, 1989.

Harper, K. *A Terminal Interface, Printer Interface, and Background Printing for an MC68000-Based System Using the 68681 DUART*. Application Note AN899, Motorola Inc., 1984.

Heath, W.S. *Real-Time Software Techniques*. New York: Van Nostrand Reinhold, 1991.

Hilf, W., and Nausch, A. *MC68000 Familie: Teil 1, Grundlagen und Architektur*. Munich, Germany: Te-wi Verlag, 1984.

Jaulent, P. *Circuits Periphériques de la Famille 68000*. Paris, France: Editions Eyrolles, 1985.

Jaulent, P. *The 68000 Hardware and Software*. London, England: Macmillan, 1985.

Kane, G., et al. *68000 Assembly Language Programming*. New York: Osborne/McGraw-Hill, 1986.

Kelly-Bootle, S. *68000 Programming by Example*. Carmel, Ind.: Howard W. Sams, 1988.

King, T., and Knight, B. *Programming the M68000*. Reading, Mass.: Addison-Wesley, 1986.

Krutz, R.L. *Interfacing Techniques in Digital Design with Emphasis on Microprocessors*. New York: Wiley, 1988.

Laws, D.A., and Levy, R.J. *Use of the Am26LS29, 30, 31 and 32 Quad Driver/Receiver Family in EIA RS-422 and 423 Applications*. Advanced Micro Devices Application Note, June 1978.

Lenk, J.D. *How to Troubleshoot and Repair Microcomputers*. Reston, Va.: Reston Publishing, 1980.

Leventhal, L., and Cordes, F. *Assembly Language Subroutines for the 68000*. New York: McGraw-Hill, 1989.

Lipovski, G.J. *16- and 32-bit Microcomputer Interfacing: Programming Examples in C and M68000 Family Assembly Language*. Englewood Cliffs, N.J.: Prentice-Hall, 1990.

MacGregor, D., and Mothersole, D.S. "Virtual Memory and the MC68010." *IEEE Micro*, Vol. 3 (June 1983): 24–39.

MacGregor, D.; Mothersole, D.S.; and Moyer, B. *The Motorola MC68020*. Motorola Inc. AR217 [reprinted from *IEEE Micro*, Vol. 4, No. 4 (August 1984): 101–118].

Mimar, T. *Programming and Designing with the 68000 Family*. Englewood Cliffs, N.J.: Prentice-Hall, 1991.

Morton, M. "68000 Tricks and Traps." *Byte*, Vol. 11, No. 9 (September 1986): 163–172.

Motorola Inc. *A Discussion of Interrupts for the MC68000*. Application Note AN1012, Motorola Inc.

Motorola Inc. *Educational Computer Board User's Manual*. Austin, Tex.: Motorola Inc., 1982.

Motorola Inc. *High Performance Memory Design Technique for the MC68000*. Application Note AN838, Motorola Inc., 1982.

Motorola Inc. *The Interrupt Controlling Capabilities of the MC68901 and the MC68230*. Application Note AN975, Motorola Inc., 1988.

Motorola Inc. *MC68000 16/32-bit Microprocessor*. Note AD1814R6, Motorola Inc., 1985.

Motorola Inc. *MC68000 16/32-bit Microprocessors Reference Manual*. Engelwood Cliffs, N.J.: Prentice-Hall, 1986.

Motorola Inc. *MC68010 Microprocessor Prototype Board*. Application Note AN996, Motorola Inc., 1988.

Motorola Inc. *MC68020 and MC68881 Platform Board for Evaluation in a 16-bit System*. Application Note AN944, Motorola Inc. 1987.

Motorola Inc. *MC68020 Minimum System Configuration*. Application Note AN1015, Motorola Inc., 1989.

Motorola Inc. *An MC68030 32-bit High Performance Minimum System*. Application Note ANE426, Motorola Inc., 1989.

Motorola Inc. *MC68040 Benchmark Board*. Application Note AN435, Motorola Inc., 1990.

Motorola Inc. *MC68230 Parallel Interface/Timer*. AD1860R2, Motorola Inc., 1983.

Motorola Inc. *MC68451 Memory Management Unit*. Motorola Inc., April 1983.

Motorola Inc. *MC68881 Floating-Point Coprocessor as a Peripheral in an M68000 System*. Application Note AN947, Motorola Inc., 1987.

Motorola Inc. *MC68HC000-Based CMOS Computer*. Application Note AN988, Motorola Inc., 1987.

Motorola Inc. *M68000 vs. iAPX86 Benchmark Performance*. Note BR150, Motorola Inc.

Motorola Inc. *A Microcomputer System Bus Technical Comparison*. Note BR172, Motorola Inc., 1984.

Motorola Inc. *Platform Boards for Development of MC68030-Based Systems*. Application Note AN972, Motorola Inc. 1987.

Motorola Inc. *32-bit Computer Design Using 68020/68881/68851*. Application Note AN994RE, Motorola Inc., 1987.

Motorola Inc. *Transmission Line Effects in PCB Applications*. Application Note AN1051, Motorola Inc.

Peterson, W.D. *The VMEbus Handbook: A User's Guide to the IEEE 1014 and IEC 821 Microcomputer Bus*. Scottsdale, Ariz.: VMEbus International Trade Association, 1989.

Protopapas, D.A. *Microcomputer Hardware Design*. Englewood Cliffs, N.J.: Prentice-Hall, 1988.

Ripps, D., and Mushinsky, B. "Benchmarks Contrast 68020 Cache-Memory Operations." *EDN* (August 8, 1985): 177–202.

Scanlon, L.J. *The 68000: Principles and Programming*. Carmel, Ind.: Howard W. Sams, 1981.

Scherer, V.A., and Peterson, W.G. *The MC68230 Parallel Interface/Timer Provides an Effective Printer Interface*. Application Note AN854, Motorola Inc.

Shooman, M.L. *Software Engineering*. New York: McGraw-Hill, 1983.

Starnes, T.W. *Design Philosophy behind Motorola's MC68000*. Note AR208, Motorola Inc.

Stone, H.S. *Microprocessor Interfacing*. Reading, Mass.: Addison-Wesley, 1982.

Treibel, S., and Singh, A. *The 68000 Microprocessor: Architecture, Software and Interfacing Techniques*. Englewood Cliffs, N.J.: Prentice-Hall, 1986.

Treibel, W.A., and Singh, A. *The 68000 and 68020 Microprocessors: Architecture, Software and Interfacing Techniques*. Englewood Cliffs, N.J.: Prentice-Hall, 1991.

Veronis, A. *The 68000 Microprocessor*. Van Nostrand Reinhold, 1988.

VMEbus International Trade Association. *The VMEbus Specification*. Scottsdale, Ariz.: VMEbus International Trade Association, 1989.

Voelzke, H. "Der mc-68000 Computer." *MC Magazine* (November 1984): 116–128; (December 1985): 50–72; (January 1985): 50–53.

Wakerly, J.F. *Microprocessor Architecture and Programming*. New York: Wiley, 1981.

Wakerly, J.F. *Microcomputer Architecture and Programming: The 68000 Family*. New York: Wiley 1989.

West, T. *Dual-Ported RAM for the MC68000 Microprocessor*. Application Note AN881, Motorola Inc.

West, T. *A High Performance MC68000L12 System with No Wait States*. Application Note AN868, Motorola Inc.

Wilcox, A.D. *68000 Microcomputer Systems: Designing and Troubleshooting*. Englewood Cliffs, N.J.: Prentice-Hall, 1987.

Williams, S. *68030 Assembly Language Reference*. Reading, Mass.: Addison-Wesley, 1988.

Witten, I.H. "The New Microprocessors." *IEE Proceedings*, Vol. 128, Part E, No. 5 (September 1981): 197–204.

Yu-Cheng Liu. *The M68000 Microprocessor Family: Fundamentals of Assembly Language Programming and Interface Design*. Englewood Cliffs, N.J.: Prentice-Hall, 1991.

Zehr, G. "Memory Management Units for 68000 Architectures." *Byte*, Vol. 11, No. 12 (December 1986): 127–135.

INDEX